Supramolecular Polymers and Assemblies

Supramolecular Polymers and Assemblies

From Synthesis to Properties and Applications

Ulrich S. Schubert
George R. Newkome
Andreas Winter

WILEY-VCH

Authors

Prof. Ulrich S. Schubert
Friedrich Schiller University Jena
Laboratory for Organic and Macromolecular Chemistry (IOMC)
Humboldtstraße 10
07743 Jena
Germany

Prof. George R. Newkome
Florida Atlantic University
Center for Molecular Biology and Biotechnology
Jupiter Campus, 5353 Parkside Drive, RF17/207
Jupiter, FL 33458
United States

Dr. Andreas Winter
Friedrich Schiller University Jena
Laboratory for Organic and Macromolecular Chemistry (IOMC)
Humboldtstraße 10
07743 Jena
Germany

Cover
Cover Image: © Sebestyen Balint/Shutterstock

Library of Congress Card No.:
applied for

British Library Cataloguing-in-Publication Data
A catalogue record for this book is available from the British Library.

Bibliographic information published by the Deutsche Nationalbibliothek
The Deutsche Nationalbibliothek lists this publication in the Deutsche Nationalbibliografie; detailed bibliographic data are available on the Internet at <http://dnb.d-nb.de>.

Print ISBN: 978-3-527-33356-1
ePDF ISBN: 978-3-527-83241-5
ePub ISBN: 978-3-527-83240-8
oBook ISBN: 978-3-527-68532-5

Typesetting SPi Global, Chennai, India
Printing and Binding Printed and bound by CPI Group (UK) Ltd, Croydon, CR0 4YY

Printed on acid-free paper

10 9 8 7 6 5 4 3 2 1

Contents

Preface *xi*
Abbreviations *xiii*
About the Authors *xix*

1 Supramolecular Polymers: General Considerations *1*
1.1 Introduction *1*
1.2 Classification Schemes *2*
1.3 Supramolecular Polymerization Mechanisms *4*
1.3.1 Isodesmic Supramolecular Polymerization *6*
1.3.2 Ring-Chain-Mediated Supramolecular Polymerization *12*
1.3.3 (Anti)-cooperative Supramolecular Polymerization *16*
1.4 Beyond Classical Supramolecular Polymerization *20*
1.5 Concluding Remarks *22*
References *23*

2 Supramolecular Assemblies Based on Ionic Interactions *29*
2.1 General Aspects *29*
2.2 Basic Binding Modes and Discrete Model Assemblies *30*
2.3 Supramolecular Polymers, Based on Ionic Interactions *35*
2.3.1 Ionic Interactions, as Integral Part of the Polymer Backbone *36*
2.3.1.1 Polymers, as Building Blocks *37*
2.3.1.2 Small Molecules, as Building Blocks *42*
2.3.2 Polymers Featuring Ionic Interactions in the Side Chain *46*
2.3.3 Self-assembly of Polyelectrolytes: Toward Polyelectrolyte Complexes *49*
2.4 Concluding Remarks *52*
References *52*

3 Supramolecular Polymers, Based on Hydrogen-Bonding Interactions *57*
3.1 General Aspects *57*
3.2 From H-Bonding Interactions to Supramolecular Polymers *63*
3.2.1 Supramolecular Polymers, Based on Single H-Bonding Interactions *65*
3.2.2 Supramolecular Polymers, Based on Double H-Bonding Interactions *66*

3.2.3 Supramolecular Polymers, Based on Triple H-Bonding Interactions *72*
3.2.4 Supramolecular Polymers, Based on Quadruple H-Bonding Interactions *84*
3.2.5 Supramolecular Polymers, Based on Sextuple H-Bonding Interactions *98*
3.2.6 Supramolecular Polymers, Based on Multiple H-Bonding Interactions *104*
3.3 Conclusion Remarks *107*
References *108*

4 Supramolecular Polymers, Based on Metal-to-Ligand Interactions *117*
4.1 General Aspects *117*
4.2 Synthesis and Design Principles *119*
4.2.1 Metal-Binding Sites in Metallo-supramolecular Polymers *119*
4.2.2 Synthesis of Metallo-supramolecular Polymers *121*
4.3 Linear Metallo-supramolecular Polymers *121*
4.3.1 Metallo-supramolecular Polymers, Based on Coordinative Bonding *121*
4.3.1.1 Metallopolymers Containing Ditopic Monodentate Ligands *123*
4.3.1.2 Coordination Polymers Containing Ditopic Bidentate Ligands *133*
4.3.1.3 Coordination Polymers Containing Ditopic Tridentate Ligands *137*
4.3.2 Metallo-supramolecular Polymers, Based on Covalent/Ionic Bonding *159*
4.3.3 Metallo-supramolecular Polymers Based on Metal–Arene Interactions *172*
4.4 Concluding Remarks *176*
References *177*

5 Supramolecular Polymers, Based on π-Electronic Interactions *195*
5.1 General Aspects *195*
5.2 Columnar Supramolecular Polymers, Based on π–π Stacking Interactions *197*
5.3 From π–π Stacking to Advanced Donor–Acceptor-Type Charge–Transfer Interactions *209*
5.4 From Charge–Transfer to π-Electronic Ion-Pairing Interactions *213*
5.5 Linear Supramolecular Polymers, Based on π-Electronic Interactions *218*
5.6 Conclusion and Outlook *232*
References *233*

6 Supramolecular Polymers, Based on Crown Ether Recognition *239*
6.1 General Aspects *239*
6.2 From Crown Ether Molecular Recognition Toward Supramolecular Polymers *240*

6.2.1 Linear Polypseudorotaxanes, Derived from Small-Molecule Building Blocks *241*
6.2.2 Linear Supramolecular Polymers Derived from Macromolecular Building Blocks *252*
6.2.3 Dendritic, Hyperbranched, Star-shaped, and Grafted Assemblies *253*
6.3 Mechanical Interlocking: From Pseudorotaxanes to Rotaxanes *258*
6.4 Poly(pseudo)rotaxanes, Derived from Preformed Polymers *263*
6.5 Supramolecular Amphiphiles *266*
6.6 Concluding Remarks *269*
References *269*

7 Supramolecular Polymers, Based on Cucurbiturils *275*
7.1 General Aspects *275*
7.2 Interactions of CB[*n*]s with Small Organic Guest Molecules *276*
7.3 Supramolecular Polymers Incorporating CB[*n*] Units *278*
7.3.1 Supramolecular Polymers, Based on Small Organic Molecules *278*
7.3.2 Supramolecular Polymers Incorporating (Bio)macromolecular Building Blocks *291*
7.4 Concluding Remarks *295*
References *296*

8 Supramolecular Polymers, Based on the Host–Guest Chemistry of Calixarenes *301*
8.1 General Aspects *301*
8.2 Calixarene-Based Supramolecular Polycaps *302*
8.3 Supramolecular Polymers Featuring Vacant Calixarene Scaffolds *309*
8.4 Supramolecular Polymers, Formed by Host–Guest Interactions *311*
8.4.1 Supramolecular Polymers, Derived from AB-Type Monomers *312*
8.4.2 Supramolecular Polymers, Derived from AA- and BB-Type Monomers *314*
8.5 Beyond Classical Calix[*n*]arenes: Calix[4]pyrroles *322*
8.6 Miscellaneous Supramacromolecular Assemblies, Based on Calixarenes *327*
8.7 Concluding Remarks *337*
References *338*

9 Cyclodextrins in the Field of Supramolecular Polymers *343*
9.1 General Aspects *343*
9.2 Cyclodextrins and Supramolecular Polymers *344*
9.2.1 Linear Polypseudorotaxanes, Derived from Preformed Linear Polymers *345*
9.2.1.1 Polypseudorotaxanes from of Poly(ethylene glycol) *347*
9.2.1.2 Polypseudorotaxanes from Other Polyethers *348*
9.2.1.3 Polypseudorotaxanes from Polyesters *349*
9.2.1.4 Polypseudorotaxanes from Polyamides and Polyurethanes *350*

9.2.1.5 Polypseudorotaxanes from Polyolefins *352*
9.2.1.6 Polyrotaxanes from Inorganic Polymers *353*
9.2.1.7 Polypseudorotaxanes from Conductive/Conjugated Polymers *353*
9.2.1.8 Polypseudorotaxanes from Polyamines *355*
9.2.2 Polypseudorotaxanes, Derived from Other Polymeric Architectures *356*
9.2.2.1 Site-Selective Complexation of Linear Block Copolymers *356*
9.2.2.2 Polypseudorotaxanes from Star-Shaped Polymers *359*
9.2.2.3 Polypseudorotaxanes from Side-Chain Polymers *360*
9.3 End-Capping: From Polypseudorotaxanes to Polyrotaxanes *362*
9.4 Polymerization of Pre-assembled Pseudorotaxanes *364*
9.5 Supramolecular Polymerization, Based on CD Recognition *368*
9.5.1 Supramolecular Polymerization of AB-Type Monomers *369*
9.5.2 Supramolecular Polymers, Derived from AA- and BB-Type Monomers *373*
9.5.3 Supramolecular Polymers, Derived from AA-Type Monomers and CDs *374*
9.6 Amphiphilic Supramolecular Diblock Copolymers *375*
9.7 Concluding Remarks *379*
References *380*

10 Supramolecular Polymers, Based on the Host–Guest Chemistry of Pillarenes *387*
10.1 General Aspects *387*
10.2 Host–Guest Complexation Between Pillarenes and Linear Polymers *389*
10.3 Supramolecular Polymers, Derived from Pillarene-based Host–Guest Interactions *391*
10.4 Hyperbranched and Cross-linked Assemblies *401*
10.5 Supramolecular Assemblies, Based on Amphiphilic Pillar[5]arenes *405*
10.6 Concluding Remarks *410*
References *411*

11 Supramolecular Polymers, Formed by Orthogonal Non-covalent Interactions *415*
11.1 Introduction *415*
11.2 Orthogonal Combinations of Supramolecular Interactions Involving Metal-to-Ligand Coordination *417*
11.3 Orthogonal Combinations of Supramolecular Interactions Involving H-Bonding *428*
11.4 Miscellaneous Orthogonal Combinations of Supramolecular Interactions *438*
11.5 Biomimetic Orthogonal Self-Assembly: Protein Recognition *443*
11.6 Concluding Remarks *447*
References *448*

12 Characterization of Supramolecular Polymers *453*
12.1 Introduction *453*
12.2 Estimation of the Molar Mass from the Theories of Supramolecular Polymer Science *454*
12.3 Size-Exclusion Chromatography *460*
12.4 Viscometry *463*
12.5 Light Scattering *465*
12.6 Vapor Pressure Osmometry *467*
12.7 Analytical Ultracentrifugation *468*
12.8 NMR Spectroscopy *470*
12.9 Mass Spectrometry *475*
12.10 Microscopy Imaging *476*
12.10.1 Scanning Probe Microscopy *476*
12.10.2 Electron Microscopy *482*
12.11 Small/Wide-Angle X-Ray Scattering *484*
12.12 X-Ray Crystallography *486*
12.13 Small-Angle Neutron Scattering *487*
12.14 Asymmetric Flow Field-Flow Fractionation *490*
12.15 Taylor Dispersion Analysis *492*
12.16 They Are Very Complex Structures but Totally Timely … *494*
References *495*

Index *503*

Preface

There is a long history of the use of both naturally occurring polymers and synthetic polymers culminating in our current deep understanding of supramolecular polymers. In the 1500s, British explorers discovered that Mayan children were playing with rubber balls made from local trees and, 150 years ago, the first synthetic polymer was made by Wesley Hyatt. He treated cellulose with camphor to create a synthetic ivory to meet the needs of the then rapidly growing billiard enterprise. This year, synthetic polymer chemistry celebrates its 100th birthday, marked by when Hermann Staudinger published his then highly controversial proposal that polymers are indeed long chains, which are formed from repeating molecular units by covalent bonds. Throughout the last century, polymer chemistry has evolved tremendously not only with respect to the design and synthesis of tailor-made architectures but also concerning the wide range of utilitarian applications to be found in our daily lives. By the 1970s, the use of polymer/plastic surpassed that of steel, aluminum, and copper – combined.

In the 1960s, supramolecular polymers were created in which two or more ions or molecules are held together by non-covalent interactions, such as ionic/Coulombic, hydrogen-bonding, and π–π-stacking interactions as well as metal-to-ligand coordination. A wide, diverse group of host–guest (inclusion) complexes was named in this context. Despite their chemically, highly different nature, they offer common characteristics, such as the unique ability to assemble linear polymer chains due to the mostly high directionality of these interactions and, even more importantly, their reversibility of binding. Thus, when incorporated into a polymer backbone, materials are obtained that exhibit properties that cannot be realized by traditional, i.e. covalent polymers.

Today, the fundamental theories that govern supramolecular polymerization reactions are well understood using a broad range of traditional and new analytical techniques, allowing their in-depth characterization both in solution and in the solid state. This overview of the different types of supramolecular polymers is organized according to the non-covalent interactions from which they have been assembled. In each case, the fundamental aspects as well as examples of supramolecular polymers with respect to their synthesis, properties, and potential applications are summarized. We have attempted to cover each field as detailed as possible in order to assist future researchers in this rapidly expanding arena.

Future research on supramolecular polymers will continue to develop and thus, in part, replace traditional historic polymeric materials. It will be important to also make forthcoming materials more biodegradable, lighter, safer, and yet still inexpensive. Hopefully, this overview will help researchers open new pathways to supramolecular systems.

June 2020 *Jena & Jupiter*

Abbreviations

18C6	18-crown-6
A	adenine **or** absorbance
A_2	second virial coefficient
acac	acetoacetate
ACQ	aggregation-caused quenching
AFM	atomic force microscopy
AF4	asymmetric flow field flow fractionation
AIE	aggregation-induced emission
Alq3	tris(8-hydroxyquinolinato)aluminium
ATP	adenosine triphosphate
ATRP	atom transfer radical polymerization
AUC	analytical ultracentrifugation
bdt	1,2-benzenedithiolate
bFGF	basic fibroblast growth factor
Bn	benzyl
BODIPY	boron-dipyrromethene
BMP32C10	bis(*m*-phenylene)-32-crown-10
BPP34C10	bis(*p*-phenylene)-34-crown-10
BSA	Bovine serum albumin
B21C7	benzo-21-crown-7
cac	critical aggregation concentration
CB[*n*]	cucurbit[*n*]uril; *n* = number of glycuril units
CD	circular dichroism **or** cyclodextrin
CDSA	crystallization-driven self-assembly
c_{eff}	effective concentration
Ce6	chlorin-e6
cgc	critical gelation concentration
Ch^+	cycloheptatrienyl cation
ChE	cholinesterase
CIA	calixarene-induced aggregation
CIE	commission Internationale de l'Éclairage
CLSM	confocal laser scanning microscopy
cmc	critical micellar concentration

CNT	carbon nanotube
CONASH	coordination nanosheet
cpc	critical polymerization concentration
CPK (models)	3D molecular models
CS	cold spray
CT	charge transfer
CuCAAC	Cu(I)-catalyzed azide-to-alkyne cycloaddition
CV	cyclic voltammetry
CVD	chemical vapor deposition
DABCO	1,4-diazabicyclo[2.2.2]octane
DBU	1,8-diazabicyclo[5.4.0]undec-7-ene
DB24C8	Dibenzo-24-crown-8
DCC	dynamic covalent chemistry
DEB	diethyl barbiturate
DHP	dihexyldecylphosphonate
DLS	dynamic light scattering
DMAc	dimethyl acetamide
DMF	dimethyl formamide
DMSO	dimethylsulfoxide
DNA	deoxyribonucleic acid
DP	degree of polymerization
DPP	diphenylphenanthrene **or** diketopyrrolopyrrole
DOSY	diffusion-ordered (NMR) spectroscopy
DOX	doxorubicin
DQ	(^{1}H) double quantum
DSC	differential scanning calorimetry
Đ	dispersity
E	molar absorptivity
EDTA	ethylenediamine tetraacetic acid
EPR	electron paramagnetic resonance
EM	effective molarity
ESI	electrospray ionization
exTTF	π-extended tetrathiafulvalene
f	packing factor
f_H	host molar fraction
FAB	fast atom bombardment
Fc^+	ferrocenium cation
FDA	U.S. Food and Drug Administration
FE	field emission
FEB	frequency-domain electric birefringence
FF	fill factor
FRET	Föster-type resonance energy transfer
FRP	free-radical polymerization
FTICR	Fourier-transform ion cyclotron resonance
FT-IR	Fourier-transform infrared

GAL1	Galectin-1
gMS^2	gradient tandem MS^2
ΔG^0	Gibbs free energy
HBC	hexabenzocoronene
HDPE	high-density polyethylene
HEEDTA	hydroxyethylethylenediaminetriacetic acid
HER	hydrogen-evolution reaction
HETPHEN	heteroleptic phenanthroline
HFIP	1,1,1,3,3,3-hexafluoroisopropanol
HG	host–guest complex
HPLC	high-performance liquid chromatography
HOMO	highest occupied molecular orbital
HOPG	highly-ordered pyrolytic graphite
HSAB	hard and soft acid and bases
HSCT	host-stabilized charge transfer
I	scattered intensity
IDP	isodesmic supramolecular polymerization
IM	ion mobility
IR	infrared
ISA	ionic self-assembly
ITC	isothermal titration calorimetry
ITO	indium tin oxide
K_a	association constant
K_d	dissociation constant
L	persistence length of polymer
LB	Langmuir–Blodgett
LC	liquid crystal (or liquid crystalline)
LCD	liquid crystal display
LCST	lower critical solution temperature
LED	light-emitting diode
LSI	liquid secondary ion
LT	low temperature
LUMO	lowest unoccupied molecular orbital
M_n	molar mass
M_w	molecular weight
MALDI	matrix-assisted laser desorption/ionization
MALS	multi-angle light scattering
MAS	magic angle spinning
Mb	myglobin
MDI	methylenediphenyl-4,4′-diisocyanate
mebip	2,6-bis(1-methyl-1H-benzo[*d*]imidazole-2-yl)pyridine
MFP	molecular force probe
MLCT	metal-to-ligand charge transfer
MMLCT	metal-metal-to-ligand charge transfer
MM2	molecular modeling 2

MPEG	PEM monomethyl ether
MOF	metal-organic frameworks
MS	mass spectrometry
MTZ	mitoxantrone
MV^{2+}	methylviologen (i.e. *N*,*N′*-Dimethyl-4,4′-bipyridinium cation)
NAND	"NAND" logic gate
Napy	2,7-diamido-1,8-naphthyridine
NBI	naphthalene bisimide
NEP	nucleation–elongation polymerization
NHC	*N*-heterocyclic carbene
NMP	nitroxide-mediated polymerization **or** *N*-methylpyrrolidone
NMR	nuclear magnetic resonance
Np	naphthalene
ODT	order–disorder transition
OEG	oligo(ethylene glycol)
OF	oligofluorene
OFET	organic field-effect transistor
OPE	oligo(phenylene-ethynylene)
OPV	oligo(phenylene-vinylene)
oxSWCNT	oxidized single-walled carbon nanotubes
P	form factor
P	probability of binding
PAA	poly(acrylic acid)
PAC	polyelectrolyte–amphiphile complex
PAH	polycyclic aromatic hydrocarbon
PANI	polyaniline
PBD	polybutadiene
PBH	poly[(*R*)-3-hydroxybutryic acid]
PBI	perylene bisimide
PBS	poly(butylene succinate)
PC	polycarbonate
PCBA	phenyl(C_{61})butyric acid
PCBM	phenyl[6.6]-C_{61}-butyric acid methyl ester
PCE	power conversion efficiency
PCl	poly(ε-caprolactone)
PDMA	poly(*N*,*N*-dimethylacrylamide)
PDMAEMA	poly(dimethylaminoethyl methacrylate)
PDMS	poly(dimethylsiloxane)
PDT	photodynamic therapy
PE	polyethylene
PEB	poly(ethylene-*co*-butylene)
PEC	polyelectrolyte complex
PEG	poly(ethylene glycol)
PEI	polyethyleneimine
PEK	poly(etherketone)

PEP	poly(ethylene-*co*-propylene)
PET	poly(ethylene terephthalate)
PFGSE	pulse field gradient spin-echo
PFSD	poly(ferrocenyldimethylsilane)
PI	polyisoprene
PIB	polyisobutylene
PIPS	polymerization-induced phase separation
PLA	poly(D,L-lactide)
PLED	polymer-based light-emitting diode
PMA	poly(methyl acrylate)
PMMA	poly(methyl methacrylate)
PMVE	poly(vinylmethyl ether)
PNIPAM	poly(*N*-isopropyl acrylamide)
POM	polarized optical microscopy
PP	polypropylene
PPA	polyphenylacetylene
PPE	poly(phenylene-ethynylene)
PPI	poly(propylene imine)
PPO	poly(propylene oxide)
PS	polystyrene **or** photosensitizer
PSS	poly(styrene sulfonate)
PT	polythiophene
PTBA	poly(tert-butyl acrylate)
PTFMS	poly(*p*-trifluoromethylstyrene)
PTHF	poly(tetrahydrofuran)
Pybox	2,6-bis(oxazol-2-yl)pyridine
P2VP	poly(2-vinylpyridine)
P3HT	poly(3-hexylthiophene)
P3MT	poly(3-methylthiophene)
P3OM	poly(oxotrimethylene)
P4HB	poly(4-hydroxybutyrate)
P4VP	poly(4-vinylpyridine)
Q	heat **or** scattering vector
QTOF	quadrupole time of flight
R_g	radius of gyration
R_h	hydrodynamic radius
RAFT	reversible addition–fragmentation chain transfer
ROMP	ring-opening metathesis polymerization
ROP	ring-opening polymerizations
SAM	self-assembled monolayer
SANS	small-angle neutron scattering
Sav	streptavidin
SAXS	small-angle X-ray scattering
SCNP	single-chain nanoparticle
SEC	size-exclusion chromatography

SEM	scanning electron microscopy
SLS	static light scattering
SMFS	single-molecule force spectroscopy
SPM	scanning probe microscopy
SQUID	superconducting quantum interferase device
STM	scanning tunneling microscopy
SWCNTs	single-walled carbon nanotubes
T	thymine
T_c	critical temperature
T_g	glass transition temperature
T_m	melting temperature
t-Boc	*tert*-butoxycarbonyl
$TATA^+$	4,8,12-triazatriagulenium cation
TBA	tetra(*n*-butyl)ammonium cation
TBN	1,3,5-trinitrobenzene
TCNQ	7,7,8,8-tetracyanoquinodimethane
TDA	Taylor dispersion analysis
TEG	triethylene glycol
TEM	transmission electron microscopy
TFA	trifluoroacetic acid
THF	tetrahydrofurane
TM	tapping mode
TNF	2,4,6-trinitrofluorene
TNT	2,4,6-trinitrotoluene
ToF	time of flight
$TOTA^+$	4,8,12-trioxa-4,8,12,12c-tetrahydrodibenzo[*cd*,*mn*]pyrenium cation
TP	tris(pyrazol-1-yl)borate
TPE	tetraphenylethylene
tpy	2,2′:6′,2″-terpyridine
TTF	tetrathiafulvalene
TWIM	travelling wave ion-mobility
UCST	upper critical solution temperature
Ug	ureidoguanosine
UHV	ultrahigh vacuum
Upy	2-ureido-4-1*H*-pyrimidinone
UV/vis	ultraviolet/visible
VPO	vapor pressure osmometry
WAXD	wide-angle X-ray diffraction
WAXS	wide-angle X-ray scattering
XPS	X-ray photoelectron spectroscopy
XRD	X-ray diffraction
XXR	X-ray reflectivity

About the Authors

Ulrich S. Schubert was born in Tübingen (Germany) in 1969. He studied chemistry in Frankfurt and Bayreuth (both Germany) and at the Virginia Commonwealth University, Richmond (USA). His PhD studies were performed at the Universities of Bayreuth and South Florida. After postdoctoral training with J.-M. Lehn at the University of Strasbourg (France), he moved to the TU Munich (Germany) and obtained his Habilitation in 1999. During 1999–2000, he was Professor at the University of Munich, and during 2000–2007 Full Professor at the TU Eindhoven (the Netherlands). Since 2007, he is a Full Professor at the Friedrich Schiller University Jena, Germany. Prof. Schubert is founder and director of the Jena Center for Soft Matter (JCSM) and the Center for Energy and Environmental Chemistry Jena (CEEC Jena). He received a VICI award of the Netherland Organization for Scientific Research (NWO) and is a fellow of the ACS Division of Polymer Chemistry, Polymer Division (USA), the Royal Society of Chemistry (FRSC, UK), and the National Academy of Inventors (NAI, USA). Prof. Schubert is elected member of the German National Academy of Science and Engineering (acatech) and external scientific member of the Max-Planck-Gesellschaft (MPI for Colloid & Interfaces, Golm). He has currently an h-Index of 102 and since 2014 is listed as ISI "Highly cited researcher."

Andreas Winter was born in Herne (Germany) in 1972 and studied chemistry at the University of Dortmund (Germany) where he graduated in organic chemistry in 1999. In 2003, he received his PhD in chemistry (University of Paderborn, Germany) for work on applications of the Mannich reaction in the synthesis of pyridine derivatives under supervision of Prof. N. Risch and stayed on as a postdoc. Subsequently, in 2005 he joined the group of Prof. U.S. Schubert at the TU Eindhoven, the Netherlands. In 2010, he moved with group from there to the Friedrich Schiller University Jena, Germany. His research is focused on the synthesis of metallo-supramolecular assemblies for opto-electronic and energy-conversion applications.

George R. Newkome received his BS and PhD in Chemistry from Kent State University, then joined Louisiana State University in 1968, and became a full professor in 1978, then Distinguished Research Master in 1982 and Executive Director of the LSU Center for Energy Studies. In 1986, Dr Newkome moved to the University of South Florida as their Vice President for Research and Professor of Chemistry, then President of the USF Research Foundation, and was named Distinguished Research Professor in 1992. In 1995–2000, he was the host and commentator on "Beyond Science," a PBS TV show, and in 2000, he received an Aurora Award for Best of Shows. In 2001, he moved back to Akron, Ohio, as Vice President for Research, Dean of the Graduate School and Professor of Chemistry and Polymer Science, the Oelschlager Professor of Science and Technology at The University of Akron. In 2015, he was awarded the Docteur Honoris Causa from the University of Bordeaux, and in 2016, he was named the Distinguished Alumni Award for Kent State University. He is currently an Affiliate Research Professor in the Center for Molecular Biology and Biotechnology at Florida Atlantic University in Jupiter, Florida. He is a fellow of the AAAS, Royal Society of Chemistry, and National Academy of Inventors as well as on numerous editorial and scientific advisory boards.

1

Supramolecular Polymers: General Considerations*

1.1 Introduction

In 1920, Staudinger proposed that polymers are indeed long chains built from repeating molecular subunits (the so-called monomer), which are linked by covalent bonds [1]. Only a few years later, Carothers developed a classification of macromolecules taking into account both types of compounds that one can polymerize and the polymerization techniques themselves [2]. About 80 years later, this classification is still contemporary and widely utilized to explain the differences between macromolecules made via different polymerization mechanisms (e.g. step-/chain-growth or ring-opening polymerization [ROP]). A broad range of polymers has been prepared over the last few decades, generally following these three basic mechanisms. More recently, the actual controlled/living techniques have widened the synthetic toolbox, but nevertheless, these metathesis, ionic, and radical polymerizations obey Carothers' taxonomy.

The interactions of macromolecules on the molecular level give rise to the materials' properties on the macroscopic scale. Such secondary or non-covalent interactions, commonly referred to as supramolecular interactions, are responsible for the programmed function of natural (e.g. DNA and enzymes) as well as synthetic polymers [3–20]. Moreover, the emerging field of supramolecular chemistry gave answers to the question if covalently linked macromolecules are necessarily required to generate polymeric materials. Considering a supramolecular polymer, i.e. a polymer whose monomeric building blocks are connected by directional and reversible non-covalent bonds, the field of polymer chemistry is no longer limited to conventional (i.e. covalent) macromolecules. The self-assembly of appropriate monomers via moderately strong, reversible, and highly directional non-covalent interactions into linear polymers of high molar mass is known as supramolecular polymerization.

The field of supramolecular chemistry was pioneered by Cram, Pedersen, and Lehn, who received the Nobel prize in chemistry in 1987 for their fundamental

* Parts of this chapter are reproduced from *Chem. Soc. Rev.* **2016**, *45*, 5311-5357 (DOI: 10.1039/c6cs00182c) and *Polymer Science: A Comprehensive Review*, Vol. 5 (Eds.: H.-W. Schmidt, M. Ueda), Elsevier BV, Amsterdam, **2012**, 269-310 by permission from The Royal Chemical Society and Elsevier B.V., respectively.

Supramolecular Polymers and Assemblies: From Synthesis to Properties and Applications, First Edition.
Ulrich S. Schubert, George R. Newkome, and Andreas Winter.

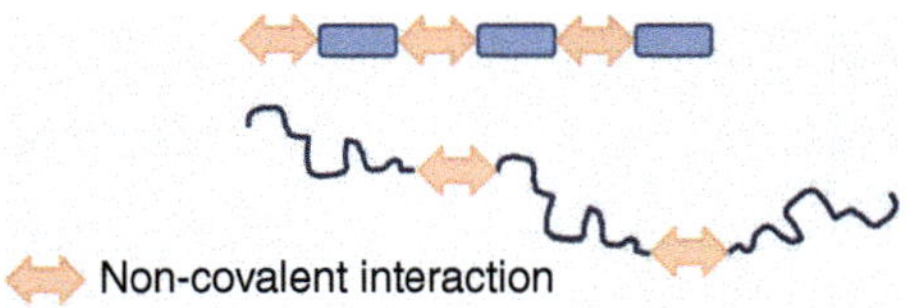

Figure 1.1 Schematic representation of a polymer based on non-covalent interactions.

and groundbreaking studies in this respect. Today, supramolecular chemistry also deals with the formation of larger entities that are beyond the classical molecular covalent chemistry but still involve intermolecular non-covalent interactions [21–23]. Several types of weak interactions are known that can keep the whole assembly together: metal-to-ligand coordination, hydrogen-bonding (H-bonding), halogen-bonding (X-bonding), ionic interactions, π–π-stacking interactions, and host–guest complexation. In principle, two different types of these assemblies can be distinguished: (i) supermolecules, which are well-defined oligomeric structures and (ii) polymolecular assemblies, which consist of a large number of monomeric species. The latter could be covalent molecules, even macromolecules [24]. Therefore, it is apparently difficult to precisely define supramolecular polymers. If the only criterion is that a supramolecular polymer "consists of non-covalently linked monomers," polymeric micelles, even a simple crystal formed by an organic compound, would already be a "supramolecular polymer" [25]. In contrast, Meijer and coworkers coined a new definition [26, 27]: "Supramolecular polymers are defined as polymeric arrays of monomeric units that are brought together by reversible and highly directional secondary interactions, resulting in polymeric properties in dilute and concentrated solution as well as in the bulk. The directionality and strength of the supramolecular bonding are important features of systems that can be regarded as polymers and that behave according to well-established theories of polymer physics."

In simple terms, a long sequence of units (i.e. monomers or polymers) connected by secondary (non-covalent) interactions can be considered as a supramolecular polymer when the overall (macromolecular) structure can also exist in solution or in the melt (Figure 1.1) [28]. The covalent counterparts of the depicted supramolecular polymers are macromolecules, which are formed by polyaddition and polycondensation reactions, respectively.

1.2 Classification Schemes

According to Meijer and coworkers, three main different aspects have to be considered when classifying supramolecular polymers [27]: first, the physical nature of the non-covalent interaction (i.e. the origin/type of the reversible interaction, "physical origin classification"), the type of molecular component or monomers utilized ("structural monomer classification"), and, finally, the Gibbs free energy of the polymer expressed as a function of the monomer conversion ("thermodynamic classification"). It has been under discussion if even a fourth classification scheme can be applied taking into account the dimensionality of the resultant assembly. However,

Table 1.1 The strength and directionality of non-covalent interactions.

Type of non-covalent interaction	Strength ($kJ\,mol^{-1}$)	Directionality
van der Waals interactions	1–5	Low
Charge–transfer interactions	7–20	Medium
Hydrogen-bonding interactions	10–20	High
Hydrophobic–hydrophilic interaction	12–15	Low
Ion-pairing interactions	12–20	Medium
Metal-to-ligand coordination	40–120	High
Covalent bond	150–1000	High

Source: Friese and Kurth [38]. © 2009 Elsevier.

it needs to be emphasized that the addition of a second (or even third) dimension is expected to introduce additional interaction energies which, in turn, will have a considerable influence on the Gibbs free energy (ΔG^0) of the supramolecular materials [26]. Thus, the understanding of how changes in certain parameters, such as concentration and/or temperature, impact the 3D structure of the material is no longer straightforward.

However, supramolecular polymers are typically classified according to the nature of the non-covalent interaction(s) on whose basis they have been assembled from, thus taking into account the main chemical driving force for the formation of large assemblies [26]; several reviews have dealt with these types of materials [29–37]. Non-covalent interactions are intrinsically much weaker (i.e. ranging from c. 1 to 120 $kJ\,mol^{-1}$), in particular when compared to the strong covalent bonds (150–1000 $kJ\,mol^{-1}$). Table 1.1 summarizes the strength of the different interactions, ranging from rather weak (van der Waals and charge–transfer) through medium (H-bonding, hydrophobic–hydrophilic, and ion-pairing interactions) to strong (metal-to-ligand coordination). Accordingly, one can correlate the strength of the supramolecular interaction(s) to the virtual molar mass and, thereby, to the degree of polymerization (DP, *vide infra*).

As a result, from their comparably low strength and non-covalent character, most of these interactions feature a certain degree of reversibility. In other words, the monomeric units of a supramolecular polymer can readily be assembled and disassembled resulting in new, interesting properties and potential applications that are not (or only hardly) accessible for covalent polymers. Utilizing all of these different types of non-covalent interactions allows one to assemble "tailor-made" supramolecular polymers, which reveal the key property, i.e. reversibility of binding [29].

Moreover, with respect to the synthesis of any type of supramolecular polymer, the directionality of the utilized non-covalent bond is of utmost relevance. In classical polymers, the covalent bonds with their perfect directionality represent the linkage of two monomeric units. On the contrary, supramolecular polymers often rely on interactions, which feature a much lower degree of directionality (e.g. ion-pairing or charge–transfer interactions). Only hydrogen-bonding interactions, certain types of metal-to-ligand coordination, and host–guest inclusion complexes offer very high directionalities, which are comparable to those of conventional covalent bonds. In summary, the directionality and strength of the non-covalent interactions represent crucial parameters when dealing with (linear) supramolecular polymers.

Taking the aforementioned "structural monomer classification" scheme, one can distinguish between two cases: first, a single molecular component is equipped with supramolecular binding site of either a self-complementary or a complementary nature. For instance, the reversible association into an $\mathbf{A}_2$-system in solution is based on the self-complementary **A**:**A** interactions (accordingly, an **A**–**B**-type system is formed by reversible complementary **A**:**B** interactions). Second, the two different homoditopic monomers **A–A** and **B–B** may exhibit only one type of interaction; in this case, the supramolecular polymerization process is driven solely by the complementary interactions between **A** and **B**.

Finally, the classification of supramolecular polymerizations according to their thermodynamics has to be considered: for supramolecular polymers, ΔG^0 can be expressed as a function of the monomer conversion from zero ($p = 0$) to full conversion ($p = 1$). Here, the particular mechanism by which a supramolecular polymer is formed from monomeric components has to be considered (this process is dependent on various parameters, e.g. concentration, solvent, and temperature).

1.3 Supramolecular Polymerization Mechanisms

The conventional polycondensation and polyaddition reactions require monomers with (at least) two functional groups by which the covalent bonds are formed. Considering linear polymers, either a single type of monomer bearing two different functionalities (the so-called AB-type monomer) or two different monomers each equipped with two functionalities of the same kind (the so-called homoditopic AA- and BB-type monomers) can basically be employed in this context. Accordingly, supramolecular polymers can be formed by the polyassociation of monomers, which possess (at least) two appropriate binding sites to undergo non-covalent interactions (Figure 1.2) [39]. As for their covalent counterparts, AB-type monomers (e.g. Ic) as well as pairs of AA- and BB-type monomers (Ib) can be utilized for this purpose. In principle, one can further distinguish between two different monomer classes: directly interacting monomers (Ia–c) and those that are connected via metal ions (IIa–c). Supramolecular polymerizations according to type-I typically involve hydrogen-bonding or ionic interactions as well as host–guest complexation (however, ionic interactions can also be involved in type-II polymers where coordinative interactions are involved). The self-complementary monomers (e.g. Ia and IIa) give supramolecular polymers with only a single type of monomeric unit. Accordingly, complementary binding units can be employed to assemble homopolymers (i.e. Ic and IIc) as well as alternating copolymers (i.e. IIb and IIc).

In general, the formation of polymers, via a polycondensation or polyaddition, represents a kinetically controlled process [26]: the potential barrier for the forward

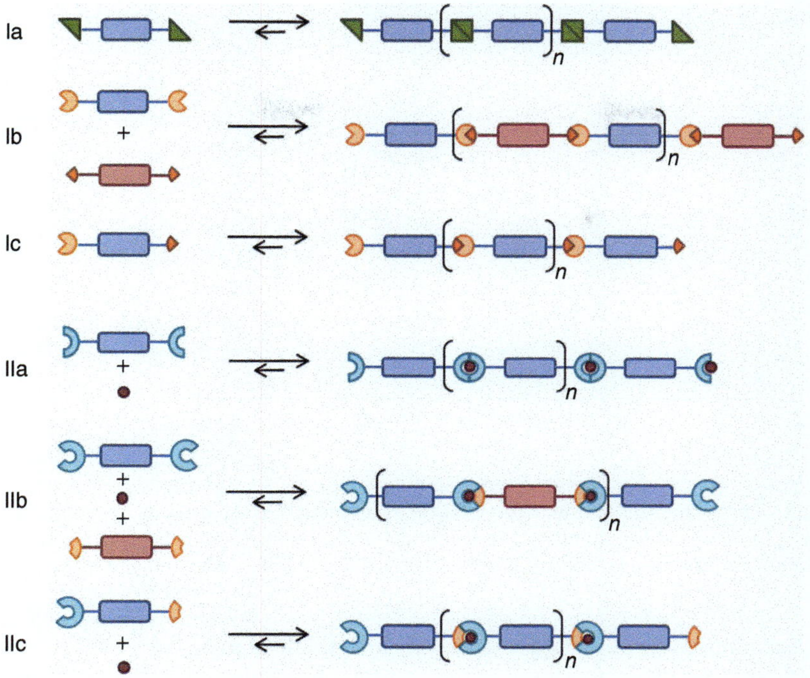

Figure 1.2 Schematic representation of the different types of supramolecular polymerization. Source: Winter et al. [39]. © 2012 Elsevier B.V.

reaction (i.e. the polymerization) is much lower than for the backward reaction (i.e. depolymerization). As a consequence, neither changes in concentration (i.e. dilution) nor temperature (i.e. heating) will affect the molar mass or the degree of conversion. In supramolecular polymerizations, the situation is much more complicated – due to the (partially) reversible character, depolymerization must be considered. Thus, heating or dilution of a kinetically unstable supramolecular polymer will have a tremendous effect with respect to the molar mass – unlike for a common macromolecule, a decrease of M_n will be the result. According to Carothers' equation (Eq. (1.1)), the molar mass of polymers, which have been prepared by either a polyaddition or a polycondensation, can be correlated to the monomer conversion. However, for supramolecular polymers, featuring a certain degree of reversibility of the non-covalent interactions, the molar mass will mainly depend on the strength of these interactions, i.e. their association constants (K_a values). As depicted in Figure 1.3, both high K_a values and high monomer concentrations are required to obtain the supramolecular polymers with high DPs. Besides these considerations, other parameters, such as concentration, temperature, and the particular reaction conditions (i.e. presence of additives or reagents, etc.), will also have an impact on the entire assembly process.

$$DP = 1\,(1-p) \tag{1.1}$$

where DP: degree of polymerization, p: monomer conversion.

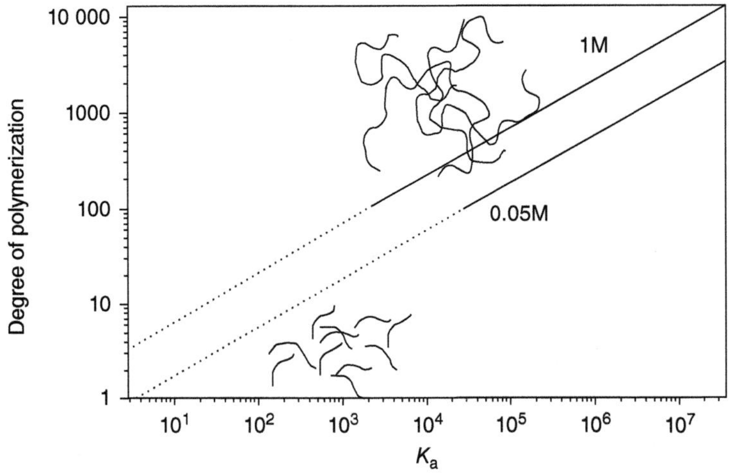

Figure 1.3 Representation of the theoretical DP as a function of the association constant (K_a in M^{-1}) for a typical supramolecular polymerization according to an isodesmic model at two different concentrations. Source: Brunsveld et al. [27]. © 2001 American Chemical Society.

This assumption, however, represents only a simplification of the actual mechanism of a supramolecular polymerization – even when only considering the thermodynamic equilibrium state, where assemblies are formed that, ideally, retain their polymeric structure over time. In this respect, three main mechanisms can be listed (Figure 1.4) [26]:

1. The isodesmic supramolecular polymerization (IDP) is comparable to the step-growth polymerization of esters (i.e. showing a broad dispersity, Đ) and the DP is strongly dependent on the K_a value of underlying supramolecular interaction.
2. The ring-chain-mediated supramolecular polymerization features an equilibrium between supramolecular cyclic species and linear polymer chains.
3. The cooperative supramolecular polymerization shows a nonlinear growth of the polymer chains and is typically nucleated.

Going beyond the "traditional" equilibrium state, supramolecular polymerizations leading to assemblies in either dissipative nonequilibrium or even non-dissipative nonequilibrium states (i.e. kinetically trapped or metastable ones) have recently attracted substantial interest. In these cases, the progress of the polymerization is heavily dependent on the applied preparative method (the so-called pathway selection), and an in-depth knowledge of the kinetics is required. In particular, the association rates for each single step become more important. These issues have recently been discussed by Sorrenti et al. in a tutorial review [40].

1.3.1 Isodesmic Supramolecular Polymerization

The isodesmic supramolecular polymerization (IDP, isos: equal, desmos: bond), also often referred to as the "multi-stage open association" mechanism [28, 41, 42],

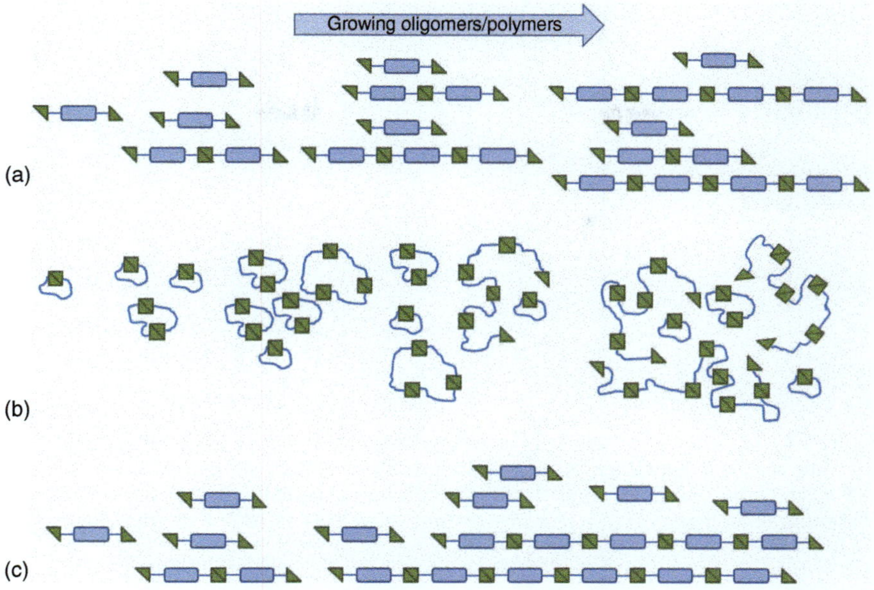

Figure 1.4 Schematic representation of the three main mechanisms known for the supramolecular polymerization processes: (a) isodesmic, (b) ring-chain mediated, and (c) cooperative supramolecular polymerization. Source: Winter et al. [39]. © 2012 Elsevier B.V.

involves the formation of one type of reversible, non-covalent interaction between monomers, oligomers, and eventually even polymer chains (Figure 1.5). All supramolecular bonds, which are formed throughout the entire process, are considered to be identical, and thus, the reactivity of all species present is considered to have the same reactivity (i.e. monomers, oligomers, and polymers). Thereby, the neighboring group effects or additional interactions with non-adjacent sites are neglected. Each single step of the process is characterized by the intermolecular

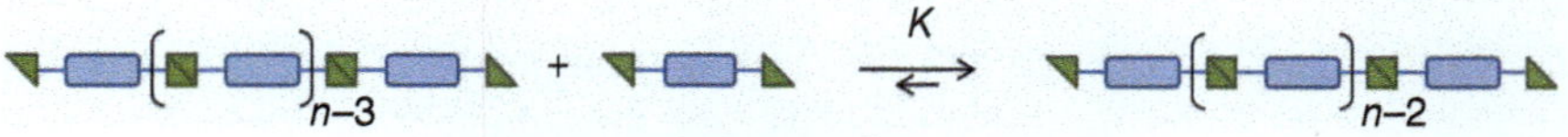

$$[\text{dimer}] = K.[\text{monomer}]^2$$
$$[\text{trimer}] = K.[\text{dimer}].[\text{monomer}] = K^2 \cdot [\text{monomer}]^3$$
$$[n\text{-mer}] = K^{n-1}.[\text{monomer}]^n$$

Figure 1.5 Schematic representation of the IDP in which the intermolecular equilibrium constant (*K*) is independent of the length of the assembly (the mechanism is shown for a bifunctional monomer of the Ia-type, see also Figure 1.2). Source: Winter et al. [39]. © 2012 Elsevier B.V.

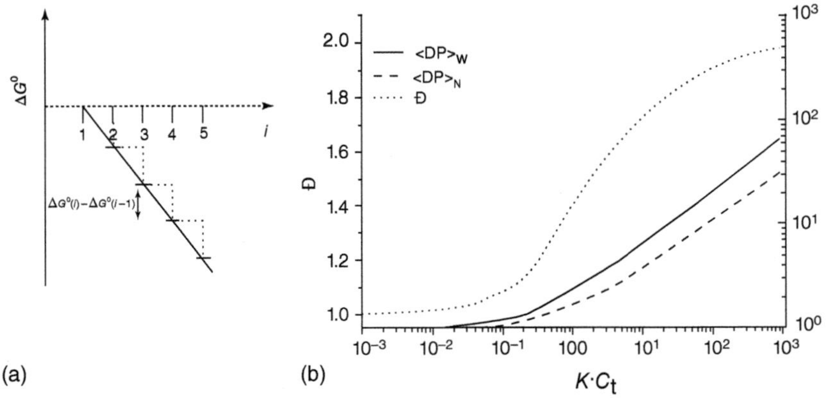

Figure 1.6 (a) Schematic drawing of an energy diagram for an IDP (i: size of the oligomer, ΔG^0: free energy in arbitrary units). (b) Evolution of the number- and weight-averaged DP ($<\mathrm{DP}>_{\mathrm{N}}$ and $<\mathrm{DP}>_{\mathrm{W}}$) and the dispersity (*Đ*) as a function of equilibrium constant and total concentration of monomer ($K{\cdot}c_t$). Source: de Greef et al. [26]. © 2009 American Chemical Society.

equilibrium constant K (Figure 1.5) – regardless of the chain length. As a result, from the equivalence of each individual polymerization step, IDPs do not exhibit any critical values for the concentration or temperature of the supramolecular polymerization (cpc: critical polymerization concentration, cpt: critical polymerization temperature) [41, 43]. Unlike for the ring-chain-mediated polymerization (*vide supra*), no cyclic species can be found during the self-assembly process. The counterpart to IDP in "traditional" polymer science is the step-by-step reversible polycondensation where intramolecular cyclizations are absent and Flory's "principle of equal reactivity" is obeyed [44, 45]. Detailed investigations have shown that, for example, the polycondensation of decanedioyl chloride with 1,10-decamethylene glycol in dioxane meets these requirements [46].

According to the rules of thermodynamics, the free energy of the system constantly decreases when the monomeric units are successively added to the growing polymer chain; this, in turn, further supports the assumption that binding of a monomer to the terminus of a polymer chain is independent of its length (an idealized energy diagram, in which kinetic barriers within the self-assembly process are neglected, is depicted in Figure 1.6a) [26].

The number- and weight-averaged DPs (i.e. $<\mathrm{DP}>_{\mathrm{N}}$ and $<\mathrm{DP}>_{\mathrm{W}}$, respectively) can be derived from the monomer concentration and equilibrium constant K according to Eq. (1.2) (though only valid for $K{\cdot}[\mathrm{monomer}] < 1$) [43]. In the ideal case, *Đ* converges to the limiting value of 2.0, and thus, the monomer concentration approaches $1/K$ (Eq. (1.2)); this scenario is comparable to a standard step-growth polymerization as known from traditional polymer chemistry [27, 33]. The correlation of these parameters with the dimensionless concentration $K{\cdot}c_t$, where K represents the equilibrium constant and c_t the total monomer concentration, is shown in Figure 1.6b. Apparently, high DPs can only be reached for high $K{\cdot}c_t$ values; thus, high monomer concentrations and high K values are both required.

Disadvantageously, the intrinsically poor solubility of monomers often excludes high concentrations, and thus, the equilibrium constants must be very high ($K > 10^6$) to compensate for this when aiming for supramolecular polymers with high molar masses. As a typical feature of IDP-type processes, increasing c_t automatically leads to a gradual and simultaneous increase of the concentration of monomers and polymer chains; thus, the monomer and polymer chains of various length coexist in solution. Finally, the equilibrium concentration of monomers converges to its maximum value, corresponding to K^{-1}, when increasing the concentration further. Thereby, the monomer remains the most abundant species in solution, independent of the values of K and c_t. As for their covalent counterparts, the precise stoichiometry of the functional groups in an IDP represents a prerequisite to obtain polymers with high molar masses: self-complementary AB-type monomers inherently bear the ideal stoichiometry, whereas complementary monomers (i.e. using a combination of AA and BB) require an exact 1 : 1 ratio. Moreover, the molar masses of the resultant polymers can be adjusted by the addition of appropriate chain-stopping agents [47–49].

$$\langle \mathrm{DP} \rangle_\mathrm{N} = \frac{1}{1 - K \cdot [\mathrm{monomer}]} \langle \mathrm{DP} \rangle_\mathrm{W} = \frac{1 + K \cdot [\mathrm{monomer}]}{1 - K \cdot [\mathrm{monomer}]} Đ = 1 + K \cdot [\mathrm{monomer}] \quad (1.2)$$

where <DP>$_\mathrm{N}$: number-averaged DP, <DP>$_\mathrm{W}$: weight-averaged DP, $Đ$: dispersity, K: equilibrium constant, [monomer]: monomer concentration.

Besides the concentration dependency, the influence of the temperature on the IDP also needs to be addressed. Basically, any type of supramolecular polymerization using a bifunctional monomer represents the polymerization of monomers by equilibrium bond formation and features an ideal polymerization temperature ($T_\mathrm{p}{}^0$) [50–54]. The Dainton–Ivin equation, initially introduced to describe the thermodynamics of ROP and polyaddition reactions, correlates the enthalpy and entropy of propagation (ΔH_pr and ΔS_pr) as well as the initial monomer mole fraction to $T_\mathrm{p}{}^0$ (Eq. (1.3)) [55, 56]. There are two fundamental cases that one must distinguish:

1. The polymerization only occurs at a temperature so high that the entropy term exceeds the enthalpy term and the system exhibits a floor temperature (ΔH_pr, $\Delta S_\mathrm{pr} > 0$).
2. The polymerization represents an enthalpically driven process, which is only allowed below a certain ceiling temperature (ΔH_pr, $\Delta S_\mathrm{pr} < 0$).

The so-called polymerization transition line, separating monomer-rich phases from polymer-rich ones, can be constructed by plotting $[M_\mathrm{i}]$ vs. the polymerization temperature, which can be determined experimentally. However, this model is only valid in those cases where a sharp monomer-to-polymer transition can be found (in general, applicable only for ring-opening, living, or cooperative polymerizations) [50]. For most of the reported IDPs, this transition is, however, very broad and the two phases rather coexist. Thus, for such a supramolecular polymerization, the polymerization transition line as a boundary appears less appropriate.

$$T_\mathrm{p}^0 = {}^{\Delta H_\mathrm{pr}}/_{\Delta S_\mathrm{pr}} + R \cdot \ln\left[M_\mathrm{i}\right] \quad (1.3)$$

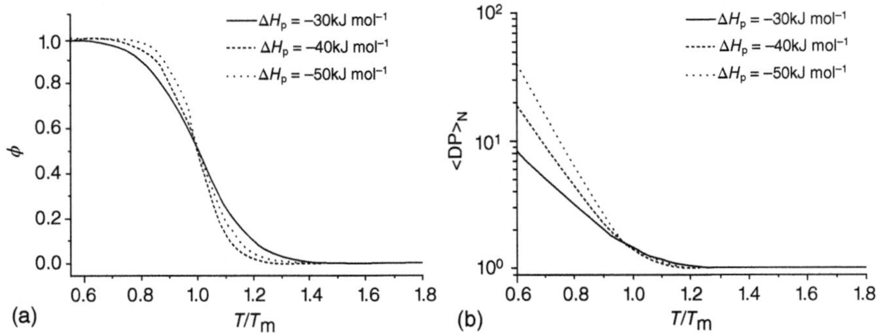

Figure 1.7 Illustration of the characteristic properties of a temperature-dependent IDP according to van der Schoot's model: (a) fraction of polymerized material (ϕ) vs. the dimensionless temperature T/T_m; (b) <DP>$_N$ vs. T/T_m. In both plots, the curves obtained for different enthalpies are shown (ΔH_p = −30, −40, and −50 kJ mol^{-1}, respectively). Source: van der Schoot et al. [57]. © 2005 Taylor & Francis.

where $T_p{}^0$: ideal polymerization temperature, ΔH_{pr}: enthalpy of propagation, ΔS_{pr}: entropy of propagation, R: gas constant, $[M_i]$: initial mole fraction of a monomer.

Historically, the temperature dependency in isodesmic self-assembly processes has been explained by means of statistical mechanics [52]. More recently, mean-field models that are free of restrictions concerning the actual mechanism of chain have been applied for the same purpose. In such models, the chain growth can occur by either the addition of a single monomer or the linkage of two existing chains. van der Schoot proposed a model, where the temperature-dependent melting temperature (T_m; in essence, the temperature at which the monomer mole fraction in the supramolecular polymer is 0.5) and the temperature-independent polymerization enthalpy (ΔH_p) were considered [57]. As one example, a system that polymerizes upon cooling is analysed by plotting the fraction of the already polymerized material (ϕ) against T/T_m for various ΔH_p values; (Figure 1.7a) in such an IDP, the steepness of the transitions of the curves only depends on ΔH_p and contributions arising from cooperativity effects can be excluded. Moreover, a gradual increase of <DP>$_N$ with decreasing temperature can typically be observed (Figure 1.7b).

Dudovich et al. introduced an alternative approach, commonly referred to as the "free association model," which is based on a mean-field incompressible lattice model derived from the Flory–Huggins theory (for the Flory–Huggins model, see [58, 59]). In this approach, the flexibility of the polymer chains and the van der Waals interactions between the monomer and solvent molecules (quantified by the parameter χ) are taken into account [50, 51]. A variety of temperature-dependent properties can be calculated from the lattice model (e.g. <DP>$_N$ and the specific heat at constant volume [C_V]). It has been shown that neither of these (as well as the Đ value) is sensitive to χ when the temperature is changed; however, the situation is different if the χ value for the polymer–solvent interaction is different from the one for the monomer–solvent interaction [50]. On the other hand, a variety of thermodynamic properties do show a strong temperature dependency of χ; these include the osmotic pressure and the critical temperature at which phase

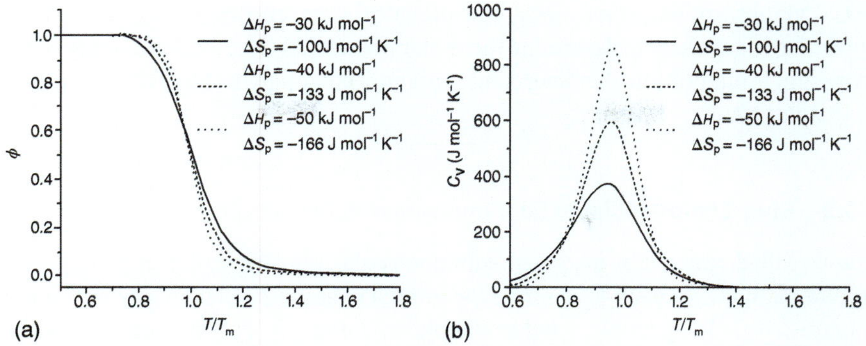

Figure 1.8 Illustration of the characteristic properties of a temperature-dependent IDP according to the "free association" model: (a) fraction of polymerized monomers (ϕ) vs. T/T_m (assuming fully flexible polymer chains and a cubic lattice); (b) heat capacity at constant volume (C_V) vs. T/T_m. In both plots, the curves obtained for various enthalpy ($\Delta H_p = -30$, -40, and $-50\,kJ\,mol^{-1}$, respectively) and entropy values ($\Delta S_p = -100$, -133, and $-166\,J\,mol^{-1}\,K^{-1}$, respectively) are shown; in all cases, the initial volume fraction of the monomers has been set to 0.1. Source: Modified from Dudowicz et al. [50]; Douglas et al. [51].

separation between monomer and solvent occurs. Two free energy parameters describe the reversibility of the supramolecular polymerization: the polymerization enthalpy (ΔH_p) and entropy (ΔS_p), which are both temperature independent. Representatively, the fraction of polymerized monomers (ϕ), as a function of the dimensionless temperature T/T_m, for a system that reversibly polymerizes upon cooling, is shown in Figure 1.8a [51]. In accordance with van der Schoot's model (*vide supra*), the curve is of sigmoidal shape and, with the values of ΔH_p and ΔS_p becoming more negative, the steepness of the curve becomes more pronounced. For fixed monomer concentrations, the C_V vs. T/T_m plots show broad and highly symmetric transition (Figure 1.8b). This feature is indicative of an IDP in which the equilibrium constant K for the addition of each monomer to the growing polymer chain has always the same value. On the other hand, the temperature dependency of C_V in ring-chain or cooperative supramolecular polymerizations shows a much sharper transition (see also Sections 1.3.2 and 1.3.3).

In the field of supramolecular polymers, independent of the nature of the involved non-covalent linkage, their formation via IDP is by far the most common mechanism. Many examples involving hydrogen-bonding (Chapter 3) or host–guest interactions (e.g. by crown ether or calixarene recognition; Chapters 6–10) as well as metal-to-ligand coordination (Chapter 4) are discussed there. It has to be pointed out that the determination of the molar mass of all these supramolecular polymers is generally nontrivial, since the established direct analytical methods commonly used for traditional, i.e. covalent, macromolecules (e.g. size-exclusion chromatography [SEC] or mass spectrometry) can often not be applied due to the weak nature of the supramolecular bonds: already small changes in temperature, solvent composition, and concentration might lead to significant changes of the DP [60, 61]. However, several spectroscopic techniques (e.g. nuclear magnetic resonance (NMR) or UV/vis absorption), calorimetry, and analytical ultracentrifugation (AUC) can be applied in many cases to determine the molar masses [33, 36, 37]. A summary of the scope and limitations in characterizing supramolecular polymers is given separately in Chapter 12.

1.3.2 Ring-Chain-Mediated Supramolecular Polymerization

The so-called ring-chain-mediated supramolecular polymerization represents the second main mechanism to describe the growth of supramolecular polymer chains (Figure 1.4b). In general, a heteroditopic monomer is polymerized reversibly; this monomer as well as its oligomers and, eventually, polymer chains feature an equilibrium between a linear and a cyclic species (Figure 1.9). Ring formation occurs via the intramolecular reaction of the end groups, whereas intermolecular reactions will accordingly give longer chains. Flexibility of the monomer represents, thus, a prerequisite for this type of mechanism; for instance, flexible alkyl or even polymer chains can be used to link the terminal supramolecular binding sites of such a monomer [62]. It is generally accepted that the covalent step-growth polymerization of such monomers typically gives some wt% of macrocyclic oligomers (thereby, the polymerization can be performed under kinetic or thermodynamic control) [44, 63, 64]. Representatively, two classic cases in which also macrocyclic species are formed shall be named briefly: the bulk polymerization of triethylene glycol with hexamethylene–diisocyanate (polycondensation under kinetic control) [65] and, as an example for a thermodynamically controlled process, the catalyzed equilibrium polymerization of α,ω-disubstituted siloxanes (in particular, the later system was widely investigated by Scott [66], Brown and Slusarczuk [67], Carmichael and Winger [68], as well as Flory and Semlyen [69]). The entropically driven ring-opening metathesis polymerization (ROMP) of cyclic olefins [70] and the ring-chain polymerization of liquid sulfur [71–73] are further representatives for covalent ring-chain polymerizations under thermodynamic control. As a general characteristic for a step-growth polymer, the reversibility of bond formation establishes an equilibrium between macrocyclic and linear species. With respect to supramolecular polymers, where the formation/cleavage

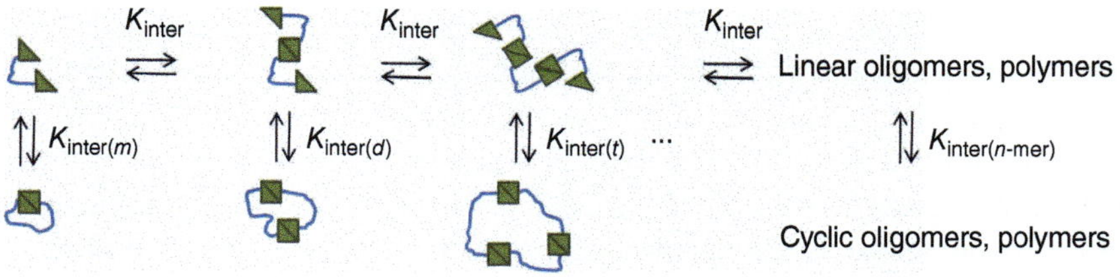

Figure 1.9 Schematic representation of the generalized mechanism of a ring-chain-mediated supramolecular polymerization. The intermolecular binding constants (K_{inter}) are related to the intermolecular association of molecules, whereas the intramolecular binding constant $K_{intra(n\text{-}mer)}$ is assigned to the ring closure of monomers, oligomers, and polymers. Source: Winter et al. [39]. © 2012 Elsevier B.V.

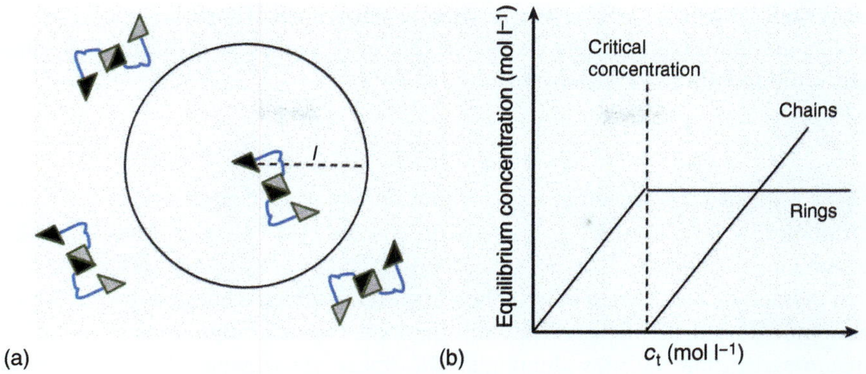

Figure 1.10 (a) Schematic representation of Kuhn's concept of effective concentration (c_{eff}) for a heteroditopic oligomer (i.e. having two different end groups, A and B) [74]. In solution, the end group A will experience an effective concentration of B, if the latter one cannot escape from the sphere of radius *l*, which is identical to the length of the stretched chain. Thus, the intramolecular association between the termini becomes favored for c_{eff} values higher than the actual concentration of B end groups. (b) Illustration of how the equilibrium concentration of chains and macrocycles can be correlated to the total concentration (c_t) of a ditopic monomer in dilute solution; such a ring-chain supramolecular polymerization typically features a critical concentration. Source: de Greef et al. [26]. © 2009 American Chemical Society.

of non-covalent bonds is typically fast, the macrocyclization pathway occurs always under thermodynamic control.

The first model to theoretically quantify the ratio of intra- and intermolecular association was provided by Kuhn already in the 1930s [74]: the effective concentration (c_{eff}) correlates the length of a polymer chain (thereby, taking the mean squared end-to-end distance and assuming Gaussian statistics) with the probability of the end groups to react, i.e. to undergo macrocyclization; the latter one was predicted to decrease by $N^{-3/2}$ (N denotes the number of bonds along the polymer chain [Figure 1.10]).

The toolbox of polymer physics, in particular utilizing random-flight statistics, enables one to calculate c_{eff} as a function of the length of the polymer chain [75]. In reasonably good approximation, the distribution function for random-coil polymers is of Gaussian shape [62]; however, this model only holds true for long, flexible chains [76]. In the same context, a particle-in-a-sphere model was utilized by Crothers and Metzger [77]. In a more realistic approach, Zhou employed a worm-like chain model to determine c_{eff} for short and, thus, semi-flexible polypeptides [78, 79].

For practical reasons, the rather theoretical concept of effective concentration, which basically relies on concentrations calculated from the physical properties of the terminal functionalities, is often replaced by a more empirical concept using effective molarities [80–85]. The effective molarity (EM) is defined as the ratio of intra- and intermolecular equilibrium constants (i.e. K_{inter} and K_{intra}, Eq. (1.4), see also Figure 1.9): cyclization is basically preferred for EM > 1, whereas linear chains are obtained for EM < 1. In addition, *EM can be considered as* a pure entropic correction, which becomes relevant when an intramolecular process replaces the analogous intermolecular one (however, this only applies to unstrained, flexible chains linking the end groups) [86].

$$\mathrm{EM} = \frac{K_{\mathrm{intra}}}{K_{\mathrm{inter}}} \tag{1.4}$$

where EM: effective molarity, K_{intra}: dimensionless equilibrium constant for the intramolecular reaction, K_{inter}: association constant (M^{-1}) for an intermolecular reaction.

In the case of a supramolecular polymerization in which a heteroditopic AB-type monomer is used, *EM* defines the limit monomer concentration below which the (macro)cyclization pathway dominates the linear chain growth. This empirical approach allows one to predict the different cyclization reactions and, even more importantly, gives an absolute measure for a monomer's cyclization ability at the cost of its polymerization (valid only for reversible, non-covalent interactions).

For thermodynamically controlled step-growth polymerizations, Jacobsen and Stockmayer predicted a critical concentration limit [87]: the system is exclusively composed of cyclic species below this value; above this value, an excess of monomer exclusively gives linear chains while the concentration of cyclic species stays constant (Figure 1.10b). These authors related the equilibrium constant for the cyclization to the probability for, thus directly connecting EM and c_{eff}. It was additionally shown that this constant would decrease with $N^{-5/2}$; in other words, a macrocycle composed of N subunits can reopen in N different ways. This study was extended by Ercolani et al., who also considered the size distribution of macrocycles under dilute conditions; thereby, a broad range of K_a values for the supramolecular macrocyclization were taken into account [83]. According to this, only for high K_a values (>10^5 M^{-1}) can a critical concentration limit be observed.

The later model is particularly suited for describing the equilibrium, which is established between cyclic and linear species during a supramolecular polymerization (they are typically conducted in relatively dilute solutions). In contrast to an IDP (*vide infra*), which commonly features K as the only thermodynamic constant, Ercolani's ring-chain model involves two such constants (Figure 1.9): K_{inter} and $K_{\mathrm{intra}(n\text{-mer})}$ (the latter represents the intramolecular binding constant for the n-th ring closure). Considering all cycles as unstrained and obeying Gaussian statistics, the $\mathrm{EM}_{n\text{-mer}}$ values can simply be expressed as a function of EM_1 (EM_1: effective molarity of the bifunctional monomer itself; Eq. (1.5)). An additional aspect that needs to be briefly mentioned is the role of the solvent: thus, volume effects cannot be neglected, and the exponent in Eq. (1.5) needs to be adjusted [62, 88–90].

$$\mathrm{EM}_{n\text{-mer}} = \frac{K_{\mathrm{intra}(n\text{-mer})}}{K_{\mathrm{inter}}} = \mathrm{EM}_1 \cdot \mathrm{DP}^{-5/2} \tag{1.5}$$

where $\mathrm{EM}_{n\text{-mer}}$: effective molarity of the n-mer, $K_{\mathrm{intra}(n\text{-mer})}$: intermolecular binding constant for the n-th ring closure, K_{inter}: association constant (M^{-1}) for an intermolecular reaction, EM_1: effective molarity of the bifunctional monomer.

Due to an additional parameter, which is the critical concentration, the situation becomes even more complex when compared to the previously discussed IDP. To

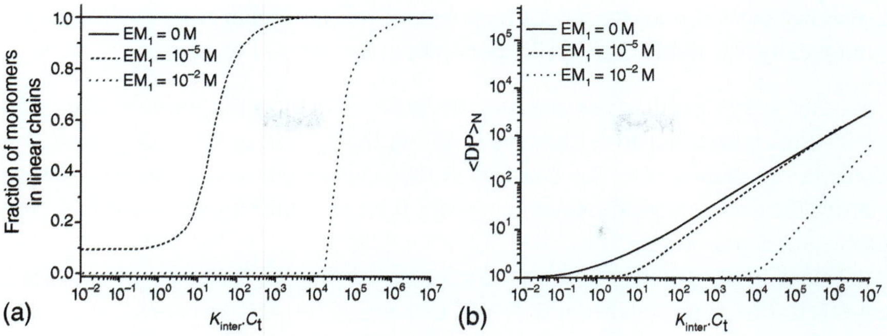

Figure 1.11 (a) Illustration of the fraction of polymerized monomer as a function of $K_{inter} \cdot c_t$ for three different EM_1 values and a fixed value of K_{inter} (10^6 M^{-1}). (b) Illustration of the evolution of $<DP>_N$ as a function of $K_{inter} \cdot c_t$ for various EM_1 values. Source: Flory and Suter [91].

account for this, the monomer fraction in linear species as well as the $<DP>_N$ and $<DP>_W$ values were calculated for a general ring-chain equilibrium, which only involved unstrained macrocycles (i.e. various $K_{intra(1)}$ values were considered, $K_{inter} = 10^6$ M^{-1}) [26, 83]. As shown in Figure 1.11a, the transition between cyclic and linear species at the critical concentration becomes much sharper when $K_{intra(1)}$ is increased. In addition, both $<DP>_N$ and $<DP>_W$ exhibit a steep increase for $c_t > EM_1$ (the sharpness of the transition still depends on $K_{intra(1)}$). In contrast, in an IDP, the DP gradually rises with increasing concentration. However, at high total concentrations, it is no longer possible to distinguish between the different modes of polymerization (i.e. the IDP or the ring-chain equilibrium polymerization), and the obtained DPs are almost identical at given concentrations that are much higher than the EM_1 value (Figure 1.11b).

Dormidontova and coworker addressed the issue of the spacer's rigidity with respect to the ring-chain equilibrium of supramolecular polymers [92]. Applying Monte Carlo simulations on such supramolecular polymerizations, these authors showed that the critical concentration was strongly dependent on the rigidity of the spacer (in these modeling studies, H-bonding interactions were representatively studied). Keeping all further parameters constant (e.g. the length of the spacer or the energy for the interaction of the end groups), the critical concentration decreased in the following order: rigid > semi-flexible > flexible. Thus, for rigid and semi-flexible systems, the probability of their end groups meeting within a bonding distance and, thus, the formation of rings, is much smaller as for flexible systems.

Various groups have reported on critical temperatures in ring-chain equilibria (T_c). These values define the transition between macrocyclic and linear species of high molar mass [71, 72, 93]. Like the supramolecular IDP elaborated in Section 1.3.1, one has to also distinguish two limiting cases for the ring-chain equilibrium polymerization [56]:

1. Above a certain ceiling temperature, polymers of high molar mass are thermodynamically less stable than cyclic monomers or oligomers.

2. Below a certain floor temperature, polymers of high molar mass are thermodynamically less stable than cyclic monomers/oligomers.

In other words, a ceiling temperature can be found in those (supramolecular) polymerizations where negative changes in the enthalpy and entropy of propagation occur; in the second case, the changes in these measures are positive and, consequently, the floor temperature defines the limit below which (supramolecular) polymerization cannot occur.

Covalent ROPs typically involve the opening of strained cycles (e.g. the cationic polymerization of tetrahydrofuran [THF] and dioxolane [42]). In general, such polymerizations represent enthalpy-driven processes for which ceiling temperatures can be observed (basically, all species are of cyclic nature above this value). Very few examples are known for ROPs exhibiting a floor temperature [94]. Examples for such processes that are characterized by a gain in entropy are the ROP of cyclic S_8 in liquid sulfur [93] and the ROMP of unstrained, macrocyclic olefins [70].

Also in the "supramolecular world," the ring-chain equilibrium polymerization is a common feature, independent of the type of employed non-covalent interactions. Representative examples in this respect are the formation of pseudorotaxanes (i.e. the supramolecular polymerization of crown ether derivatives equipped with a pending positively charged amine; Figure 1.12, see Chapter 6) [95–97], the polymerization of poly(dimethylsiloxane)s functionalized in α,ω-position with carboxylic acids (see Chapter 2) [98], and the equilibrium between linear, tape-like, and cyclic structures that can be observed in stoichiometric mixtures of cyanuric acid and melamine derivatives (see Chapter 3) [99].

1.3.3 (Anti)-cooperative Supramolecular Polymerization

The third and last mechanism for supramolecular polymerization to be discussed herein involves (at least) two distinct stages, resulting in a cooperative or an anti-cooperative growth of the polymer chains. At first glance, the mechanism of the cooperative supramolecular polymerization is reminiscent to the one for the IDP; however, the polymerization initially occurs via the reversible binding of monomers to the growing chain (as for the IDP, all these steps basically possess the same equilibrium constant K_n). At a certain DP, a nucleus is formed and, from this point on, the binding of monomers to the polymer chain features an association constant K_e, which is higher than K_n (Figure 1.13). In such a nucleation-elongation polymerization (NEP) model, the supramolecular polymerization proceeds via a

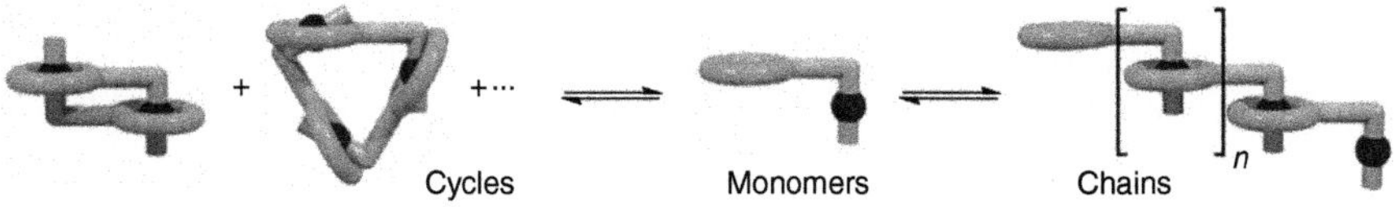

Figure 1.12 Schematic representation of the formation of a poly(pseudorotaxane) via a ring-chain equilibrium. Source: Cantrill et al. [95]. © 2001 American Chemical Society.

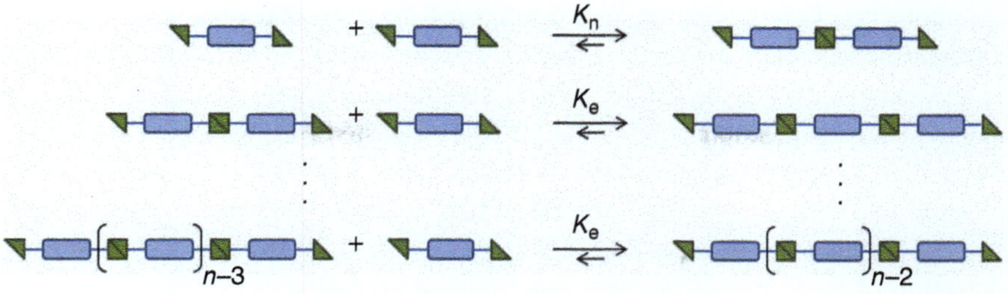

Figure 1.13 Schematic representation of a typical cooperative supramolecular polymerization reaction (nucleation-elongation mechanism). K_n and K_e represent the association constants for the nucleation and the elongation phase, respectively ($K_n < K_e$). Source: Winter et al. [39]. © 2012 Elsevier B.V.

linear IDP. In this elongation phase, the actual association constant is now K_e rather than K_n [26, 43, 100, 101].

The complex thermodynamics of the (anti-)cooperative supramolecular polymerization have already been summarized by de Greef et al.; the reader is referred to this review for a more in-depth discussion [26]. In the following, a few general aspects concerning the different types of cooperative supramolecular polymerization shall be named. First, one can distinguish between the nucleated and the downhill cooperative supramolecular polymerizations. Ferrone defined a nucleated supramolecular polymerization as a process wherein the initial steps of the chain growth are characterized by an increase of ΔG^0 of the oligomers relative to the monomer (Figure 1.14a) [102]. Beyond the point of nucleation, characterized by a maximum in ΔG^0, polymerization becomes energetically favorable. Now, the nucleus represents the least stable and, thus, the least abundant species in the

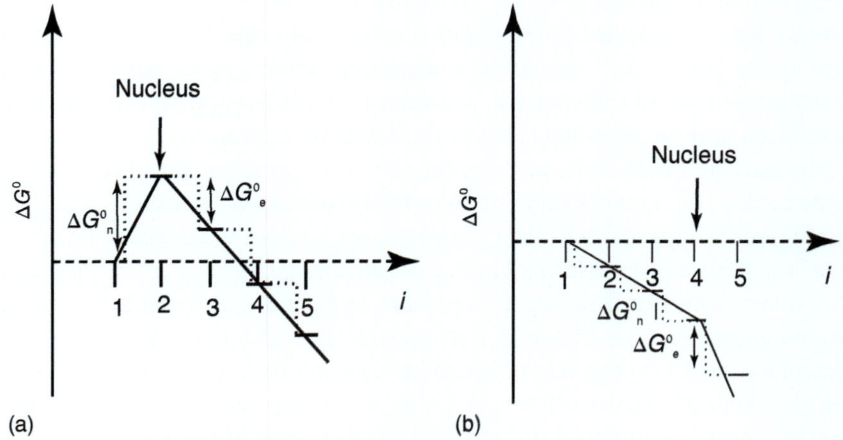

Figure 1.14 Schematic illustration of the energy diagrams of a cooperative nucleated (a) and a cooperative downhill supramolecular polymerization (b). In both plots, the axis of abscissae represents the oligomer's size (*i*), whereas the ordinate measures the ΔG^0 in arbitrary units. In diagram (a), the size of the nucleus is 2 (i.e. dimeric nucleus); in diagram (b), a tetrameric nucleus is depicted. Source: de Greef et al. [26]. © 2009 American Chemical Society.

supramolecular polymerization; as a result, the formation of new polymer chains is retarded (the so-called bottleneck effect). It is widely accepted that formation of the nucleus occurs via homogeneous nucleation (the analysis of various examples of nucleated supramolecular polymerization has suggested this feature). Noteworthy, however, heterogeneous nucleation is also known, but commonly refers to the nucleation processes on foreign substrates [103]: foreign molecules (e.g. impurities) [104–106], external surfaces (e.g. substrates) [107], dust particles, or secondary nucleation of monomers. The latter eventually give a supramolecular polymer on an already existing one. In particular, the latter has been reported to be dominant in various bio-supramolecular polymerizations [26].

In summary, three key criteria can be listed according to Frieden to distinguish between an NEP and an IDP [280]:

1. The supramolecular polymerization process is retarded time dependently;
2. This delay of the polymerization can be compensated by adding a preformed nucleus (i.e. seeding); and
3. An equilibrium between the monomer and the supramolecular polymer is established at a certain critical concentration (or temperature).

In contrast to the cooperative nucleated supramolecular polymerization, the cooperative downhill counterpart does not exhibit any increase of ΔG^0 in the initial steps. Instead, the initial growth of the polymer is characterized by a lower association constant than the following elongation (i.e. $K_n > K_e$; Figure 1.14b). Thus, the monomer is always the species of highest energy in such a cooperative polymerization for which Powers and Powers defined the "nucleus" as the critical chain length at which the absolute ($\mathrm{d}\Delta G^0/\mathrm{d}i$)-increment steeply increases [108]. The distinction between the two aforementioned possibilities for cooperative polymerization is associated to the concentration, and, at high total monomer concentrations, a nucleated polymerization process can even be converted into a downhill one [108, 109]. Concentration-dependent kinetic measurements might, for example, be utilized to distinguish between the two different types of cooperative supramolecular polymerization [108]: in the downhill supramolecular polymerization, the nucleus will be different from the one to be found in a nucleated process (i.e. the nucleus represents a stable or an unstable species, respectively).

For the second type of mechanism, the anti-cooperative supramolecular polymerization, the initial oligomer formation features an association constant that is much higher than the one for the elongation process. So far, the anti-cooperative growth in supramolecular polymerizations has attracted less attention, though discrete objects of low dispersity might be obtained (on the contrary, cooperative growth typically gives supramolecular polymers with high Đ values). For example, Mukerjee [110–114] as well as Tanford [115] reported the formation of large aggregates due to the self-assembly of the surfactants. Due to a high degree of cooperativity in the early stages of the micellar growth, the formation of molecular clusters (i.e. dimers and trimers) was almost fully suppressed. Moreover, (electro)static interactions between the polar head groups of the molecules were identified as the origin of the anti-cooperative effects, affording micelles of finite size.

What is the covalent counterpart to cooperative supramolecular polymerization? In typical chain-growth polymerizations (either ionic or radical), the initiation step is analogous to the formation of the nucleus in the NEP; both systems also feature a sequence of propagation steps; however, a conventional termination is generally absent in supramolecular polymerizations. One can still utilize that numerous features of the living ionic or the radical polymerization explain the cooperative supramolecular polymerization: e.g. in both cases, monomers and polymer chains are present halfway through the polymerization [26]. Furthermore, cooperative effects have also been observed in conventional polycondensation reactions: Flory's "principle of equal reactivity" [44] is disobeyed if the electronic properties of the termini are changed in course of the chain growth, in particular when the reactivity of the polymer end groups becomes higher as for the monomer (i.e. the reaction of monomers becomes retarded). The polycondensation of phenyl 4-(*N*-alkylamino)benzoate in THF under basic conditions, initiated by phenyl 4-nitrobenzoate, represents one example in this respect [116]: the reactivity of the remote phenyl ester moiety is strongly decreased as soon as the monomer is deprotonated; thus, individual monomers will not react with each other. The monomer anion will then only react with the initiator, thus generating an activated species which, in turn, contains phenyl ester moiety with increased reactivity (relative to the anionic monomer). Consequently, only the activated monomers will react, resulting in a polycondensation which shows a chain growth rather than a step-growth behavior [117, 118].

The cooperative supramolecular polymerization can be induced by electronic (as shown in the previous example), hydrophobic, as well as structural effects (e.g. the formation of helical structures). In the latter case, the initial polymerization is thermodynamically rather unfavored when compared to the elongation phase, and thus, polymerization is preferred only when a critical length of the growing polymer chain is reached; then, due to conformational and/or structural changes, the growth of the polymer chain eventually becomes more favorable. As pointed out above, the nucleus is defined as the critical oligomer at which length the elongation of the polymer chain becomes more favored compared to dissociation. If cooperativity is arising (mainly) from structural changes, the nucleus can be regarded as the smallest possible species at which an unstructured, disordered assembly is transformed into an ordered one. A typical example for such a covalent polymerization reaction is the acid-initiated polymerization of isocyanides, generating helical polymers. It has been proposed that the formation of an initial helical oligomer is required that, subsequently, serves as template for the attachment of further monomer units [119]. Remarkably, the anionic polymerization of triphenylmethyl methacrylate, using 9-fluorenyllithium, as initiator, can be conducted asymmetrically when additional chiral ligands are present. Nakano et al. reported the synthesis of a one-handed helical polymer [120]; the reactivity of each anionic oligomer was attributed to the DP and, thereby, to the specific chain conformation. The addition of further monomers to the chain occurred more readily when a stable helical conformation of the oligomer could be adopted (in this case, at a DP of 7–9).

Similarly, the Oya group reported on the cooperative chain growth of synthetic polypeptides via heterogeneous polymerization of amino acid anhydrides [121, 122]. Presumably due to the very short chains, preventing the formation of the anticipated *R*-helical structure, antiparallel β-sheet-type oligopeptide species were initially observed; however, at a DP of c. 8, a conformational change into the *R*-helix induced a more favorable chain growth via the addition of monomers to the active end of the helical structure.

Besides these representatives from the "covalent world," various other examples for cooperative supramolecular polymerizations can be found. In particular, the structural cooperativity of H-bonding interactions has been addressed by the Lehn [123], Meijer [124], and the Würthner groups [125] (see also Chapter 3). Other examples include the aggregation of cationic and anionic porphyrins (see Chapter 2) [126] as well as the stacking of oligo(*p*-phenylene)s bearing dendritic ethylene glycol substituents [127], both driven by strong hydrophobic cooperativity effects. Moreover, metallophilic interactions have recently been identified to be the driving force in the supramolecular polymerization of oligo(phenylene ethynylene)-based Pd(II)–pyridyl complexes [128] or linear Pt(II)–acetylide complexes (Chapter 4) [129]. In contrast, anti-cooperative supramolecular polymerizations are rare: for instance, the polymerization of amphiphilic perylene derivatives in water [130] or cyclic peptides based on α- and ε-amino acids [131].

1.4 Beyond Classical Supramolecular Polymerization

The three aforementioned mechanisms represent the classical examples for how supramolecular polymerization might proceed. Going beyond these, various strategies have been developed to circumvent some limitations arising from the traditional approaches [37]. In particular, the control over the molar mass and the dispersity remain as major challenges. The living supramolecular polymerization aims to adopt the key features from covalent living polymerizations, i.e. good kinetic control over the initiation and propagation steps, and to transfer these into self-assembly processes [132]. Meijer and coworkers identified the NEP process (see Section 1.3.3) as the most appropriate mechanism for this purpose. It was shown that a finely tuned balance between attractive (i.e. a combination of various non-covalent interactions) and repulsive forces (i.e. electrostatic interactions) was crucial to enable control over the supramolecular polymerization [133]. Based on an interplay between isodesmic and cooperative pathways, Ogi et al. realized that the supramolecular self-assembly of a porphyrin monomer equipped with H-bonding entities and hydrophobic alkyl chains into nanofibers of narrow length dispersities ($Đ$ value of 1.10) [134]. In the same context, the "living crystallization," i.e. the seeded growth of block copolymers into micrometer-sized micelles, needs also to be mentioned [135]. In particular, the Manners and Winnick groups employed this strategy to assemble block copolymers in a highly controlled fashion.

A "supramonomer," i.e. a monomer that was formed via non-covalent interactions, can be polymerized using covalent or supramolecular polymerization techniques [37]. The covalent polymerization of preformed metal complexes into metallopolymers represents one of the most common applications in this respect (see also Chapter 4) [36, 136]. However, these polymerizations (as well as other examples, e.g. [137, 138]) do not offer sufficient control over the molar mass (distribution) since the reactions typically follow a step-growth mechanism. Nonetheless, this approach provides an alternative to the direct supramolecular polymerizations and, thus, widens the options to prepare tailor-made polymers that incorporate supramolecular units.

The self-sorting of molecular components due to the selectivity of non-covalent binding represents an established approach to assemble alternating supramolecular polymers, in particular when making use of orthogonal types of secondary interactions (see Chapter 11) [139]. Huang et al. demonstrated that supramolecular self-sorting of two different cucurbit[n]uril derivatives ($n = 7$ and 8), in the presence of an appropriate homoditopic guest, can even proceed with reasonable control (see also Chapter 7) [140]. At an equimolar ratio of all three components, a maximum molar mass (M_w) of 9.7×10^4 g mol^{-1} was obtained; the dispersity of the polymer was c. 1.5 and thus remarkably low for such a supramolecular polymerization.

As one further example, the stimuli-controlled supramolecular polymerization needs to be mentioned. Yang et al. proposed that supramolecular polymerizations, driven by external stimuli, might proceed with good control (assuming optimized reaction conditions) [37]. However, the broad range of stimuli-responsive polymers reported so far basically rely on the ability to reversibly polymerize/depolymerize when, e.g. the pH-switchability of ionic interactions (Chapter 2) or the redox-switchability of host-guest interactions (Chapter 7).

Not only supramolecular polymers are typically assembled under thermodynamic control, but also materials generated in kinetically trapped or metastable states have also been reported [40, 141]. For supramolecular polymers, which were assembled far from equilibrium, unique possibilities have been suggested: inherent dynamic nature, adaptivity, and spatiotemporal controllability [142–144]. Noteworthy, many natural self-assembled systems operate in such states and fulfill crucial functions [145]. Kinetic control of a supramolecular polymerization can be achieved utilizing strong non-covalent interactions; thus, materials in non-dissipative, nonequilibrium states can be formed (Figure 1.15). In such processes, the experimental details are of significant importance since nonlinear phenomena (e.g. nucleation or multiple competitive growth) are known to contribute to the outcome of the supramolecular self-assembly (e.g. the nanoscale morphology of the obtained materials) [40]. Remarkable progress has been made to understand the kinetics and pathway complexity, and, today, the selection of a specific self-assembly pathway is possible by cautious optimization of the experimental details. As a result, one can generate supramolecular materials with different functional properties, even from a single monomer. The fourth case, which is the dissipative nonequilibrium state, is known from living systems. In these supramolecular assemblies, energy is continuously consumed to account for their stability. In Nature, nucleobase triphosphates represent the main energy resources, which can control the nonequilibrium self-assembly process in a highly spatiotemporal fashion – a prerequisite for cells

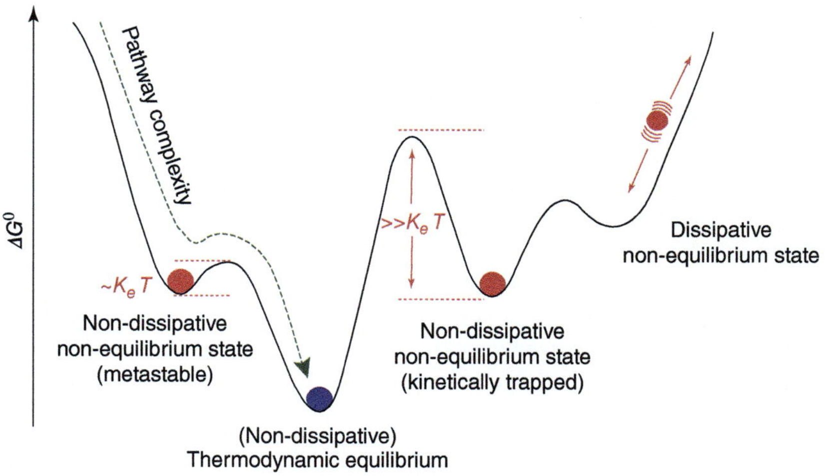

Figure 1.15 Illustration of the various thermodynamic states in supramolecular polymerizations on Gibbs free energy landscape. Source: Sorrenti et al. [40]. Licenced under CC BY 3.0.

to perform their complex functions (e.g. cell division, motility, and intracellular transport) [146]. The evolution of this field of research, an in-depth discussion of the underlying mechanisms, and selection criteria to assemble supramolecular polymers in nonequilibrium states can be found in various recent reviews [40, 145]. In a recent example, Xu and coworkers demonstrated that only a photo-reduced homoditopic viologen monomer gave a supramolecular polymer due to the formation of host–guest complexes with cucurbit[8]uril; whereas, under ambient conditions, depolymerization occurred due to a reoxidative process (see Chapter 7) [147].

1.5 Concluding Remarks

In polymer science, two types of materials are in the focus of research: conventional polymers and, with increasing interest, supramolecular polymers. Whereas the former ones are based on covalent bonds, the latter are formed via assembly of smaller entities by specific directional secondary interactions. Supramolecular polymers exhibit properties that are comparable to those of well-known traditional macromolecules; however, reversibility of the secondary interaction represents an additional feature that gives rise to new applications: supramolecular polymers typically represent species in their thermodynamic equilibrium and their properties can be adjusted by applying external stimuli (e.g. changes in temperature, concentration, or solvent). Moreover, supramolecular polymers in non-dissipative and/or nonequilibrium states also have to be considered and might be important in the future for the fine-tuning of, e.g. shape, molecular organization, chirality, and/or dispersity of supramolecular polymers [40]. These aspects have been proposed to be crucial for utilitarian applications, as in energy conversion or biomedicine areas [148].

The broad range of supramolecular polymers that has been published so far can be classified by mainly two approaches: the type of the secondary interaction involved (as noted in the following chapters) or the mechanism by which they have been formed (in accordance with Carothers' classification from the 1930s [2]). From the three main mechanisms discussed in this chapter, particularly the isodesmic and ring-chain-mediated supramolecular polymerizations are now well understood; however, various effects (e.g. hysteresis and heterogeneous nucleation) make cooperative supramolecular polymerizations much more difficult to understand. Insight into the kinetics and thermodynamics of supramolecular polymers can be gained by comparing these artificial systems to the well-documented protein aggregation. Moreover, one may compare these three mechanisms for supramolecular polymerizations to the three classes of covalent polymerization: step-growth, chain-growth, and ring-opening polymerizations. For covalent polymers, the field of application often dictates the way by which mechanism the polymer may be prepared. The same also holds basically true for the supramolecular polymers.

In recent years, new concepts have evolved addressing the issue on how to control supramolecular polymerizations regarding the molecular structure and even the dispersity of the self-assembled materials. From these, the so-called living supramolecular polymerization represents a highly promising approach to prepare novel, designer, supramolecular materials via control over, e.g. their shape, size, and dispersity.

Going beyond a rather theoretical discussion, various types of supramolecular polymers will be introduced in the subsequent chapters – differentiated by the nature of their underlying non-covalent/supramolecular interactions. Nonetheless, a detailed knowledge of the kinetic and thermodynamic driving forces for the formation of these materials remains a fundamental requirement.

References

1 Staudinger, H. (1920). *Ber. Dtsch. Chem. Ges.* 53: 1073–1085.
2 Carothers, W.H. (1931). *Chem. Rev.* 8: 353–426.
3 Schneider, H.-J. and Strongin, R.M. (2009). *Acc. Chem. Res.* 42: 1489–1500.
4 Shunmugam, R., Gabriel, G.J., Aamer, K.A., and Tew, G.N. (2010). *Macromol. Rapid Commun.* 31: 784–793.
5 Wang, X.-S., Guerin, G., Wang, H. et al. (2007). *Science* 317: 644–647.
6 Schappacher, M. and Deffieux, A. (2008). *Science* 319: 512–1515.
7 Tang, C.-B., Lennon, E.M., Fredrickson, G.H. et al. (2008). *Science* 322: 429–432.
8 Ruokolainen, J., Mäkinen, R., Torkkeli, M. et al. (1998). *Science* 280: 557–560.
9 Pochan, D.J., Chen, Z.-Y., Cui, H.-G. et al. (2004). *Science* 306: 94–97.
10 Cui, H.-G., Chen, Z.-Y., Zhong, S. et al. (2007). *Science* 317: 647–650.
11 Cornelissen, J.J.L.M., Fischer, M., Sommerdijk, N.A.J.M., and Nolte, R.J.M. (1998). *Science* 280: 1427–1430.
12 Fréchet, J.M.J. (2002). *Proc. Natl. Acad. Sci. U.S.A.* 99: 4782–4787.

13 Kato, T., Mizoshita, N., and Kishimoto, K. (2006). *Angew. Chem. Int. Ed.* 45: 36–68.

14 Kato, T., Hirai, Y., Nakaso, S., and Moriyama, M. (2007). *Chem. Soc. Rev.* 36: 1857–1867.

15 Sivakova, S. and Rowan, S.J. (2005). *Chem. Soc. Rev.* 34: 9–21.

16 Pollino, J.M. and Weck, M. (2005). *Chem. Soc. Rev.* 34: 193–207.

17 Weck, M. (2007). *Polym. Int.* 56: 453–460.

18 Percec, V., Dulcey, A.E., Balagurusamy, V.S.K. et al. (2004). *Nature* 430: 764–768.

19 Feldman, K.E., Kade, M.J., de Greef, T.F.A. et al. (2008). *Macromolecules* 41: 4694–4700.

20 Gohy, J.-F. (2009). *Coord. Chem. Rev.* 253: 2214–2225.

21 Lehn, J.-M. (1988). *Angew. Chem. Int. Ed. Engl.* 27: 89–112.

22 Lehn, J.-M. (1990). *Angew. Chem. Int. Ed. Engl.* 29: 1304–1319.

23 Lehn, J.-M. (1993). *Science* 260: 1762–1763.

24 Lehn, J.-M. (2002). *Polym. Int.* 51: 825–839.

25 Ciferri, A. (2005). *Supramolecular Polymers*. New York, NY: Taylor & Francis.

26 de Greef, T.F.A., Smulders, M.M.J., Wolffs, M. et al. (2009). *Chem. Rev.* 109: 5687–5754.

27 Brunsveld, L., Folmer, B.J.B., Meijer, E.W., and Sijbesma, R.P. (2001). *Chem. Rev.* 101: 4071–4098.

28 Ciferri, A. (2000). *Supramolecular Polymers*. New York, NY: Marcel Dekker.

29 Fox, J.D. and Rowan, S.J. (2009). *Macromolecules* 42: 6823–6835.

30 Binder, W.H. and Zirbs, R. (2007). *Adv. Polym. Sci.* 207: 1–78.

31 Bouteiller, L. (2007). *Adv. Polym. Sci.* 207: 79–112.

32 Rieth, S., Baddeley, C., and Badjić, J.D. (2007). *Soft Matter* 3: 137–154.

33 de Greef, T.F.A. and Meijer, E.W. (2010). *Aust. J. Chem.* 63: 596–598.

34 Hofmeier, H. and Schubert, U.S. (2005). *Chem. Commun.*: 2423–2432.

35 Hoogenboom, R., Fournier, D., and Schubert, U.S. (2007). *Chem. Commun.*: 155–162.

36 Winter, A. and Schubert, U.S. (2016). *Chem. Soc. Rev.* 45: 5311–5357.

37 Yang, L., Tan, X., Wang, Z., and Zhang, X. (2015). *Chem. Rev.* 115: 7196–7239.

38 Friese, V.A. and Kurth, D.G. (2009). *Curr. Opin. Colloid Interface Sci.* 14: 81–93.

39 Winter, A., Hager, M.D., and Schubert, U.S. (2012). Supramolecular polymers. In: *Polymer Science: A Comprehensive Review*, vol. 5 (eds. H.-W. Schmidt and M. Ueda), 269–310. Amsterdam: Elsevier BV.

40 Sorrenti, A., Leira-Iglesias, J., Markvoort, A.J. et al. (2017). *Chem. Soc. Rev.* 46: 5476–5490.

41 Ciferri, A. (2002). *Macromol. Rapid Commun.* 23: 511–529.

42 Odian, G.G. (2004). *Principles of Polymerization*, 4e. Hoboken, NJ: Wiley-Interscience.

43 Zhao, D.-H. and Moore, J.S. (2003). *Org. Biomol. Chem.* 1: 3471–3491.

44 Flory, P.J. (1953). *Principles of Polymer Chemistry*. Ithaca, NY: Cornell University Press.

45 Flory, P.J. (1946). *Chem. Rev.* 39: 137–197.

46 Ueberreier, K. and Engel, M. (1977). *Makromol. Chem.* 178: 2257–2260.

47 Chiper, M., Meier, M.A.R., Wouters, D. et al. (2008). *Macromolecules* 41: 2771–2777.

48 Knoben, W., Besseling, N.A.M., and Cohen Stuart, M.A. (2006). *Macromolecules* 39: 2643–2653.

49 Knoben, W., Besseling, N.A.M., and Cohen Stuart, M.A. (2007). *J. Chem. Phys.* 126, Article ID: 2409296.

50 Dudowicz, J., Freed, K.F., and Douglas, J.F. (2003). *J. Chem. Phys.* 119, Article ID: 12645.

51 Douglas, J.F., Dudowicz, J., and Freed, K.F. (2008). *J. Chem. Phys.* 128, Article ID: 224901.

52 Greer, S.C. (2002). *Annu. Rev. Phys. Chem.* 53: 173–200.

53 Greer, S.C. (1998). *J. Phys. Chem. B* 102: 5413–5422.

54 Greer, S.C. (1996). *Adv. Chem. Phys.* 94: 261–296.

55 Dainton, F.S. and Ivin, K.J. (1948). *Nature* 162: 705–707.

56 Dainton, F.S. and Ivin, K.J. (1958). *Q. Rev. Chem. Soc.* 12: 61–92.

57 van der Schoot, P. (2005). Theory of supramolecular polymerization. In: *Supramolecular Polymers* (ed. A. Ciferri), 77–106. London: Taylor & Francis.

58 Flory, P.J. (1942). *J. Chem. Phys.* 10: 51–61.

59 Huggins, M.L. (1942). *Ann. N.Y. Acad. Sci.* 43: 1–32.

60 Lou, X.-W., Zhu, Q.-S., van Dongeren, J.L.J., and Meijer, E.W. (2004). *J. Chromatogr. A* 1029: 67–75.

61 Meier, M.A.R., Hofmeier, H., Abeln, C.H. et al. (2006). *e-Polymers* 6, Article ID: 016.

62 Winnik, M.A. (1981). *Chem. Rev.* 81: 491–524.

63 Semlyen, J.A. (2000). *Cyclic Polymers*. Dordrecht: Kluwer Academic.

64 Kuchanov, S., Slot, H., and Stroeks, A. (2004). *Prog. Polym. Sci.* 29: 563–633.

65 Kricheldorf, H.R. and Schwarz, G. (2003). *Macromol. Rapid Commun.* 24: 359–381.

66 Scott, D.W. (1946). *J. Am. Chem. Soc.* 68: 2294–2298.

67 Brown, J.F. Jr., and Slusarczuk, G.M.J. (1965). *J. Am. Chem. Soc.* 87: 931–932.

68 Carmichael, J.B. and Winger, R. (1965). *J. Polym. Sci., Part A: Gen. Pap.* 3: 971–984.

69 Flory, P.J. and Semlyen, J.A. (1966). *J. Am. Chem. Soc.* 88: 3209–3212.

70 Hodge, P. and Kamau, S.D. (2003). *Angew. Chem. Int. Ed.* 42: 2412–2414.

71 Gee, G. (1952). *Trans. Faraday Soc.* 48: 515–526.

72 Tobolsky, A.V. and Eisenberg, A. (1959). *J. Am. Chem. Soc.* 81: 780–782.

73 Steudel, R., Mäusle, H.-J., Rosenbauer, D. et al. (1981). *Angew. Chem. Int. Ed. Engl.* 20: 394–395.

74 Kuhn, W. (1934). *Colloid. Polym. Sci.* 68: 2–15.

75 Flory, P.J. (1969). *Statistical Mechanics of Chain Molecules*. New York: Wiley-Interscience.

76 Morawetz, H. and Goodman, N. (1970). *Macromolecules* 3: 699–700.

77 Crothers, D.M. and Metzger, H. (1972). *Immunochemistry* 9: 341–357.

78 Zhou, H.-X. (2001). *J. Phys. Chem. B* 105: 6763–6766.

79 Zhou, H.-X. (2003). *J. Mol. Biol.* 329: 1–8.

80 Mandolini, L. (1987). *Adv. Phys. Org. Chem.* 22: 1–111.

81 Page, M.I. and Jencks, W.P. (1971). *Proc. Natl. Acad. Sci. U.S.A.* 68: 1678–1683.

82 Page, M.I. (1973). *Chem. Soc. Rev.* 2: 295–323.

83 Ercolani, G., Mandolini, L., Mencarelli, P., and Roelens, S. (1993). *J. Am. Chem. Soc.* 115: 3901–3908.

84 Galli, C. and Mandolini, L. (2000). *Eur. J. Org. Chem.* 2000: 3117–3125.

85 Kirby, A.J. (2008). *Adv. Phys. Org. Chem.* 17: 183–278.

86 Hamacek, J., Borkovec, M., and Piguet, C. (2006). *Dalton Trans.*: 1473–1490.

87 Jacobsen, H. and Stockmayer, W.H. (1950). *J. Chem. Phys.* 18: 1600–1606.

88 Chan, H.-S. and Dill, K.A. (1989). *J. Chem. Phys.* 90: 492–509.

89 Hiley, B.J. and Sykes, M.F. (1961). *J. Chem. Phys.* 34: 1531–1537.

90 Martin, J.L., Sykes, M.F., and Hioe, F.T. (1967). *J. Chem. Phys.* 46: 3478–3481.

91 Flory, P.J., Suter, U.W., and Mutter, M. (1976). *J. Am. Chem. Soc.* 98: 5733–5739.

92 Chen, C.-C. and Dormidontova, E.E. (2004). *Macromolecules* 37: 3905–3917.

93 Harris, R.E. (1970). *J. Phys. Chem.* 74: 3102–3111.

94 Hodge, P. and Colquhoun, H.M. (2005). *Polym. Adv. Technol.* 16: 84–94.

95 Cantrill, S.J., Youn, G.J., Stoddard, J.F., and Williams, D.J. (2001). *J. Org. Chem.* 66: 6857–6872.

96 Ashton, P.R., Baxter, I., Cantrill, S.J. et al. (1998). *Angew. Chem. Int. Ed. Engl.* 37: 1294–1297.

97 Ashton, P.R., Parsons, I.W., Raymo, F.M. et al. (1998). *Angew. Chem. Int. Ed. Engl.* 37: 1913–1916.

98 Abed, S., Boileau, S., and Bouteiller, L. (2000). *Macromolecules* 33: 8479–8487.

99 Bielejewska, A.G., Marjo, C.E., Prins, L.J. et al. (2001). *J. Am. Chem. Soc.* 123: 7518–7533.

100 Zhao, D.-H. and Moore, J.S. (2003). *J. Am. Chem. Soc.* 125: 16294–16299.

101 Zhao, D.-H. and Yue, K. (2008). *Macromolecules* 41: 4029–4036.

102 Ferrone, F.A. (1999). Analysis of protein aggregation kinetics. In: *Methods of Enzymology*, vol. 309 (ed. R. Wetzel), 256–273. New York, NY: Academic Press.

103 Katshchiev, D. (2000). *Nucleation: Basic Theory with Applications*. Oxford: Butterworth-Heinemann.

104 Wolffs, M., Korevaar, P.A., Jonkheijm, P. et al. (2008). *Chem. Commun.*: 4613–4615.

105 Cabaleiro-Lago, C., Quinlan-Pluck, F., Lynch, I. et al. (2008). *J. Am. Chem. Soc.* 130: 15437–15443.

106 Linse, S., Cabaleiro-Lago, C., Xue, W.-F. et al. (2007). *Proc. Natl. Acad. Sci. U.S.A.* 104: 8691–8696.

107 Rogers, S.S., Krebs, M.R.H., Bromley, E.H.C. et al. (2006). *Biophys. J.* 90: 1043–1054.

108 Powers, E.T. and Powers, D.L. (2006). *Biophys. J.* 91: 122–132.

109 Firestone, M.P., De Levie, R., and Rangarajan, S.K. (1983). *J. Theor. Biol.* 104: 535–552.

110 Mukerjee, P. (1972). *J. Phys. Chem.* 76: 565–570.

111 Mukerjee, P. (1967). *Adv. Colloid Interface Sci.* 1: 242–275.

112 Mukerjee, P., Mysels, K., and Kapauan, P. (1967). *J. Phys. Chem.* 71: 4166–4175.
113 Mukerjee, P. (1969). *J. Phys. Chem.* 73: 2054–2056.
114 Mukerjee, P. (1974). *J. Pharm. Sci.* 63: 972–981.
115 Tanford, C. (1974). *J. Phys. Chem.* 78: 2469–2479.
116 Yokozawa, T., Asai, T., Sugi, R. et al. (2000). *J. Am. Chem. Soc.* 122: 8313–8314.
117 Yokozawa, T. and Yokoyama, A. (2007). *Prog. Polym. Sci.* 32: 147–172.
118 Yokoyama, A. and Yokozawa, T. (2007). *Macromolecules* 40: 4093–4101.
119 Metselaar, G.A., Cornelissen, J.J.L.M., Rowan, A.E., and Nolte, R.J.M. (2005). *Angew. Chem. Int. Ed.* 44: 1990–1993.
120 Nakano, T., Okamoto, Y., and Hatada, K. (1992). *J. Am. Chem. Soc.* 114: 1318–1329.
121 Iwakura, Y., Uno, K., and Oya, M. (1967). *J. Polym. Sci., Part A1: Polym. Chem.* 5: 2867–2874.
122 Komoto, T., Akaishi, T., Oya, M., and Kawai, T. (1972). *Makromol. Chem.* 154: 151–159.
123 Ikeda, M., Nobori, T., Schmutz, M., and Lehn, J.-M. (2005). *Chem. Eur. J.* 11: 662–668.
124 Tomović, Ž., van Dongen, J., George, S.J. et al. (2007). *J. Am. Chem. Soc.* 129: 16190–16196.
125 Kaiser, T.E., Stepanenko, V., and Würthner, F. (2009). *J. Am. Chem. Soc.* 131: 6719–6732.
126 Kano, K., Fukuda, K., Wakami, H. et al. (2000). *J. Am. Chem. Soc.* 122: 7494–7502.
127 Hong, D.-J., Lee, E., and Lee, M. (2007). *Chem. Commun.*: 1801–1803.
128 Mayoral, M.J., Rest, C., Stepanenko, V. et al. (2013). *J. Am. Chem. Soc.* 135: 2148–2151.
129 Wang, X., Han, Y., Liu, Y. et al. (2017). *Angew. Chem. Int. Ed.* 56: 12466–12470.
130 Arnaud, A., Belleney, J., Boué, F. et al. (2004). *Angew. Chem. Int. Ed.* 43: 1718–1721.
131 Horne, W.S., Stout, C.D., and Ghadiri, M.R. (2003). *J. Am. Chem. Soc.* 125: 9372–9376.
132 Mukhopadhyay, R.D. and Ajayaghosh, A. (2015). *Science* 349: 241–242.
133 Besenius, P., Portale, G., Bomans, P.H.H. et al. (2010). *Proc. Natl. Acad. Sci. U.S.A.* 107: 17888–17893.
134 Ogi, S., Sugiyasu, K., Manna, S. et al. (2014). *Nat. Chem.* 6: 188.
135 Gilroy, J.B., Gädt, T., Whittle, G.R. et al. (2010). *Nat. Chem.* 2: 566–570.
136 Ulbricht, C., Beyer, B., Friebe, C. et al. (2009). *Adv. Mater.* 21: 4418–4441.
137 Roy, N., Buhler, E., and Lehn, J.-M. (2013). *Polym. Chem.* 4: 2949–2957.
138 Yang, L., Liu, X., Tan, X. et al. (2014). *Polym. Chem.* 5: 323–326.
139 Li, S.-L., Xiao, T.-X., Lin, C., and Wang, L.-Y. (2012). *Chem. Soc. Rev.* 41: 5950–5968.
140 Huang, Z., Yang, L., Liu, Y. et al. (2014). *Angew. Chem. Int. Ed.* 53: 5351–5355.
141 Chen, H., Huang, Z., Wu, H. et al. (2017). *Angew. Chem. Int. Ed.* 56: 16575–16578.

142 Sorrenti, A., Leira-Iglesias, J., Sato, A., and Hermans, T.M. (2017). *Nat. Commun.* 8: 15899.

143 Mishra, A., Korlepara, D.B., Kumar, M. et al. (2018). *Nat. Commun.* 9: 1295.

144 Leira-Iglesias, J., Tassoni, A., Adachi, T. et al. (2018). *Nat. Nanotechnol.* 13: 1021–1027.

145 van Rossum, S.A.P., Tena-Solsona, M., van Esch, J.H. et al. (2017). *Chem. Soc. Rev.* 46: 5519–5535.

146 Alberts, B., Johnson, A., Lewis, J. et al. (2014). *Molecular Biology of the Cell*, 6e. New York: Norton & Company.

147 Yin, Z., Song, G., Jiao, Y. et al. (2019). *CCS Chem.* 1: 335–342.

148 Aida, T., Meijer, E.W., and Stupp, S.I. (2012). *Science* 335: 813–817.

2

Supramolecular Assemblies Based on Ionic Interactions

2.1 General Aspects

The self-assembly of (macro)molecules into supramolecular architectures of higher order on the basis of ionic (or Coulombic) interactions is intrinsically always an asymmetric process – due to the intrinsically nondirectional attraction between species of opposite charges, either a permanent one or a dipole. Besides the classical combination of ion-to-ion (i.e. ion pairing), ion-to-dipole, and ion-to-quadrupole interactions is also known (e.g. host–guest complexes, such as the coordination of cations to crown ether macrocycles, represent prominent examples in this respect, see Chapter 6) [1]. Besides the nondirectionality of ionic interactions, in particular when compared, for example, to hydrogen-bonding (H-bonding, see Chapter 3) or host–guest interactions (see Chapters 6–10), the scope of this type of non-covalent interaction in the field of supramolecular polymers is limited further by relatively weak binding strength in solution, which are typically in the range of 12–20 kJ mol^{-1} (as noted in Chapter 1, coordinative bonds are significantly stronger, i.e. 40–120 kJ mol^{-1}) [2]. Why might ionic interactions still be of interest for supramolecular chemists? H-bonding interactions, for comparison, are limited in strength: as a rule of thumb, their strength decreases with increasing solvent polarity [3, 4]. Polar solvent molecules are well known to competitively bind to the donor and/or acceptor sites; thus, supramolecular assemblies, based exclusively on H-bonding interactions, only feature reasonably high binding energies in aprotic, barely polar solvents and can scarcely be preserved in water. Moreover, the self-assembly, based on this type of interaction, is highly reversible and can be stimulated by changing the solvent polarity: ionic groups are readily solvated in a polar (or even protic) environment; thus, decreased binding strengths or even dissociation will occur [5]. This limitation can, however, be overcome by two approaches:

1) Combination of ionic interactions with additional (orthogonal) non-covalent interactions (e.g. metal-to-ligand coordination, π–π stacking, and H-bonding interactions) or
2) Utilizing building blocks that show an increased binding strength and directionality due to cooperativity of multiple-ionic interactions.

Supramolecular Polymers and Assemblies: From Synthesis to Properties and Applications, First Edition.
Ulrich S. Schubert, George R. Newkome, and Andreas Winter.

The first approach is discussed separately later; for example, the dimension-controlled self-assembly of π-electronic ions into highly ordered arrays is worth mentioning in this context (see Chapter 5) [6]. Later in this chapter, supramolecular assemblies, based solely on ionic interactions, are summarized [7]. The theory of ionic interactions, in particular with respect to macromolecules, has been reviewed extensively by Ciferri in 2010 [8]; earlier, Faul and Antonietti highlighted the formation of nanostructured supramolecular materials based on this type of binding [9].

2.2 Basic Binding Modes and Discrete Model Assemblies

Natural anion binding systems typically feature a remarkably high substrate affinity and selectivity. Thus, a broad range of synthetic receptors has been developed to mimic such binding modes for the formation of supramolecular assemblies that are stable even in protic environments. In general, N-containing systems are commonly used for this purpose (e.g. pyrroles, protonated or quaternary amines, guanidinium or amidinium salts) [10]. In particular, guanidinium salts are of interest since they remain protonated over a wide pH regime especially when compared to ammonium salts (the pK_a value of the guanidinium cation is 13.5). In the presence of chelating anions, such as carboxylates, a bidentate-like electrostatic interaction is established (Figure 2.1) [13]. The binding constants of such systems were determined by Lehn and coworkers already in the late 1970s [11]: bis-(**1**) and tris-guanidinium (**2**) salts were found to strongly coordinate carboxylates, such as acetate or maleate; whereas, only weak binding was observed for the simple 1,3-dibenzylguanidinium salt ($K_a < 100\,M^{-1}$). Later, Linton and Hamilton determined the association constants for the binding of acetate to various acyclic, mono-, and bicyclic guanidinium salts (**3–5**) by isothermal titration calorimetry (ITC) and ^{1}H NMR (nuclear magnetic resonance) spectroscopy [12]; in general, a decrease of K_a with increasing degree of substitution was observed.

Similar binding studies were also conducted for other systems [13]: protonated amines, quaternary ammonium salts, amidinium salts, etc. On the basis of these fundamental studies, various types of supramolecular architectures have been created. Schmuck reported that the binding of a carboxylate anion to a guanidinium moiety can be enhanced by additional H-bonding units [14, 15]. For example, the derivative **6** featured a strong fixation of *N*-protected α-amino acids in aqueous dimethylsulfoxide (DMSO) solutions (i.e. K_a values in the range of 3.5×10^2 to $5.3 \times 10^3\,M^{-1}$); thus, the side chain of the amino acid determined the binding strength for steric and/or electronic reasons (Figure 2.2).

If the zwitterionic pyrrole derivative **7** was utilized instead, self-assembly into the dimeric species $(\mathbf{7})_2$ occurred – revealing the self-complementary character of the supramolecular building block (Figure 2.3a) [16, 18, 19]. A remarkably high binding strength (K_a values of 10^{12} and $170\,M^{-1}$ in DMSO and water, respectively), resulting from the combination of various weak ionic interactions, was estimated from ^{1}H NMR titration experiments. The dimerization process was found to be

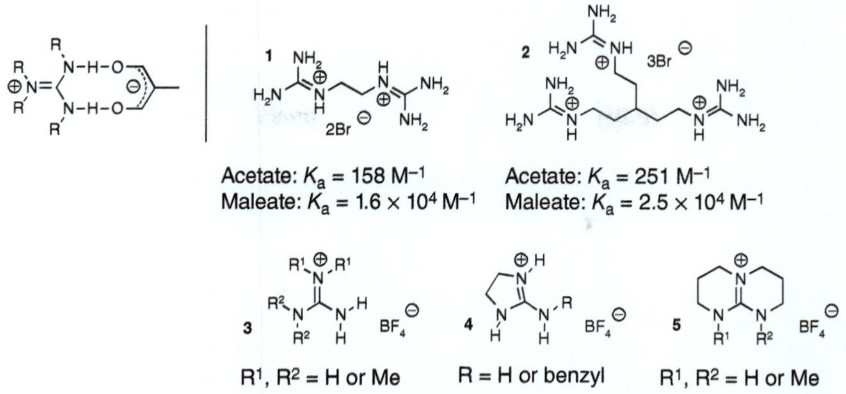

Figure 2.1 Schematic representation of the basic guanidinium–carboxylate interaction as well as of several guanidinium receptor. Source: Dietrich et al. [11]; Linton and Hamilton [12].

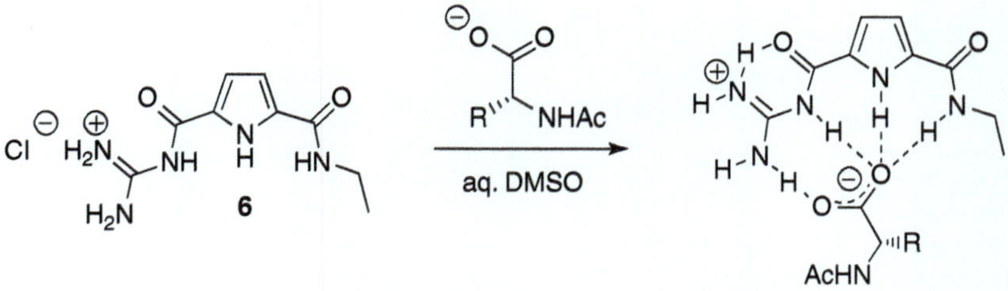

Figure 2.2 Schematic representation of the binding of *N*-acetyl-protected α-amino carboxylates to the receptor **6**. Source: Schmuck [14].

sensitive toward the pH value of the environment; thus, neither the cationic nor the anionic form gave a stable dimer. The same type of intramolecular folding also generated the dimer of the alkyl-bridged derivative **8a** [17]. The alternative intermolecular association was excluded on the basis of detailed temperature- and concentration-dependent ^{1}H NMR spectroscopy measurements. In contrast, using shorter linkers between the terminal binding sites prevented the intramolecular interaction (**8b**, Figure 2.3b). The association mode could be further modulated by simply changing from a 2,5- to a 2,4-substitution pattern on the central pyrrole entity [20]; in the latter case, no intramolecular association could be observed, instead intermolecular assembly into supramolecular oligomers/polymers was detected (*vide infra*).

In general, with an increasing number of such zwitterionic moieties within a molecule, the self-assembly behavior will become more complicated and less predictable. Thereby, the type of nanostructure formed is influenced *inter alia* by the chemical structure of the linker (i.e. its length, flexibility, and chemical nature). For example, a hierarchical self-assembly was observed for a flexible tripodal triple-zwitterion, featuring the structural binding motif of **7**; initially, dimers were

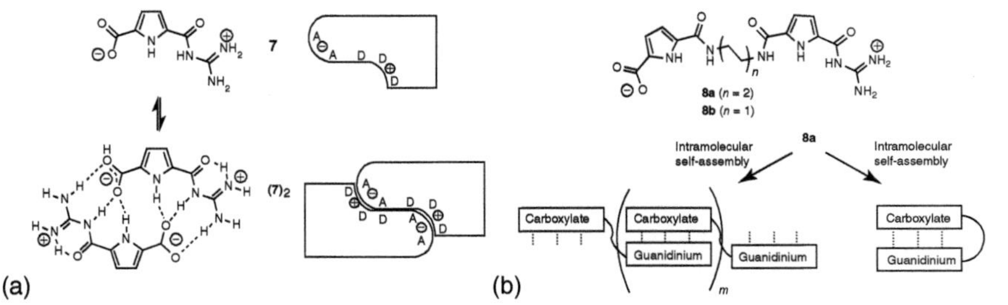

Figure 2.3 (a) Schematic representation of the dimerization of the self-complementary guanidinium derivative **7** [16]. (b) Schematic representation of the supramolecular self-assembly of the heteroditopic derivatives **8**. Source: Schmuck et al. [17]. Figure reproduced with kind permission. © 1999 Wiley-VCH and 2000 American Chemical Society, respectively.

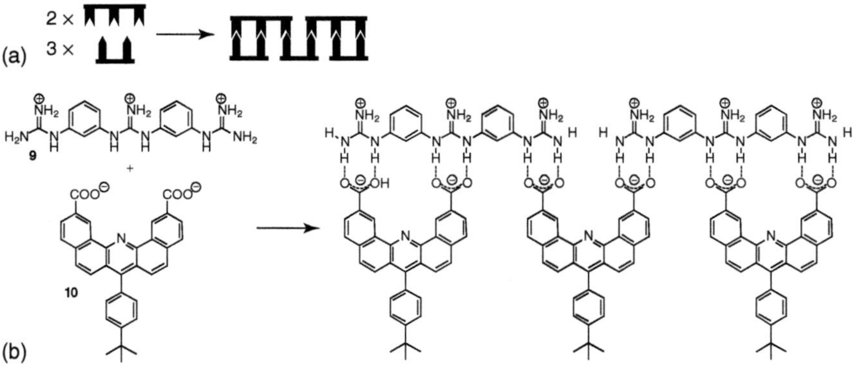

Figure 2.4 (a) Schematic representation of the self-assembly of two complementary components in a Vernier-type fashion (the most straightforward case, i.e. combining a ditopic and a tritopic building block is depicted). (b) Schematic representation of the self-assembly of **9** and **10** into a molecular [2×3]-Vernier motif. Source: Kelly et al. [21].

formed that assembled into 2D ribbons when the concentration was increased. The ribbons, in turn, slowly aggregated into the final supramolecular nanospheres [7].

The intermolecular carboxylate-to-guanidinium binding was employed by Kelly et al. to construct a stable molecular [2×3]-Vernier motif (Figure 2.4) [21]. The spontaneous self-assembly of the linear tris(guanidinium salt) **9** and the rigid biscarboxylate **10** afforded the supramolecular system ($\mathbf{9_2 10_3}$) in high yield (95%) and purity (according to elemental analysis).

Amidinium salts offer a similar binding site, e.g. for carboxylates, as the structurally related guanidinium salts – though, they are less basic and might serve as a stronger H-bond donor upon protonation [13]. A remarkable example for the formation of a supramolecular architecture, based on the interaction of amidinium salts with carboxylates, was reported by von Kiedrowski and coworker in which the strong affinity of acetate to the benzamidinium cation (K_a = 350 M^{-1} in DMSO at 35 °C) was applied as the templating unit in the self-replicating imino condensation of 3-aminobenzamidines **11** with 2-formylphenoxyacetic acids **12** (Figure 2.5)

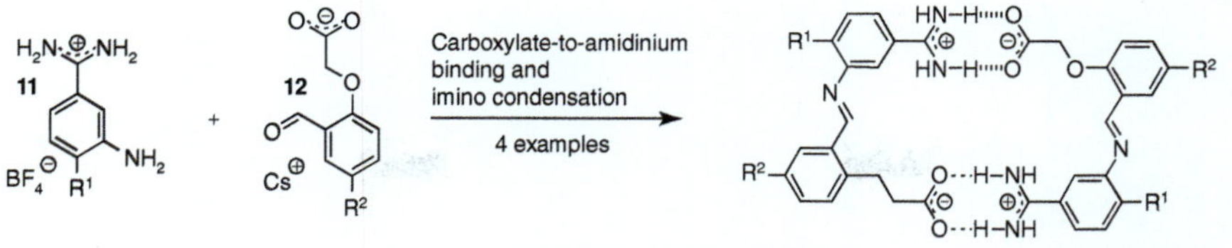

Figure 2.5 Schematic representation of the template-driven self-assembly of **11** and **12** into a supramolecular rectangle. Source: Terfort and von Kiedrowski [22].

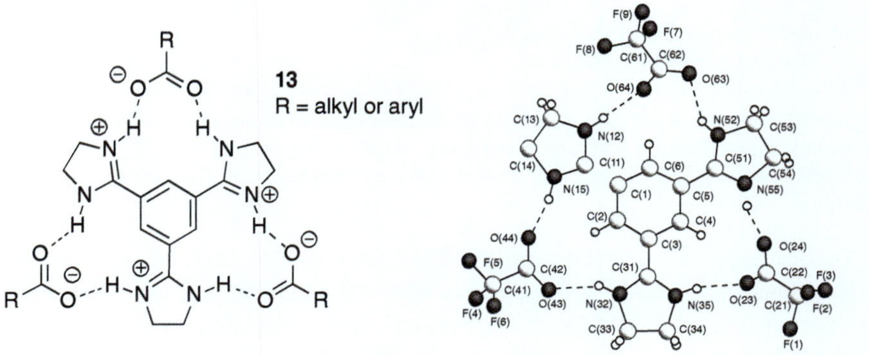

Figure 2.6 Schematic representation of star-shaped assembly **13** and the corresponding X-ray single crystal structure (R = CF_3). Source: Kraft and Fröhlich [23]. © 1998 Royal Chemical Society.

[22]. A star-shaped architecture (**13**) was derived from the binding of three carboxylate molecules to a benzene-centered tris-imidazolinium derivative (Figure 2.6). A K_a value of c. 1000 M^{-1} was calculated on the basis of ^1H NMR spectroscopy in $CD_3OD/CHCl_3$ solutions (R = phenyl). Moreover, single crystals suited for X-ray structure analysis were obtained; thus, the formation of **13** (R = CF_3) was confirmed in the solid state.

An example for the ion pairing between a dicarboxylate and a bis(ammonium salt) was reported by Breslow et al., in which the conformational freedom of the dicarboxylate was reduced upon interaction with the cationic centers of the benzophenone derivative (**14**) [24]. Subsequent photolysis led to an intramolecular reaction of the carbonyl group with one of the CH_2 units in an almost quantitative fashion and with high regioselectivity (according to ^{13}C NMR spectroscopy, 72% of the central CH_2 moieties of the C7 chains were reacting; Figure 2.7).

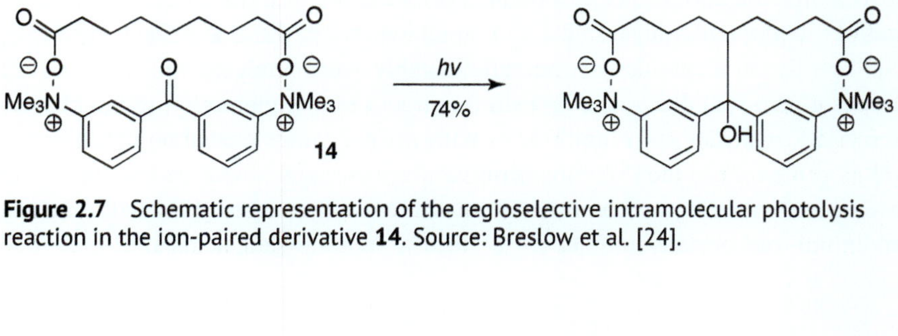

Figure 2.7 Schematic representation of the regioselective intramolecular photolysis reaction in the ion-paired derivative **14**. Source: Breslow et al. [24].

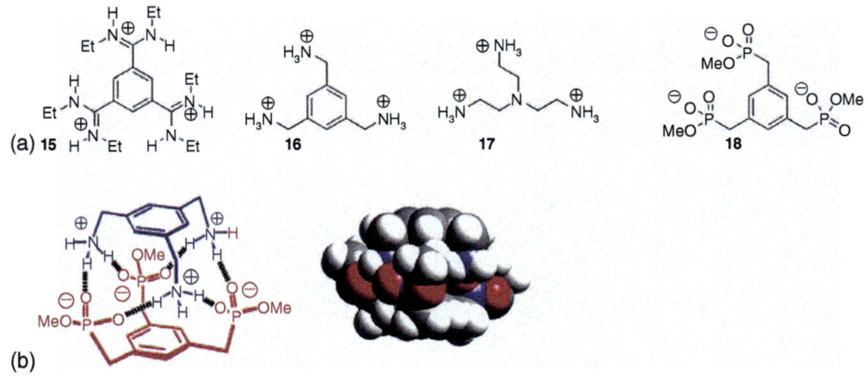

Figure 2.8 (a) Schematic representation of the bowl-shaped triple-ions **15–18**; (b) schematic representation (left) and space-filling model of the ion pair **16 × 18** (right). Source: Grawe et al. [27]. Figure reproduced with kind permission. © 2002 American Chemical Society.

The formation of spherical capsules, based on reversible multiple non-covalent interactions, is of interest with respect to potential applications, e.g. as drug delivery carriers [25, 26]. To realize such systems, at least two complementary half-sphere molecules are required. The requirements that these molecules have to meet are sufficient rigidity, pre-organization of the binding sites, as well as a reasonable binding strength in water [7]. The first aspect facilitates the formation of capsules of defined interiors, whereby a pre-organization of the binding sites enhances the stability of the aggregates (due to a reduced conformational freedom).

Grawe et al. investigated the capsule formation via the hetero-association of the bowl-shaped triple or quadruple cations with appropriate anions [27]. For instance, capsules of 1 : 1 stoichiometry were obtained from the ion-pairing reactions of cations **15–17** with the tris-phosphonate **18** (Figure 2.8). The rigid, pre-organized tris-cations **15** and **16** were binding stronger to **18** than the more flexible derivative **17** (**15**···**18**: $K_a = 1.1 \times 10^6\ M^{-1}$, **16**···**18**: $K_a = 1.4 \times 10^5\ M^{-1}$, **17**···**18**: $K_a = 4.8 \times 10^4\ M^{-1}$, ^1H NMR titration experiments in CD_3OD). However, the same experiments in D_2O revealed a stronger binding of the harder ammonium cations (**16** and **17**) to **18** (the **15**···**18** interaction was weaker by a factor of 4, $K_a = 1.0 \times 10^3\ M^{-1}$).

The Reinhoudt group reported on the hetero-association of two tetra-ionic species – a *meso*-functionalized Zn^{II}-porphyrin derivative and a calix[4]arene-sulfonate system (**19** and **20**, Figure 2.9a/b) [28, 29]. An association constant in the range of $K_a = 10^7\ M^{-1}$ was extracted from UV/vis absorption titration experiments; the appearance of a defined isobestic point in the characteristic Soret band of the porphyrin suggested a 1 : 1 stoichiometry of the assembly (Figure 2.9c) [28]. The thermodynamics of the self-assembly were analyzed via a van't Hoff analysis of the ITC data – the process featured a strongly entropy-driven character (i.e. $\Delta S = (129 \pm 24)\ J\,mol^{-1}\ K^{-1}$) with only a minor enthalpic contribution [$\Delta H = (-8 \pm 6.7)\ kJ\,mol^{-1}$]. This observation was rationalized by the release of solvent molecules upon binding. These same authors also utilized the strong amidinium–carboxylate interaction to form a supramolecular capsule from two

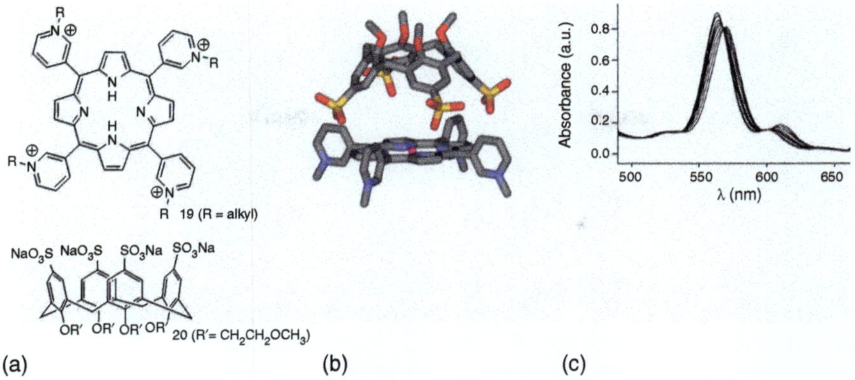

Figure 2.9 (a) Schematic representation of metalloporphyrin **19** and calix[4]arene **20**, as building blocks for supramolecular capsule formation. (b) Illustration of the simulated structure of the capsule (CHARMn 24.0). Source: Rehm and Schmuck [7]. © 2010 Royal society of chemistry. (c) Representative UV/vis absorption titration of **19** with **20** (the evolution of the Soret band of **19** is shown). Source: Fiammengo et al. [28]. © 2002 Elsevier B.V.

complementary half-sphere calix[4]arene systems, functionalized on their upper rims [30]. In contrast to the previous example, the self-assembly process with $K_a = 3.3 \times 10^4\,M^{-1}$ was exothermic ($\Delta H = -13.8\,kJ\,mol^{-1}$, $T\Delta S = 12.1\,kJ\,mol^{-1}$). ^1H NMR spectroscopy and mass spectrometry were further applied to prove that the encapsulation of guest molecules, such as the *N*-methyl-quinuclidinium cation, into the void was possible. The formation of supramolecular capsules due to the dimerization of calix[*n*]arenes, functionalized with ion-bonding moieties on the upper rim, represents an important approach; the utilization of such "supercaps" with respect to supramolecular polymers is detailed in Chapter 8.

A different type of architecture was published by Oshovsky et al., in which two molecules of the cavitand **21**, bearing four pending pyridinium cations, were self-assembled into a capsule via triple-ion interactions with four monovalent anions (e.g. Br^-, NO_3^-, acetate, $TosO^-$; Figure 2.10) [31, 32]. In solution, fast exchange processes between the participating molecules occurred; however, nuclear Overhauser enhancement spectroscopy (NOESY) could be applied to determine the position of the anions (i.e. through-space coupling of the acetates' or tosylates' CH_3-groups with the α,β-protons of the pyridine rings; Figure 2.10). Electrospray ionization mass spectrometry (ESI MS) reflected fragmentation of the supramolecular assembly as well as the presence of host–guest complexes of the stoichiometry $[\mathbf{21}_2Br_6]^{2-}$, where two Br^- anions were encapsulated.

2.3 Supramolecular Polymers, Based on Ionic Interactions

Ciferri suggested a classification scheme for supra(macro)molecular materials, based on ionic interactions, in which the molecular mechanisms by which they

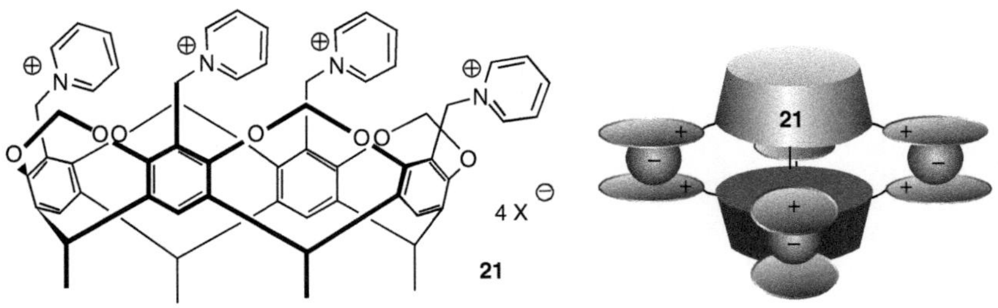

Figure 2.10 Schematic representation of cavitand **21** and its anion-supported self-assembly into a ($\mathbf{21}_2X_4$) capsule (X denotes as monovalent anion). Source: Oshovsky et al. [31]. © 2006 American Chemical Society.

are formed were taken into account [8]. The first class includes materials in which the repulsion (or the attraction) of fixed charges as well as their migration involving mobile ions is responsible for their formation. Polyelectrolytes and polyelectrolyte–surfactant complexes are examples for the repulsive case, whereas polyampholytes, proteins, layered polyelectrolytes, and host–guest complexes are representative for the attractive case. Second, in hydrophobic polyelectrolytes, for example, the charge-independent hydrophobic attraction is increased and can be identified as the major driving force for the supramolecular assembly. The third case, in which the self-assembly process is supported by multivalent mobile ions, as in metallo-supramolecular polymers or metalloproteins, has to be distinguished from the former schemes. The so-called Hofmeister effect, i.e. the ability of simple mono- and divalent ions to coordinate to a polymer via its polar entities, is known to contribute to the overall structure in solution [33]. Finally, charged systems, where low-melting salts represent the charge-carrying species, have to be listed – solid polymeric electrolytes applied for energy sources are representative of this type of material.

However, not all of these systems are of relevance with respect to the scope of this chapter. In the following, three different types of polymeric architectures will be evaluated: supramolecular polymers comprising ionic interactions as integral part of the main chain and, briefly, covalent polymers where molecules have been grafted onto via ionic interactions. Within the scope of this chapter, only materials based on organic molecules will be discussed – the "halatopolymers," which rely on ionic interactions of homotelechelics (e.g. monomers carrying terminal carboxylate moieties) and (divalent) transition metal ions will not be considered herein [34]. They can also be defined as supramolecular polymers based on metal-to-ligand coordination and are, thus, rather a subject of Chapter 4.

2.3.1 Ionic Interactions, as Integral Part of the Polymer Backbone

When aiming for supramolecular polymers where the polymer backbone is held together by non-covalent bonds, i.e. ionic interactions in this particular case, two limitations have to be overcome: (i) directionality and (ii) kinetic lability in polar environments. These issues can, to some extent, be solved by supporting the ionic interaction by further, ideally orthogonal types of non-covalent interactions – this approach is now well established in the field of supramolecular polymers and will be discussed separately (Chapter 11) [35]. Following Ibidapo's classification of ionic polymers, the two main classes of organic building blocks to deal with are (i) polymers and (ii) small molecules [36].

2.3.1.1 Polymers, as Building Blocks

Remarkably, Coulombic interactions have already been considered in polymer blends in the 1950s and, together with the broad field of ionomers, predate the supramolecular polymers – research on this emerged much later in the early 1990s [37–40]. Basically, the pioneering work aimed to enhance the miscibility of components in polymer blends via the transfer of protons from acidic to basic sites that are present in the different polymers [41]. From today's point of view, the materials referred to therein have to be classified as supramolecular graft-block copolymers or polymer networks.

The linking of polymer chains of different chemical structures generates block copolymers. The range of possible architectures includes linear systems (e.g. AB-diblock, ABA- or ABC-triblock or multiblock polymers) as well as nonlinear systems (e.g. graft/star polymers as well as dendritic polymers). Block copolymers are well known for their ability to assemble into nanostructures with well-defined geometries of long-range order, in which the length and the chemical structure of the constituent blocks contribute to the observed morphology: body-centered cubic, hexagonally packed cylinders, and lamellae [42]. Besides covalently linking the polymer blocks, supramolecular interactions have also been applied for the very same purpose [43], and the resulting supramolecular materials are also referred to as pseudo-block copolymers [44] or block-type supramacromolecules [45].

Regarding ionic interactions, pioneering work in the field was performed by Jérôme and coworkers who reported on supramolecular $(AB)_n$-type multiblock copolymer chains in the late 1980s [46–50]. These authors investigated telechelic polystyrenes (PSs) with –COOH or $–SO_3H$ end groups and polyisoprene (PI) end functionalized with two tertiary amine moieties. The solution blending of these polymers allowed for a Brønsted-type acid–base reaction, thereby generating a multiblock copolymer structure. Small-angle X-ray scattering (SAXS) and oscillator shear rheology were applied to study the phase separation of these materials [47, 49]: the binary mixtures of R_2N–PI–NR_2 and PS–COOH exhibited macrophase separation, whereas the combination of R_2N–PI–NR_2 and HOOC–PS–COOH showed microphase separation at room temperature. Upon heating to 130 °C, the latter system underwent macrophase separation attributed to the dissociation of the stabilizing ion pairs (UCST-like behavior, UCST: upper critical solution temperature). Utilizing the bis-sulfonated polymers instead led to a significantly stronger ion pairing as revealed by an increased UCST temperature (c. 170 °C). Moreover, these supramolecular polymers showed order–disorder transitions (ODTs) upon heating prior to macrophase separation; these processes were found to depend mainly on the molecular mass of the block copolymer itself. The proposed phase diagram for the combination of HOOC–PαMS–COOH and H_2N–PI–NH_2, aiming to

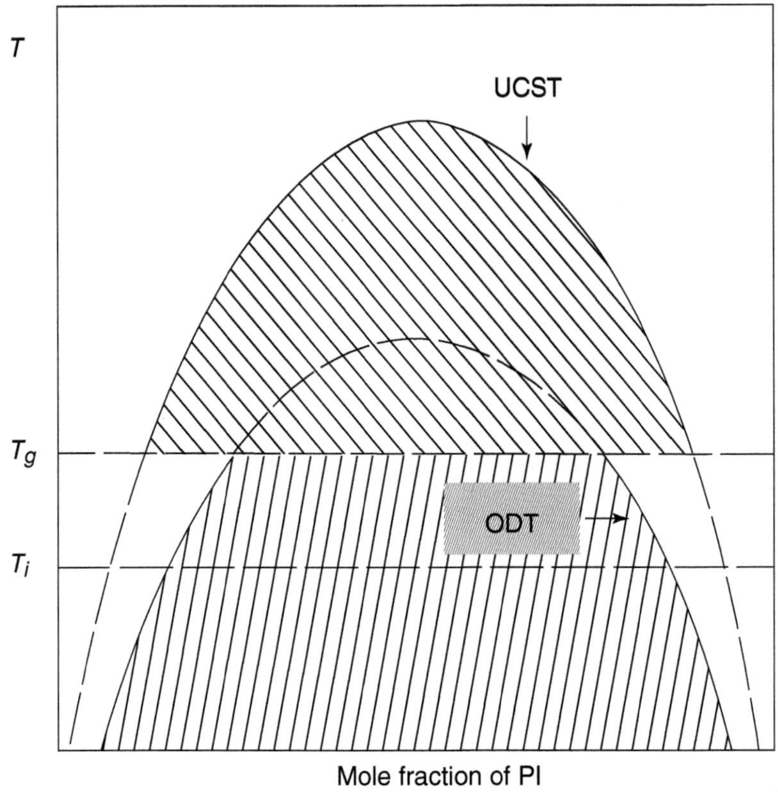

Figure 2.11 Proposed phase diagram for the supramolecular polymer formed by HOOC–PαMS–COOH (M_w = 10 kDa) and H_2N–PI–NH_2PIP (M_w = 18 kDa). The constituent blocks phase separated at the UCST. ODT denotes the regime for the order–disorder transition of the block copolymer structure. T_g and T_i define the glass transition and dissociation temperature, respectively. Microphase separation can be observed in the left-to-right diagonally hatched area; the mixture of telechelic polymers is macroscopically phase separated in the right-to-left diagonally hatched area. In the stippled regime, the copolymer is disordered phase, and finally, the clear area represents a homogeneous mixture of the two constituent polymers. The dashed lines represent the proposed continuations of the curves, which were experimentally not accessible due to ionic aggregation and/or cleavage of the supramolecular bonds. Source: Russell et al. [47]. © 1988 American Chemical Society.

rationalize the experimental results, is shown in Figure 2.11 [47]. Later, rheological studies on the supramolecular block copolymers gave experimental proof for the proposed model [49].

Based on the same binding mode, i.e. proton transfer, a range of further supramolecular block copolymers has been reported. These include the combinations of PS and poly(ethylene glycol) (PEG) [51–54], PS and polyisobutylene (PIB) [55], PS and poly(methyl acrylate) (PMA) [56], as well as PS and PI [45, 57]. Further, proton-transfer processes have been applied for the interfacial modification of immiscible polymer blends [58] or to prepare anti-reflective

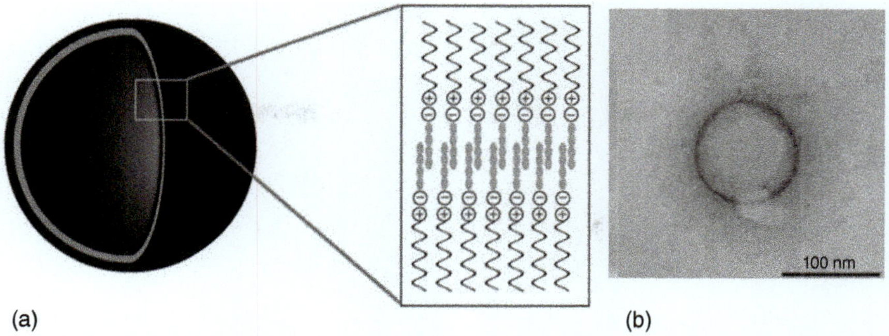

Figure 2.12 (a) Scheme representation of the vesicles formed by the self-assembly of PS-COOH and PNIPAM-NH_2 in aq. dioxane. (b) Representative TEM image of the thusly obtained vesicles. Source: Qian and Wu [64]. Figure reproduced with kind permission. © 2008 American Chemical Society.

coatings [59], structured nanoparticles [60], nanoporous films [61–63], as well as vesicles [64]. The latter were obtained from the self-assembly of PS–COOH and NH_2-functionalized poly(*N*-isopropyl acrylamide) (PNIPAM–NH_2) in aqueous dioxane (Figure 2.12).

A different approach was followed by Wiesner and coworkers who investigated the self-assembly of heterotelechelic block copolymers [65–67]. For this purpose, PS-*b*-PI was end-functionalized with a sulfonate and a trimethylammonium unit. Due to the presence of the ionic substituents, a pronounced change in the material's morphology and phase-separation behavior was observed: the parent PS-*b*-PI system is well known to phase separate as a result of the immiscibility of its constituent blocks in the solid state. However, the salt-free betainic Me_3N^+–PS-*b*-PI–SO_3^- analog did not show this clear behavior. The supramolecular interaction of the terminal ionic moieties enabled the mixing of the intrinsically nonmiscible segments, thereby suppressing any solid-state ordering. In contrast, the monofunctionalized block copolymers with either of the ionic moieties, as end groups, exhibited an even pronounced segregation arising from the aggregation of copolymer chains into a quasi-star block copolymer architecture (Figure 2.13).

Beyond the aforementioned studies dealing with the miscibility/phase separation of block copolymers, ionic interactions between macromolecules have been employed in self-healing or shape-memory materials, where the non-covalent interactions account for the macroscopic properties of the materials. However, in most cases, side-chain polymers are employed in this context. As one example, the "intelligent" supramolecular rubber, as reported by Zhao, Xu, and coworkers is worth mentioning [68]. This polymer network contained covalent as well as non-covalent cross-links (i.e. ionic interactions between sulfonate and ammonium sites) and exhibited such self-healing and shape-memory properties (Figure 2.14).

As an alternative approach, polyoxometalates (POMs), as discrete metal oxide clusters, have been utilized to assemble polymers bearing cationic end groups. In such supramolecular hybrid assemblies, the polyanionic POM serves as a template to organize the cationic polymers into ordered structures via ionic

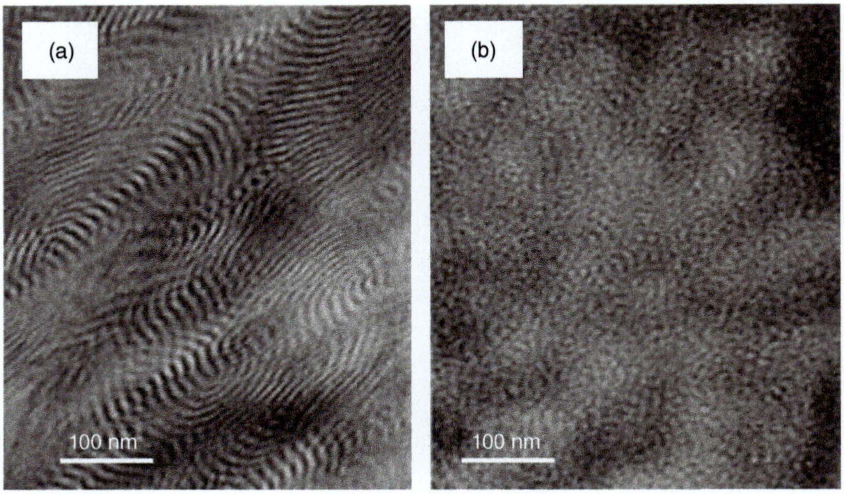

Figure 2.13 Transmission electron microscopy (TEM) images of ionically end-capped PS-*b*-PI: (a) Me_3N^+-PS-*b*-PI; (b) Me_3N^+ - PS - *b* - PI - SO_3^-. Source: Schädler et al. [67]. Figure reproduced with kind permission. © American Chemical Society.

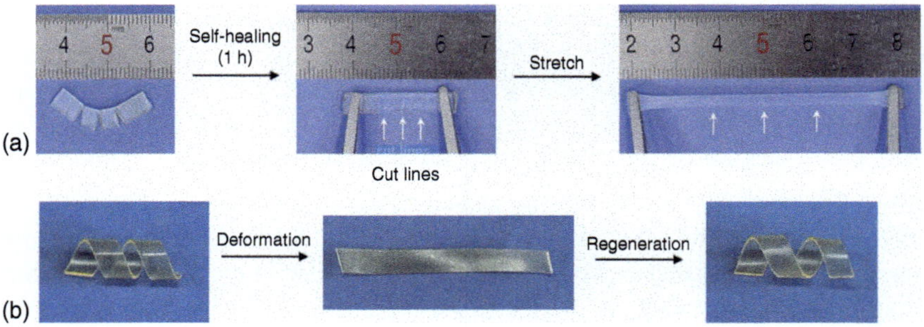

Figure 2.14 Pictures of an "intelligent" supramolecular rubber, which exhibited (a) self-healing and (b) shape-memory properties. Source: Wang et al. [68]. Figure reproduced with kind permission. © 2015 The Royal Chemical Society.

interactions. The first example in this respect was reported by Bu and coworkers who employed the giant Keplerate-type POM and PS, end-functionalized with an 11′-(*N*,*N*,*N*-trimethylammonium)-undecanylisobutyrate group (Figure 2.15) [69]. The electrostatic interactions gave star-shaped polymers in which the number of arms could be controlled via the size of the PS, i.e. in the case of shortest PS (degree-of-polymerization [DP] of 124), c. 39 of such chains were bound to the POM core, according to elemental analysis. The multi-arm, star-shaped polymers gave, in $CHCl_3$/MeOH solutions, vesicles, which were analyzed by microscopy imaging and dynamic light-scattering (DLS) measurements (Figure 2.15). In a similar fashion, the $\{EuW_{10}O_{36}\}^{9-}$ POM was used, as a polyanionic core, to assemble PSs bearing a cationic end group [70]. Langmuir monolayers and drop-cast films for the hybrid materials were produced in which the distance between the POM cores could be adjusted via the size of the PS arms. As a result, from the well-separated ordering of

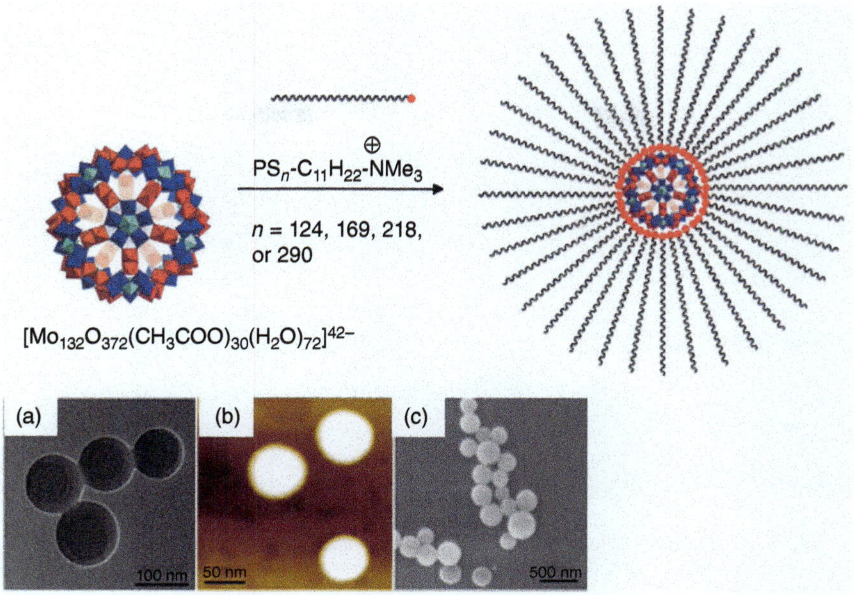

Figure 2.15 Schematic representation of the self-assembly of a multi-arm, star-shaped polymers containing a POM, as polyanionic core (counterions omitted for clarity). Representative (a) TEM, (b) AFM, and (c) Scanning electron microscopy (SEM) images of the vesicles formed by the star-shaped polymers are also shown ($n = 169$). Source: Zhang et al. [69]. Figure reproduced with kind permission. © 2012 The Royal Chemical Society.

the POMs, bright luminescence of high color purity was observed. These authors also assembled star-shaped architectures from the same luminescent POM core and neutral–cationic diblock copolymers, comprising a hydrophilic PEG and a polycationic poly(*N*,*N*-dimethylaminoethyl methacrylate) (PDMAEMA) block [71]. The emission intensity was found to strongly depend on the applied charge ratio between the POM and the copolymer; compared to the emission of the parent POM in aqueous solution, the emission intensity was increased by a factor of 20 at a proper POM-to-polymer ratio.

A different synthetic strategy for the synthesis of POM-centered supramolecular polymers was followed by Li and coworkers [72]. These authors assembled two cationic chain-transfer agents to the Lindqvist-type hexatungstate anion and employed this in the reversible addition–fragmentation chain-transfer (RAFT) polymerization of styrene. Thus, two PS chains were grown from the surface of the POM anion in a highly controlled fashion, as concluded from the analysis of the as-grown polymer chains by size-exclusion chromatography (SEC) and mass spectrometry (remarkably, the dispersities [$Đ$] of the polymers were <1.1). Due to their amphiphilic nature, the supramolecular polymers assembled into spherical nanostructures in water, in which the molecular weight of the PS chains governed the final morphologies: at a DP of 26, vesicles with an average hydrodynamic radius (R_h) of 90 nm were found, whereas for the larger PS chains (DP of 67), the formation of micelles was favored (R_h of 6 nm) (Figure 2.16).

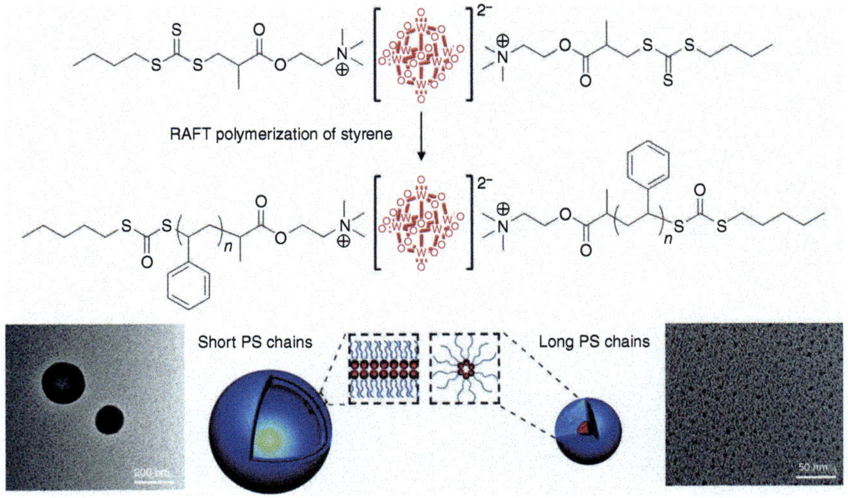

Figure 2.16 Schematic representation of POM-centered supramolecular polymers via a surface-started RAFT polymerization. As shown by representative TEM images, the morphology of the self-assembled nanostructure depended on the length of the PS chains. Source: Cao et al. [72]. Figure reproduced with kind permission. © 2016 The Royal Chemical Society.

Besides these highly ordered supramolecular architectures, POMs have also combined with block copolymers to generate nanostructures, such as nanosheets, nanorods, micelles, and vesicles [73–75]. For example, core-corona assemblies were obtained from a mixture of PS-*b*-poly(4-vinyl-*N*-methylpyridinium iodide) and the Keggin-type $\{\alpha\text{-}PW_{12}O_{40}\}^{3-}$ anion [73]. Moreover, the solid-state morphology of block copolymers can be modulated due to electrostatic interactions with added POMs. This was demonstrated by Wu, Li, and coworkers who investigated the changes in the morphology of PS-*b*-P4VP diblock copolymers when the Keggin-type $H_4SiW_{12}O_{40}$ polyoxoacid was added (P4VP: poly(4-vinylpyridine)) [76]. The initially hexagonally packed cylindrical morphology of PS_{488}-*b*-$P4VP_{95}$ was converted into a disordered bicontinuous morphology due to the protonation of the P4VP block and electrostatic interactions (Figure 2.17). As a result, the glass transition of the P4VP block was absent, and the nanocomposites showed a reinforcement of their mechanical properties. However, PS_{1057}-*b*-$P4VP_{95}$, showing a disordered spherical morphology, was not transformed by the polyanion, presumably due to its insulating spherical microstructure. The concept was applied also to other diblock copolymers, such as PS-*b*-P2VP (P2VP: poly(2-vinylpyridine)) [77, 78].

2.3.1.2 Small Molecules, as Building Blocks

In contrast to the utilization of ionic interactions to link appropriately functionalized polymers in a supramolecular fashion, the supramolecular polymerization of small organic molecules via this type of secondary interactions has up-to-now attracted only little attention. On the other hand, many examples for stable discrete assemblies, based on ion pairing, have been reported in literature (e.g. see Section 2.2).

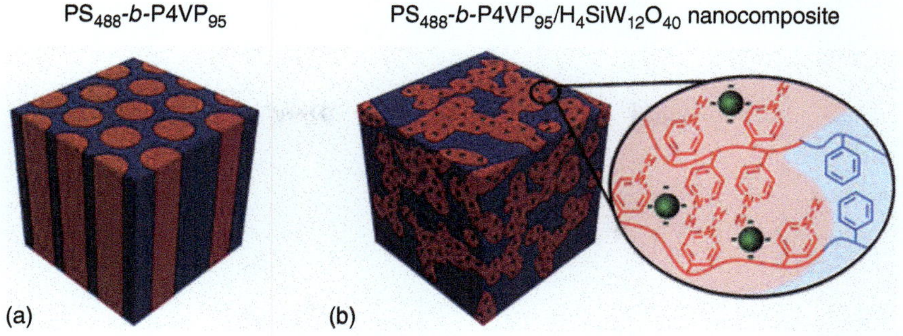

Figure 2.17 Schematic representation of the morphologies of pristine PS_{488}-*b*-$P4VP_{95}$ (a) and the nanocomposite with added $H_4SiW_{12}O_{40}$ (b). Source: Zhang et al. [76]. Figure reproduced with kind permission. © 2016 Elsevier B.V.

Presumably, stability and/or characterization issues represent the main reasons for the sporadic investigation of ion-paired supramolecular polymers based on small molecules.

An unique example in this respect was reported by Schmuck, who investigated the self-assembly behavior of the zwitterion **22** [20]. In contrast to **7**, regioisomeric **22** featured a 2,4-substitution pattern, and thus, a change in the association mode, i.e. assembly in a head-to-tail fashion, was observed exclusively. The authors applied concentration-dependent ^1H NMR spectroscopy in DMSO-D_6 and observed a broadening of the signal assigned to guanidinium's NH proton. The concentration-dependency of the chemical shift was rationalized by a dimerization for which a binding constant (K_{dim}) of 22.2 M^{-1} was determined (Figure 2.18). Taking into account that the self-assembly process was not a simple dimerization but (at least) an oligomerization of **22**, the isodesmic model (see also Section 1.3.1) had to be applied instead (corresponding to a K value of 44.4 M^{-1}) [1]. In view of the low K_a value, high concentrations (significantly above 0.1 M) were required to finally achieve oligomerization of **22** (a DP of 2.5 was determined in DMSO at a concentration of 0.1 M) [20]. Temperature-variable NMR experiments afforded one to conclude on the thermodynamics of the self-assembly process: both ΔH and ΔS were positive, suggesting an entropy-driven process.

In line with the observations made in the self-assembly of **8**, where the intermolecular association was due to steric constraints, only found for the derivative **8b** with the shorter spacer (Figure 2.3b) [17], that the zwitterion **23** also exhibited a spacer-dependent self-assembly behavior (Figure 2.19) [79]. For the more flexible representatives (i.e. $n \geq 3$), NMR dilution experiments indicated the exclusive formation of dimers (K_{dim} values in the range of 30–50 M^{-1}). In particular, for the zwitterion with the shortest chain ($n = 1$), a completely different behavior, according to the mechanism of a ring-chain-mediated supramolecular polymerization (see Section 1.3.2), was found. Below a certain critical concentration, only small cycles were formed that experienced a ring-chain transition above this value. At high concentration, the amount of linear assemblies exceeded that for the cyclic ones. At the concentration of 5 mM in DMSO, a DP of 8 could be estimated by NMR spectroscopy.

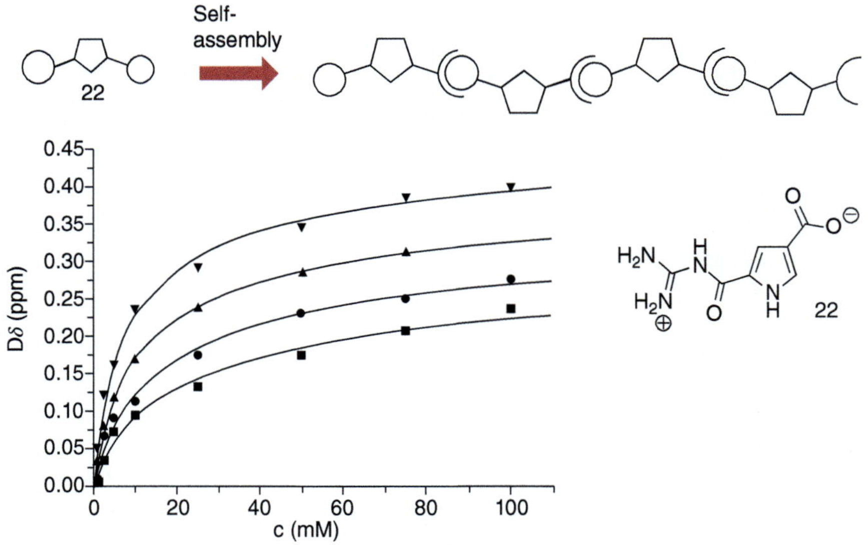

Figure 2.18 Schematic representation of the self-assembly of the heteroditopic monomer **22** in a head-to-tail fashion. The concentration-dependency of the self-assembly process, as studied by ^{1}H NMR spectroscopy, is also shown. Source: Schmuck [20]. © 2001 Elsevier B.V.

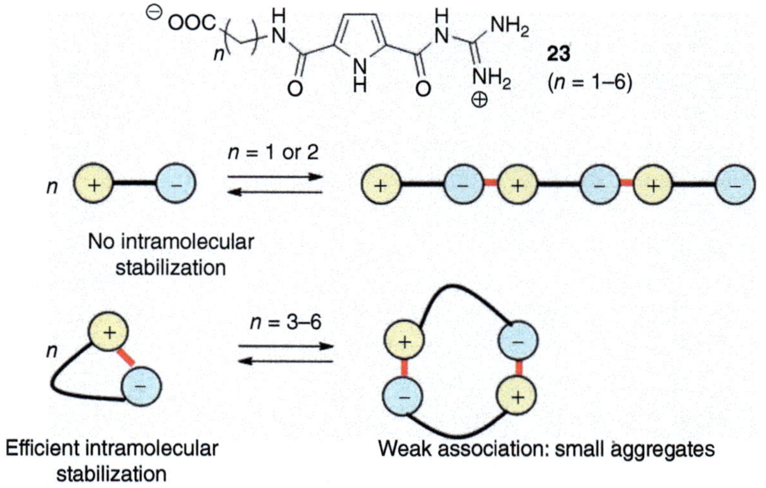

Figure 2.19 Schematic representation of the different self-assembly modes of zwitterion **23** dictated by the length of the alkyl chain. Source: Schmuck et al. [79]. © 2007 American Chemical Society.

Due to the precipitation at higher concentrations, a quantitative analysis could not be performed; the size of the aggregates, up to 100 nm, was determined by DLS measurements.

A series of zwitterions that was prepared from amino acids (**24**) has also been investigated by Schmuck and coworkers. In contrast to the glycine derivative

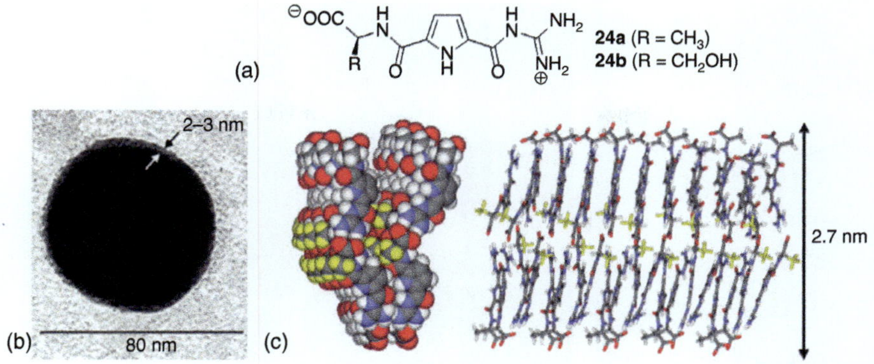

Figure 2.20 (a) Schematic representation of zwitterions **24** functionalized with amino-acid residues. (b) Representative TEM image of a vesicle formed by the self-assembly of **24a** in DMSO (after staining with uranyl acetate). (c) Calculated structure of a membrane segment (the Me-groups of the alanine moieties, essential for the vesicle formation are depicted in yellow); a view along the rows of stacked dimers showing the vesicle curvature as well as side view of the stacked dimers showing their alternating antiparallel orientation. Source: Rehm et al. [80]. Figure reproduced with kind permission. © 2008 American Chemical Society.

(**23**, $n = 1$), the alanine analog **24a** generated vesicles for which a diameter in the range of 25–50 nm was determined by DLS measurements as well as atomic force microscopy (AFM) and transmission electron microscopy (TEM) imaging (the latter two methods suggested a bilayer-type membrane structure) [80]. The authors proposed a model for this self-assembly on the basis of molecular modeling calculations: according to these, linear dimers of **24a** were initially formed, which assembled in an antiparallel fashion, thus allowing the electrostatic attraction between dimers due to oppositely charged groups in concert with additional π–π stacking and hydrophobic interactions between the aromatic residues (Figure 2.20). Vesicles of similar size could also be obtained from the corresponding serine derivative **24b** [81]. In contrast to the assemblies based on **24a**, vesicle formation and disassembly were found to be reversible and could be modulated by adjusting the pH value (i.e. (de)protonation of the zwitterion).

The next level of complexity is reached when multitopic building blocks are used in supramolecular polymerizations. The topology of the formed nanostructures is dictated *inter alia* by the chemical nature of the linker between the binding sites (e.g. its length and flexibility), and, in these cases, the self-assembly processes become much more complicated and, consequently, also less predictable. For example, Rehm et al. observed the assembly of the tris-zwitterion **25** into dimers that, when the concentration was further increased, generated flexible twisted 2D ribbons [82]. The latter ones eventually coiled into soft nanospheres. The authors applied diffusion-ordered (NMR) spectroscopy (DOSY), small-angle neutron scattering (SANS), and AFM imaging, supported by molecular modeling studies, to develop this model for the hierarchical self-assembly process of **25** in DMSO (Figure 2.21). One further representative example in this respect is the two-component self-assembly of a tetrakis(guanidiniumcarbonyl

Figure 2.21 Schematic representation of the tris-zwitterion **25**. The AFM images of the assemblies after spin-coating onto mica substrates are also depicted: large plates were obtained on the surface at high rotational speed (7000 rpm, a and b); the plates merged at lower rotational speed (5000 rpm, c) and, finally, fully disintegrate into the 2D network of ribbons (d). A cross-sectional plot of three representative plates is also shown (e, also depicted as yellow line in b) exhibited uniform heights and diameters of c. 450 and 2.5 nm, respectively. Source: Rehm et al. [82]. Figure reproduced with kind permission. © 2012 The Royal Chemical Society.

pyrrole) cation and a tetracarboxylate, i.e. the sodium salt of EDTA (EDTA: ethylenediamine-tetraacetic acid) [83]. In the pH range of 5–8 and using an equimolar ratio of the two components, the formation of a supramolecular 3D network was observed that reversibly dissociated into the single building blocks by adding either an acid or a base. Viscosity measurements, DLS, as well as microscopy imaging of the drop- or spin-coated films were applied for the characterization of this pH-sensitive supramolecular network.

2.3.2 Polymers Featuring Ionic Interactions in the Side Chain

The interaction of ionic molecules (might even be a polymer) with an oppositely charged polymer gives the so-called polyelectrolyte–surfactant complexes, which, by far, represent the most common type of supramolecular polymer, based on ionic interactions [8, 84–86]. The main general synthetic approaches toward such materials can be summarized as follows: ion metathesis, acid–base reaction, redox chemistry, and side-chain functionalization (e.g. quaternization of amines) [84]. The fundamental aspects of the formation of polyelectrolyte–surfactant complexes were identified by Antonietti et al. almost 25 years ago:

1) Already at very low concentrations, i.e. well below the critical micellar concentration (cmc) of the surfactant, polyelectrolytes and surfactant molecules will start to assemble. The binding process is typically characterized by a high degree-of-cooperativity (for details on this type of supramolecular

polymerization, see Section 1.3.3): the binding of a surfactant molecule next to an already bound one is significantly faster (by a factor of c. 2000) than the initial binding step.
2) When a sub-stoichiometric amount of a surfactant is added to the polyelectrolyte, often the so-called string of pearls morphology is obtained, in which surfactant micelles are aligned along the polymer backbone.
3) Polyelectrolytes with highly hydrophobic backbones, such as poly(styrene sulfonate) (PSS), exhibit a two-step binding mechanism: in the first cooperative step, about 50–70% of the ionic sites along the polymer backbone are occupied. To achieve higher degrees-of-binding, either much higher surfactant concentrations or more forcing binding conditions are required.

In particular, two types of combination have been widely employed for the formation of polyelectrolyte–surfactant complexes: PSS and *n*-alkyltrimethylammonium surfactants, on the one hand, as well as P4VP and anionic surfactants, on the other. Concerning the former one, Antonietti et al. observed the formation of highly ordered mesomorphous phases; the periodicities of the layered phase modifications in the range of 2.5–5 nm clearly reflected the lengths of the employed surfactants [86]. Poly(acrylic acid) (PAA), as one further anionic polyelectrolyte, was complexed by dodecyltrimethylammonium cations, and a highly ordered liquid crystalline (LC) morphology was formed (Figure 2.22) [87]. The cylindrical arrangement of the polymer chains on the nanoscale led to overall tendon-like mechanical properties of the material. Remarkably, mesostructures of similar morphology could also be obtained from the self-assembly of polar polymers with alkylphenols via H-bonding interactions (see Chapter 3) [88, 89]. Thus, the formation of nanosized structures of the supramolecular materials does occur almost independent of the nature of the underlying non-covalent interactions as long as the binding strength is sufficiently high and physical character of the (macro)molecular subunits is different enough (i.e. allowing for phase separation).

The properties of such supramolecular polymers in solution were analyzed by the methods of molecular hydrodynamics. For example, Tsvetkov et al. prepared poly(2-acrylamido-2-methylpropane sulfonate) with grafted cetyltrimethylammonium side chains by micellar polymerization and determined the molecular, conformational, and optical characteristics of the resultant materials [90, 91]. These data were almost identical to those obtained for covalent acrylic and methacrylic comb polymers with aliphatic side chains of similar lengths. Thus, the conformation of the polymers in solution was found to be independent from the nature of the connection of the side chains to the backbone (i.e. supramolecular vs. covalent) [92].

The second major class of materials is represented by cationic templates, such as P4VP (**23**) [93] or other pyridine- [94, 95] and related *N*-heteroaromatic-containing polymers [96], that can interact with surfactant-like molecules containing anionic head groups (e.g. sulfonates, carboxylates, or phenolates), thereby forming supramolecular graft copolymers, such as **24** (Figure 2.23). Similar architectures were obtained, for instance, by combining linear poly(ethylene imine) (l-PEI) with sulfonic acids [97] as well as PSS or PAA with quaternary ammonium salts

Figure 2.22 Polarizing microscopy image of the PAA–dodecyltrimethylammonium complex (extension of $\lambda = 1.5$ nm). Source: Antonietti and Conrad [87]. Figure reproduced with kind permission. © 1994 Wiley-VCH.

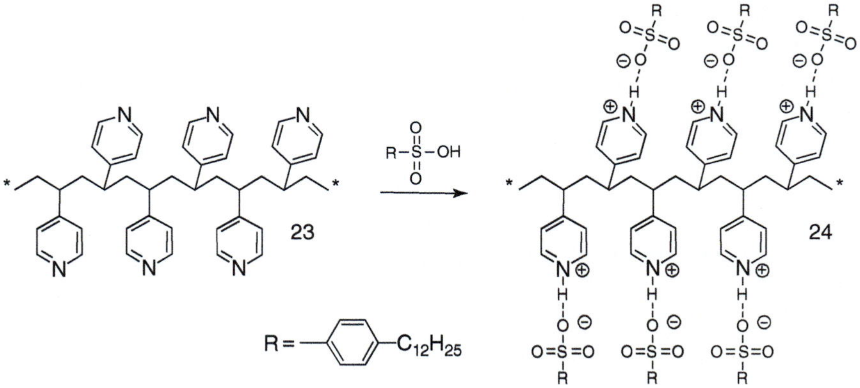

Figure 2.23 Schematic representation of the supramolecular grafting of sulfonates onto P4VP. Source: Ikkala et al. [93].

[87, 98]. Ionic interactions between the pyridine moieties of a PS-*b*-P4VP diblock copolymer and oligo(ethylene oxide) with a sulfonic acid head group gave coil-comb supramolecular block copolymer architectures; these systems phase separated into lamellae, and, after the addition of $LiClO_4$, as supporting electrolyte, the material became conductive at an ambient temperature [98].

The decoration of P4VP with sulfonic-acid-functionalized dendritic building blocks was investigated by Zhu et al. [99]. The self-assembly of the components in bulk was found to be strongly dependent on the applied ratio between the dendrons and pyridine moieties of the polymer: the authors reported on the transition from a lamellar to a hexagonal columnar phase when the ratio of dendrons was increased. The dendrons exhibited a certain degree-of-mobility, i.e. their inter- as well as intra-chain movement could be observed. Not only graft- and comb-like copolymers can be formed by ionic interactions, a variety of bis(sulfonic acids) was employed to also assemble supramolecular networks with P4VP. The supramolecular cross-linking of polymers and the origin of the network formation allowed one to adjust both the glass transition and the melting temperature over a broad range, while a high thermal stability of the material was retained [100].

It was demonstrated that "information" can be transferred from small chiral molecules to polymer chains in a random-coil conformation: combining the regioregular but achiral 4-carboxyphenyl-substituted *cis*-transoid polyacetylene (**25**) with (*R*)-1-naphthylethylamine (**26**) in DMSO induced a strong circular dichroism (CD) effect [101]. Thereby, the chiral information was translated from the substrate to the polymer chain via ion-pairing interactions that, in turn, adopted a helical conformation (Figure 2.24). In the presence of an even stronger base of opposite chirality [e.g. (*S*)-2-aminopropanol, **27**], pushing (*R*)-**26** aside, the mirror image of the initial CD spectrum was obtained. Thus, the chirality of the helix could simply be controlled via the absolute configuration of the amine, which was ionically bound to the polymer; the stability of the entire supramolecular architecture was attributed to an efficient charge balancing between the negative charges along the polymer chain and the positive charges of the amines [102]. Remarkably, the helicity was memorized even upon the addition of 2-aminoethanol as a highly competitive but achiral base. Apparently, the amine-exchange reaction was faster than the backbone equilibration and the resultant loss of the superstructure.

Related to this, a strategy to self-assemble π-conjugated polymer chains into a double-stranded helix was reported by the Maeda group [103]: two poly(phenylene-ethynylene)s (PPEs) were prepared, bearing either chiral amidine (*R*-**28**) or achiral carboxylic acid groups (**29**) on the *meta*-constituted terphenyl units of their backbones (M_n values in the range of 3×10^4 g mol^{-1}, $Đ$ of 1.9–3.4; Figure 2.25a). Mixing these polymers in tetrahydrofuran (THF) gave a right-handed helical structure, as confirmed by CD spectroscopy. Thereby, the helicity was induced by the *R*-configuration of the substituents on the amidinium moieties. AFM imaging as well as X-ray diffraction (XRD) measurements revealed a helical pitch of c. 1.70 nm – a value that was in fair agreement with the one predicted from molecular modeling studies (Figure 2.25b). It was further shown that in less polar solvents, such as $CHCl_3$, kinetic control supported the formation of imperfectly hybridized double-stranded helices (strong ion-pairing interactions resulted in random cross-linking of the polymer strands). Rearrangement of these into the right-handed double-strands was achieved by treatment with CF_3CO_2H and subsequent neutralization with diisopropylamine. A similar polymer featuring additional metal-to-ligand coordination, as an orthogonal type of non-covalent interaction, was also prepared – this example is discussed later in Section 11.2 [104].

2.3.3 Self-assembly of Polyelectrolytes: Toward Polyelectrolyte Complexes

The cooperative electrostatic interactions between polycations and polyanions in water result in the so-called polyelectrolyte complexes (PECs) in which the

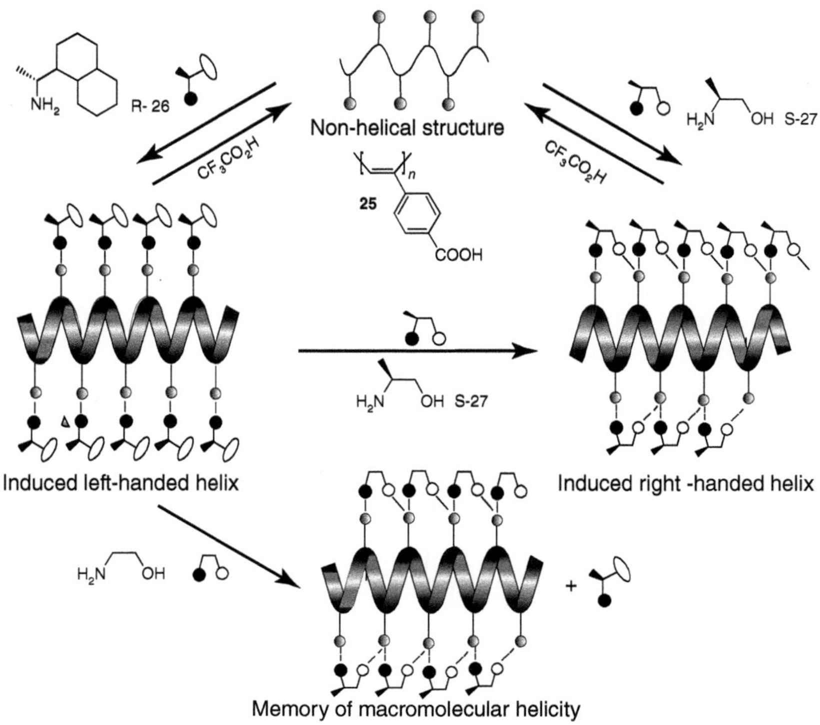

Figure 2.24 Schematic representation of the formation of helical strands from the achiral polyacetylene derivative **25** induced by ion pairing with chiral amines. Source: Yashima et al. [101]. © 1999 Springer Nature.

oppositely charged polymer chains ideally pack in a dense fashion, thus forming a phase well separated from the solvent [105]. PECs are already known since the late 1920s when Bungenberg de Jong and Kruyt observed the occurrence of colloidal complexes of natural polyelectrolytes in water (referred to as "coacervates") [106]. The main driving force in the phase-separation process, which is the electrostatic interactions between inversely charged polymer chains, was already identified by Kossel in 1897 for proteins and carbohydrates [107]. Later, the insoluble PECs were also obtained from linear synthetic polymers featuring a high charge density [108]. For instance, Michaels and Miekka prepared stoichiometric complexes of poly(4-vinylbenzyltrimethylammonium chloride) with the sodium salt of PSS [109]. In recent years, a broad range of PECs has been prepared by combining synthetic and natural charge-carrying polymers. These materials can be classified according to the charge ratio, the DPs, and the structure of the initial polyelectrolytes: both types of structures can basically be formed – non-stoichiometric, yet water-soluble PECs as well as stoichiometric but insoluble PECs can be formed [110]. The former ones can be obtained by mixing the polyelectrolytes in a non-stoichiometric fashion (thereby, the molar masses of the polycation and the polyanion should differ significantly). Stoichiometric PECs are much more common and arise from the self-assembly of polycations and polyanions in a 1 : 1 ratio. Taking the nature of the

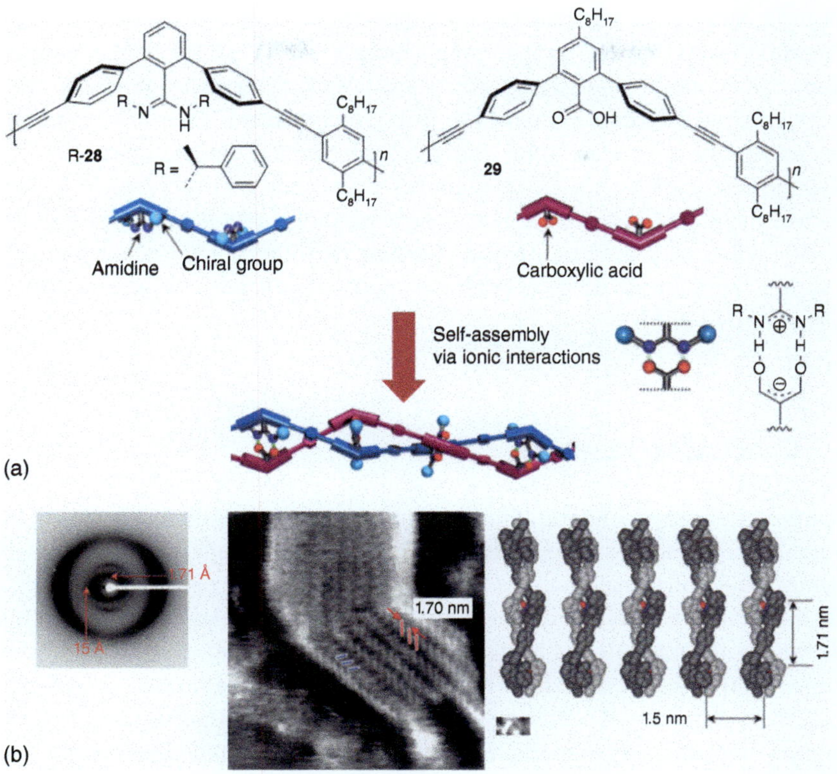

Figure 2.25 (a) Schematic representation of polymers *R*-**28** and **29** that self-assemble into right-handed double helices in polar-aprotic solvents. (b) Wide-angle XRD pattern and AFM phase image (40 nm × 40 nm) of *R*-**28** × **29** along with a molecular model of the double-stranded helical structure. Source: Maeda et al. [103]. Figure reproduced with kind permission. © 2008 American Chemical Society.

Figure 2.26 Schematic representation of the two different structural modes for stoichiometric PECs. (a) Ladder-type structure and (b) "Scrambled egg" structure.

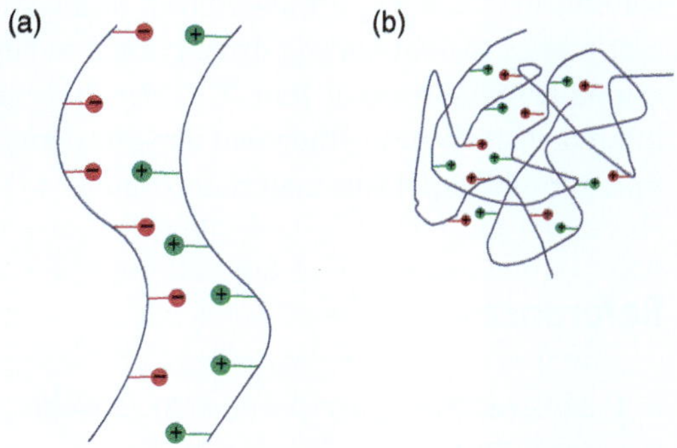

ionic groups, the molar mass of the polymer chains, and external parameters into account (e.g. concentration, pH, temperature, presence of additives), two structural motives can be proposed – the ladder-type and the "scrambled egg" structures (Figure 2.26).

Recently, Meka et al. summarized the range of potential applications of PECs, in particular with respect to their biomedicinal use [111]. From the current cutting-edge applications, e.g. gene delivery, site-specific or gastro-retentive drug delivery and vaccine development have to be foregrounded. From the polymers discussed in this context, natural chitosan and synthetic poly(ethylene imine) (PEI) represent highly versatile carriers for drug- and gene-delivery purposes, respectively [112, 113]. Due to the diversity of structures and potential applications, an in-depth discussion of PECs is beyond the scope of this chapter; a detailed survey also addressing the wide range of potential applications is given in several review articles [111, 114–116].

2.4 Concluding Remarks

Ionic interactions represent one of the fundamental types of secondary interactions commonly employed in supramolecular chemistry. Accordingly, a broad range of supramacromolecular materials has been derived making use of the ionic (electrostatic) interaction between the charge-carrying organic molecules. In particular, this holds true for the formation of block copolymers, chain-extended and cross-linked polymers – ionic interactions have been identified as a versatile tool to manipulate the bulk properties of polymers, e.g. to overcome (or manipulate) macrophase separation of intrinsically immiscible constituent blocks. Even more important, the PECs – associates between polycationic and polyanionic species – have attracted considerable interest for biomedical applications, including drug and gene delivery. However, supramolecular polymers – where polymers result from the non-covalent interaction between small-molecule monomers – have hardly been realized on the basis of ionic interactions. This limitation can be attributed to the low directionality of the interaction and the poor solubility of the resultant materials in the polar solvents in which they are assembled. Nonetheless, ionic interactions play an important role as the supporting driving force in supramolecular chemistry, e.g. for the assembly of π-electronic ions (Chapter 5) or calix[*n*]arenes (Chapter 8); also, their interplay with other, orthogonal non-covalent forces is of importance to create multiple stimuli-responsive materials (Chapter 11).

References

1 de Greef, T.F.A., Smulders, M.M.J., Wolffs, M. et al. (2009). *Chem. Rev.* 109: 5687–5754.

2 Winter, A., Hager, M.D., and Schubert, U.S. (2012). Supramolecular polymers. In: *Polymer Science: A Comprehensive Review*, vol. 5 (eds. H.-W. Schmidt and M. Ueda), 269–310. Amsterdam: Elsevier BV.

3 Huggins, M.L. (1971). *Angew. Chem. Int. Ed. Engl.* 10: 147–152.

4 Kollman, P.A. and Allen, L.C. (1972). *Chem. Rev.* 72: 283–303.

5 Hoogenboom, R., Fournier, D., and Schubert, U.S. (2008). *Chem. Commun.*: 155–162.
6 Haketa, Y. and Maeda, H. (2017). *Chem. Commun.* 53: 2894–2909.
7 Rehm, T.H. and Schmuck, C. (2010). *Chem. Soc. Rev.* 39: 3597–3611.
8 Ciferri, A. (2010). *Chem. Eur. J.* 16: 10930–10945.
9 Faul, C.F.J. and Antonietti, M. (2003). *Adv. Mater.* 15: 673–683.
10 Kubik, S., Reyheller, C., and Stüwe, S. (2005). *J. Incl. Phenom.* 52: 137–187.
11 Dietrich, B., Fyles, D.L., Flyes, T.M., and Lehn, J.-M. (1979). *Helv. Chim. Acta* 62: 2763–2787.
12 Linton, B. and Hamilton, A.D. (1999). *Tetrahedron* 55: 6027–6038.
13 Fitzmaurice, R.J., Kyne, G.M., Douheret, D., and Kilburn, J.D. (2002). *J. Chem. Soc., Perkin Trans.* 1: 841–864.
14 Schmuck, C. (1999). *Chem. Commun.*: 843–844.
15 Schmuck, C. (2000). *Chem. Eur. J.* 6: 709–718.
16 Schmuck, C. (1999). *Eur. J. Org. Chem.* 1999: 2397–2403.
17 Schmuck, C. (2000). *J. Org. Chem.* 65: 2432–2437.
18 Schmuck, C. and Wienand, W. (2003). *J. Am. Chem. Soc.* 125: 452–459.
19 Schmuck, C. and Wienand, W. (2001). *Angew. Chem. Int. Ed.* 40: 4363–4369.
20 Schmuck, C. (2001). *Tetrahedron* 57: 3063–3067.
21 Kelly, T.R., Xie, R.L., Weinreb, C.K., and Bregant, T. (1998). *Tetrahedron Lett.* 39: 3675–3678.
22 Terfort, A. and von Kiedrowski, G. (1992). *Angew. Chem. Int. Ed. Engl.* 31: 654–656.
23 Kraft, A. and Fröhlich, R. (1998). *Chem. Commun.*: 1085–1086.
24 Breslow, R., Rajagopalan, R., and Schwarz, J. (1981). *J. Am. Chem. Soc.* 103: 2905–2907.
25 Liu, R., Lai, Y., He, B. et al. (2012). *Int. J. Nanomed.* 7: 5249–5258.
26 Couvreur, P. (2013). *Adv. Drug Delivery Rev.* 65: 21–23.
27 Grawe, T., Schrader, T., Zadmard, R., and Kraft, A. (2002). *J. Org. Chem.* 67: 3755–3763.
28 Fiammengo, R., Timmerman, P., Huskens, J. et al. (2002). *Tetrahedron* 58: 757–764.
29 Fiammengo, R., Timmermman, P., de Jong, F., and Reinhoudt, D.N. (2000). *Chem. Commun.*: 2313–2314.
30 Corbellini, F., Di Costanzo, L., Crego-Calama, M. et al. (2003). *J. Am. Chem. Soc.* 125: 9946–9947.
31 Oshovsky, G.V., Reinhoudt, D.N., and Verboom, W. (2006). *J. Am. Chem. Soc.* 128: 5270–5278.
32 Oshovsky, G.V., Reinhoudt, D.N., and Verboom, W. (2006). *J. Org. Chem.* 71: 7441–7448.
33 Zhang, Y. and Cremer, P.S. (2006). *Curr. Opin. Chem. Biol.* 10: 658–663.
34 Ibidapo, T.A. (1996). *Polym. Eng. Sci.* 36: 1920–1927.
35 Li, S.-L., Xiao, T.-X., Lin, C., and Wang, L.-Y. (2012). *Chem. Soc. Rev.* 41: 5950–5968.
36 Ibidapo, T.A. (1988). *Polym. Eng. Sci.* 28: 1473–1476.

37 Lehn, J.-M. (1990). *Angew. Chem. Int. Ed. Engl.* 29: 1304–1319.
38 Lehn, J.-M. (1995). *Supramolecular Chemistry – Concepts and Perspectives*. Weinheim: VCH-Wiley.
39 Kato, T. and Frechet, J.M.J. (1989). *Macromolecules* 22: 3818–3819.
40 Lange, R.F.M. and Meijer, E.W. (1995). *Macromolecules* 28: 782–783.
41 Gao, Z., Molnár, A., and Eisenberg, A. (1997). Supramolecular polymers. In: *Ionomers: Synthesis, Structure, Properties and Applications* (eds. M.R. Tant, K.A. Mauritz and G.L. Wilkes), 390–443. London: Chapman & Hall.
42 Hamley, I.W. (1998). *The Physics of Block Copolymers*. Oxford: University Press.
43 Yang, S.K., Ambade, A.V., and Weck, M. (2011). *Chem. Soc. Rev.* 40: 129–137.
44 Binder, W.H., Bernstorff, S., Kluger, C. et al. (2005). *Adv. Mater.* 17: 2824–2828.
45 Noro, A., Tamura, A., Wakao, S. et al. (2008). *Macromolecules* 41: 9277–9283.
46 Horrion, J., Jerome, R., and Teyssie, P. (1986). *J. Polym. Sci., Part C: Polym. Lett.* 24: 69–76.
47 Russell, T.P., Jerome, R., Charlier, P., and Foucart, M. (1988). *Macromolecules* 21: 1709–1717.
48 Horrion, J., Jérôme, R., and Teyssié, P. (1990). *J. Polym. Sci., Part A: Polym. Chem.* 28: 153–171.
49 Charlier, P., Jerome, R., Teyssie, P., and Utracki, L.A. (1992). *Macromolecules* 25: 2651–2656.
50 Horrion, J., Jerome, R., Teyssie, P. et al. (1989). *Polym. Bull.* 21: 627–633.
51 Iwasaki, K., Hirao, A., and Nakahama, S. (1993). *Macromolecules* 26: 2126–2131.
52 Haraguchi, M., Nakagawa, T., and Nose, T. (1995). *Polymer* 36: 2567–2572.
53 Inomata, K., Haraguchi, M., and Nose, T. (1996). *Polymer* 37: 4223–4228.
54 Inomata, K., Liu, L.-Z., Nose, T., and Chu, B. (1999). *Macromolecules* 32: 1554–1558.
55 Zhang, L., Kucera, L.R., Ummadisetty, S. et al. (2014). *Macromolecules* 47: 4387–4396.
56 Zhang, L., Brostowitz, N.R., Cavicchi, K.A., and Weiss, R.A. (2014). *Macromol. React. Eng.* 8: 81–99.
57 Huh, J., Park, H.J., Kim, K.H. et al. (2006). *Adv. Mater.* 18: 624–629.
58 Fleischer, C.A., Morales, A.R., and Koberstein, J.T. (1994). *Macromolecules* 27: 379–385.
59 Gao, J., Li, X., Li, B., and Han, Y. (2010). *Polymer* 51: 2683–2689.
60 Huh, J., Jung, J.Y., Lee, J.U. et al. (2011). *ACS Nano* 5: 115–122.
61 Takahashi, A., Rho, Y., Higashihara, T. et al. (2010). *Macromolecules* 43: 4843–4852.
62 Yu, H., Stoffelbach, F., Detrembleur, C. et al. (2012). *Eur. Polym. J.* 48: 940–944.
63 Song, G., Cho, S.M., Jung, H.J. et al. (2012). *Chem. Eur. J.* 18: 15662–15668.
64 Qian, J. and Wu, F. (2008). *Macromolecules* 41: 8921–8926.
65 Schädler, V., Spickermann, J., and Räder, H.J. (1996). *Macromolecules* 29: 4865–4870.
66 Schädler, V. and Wiesner, U. (1997). *Macromolecules* 30: 6698–6701.

67 Schädler, V., Kniese, V., Thurn-Albrecht, T. et al. (1998). *Macromolecules* 31: 4828–4837.

68 Wang, D., Guo, J., Zhang, H. et al. (2015). *J. Mater. Chem. A* 3: 12864–12872.

69 Zhang, Q., He, L., Wang, H. et al. (2012). *Chem. Commun.* 48: 7067–7069.

70 Li, D., Li, H., and Wu, L. (2014). *Polym. Chem.* 5: 1930–1937.

71 Zhang, J., Liu, Y., Li, Y. et al. (2012). *Angew. Chem. Int. Ed.* 51: 4598–4602.

72 Cao, X., Zhang, L., Xu, T. et al. (2016). *Polym. Chem.* 7: 3216–3220.

73 Bu, W., Uchida, S., and Mizuno, N. (2009). *Angew. Chem. Int. Ed.* 48: 8281–8284.

74 Yelamanchili, R.S., Walther, A., Müller, A.H.E., and Breu, J. (2008). *Chem. Commun.*: 489–491.

75 Liu, N., Wang, H., and Bu, W. (2014). *J. Mater. Chem. C* 2: 5271–5274.

76 Zhang, L., Liu, C., Shang, H. et al. (2016). *Polymer* 106: 53–61.

77 Zhang, L., Cui, T., Cao, X. et al. (2017). *Angew. Chem. Int. Ed.* 56: 9013–9017.

78 Tan, J., Liu, Z., Wang, R. et al. (2019). *Soft Matter* 15: 6988–6993.

79 Schmuck, C., Rehm, T., Geiger, L., and Schäfer, M. (2007). *J. Org. Chem.* 72: 6162–6170.

80 Rehm, T., Stepanenko, V., Zhang, X. et al. (2008). *Org. Lett.* 10: 1469–1472.

81 Rodler, F., Linders, J., Fenske, T. et al. (2010). *Angew. Chem. Int. Ed.* 49: 8747–8750.

82 Rehm, T.H., Grohn, F., and Schmuck, C. (2012). *Soft Matter* 8: 3154–3162.

83 Samanta, K. and Schmuck, C. (2015). *Chem. Commun.* 51: 16065–16067.

84 Ober, C.K. and Wegner, G. (1997). *Adv. Mater.* 9: 17–31.

85 Ikkala, O. and ten Brinke, G. (2004). *Chem. Commun.*: 2131–2137.

86 Antonietti, M., Conrad, J., and Thuenemann, A. (1994). *Macromolecules* 27: 6007–6011.

87 Antonietti, M. and Conrad, J. (1994). *Angew. Chem. Int. Ed. Engl.* 33: 1869–1870.

88 Ruokolainen, J., Torkkeli, M., Serimaa, R. et al. (1996). *Macromolecules* 29: 6621–6628.

89 Ruokolainen, J., ten Brinke, G., Ikkala, O. et al. (1996). *Macromolecules* 29: 3409–3415.

90 Tsvetkov, N.V., Mikhailova, M.E., Mikusheva, N.G. et al. (2017). *Int. J. Polym. Anal. Charact.* 22: 27–34.

91 Tsvetkov, N.V., Mikusheva, N.G., Lezov, A.A. et al. (2016). *Eur. Polym. J.* 75: 251–263.

92 Andreeva, L.N., Shcherbinina, T.M., Zorin, I.M. et al. (2013). *Polym. Sci. Ser. A Polym. Phys.* 55: 289–294.

93 Ikkala, O., Ruokolainen, J., ten Brinke, G. et al. (1995). *Macromolecules* 28: 7088–7094.

94 Huang, W.-Y. and Han, C.-D. (2006). *Macromolecules* 39: 4735–4745.

95 Vuillaume, P.Y. and Bazuin, C.G. (2003). *Macromolecules* 36: 6378–6388.

96 Green, O., Grubjesic, L., Lee, S., and Firestone, M.A. (2009). *Polym. Rev.* 49: 339–360.

97 Chen, H.-L. and Hsiao, M.-S. (1999). *Macromolecules* 32: 2967–2973.

98 Kosonen, H., Valkama, S., Hartikainen, J. et al. (2002). *Macromolecules* 35: 10149–10154.

99 Zhu, X.-M., Beginn, U., Möller, M. et al. (2006). *J. Am. Chem. Soc.* 128: 16928–16937.

100 Shibata, M., Kimura, Y., and Yaginuma, D. (2004). *Polymer* 45: 7571–7577.

101 Yashima, E., Maeda, H., and Okamoto, Y. (1999). *Nature* 399: 449–451.

102 Maeda, H., Morino, K., Okamoto, Y. et al. (2004). *J. Am. Chem. Soc.* 126: 4329–4342.

103 Maeda, T., Furusho, Y., Sakurai, S.-I. et al. (2008). *J. Am. Chem. Soc.* 130: 7938–7945.

104 Ikeda, M., Tanaka, Y., Hasegawa, T. et al. (2006). *J. Am. Chem. Soc.* 128: 6806–6807.

105 Jha, P., Desai, P., Li, J., and Larson, R. (2014). *Polymers* 6: 1414.

106 Bungenberg de Jong, H.G. and Kruyt, H.R. (1929). *Proc. Kon. Ned. Akad. Wetensch.* 32: 849–856.

107 Kossel, A. (1897). *Z. Physiol. Chem.* 22: 176–187.

108 Michaels, A.S. (1965). *Ind. Eng. Chem.* 57: 32–40.

109 Michaels, A.S. and Miekka, R.G. (1961). *J. Phys. Chem.* 65: 1765–1773.

110 Buriuli, M. and Verma, D. (2017). Supramolecular polymers. In: *Advances in Biomaterials for Biomedical Applications* (eds. A. Tripathi and J.S. Melo), 45–93. Heidelberg: Springer.

111 Meka, V.S., Sing, M.K.G., Pichika, M.R. et al. (2017). *Drug Discovery Today* 22: 1697–1706.

112 Hamman, J.H. (2010). *Marine Drugs* 8: 1305.

113 Patrick, M., Gilles, B., Jean Pierre, G., and Chantal, P. (2008). *Curr. Gene Ther.* 8: 335–352.

114 van der Gucht, J., Spruijt, E., Lemmers, M., and Cohen Stuart, M.A. (2011). *J. Colloid Interface Sci.* 361: 407–422.

115 Thünemann, A.F., Müller, M., Dautzenberg, H. et al. (2004). Supramolecular polymers. In: *Polyelectrolytes with Defined Molecular Architecture II* (ed. M. Schmidt), 113–171. Berlin, Heidelberg: Springer.

116 Cooper, C.L., Dubin, P.L., Kayitmazer, A.B., and Turksen, S. (2005). *Curr. Opin. Colloid Interface Sci.* 10: 52–78.

3

Supramolecular Polymers, Based on Hydrogen-Bonding Interactions*

3.1 General Aspects

Over a century ago, Moore and Winmill introduced the concept of hydrogen-bonding (H-bonding) that basically originates from the dipole–dipole-type interaction between an electronegative atom and a hydrogen atom which, in turn, is located on an electronegative atom (e.g. nitrogen, oxygen, or fluorine) [1]. The utilization of H-bonding interactions, as non-covalent type of interaction in the field of supramolecular polymers, was pioneered by Jean-Marie Lehn and coworkers [2]. The first main-chain supramolecular polymer, based on the directed self-assembly of tartaric acids with either two diaminopyridines or two uracil derivatives as telechelic, complementary end groups, was reported in 1990, a triple H-bonding interaction afforded a polymer with liquid crystalline behavior.

In general, H-bonds can – when placed at defined positions of a building block, in concert with its initial geometry and size – be applied for the molecular self-assembly of various types of supramolecular architectures: network-, rod-, fibrous-, disc-, helical-, lamellar-, and chain-like structures (Figure 3.1) [3, 4].

In comparison to other non-covalent interactions (i.e. metal-to-ligand coordination or certain host–guest complexes), H-bonding between two neutral organic molecules is still relatively weak (c. 10–20 kJ mol^{-1}, see also Section 1.2). Nevertheless, this particular type of non-covalent interaction plays a pivotal role in today's supra(macro)molecular chemistry. The reason for this can be found in directionality of binding and the versatility to assemble complex supramolecular structures. Accordingly, various review articles have appeared [3, 5–15]. Jeffrey and Saenger classified H-bonding interactions on the basis of their strength and chemical nature (Figure 3.3a) [16]: strong, medium (or weak) as well as nonclassical H-bonds. Strong H-bonding interactions exhibit short distances between the interacting centers. Furthermore, they are highly directional and their association energies typically exceed 40 kJ mol^{-1}. The weaker H-bonds, generally described as "D–H···A" interactions, have low directionality in binding and the bond energies are decreased (typically between 20 and 40 kJ mol^{-1}). In such a system, the electronegative group

* Parts of this chapter are reproduced from *Polymer Science: A Comprehensive Review*, Vol. 5 (Eds.: H.-W. Schmidt, M. Ueda), Elsevier BV, Amsterdam, 2012, 269–310 by permission from Elsevier B.V.

Supramolecular Polymers and Assemblies: From Synthesis to Properties and Applications, First Edition.
Ulrich S. Schubert, George R. Newkome, and Andreas Winter.

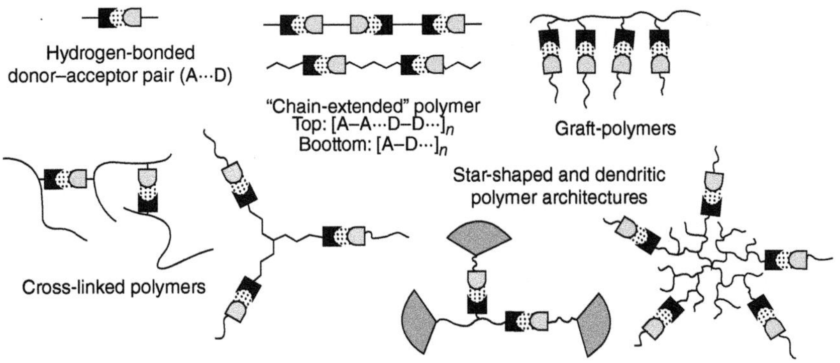

Figure 3.1 Schematic representation of the various types of architectures accessible via H-bonding interactions (A: H-bonding acceptor, D: H-bonding donor) [3]. Source: Redrawn from Binder and Zirbs [3]. © 2007 Springer Nature.

(A) interacts with a hydrogen atom attached to an electronegative or even carbon atom (D). The last case, which is the nonclassical H-bonding interactions, involves the binding of D–H to acceptor moieties, which can be π-conjugated systems, transition metal ions, or boron hydrides. To briefly illuminate this unconventional class of hydrogen bonds, a correlation between the bonding enthalpy (ΔH) and the length of the hydrogen bond (r) for various (semi-)metal complexes, as H-bond acceptors, is depicted in Figure 3.2b [17].

It has to be pointed out that solvent effects will have a tremendous impact on the strength of a H-bond – independent of its chemical nature. This holds particularly true for highly polar or even protic solvents, which might act as an efficient competitor for H-bonding. In general, the addition of a polar solvent will significantly weaken H-bonds, often over several orders of magnitude. Thus, the formation of supramolecular polymers by H-bonding interactions is intrinsically restricted to the use of aprotic and nonpolar solvents (e.g. alkanes, toluene, and halogenated solvents) [3].

The strength of H-bonds represents one of the most crucial parameters when dealing with advanced supramolecular materials. Basically, the bond strength can be strongly enhanced via the number of involved bonds: the synergy of multiple H-bonding offers an overall enhanced binding strength with a contribution of c. 7.4 $kJ\,mol^{-1}$ per single hydrogen bond. The most relevant types of H-bonding arrays, according to the number of H-bonds involved in the interaction, are shown in Figures 3.3–3.5 and 3.7 [3]. The complex interplay of multiple H-bonding interactions in natural polymers, such as the nucleobase pairing in DNA [22], has paved the way to develop artificial systems to apply this type of supramolecular interaction also for the preparation of synthetic polymers. In particular, Sivakova and Rowan emphasized the versatility of nucleobase pairing in the assembly of supramolecular polymers [23].

The particular arrangement of adjacent donor-(D) and acceptor-type sites (A) is also of significant importance concerning the strength of the supramolecular array. This aspect was elaborated by the Jorgensen group with the example of linear

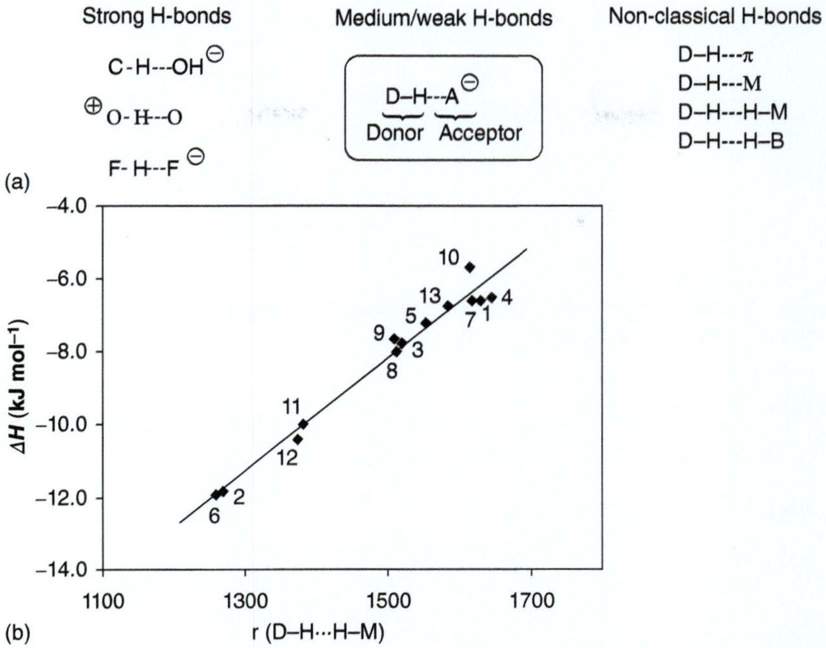

Figure 3.2 (a) Classification of H-bonds according to Jeffrey and Saenger (D: H-bonding donor, A: H-bonding acceptor, M: metal ion). Source: Jeffrey and Saenger [16]. © 1991 Springer Nature. (b) Correlation between the binding energy of H-bonding and the bond lengths for some nonclassical complexes: AlH_4^- with various solvents [CH_3OH (1), CF_3OH (2), CF_3CH_2OH (3)], BH_4^- with various solvents [CH_3OH (4), CF_3CH_2OH (5)], GaH_4^- with various solvents [CF_3OH (6), CH_3OH (7), CF_3CH_2OH (8)], Cp*FeH(1,2-diphosphinoethane) with various solvents [$(CF_3)_2CHOH$ (9), FCH_2CH_2OH (10), $(CF_3)_3COH$ (11), CF_3COOH (12), and CF_3CH_2OH (13)]. Source: Belkova et al. [17]. © 2005 American Chemical Society.

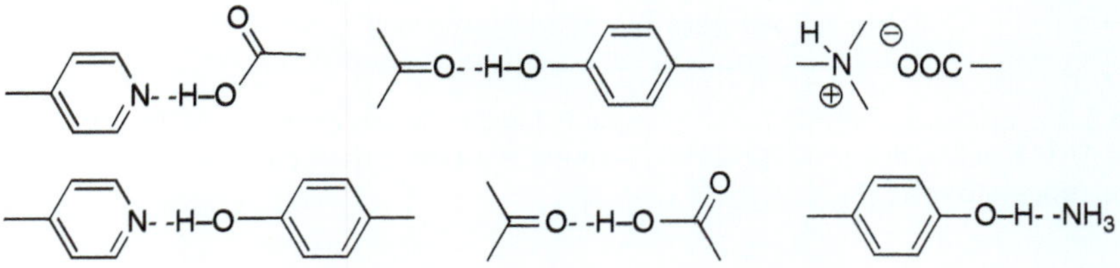

Figure 3.3 Schematic representation of various single H-bonding motifs.

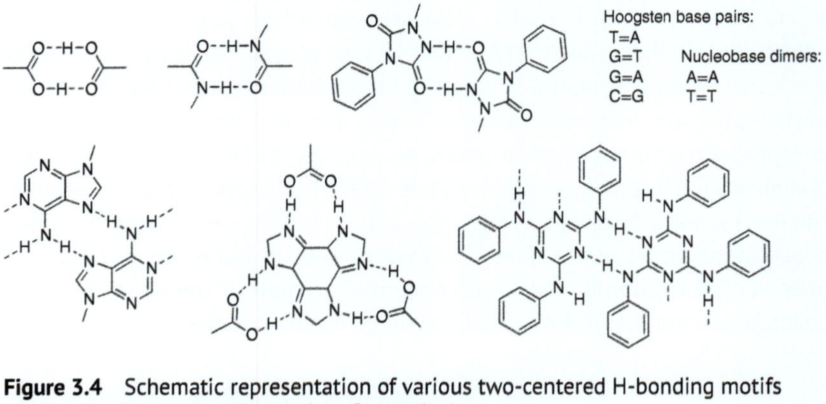

Figure 3.4 Schematic representation of various two-centered H-bonding motifs (A: adenine, T: thymine, G: guanine, C: cytosine).

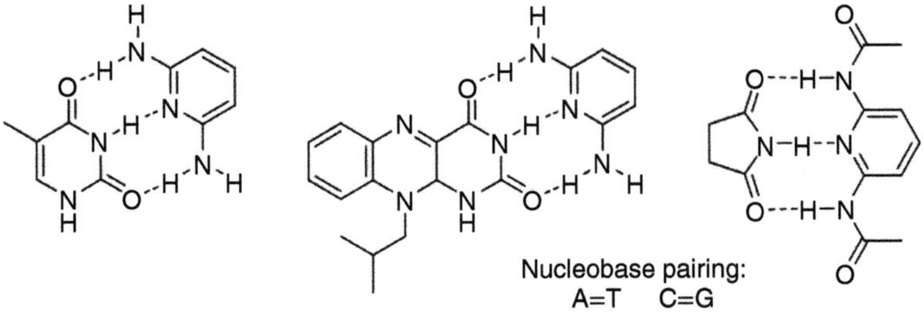

Figure 3.5 Schematic representation of various triple H-bonding motifs (A: adenine, T: thymine, G: guanine, C: cytosine).

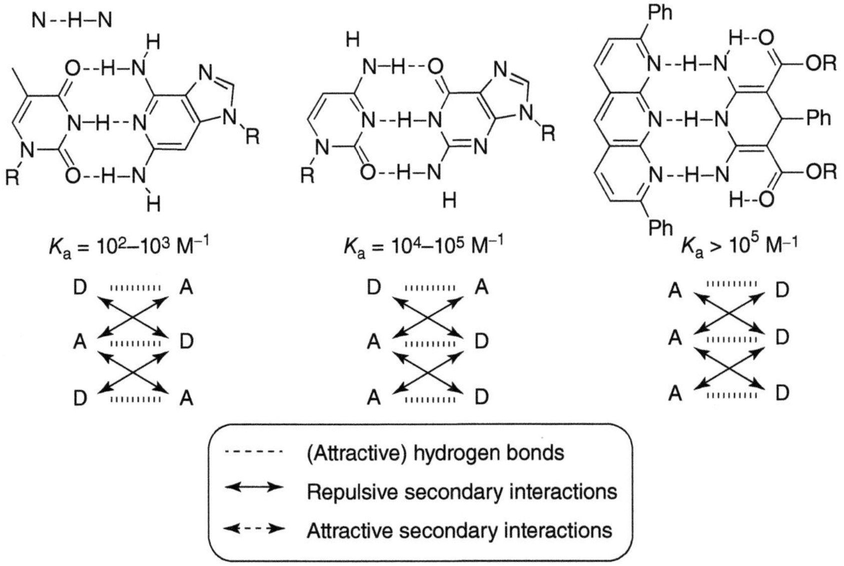

Figure 3.6 Schematic representation of triple H-bonding arrays exhibiting different K_a values (D: H-bonding donor; A: H-bonding acceptor, K_A: association constant in $CHCl_3$). Source: Brunsveld et al. [5].

supramolecular assemblies, based on triple H-bonding interactions (Figure 3.6) [24, 25]. In detail, the system featuring a DAD···ADA interaction had a binding strength (K_A) of c. $10^2\ M^{-1}$ in $CHCl_3$. Exclusively, repulsive forces were present in the former case; whereas, additional attractive forces, as in the DAA···ADD array, increased the binding strength. A general consideration concerning diagonally located D and A sites can be made: two D or A sites will repel each other when they are placed diagonally to each other; accordingly, unlike diagonal pairs will attract each other. Thus, the assemblies of the AAA···DDD-type feature a maximum in attractive forces and, thus, the K_A value will be the highest within the series. Secondary electrostatic interactions (i.e. making use of additional π–π stacking interactions of, for example, extended condensed aromatics) were found to raise the association constant in AAA···DDD systems even up to $2 \times 10^7\ M^{-1}$ [26, 27].

Schneider and coworker suggested an incremental system to describe the role of such secondary interactions with respect to the strength of H-bonding interactions [28]: a secondary interaction energy of ±2.9 kJ mol^{-1} is added to (subtracted from) the initial bond strength for an enhancement (destabilization) of the H-bond. Another aspect that needs to be considered is the impact of electronic effects on the strength of the non-covalent interaction, in particular for aromatic systems [29].

Kinetically and thermodynamically highly stable supramolecular assemblies can be formed on the basis of quadruple [18, 19, 21, 30–33] or even more extended H-bonding entities [34, 35]. From these, the synthetic quadruple H-bonding moieties combine high directionalities with very high K_a values in apolar environments ($K_A > 10^6$ mol^{-1}). In particular, the self-complimentary DADA- (e.g. aminotriazines [18] or ureidotriazines [36]) and DDAA-type units (e.g. 2-ureido-4-1*H*-pyrimidinone (UPy) [18]) play a central role in supramolecular polymer science (Figure 3.7a) [14]. For the latter system, commonly referred to as UPy, a K_a value of 6×10^7 M^{-1} was determined. These moieties, however, might show some tautomerism by which the self-complementary nature is partially lost: for example, a self-complementary DDAA-type unit tautomerizes into a DADA-type one, which possesses an increased number of repulsive forces and, thus, an overall reduced K_a value [5]. To overcome this limitation, the Zimmerman group utilized a highly stable self-complementary H-bonding array for which all four possible tautomers could dimerize via quadruple

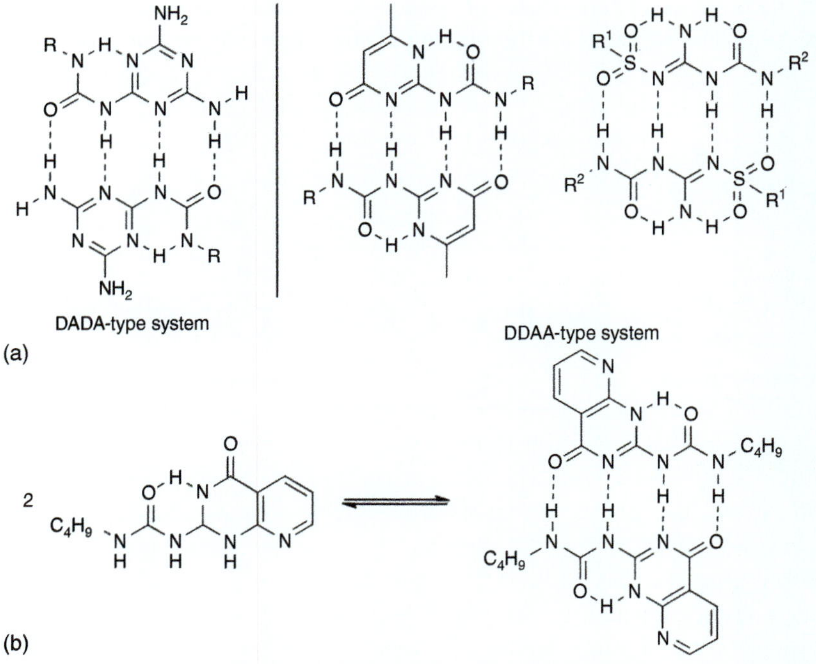

Figure 3.7 (a) Schematic representation of various quadruple H-bonding motifs (from left to right). Source: Refs. [18, 19] and Prabhakaran et al. [20]. (b) Schematic representation of a H-bonding monomer that dimerizes via quadruple H-bonding into four different tautomers (only one of these is depicted). Source: Corbin and Zimmerman [21].

H-bonding interactions (one selected example for this dimerization is given in Figure 3.7b) [21]. Later, Prabhakaran et al. showed that tautomerism of the binding site could be suppressed by preventing rotational freedom due to internal non-covalent fixation [20].

However, the intrinsic self-complementary nature of these quadruple H-bonding arrays prevents, in general, the controlled self-assembly of two different components, e.g. in a strictly alternating fashion. Corbin and Zimmerman reported on the directional conformational transition of a self-complementary AADD-type array into a non-self-complementary DAAD-type one in the presence of a 2,7-diamido-1,8-naphthyridine unit (Napy), itself representing a strong ADDA-type system; as a consequence, highly stable heterodimers were formed (Figure 3.8a) [21]. This approach offers the possibility to overcome the aforementioned restriction when using quadruple H-bonding systems for the formation of heteroassemblies. In a similar fashion, Chen and coworkers showed that the UPy-moiety is also able to form stable heterodimers in the presence of diamidonaphthyridine, as the complementary unit (Figure 3.8b) [37].

In addition, the use of multiple H-bonding interactions represents a further extension of the general idea to increase the binding strength via the number of connected H-bonding sites within a molecular system. From these extended systems, the DADDAD-type array, as initially reported by the Hamilton group [38–40], is the most prominent example featuring the formation of heteroassemblies in the presence of barbiturates or N-alkyl-cyanurates via sextuple H-bonding (Figure 3.9a). The K_a value for the binding of such molecules to this receptor in

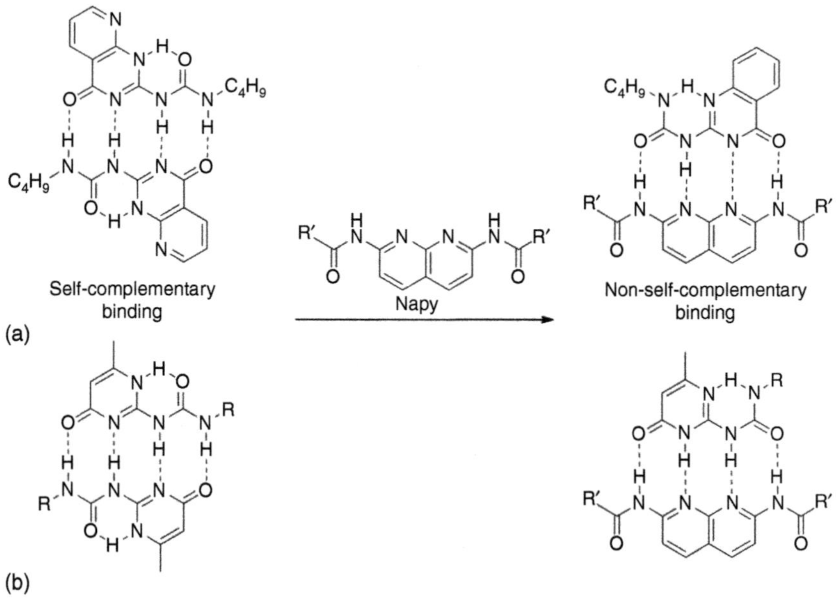

Figure 3.8 Schematic representation of Napy-induced translation of homodimeric into heterodimeric assemblies via quadruple H-bonding. Two examples according to Corbin and Zimmerman (a) and Chen (b) are shown in [21, 37].

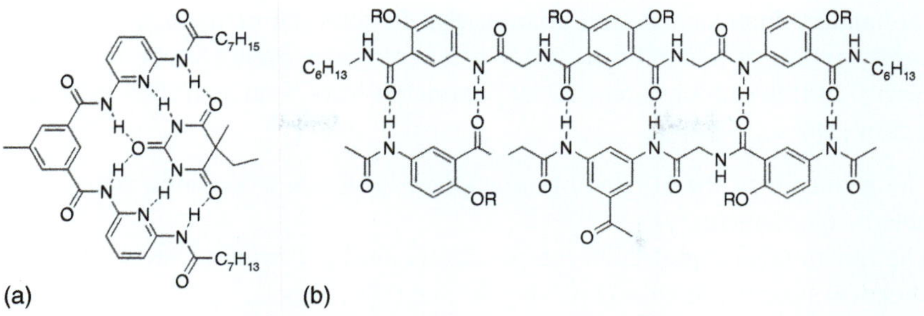

Figure 3.9 Schematic representation of heterodimers based on sextuple H-bonding systems. Source: (a) Chang and Hamilton [38] and (b) Yang et al. [41].

a hetero-complementary fashion is the same range as the one reported for the homodimerization of UPy-type derivatives (K_a value of c. $10^5\ M^{-1}$). In a related approach, short oligoamide chains were employed as non-self-complementary templates for the assembly of heterodimers via sextuple H-bonding interactions (Figure 3.9b); for this particular type of supramolecular array, remarkably high K_a values of c. $10^9\ M^{-1}$ were reported [41].

When dealing with supramolecular materials, in this particular case with those formed via H-bonding interactions, the characterization, for instance, with respect to binding energies, the K_a values, or the bond dynamics, is not straightforward [7]. The equilibria underlying the association of molecules via H-bonding into advanced supramolecular arrays, either small molecules or even macromolecules, are often complex; the mechanism involved in supramolecular polymers, as detailed in Chapter 1, can be studied by nuclear magnetic resonance (NMR) spectroscopy and supported by mathematical fitting procedures [42, 43]. Moreover, isothermal titration calorimetry (ITC) represents a highly efficient method to study the thermodynamics of the binding process – the association energy as well as free enthalpy, entropy, and enthalpy can thusly be determined [44, 45]. Infrared (in solution and in the solid state) and solid-state NMR spectroscopy are commonly applied to investigate the nature and the dynamics of H-bonding interactions [46, 47]. Furthermore, the kinetics, i.e. the time scale on which H-bonds have been broken/reformed, have been extensively studied [48]. For a more detailed evaluation of the range of possible characterization techniques (and their limitations) with respect to supramolecular polymers, the reader is referred to Chapter 12.

3.2 From H-Bonding Interactions to Supramolecular Polymers

The broad range of supra(macro)molecular materials that can be assembled via H-bonding interactions has already been summarized (see Figure 3.1). The linear polymers, such as block copolymers or "chain-extended" polymers, feature H-bonds within the polymers' main chains. These structurally most straightforward architectures are basically formed by the linking of homo- or heterotelechelic monomers via H-bonding interactions and, accordingly, the strength of these non-covalent linkages represents a significant factor concerning the overall properties of the materials. In this respect, three crucial parameters have been identified by Binder and Zirbs [3]:

1. The character of the H-bonding interaction itself (i.e. self-complementary vs. hetero-complementary),
2. The total number of molecular components linked by H-bonding interactions (i.e. chemical structure, degree of polymerization [DP], dispersity [Đ]), and
3. At least with respect to block copolymers, the compatibility or miscibility of the constituent segments/units (i.e. (micro)phase separation, micellization).

Concerning the later one, the association of monomers (as expressed by the K_a value), the polymers' mobility and, if applicable, the phase-separation behavior have to be correlated. Several molecular modeling [49, 50] and extensive experimental studies have dealt with this issue. As outlined in Chapter 1, supramolecular polymers of high molar mass will only be formed when high K_a values for the interaction are given [12]. The assembly of single H-bonding systems, supported by additional non-covalent interactions (e.g. π–π stacking, hydrophobic, or dipolar interactions), often gives materials with liquid crystalline (LC) behavior. Recently, halogen-bonding (X-bonding) and metallophilic interactions have been reported as alternative non-covalent interactions in the same respect [51, 52]. In other words, H-bonding can even be seen as a structural tool to stabilize liquid crystallinity [5]. The strong effect of single (or isolated) H-bonding interactions is typically restricted to the solid state; on the other hand, multiple H-bonding interactions may even give rise to miscibility effects in polymer blends. The higher the order of H-bonding interactions are, the more feasible the formation of well-defined (linear) supramolecular polymeric systems becomes: already two-centered H-bonding moieties can be utilized for this purpose. This trend holds even truer when the formation of polymer networks is considered: the reversible cross-linking via H-bonds results in significant changes in both the (melt-)rheological properties and the elastomeric behavior. Even more important, the materials exhibit a strong temperature dependency of their properties, which can be utilized to realize stimuli-responsive and self-healing materials [53–55]. As expected, these effects are more pronounced when effective (i.e. multivalent) H-bonding motifs are used. In addition, miscibility effects in terms of microphase separation become more pronounced. In particular, with respect to networks, based on H-bonding interactions, their properties can be adjusted simply via the number of such units (even when taking only a few of H-bonding sites per some thousands of monomers) [3]. The utilization of these effects represented the driving force for the optimization/tuning of the H-bonding systems: the Meijer and Hamilton groups developed highly efficient quadruple and sextuple systems ($K_a > 10^5\ M^{-1}$) widely utilized today in the field of supramolecular polymers.

H-Bonding arrays have been widely implemented into functional polymers and identified as beneficial with respect to properties of the material/device that they have been used for; representatively, self-healing materials [56] and bulk-heterojunction solar cells [57] are named in this context.

3.2.1 Supramolecular Polymers, Based on Single H-Bonding Interactions

From a synthetic chemist's point of view, the incorporation of a single H-bonding unit into a molecule is much easier to realize when compared to, for example, triple or quadruple H-bonding moieties. However, the intrinsic weakness of such arrays limits the range of potential applications. Supramolecular polymers with LC-like properties were developed, but this property was mainly observed for materials in which side chains were grafted to a polymer backbone, e.g. via the interaction between carboxylic acids and pyridine moieties [58–64]. Also, the assembly of polymer chains via multiple, yet isolated H-bonding sites was reported (the enhancement of miscibility of intrinsically immiscible polymers represented the main target of this research, Figure 3.10) [65]. These particular supramolecular systems can be realized, since the K_a value for the binding of pyridine to a carboxylic acid is much higher (K_a of c. 500 M^{-1}) than that for the dimerization of two acid molecules [59].

However, numerous publications are known where linear supramolecular polymers were formed due to the interaction between molecular components equipped with two carboxylic acids and two pyridine moieties, respectively, at their termini [66–70]. For instance, supramolecular polymers of rod-coil conformation were prepared by the self-assembly of 4,4′-bipyridine derivatives with telechelic poly(propylene oxide) equipped with benzoic acid end groups; highly ordered LC phases were observed for these materials by polarized optical microscopy [70]. H-bonding interactions between derivatives of tartaric acid and 4,4′-bipyridine units afforded chiral main-chain LC polymers, which were analyzed by circular dichroism (CD) spectroscopy, optical microscopy, and X-ray diffraction (XRD) measurements [71, 72].

The self-assembly of small molecules via single H-bonding interactions was also tested but gave only oligomers rather than the targeted supramolecular polymers [7]. As a notable exception, St. Pourcain and Griffin observed materials for which a polymer-like behavior was disclosed [73]: the supramolecular materials were assembled from the star-shaped tetrakispyridine derivative **1** and the linear dicarboxylic acid **2** (Figure 3.11). The ladder-type polymeric strands with thermoreversible behavior as well as polymer networks with LC properties were found; above the isotropization temperature, a large drop in material properties occurred. Thus, the interplay between association and LC behavior was identified to be crucial for

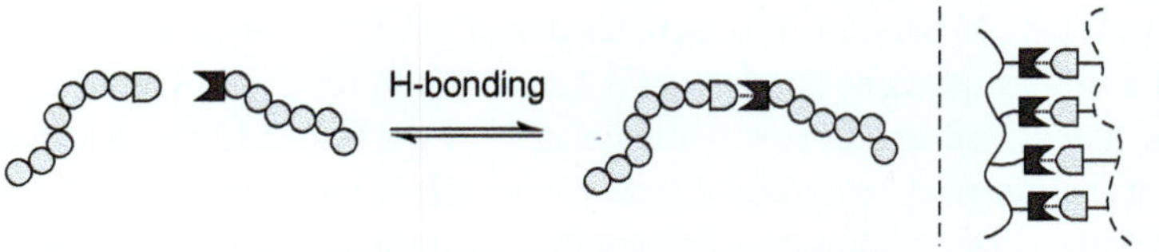

Figure 3.10 The utilization of single H-bonding for the formation of main-chain supramolecular materials (left) and inter-chain connection polymer blends (right). Source: Redrawn from Binder and Zirbs [3]. © 2007 Springer Nature.

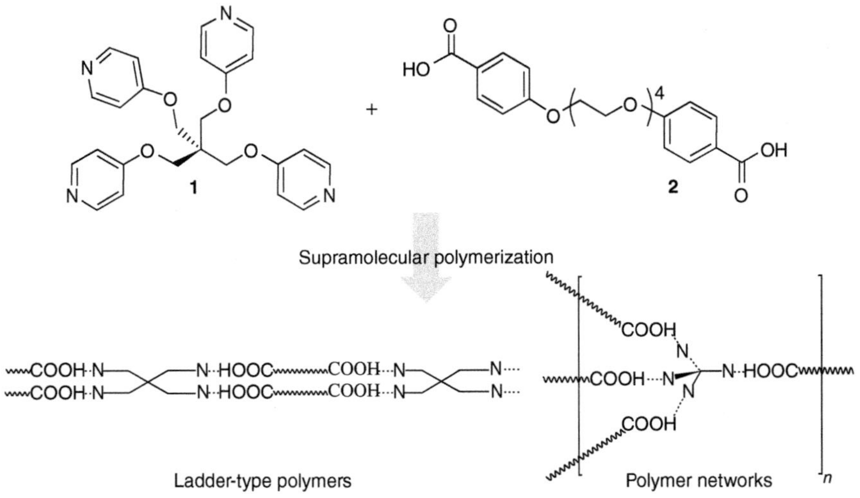

Figure 3.11 The formation of supramolecular ladder-type polymers and networks based on monomers **1** and **2**. Source: St.Pourcain and Griffin [73].

the polymerization process. Differential scanning calorimetry (DSC) measurements suggested a "memory effect": the crystallinity decreased with the time and that the materials were held in their isotropic state [74].

3.2.2 Supramolecular Polymers, Based on Double H-Bonding Interactions

In general, the interaction of molecules via double H-bonding interactions is stronger as in the case of isolated ones; however, their utility in the field of supramolecular polymers is still rather poor.

Uradiazoles (i.e. representing a H-bonding motif of the AD-type), for example, were employed by Stadler et al. for the formation of reversible cross-links within polymers (Figure 3.12) [75–93]. In general, H-bonding interactions enforced by (micro)phase separation, i.e. formation of crystalline domains, are similar to the enforcement due to liquid crystallinity (as in the case of single H-bonding arrays, see Section 3.2.1) [5]. Commercial polybutadiene (butadiene rubber, BR) was functionalized with H-bonding sites (degree-of-functionalization of c. 2%); the resultant material exhibited significant changes in the thermal and mechanical properties when compared to the parent BR (making it similar to thermoplastic elastomers): the intact H-bonding interactions contributed to the properties in almost the same way as covalent bonds would do (although only at low temperatures); whereas, at elevated temperatures, dissociation of the H-bonds resulted in a flow behavior of the material (as typical for polymers of low molar mass). Various types of characterization techniques were applied to understand the reversible thermoresponsive properties of the polymers: DSC [78, 79], light scattering [80], X-ray scattering [77], rheology [81–87], dielectric spectroscopy [88, 89], as well as ^{2}H NMR [90] and infrared (IR) spectroscopy [91–93]. Brunsveld et al. suggested a range

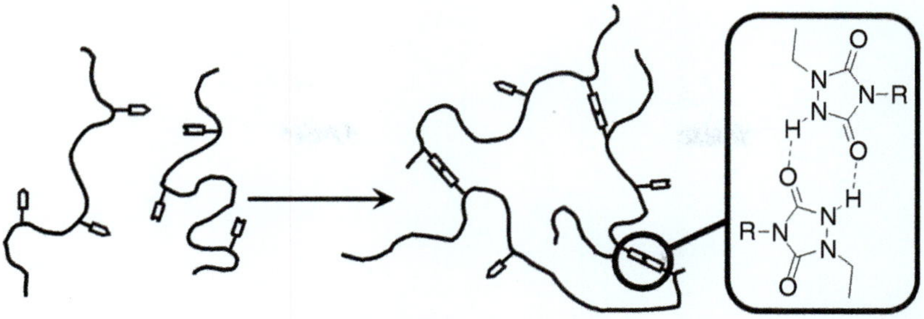

Figure 3.12 Schematic representation of the formation of a supramolecular polymeric network using double H-bonding interactions of the self-complementary uradiazole units.

of possible applications of these "associative polymers" [5]: rheology modifiers, adhesives, adsorbents, coatings, surfactants, and stabilizers. A summary of further potential applications for supramolecular polymers, where the double H-bonding interactions are placed in the side chains (e.g. the bivalent association of amides), can be found elsewhere [3].

Besides the reversible cross-linking of polymer chains via side-chain functionalities, examples for the formation of linear supramolecular polymers are also known. The dimerization of carboxylic acids has been mainly utilized for this particular purpose. In this respect, the telechelic poly(dimethylsiloxane)s (**3**, PDMSs), bearing two benzoic acid end groups (Figure 3.13a), were investigated by Bouteiller and coworkers [4, 94, 95]. Upon supramolecular polymerization, the formation of cyclic oligomers was observed – this behavior is prototypal for ring-chain equilibrium supramolecular polymerizations (for a detailed discussion of the mechanism, see Section 1.3.2). The polymerization of monomers of different chain lengths was studied by Fourier-transform infrared (FT-IR) spectroscopy and viscometry (Figure 3.13b/c). The later technique allowed one to determine a critical concentration ($c = 10\,\mathrm{g\,l^{-1}}$) for the crossover from supramolecular (macro)cycles to linear polymer chains (in control experiments; however, the corresponding dibenzyl ester did not exhibit such a behavior). Applying the theoretical models according to Jacobsen and Stockmeyer [96] as well as Ercolani et al. [97] allowed the quantification of the equilibrium between the two different species: concentration-dependent FT-IR absorption measurements were conducted to determine a K_a value of $260 \pm 50\,\mathrm{M^{-1}}$ for the dimerization of two acid units as well as the Jacobson–Stockmayer parameters B_1 and B for cyclization (B_1 and B denote the constants for the cyclization of the monomer and the cyclization of all linear oligomers, respectively). The parameter B_1 was found to increase with the length of the linker, implying a larger fraction of cyclic species for larger macromonomers. However, no clear trend was found for the second parameter, since opposite effects contribute to B (i.e. the distance between the binding sites and the stiffness of this linker). For diacids with long linkers, representing unstrained cyclic monomers, the value of the B/B_1-ratio was close to unity. As a result, from their studies, the authors could describe the $Đ$ values of cyclic and linear species in solution to be concentration dependent. ^{13}C NMR relaxation studies were conducted revealing

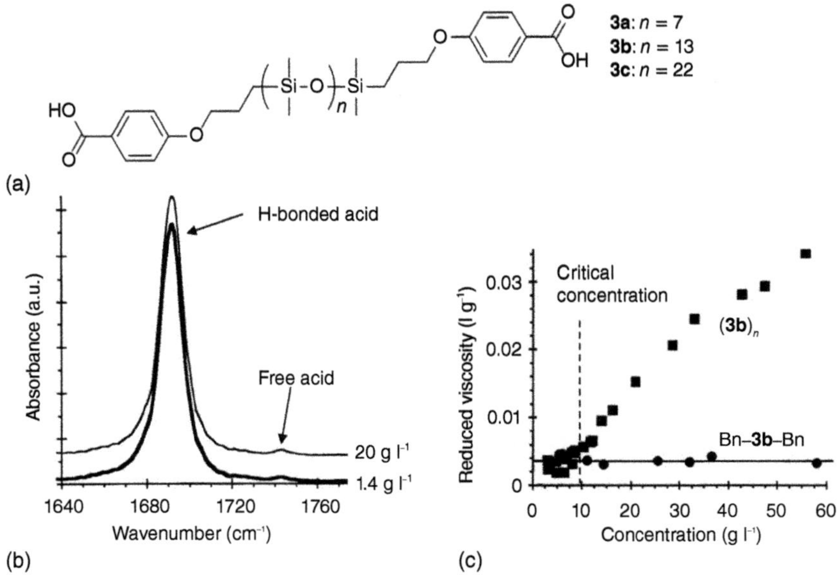

Figure 3.13 (a) Schematic representation of the telechelic PDMSs **3** as monomers for the supramolecular ring-chain equilibrium polymerization. (b) FT-IR spectra of $(\mathbf{3b})_n$ at two different concentrations (c = 1.4 and 20 g l^{-1}). (c) Concentration dependence of the reduced specific viscosity of $(\mathbf{3b})_n$ and the corresponding dibenzyl ester Bn−**3b**−Bn (Bn = benzyl, hexane, 25 °C). Source: Abed et al. [4]. Figure reproduced with kind permission. © 2000 American Chemical Society.

a polymer-like behavior, such as reduced chain-mobility, which was rationalized by the presence of sufficiently stable H-bonding interactions. According to this analytical method, a "virtual" molar mass in the order of c. 200 000 g mol^{-1}, corresponding to a DP of c. 100, was determined [94].

In a different approach, the Sleiman group investigated the self-assembly of diacids having an azobenzene moiety within the linker [98]. The formation of either linear or cyclic species could be triggered via the configuration of the central N=N-linkage, which could be switched via photoisomerization. The former were preferred for the trans-isomer; whereas, the cis-isomer favored the latter. For the cis-isomer, the formation of fibers upon deposition onto carbon grids was observed by transmission electron microscopy (TEM) imaging. An example for isodesmic supramolecular polymerization (IDP) (for the mechanism of this type of supramolecular polymerization, see Section 1.3.1) was reported by Wärnmark and coworkers [99]: the chiral two-centered self-complementary H-bonding unit **4** (Figure 3.14a) self-assembled into a helical, tubular supramolecular polymer. The polymerization was monitored by ^{1}H NMR, where the resonance of the amido-proton showed a clear dependency on the applied concentration. Modeling of the isodesmic binding isotherm for different solvents gave K_a values of $1.1 \times 10^3\ M^{-1}$ and $1.1 \times 10^2\ M^{-1}$ (in CD_2Cl_2 and $CDCl_3$, respectively); thus, $CHCl_3$, as the slightly more acidic solvent, suppressed the H-bonding interaction to a larger extent than CH_2Cl_2. Moreover, the thermodynamics of the polymerization were analyzed by temperature-dependent ^{1}H NMR.

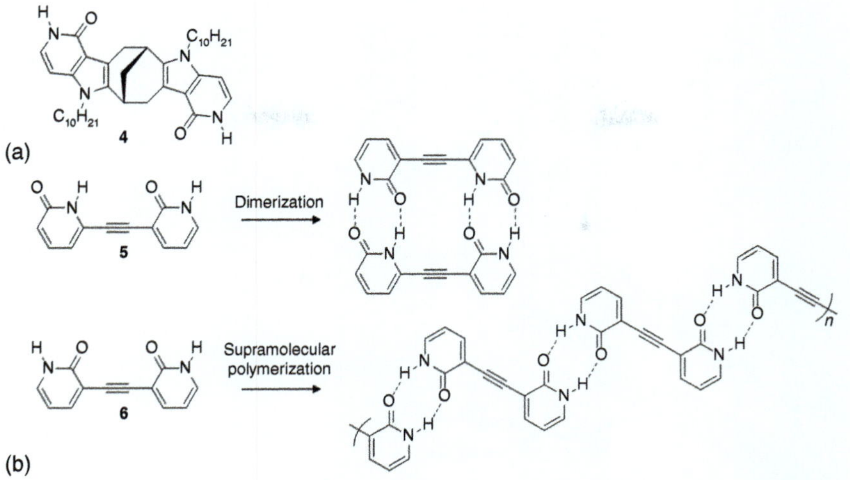

Figure 3.14 (a) Schematic representation of the chiral two-centered H-bonding unit **4** with its bicyclo[3.3.1]nonane core. (b) Schematic representation of the structure-dependent supramolecular self-assembly of monomers **5** and **6**.

As a characteristic feature for the enthalpy-driven polymerization, negative values were determined for both the enthalpy (ΔH_p) and the entropy of association (ΔS_p). Additionally, vapor pressure osmometry (VPO) experiments allowed one to exclude that only an equilibrium between monomers and dimers was present, similar K_a values to those determined via NMR were obtained. In $CHCl_3$ at 313 K, a very low DP of c. 4 was observed; according to the mechanism of an IDP, low K_a values always result in low DPs. Moreover, π–π stacking interactions, as a contributor to the aggregation, were excluded on the basis of UV/vis absorption spectroscopy, the helical nature of the polymer was confirmed by CD measurements.

The formation of cyclic and linear assemblies can also be controlled via the molecular structure of the monomer. For instance, Ducharme and Wuest reported on the self-assembly of the two isomeric dipyridones **5** and **6** (Figure 3.14b) [100]. The asymmetric isomer **5** was found to dimerize in solution (the Gibbs free energy for this process was estimated to be $-27.2\,\text{kJ mol}^{-1}$); whereas, the symmetric isomer **6** polymerized into planar, linear structures. X-ray single crystal analyses of the two monomers showed that different modes of interaction were retained also in solid state.

The perylene bisimides (PBIs) represent an important class of organic dyes that has been widely studied due to their self-assembly behavior in solution as well as in the solid state. Though these processes are mainly governed by π–π stacking interactions (see Chapter 5), H-bonding interactions have been employed to organize such molecules into supramolecular structures of higher order [101]. In this respect, the Würthner group designed the *N,N′*-unsubstituted PBI derivative **7** featuring four bulky tris-alkoxyphenyl substituents in the so-called bay-positions [102]. Due to the steric crowding, a rotational twist of c. 25° between the outer naphthalene imide subunits was achieved that efficiently suppressed the formation of columnar stacks

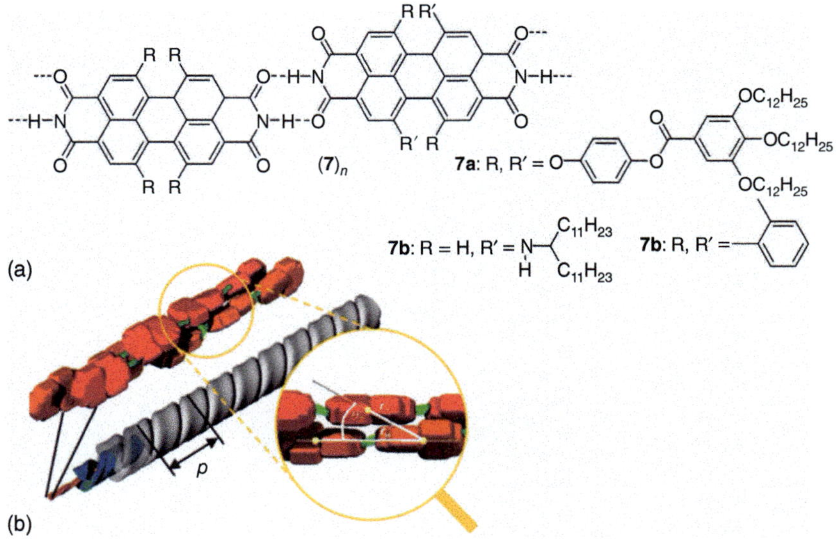

Figure 3.15 (a) Schematic representation of the supramolecular polymer $(\mathbf{7})_n$. (b) Schematic representation of the dense 1D packing of **7a** in a slipped fashion (the red twisted blocks represent the PBI core, the bay substituents are shown in gray cones with a blue apex and H-bonding interactions are indicated as green lines). Source: Würthner et al. [101]. Figure reproduced with kind permission. © 2016 American Chemical Society.

via π–π stacking interactions and, thereby, crystallization in solution. Whereas the interplay of π–π stacking interactions between PBIs, stacked in a slipped fashion, and H-bonding interaction between the imide groups of adjacent PBIs allowed the molecules to assemble into the helical double-stringed nanofibers $(\mathbf{7})_n$ in which a selective head-to-tail orientation of the molecules was found (Figure 3.15). The material exhibited an intense fluorescence, which was attributed to the dense packing and, thus, strong excitonic coupling of the chromophores (J-type aggregates) [103]. Single-molecule emission spectroscopy measurements of the polymer revealed a strong fluctuation of the emission intensity (the so-called "blinking"); this finding was rationalized by exciton-transport processes along the chains (operable over a distance of c. 70 nm) [104].

The Wasielewski group introduced 1,7-diamino-substituted PBIs as an "artificial chlorophyll" due to their bright green color [105]. The same binding mode as in the previous example was used to assemble **7b** into J-aggregate fibers (Figure 3.15) [106]. The supramolecular material showed a bathochromically shifted J-type absorption in the near-infrared (NIR) regime (accompanied by an absorption band at higher energies, i.e. c. 450 nm). Due to the particular stacking mode and the resultant spectral features of $(\mathbf{7b})_n$, a potential use of this polymer in NIR-sensitizing applications was proposed. The tetra-substituted derivative **7c** (Figure 3.15) was found to function as an efficient organo-gelator [107]; its self-assembly into highly fluorescent 1D J-type aggregates via a nucleation–elongation chain-growth mechanism involved a cooperativity of H-bonding and π–π stacking interactions. Its gelation at very low critical gelation concentrations (cgc) was observed in toluene (cgc as low as 0.05 wt%). Noteworthy, the methyl groups in the substituents were identified as crucial to retain solubility and, even more importantly, induced a torsional twist, which prevented columnar association.

Diketopyrrolopyrroles (DPPs) represent one further class of organic dyes, which have been assembled into supramolecular polymers by π–π stacking interactions (also see Section 5.2) [108]. As for the aforementioned PBIs, H-bonding interactions have been employed for the self-assembly of DPPs. Dyes of the DPP family feature remarkable optical and electronic properties, which make them interesting candidates or applications, even on the industrial scale (the famous "Ferrari" red pigment is a DPP derivative). However, due to the presence of complementary H-bonding arrays, structurally simple DPPs are nearly insoluble in common organic solvents. This behavior makes a processing of DPPs from solution demanding and chemical-vapor deposition (CVD) has often to be applied. To overcome this limitation, Zambounis et al. utilized *N*-*t*-Boc-protected DPP **8**, as a "latent" and organo-soluble pigment, which could readily be processed from organic solutions (*t*-Boc: *tert*-butoxycarbonyl) [109]. Thermal annealing of the resultant films cleaved the protecting group and generated DPP which, in turn, organized itself into a supramolecular polymer (Figure 3.16). However, due to the rather rough and inhomogeneous morphology, such DPP films showed a much lower hole conductivity when compared to a film fabricated by CVD [110].

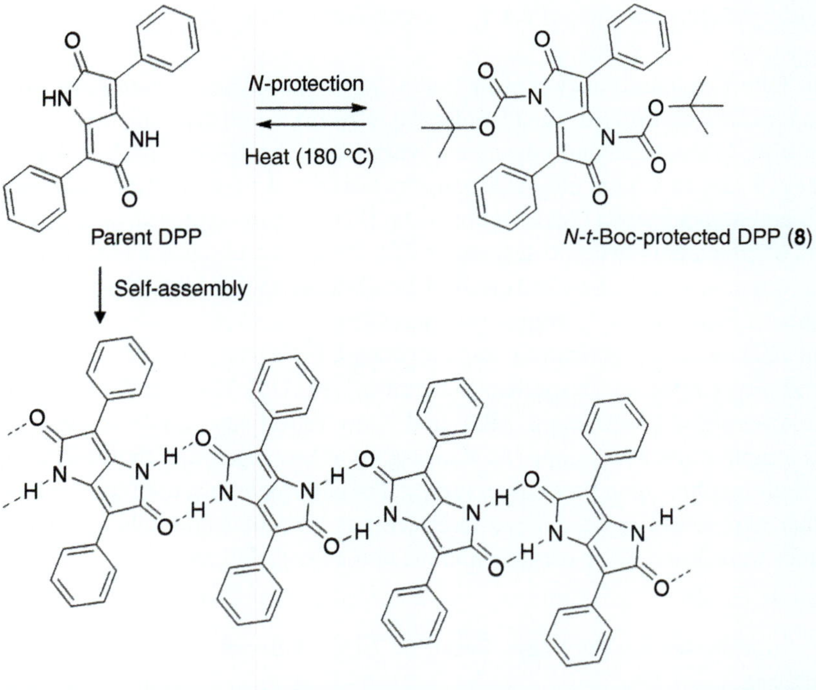

Figure 3.16 Schematic representation of the formation of parent DPP from its *N*-*t*-Boc-protected derivative **8**; DPP self-assembles into linear supramolecular polymer due to double H-bonding interactions.

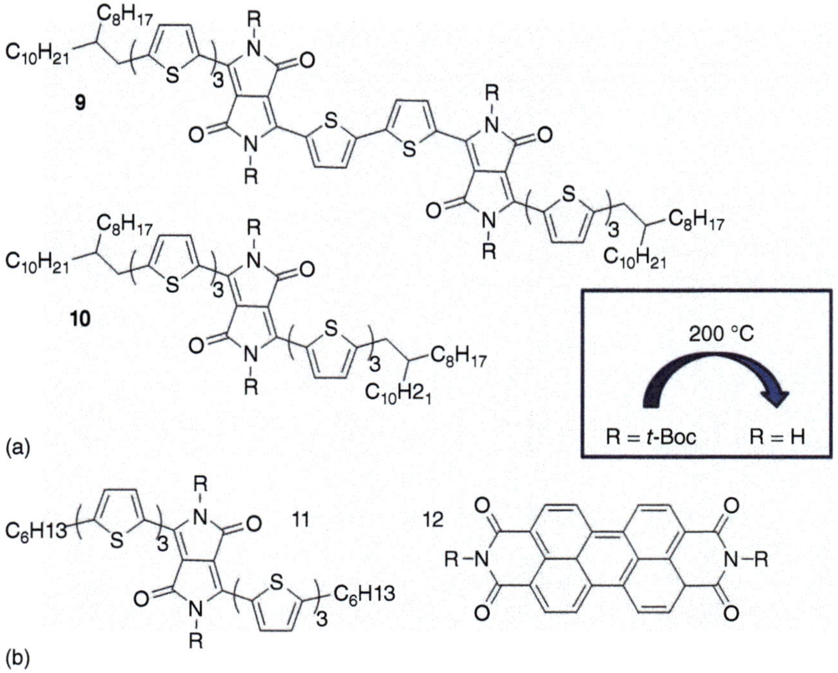

Figure 3.17 (a) Schematic representation of the *N-t*-Boc-protected DPP oligomers **9** and **10**. (b) Schematic representation of the *N-t*-Boc-protected DPP and PBI dyes **11** and **12**.

In the following, the strategy of using "latent" dyes was applied by various authors to a range of DPP derivatives – small molecules, oligomers, and even polymers [108]. For example, Yamashita and coworkers synthesized the *N-t*-Boc-protected DPP oligomers **9** and **10** which, after solution processing and thermal treatment, gave smooth and homogeneous films (Figure 3.17a) [111]. When employed, as an active layer, in organic field-effect transistors (OFETs), efficient ambipolar charge-carrier transfer was observed, i.e. hole and electron mobilities of up to 6.7×10^{-3} cm^2 V^{-1} s^{-1} and 5.6×10^{-3} cm^2 V^{-1} s^{-1}, respectively, were found. In another study, Beverina, Caironi and coworkers fabricated complementary OFETs using DPP **11** and PBI **12** as *p*- and *n*-type materials, respectively (Figure 3.17b) [112]. These authors applied UV/vis absorption spectroscopy, XRD, and X-ray reflectivity (XRR) analyses as well as atomic force microscopy (AFM) imaging to investigate the morphological changes of the films upon thermal annealing. For both materials, the films became smoother and denser upon thermal treatment; as a result, higher charge-carrier mobilities were found when compared to the non-annealed films.

3.2.3 Supramolecular Polymers, Based on Triple H-Bonding Interactions

In contrast to the examples listed in Sections 3.2.1 and 3.2.2, triple H-bonding interactions represent a widely used motif in the field of supramolecular

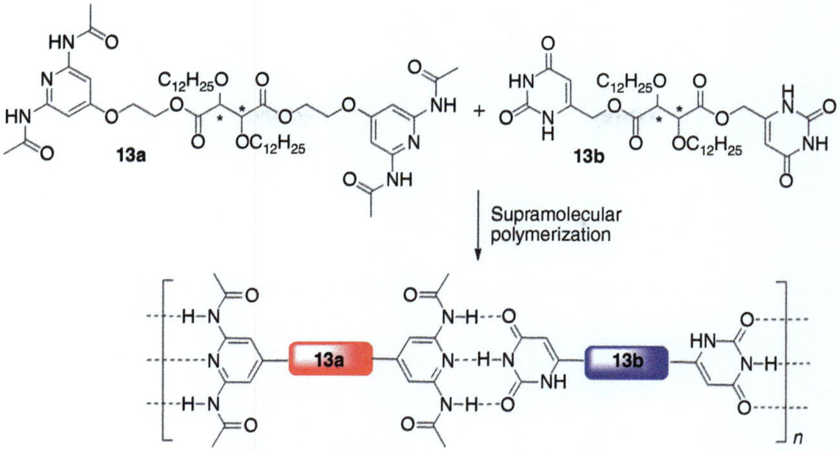

Figure 3.18 Schematic representation of the supramolecular polymerization of the homotelechelic monomers **13a** and **13b** into an alternating copolymer.
Source: Fouquey et al. [2].

polymer chemistry. In this context, combinations of 2,6-diaminopyridines and 2,6-diamino-1,3,5-trizines with derivatives of flavine, thymine/uracil, and succinimide are of particular importance (Figure 3.5); moreover, the natural nucleobase-pairing modes in a Watson–Crick- or Hogsteen-type fashion have also been employed in this context [23].

Outstanding credit has to be given to Lehn and coworkers for recognizing the importance of the interaction between 2,6-diaminopyridine and uracil; this particular interaction was used to assemble the first linear supramolecular polymer homotelechelic monomers via triple H-bonding interactions (**13a/b**, Figure 3.18) [2, 113]. At an ideal 1 : 1 stoichiometry of the molecular components, superstructures of columnar shape were formed showing a LC-type behavior up to c. 200 °C (the pure monomers instead melted into isotropic liquids not showing any LC properties). In the presence of a chiral tartrate spacer helicity was induced and fibrous superstructures were found for the supramolecular polymer; the self-assembly process was studied by electron microscopy imaging from nano- to micrometer scales [114]. Further, using planar 9,10-dialkoxyanthracene moieties within the monomers, as in **14**, gave rigid-rod supramolecular polymers (Figure 3.19) [115]. As a result, from the increased rigidity of the polymer chain, the LC properties were changed – instead of the thermotropic behavior, a lyotropic LC phase was identified in apolar solvents, which was birefringent and possessing high viscosity [116].

Kitamura and coworkers investigated the H-bonding between melamine derivatives and complementary cyanurates or barbiturates [117–121]. For example, the amido-bridged melamine dimers, having a diazobenzene core, aggregated into the binary organogels **15** [119]. In liquid hydrocarbon solvents, the formation of entangled nanofibers was observed (Figure 3.20a). As shown by UV/vis and IR spectroscopy, packing of the polymer chains was enhanced by both π–π stacking interactions of the rigid diazobenzene tethers as well as by additional interchain

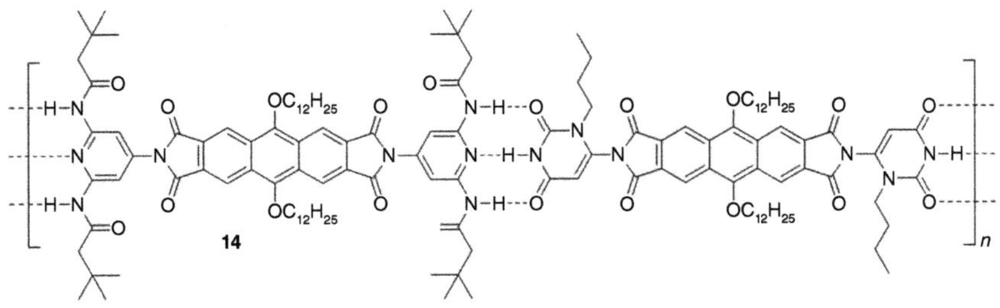

Figure 3.19 Schematic representation of the LC supramolecular polymer **14**. Source: Kotera et al. [115].

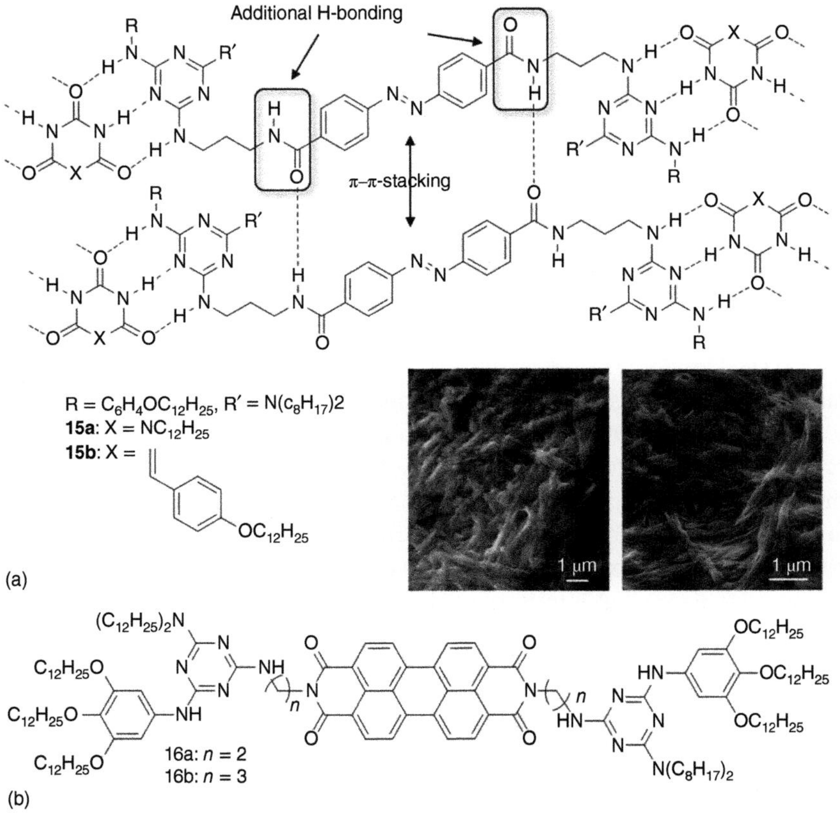

Figure 3.20 (a) Schematic representation of the formation of supramolecular gels based on triple H-bonding motifs in concert with additional non-covalent interchain interactions. Representative field-emission scanning electron microscopy (FE-SEM) images of the fibers obtained from the dried benzene gels are also shown (left: **15a**, right: **15b**). (b) Schematic representation of monomer **16** used for the supramolecular polymerization with *N*-dodecyl cyanuric acid. Source: Yagai et al. [119]. Figure reproduced with kind permission; © 2005 American Chemical Society.

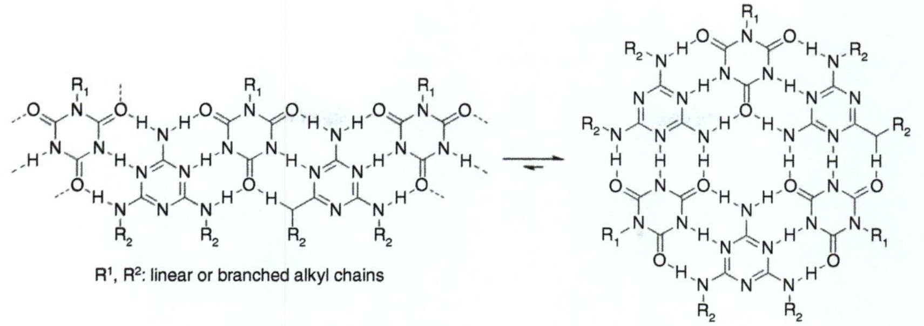

Figure 3.21 Schematic representation of the equilibrium between supramolecular structures of linear tape-like and cyclic rosette-like shapes.

H-bonding of the amide functionalities. Moreover, the well-defined self-assembly in methylcyclohexane, according to the mechanism for an IDP (see Section 1.3.1), was reported for **16b** with *N*-dodecyl cyanuric acid, as the complementary substrate (in contrast, **16a** generated only a dimeric species under the same conditions, Figure 3.20b) [122, 123]. Thus, the structure of the supramolecular architecture, which was analyzed by dynamic light scattering (DLS) measurements and AFM imaging, was mainly governed by the length of the alkyl spacer of **16**. According to temperature-dependent UV/vis absorption spectroscopy, secondary π–π stacking interactions could also have contributed to the structure of the supramolecular polymer.

The complementary couple melamine/cyanurate was also studied by Reinhoudt and coworkers [124]; in their work, the parameters influencing the equilibrium between tape-like linear and rosette-like cyclic assemblies were in the focus of interest (Figure 3.21). In their proposed model, the presence of linear structures consisting of up to eight components was assumed to be present under the experimental conditions (i.e. in total 270 stereoisomers may have formed). Moreover, various steric parameters (describing the steric interactions within the assemblies), a bimolecular K_a value for the cyanurate–melamine interaction as well as the equilibrium constant for cyclization of linear hexamers, were taken into consideration: due to a higher equilibrium constant for cyclization, the formation of rosette-like structures should be preferred over tape-like ones. Experimentally, the ratio of linear to cyclic species was not to be affected by steric repulsions in the tape-like structure (though, the model predicted a pronounced sensitivity). According to Whitesides' concept of peripheral crowding [125–127], with increasing bulkiness of the substituents on the melamine (leading to a nonplanar structure of the linear assembly) the tendency of the cyclic motif to crystallize increased. Besides linear or branched alkyl chains as substituents, more complex moieties, such as (metallo-)porphyrins on the melamine units [128] or metal complexes on both building blocks [129], were studied against the same background.

The directional imide–melamine H-bonding interaction was utilized by Würthner et al. for the self-assembly of *N*,*N*′-unsubstituted PBIs dyes and melamine derivatives into linear supramolecular polymers, such as **17** (Figure 3.22) [130, 131].

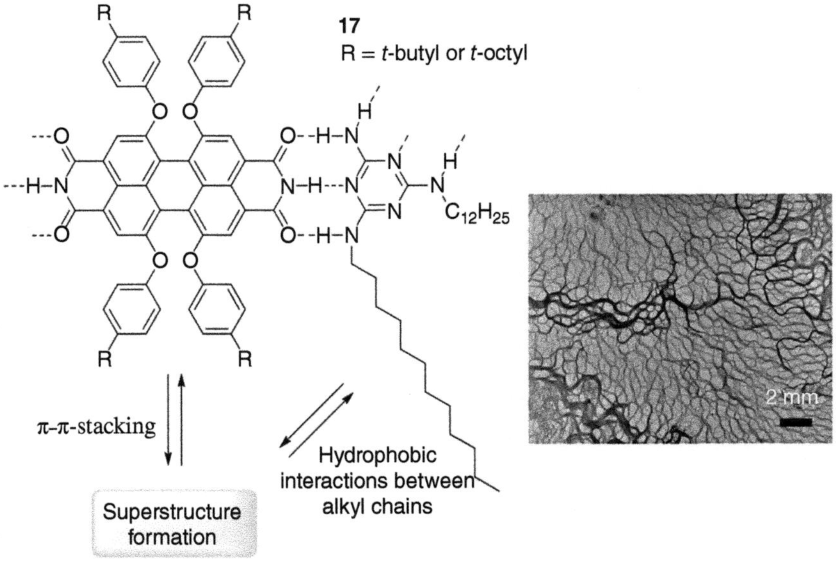

Figure 3.22 Schematic representation of the supramolecular polymer **17** based on triple H-bonding interactions, supported by π-π stacking hydrophobic interactions and a PBI core. A representative TEM image of the mesoscopic superstructures that were fabricated by evaporation of a methylcyclohexane solutions of **17** ($c = 5 \times 10^{-5}$ mol l^{-1}) on a carbon-Formvar-coated nickel grid (200 mesh) and subsequent staining with uranyl acetate is also shown. Source: Würthner et al. [130]. Figure reproduced with kind permission; © 1999 Wiley-VCH.

This approach is complementary to the assembly of ligand-equipped PBIs via metal-to-ligand coordination (see Chapter 4). Due to the low association constants ($K_a < 100\,M^{-1}$) in concert with the low solubility, supramolecular polymerization was not observed in halogenated solvents (e.g. $CHCl_3$ or CCl_4). However, in methylcyclohexane (where the solubility of both monomers was still sufficient), the two monomers gave well-defined mesoscopic structures of high photostability and a bright fluorescence behavior. The self-assembly process was found to be of complex nature: not only H-bonding interactions but also π–π stacking and hydrophobic interactions contributed to the resulting supramolecular superstructure, even at very low concentrations (10^{-5} to 10^{-6} mol l^{-1}). TEM and scanning electron microscopy (SEM) imaging were used to determine the dimensions of the obtained cylindrical strands (diameters in the range of 200–300 nm were found, Figure 3.22).

Combining the same PBI chromophores with melamines, derived from amino acids, allowed one to induce chiroptical properties into the materials [132]. In the case of the L-alanine-derived melamine, exciton-coupled CD spectroscopy showed a positive Cotton effect; whereas, the supramolecular polymer incorporating a D-alanine-based melamine monomer had a negative Cotton effect – these findings were assigned to M- and P-helical stacks of the PBI molecules within the supramolecular material, respectively. Similar results were obtained when using melamines based on L- or D-phenylalanine.

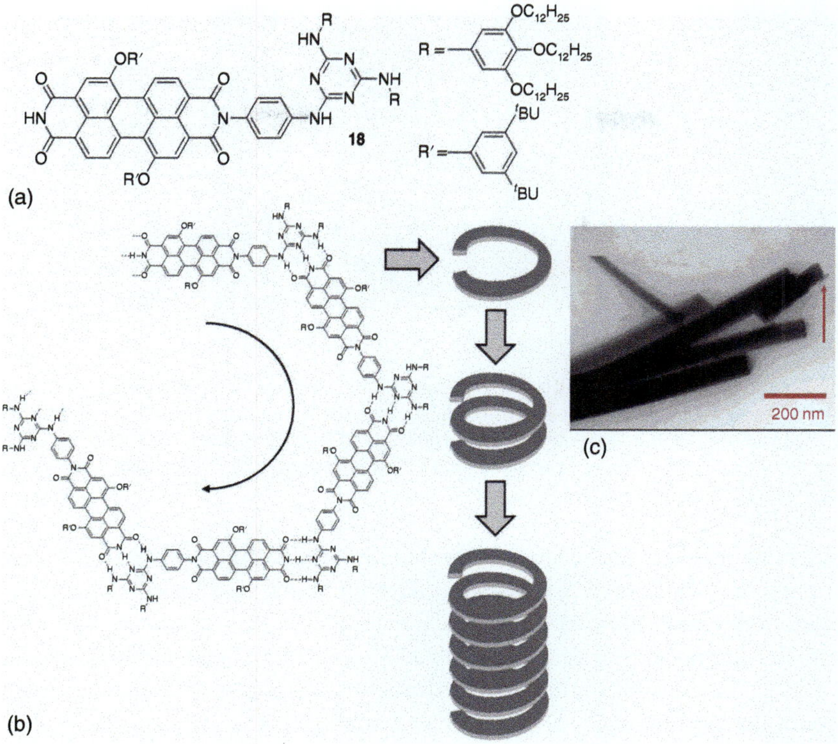

Figure 3.23 (a) Schematic representation of the heteroditopic PBI **18**. (b) Schematic representation of the supramolecular polymerization of **18** into cylindrical nanostructures. (c) Representative TEM image of the nanocylinders. Source: Sinks et al. [133]. Figure reproduced with kind permission. © 2005 American Chemical Society.

Sinks et al. studied the supramolecular polymerization of the heterotelechelic PBI **18** (Figure 3.23a) through an interplay of H-bonding and π–π stacking interactions as well as microsegregation into cylindrical nanostructures with large diameters (both in solution and in the solid phase) [133]. The structure of the supramolecular assembly was investigated by UV/vis absorption and ^{1}H NMR spectroscopy as well as small-angle X-ray scattering (SAXS) measurements; in solution, the highly ordered H-bonded hexagonal motif possessing a two-turn helical shape was found (each comprising 12 monomers, Figure 3.23b). These arrays, in turn, grew up to c. 1 μm in length and eventually generated bundled cylindrical structures in the solid state (as evidenced by XRD measurements and microscopy imaging, see Figure 3.23c). The photophysics of the assembly in solution was studied by UV/vis absorption, time-resolved fluorescence, and fs-transient absorption spectroscopy; on the basis of these measurements, the initial population of an excimer-type state and the subsequent ultrafast transfer of an electron from the donor-type tris-alkoxybenzene substituents to the acceptor-type PBI cores was concluded.

Besides formation of supramolecular polymers in solution, the self-assembly of PBIs on surfaces, driven by directed H-bonding interactions, has attracted

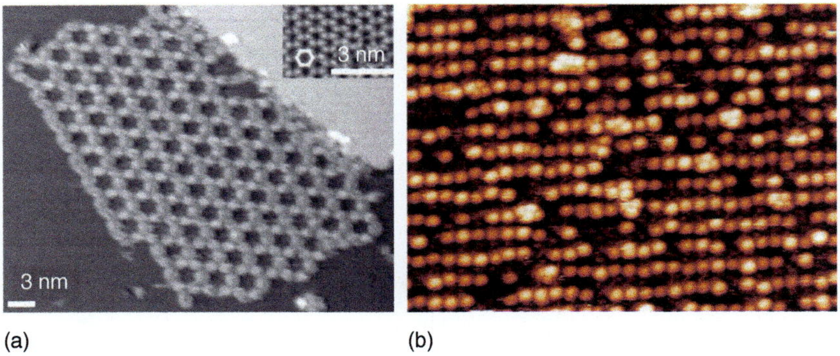

Figure 3.24 (a) Scanning tunnelling microscopy (STM) image of the honeycomb structure on a Ag-layered Si substrate (inset: high-resolution view of the substrate surface) [134]. (b) STM image of the Lu@C_{82}-loaded network on the Au(111) surface (area of $55 \times 40\ nm^2$). Source: Silly et al. [135]. Figure reproduced with kind permission; © 2003 Nature Publishing Group and 2008 The Royal Chemical Society.

considerable interest. For example, Theobald et al. assembled pristine PBI and melamine on a Ag-layered Si substrate under ultrahigh vacuum conditions into an open honeycomb network (Figure 3.24) [134, 136]. In a similar fashion, the supramolecular network was also assembled on transition metal surfaces, such as gold [137]. The pores of the honeycomb structure were utilized as templates for the controlled capture of C_{60} clusters (up to heptamers). Moreover, single Lu@C_{82} metallofullerenes could be assembled onto the pores of the surface-deposited supramolecular network, e.g. on a Au(111) surface (Figure 3.24) [135].

The solution-based deposition of PBI and melamine onto a gold surface also afforded a honeycomb-type arrangement, which allowed the subsequent formation of a self-assembled monolayer (SAM) of thiol **19** within the pores of the network [138]. This hybrid material allowed further modification of the surface via the underpotential deposition of Cu atoms selectively onto the SAM-filled pores (Figure 3.25) [139].

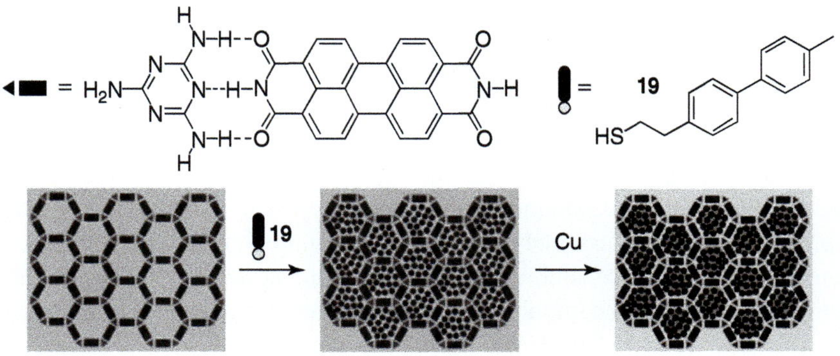

Figure 3.25 Schematic representation of the stepwise fabrication of a hybrid structure starting from a supramolecular honeycomb PBI–melamine network. Source: Silien et al. [139]. Figure reproduced with kind permission; © 2009 Wiley-VCH.

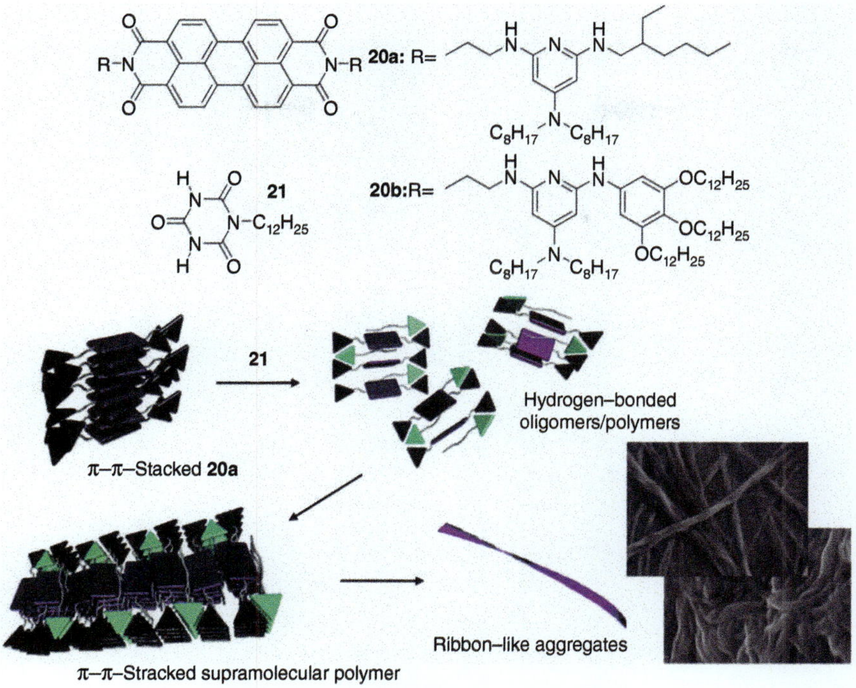

Figure 3.26 Schematic representation of the supramolecular self-assembly of **20a** and **21** into ribbon-like aggregates. Representatively, a field-emission SEM image of the aggregates is also shown. Source: Yagai et al. [123]. Figure reproduced with kind permission. © 2007 American Chemical Society.

The Yagai group employed melamine-bearing PBIs in combination with complementary cyanurate derivatives to assemble supramolecular architectures, based on triple H-bonding interactions [101, 140]. For example, the PBI–melamine conjugate **20a** showed the characteristic π–π stacking behavior in a face-to-face manner (UV/vis absorption spectroscopy revealed the typical spectral features for an excitonically coupling of PBIs in such an array) [123]. In the presence of 1 equiv of cyanurate **21**, however, the formation of a H-bonded supramolecular polymer occurred, even at the low concentration of 10 μmol in aliphatic solvents. At higher concentrations, ribbon- or rope-like aggregates were formed as a result of a hierarchically self-organization driven by π–π stacking of the PBI units along the polymer strands (Figure 3.26). A similar behavior was observed for the supramolecular material based on **20b** and **21**; whereas, a shorter spacer in the PBI–melamine conjugates generated discrete dimers when combined with **21** [122]. The complementary H-bonding interactions between melamine derivatives and barbiturates/cyanurates have also been applied to other π-conjugated systems, such as azobenzenes, diarylethenes, merocyanines, and short π-conjugated oligomers [e.g. oligo(*p*-phenylenevinylene)s, oligo(*p*-phenyleneethynylene)s, and oligothiophenes]. For an in-depth discussion of these supramolecular assemblies, the reader is referred to Yagai's review on this topic [140].

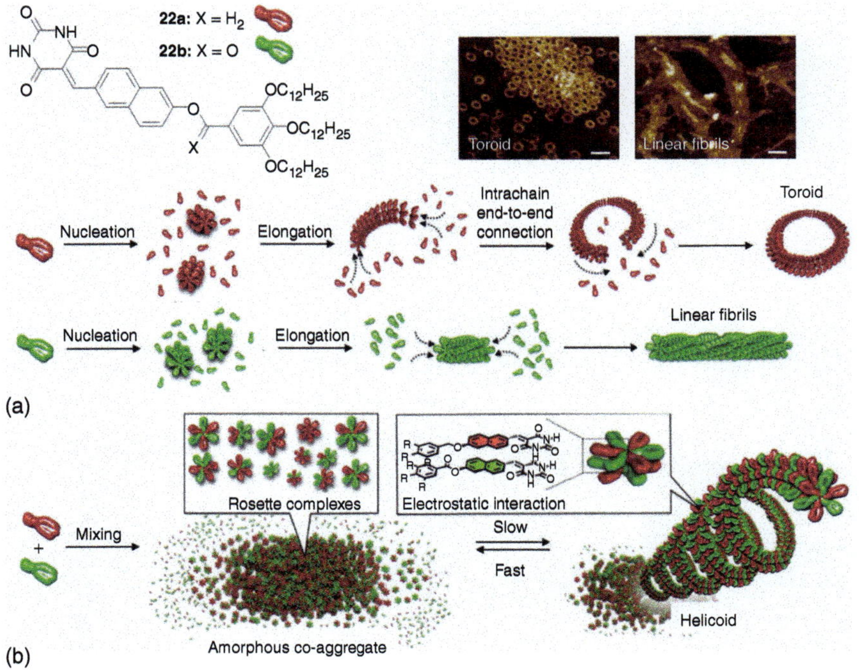

Figure 3.27 Schematic representation of the monomers **22**, which self-assembled into toroids and linear fibrils (a) or, upon mixing, generated a supramolecular helical polymer (b). Source: Aratsu et al. [141]. Figure reproduced with kind permission. © 2020 Springer Nature.

Very recently, Yagai and coworkers reported on a supramolecular polymerization, which was driven by the self-sorting of H-bonded, rosette-shaped precursors [141]. The individual monomers **22a** and **22b** assembled, following a cooperative nucleation–elongation mechanism, into toroids and linear fibrils, respectively (Figure 3.27a). The mixture of **22a** and **22b**, however, gave an amorphous co-aggregate of diverse rosettes, which were formed by a kinetic co-assembly process. The co-aggregate slowly transformed into a helical architecture (self-sorting of the rosettes via electrostatic interactions was proposed as the origin for this process, Figure 3.27b). Upon heating, the helical structure collapsed rapidly, thus regenerating an amorphous co-aggregate of rosettes. It was shown further that the initial rosettes formed from **22b** were only metastable and could be transformed into nanosheets due to a rearrangement of the H-bonding pattern [142]. In a related study, two isomeric monomers were employed to assemble thermoresponsive nanotubes (i.e. the nanotubes were stable at 0 °C, but dissociated into the monomers upon heating to 90 °C) [143]. As in the previous example, the supramolecular polymerization involved intramolecular H-bonding and π–π stacking interactions and followed a cooperative nucleation–elongation pathway.

With respect to more classical polymer science, Binder et al. combined homotelechelic poly(etherketone) (PEK) and poly(iso-butylene) (PIB) into a chain-extended copolymer [144–146]. As depicted in Figure 3.28, the alternate

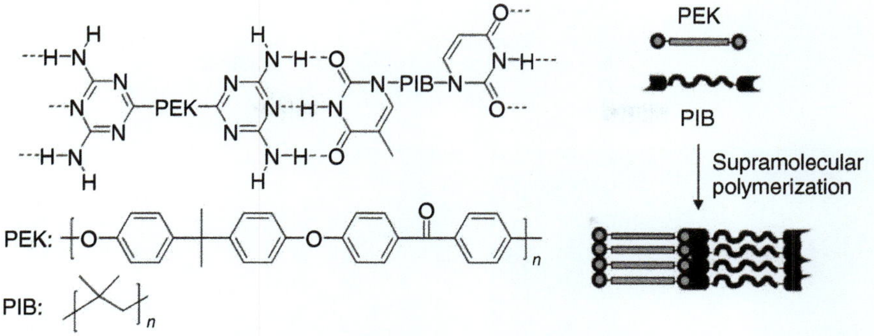

Figure 3.28 A chain-extended supramolecular block-copolymer based on triple H-bonding arrays. PIB: poly(iso-butylene), PEK: poly(etherketone). Source: Binder et al. [146]. Figure reproduced with kind permission. © 2005 Wiley-VCH.

order of the blocks within the supramolecular polymers was achieved via complementary pairs of triple H-bonding entities – either a strong or a weak one. For this purpose, combinations of either 2,6-diamino-1,3,5-triazine and thymine or 2,6-diamino-1,3,5-triazine and cytosine were utilized. The directional H-bonding interactions improved the miscibility of the two intrinsically incompatible polymers, which typically feature a pronounced phase separation in the solid state. NMR spectroscopy was used to study the self-assembly of the polymers in solution and gave K_A values that were similar to those measured for the analogous small molecules (K_a values of 400–1000 M^{-1} and c. 8 M^{-1} for the couples triazine/thymine and triazine/cytosine, respectively). According to NMR data (i.e. ^{13}C NMR relaxation studies), TEM imaging, and thermogravimetric analysis, sheet-like structures were proposed to be formed in the solid state; DSC measurements further support the assumption of phase separation in the solid state on the nano- to the micrometer scale [144].

Moreover, the PIBs were functionalized with either thymine or 2,6-diaminotriazine moieties at both chain ends. The stoichiometric mixture of these polymers gave a supramolecular polymer composed of strong H-bonded thymine–diaminotriazine pairs along the chains (K_a of 10^3 M^{-1}) [147]. Rheological measurements of solid-state samples were conducted to identify the temperature-dependent dynamics within the material: at temperatures from 20 to 60 °C, aggregation into sheet-type or even cross-linked structures occurred; however, at elevated temperatures, disaggregation of the chains generated a polymer melt of low viscosity. The thermoreversibility of the process suggested a possible application of such H-bonded polymers in self-healing materials [54].

Related to this, Leibler and coworkers studied the self-assembly behavior of unentangled poly(propylene oxide) (PPO) of low molar mass (M_n of 2200 g mol^{-1}), equipped with thymine or diaminotriazine units at the chain ends (**23a/b**, Figure 3.29) [148]. In contrast to **23a**, which was a liquid of low viscosity, **23b** was a solid material. According to DSC and SAXS measurements, crystallization of the H-bonded thymine-dimers led to a crystallization of the H-bonding moieties (presumably, this process was driven by secondary weak H-bonding interactions

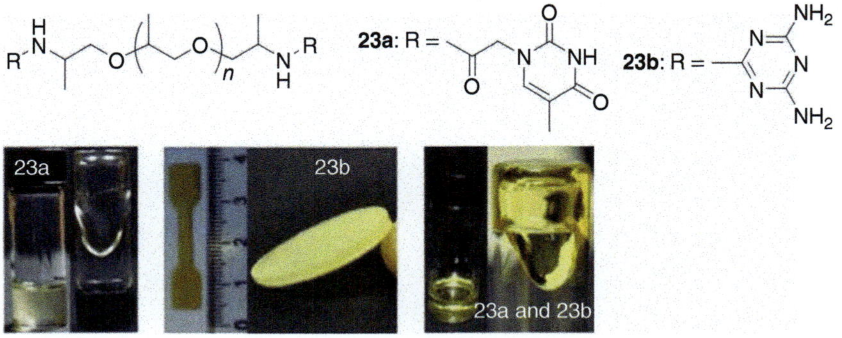

Figure 3.29 Schematic representation of the homotelechelic PPOs **23**. The pictures show pure **23a**, pure **23b**, and an equimolar mixture of both (from left to right). Source: Cortese et al. [148]. Figure reproduced with kind permission. © 2012 American Chemical Society.

involving the amide groups that were in a close proximity to the thymine sites). However, an equimolar mixture of **23a** and **23b** again gave a liquid – since the strong hetero-association of thymine and diaminotriazine units suppressed the crystallization of thymine-dimers. The viscosity of this **23a/23b** mixture was significantly higher when compared to that of **23a** due to the higher K_a value and, thus, DP of the resultant supramolecular polymer.

A similar approach was followed by the Rowan group, who assembled rigid π-conjugated OPE fragments [OPE: oligo(phenylene-ethynylene)], equipped with complementary nucleobases (i.e. adenine [A] and thymine [T] derivatives), into main-chain LC supramolecular polymers (Figure 3.30) [149, 150]. The melt mixing of appropriate components **24** (i.e. heterotelechelic **24a** alone or a 1 : 1 mixture of the homotelechelic components **24b** and **24c**, respectively) provided polymers featuring a broad range of LC behavior – much broader than for the individual components. These authors rationalized this finding by a favored Hogsteen-type base-pairing interaction of the utilized A- and T-derivatives (thereby circumventing the more common pairing according to a Watson–Crick-type binding); as a result, from this association mode, a polymer network was formed in which the overall aspect ratio was strongly increased (compared to that of 1D polymer chains) and the LC phase was stabilized.

The concept of supramolecular polymerization, based on nucleobase-pairing interactions, was further applied, e.g. in the self-assembly of homo- and heteroditopic bolaamphiphiles [151] as well as poly(tetrahydrofuran)s (PTHFs) [152] equipped with adenine- and thymine-moieties (for further examples, see Ref. [23]). With respect to the former, the bolaamphiphiles **25**, equipped with the nucleobases T and/or A, were self-assembled from solution [151]. Thus, the stoichiometric combination of **25b** and **25c** afforded nanometer-sized fibers and not the twisted or helical motifs that were found for pure **25c** (Figure 3.31). The authors rationalized the formation of these superstructures by the individual molecular packing of the hydrophobic chains supported by additional interchain H-bonding interactions.

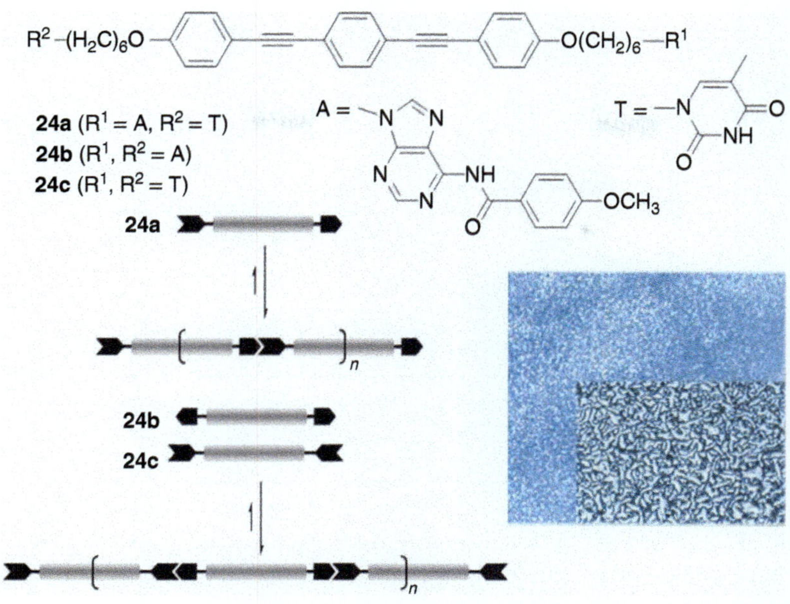

Figure 3.30 Schematic representation of the reversible self-assembly of the nucleobase-functionalized oligo(phenylene-ethynylene)s **24** (A: adenine, T: thymine). The optical micrograph of an annealed film of the polymer $(\mathbf{24a}\cdots\mathbf{24b}\cdots)_n$ at 130 °C is also shown (100× magnification, the inset shows a section with a 500× magnification). Source: Figure reproduced with kind permission; © 2003 The Royal Society of Chemistry.

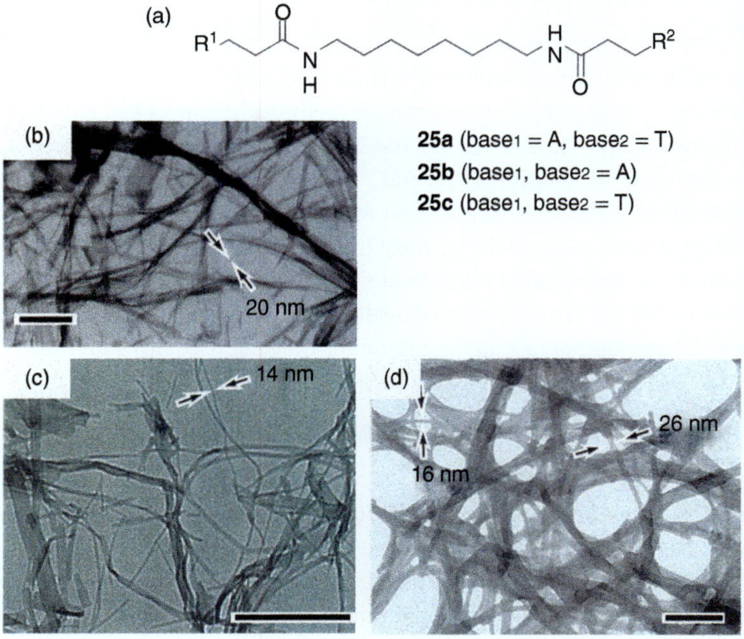

Figure 3.31 (a) Schematic representation of the bolaamphiphiles **25** (for the structures of adenine [A] and thymine [T], see Figure 3.30). (b–d) Energy-filtered transmission electron microscopy (EF-TEM) images of the fibers formed by self-assembly in 10% ethanolic aqueous solution at 25 °C (b: **25a**, c: **25c**, d: 1 : 1 ratio of **25b** and **25c**; the scale bar represents always 1 μm). Source: Modified from Shimizu et al. [151]. Figure reproduced with kind permission. © 2001 American Chemical Society.

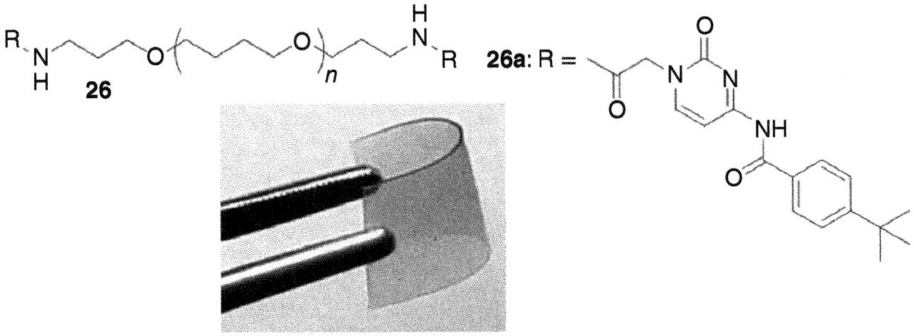

Figure 3.32 A highly flexible material was formed via the self-assembly of α,ω-dinucleobase-functionalized PTHF (e.g. **26a** with two N^4-(4-*tert*-butylbenzoyl)-cytosine moieties). Source: Modified from Sivakova et al. [152]. Figure reproduced with kind permission. © 2005 American Chemical Society.

> **Bolaamphiphiles**, which are also referred to as "bolaform surfactants," "bolaphiles," or "α,ω-type surfactants," represent amphiphilic molecules with two highly hydrophilic end groups on a hydrophobic chain. The additional polar head groups improve the surfactants' solubility in water (when compared to a single-headed amphiphile) and, thereby, raise the critical micelle concentration (cmc) as well as lower the aggregation number. Various types of morphologies have been reported for their aggregates, including spheres, cylinders, disks, and vesicles. The fundamental properties and applications of bolaamphiphiles have been summarized by Huang and coworkers [153].

Finally, the nucleobase-pairing approach was also utilized for the association of PTHF segments ($M_n < 2000\,\text{g mol}^{-1}$) via terminally bound adenine- or cytosine-type moieties (**26**, Figure 3.32) [152]. Although the H-bonding interaction between A and C was rather weak in solution ($K_a < 5\,\text{M}^{-1}$), the properties of the resultant material were changed significantly compared to pristine PTHF as a typically soft polymer: PTHF represents a low-melting, waxy material; whereas, the supramolecular polymer exhibited high stability and flexibility. Upon phase separation, hard domains – arising from the cross-linking of nucleobases – and soft PTHF domains could be observed – presumably being the reason for the macroscopic properties.

3.2.4 Supramolecular Polymers, Based on Quadruple H-Bonding Interactions

Within the last two decades, the quadruple H-bonding interactions have generated highly versatile supramolecular arrays and numerous examples of linear supramolecular polymers incorporating this particular motif have been reported [3, 6, 154, 155]. In line with the theories of supramolecular polymerization (Chapter 1), the self-assembly of small-molecule components into materials, which exhibit polymer-like properties in solution, is limited by the binding strength, i.e. the K_a value [6, 12]: at a monomer concentration of about 1 M, a K_a value of $10^5\,\text{M}^{-1}$

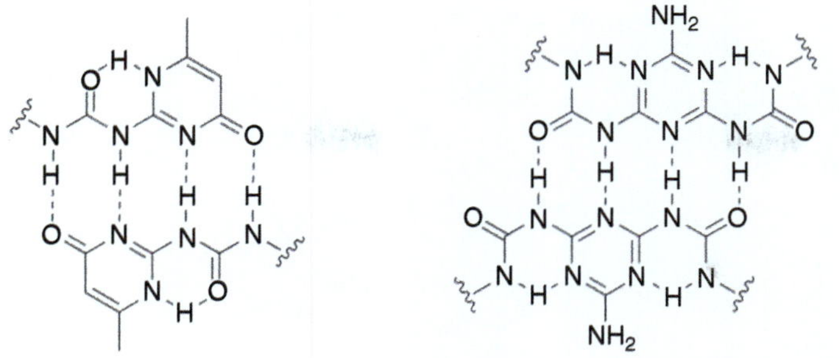

2-Ureido-4[1*H*]-pyrimidione $k_a = 6 \times 10^7\ M^{-1}$

1,1′-(6-Amino-1,3,5-triazine-2,4-diyl)diurea $k_a = 2 \times 10^4\ M^{-1}$

Figure 3.33 Schematic representation of two important self-complementary quadruple H-bonding arrays with high K_a values.

will offer a DP of c. 100 (Figure 1.3). Thus, when aiming for supramolecular polymers of high molar masses, much higher K_a values are a necessary prerequisite. Various types of quadruple H-bonding systems, such as the famous UPy [18] or the 1,1′-(6-amino-1,3,5-triazine-2,4-diyl)diurea unit [36], both featuring K_a values in the range of 10^4–$10^7\ M^{-1}$, meet this requirement [Figure 3.33; see Refs. [5, 6, 10, 11, 156] for reviews dealing with quadruple H-bonding in supra(macro)-molecular science].

Meijer and coworkers reported on the first application of such systems with respect to supramolecular polymers in the late 1990s [18, 32, 35]: dilute solutions of the telechelic derivative **27a** in $CHCl_3$ exhibited high viscosities (Figure 3.34a), even at low concentrations (i.e. down to 40 mM); this behavior was attributed to the self-assembly into long supramolecular chains via quadruple H-bonding interactions. The titration of this polymer $(\mathbf{27a})_n$ with the monotopic counterpart **28** led to a steady decrease of the specific viscosity, attributed to the decreasing DP [32]. The authors assumed an IDP (IDP mechanism) and calculated a number-averaged DP of 700 at a concentration of 40 mmol l^{-1} (estimated K_{dim} value of $2.2 \times 10^6\ M^{-1}$); however, in some cases, the formation of macrocyclic assemblies, in equilibrium with linear ones, was concluded from ^1H NMR spectroscopy and viscosity measurements [11]. The conformational preorganization of **27**, via an appropriate spacer structure, facilitated the formation of cyclic species. The critical concentration for all UPy derivatives **27** was determined on the basis of ^1H NMR spectroscopy due to the slow exchange (on the NMR timescale) of resonances assigned to the UPy-dimers in cycles and linear species (Figure 3.34b); the existence and magnitude of this value could be further confirmed by Ubbelohde viscometer studies. Due to the high K_a value for the self-complementary H-bonding of the UPy unit, the true critical concentration – as predicted by the theory of supramolecular ring-chain equilibrium polymerizations (Section 1.3.2) – could be derived [11]. In general, a preference for cyclization into dimers was observed at low concentrations; whereas, linear polymers were formed at higher concentrations. In comparison, **27a** (having a simple *n*-alkyl spacer) exhibited very high viscosities at concentrations of c. 75 mM

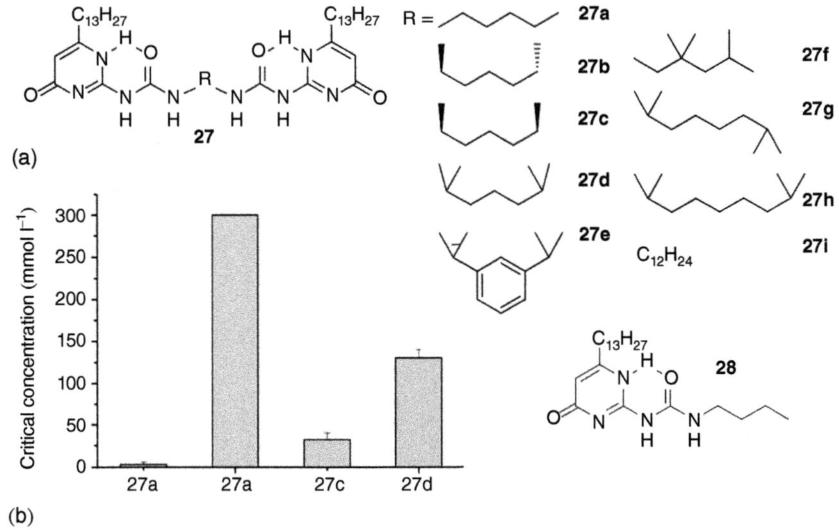

Figure 3.34 (a) Schematic representation of the UPy-functionalized homotelechelic monomers **27**. (b) Representation of the critical concentrations of **20** in supramolecular ring-chain equilibrium polymerizations. Source: ten Cate and Sijbesma [11]. Figure reproduced with kind permission. © 2003 Wiley-VCH.

in $CHCl_3$; whereas, the branched systems (**27b/d**) showed only low viscosities under the same conditions as small cycles were the dominate species. This increase in the critical concentration was attributed to the presence of the methyl-substituents, which shifted the equilibrium of the *anti-/gauche*-conformations of the linker (Figure 3.34b): if the CH_3-groups were oriented in a strict *anti*-conformation relative to the alkyl chain, the UPy sites were preorganized for cyclization/dimerization (thus, increasing the critical concentration). Consequently, high spacer rigidities, as in **27e**, forced the system to dimerize, almost independent of the concentrations [35]. From a thermodynamic point of view, **27g** resembles an exception from the generally enthalpically driven supramolecular polymerizations of monomers **27**: a significant increase in viscosity upon heating solutions of **27g** in $CHCl_3$, attributed to the formation of linear chains at the cost of cyclic species, is characteristic for such a process [157].

In this respect, the work of Hailes and coworkers has to be mentioned: secondary non-covalent interactions might facilitate cyclization [33]. For example, the formation of cyclic structures became strongly favored due to additional H-bonding interactions involving the carbamate and carbonyl groups of the pyrimidinone unit of **29** (Figure 3.35). In absence of the carbamate unit within the spacer geometry, intermolecular H-bonding was excluded, and thus, the molecule's rigidity was reduced. As a consequence, **30** gave linear supramolecular polymers at high concentrations (Figure 3.35). For the homopolymerization of either **27a** or **31**, the presumably very low critical polymerization concentration (cpc) could not be determined via viscometry, since the flexibility of the molecules strongly hampered the cyclization [158]. However, the antiparallel arrangement of **31** (Figure 3.35) favored the co-assembly

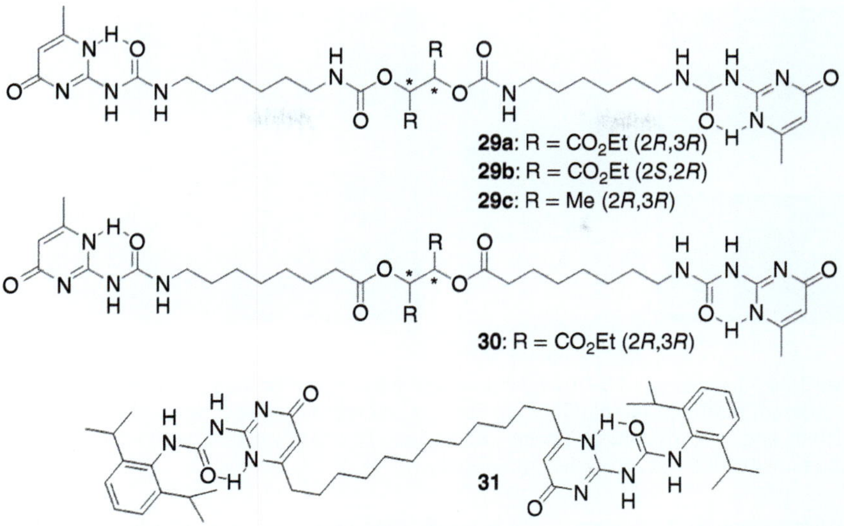

Figure 3.35 Schematic representation of the UPy-functionalized homotelechelic monomers **29–31**.

with **27a** into cyclic heterodimers, as shown by viscosity and diffusion-ordered spectroscopy (DOSY) measurements of their mixtures.

Besides these examples for the self-assembly of small molecules, polymers, functionalized with UPy moieties at their chain ends, were utilized for the same purpose. In 1999, Meijer and coworkers reported on the association of star-shaped polyethers into stable polymer networks that did not require any further stabilization due to crystallization or phase separation [159]. The reversibility of the supramolecular binding was shown by either adding water (i.e. c. 30 equiv per UPy unit, to significantly reduce the viscosity of the solution) or an excess of UPy monomer (thus, achieving a quantitative depolymerization of the material). The self-assembly of a range of homotelechelic bis-UPy-functionalized polymers [e.g. poly(ethylene-*co*-butylene), polyethers, polyesters, and polycarbonates] with molar masses of 2000–3500 $g\,mol^{-1}$ was investigated by Sijbesma and coworkers [160–163]. These supramolecular polymers featured the mechanical properties similar to those of analogous covalent polymers as well as a low melt viscosity typical for small organic molecules. The remarkable ease-in-processability, thus, offered a potential use in applications, e.g. adhesives, inks, coatings, and cosmetics [164].

Studies in the solid state as well as in the molten state revealed significant changes of the materials' properties when compared to the non-modified polymers, i.e. those missing the terminal UPy moieties. The nontrivial thermo-rheological behavior was attributed to the reversibility of the supramolecular interactions. The formation of clusters – based on H-bonded structures – was observed. This "physical cross-linking" was reminiscent of that known from conventional thermoplastic elastomers, i.e. a strong decrease of the viscosity with the temperature (here, the melt viscosity decreased by a factor of c. 4 as the temperature was increased by 10 °C).

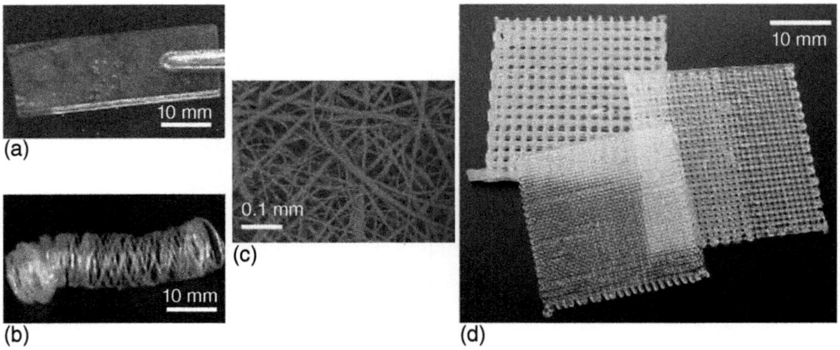

Figure 3.36 Images of a *bis*-UPy-functionalized poly(ε-caprolactam) that was processed into different scaffolds: (a) thin films, (b) fibers, (c) meshes (a representative SEM image is depicted), and (d) grids. Source: Dankers et al. [165]. Figure reproduced with kind permission. © 2005 Nature Publishing Group.

Supramolecular polymers, which were obtained from quadruple H-bonding interactions, were also processed from solution into thin films (applying solvent casting or compression molding), fiber- or mesh-like structures (via melt- or electrospinning) as well as grids (by fused deposition modeling, Figure 3.36). In this respect, Dankers et al. used mixtures of a bis-UPy-functionalized biocompatible poly(ε-caprolactam) and UPy-functionalized peptide recognition sequences (e.g. the synthetic peptide with the sequence Gly-Arg-Gly-Asp-Ser, as a mimic for the cellular binding site of many adhesive proteins) for the fabrication of bioactive scaffolds for tissue engineering applications [165]. In more detail, the spun polymer was incubated into its final shape and subsequently treated with cells. After cellular binding and growth, the supramolecular polymer was found to slowly dissolve in the aqueous medium – either due to proteolytic cleavage or dissociation of the supramolecular bonds. The authors pointed out that the material represented a highly versatile cellular scaffold with a remarkable resorption behavior.

More recently, these authors synthesized a chain-extended UPy-containing poly(ε-caprolactone) (PCl) and a homotelechelic bis-UPy-functionalized poly(ethylene glycol) (PEG). The self-assembled material, at a macromonomer ratio of 9 : 1, was electrospun into a vascular graft (Figure 3.37). It was demonstrated that the *in vivo* cell adhesion was strongly reduced. A similar bilayer scaffold, comprising a cell-adhesive and a non-cell-adhesive side, was also produced using appropriate UPy-containing polymers [167]. The nonadhesive layer could be bioactivated and made adhesive for kidney epithelial cells by adding 4 mol% of a short UPy-terminated oligopeptide to the electrospinning solution. These experiments show how the interplay of directed H-bonding interactions, mainly based on UPy, and thermoplastic elastomers can be utilized to generate smart biomaterials, which are of significant interest due to a possible implementation in medicinal applications [155]. In this context, various reports by Dankers et al. are worth mentioning in which biomaterials were fabricated from UPy–PCl conjugates [168–173].

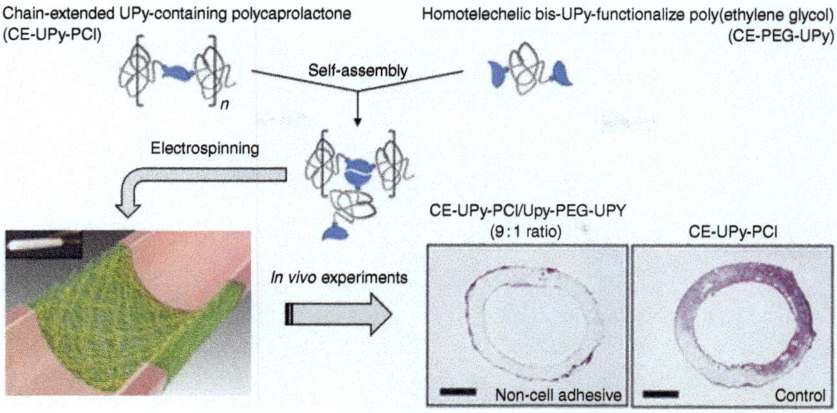

Figure 3.37 Schematic representation of the fabrication of vascular graft from UPy-equipped macromolecular components. The pictures show the results of the *in-vivo* cell-adhesion experiments. Source: van Alme et al. [166]. Figure reproduced with kind permission. © 2016 Wiley-VCH.

UPy-functionalized amphiphiles represent one further class of materials that has been employed to assemble hierarchical supramolecular architectures. For example, Meijer and coworkers used such macromonomers to generate supramolecular fibers, which featured the exchange of molecules when different monomers were mixed [174].

The process of polymerization-induced phase separation (PIPS), i.e. the phase separation in a mixture of various components due to the polymerization of (at least) one of these, is well known to give materials with well-defined morphologies. For instance, Keizer et al. reported on the formation of "supramolecular droplets" within acrylate matrices. Initially, supramolecular polymers, miscible in bulk acrylates, were formed by the self-assembly of bis-UPy-functionalized PTHF or polycarbonates [162]. However, the miscibility of such a supramolecular polymer was reduced significantly in the course of acrylate polymerization, eventually generating "droplets" of the supramolecular polymer finely dispersed within the polyacrylate matrix. The droplet size could be varied over a broad range, i.e. from below 100 nm up to some micrometers in diameter, by adjusting the polymerization conditions or the ratio between the acrylate monomers and the supramolecular polymer.

The roughness of surfaces can be modified by immobilizing H-bonding arrays equipped with trialkoxysilyl functionalities onto surfaces (Figure 3.38) [175]. The H-bonded dimer **32** was cross-linked via its siloxane moieties in a sol–gel-mediated process to afford a 3D polymer network. Due to the phase separation within this material, granules and holes of defined sizes were formed on the substrate surface. The surface roughness was studied by contact-angle measurements revealing values in the superhydrophobic regime of c. 150°; the properties of the hybrid material were reminiscent of those of lotus leaves featuring the famous self-cleaning behavior

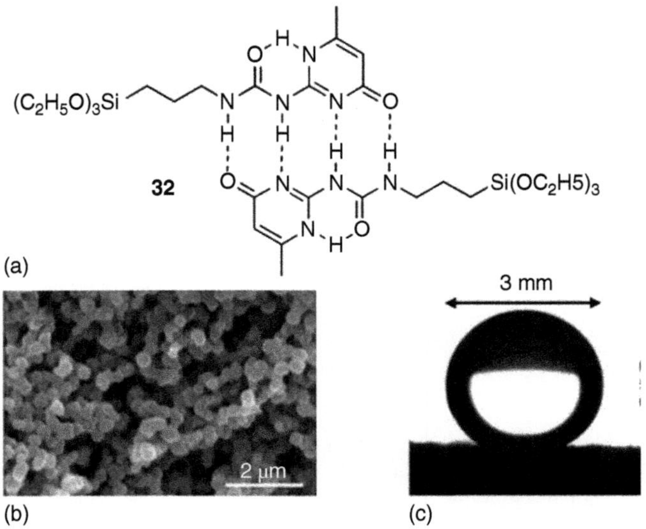

Figure 3.38 (a) Schematic representation of the H-bonded dimer **32**. (b) SEM image of the surface roughness (scale bar: 2 μm). (c) Representation of the superhydrophobicity of the resulting surface: Water contact angle of c. 150°. Source: Han et al. [175]. Figure reproduced with kind permission. © 2004 American Chemical Society.

[176, 177]. It was also shown that the non-covalent pre-organization, arising from the H-bonding interactions, gave phase effects, which could not be reached when utilizing analogous covalent polymers (for comparison, on smooth PDMS surfaces the water contact angle is only in the range of 11°).

In general, AB-type diblock copolymers can be formed by linking two appropriately functionalized polymer chains in a covalent or non-covalent fashion. In the latter case, diblock copolymers are obtained in which a supramolecular bonding motive connects the two chemically different segments (Figure 3.39). Such materials have attracted considerable interest due to their highly ordered morphologies on the nano- and micrometer scales (many examples for this can be found throughout the subsequent chapters) [164]. To assemble such systems based on H-bonding, a high directionality of the interaction as well as a selectivity for hetero-association must be given (i.e. formation of the AB-type diblock copolymer must be favored compared to the AA- and BB-type homopolymers). These requirements are met by the UPy–Napy recognition pair in which one tautomer of UPy and Napy selectively forms a strong ADDA-DAAD-type hetero-associate (Figure 3.8) [37]. Accordingly, this complementary H-bonding has been utilized to assemble supramolecular diblock copolymers.

For example, Feldman et al. synthesized the UPy- and Napy-functionalized polyacrylates and polymethacrylates **33** and **34** by an atom-transfer radical polymerization (ATRP, Figure 3.40) [178, 179]. The physical properties of the materials were correlated directly with the polymers' terminal H-bonding sites. Bulk-blending experiments were performed and revealed a strong influence of the chemical structure of the H-bonding units on the phase-separation behavior: in detail, the

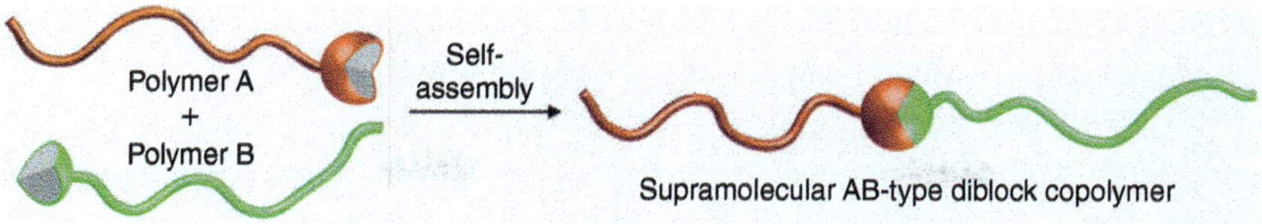

Figure 3.39 Schematic representation of the formation of supramolecular AB-type diblock copolymers due to complementary H-bonding interactions. Source: Feldman et al. [178]. © 2008 American Chemical Society.

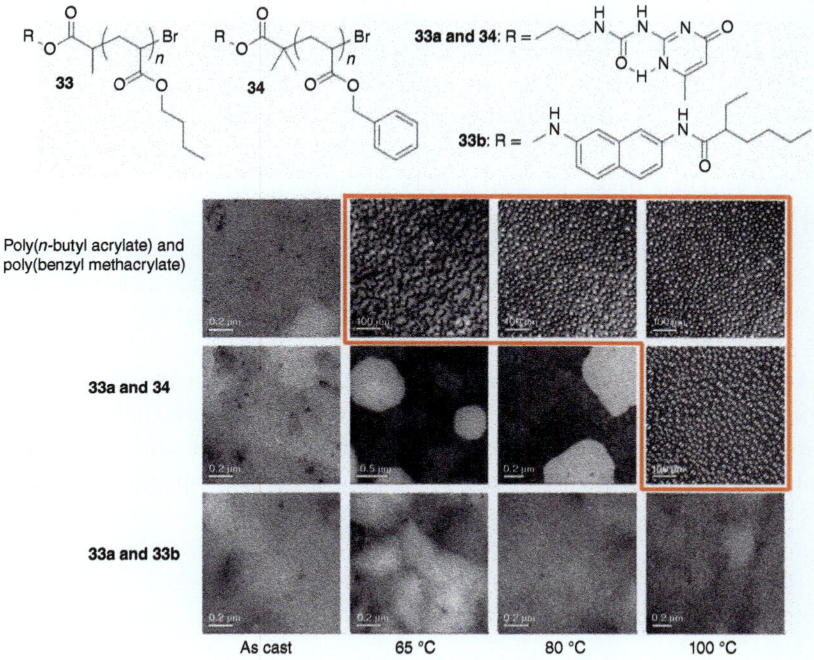

Figure 3.40 Schematic representation of the UPy- and Napy-functionalized homopolymers **33** and **34**. The microscopy images show the microstructure of the blends after annealing at various temperatures. Source: Modified from [178]. Figure reproduced with kind permission. © 2008 American Chemical Society.

complementary UPy–Napy pairs strongly reduced the phase separation of the components; whereas, the self-complementary UPy–UPy pairs even enabled a partial miscibility of **33a** and **34** within the blend. In the blend of **33a** and **33b**, the UPy–Napy interaction gave some diblock copolymers at the interfaces of the blends, which served as compatibilizers and stabilized the nanostructured materials (i.e. no phase separation could be achieved, not even after thermal annealing at 100 °C, Figure 3.40). More recently, Nojiri et al. combined isotactic polypropylene (iPP) with poly(ethylene-*co*-propylene) (PEP), both end-functionalized with a UPy and a Napy moiety, respectively, into a pseudo-diblock copolymer by melt mixing [180]. TEM imaging of the blend revealed the elastic PEP domains to be dispersed in the iPP matrix (almost comparable to the behavior known for the iPP/PEP impact copolymer). Thus, the quadruple UPy–Napy H-bonding acted as a compatibilizer and improved the overall mechanical properties of the material.

Further, Sijbesma, Meijer, and coworkers utilized this approach to assemble a chain-extended supramolecular polymer from two homotelechelic monomers – bis-UPy-functionalized PTHF and an alkyl-bridged bis-Napy derivative [181]. In a similar fashion, Long and coworkers applied the self-complementary UPy moiety for the preparation of microphase-separated block copolymers; for this purpose, a UPy-terminated diblock copolymer, i.e. PS-*b*-PI [PI: poly(*cis*-isoprene)], prepared via anionic polymerization, was dimerized [182].

From the broad range of (dynamic) supramolecular copolymers that exhibit phase separation in solution as well as in the solid state [183], the pseudo-diblock copolymers **35** and **36** as reported by Rao et al. also need to be mentioned (Figure 3.41). In contrast to the previous examples, where the UPy–Napy interaction was utilized to link two polymer chains, there the combination of Napy with ureidoguanosine (UG) was employed. The non-self-complementary UG moiety, offering a similar ability to bind to Napy as UPy does, was initially reported by Zimmerman and coworkers (K_a of c. $10^7\ M^{-1}$ in $CHCl_3$); due to the negligible ability of UG to dimerize (K_{dim} of $230\ M^{-1}$ in $CHCl_3$) the homo-association can be suppressed efficiently [184, 185]. The formation of the supramolecular diblock copolymers mimicking PS-*b*-PI (**35**) and PS-*b*-PEG (**36**) was concluded from ^1H NMR and UV/vis spectroscopic studies [186]. According to AFM and TEM imaging as well as SAXS measurements, **35** and **36** showed a phase-separation behavior in thin spin-coated films and in bulk to give cylindrical morphology with domain sizes in the range of 20–25 nm (Figure 3.41). These features were similar to those observed for the corresponding covalent diblock copolymers of the same chain lengths. In contrast to these, a long-range

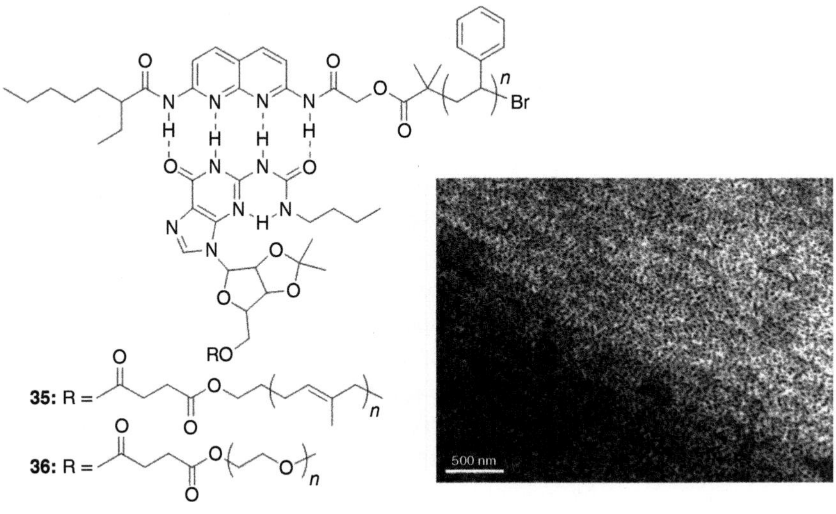

Figure 3.41 Schematic representation of the supramolecular diblock copolymers **35** and **36**. A representative TEM image of **35** showing the microphase separation in the solid state is also shown. Source: Rao et al. [186]. Figure reproduced with kind permission. © 2012 The Royal Chemical Society.

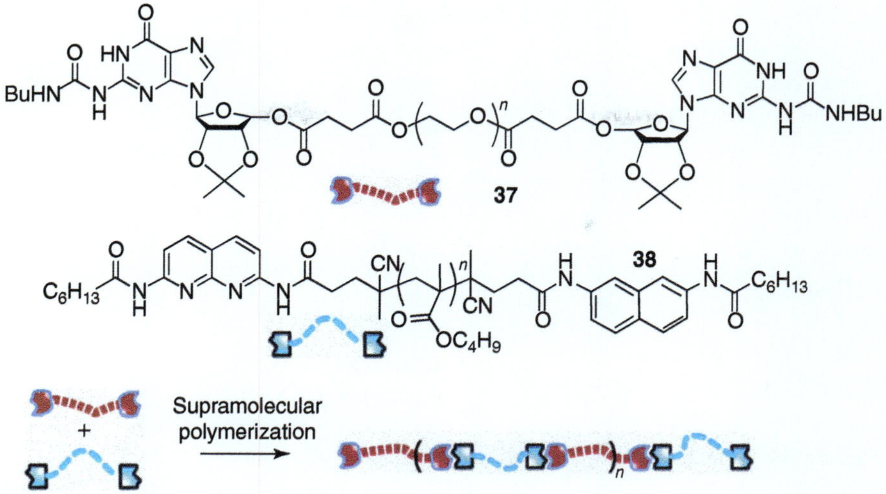

Figure 3.42 Schematic representation of the formation of a strictly alternating supramolecular copolymer using the non-self-complementary monomers **37** and **38**. Source: Based on Park and Zimmerman [187]. © 2006 American Chemical Society.

lateral order could not be realized due to incompatibility of the supramolecular materials with the required annealing conditions, i.e. high humidity. The same type of interaction was applied by Park and Zimmerman to assemble a strictly alternating supramolecular copolymer. The combination of the homoditopic macromonomers **37** and **38**, in equimolar amounts, afforded this particular material; whereas, none of the monomers alone gave any supramolecular polymer (Figure 3.42) [187].

Besides the previous examples in which either the homo- or hetero-association of mono-functionalized building blocks (e.g. giving pseudo-diblock copolymers) as well as the supramolecular association of homoditopic, self-complementary components or of two homoditopic, complementary ones into chain-extended polymers have been discussed, also heteroditopic derivatives, e.g. **39–41** (Figure 3.43), have been utilized in supramolecular polymerization reactions. For example, the self-assembly of **39** in dilute $CHCl_3$ solutions, due to the strong quadruple H-bonding interactions in the UPy–Napy pairs, was reported by Scherman et al. [188]; the K_a value for this system was $6 \times 10^6\,M^{-1}$ (see also Section 3.1 and Figure 3.1) [21, 37]. ^1H NMR spectroscopy and viscometry measurements revealed a pronounced transition from cyclic species (present at low concentrations) to linear ones (preferred at high concentrations). Due to the significantly lower K_a value for hetero-association than for homo-association (by a factor of >20), the Napy derivatives acted as chain-stopping agents, thereby limiting the resultant DP (the so-called "self-stoppered" effect). The introduction of a $-N(C_4H_9)_2$ group at the UPy moiety (**41**) stabilized its enol form and, thus, reduced the tendency for the homo-association relative to **40**, though not affecting the K_a value for this process [189]. Thus, the "self-stoppered" effect at high concentrations was efficiently suppressed and cyclization became even more favored at lower concentrations.

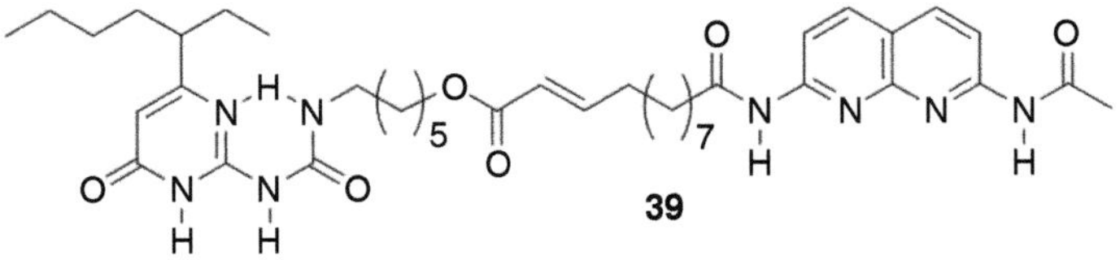

39

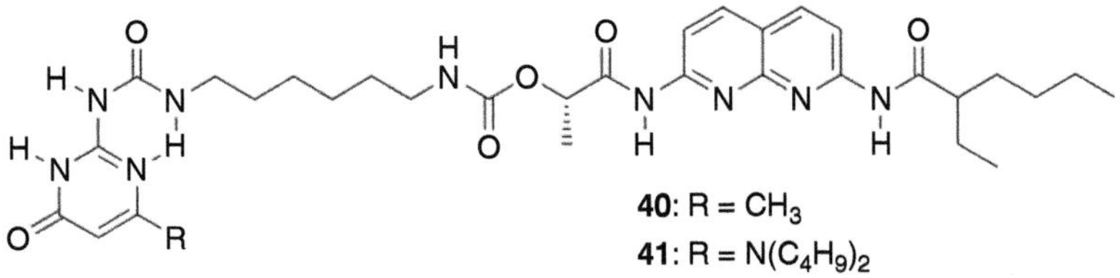

40: R = CH_3
41: R = $N(C_4H_9)_2$

Figure 3.43 Schematic representation of the heteroditopic H-bonding units **39–41**.

The photoreversible supramolecular polymerization was investigated by Takeshita et al. [190]. These authors used a bis-UPy-functionalized dithienylethylene derivative, as a photochromic compound, which can be switched between an open (*o*-**42**) and a closed structure (*c*-**42**) via a reversible [3,3]-sigmatropic Cope-type reaction (Figure 3.44). The irradiation at 366 nm generated the closed form; whereas, the open species was formed by applying light of lower energy (λ_{ex} > 540 nm). The equilibrium between a transoid- and cisoid-conformer strongly decreased the self-assembly tendency of *o*-**42** via H-bonding interactions. In contrast, *c*-**42** readily gave a supramolecular polymer. The ability to photochemically switch between an assembled and a disassembled state was controlled by viscosity as well as DLS measurements. Utilizing the later technique, particle sizes of c. 650 nm were estimated – corresponding to a DP of 200 and a M_n value of 2.5×10^5 g mol^{-1} (assuming a monomer length of c. 3 nm and a molar mass of 1200 g mol^{-1}). In the same context, the photochromic diazobenzene moiety was incorporated into a homoditopic UPy-based monomer [191]. The E-isomer generated a linear supramolecular polymer in $CHCl_3$ due to dimerization of the self-complementary UPy sites; the Z-isomer, however, formed much shorter chains at the same concentration (attributed to steric constraints). Moreover, the thin-solid films of the material also showed a highly reversible photo-induced trans/cis-isomerization upon irradiation with light of the appropriate wavelength.

With respect to opto-electronic applications, Meijer and coworkers reported on the supramolecular self-assembly of combinations of various UPy-functionalized π-conjugated chromophores {i.e. an oligofluorene [OF], a PBI, and an oligo (phenylene-vinylene) [OPV]}. The supramolecular polymerization in solution via H-bonding interactions enabled, for example, the energy transfer from the blue-absorbing OF moieties to the green-emissive OPV ones. The enhancement of this process was rationalized by the ordering of the chromophores within the supramolecular material [192]; in control experiments, the energy transfer was less than 25% as efficient if either the donor or the acceptor component contained no H-bonding groups. Furthermore, OPV dimers were assembled in solution into

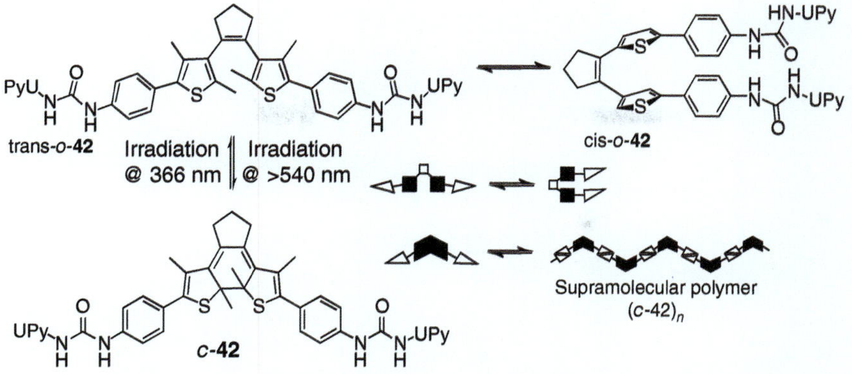

Figure 3.44 Schematic representation of the photoreversible formation of the supramolecular polymer (*c*-**42**)$_n$.

disc- [193] and ladder-type superstructures [194] for which a strong excitonic coupling was observed. Methanofullerene phenyl-[6,6]-C_{61}-butyric acid methyl ester (PCBM), a common acceptor-type material in bulk-heterojunction solar cells, was added to these solutions; the resultant materials were subsequently processed into the solid state. Spin-coating offered good control over the morphology of the solid films for which thicknesses of c. 100 nm were achieved. The morphology of these films was investigated further by neutron-scattering experiments and revealed the presence of rod-like structures in which the PCBM molecules were found in close proximity to the π-conjugated units. The authors also tested the applicability of the material in photovoltaic devices; the high efficiency (c. 12%) for the photo-to-current transformation was explained by the phase-separated morphology of the material in the solid state.

Combining three different π-conjugated chromophores, all having two self-complementary UPy-units, afforded a supramolecular random copolymer incorporating the blue-emitting OF, the green-emitting OPV, and the red-emitting PBI derivative (**43–45**, Figure 3.45a) [195]. The supramolecular polymerization of these three components could be performed in solution as well as in bulk. The random supramolecular copolymers containing all three types of chromophores (in different stoichiometric ratios) exhibited energy transfer when the donor-type OF unit became excited (Figure 3.45b). Emission of white light was found in solution for the polymer containing the monomers **43–45** in a 59 : 33 : 8 ratio. Homogeneous films on solid substrates with no indication of phase separation were obtained by spin-coating of the supramolecular copolymers from solution (in the absence of the UPy moieties, no such films were formed). Furthermore, the energy transfer in the solid state was by far more efficient than in solution: thus, emission of white light could be realized at much lower acceptor-to-donor ratios (i.e. at ratio of 84 : 10 : 6 for **43–45**). The authors also fabricated light-emitting diodes (LEDs) from the supramolecular copolymers: all three supramolecular homopolymers, derived from **43–45**, gave blue-, green-, and red-emitting LEDs, respectively (Figure 3.45c); cautious optimization of the chromophores' ratio afforded even

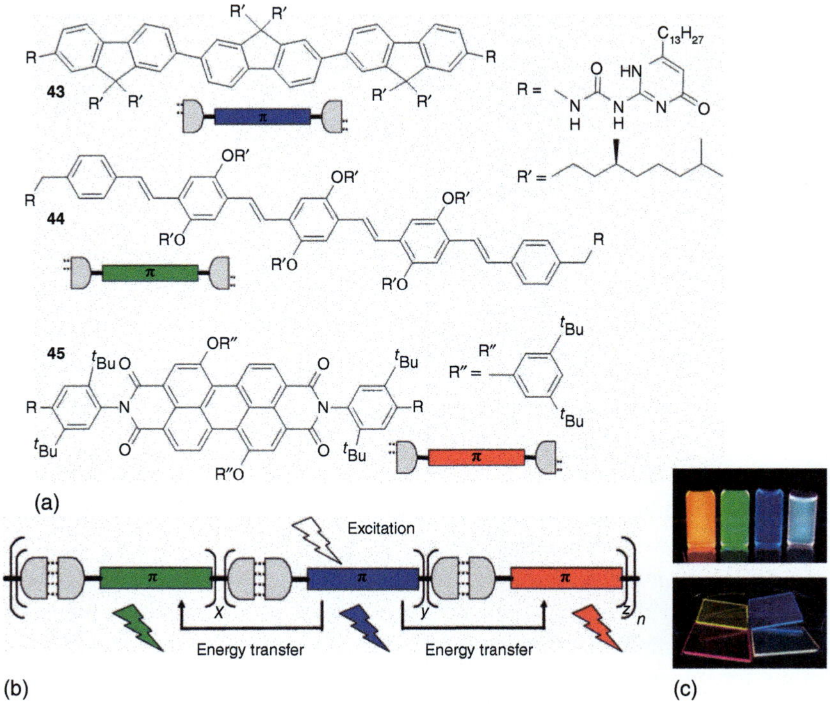

Figure 3.45 (a) Schematic representation of the various bis-UPy-functionalized π-conjugated chromophores **43–45**. (b) Schematic representation of the proposed energy-transfer process within the supramolecular copolymers. (c) Representation of the emission colors of the three individual homopolymers and the three-component copolymer in solution (top) as well as in the solid state (bottom). Source: Abbel et al. [195]. Figure reproduced with kind permission. © 2009 American Chemical Society.

white electroluminescence (according to the CIE chromaticity diagram, CIE: Commission Internationale de l'Éclairage).

As previously pointed out, the UPy moiety represents, by far, the most common structural unit among all quadruple H-bonding arrays; however, ureidotriazines that assemble in a similar fashion as the UPy system have also been employed. As shown in Figure 3.7a, this example for quadruple H-bonding is also of self-complementary nature with a K_a value of c. 2×10^4 M^{-1} [196]. The stacking of the dimers **46** and **47** into columnar architectures, driven by solvophobic interactions, was demonstrated (Figure 3.46), thereby, the overall structure and the helicity could be tuned via the chemical nature of the solubilizing side chains (e.g. by using chiral or achiral alkyl chains). Well-defined helices were formed from these associates [161]. This type of H-bonding unit can easily be attached to, for instance, isocyanate-terminated (macro)molecules by reaction with 1,3,5-triamino-2,4,6-triazine and might, therefore, be also of interest for large-scale, industrial applications. In this respect, the chain-extension of homotelechelic PTHF (M_n = 2000 g mol^{-1}) into a supramolecular polymer of high molar mass featuring a pronounced temperature-dependent viscosity has to be mentioned [36].

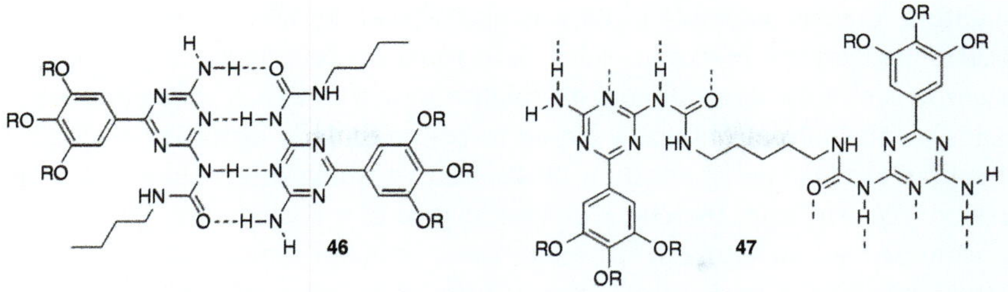

Figure 3.46 Schematic representation of the quadruple H-bonding arrays **46** and **47** based on self-complementary 1,3,5-triamino-2,4,6-triazine derivatives (R denotes aliphatic and oligo(ethylene glycol) chains – chiral and non-chiral ones).

To the end of this section, a typical example for a cooperative supramolecular polymerization, driven mainly by electronic effects, as published by Lehn and coworkers, will be discussed [197]. The supramolecular polymerization of the self-complementary H-bonding array **48** (acting as a heterodifunctional ADDA-DAAD-type unit, Figure 3.47a/b) was analyzed by concentration-dependent ^{1}H NMR spectroscopy in $CDCl_3$ solutions. The equilibrium between the monomer and dimer was the only process to be identified in this environment; a K_a value of c. 20 M^{-1} for the dimerization was estimated from the chemical shifts (i.e. the resonances of the amide- and amino-protons in the free vs. the H-bonded ones).

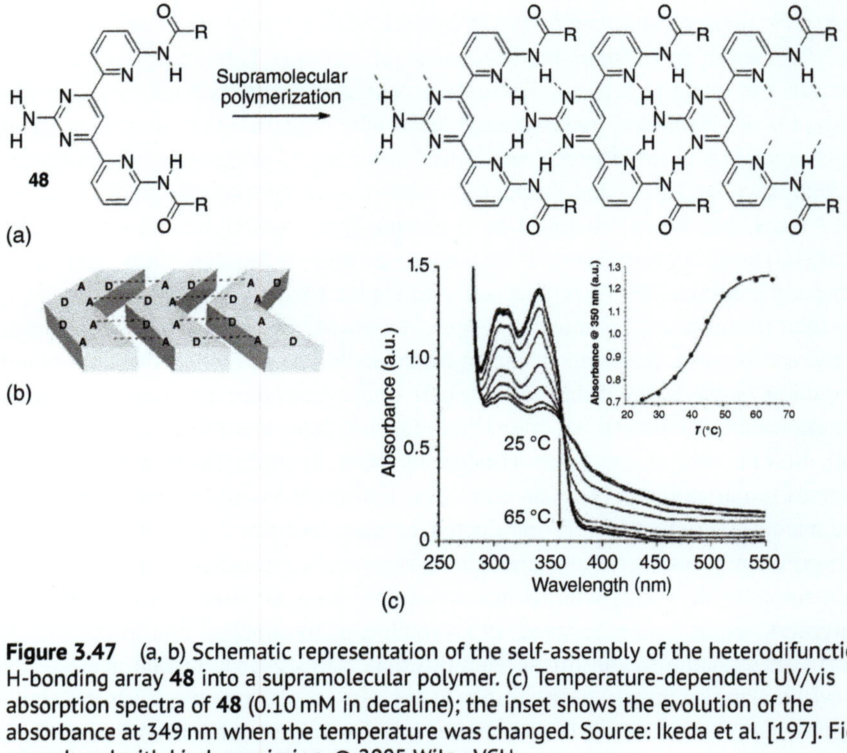

Figure 3.47 (a, b) Schematic representation of the self-assembly of the heterodifunctional H-bonding array **48** into a supramolecular polymer. (c) Temperature-dependent UV/vis absorption spectra of **48** (0.10 mM in decaline); the inset shows the evolution of the absorbance at 349 nm when the temperature was changed. Source: Ikeda et al. [197]. Figure reproduced with kind permission. © 2005 Wiley-VCH.

In addition, the self-assembly of **48** was also studied by UV/vis absorption and emission spectroscopy in various solvents at equal concentrations; a pronounced hypochromism of the absorption of **48**, relative to that in $CHCl_3$, was observed in decaline, as the solvent (**48** was believed to be molecularly dissolved in $CHCl_3$). The supramolecular polymerization of **48** via H-bonding interactions led to an increased planarity and, thereby, increased degree of π conjugation. The process was identified as an enthalpy-driven process: variable temperature, ^{1}H NMR spectroscopy (10^{-2} M in decaline), revealed the dissociation of the assemblies at elevated temperatures. Raising the temperature of the decaline solution of **48** to 65 °C, at a concentration of 10^{-4} M, a bathochromic shift of the UV/vis absorption band was observed which was attributed to the disassembly of $(\mathbf{48})_n$ back to the monomeric state (Figure 3.47c). This process could be followed by monitoring the evolution of the UV/vis absorption intensity at 349 nm. The non-sigmoidal shape of the curve pointed to a cooperative supramolecular polymerization process [6]. The authors correlated the cooperativity of this process to electronic effects, e.g. long-range dipole–dipole interactions in concert with polarization effects (additionally, allosteric effects resulting from the planarization of the π-conjugated system in course of the supramolecular polymerization could be involved).

3.2.5 Supramolecular Polymers, Based on Sextuple H-Bonding Interactions

The UPy moiety (along with its structural analogs) may represent the most powerful system to be used in the context of multiple H-bonding interactions; however, the intrinsic self-complementary character limits its applicability when aiming for the formation heteroassemblies in which a strictly alternating order of two components is anticipated (*vide supra*). A variety of extended complementary H-bonding arrays, increasing the stability when compared to quadruple H-bonding interactions, has been developed to overcome this restriction. From these, the so-called Hamilton-receptor [38–40], as a sextuple H-bonding unit, has been frequently employed in this respect (see also Figure 3.9a).

In their initial work, Lehn and coworkers employed two homoditopic molecules for the self-organization into linear supramolecular polymers via the directional interaction between the Hamilton-receptor and cyanurates; the thusly obtained supramolecular polymers **49** assembled further into nanofibers (Figure 3.48) [198]. Making use of the same association mode, tritopic building blocks gave supramolecular polymer gels or networks. Extensive viscosity and NMR spectroscopic measurements were conducted to elucidate the fiber formation. The aggregation process, in the presence and absence of appropriate mono-functional chain-stopping agents as well as at imbalanced stoichiometric ratios of the two monomeric units, was monitored as a function of the applied concentration. In agreement with the model for an IDP (Section 1.3.1), a pronounced decrease of the (virtual) molar masses of the resultant supramolecular polymers was generally observed. To further verify this, a model for the polymerization was suggested, assuming only a single equilibrium constant ($K = 4 \times 10^4\ M^{-1}$) and neglecting

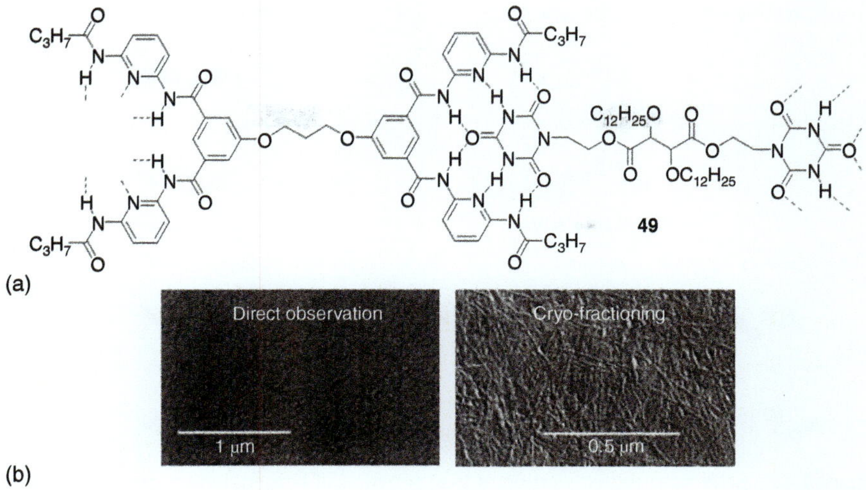

Figure 3.48 (a) Schematic representation of the linear supramolecular polymer **49**. (b) Representative electron-microscopy images of the extended, densely packed fiber networks obtained from the solution of **49** (2.5 mmol l^{-1} in $CHCl_3$/heptane, 1 : 4 ratio). Source: Berl et al. [198]. Figure reproduced with kind permission. © 2002 Wiley-VCH.

all chain-length and cyclization effects. The modeling studies, in particular the calculation of the DP, were in good agreement with the experimental data. Microscopy imaging as well as neutron-scattering measurements confirmed the aggregation of the supramolecular polymer chains into bundles of fibers in nonpolar solvents (e.g. decane or toluene) [199].

Related to this, Binder and coworkers combined PEK and PIB blocks, as two highly different systems that are well known for their microphase separation in the solid state (see also Figure 3.28), via the sextuple H-bonding interaction between the Hamilton-receptor and barbiturates (this work can be seen as a continuation of the authors' previous work focusing on triple H-bonding arrays in the field of polymer science [54]). In contrast to the chain-extended supramolecular block copolymers based on those triple H-bonding interactions ($K_a = 800\,M^{-1}$), the sextuple H-bonding interaction increased the stability of the material significantly – even up to 80 °C, well above the glass-transition temperature (T_g) of the PEK block [146]. Later, the concept was extended further toward supramolecular reversibly thermoresponsive gels [200]; this approach represents an important contribution with respect to the development of efficient self-healing materials [53, 54].

To assemble supramolecular block copolymers on the basis of the same sextuple H-bonding motive, Bernard and coworkers synthesized a variety of homo and diblock copolymers from RAFT agents functionalized with a Hamilton wedge or a barbiturate moiety (RAFT: reversible addition-fragmentation chain-transfer) [201]. These polymers, in turn, were assembled into supramolecular pseudo-block copolymers of the general types A–B, AB–C, A–BC, and AB–CD via sextuple H-bonding interactions (Figure 3.49). A different polymer topology was reported by Altintas et al., who combined a Hamilton-receptor carrying one PS and one

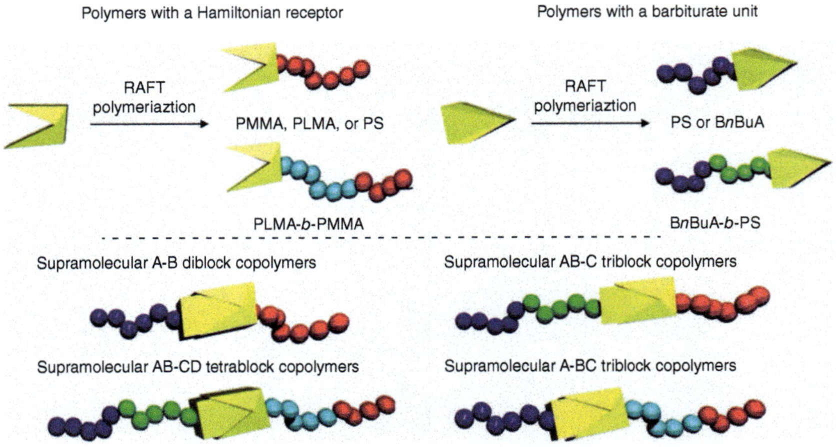

Figure 3.49 Schematic representation of the formation of homopolymers and diblock copolymers equipped with Hamiltonian receptors or barbiturate units. PS: polystyrene, PMMA: poly(methyl methacrylate), PLMA: poly(lauryl methacrylate), P*n*BuA: poly(*n*-butyl acrylate). These macromonomers were assembled into supramolecular pseudo-block copolymers via sextuple H-bonding interactions. Source: Chen et al. [201]. Figure reproduced with kind permission. © 2012 The Royal Chemical Society.

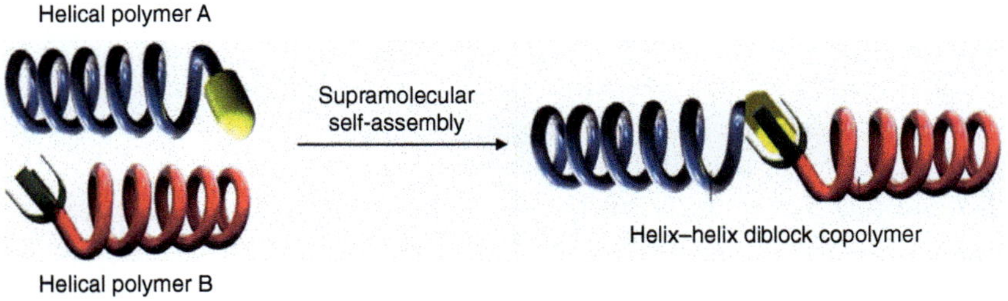

Figure 3.50 Schematic representation of the formation of a supramolecular helix–helix diblock copolymer. Source: Croom et al. [204]. Figure reproduced with kind permission. © 2016 American Chemical Society.

PEG arm with a poly(*n*-butyl acrylate) (PnBA) chain end-functionalized with two cyanuric acid residues; the self-assembly of these macromolecular components, in a 2 : 1 ratio, gave a H-shaped terpolymer, which was analyzed by 1D- and 2D-NMR spectroscopy as well as DLS measurements [202]. Binder and coworkers analyzed the phase-separation behavior of various AB-type diblock copolymers by temperature-dependent SAXS measurements [203]: the diblock copolymers with high phase-segregation strength exhibited an earlier dissociation of the H-bonds driven by the demixing of the incompatible blocks. The secondary structures of two different helical polymer strands in solution were retained upon linking them via sextuple H-bonding (CD and IR spectroscopy were applied to confirm this, Figure 3.50). However, the absence of such microstructures in the solid state could be excluded on the basis of wide-angle X-ray scattering (WAXS) and DSC measurements [204].

A chain-extended supramolecular polymer was assembled from heteroditopic PnBA equipped with a Hamilton wedge and barbiturate moiety. Chen et al. prepared the bifunctional macromonomer by a RAFT polymerization and subsequent chain-end functionalization via a thiol-bromo substitution reaction [205]. Formation of the supramolecular polymer was concluded from temperature-dependent ^{1}H NMR and DOSY spectroscopy as well as rheology measurements. Barner-Kowollik and coworkers prepared the heterotelechelic PS derivative **50** by combining ATRP and end-group functionalization by the Cu(I)-catalyzed azide-to-alkyne cycloaddition (CuAAC, a classic "click" reaction, Figure 3.51) [206]. At low concentrations (i.e. $c \leq 0.5\,\text{mM}$), polymer $(\mathbf{50})_n$ showed a single-chain self-assembly via sextuple H-bonding between the terminal binding sites. DLS measurements suggested that the entropically favored cyclic species reversibly reorganized into a chain-extended supramolecular polymer at higher concentrations (Figure 3.51). It was further shown that the single-chain self-folded structure could be stabilized by additional intramolecular H-bonding interactions [207]. For this purpose, a modified polymer

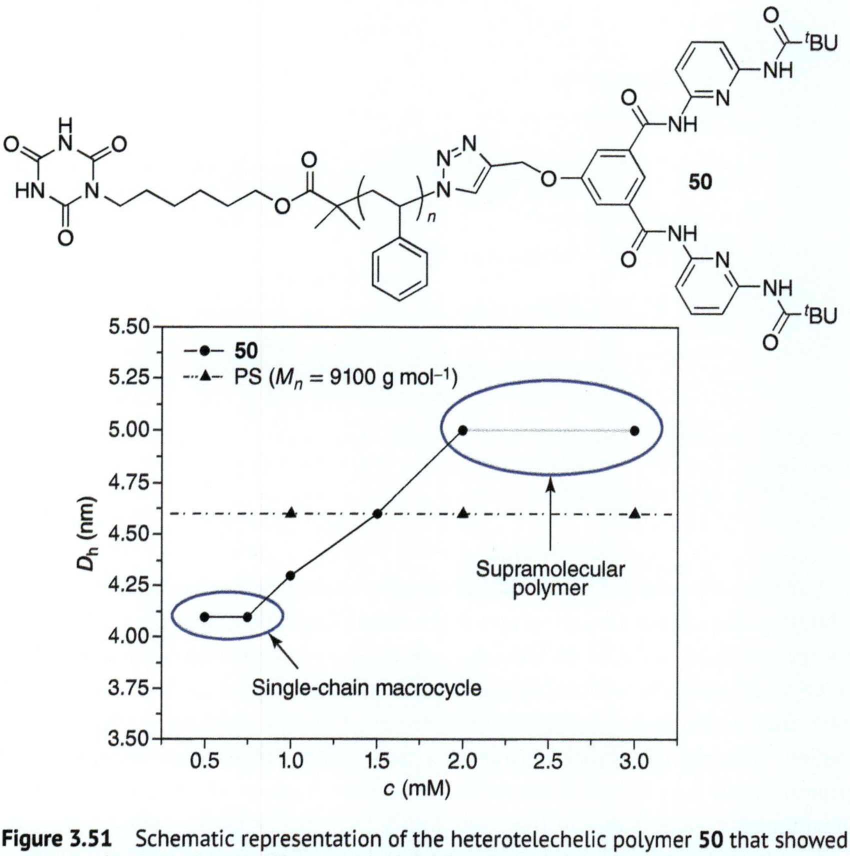

Figure 3.51 Schematic representation of the heterotelechelic polymer **50** that showed a concentration-dependent self-assembly behavior (i.e. formation of single-chain macrocycles vs. chain-extended polymers). Source: Altintas et al. [206]. Figure reproduced with kind permission. © 2010 The Royal Chemical Society.

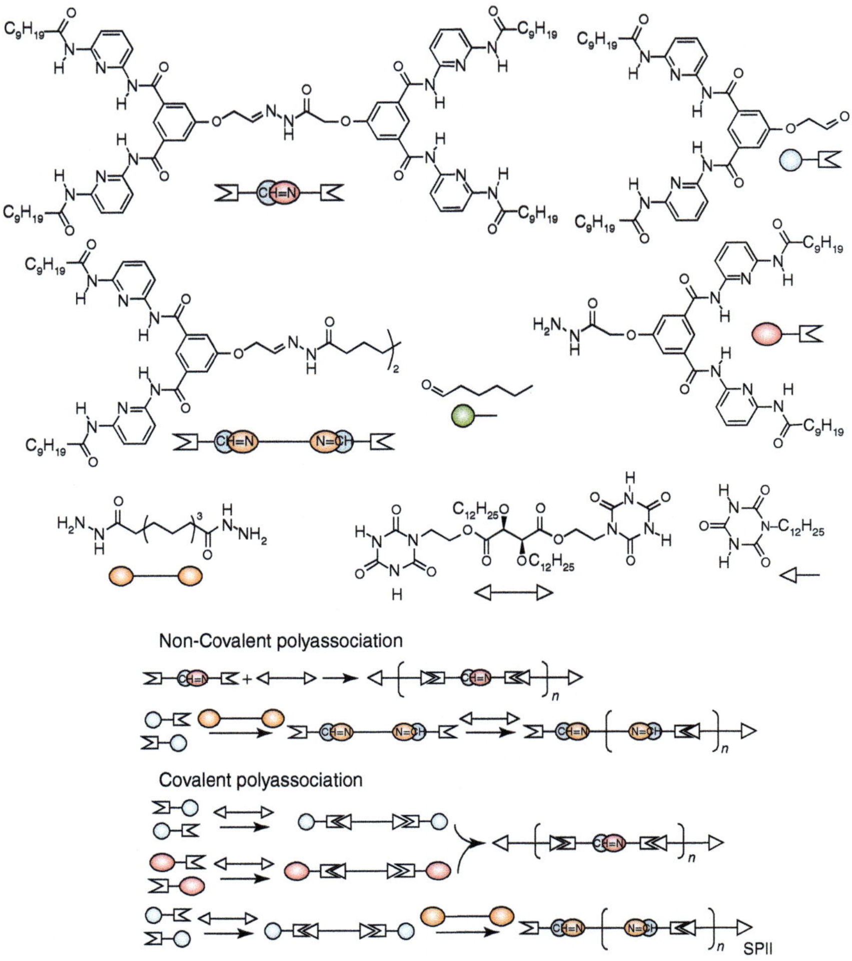

Figure 3.52 Schematic representation of Lehn's concept of double dynamic supramolecular polymers ("double dynamers"). Source: Kolomiets et al. [209]. © 2005 The Royal Chemical Society.

was used which contained an additional complementary pair of H-bonding entities, as pending side chains (i.e. a thymine and a diaminopyridine).

The concept of multiple H-bonding interactions was also combined with the formation of (partially) reversible covalent bonds (e.g. imines [208] or hydrazones [209]), thus forming a combinatorial libraries of double dynamic polymers, the so-called "double dynamers." Kolomiets and Lehn coined this definition for supramolecular polymers that exhibited a dynamic character (i.e. exchange of their molecular components due to the reversibility of bond formation) simultaneously on the molecular as well as supramolecular level (Figure 3.52) [209]. In these particular materials, the self-assembly processes were governed by an interplay between the H-bonding interactions and the covalent-bonding system of a partially reversible nature. It was envisioned that these double dynamic polymers might have potential applications as adaptive or self-healing materials [55].

Already in 1990, Lehn et al. studied the self-assembly of ring-substituted 1,3,5-triamino-2,4,6-triazine and barbituric acid into supramolecular, quasi-polymeric strands [210]. The level-of-order could be controlled via the nature of the substituents (i.e. hydrophobicity and solubility effects). The pronounced cooperativity of the supramolecular polymerization facilitated a high level of association and, thus, a high virtual molar mass of the materials. Reinhoudt and coworkers utilized this particular system to assemble calix[4]arenes in a sheet-type fashion (see also Chapter 8) [211]. For this purpose, calix[4]arenes bearing either two melamine (**51**) or two cyanurate moieties (**52**), as designated H-bonding units, were prepared. The location of these entities on the upper rims of the calix[4]arene scaffolds enabled the formation of polymeric aggregates and, at the same time, prohibited the formation of cyclic structures (Figure 3.53). In the solid state, the two components, at a

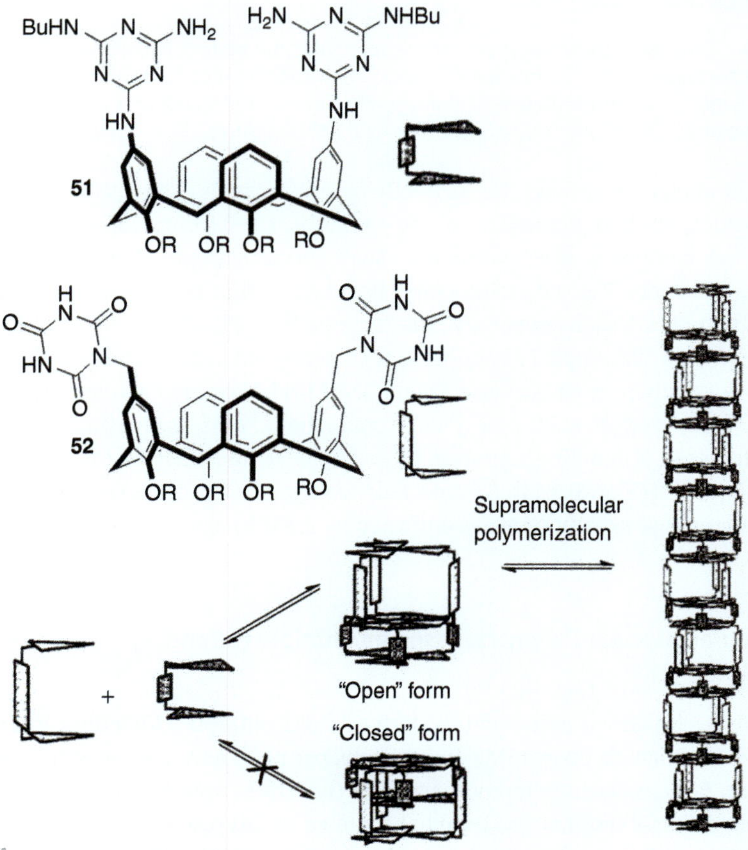

Figure 3.53 Schematic representation of the directed self-assembly of the calix[4]arenes **51** and **52** into columnar structures via tris-aminotriazine/barbiturate interactions. Source: Klok et al. [211]. Figure reproduced with kind permission. © 1999 American Chemical Society.

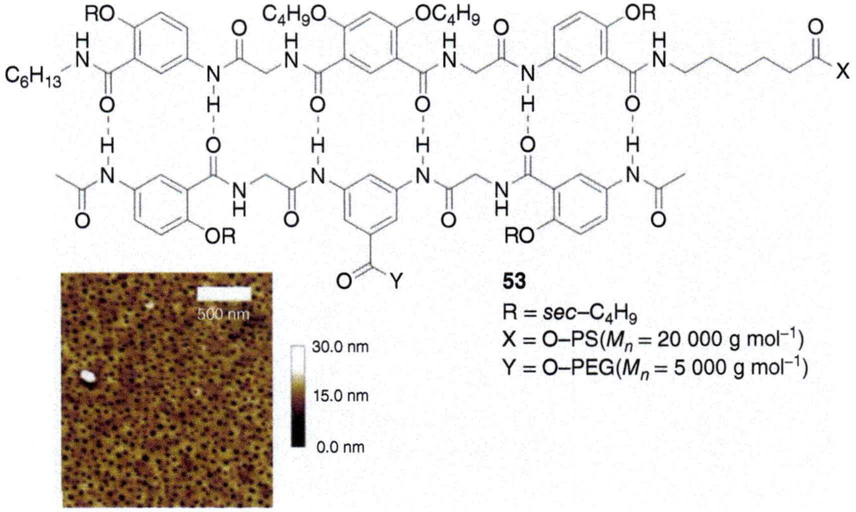

Figure 3.54 Schematic representation of the supramolecular diblock copolymer **53** featuring a sextuple H-bonding linkage (PS block: M_n = 20 000 g mol^{-1}, PEG block: M_n = 5000 g mol^{-1}). A representative AFM image of a spin-coated thin film is also depicted. Source: Yang et al. [41]. Figure reproduced with kind permission. © 2004 Wiley-VCH.

1 : 1 stoichiometry, assembled into perfectly ordered strands. These non-covalent, rod-like structures had diameters in the range of 3.5–5.5 nm; the amphiphilic structure was composed of a hydrophobic shell and a hydrophilic core, based on H-bonding networks. Related architectures based on similar H-bonding arrays also generated strands via such cooperative interactions [120, 212–214].

A different type of sextuple H-bonding array was applied to assemble supramolecular diblock copolymers **53** (see also Figure 3.9b) [41]. Two incompatible polymer segments, composed of a PS and a PEG component, were each coupled to an oligoamide-based H-bonding template ($K_a > 10^9$ M^{-1}) and merged in a solution blending approach (Figure 3.54). As expected, two separate T_g values were observed, and the microphase separation was confirmed by AFM imaging of the spin-coated thin films.

3.2.6 Supramolecular Polymers, Based on Multiple H-Bonding Interactions

Structurally well-defined architectures derived from multiple H-bonding interactions (i.e. more than six cooperative interactions per unit) have also been reported, but are rare. Besides Nature's famous examples (i.e. DNA, proteins, or oligosaccharides) some artificial systems also have to be named in this respect.

Ghadiri et al. utilized cyclic peptide scaffolds for the assembly of large supramolecular nanotubes in aqueous media [215–225]. Initially, cyclic peptides **54** with eight amino acid residues of alternating L- and D-configurations were investigated (Figure 3.55a) [215, 216, 218, 221]. It was shown by FT-IR spectroscopy that intermolecular H-bonding interactions represented the main driving force for

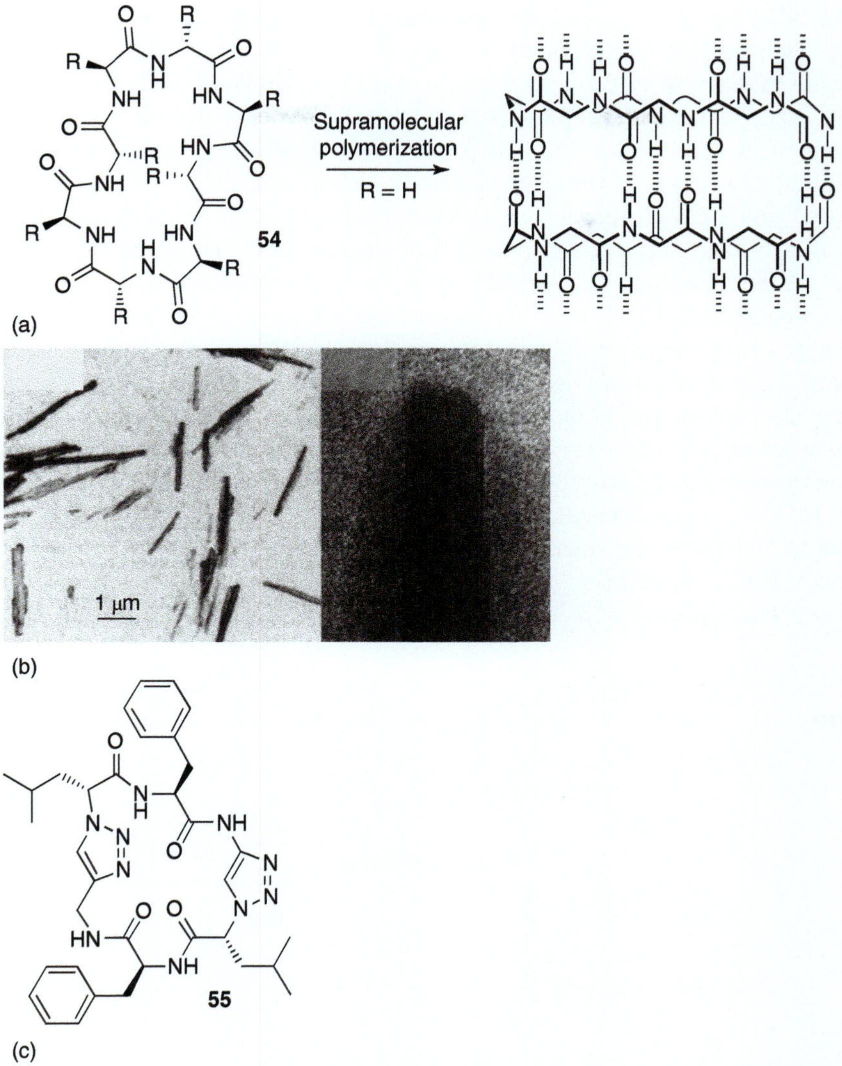

Figure 3.55 (a) Schematic representation of the cyclic eight-residue peptide **54** and of its self-assembly into supramolecular nanotubes. (b) b1: TEM image of the supramolecular nanotubes adsorbed onto a carbon support film; b2: Low-dose cryo-microscopy image of a single nanotube. (c) Schematic representation of the cyclic four-residue peptide **55**. Source: Khazanovich et al. [217]. Figure reproduced with kind permission. © 1994 American Chemical Society.

the tubular growth of the supramolecular assembly [215, 221]. An antiparallel β-sheet-type stacking of the cyclic peptides was concluded from the FT-IR data and electron diffraction measurements on crystalline samples [221]. The low solubility as well as the high stability of the nanostructures (Figure 3.55b) was explained by the high cooperativity of the supramolecular polymerization process in solution. The authors proposed a templating effect (due to the preorganization of the cyclic peptide) and the concerted formation of multiple non-covalent bonds to be responsible for the observed allosteric cooperativity [221, 225]. It has to be noted in this context that computational modeling studies confirmed the presence of the suggested antiparallel β-sheets as a result of the cooperative supramolecular polymerization mechanism [226–228]. This process could be partially suppressed by alkylation of the amino groups – as a result, only dimers and not polymers were formed. This approach allowed one to determine the K_a value for the dimerization in $CHCl_3$ by concentration-dependent ^1H NMR [223, 225]. The structural change from the eight-residue cyclic peptide **54** to the four-residue one **55** (comprising two 1,4-disubstituted [1*H*]-1,2,3-triazole ε-amino acid units, Figure 3.55c) induced a change of the mechanism for the supramolecular polymerization: in this case, an anti-cooperative supramolecular polymerization was observed (for a discussion of the mechanism, see Section 1.3.3 and Ref. [6]) [229]. ^1H NMR spectroscopy, mass spectrometry, and X-ray crystallography were used to elucidate the structure of the supramolecular assembly (i.e. a nanotubular superstructure): a K_a value of $8.6 \times 10^4\ M^{-1}$ for the dimerization of **54** and an elongation equilibrium constant (K) of $3.8 \times 10^4\ M^{-1}$ were determined by concentration-dependent ^1H NMR experiments in $CHCl_3$. Since the K_{dim}/K_a fraction was larger than unity, a small degree of anti-cooperativity was assumed – however, the origin of this could not be clarified.

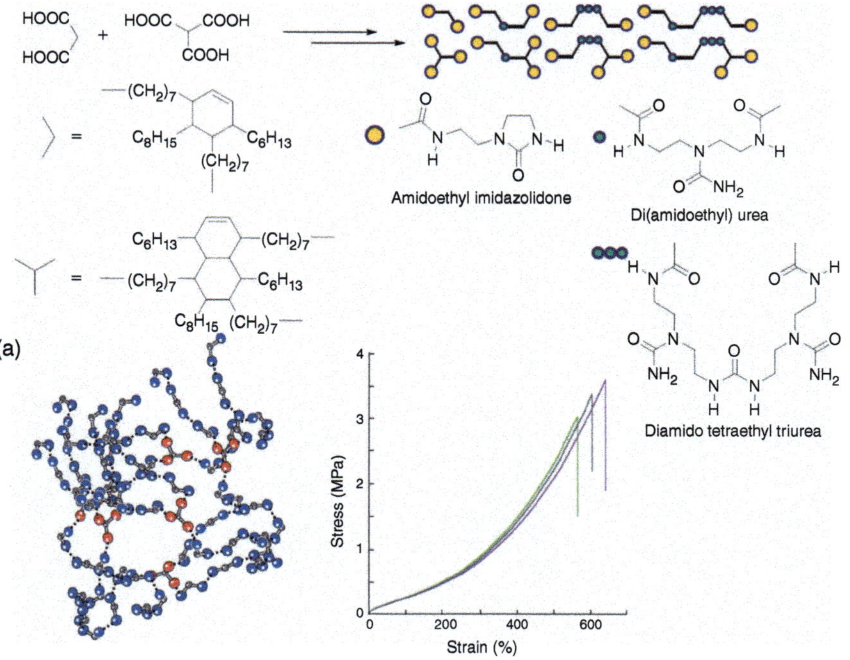

(a)
(b)
(c)

Figure 3.56 (a) Schematic representation of the formation of a "supramolecular rubber," based on multiple H-bonding. (b) Representation of a polymer network formed by mixtures of ditopic (blue) and tritopic (red) building blocks, self-assembled by directional interactions (dotted lines). (c) Stress–strain curve of the supramolecular rubber. Source: Cordier et al. [230]. Figure reproduced with kind permission. © 2008 Springer Nature.

Finally, the enormous potential of these supramolecular polymers, derived from multiple H-bonding interactions, is highlighted by their application as self-healing materials and thermoreversible rubbers. For this purpose, fatty acid dimers, obtained from natural feedstock (i.e. vegetable oil), were modified with multiple H-bonding moieties, such as amidoethyl imidazolidone, di(amidoethyl) urea, and diaminotetraethyl diurea (Figure 3.56a) [230]. The synthesis of these monomers involved the condensation of the diacids with diethylene triamine; the resultant products were than treated with urea. The initial products resembled translucent plastics, which behaved as soft rubbers at 90 °C (thus, significantly above the T_g of 28 °C). According to size-exclusion chromatography (SEC) and viscosity measurements, oligomers were present under these conditions rather than polymers. When the temperature was increased (>160 °C), the materials' properties changed significantly and the material could, as typical for covalent viscoelastic polymers, even be molded, extruded, and (re)shaped. The cross-linking of the supramolecular material via long alkyl chains lowered the T_g value and the material behaved like a nonsticky rubber (Figure 3.56b). The rubber-type properties were shown by rheological and mechanical measurements: the stress–strain curves, for instance, resembled those of covalent soft rubbers (Figure 3.56c). The process of mechanically breaking the non-covalent bonds and re-assembling them by thermal curing could be repeated many times without loss of efficiency.

3.3 Conclusion Remarks

Within the wide range of supramolecular polymers, the systems that are formed via (directional) H-bonding interactions share a key role (together with metal-to-ligand coordination), due to the broad diversity of chemical structures can be utilized to function as H-bonding systems in combination with their high degree-of-directionality and specificity in recognition. As demonstrated by Nature, the stability of this type of non-covalent interactions can easily be varied with respect to the binding strength and this, in turn, can be utilized to fine-tune their properties. Thus, supramolecular polymers with elastomeric or thermoplastic properties as well as highly cross-linked duroplastic structures and networks can be generated and these properties often resemble those as known for analogous covalent polymers.

With respect to the remarkable thermoreversibility of H-bonding interactions, materials exhibiting highly tunable properties can be designed and constructed. The mechanical and/or opto-electronic properties can be reversibly changed readily by applying small external stimuli. Thus, H-bonded systems are of utmost relevance for the realization of future's addressable and adaptable materials. This even includes biomaterials for advanced medicinal applications.

In addition, the high specificity and reversibility of H-bonding interactions have paved the way to a new field in modern supra(macro)molecular polymer science: dynamic supramolecular materials exhibiting self-selection or self-sorting abilities. This field of research, as pioneered by the work of Jean-Marie Lehn on his "dynamers," is of significant importance with respect to the preparation of new supra(macro)molecular materials with properties that cannot be realized with conventional monomers and polymers where only covalent bonds are involved.

References

1 Moore, T.S. and Winmill, T.F. (1912). *J. Chem. Soc. Trans.* 101: 1635–1676.
2 Fouquey, C., Lehn, J.-M., and Levelut, A.-M. (1990). *Adv. Mater.* 2: 254–257.
3 Binder, W.H. and Zirbs, R. (2007). *Adv. Polym. Sci.* 207: 1–78.
4 Abed, S., Boileau, S., and Bouteiller, L. (2000). *Macromolecules* 33: 8479–8487.
5 Brunsveld, L., Folmer, B.J.B., Meijer, E.W., and Sijbesma, R.P. (2001). *Chem. Rev.* 101: 4071–4098.
6 de Greef, T.F.A., Smulders, M.M.J., Wolffs, M. et al. (2009). *Chem. Rev.* 109: 5687–5754.
7 Bouteiller, L. (2007). *Adv. Polym. Sci.* 207: 79–112.
8 Calhorda, M.J. (2000). *Chem. Commun.*: 801–809.
9 Sherrington, D.C. and Taskinen, K.A. (2001). *Chem. Soc. Rev.* 30: 83–93.
10 Schmuck, C. and Wienand, W. (2001). *Angew. Chem. Int. Ed.* 40: 4363–4369.
11 ten Cate, T. and Sijbesma, R.P. (2003). *Macromol. Rapid Commun.* 23: 1094–1112.
12 Brunsveld, L., Folmer, B.J.B., and Meijer, E.W. (2000). *MRS Bull.* 25: 49–53.
13 Armstrong, G. and Buggy, M. (2005). *J. Mater. Sci.* 40: 547–559.
14 Krieg, E., Bastings, M.M.C., Besenius, P., and Rybtchinski, B. (2016). *Chem. Rev.* 116: 2414–2477.
15 Bosman, A.W., Sijbesma, R.P., and Meijer, E.W. (2004). *Mater. Today* 7: 34–39.
16 Jeffrey, G.A. and Saenger, W. (1991). *Hydrogen Bonding in Biological Structures.* Berlin, Heidelberg: Springer-Verlag.
17 Belkova, N.V., Shubina, E.S., and Epstein, L.M. (2005). *Acc. Chem. Res.* 38: 624–631.
18 Beijer, F.H., Sijbesma, R.P., Kooijman, H. et al. (1998). *J. Am. Chem. Soc.* 120: 6761–6769.
19 Beijer, F.H., Kooijman, H., Spek, A.L. et al. (1998). *Angew. Chem. Int. Ed. Engl.* 37: 75–78.
20 Prabhakaran, P., Puranik, V.G., and Sanjayan, G.J. (2005). *J. Org. Chem.* 70: 10067–10072.
21 Corbin, P.S. and Zimmerman, S.C. (1998). *J. Am. Chem. Soc.* 120: 9710–9711.
22 Watson, J.D. and Crick, F.H.C. (1953). *Nature* 171: 737–738.
23 Sivakova, S. and Rowan, S.J. (2005). *Chem. Soc. Rev.* 34: 9–21.
24 Pranata, J., Wierschke, S.G., and Jorgensen, W.L. (1991). *J. Am. Chem. Soc.* 113: 2810–2819.
25 Jorgensen, W.L. and Pranata, J. (1990). *J. Am. Chem. Soc.* 112: 2008–2010.
26 Djurdjevic, S., Leigh, D.A., McNab, H. et al. (2007). *J. Am. Chem. Soc.* 129: 476–477.
27 Murray, T.J. and Zimmerman, S.C. (1992). *J. Am. Chem. Soc.* 114: 4010–4011.

28 Sartorius, J. and Schneider, H.-J. (1996). *Chem. Eur. J.* 2: 1446–1452.

29 Zimmerman, S.C. and Corbin, P.S. (2000). Heteroaromatic modules for self-assembly using multiple hydrogen bonds. In: *Structure and Bonding. Molecular Self-Assembly - Organic Versus Inorganic Approaches*, vol. 96 (ed. M. Fujita), 63–94. Berlin, Heidelberg: Springer-Verlag.

30 Sessler, J.L. and Wang, R.-H. (1998). *Angew. Chem. Int. Ed. Engl.* 37: 1726–1729.

31 Kolotuchin, S.V. and Zimmerman, S.C. (1998). *J. Am. Chem. Soc.* 120: 9092–9093.

32 Sijbesma, R.P., Beijer, F.H., Brunsveld, L. et al. (1997). *Science* 278: 1601–1604.

33 Lafitte, V.G.H., Aliev, A.E., Horton, P.N. et al. (2006). *J. Am. Chem. Soc.* 128: 6544–6545.

34 Zeng, H.-Q., Miller, R.S., Flowers, R.A. II, and Gong, B. (2000). *J. Am. Chem. Soc.* 122: 2635–2644.

35 Folmer, B.J.B., Sijbesma, R.P., Kooijman, H. et al. (1999). *J. Am. Chem. Soc.* 121: 9001–9007.

36 Öjelund, K., Loontjens, T., Steeman, P. et al. (2003). *Macromol. Chem. Phys.* 204: 52–60.

37 Wang, X.-Z., Li, X.-Q., Shao, X.-B. et al. (2003). *Chem. Eur. J.* 9: 2904–2913.

38 Chang, S.-K. and Hamilton, A.D. (1988). *J. Am. Chem. Soc.* 110: 1318–1319.

39 Tecilla, P. and Hamilton, A.D. (1990). *J. Chem. Soc., Chem. Commun.*: 1232–1234.

40 Chang, S.-K., Van Engen, D., Fan, E., and Hamilton, A.D. (1991). *J. Am. Chem. Soc.* 113: 7640–7645.

41 Yang, X.-W., Hua, F.-J., Yamato, K. et al. (2004). *Angew. Chem. Int. Ed.* 43: 6471–6474.

42 Macomber, R.S. (1992). *J. Chem. Educ.* 69: 375.

43 Barrans, R.E. Jr., and Dougherty, D.A. (1994). *Supramol. Chem.* 4: 121–130.

44 Bouteiller, L., Colombani, O., Lortie, F., and Terech, P. (2005). *J. Am. Chem. Soc.* 127: 8893–8898.

45 Arnaud, A. and Bouteiller, L. (2004). *Langmuir* 20: 6858–6863.

46 Schnell, I., Langer, B., Söntjens, S.H.M. et al. (2002). *Phys. Chem. Chem. Phys.* 4: 3750–3758.

47 Spiess, H.W. (2003). *Macromol. Symp.* 201: 85–88.

48 Söntjens, S.H.M., Sijbesma, R.P., van Genderen, M.H.P., and Meijer, E.W. (2000). *J. Am. Chem. Soc.* 122: 7487–7493.

49 Loverde, S.M., Ermoshkin, A.V., and De la Cruz, M.O. (2005). *J. Polym. Sci., Part B: Polym. Phys.* 43: 796–804.

50 Huh, J. and Jo, W.-H. (2004). *Macromolecules* 37: 3037–3038.

51 Gilday, L.C., Robinson, S.W., Barendt, T.A. et al. (2015). *Chem. Rev.* 115: 7118–7195.

52 Lima, J.C. and Rodríguez, L. (2015). *Inorganics* 3: 1–18.

53 Hager, M.D., Greil, P., Leyens, C. et al. (2010). *Adv. Mater.* 22: 5424–5430.

54 Herbst, F., Döhler, D., Michael, P., and Binder, W.H. (2013). *Macromol. Rapid Commun.* 34: 203–220.

55 Roy, N., Bruchmann, B., and Lehn, J.-M. (2015). *Chem. Soc. Rev.* 44: 3786–3807.
56 Campanella, A., Döhler, D., and Binder, W.H. (2018). *Macromol. Rapid Commun.* 39: 1700739.
57 Xiao, Z., Duan, T., Chen, H. et al. (2018). *Sol. Energy Mater. Sol. Cells* 182: 1–13.
58 Aoki, K., Nakagawa, M., and Ichimura, K. (1999). *Chem. Lett.* 28: 1205–1206.
59 Lee, J.Y., Painter, P.C., and Coleman, M.W. (1988). *Macromolecules* 21: 954–960.
60 Kumar, U., Kato, T., and Fréchet, J.M.J. (1992). *J. Am. Chem. Soc.* 114: 6630–6639.
61 Kato, T., Kihara, H., Kumar, U. et al. (1994). *Angew. Chem. Int. Ed. Engl.* 33: 1644–1645.
62 Ambrožič, G. and Žigon, M. (2000). *Macromol. Rapid Commun.* 21: 53–56.
63 Kato, T., Mizoshita, N., and Kanie, K. (2001). *Macromol. Rapid Commun.* 22: 797–814.
64 Kato, T., Mizoshita, N., and Kishimoto, K. (2006). *Angew. Chem. Int. Ed.* 45: 36–68.
65 Jiang, M., Li, M., Xiang, M.-L., and Zhou, H. (1999). *Adv. Polym. Sci.* 146: 121–196.
66 Lee, C.-M., Jariwala, C.P., and Griffin, A.C. (1994). *Polymer* 35: 4550–4554.
67 Lee, C.-M. and Griffin, A.C. (1997). *Macromol. Symp.* 117: 281–290.
68 Alexander, C., Jariwala, C.P., Lee, C.-M., and Griffin, A.C. (1994). *Macromol. Symp.* 77: 283–294.
69 He, C.-B., Donald, A.M., Griffin, A.C. et al. (1998). *J. Polym. Sci., Part B: Polym. Phys.* 36: 1617–1624.
70 Lee, M.-S. and Cho, B.-K. (1999). *Macromolecules* 32: 8531–8537.
71 Singh, A., Lvov, Y., and Qadri, S.B. (1999). *Chem. Mater.* 11: 3196–3200.
72 He, C.-B., Lee, C.-M., Griffin, A.C. et al. (1999). *Mol. Cryst. Liq. Cryst. Sci. Technol., Sect. A* 332: 251–258.
73 Pourcain, C.B.S. and Griffin, A.C. (1995). *Macromolecules* 28: 4116–4121.
74 Wiegel, K.N. and Griffin, A.C. (1999). *Polym. Prepr. (Am. Chem. Soc., Div. Polym. Chem.)* 40: 1136–1137.
75 Stadler, R. and De Lucca Freitas, L. (1989). *Makromol. Chem. Macromol. Symp.* 26: 451–457.
76 Müller, M., Dardin, A., Seidel, U. et al. (1996). *Macromolecules* 29: 2577–2583.
77 Schirle, M., Hoffmann, I., Pieper, T. et al. (1996). *Polym. Bull.* 36: 95–102.
78 Stadler, R. and Burgert, J. (1986). *Makromol. Chem.* 187: 1681–1690.
79 Hilger, H. and Stadler, R. (1991). *Makromol. Chem.* 192: 805–817.
80 Bica, C.I.D., Burchard, W., and Stadler, R. (1996). *Macromol. Chem. Phys.* 197: 3407–3426.
81 Stadler, R. and De Lucca Freitas, L. (1986). *Colloid. Polym. Sci.* 264: 773–778.
82 De Lucca Freitas, L., Burgert, J., and Stadler, R. (1987). *Polym. Bull.* 17: 431–438.
83 De Lucca Freitas, L. and Stadler, R. (1987). *Macromolecules* 20: 2478–2485.

84 De Lucca Freitas, L. and Stadler, R. (1988). *Colloid. Polym. Sci.* 266: 1095–1101.
85 Hilger, H., Stadler, R., and De Lucca Freitas, L. (1990). *Polymer* 31: 818–823.
86 Hilger, H. and Stadler, R. (1991). *Polymer* 32: 3244–3249.
87 Müller, M., Seidel, U., and Stadler, R. (1995). *Polymer* 36: 3143–3150.
88 Müller, M., Stadler, R., Kremer, F., and Williams, G. (1995). *Macromolecules* 28: 6942–6949.
89 Müller, M., Fischer, E.W., Kremer, F. et al. (1995). *Colloid. Polym. Sci.* 273: 38–46.
90 Dardin, A., Stadler, R., Böffel, C., and Spiess, H.W. (1993). *Makromol. Chem.* 194: 3467–3477.
91 Stadler, R. and De Lucca Freitas, L. (1986). *Polym. Bull.* 15: 173–179.
92 Seidel, U., Stadler, R., and Fuller, G.G. (1994). *Macromolecules* 27: 2066–2072.
93 Seidel, U., Stadler, R., and Fuller, G.G. (1995). *Macromolecules* 28: 3739–3740.
94 Duweltz, L., Abed, F.S., Bouteiller, L., and Boileau, S. (2003). *Polymer* 44: 2295–2302.
95 Abed, S., Boileau, S., and Bouteiller, L. (2001). *Polymer* 42: 8613–8619.
96 Jacobsen, H. and Stockmayer, W.H. (1950). *J. Chem. Phys.* 18: 1600–1606.
97 Ercolani, G., Mandolini, L., Mencarelli, P., and Roelens, S. (1993). *J. Am. Chem. Soc.* 115: 3901–3908.
98 Rakotondradany, F., Whitehead, M.A., Lebuis, A.-M., and Sleiman, H.F. (2003). *Chem. Eur. J.* 9: 4771–4780.
99 Stončius, S., Orentas, E., Butkus, E. et al. (2006). *J. Am. Chem. Soc.* 128: 8272–8285.
100 Ducharme, Y. and Wuest, J.D. (1988). *J. Org. Chem.* 53: 5787–5789.
101 Würthner, F., Saha-Möller, C.R., Fimmel, B. et al. (2016). *Chem. Rev.* 116: 962–1052.
102 Kaiser, T.E., Wang, H., Stepanenko, V., and Würthner, F. (2007). *Angew. Chem. Int. Ed.* 46: 5541–5544.
103 Kaiser, T.E., Stepanenko, V., and Würthner, F. (2009). *J. Am. Chem. Soc.* 131: 6719–6732.
104 Lin, H., Camacho, R., Tian, Y. et al. (2010). *Nano Lett.* 10: 620–626.
105 Giaimo, J.M., Gusev, A.V., and Wasielewski, M.R. (2002). *J. Am. Chem. Soc.* 124: 8530–8531.
106 Wang, H., Kaiser, T.E., Uemura, S., and Würthner, F. (2008). *Chem. Commun.*: 1181–1183.
107 Xie, Z., Stepanenko, V., Fimmel, B., and Würthner, F. (2014). *Mater. Horiz.* 1: 355–359.
108 Ghosh, S., Shankar, S., Philips, D.S., and Ajayaghosh, A. (2020). *Mater. Today Chem.* 16: 100242.
109 Zambounis, J.S., Hao, Z., and Iqbal, A. (1997). *Nature* 388: 131–132.
110 Yanagisawa, H., Mizuguchi, J., Aramaki, S., and Sakai, Y. (2008). *Jpn. J. Appl. Phys.* 47: 4728–4731.
111 Suna, Y., Nishida, J.-i., Fujisaki, Y., and Yamashita, Y. (2012). *Org. Lett.* 14: 3356–3359.

112 Maqueira-Albo, I., Ernesto Bonacchini, G., Dell'Erba, G. et al. (2017). *J. Mater. Chem. C* 5: 11522–11531.
113 Lehn, J.-M. (1993). *Makromol. Chem. Macromol. Symp.* 69: 1–17.
114 Gulik-Krzywicki, T., Fouquey, C., and Lehn, J.-M. (1993). *Proc. Natl. Acad. Sci. U.S.A.* 90: 163–167.
115 Kotera, M., Lehn, J.-M., and Vigneron, J.-P. (1995). *Tetrahedron* 51: 1953–1972.
116 Kotera, M., Lehn, J.-M., and Vigneron, J.-P. (1994). *J. Chem. Soc., Chem. Commun.*: 197, 199.
117 Yagai, S., Higashi, M., Karatsu, T., and Kitamura, A. (2006). *Chem. Commun.*: 1500–1502.
118 Yagai, S., Kinoshita, T., Higashi, M. et al. (2007). *J. Am. Chem. Soc.* 129: 13277–13287.
119 Yagai, S., Karatsu, T., and Kitamura, A. (2005). *Langmuir* 21: 11048–11052.
120 Yagai, S., Higashi, M., Karatsu, T., and Kitamura, A. (2005). *Chem. Mater.* 17: 4392–4398.
121 Yagai, S., Higashi, M., Karatsu, T., and Kitamura, A. (2004). *Chem. Mater.* 16: 3582–3585.
122 Seki, T., Yagai, S., Karatsu, T., and Kitamura, A. (2008). *J. Org. Chem.* 73: 3328–3335.
123 Yagai, S., Monma, Y., Kawauchi, N. et al. (2007). *Org. Lett.* 9: 1137–1140.
124 Bielejewska, A.G., Marjo, C.E., Prins, L.J. et al. (2001). *J. Am. Chem. Soc.* 123: 7518–7533.
125 Zerkowski, J.A., Seto, C.T., and Whitesides, G.M. (1992). *J. Am. Chem. Soc.* 114: 5473–5475.
126 Seto, C.T. and Whitesides, G.M. (1993). *J. Am. Chem. Soc.* 115: 1330–1340.
127 Mathias, J.P., Simanek, E.E., Zerkowski, J.A. et al. (1994). *J. Am. Chem. Soc.* 116: 4316–4325.
128 Chauhan, S.M.S. and Giri, N.G. (2008). *Supramol. Chem.* 20: 743–752.
129 Zhang, F. and Therrien, B. (2018). *Eur. J. Inorg. Chem.* 2018: 2399–2407.
130 Würthner, F., Thalacker, C., and Sautter, A. (1999). *Adv. Mater.* 11: 754–758.
131 Würthner, F., Thalacker, C., Sautter, A. et al. (2000). *Chem. Eur. J.* 6: 3871–3886.
132 Thalacker, C. and Würthner, F. (2002). *Adv. Funct. Mater.* 12: 209–218.
133 Sinks, L.E., Rybtchinski, B., Iimura, M. et al. (2005). *Chem. Mater.* 17: 6295–6303.
134 Theobald, J.A., Oxtoby, N.S., Phillips, M.A. et al. (2003). *Nature* 424: 1029.
135 Silly, F., Shaw, A.Q., Porfyrakis, K. et al. (2008). *Chem. Commun.*: 4616–4618.
136 Theobald, J.A., Oxtoby, N.S., Champness, N.R. et al. (2005). *Langmuir* 21: 2038–2041.
137 Perdigão, L.M.A., Perkins, E.W., Ma, J. et al. (2006). *J. Phys. Chem. B* 110: 12539–12542.
138 Madueno, R., Räisänen, M.T., Silien, C., and Buck, M. (2008). *Nature* 454: 618.
139 Silien, C., Räisänen, M.T., and Buck, M. (2009). *Angew. Chem. Int. Ed.* 48: 3349–3352.
140 Yagai, S. (2015). *Bull. Chem. Soc. Jpn.* 88: 28–58.

141 Aratsu, K., Takeya, R., Pauw, B.R. et al. (2020). *Nat. Commun.* 11: 1623.
142 Aizawa, T., Aratsu, K., Datta, S. et al. (2020). *Chem. Commun.* 56: 4280–4283.
143 Saito, T. and Yagai, S. (2020). *Org. Biomol. Chem.* 18: 3996–3999.
144 Binder, W.H., Kunz, M.J., and Ingolic, E. (2004). *J. Polym. Sci., Part A: Polym. Chem.* 42: 162–172.
145 Kunz, M.J., Hayn, G., Saf, R., and Binder, W.H. (2004). *J. Polym. Sci., Part A: Polym. Chem.* 42: 661–674.
146 Binder, W.H., Bernstorff, S., Kluger, C. et al. (2005). *Adv. Mater.* 17: 2824–2828.
147 Herbst, F., Schröter, K., Gunkel, I. et al. (2010). *Macromolecules* 43: 10006–10016.
148 Cortese, J., Soulié-Ziakovic, C., Tencé-Girault, S., and Leibler, L. (2012). *J. Am. Chem. Soc.* 134: 3671–3674.
149 Sivakova, S., Wu, J., Campo, C.J. et al. (2006). *Chem. Eur. J.* 12: 446–456.
150 Sivakova, S. and Rowan, S.J. (2003). *Chem. Commun.*: 2428, 2429.
151 Shimizu, T., Iwaura, R., Masuda, M. et al. (2001). *J. Am. Chem. Soc.* 123: 5947–5955.
152 Sivakova, S., Bohnsack, D.A., Mackay, M.E. et al. (2005). *J. Am. Chem. Soc.* 127: 18202–18211.
153 Yan, Y., Lu, T., and Huang, J.-B. (2009). *J. Colloid Interface Sci.* 337: 1–10.
154 Yang, L., Tan, X., Wang, Z., and Zhang, X. (2015). *Chem. Rev.* 115: 7196–7239.
155 Goor, O.J.G.M., Hendrikse, S.I.S., Dankers, P.Y.W., and Meijer, E.W. (2017). *Chem. Soc. Rev.* 46: 6621–6637.
156 Brunsveld, L., Vekemans, J.A.J.M., Hirschberg, J.H.K.K. et al. (2002). *Proc. Natl. Acad. Sci. U.S.A.* 99: 4977–4982.
157 Folmer, B.J.B., Sijbesma, R.P., and Meijer, E.W. (2001). *J. Am. Chem. Soc.* 123: 2093–2094.
158 Söntjens, S.H.M., Sijbesma, R.P., van Genderen, M.H.P., and Meijer, E.W. (2001). *Macromolecules* 34: 3815–3818.
159 Lange, R.F.M., Van Gurp, M., and Meijer, E.W. (1999). *J. Polym. Sci., Part A: Polym. Chem.* 37: 3657–3670.
160 Folmer, B.J.B., Sijbesma, R.P., Versteegen, J.A.J., and Meijer, E.W. (2000). *Adv. Mater.* 12: 874–878.
161 Hirschberg, J.H.K.K., Ramzi, A., Sijbesma, R.P., and Meijer, E.W. (2003). *Macromolecules* 36: 1429–1432.
162 Keizer, H.M., Sijbesma, R.P., Jansen, J.F.G.A. et al. (2003). *Macromolecules* 36: 5602–5606.
163 van Beek, D.J.M., Gillissen, M.A.J., van As, B.A.C. et al. (2007). *Macromolecules* 40: 6340–6348.
164 Yang, S.K., Ambade, A.V., and Weck, M. (2011). *Chem. Soc. Rev.* 40: 129–137.
165 Dankers, P.Y.W., Harmsen, M.C., Brouwer, L.A. et al. (2005). *Nat. Mater.* 4: 568–574.
166 van Almen, G.C., Talacua, H., Ippel, B.D. et al. (2016). *Macromol. Biosci.* 16: 350–362.

167 Mollet, B.B., Comellas-Aragonès, M., Spiering, A.J.H. et al. (2014). *J. Mater. Chem. B* 2: 2483–2493.

168 Muylaert, D.E.P., van Almen, G.C., Talacua, H. et al. (2016). *Biomaterials* 76: 187–195.

169 Dankers, P.Y.W., Boomker, J.M., der Vlag, A. H.-v. et al. (2010). *Macromol. Biosci.* 10: 1345–1354.

170 Appel, W.P.J., Meijer, E.W., and Dankers, P.Y.W. (2011). *Macromol. Biosci.* 11: 1706–1712.

171 Wisse, E., Spiering, A.J.H., Dankers, P.Y.W. et al. (2011). *J. Polym. Sci., Part A: Polym. Chem.* 49: 1764–1771.

172 Dankers, P.Y.W., van Leeuwen, E.N.M., van Gemert, G.M.L. et al. (2006). *Biomaterials* 27: 5490–5501.

173 Goor, O.J.G.M., Keizer, H.M., Bruinen, A.L. et al. (2017). *Adv. Mater.* 29: 1604652.

174 Hendrikse, S.I.S., Wijnands, S.P.W., Lafleur, R.P.M. et al. (2017). *Chem. Commun.* 53: 2279–2282.

175 Han, J.-T., Lee, D.-H., Ryu, C.-Y., and Cho, K. (2004). *J. Am. Chem. Soc.* 126: 4796–4797.

176 Feng, L., Li, S., Li, Y. et al. (2002). *Adv. Mater.* 14: 1857–1860.

177 Marmur, A. (2005). *Langmuir* 20: 3517–3519.

178 Feldman, K.E., Kade, M.J., de Greef, T.F.A. et al. (2008). *Macromolecules* 41: 4694–4700.

179 Feldman, K.E., Kade, M.J., Meijer, E.W. et al. (2010). *Macromolecules* 43: 5121–5127.

180 Nojiri, S., Kimata, S., Ikeda, K. et al. (2017). *Macromolecules* 50: 5687–5694.

181 Ligthart, G.B.W., Ohkawa, H., Sijbesma, R.P., and Meijer, E.W. (2005). *J. Am. Chem. Soc.* 127: 810–811.

182 Yamaguchi, K., Lizotte, J.R., Hercules, D.M. et al. (2002). *J. Am. Chem. Soc.* 124: 8599–8604.

183 Stuparu, M.C., Khan, A., and Hawker, C.J. (2012). *Polym. Chem.* 3: 3033–3044.

184 Park, T., Todd, E.M., Nakashima, S., and Zimmerman, S.C. (2005). *J. Am. Chem. Soc.* 127: 18133–18142.

185 Park, T., Zimmerman, S.C., and Nakashima, S. (2005). *J. Am. Chem. Soc.* 127: 6520–6521.

186 Rao, J., Păunescu, E., Mirmohades, M. et al. (2012). *Polym. Chem.* 3: 2050–2056.

187 Park, T. and Zimmerman, S.C. (2006). *J. Am. Chem. Soc.* 128: 13986–13987.

188 Scherman, O.A., Ligthart, G.B.W., Sijbesma, R.P., and Meijer, E.W. (2006). *Angew. Chem. Int. Ed.* 45: 2072–2076.

189 de Greef, T.F.A., Ercolani, G., Ligthart, G.B.W. et al. (2008). *J. Am. Chem. Soc.* 130: 13755–13764.

190 Takeshita, M., Hayashi, M., Kadota, S. et al. (2005). *Chem. Commun.*: 761–763.

191 Tan, S., Sha, Y., Zhu, T. et al. (2018). *Polym. Chem.* 9: 5395–5401.

192 Dudek, S.P., Pouderoijen, M., Abbel, R. et al. (2005). *J. Am. Chem. Soc.* 127: 11763–11768.

193 El-ghayoury, A., Schenning, P.H.J., van Hal, P.A. et al. (2001). *Angew. Chem. Int. Ed.* 40: 3660–3663.

194 El-ghayoury, A., Peeters, E., and Schenning, P.H.J. (2000). *Chem. Commun.*: 1969–1970.

195 Abbel, R., Grenier, C., Pouderoijen, M.J. et al. (2009). *J. Am. Chem. Soc.* 131: 833–843.

196 Hirschberg, J.H.K.K., Brunsveld, L., Ramzi, A. et al. (2000). *Nature* 407: 167–170.

197 Ikeda, M., Nobori, T., Schmutz, M., and Lehn, J.-M. (2005). *Chem. Eur. J.* 11: 662–668.

198 Berl, V., Schmutz, M., Krische, M.J. et al. (2002). *Chem. Eur. J.* 8: 1227–1244.

199 Kolomiets, E., Buhler, E., Candau, S.J., and Lehn, J.-M. (2006). *Macromolecules* 39: 1173–1181.

200 Binder, W.H., Petraru, L., Roth, T. et al. (2007). *Adv. Funct. Mater.* 17: 1317–1326.

201 Chen, S., Rocher, M., Ladavière, C. et al. (2012). *Polym. Chem.* 3: 3157–3165.

202 Altintas, O., Schulze-Suenninghausen, D., Luy, B., and Barner-Kowollik, C. (2013). *ACS Macro Lett.* 2: 211–216.

203 Chen, S., Yan, T., Fischer, M. et al. (2017). *Angew. Chem. Int. Ed.* 56: 13016–13020.

204 Croom, A., Manning, K.B., and Weck, M. (2016). *Macromolecules* 49: 7117–7128.

205 Chen, S., Ströhl, D., and Binder, W.H. (2015). *ACS Macro Lett.* 4: 48–52.

206 Altintas, O., Gerstel, P., Dingenouts, N., and Barner-Kowollik, C. (2010). *Chem. Commun.* 46: 6291–6293.

207 Altintas, O., Lejeune, E., Gerstel, P., and Barner-Kowollik, C. (2012). *Polym. Chem.* 3: 640–651.

208 Kanazawa, H., Higuchi, M., and Yamamoto, K. (2006). *Macromolecules* 39: 138–144.

209 Kolomiets, E. and Lehn, J.-M. (2005). *Chem. Commun.*: 1519–1521.

210 Lehn, J.-M., Mascal, M., Decian, A., and Fischer, J. (1990). *J. Chem. Soc., Chem. Commun.*: 479–481.

211 Klok, H.-A., Jolliffe, K.A., Schauer, C.L. et al. (1999). *J. Am. Chem. Soc.* 121: 7154–1755.

212 Mascal, M., Hansen, J., Fallon, P.S. et al. (1999). *Chem. Eur. J.* 5: 381–384.

213 Gong, H. and Krische, M.J. (2005). *J. Am. Chem. Soc.* 127: 1719–1725.

214 Yagai, S., Iwashima, T., Karatsu, T., and Kitamura, A. (2004). *Chem. Commun.*: 1114–1115.

215 Ghadiri, M.R., Granja, J.R., Milligan, R.A. et al. (1993). *Nature* 366: 324–326.

216 Ghadiri, M.R., Granja, J.R., and Bühler, L.K. (1994). *Nature* 369: 301–303.

217 Khazanovich, N., Granja, J.R., McRee, D.E. et al. (1994). *J. Am. Chem. Soc.* 116: 6011–6012.

218 Engels, M., Bashford, D., and Ghadiri, M.R. (1995). *J. Am. Chem. Soc.* 117: 9151–9158.

219 Ghadiri, M.R., Kobayashi, K., Granja, J.R. et al. (1995). *Angew. Chem. Int. Ed. Engl.* 34: 93–95.

220 Kobayashi, K., Granja, J.R., and Ghadiri, M.R. (1995). *Angew. Chem. Int. Ed. Engl.* 34: 95–98.

221 Hartgerink, J.D., Granja, J.R., Milligan, R.A., and Ghadiri, M.R. (1996). *J. Am. Chem. Soc.* 118: 43–50.

222 Clark, T.D., Bühler, L.K., and Ghadiri, M.R. (1998). *J. Am. Chem. Soc.* 120: 651–656.

223 Clark, T.D., Buriak, J.M., Kobayashi, K. et al. (1998). *J. Am. Chem. Soc.* 120: 8949–8962.

224 Hartgerink, J.D., Clark, T.D., and Ghadiri, M.R. (1998). *Chem. Eur. J.* 4: 1367–1372.

225 Bong, D.T., Clark, T.D., Granja, J.R., and Ghadiri, M.R. (2001). *Angew. Chem. Int. Ed.* 40: 988–1011.

226 Kortemme, T., Ramírez-Alvarado, M., and Serrano, L. (1998). *Science* 281: 253–256.

227 Schenck, H.L. and Gellman, S.H. (1998). *J. Am. Chem. Soc.* 120: 4869–4870.

228 Rossmeisl, J., Nørskov, J.K., and Jacobsen, K.W. (2004). *J. Am. Chem. Soc.* 126: 13140–13143.

229 Horne, W.S., Stout, C.D., and Ghadiri, M.R. (2003). *J. Am. Chem. Soc.* 125: 9372–9376.

230 Cordier, P., Tournilhac, F., Soulié-Ziakovic, C., and Leibler, L. (2008). *Nature* 451: 977–980.

4

Supramolecular Polymers, Based on Metal-to-Ligand Interactions*

4.1 General Aspects

The interaction of (transition) metal ions with organic ligands, i.e. metal-to-ligand coordination, represents one of the most important tools in today's supramolecular chemistry and is, for a number of reasons, of particular value for the preparation of supramolecular polymers via self-assembly strategies [1–3]:

1) Metal-to-ligand coordinative bonds exhibit a high degree-of-directionality;
2) The molecular structure of the organic ligands can easily be modified, and
3) The thermodynamic and/or kinetic stability of an assembly can be adjusted by the proper choice of metal centers and ligands.

Going beyond these fundamental aspects related to supramolecular chemistry, merging conventional organic polymers and inorganic coordination compounds enabled the preparation of new classes of functional materials possessing remarkable properties. Thereby, the mechanical properties related to traditional macromolecules are combined with the additional features caused by the presence of the metal centers: bioactivity [4–7], molecular magnetism [8–10], conductivity [11, 12], nonlinear optical behavior [13, 14], ferroelectricity [15, 16], or sensitization [17, 18]. Furthermore, the immobilization of catalytically active organometallic species gave polymer-based catalysts, homogeneous as well as heterogeneous ones, which have been used in organic and macromolecular chemistry [19–21]; cross-coupling reactions, olefin metathesis, and/or olefin polymerization have to be named in this context. In recent years, polymer photocatalysts have emerged as new materials also addressing energy-related applications, such as water splitting or CO_2 reduction [22, 23].

The development of such metal-containing polymers was, for a long period of time, hampered by limitations in the availability of good synthetic procedures. For example, the established protocols to synthesize well-defined organic polymers cannot (at least in many cases) be directly translated to metal-based systems, which maybe prone to undergo undesirable side reactions; thus, materials of low molar

* Parts of this chapter are reproduced from *Chem. Soc. Rev.* **2016**, *45*, 5311-5357 (DOI: 10.1039/c6cs00182c) by permission of The Royal Chemical Society.

Supramolecular Polymers and Assemblies: From Synthesis to Properties and Applications, First Edition.
Ulrich S. Schubert, George R. Newkome, and Andreas Winter.

masses were often obtained in the early studies in this field. Moreover, a reliable structural characterization was impeded due to the commonly poor solubility of these materials. A broad range of methodologies, tolerating the presence of (charged) metal sites, has been developed over the past couple of decades. These approaches allow the preparation of metallopolymers of high molar masses in a controlled manner and in reasonable amounts. Thus, the primary constraints related to the synthesis and characterization of metal-containing polymers have largely been overcome and, as a result, a wide range of such architectures can be realized. These materials typically feature sufficient solubility and, therefore, can be adequately characterized. In particular, the enormous progress in the development of analytical techniques applicable to non-covalent polymers has accounted for a much easier structural characterization of metal-containing materials (see also Chapter 12) [2, 24–29]. Going beyond these aspects, the improved solubility enabled an use-oriented processing further from solution, for example, by spin-coating or inkjet-printing techniques. As a result, metallopolymers could be brought into action and a broad range of applications have started to emerge [1–3, 23, 30–70].

This chapter is devoted to metallo-supramolecular coordination polymers, their syntheses, structures, properties, and potential applications will be evaluated. However, before doing so, a precise definition for "metallo-supramolecular polymers" needs to be established. In general, a metallopolymer can be defined as "any polymer containing metal atoms in the repeating monomer, either as part of the backbone or in side chains" (https://www.yourdictionary.com/metallopolymer). It must be emphasized that this simple and catchy description is highly deceptive, as the term "coordination polymer" might be used in rather different ways. In the field of inorganic chemistry, for example, coordination networks of whatever dimensionality are regarded as coordination polymers [71]. These solid-state materials are typically crystalline (or sometimes amorphous) and prone to disassemble into discrete molecular components in solution. Such assemblies are often referred to (though with some inconsistency) as "metal–organic coordination networks" [72–74] or "metal–organic frameworks" (MOFs) [75, 76]. However, IUPAC has not yet offered a precise definition of these materials (currently, a project is dealing with this issue) [77]. The utilization of MOFs, in particular with respect to gas-storage and gas-separation applications, has extensively been discussed elsewhere [78–83]. Herein, only materials that retain their macromolecular structure in solution are considered and those assemblies that exhibit polymer-like structures and properties due to metal-to-ligand coordinative interactions only in the solid state are not included.

In polymer science, however, "metallopolymers" are defined stricter and much more precisely: the metal centers represent integral parts of the polymer backbone or, alternatively, they are attached to the main chain or the pending side chains. These materials should ideally not disassemble or dissociate in the presence of a solvent; thus, the metal-to-ligand interactions must be of sufficient strengths. Moreover, in the field of supramolecular chemistry, materials that have been formed via a supramolecular approach are regarded as "coordination polymers." In other words, the polymer chain itself is assembled by forming non-covalent bonds between molecular components. Metal ions and organic ligands, for example, give Lewis-type acid–base pairs via dative bonds from the Lewis-basic organic ligands to the electropositive metal centers. As noted above, this particular type of interaction features some advantages when compared to other supramolecular systems (e.g. π–π stacking or ionic interactions). First, the coordinative bond between metal centers and organic ligands is highly directional, thus allowing the assembly of highly ordered architectures. Moreover, the structure of the supramolecular system can be adjusted via the molecular structure of the organic ligand and, finally, the combination of metal centers and organic ligands offers a good control over the thermodynamic and/or kinetic stability of the resultant assembly [3].

In an ideal case, metal-containing polymers should exhibit (macroscopic) properties similar to those known from conventional (i.e. covalent) polymers, such as an increased viscosity in comparison to the monomeric building blocks or a glassy solid state [84]. In the late 1990s, Rehahn and Ciardelli suggested a still valid classification scheme for a broad variety of materials that fulfill these fundamental requirements [55, 85]. According to these authors, three fundamental types of metal-containing polymers can be distinguished (Figure 4.1). The first class embraces metal centers bound to a polymer chain via electrostatic or coordinative interactions as well as covalent bonding (Type I). Alternatively, the metal centers might be an integral part of the polymer main chain (Type II); thereby, these metal centers are incorporated either by covalent linkages or by metal-to-ligand coordination. In both former cases, i.e. Type-I and Type-II materials, a close proximity of the metal sites to the polymer backbone will have a profound influence on the resultant properties, in particular, when dealing with π-conjugated backbones [40]. Finally, the Type-III materials are composed of metal ions (or complexes) embedded in a polymer matrix via physical interactions. The negative free energy (ΔG^0) amounting from the polymer's chelate effect represents the main driving force for the formation of these materials. Going beyond this classification scheme and in line with the supramolecular polymers obtained from other non-covalent interactions (e.g. hydrogen-bonding interactions or host–guest complexation), a further discrimination can be made according to the topology of the metallopolymers – a broad range of linear, grafted, hyperbranched, or dendritic architectures is known from literature. In this chapter, the focus will be on metallo-supramolecular polymers, which resemble Type-II materials, i.e. main-group elements or transition/rare-earth metal ions represent integral parts of the polymer main chain. Metal-containing polymers of other topologies have been summarized in various review articles elsewhere [31, 50, 69, 70].

4.2 Synthesis and Design Principles

4.2.1 Metal-Binding Sites in Metallo-supramolecular Polymers

Interactions between metal centers and organic building blocks, the ligands, represent the foundation of metallo-supramolecular polymers, and thus, these

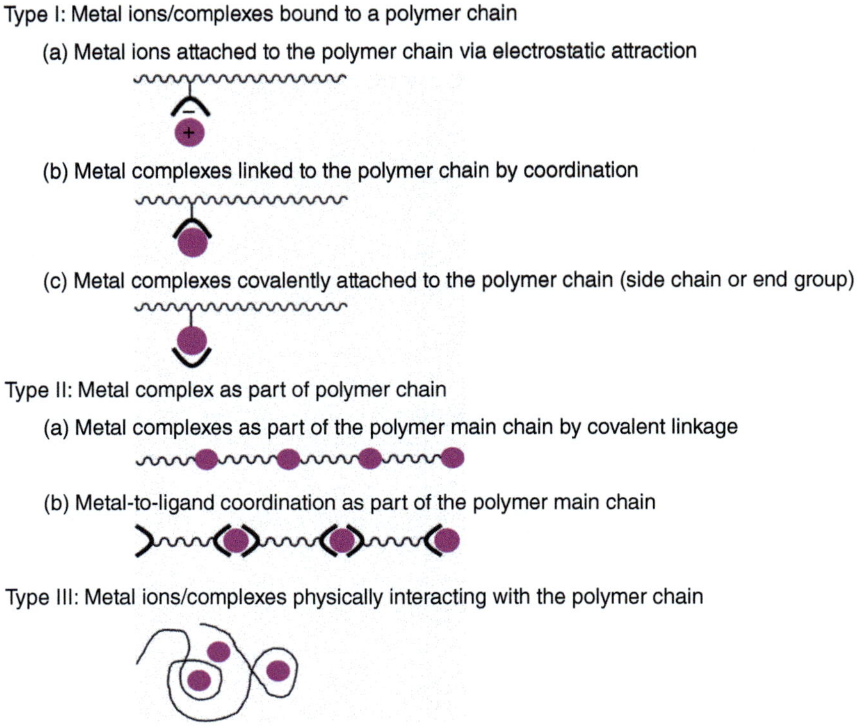

Figure 4.1 Schematic representation of the general architectures of metal-containing polymers. Source: Wild et al. [67]. © 2011 The Royal Chemical Society.

two components need to be considered. First of all, the organic ligand must be capable of doing both – allow the coordinated metal center to show a desired property (e.g. redox potential, emission, magnetism) and enable the polymer formation while the macroscopic properties of the overall material (e.g. solubility) are retained [46]. In this respect, the coordination numbers of the respective metal ion and the organic ligand also become important: the self-assembly into a linear supramolecular polymer via a bidirectional chain-growth mechanism (see Chapter 1) is only possible, if two opposing contact points are present at the metal center and the organic ligand itself is able to coordinate to two metal centers. Basically, small molecules (monomers), oligomers, and even polymers might be used as organic ligands. Despite its size, the organic ligand should exhibit a high binding affinity toward the used metal center (K_a values) to provide a high thermodynamic driving force for the supramolecular polymerization – one prerequisite to assemble polymers of high molar masses.

Preferably, two-sided, i.e. ditopic ligands are used for the synthesis of linear metallo-supramolecular polymers. The ligand's binding sites can interact with metal ions in very different ways. Basically, three fundamental types of interactions can be distinguished in this respect: coordinative bonding, ionic/covalent bonding as well as arene π-complexation (Figure 4.2) [2].

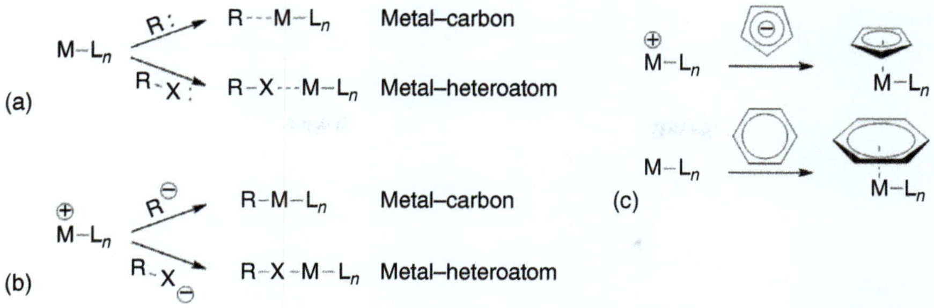

Figure 4.2 Schematic representation of the three main types of metal-to-ligand interactions. (a) Coordinative bonding; (b) ionic bonding; and (c) arene π-complexation. Source: Refs. [2, 46, 84].

4.2.2 Synthesis of Metallo-supramolecular Polymers

In the past decades, a plethora of methods for the synthesis of structurally well-defined metal-containing polymers have been developed. In principle, the whole range of protocols known from traditional polymer synthesis can also be applied to their metal-containing analogs; however, some restrictions might still apply. Overall, a rich tool box exists allowing chemists to prepare tailor-made metallopolymers; examples are known where polycondensation, ring-opening polymerization, electro-polymerization, and living ionic as well as controlled radical polymerization techniques have been applied for this purpose [2]. Furthermore, metal-containing polymers have been prepared by modifying preformed polymers, for example, by coordinating metal centers to a polymer backbone or by grafting metal-containing residues onto a functionalized polymer chain. The basic methods, which one can apply to prepare metal-containing (co)polymers, are depicted in Figure 4.3. From all these methods, only Method III and Vb are of importance when aiming for the formation of linear metallo-supramolecular polymers that feature metal centers within the polymer backbone. To further narrow down the scope of methods, only Method III applies as a typical metallo-supramolecular polymerization for which the different mechanisms, as discussed in Chapter 1, imply.

4.3 Linear Metallo-supramolecular Polymers

Linear metallo-supramolecular polymers can be assembled from appropriate combinations of ditopic organic ligands and metal centers. In line with the classification of the three fundamental modes of metal-to-ligand interactions (Figure 4.2), the different types of metallo-supramolecular polymers will be discussed on the same basis.

4.3.1 Metallo-supramolecular Polymers, Based on Coordinative Bonding

The vast majority of metallo-supramolecular polymers results from coordinative, dative-type interactions between organic ligands and metal centers. The mostly

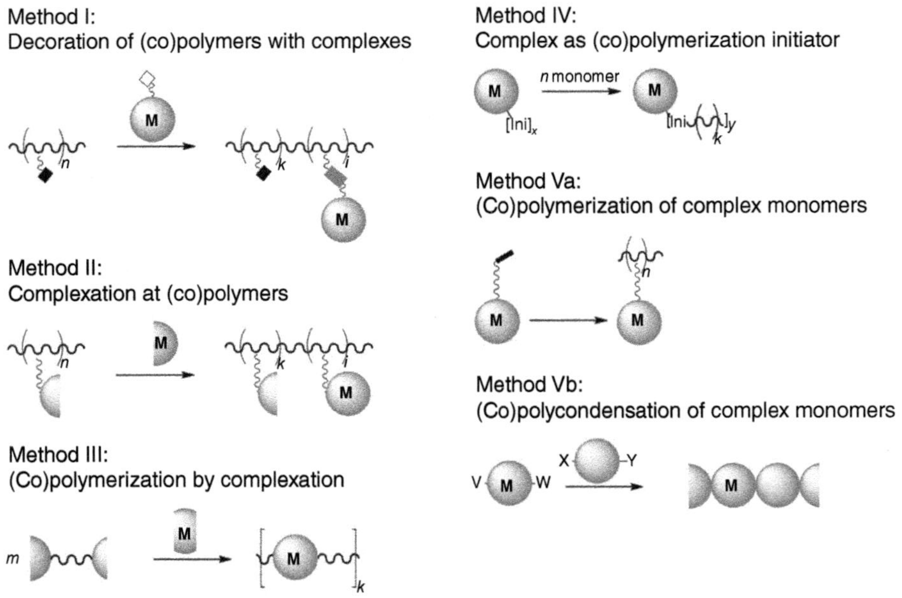

Figure 4.3 Schematic representation of the different general methods utilized for the synthesis of metal-containing (co)polymers. Source: Refs. [2, 61].

neutral binding sites include structurally highly diverse functionalities, such as phosphines, carbenes, amines, pyridines, ethers, ketones, imines, nitriles, and so on [46]. Accordingly, the K_a values for the interaction of these moieties with metal centers will cover a wide range, and thus, the resultant metallopolymers will exhibit different stabilities – polymers that easily disassemble in solution and polymers that even retain their polymeric structure when addressed by external stimuli (e.g. heat, light, ultrasound, pH value, etc.) represent the two extreme cases in this respect. To increase the stability of a metal-to-ligand interaction, two approaches can be followed: organic ligands with high affinities toward metal centers and/or organic ligands with an increased number of metal-binding sites (i.e. chelation or multidentate binding) can be used. In particular, the latter one has been widely exploited using pyridine-type ligands (Figure 4.4a). Würthner and coworker elaborated the influence of the ligand's denticity on the complex stability; the so-called chelation effect was studied as an example: polypyridyl ligands [3]. A K_a value of $10^3\ M^{-1}$ was found for the single Zn(II)–pyridine interaction, whereas the Zn(II)–tpy system exhibited a K_a value as high as $10^8\ M^{-1}$ (tpy: 2,2′:6′,2″-terpyridine). Stable bis-tridentate complexes of pseudo-octahedral geometry can be derived from many transition metal ions, though only in low oxidation states, and tpy ligands. Such complexes feature a remarkable stability, which can be rationalized by the dynamic chelate effect in combination with an efficient metal-to-ligand d–π* back-donation. In most cases, the order of the complex stability, as expressed by the stability constants [86, 87], follows the sequence predicted by Irving and Williams for the divalent first-row transition metal ions (Figure 4.4b) [88]. Beyond these

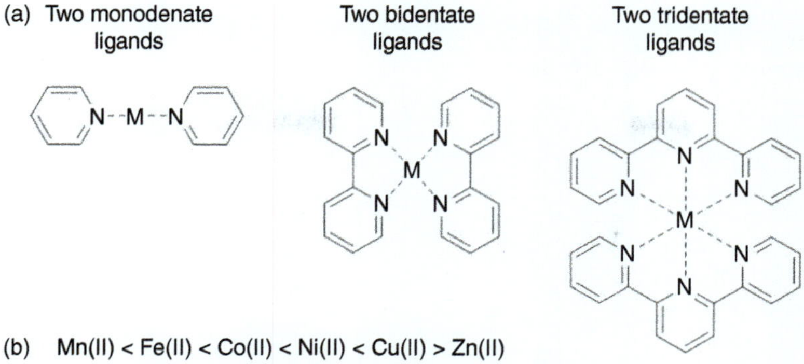

(b) Mn(II) < Fe(II) < Co(II) < Ni(II) < Cu(II) > Zn(II)

Figure 4.4 (a) Schematic representation of metal complexes containing pyridine-based ligands of increasing denticity. (b) The Irving–Williams series.

fundamental considerations, Seifert and coworkers recently identified one further aspect to be noted at this point. These authors compared the kinetic stability of various tpy bis-complexes bearing a polymer chain on the central pyridine rings of the tpy scaffold [89]. Rheology measurements revealed a decreasing stability with an increasing length of the polymer chain (due to the dynamic activity of the polymer, the dissociation occurred up to six times faster as for analogous small-molecule complexes).

4.3.1.1 Metallopolymers Containing Ditopic Monodentate Ligands

According to previous considerations, only some kinetically and thermodynamically stable metallo-supramolecular polymers, based on homoditopic monodentate organic ligands can be found in the literature. In such assemblies, the absence of any stabilizing chelate effect needs to be compensated by a strong donor character of the binding site to retain structural integrity in solution [2]. The selected examples, as depicted in Figure 4.5, represent the full stability regime – ranging from a kinetically labile metallopolymer to a kinetically inert one. Kinetically very labile metallo-supramolecular polymers, such as **1**, are only known to exist as monolayers at the water/air interface [90, 91]. UV/vis absorption and X-ray photoelectron spectroscopy (XPS) gave evidence for the proposed polymeric structure of **1** (for the XPS measurements, the monolayers were transferred as films onto quartz or indium tin oxide [ITO] slides). Due to the pronounced kinetic lability of the metal-to-ligand interactions, fast dissociation was observed even in organic solvents of moderate polarity.

A metallo-supramolecular polymer (**4**) was also formed from the interaction of trans-$(PPh_3)_2Pt(C{\equiv}CPh)_2$ with Cu(I) halides [92]. In the former case, a tetrahedral $\{Cu_2X_2\}$ fragment was identified as the bridging unit, which was bound in a monodentate fashion to each alkynyl group of the Pt(II) complex (Figure 4.6). However, this heterobimetallic polymer **4** exhibited a poor stability in solution: polar solvent molecules acting as strong competitive ligands induced a fast depolymerization (basically the same behavior was also observed for the Ag(I)-containing analog to **4**).

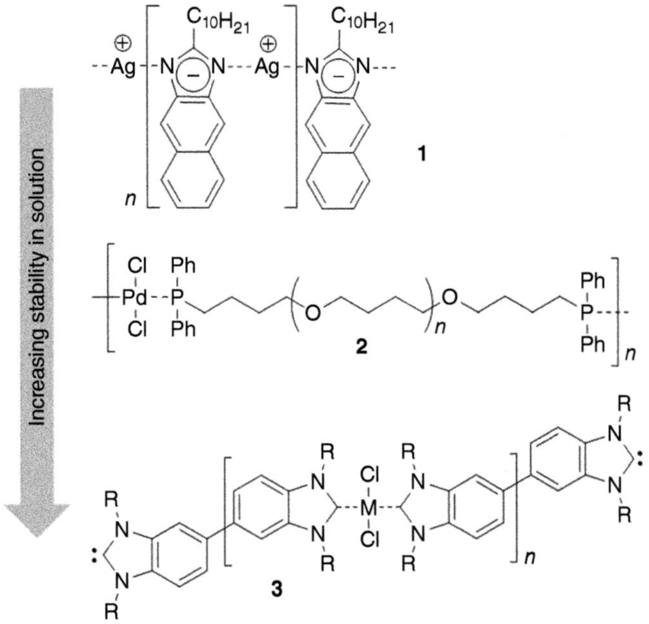

Figure 4.5 Schematic representation of the coordination polymers **1–3** showing an increased stability in solution due to enhanced metal-to-ligand interactions.

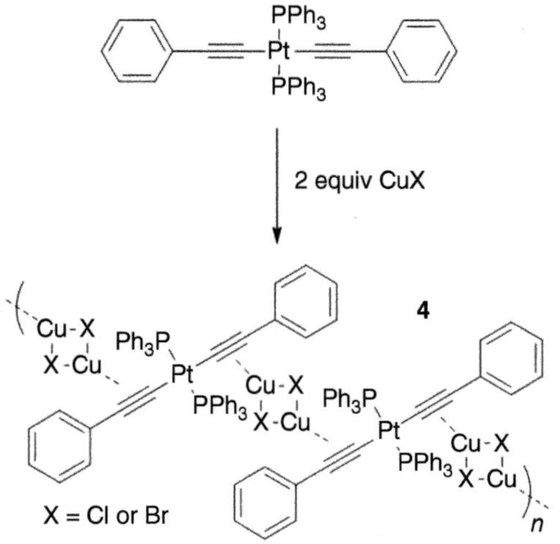

Figure 4.6 Schematic representation of the synthesis of the heterobimetallic coordination polymer **4**.

The metallopolymers, such as **2** (Figure 4.5), which were assembled from homoditopic phosphine ligands and $PdCl_2$, represent materials that retained their polymeric character even in highly coordinating solvents (e.g. MeCN) [93]. This behavior was attributed to the much stronger donor ability of the phosphine-binding sites. The stimuli-responsive behavior of **2** was demonstrated as the example of the sonication-induced dissociation; the high fidelity of the switching between an assembled and disassembled state is of relevance to create materials with reversible

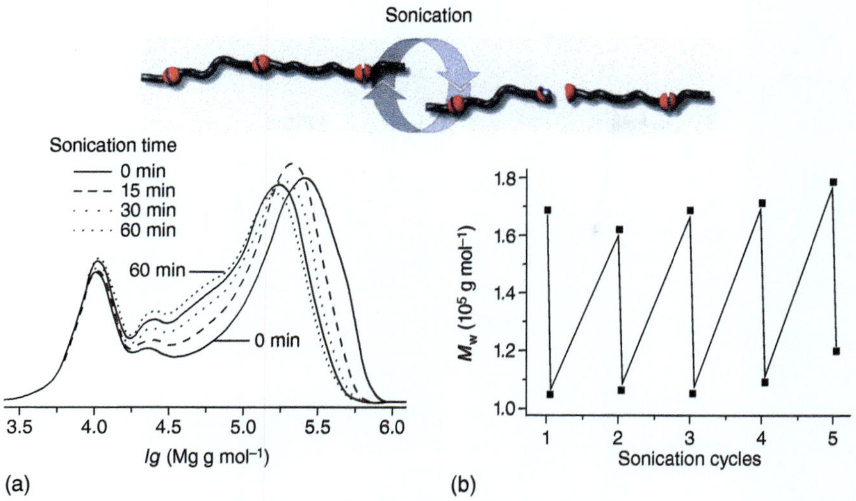

Figure 4.7 Illustration of the sonication-induced decrease in the molar mass of **2** (1.5 mM in toluene): (a) Molar mass distribution of samples taken during sonication. (b) Evolution of the M_W value over five cycles of sonication (1 hour) followed by equilibration (23 hours). Source: Paulusse and Sijbesma [94]. Figure reproduced with kind permission. © 2004 Wiley-VCH.

mechanochemical properties (Figure 4.7) [94]. As addressed in Chapter 12, a detailed analysis of these metallo-supramolecular polymers even by size-exclusion chromatography (SEC) in solution was enabled. Basically, SEC is demanding to apply for supramolecular polymers that prone to disassembly/dissociation upon dilution and/or in the presence of coordinating solvents [52]. A different type of metallopolymer, comprising carbonyl clusters as integral part of the backbone, was prepared by Johnson et al. from the $Ru_6(CO)_{17}$ precursor and a linear, alkyne-bridged diphosphine ligand; the initial metallopolymer was converted into conductive nanowires by applying an electron beam [95]. Related bi- or even trimetallic assemblies were reported earlier by Phang et al.; these presumably only oligonuclear species were derived from appropriate carbonyl precursors and bisphosphine ligands (e.g. 1,1′-bis(diphenylphosphino)ferrocene or alkyl-bridged bisphosphines) [96].

Metallo-supramolecular polymers, which are based on the interaction of isocyanides with transition metal ions, have attracted considerable attention due the (partial) multiple-bond character of the metal-to-carbon linkage. This particular bonding scenario is considered as beneficial when integrated within π-conjugated metallopolymers due to the proposed increased electronic communication along the chain. The Efraty, Feinstein-Jaffe, and Hanack groups already reported on such metal–aryldiisocyanide polymers in the 1980s: due to the poor solubility of these materials, a characterization, in solution, was obsolete; however, in the solid state, ordered as well as unordered structures could be identified [97–108]. Besides these examples, based on group-10 metal centers, metallo-supramolecular polymers were also assembled from Cu(I) or Ag(I) and diisocyanide-equipped

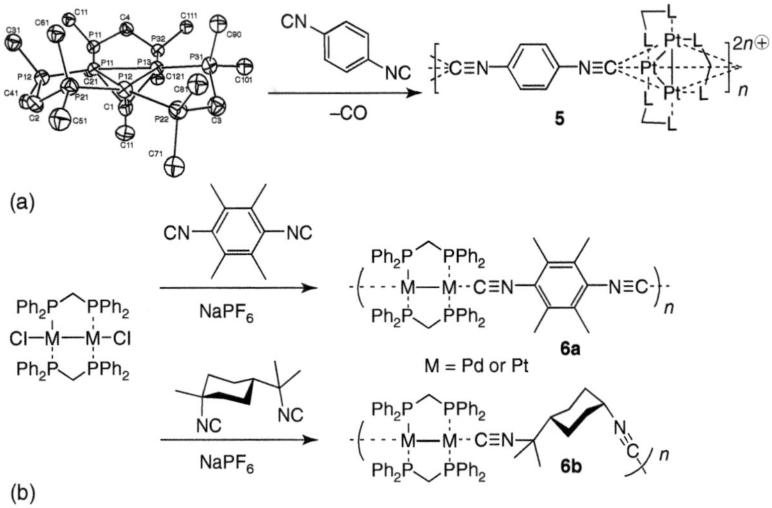

Figure 4.8 (a) Schematic representation of the synthesis of coordination polymer **5** via the coordination of 1,4-benzenediisocyanide to the $[Pt_3(\mu_3\text{-}CO)(\mu\text{-}dppm)_3]^{2+}$ cluster (the counterions are omitted for clarity, L ^ L: dppm). A representation of the X-ray single-crystal structure of the precursor cluster is also shown (here, only the phenyl C-atoms bonded to each P-atom are shown for clarity). Source: Bradford et al. [111]. Figure reproduced with kind permission. © 1994 American chemical society. (b) Schematic representation of the synthesis of coordination polymers **6**, based on diisocyanides and $[M_2(dppm)_2]^{2+}$ units.

ligands [109, 110]. Related to this, the so-called Shish–Kebab-type metallopolymers, in which diisocyanides are used to bridge metalloporphyrins (or structurally related macrocycles), have to be mentioned; these materials are detailed later in the context of porphyrin coordination.

In a similar context, the coordinatively unsaturated $[Pt_3(\mu_3\text{-}CO)(\mu\text{-}dppm)_3]^{2+}$ precursor cluster was used by Puddephatt and coworkers to assemble the linear metallopolymer **5**; thereby, the aryldiisocyanide ligands were coordinated axially to the stable, almost planar $\{Pt_3(dppm)_3\}$ framework (dppm: bis(diphenylphosphanyl)methane, Figure 4.8a) [111]; the electronically unsaturated nature of the material allowed for the binding of two-electron donors, thus increasing the sum of valence electrons from 42 to 44.

Puddephatt and coworkers also reported on the formation of polymers **6** featuring binuclear $[M_2(\mu\text{-}dppm)_2]^{2+}$ units with a d^9—d^9 metal-to-metal bond (M = Pd or Pt) [110, 112, 113]. Both rigid (e.g. aryl) and flexible (e.g. cyclohexyl) diisocyanide ligands were coordinated to the precursor complexes in the presence of a Cl^--scavenger (Figure 4.8b). Oligomeric assemblies rather than polymeric ones were identified in solution (i.e. degree-of-polymerization [DP] of 4–15); however, extended polymer chains were observed by X-ray diffraction (XRD) analysis in the solid state. Similar to this, the $Pd_2(dmb)_2Cl_2$ precursor complex (dmb: 1,8-diisocyano-*p*-menthane) was incorporated into a metallo-supramolecular polymer by replacing the chloride ligands by tertiary phosphines (ditopic bisphosphine ligands were used, as organic ligands) [114]. Moreover, a unique linear $\{Pt_4\}$

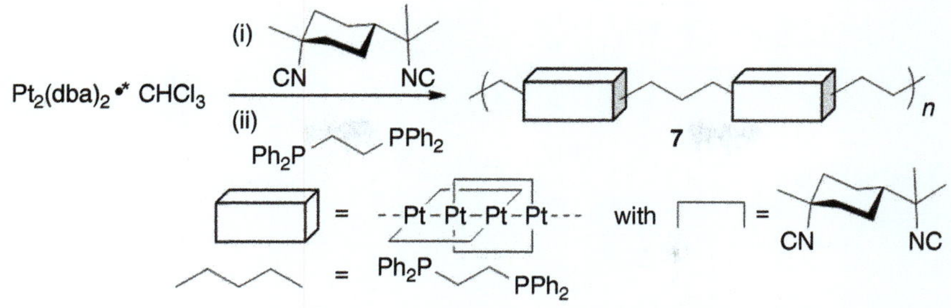

Figure 4.9 Schematic representation of the synthesis of the metallopolymer **7** with linear Pt_4 clusters within the main chain.

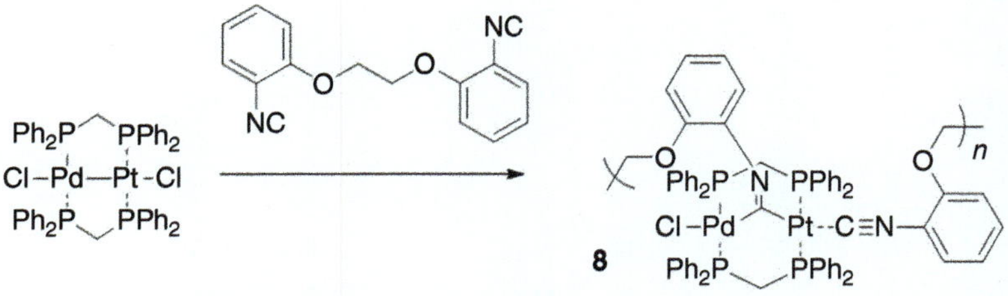

Figure 4.10 Schematic representation of the synthesis of coordination polymer **8** featuring A-frame-units within the backbone.

cluster was introduced into a metallo-supramolecular polymer in the presence of bridging bisphosphine and ancillary diisocyanide ligands (Figure 4.9) [115]. The clusters featured a total of 58 valence electrons and a (formal) Pt oxidation state of 0.5. In the supramolecular polymer, two chelating dmb molecules stabilized the bisphosphine-bridged Pt–Pt–Pt–Pt units.

The metallopolymer **8**, incorporating the so-called "A-frame" motive, was synthesized by Clément et al. [116]; the authors used the bimetallic $ClPd(\mu\text{-dppm})_2PtCl$ precursor complex and took advantage of the Pt center's site-selectivity in ligand binding to place this particular substructure into a metallopolymer's backbone (Figure 4.10). The structure of the poorly soluble **8** was verified by comparison of the available analytical data to those of analogous small-molecule model complexes. A polymer-analog to cisplatin, widely used in chemotherapy, was synthesized by the stoichiometric reaction of K_2PtX_2 (X = Cl or I) with various ditopic primary amines (**8**, Figure 4.11) [117]. This work by Carraher et al. represents one of the earliest examples for the attempts to prepare polymer-based metallodrugs for pharmaceutical purposes – a currently highly active field of research [118].

Linear metallo-supramolecular polymers can be assembled from planar metalloporphyrins (or their structurally related analogs, such as metallophthalocyanines) by coordinating ditopic ligands to the axial positions of the central metal ion of the frameworks. According to Abd-El-Aziz et al., three different types of metallo-supramolecular polymers can be formed on this basic concept (Figure 4.12) [68]. From the wide range of possible ditopic ligands, N-based ones are most

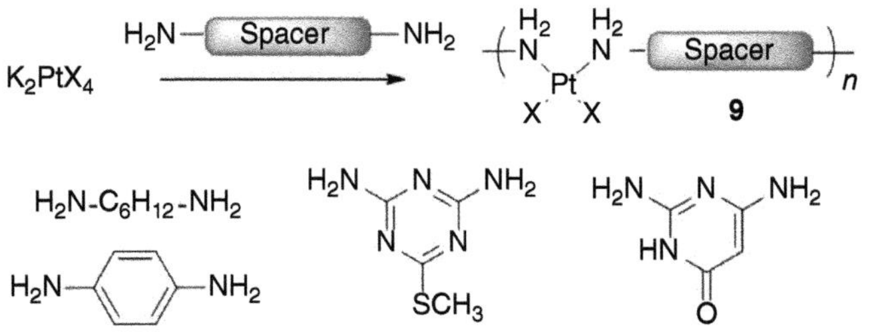

Figure 4.11 Schematic representation of the synthesis of **9**, as a polymer-analogs to cisplatin.

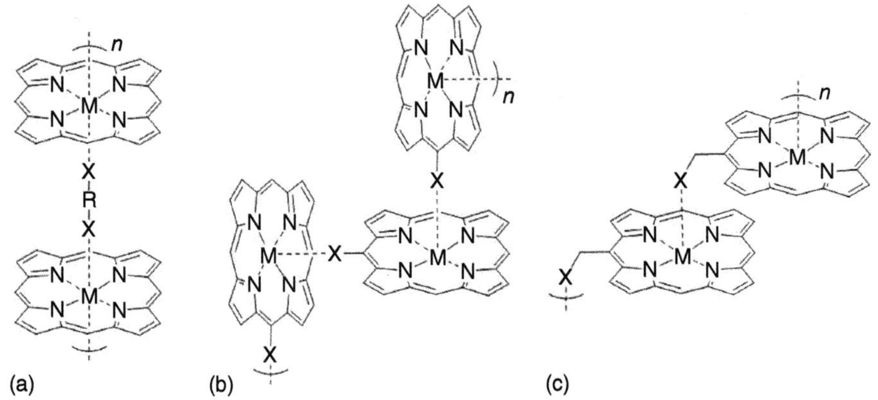

Figure 4.12 Schematic representation coordination polymers through the center of the porphyrin rings. (a) Shish-Kebab-type assembly, (b and c) assemblies based on AB-type monomers.

commonly utilized for the supramolecular linking of the rings in either of the three ways.

From these, the Shish–Kebab-type supramolecular polymerization of metalloporphyrins in the presence of π-electronic ditopic ligands (e.g. cyanide, thiocyanate, acetylide, pyrazine, 4,4′-bipyridine, etc.) is in the focus of interest already since the late 1970s. This particular interest arose from the potential electronic interaction of aligned porphyrins or phthalocyanines; such supramolecular materials can be considered as molecular wires [119]. Basically, the degree-of-electronic interaction can be modulated via the properties of the bridging ligand between the aligned macrocyclic units. Two parameters have to be considered in this respect – the distance between two such macrocycles and the character of the metal-to-ligand bonds (note that pyrazine interacts via two coordinative bonds, whereas $(C{\equiv}C)^{2-}$ binds via two formal σ-bonds). Hanack et al. assembled the metallopolymers **10** from macrocyclic building blocks (e.g. phthalocyanine or naphthalocyanine), complexed with divalent group-8 metal ions, and various *N*-heterocyclic bridging ligands L (e.g. tetrazine, pyrazine, or 1,4-diazabicyclo[2.2.2]octane [DABCO]; Figure 4.13). In all cases, thus independent from the nature of the metal ion, a remarkable increase in conductivity was recorded (e.g. 7 orders of magnitude, when using tetrazine, as

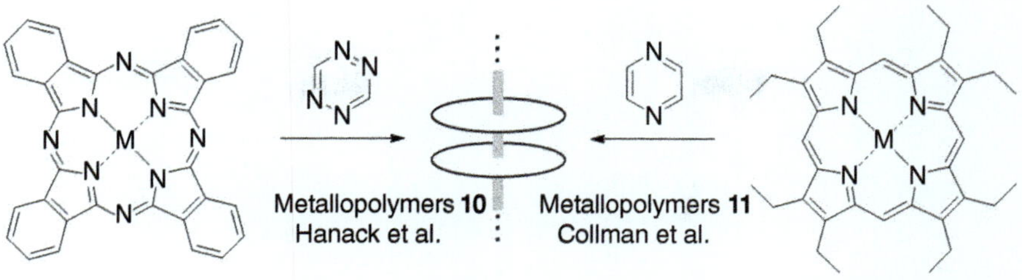

Figure 4.13 Schematic representation of Shish–Kebab-type metallo-supramolecular polymers **10** and **11** incorporating phthalocyanine.

the bridging ligand) [120]. Based on this concept, various metallo-supramolecular Shish–Kebab-type assemblies were synthesized from planar metalloporphyrin or metallophthalocyanine scaffolds [121]. As one representative example, the supramolecular polymerization of an octaethyl metalloporphyrin with pyrazine to give **11** is also shown (Figure 4.13) [122].

Basically, the same approach was employed for the supramolecular cross-linking of the metalloporphyrin-containing polymers **12** (Figure 4.14a) [124–128]. The efficient π-conjugation length of rigid-rod **12** in solution was truncated by the low-energetic rotation barrier of one porphyrin unit with respect to the neighboring ones. When 0.5 equiv of 4,4′-bpy was added to the π-conjugated metallopolymer, a double-stranded ladder-shaped supramolecular assembly was formed (Figure 4.14b). In this confined environment, the rotation of the metalloporphyrin units was precluded, and thus, the π-conjugation length was increased significantly

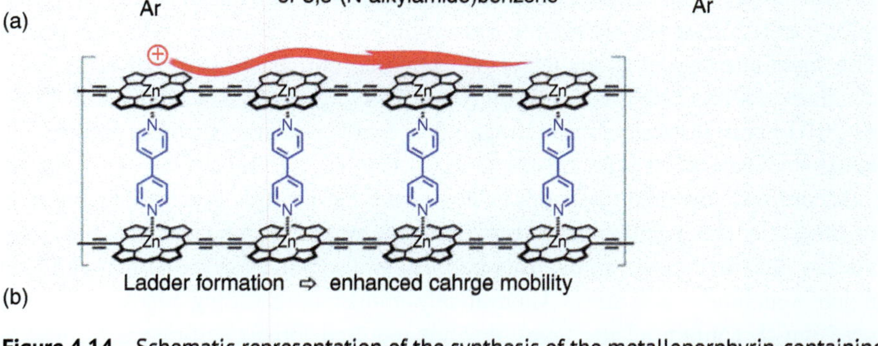

Figure 4.14 Schematic representation of the synthesis of the metalloporphyrin-containing polymer **12** via a Glaser-type polyaddition (a) and of the metallo-supramolecular 2D ladder-shaped assembly featuring enhanced charge mobility (b). Source: Taylor and Anderson [123]. © 1999 American chemical society.

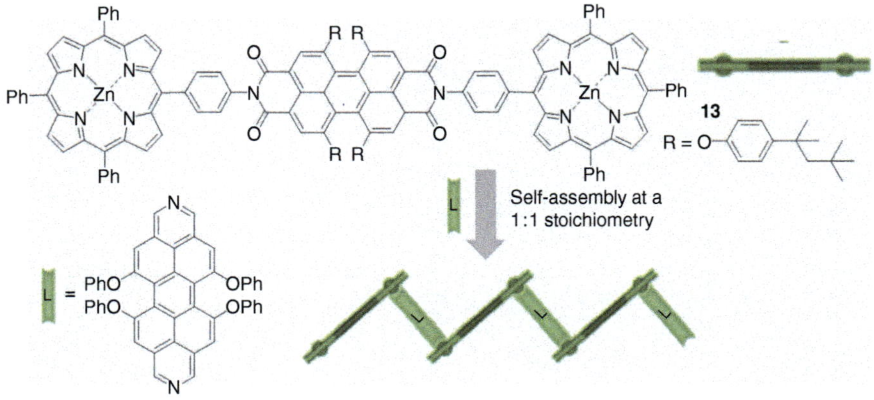

Figure 4.15 Schematic representation of the synthesis of a zigzag-shaped metallo-supramolecular coordination polymer. Source: You and Würthner [129]. © 2004 American Chemical Society.

(e.g. the charge-carrier mobility along the molecular wires was enhanced by a factor of 10). Highly atypical for metallo-supramolecular polymerizations, the cross-linking of **12** obeyed the rules for a cooperative supramolecular polymerization (see Chapter 1) [123].

The Würthner group reported on the self-assembly of a Zn(II)-containing porphyrin–PBI–porphyrin triad (**13**, PBI: perylenebisimide) in the presence of various ditopic ligands [129]. It was found that, bulky ligands favored the formation of zigzag-shaped metallo-supramolecular polymers at the cost of the entropically more preferred rectangular discrete assemblies (Figure 4.15). Not only homoditopic ligands but also heteroditopic ones can be used to assemble Shish–Kebab-type metallo-supramolecular polymers. It was shown that both a carboxylate and a N-containing ligand could be coordinated axially to an Al(III)–porphyrin derivative, thus establishing a six-coordinate geometry at the Al(III) center [130]. The addition of isonicotinic or nicotinic acid, as the simplest structures combining the aforementioned functionalities, afforded the corresponding metallo-supramolecular polymers. Moreover, a range of diisocyanides and ambient anions (e.g. CN^- and SCN^-) were also used, as bridging ligands, in Shish–Kebab-type metallopolymers. The materials derived from tetraazaporphyrins, phthalocyanines, or naphthalocyanines; various divalent transition metal ions (e.g. Fe(II), Co(II), Ru(II), Pd(II), or Pt(II)); and, in most cases, aromatic diisocyanide derivatives often featured an electroactive behavior [120, 131–145].

According to the theory of supramolecular polymerizations, precise stoichiometry of the molecular components are required to assemble polymers of high molar masses (Chapter 1). Naturally, this premise is also valid for the aforementioned Shish–Kebab-type polymers. Alternatively, molecular building blocks might be used which contain all the functionalities required for the self-assembly process (i.e. the so-called AB-type monomers). The main advantage of such monomers is the intrinsically given ideal stoichiometry. This concept was used to assemble metalloporphyrins into oligomeric and even polymeric materials. To reach this

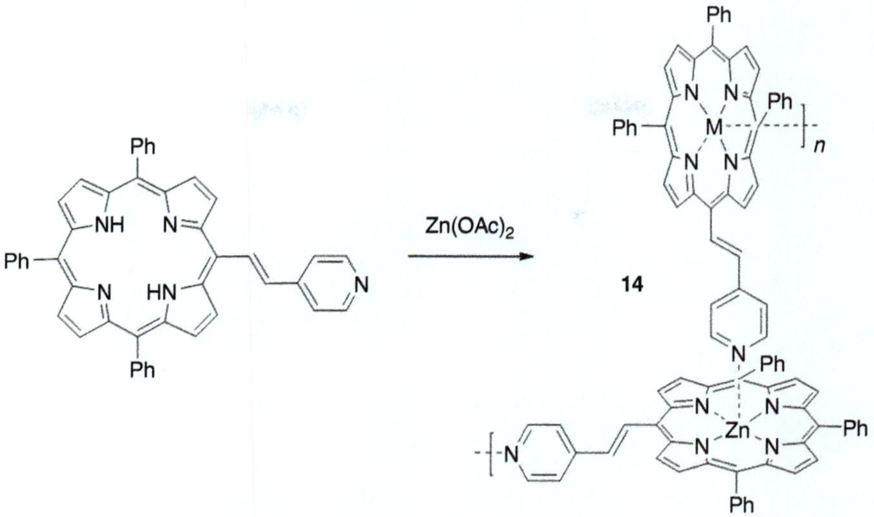

Figure 4.16 Schematic representation of the self-assembly of a 4-pyridyl-substituted porphyrin derivative with Zn(II) ions toward **14**.

goal, appropriate metal binding sites were introduced in *meso*-positions of the porphyrin rings. In a straightforward example, a tetra-*meso*-NO_2-substituted porphyrin derivative was complexed with Zn(II) ions and afforded a supramolecular polymer in which the metal centers in the porphyrin scaffold interacted with the NO_2-groups of neighboring molecules [146]. Burell et al. followed a much more defined strategy to assemble metallo-supramolecular polymers (referred to as type-B in Figure 4.12b). For this purpose, a 4-pyridyl moiety was attached to only one of the *meso*-positions by Wittig olefination. The subsequent addition of Zn(II) ions afforded the supramolecular polymer **14** (Figure 4.16) [147]. The structure of the assembly depended on the stereochemistry of the *exo*-cyclic sOC=C bond: only the *trans*-derivative gave a metallo-supramolecular polymer, whereas the *cis*-isomer exclusively gave dimers for steric reasons. In a similar fashion, a Zn(II)-containing bisporphyrin, equipped with two methylated imidazol-2-yl units in opposite *meso*-positions, also generated a metallo-supramolecular polymer. Remarkably, the material was sufficiently stable to allow its characterization by SEC (see Chapter 12) [148]. Moreover, a porphyrin derivative featuring two ligand arms in the *meso*-5 and -15 positions was complexed with Co(II) ions. Due to the preferred six-coordinate complex geometry of the metal centers, the anticipated zigzag-shaped assembly was obtained [149].

Puddephatt and coworkers investigated linear Au(I)-containing metallopolymers. These materials were prepared by coordinating a flexible bisphosphine ligand and a rigid bispyridine ligand to the metal centers in a stepwise fashion [150–154]. The conformation of the polymer exhibited a remarkable odd–even effect (Figure 4.17): a folded structure (**15a**) in the *syn*-conformation, which was reminiscent of an accordion, was obtained, if *m* was an odd number; in the other cases (*m* being an

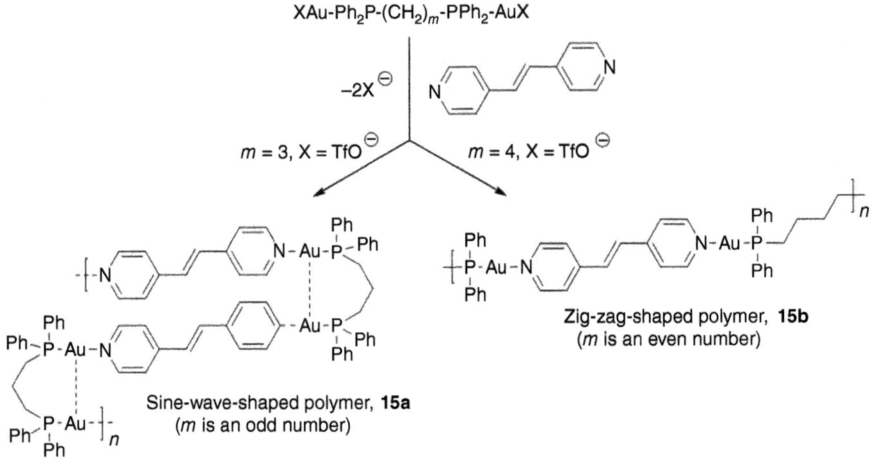

Figure 4.17 Schematic representation of the linear Au(I)-containing metallopolymers **15**, which exhibited an odd–even effect with respect to the chain conformation.

even number), a zigzag-shaped, anticonformation was observed (**15b**). The authors identified packing effects, presumably involving aurophilic intrachain interactions, as the main reason for this behavior.

The chemistry of carbene–metal complexes has developed tremendously over the past two decades and, in particular, *N*-heterocyclic carbenes (NHCs), as neutral two-electron donors, have attracted considerable interest with respect to the development of efficient photoluminescent transition metal complexes [155]. Very high K_a values have been reported for the interaction of NHCs with transition metal ions – in particular to the divalent group-10 cations. It has to be expected that the incorporation of this metal-to-ligand interaction into a polymer main chain results in kinetically highly inert materials with rich opto-electronic or magnetic properties. Surprisingly, this strategy has been applied to prepare linear metallo-supramolecular polymers only sporadically; some examples reported by the Bielawski group can be found in the literature [156–159]. The first chelating carbene ligands (the so-called pincer ligands) have been reported about 20 years ago [160]; however, the synthesis of appropriate ditopic carbene-based ligands remained a challenge and, thus, prevented any application in the field of metallo-supramolecular chemistry. These limitations have been overcome and a variety of rigid bis-NHC derivatives, as building blocks for the formation of metallo-supramolecular architectures, has been introduced [158, 161–163]. In contrast to the highly stable Pd(II)- and Pt(II)-containing metallopolymers **3** and **16–17** (Figure 4.18a), their Ni(II)-containing analogs were reported to be rather labile materials. For example, metallopolymer **3** exhibited a high chemical and thermal stability; SEC could be applied to estimate the molar mass of the polymer (M_n of c. 10^4 g mol^{-1}). The stability of the Ni(II)-based assemblies could be somehow increased by introducing stabilizing substituents, which acted as additional binding sites (**18**, Figure 4.18b) [157].

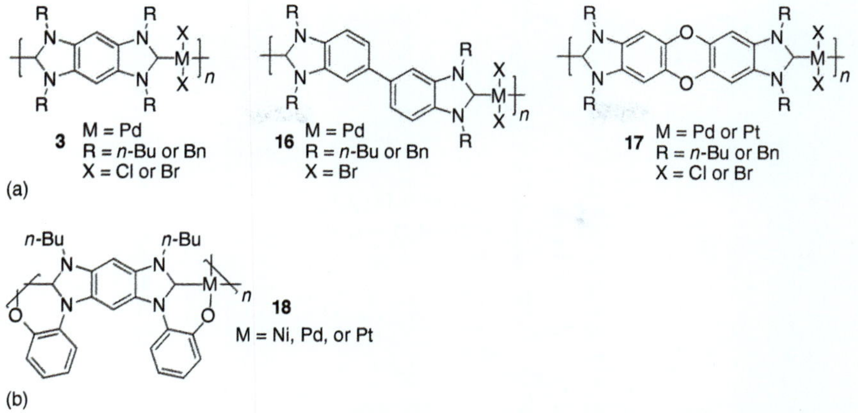

Figure 4.18 Schematic representation of various metallopolymers containing ditopic NHC-type ligands. (a) Monodentate and (b) bidentate binding.

4.3.1.2 Coordination Polymers Containing Ditopic Bidentate Ligands

The formation of metallo-supramolecular polymers from transition metal ions and neutral ditopic bidentate ligands has attracted comparably little attention. The main reason for this is that such materials can conveniently by prepared by the (metal-catalyzed) coupling of preformed complexes. These polyaddition-type reactions correspond to the general preparation method denoted as Vb (Figure 4.3); accordingly, such materials do not represent metallo-supramolecular polymers in the sense of meaning, as they have not been assembled directly by metal-to-ligand interactions, and are not discussed here. The various Ir(III)-containing metallopolymers, applied in optoelectronic devices, are presumably the most prominent examples in this respect [61, 164, 165].

However, (hyper)branched or dendritic architectures have been prepared by the multiple-step assembly of such complexes. In particular, Ru(II) ions have widely been used in this context due to the high directionality in coordinating bidentate ligands centers and the very high kinetic stability of the resultant tris-bidentate complexes. For example, the directional synthesis enabled the synthesis of dendritic architectures comprising up to 20 Ru(II) centers [42]. Recently, Higuchi and coworkers assembled a coordination nanosheet ("CONASH"), as designated electrochromic material, from Fe(II) ions and rigid ditopic bidentate ligands [166]. The first example for a Ru(II)-containing polymer, which was assembled from the $[Ru(R_2bpy)Cl_3]_x$ precursor complex ($R = H$ or C_6H_5) and tetrapyridophenazine, was reported by the Rehahn group in 1996 (**19**, Figure 4.19) [167, 168]. Cautious optimization of the reaction conditions eventually gave soluble materials of high molar masses. The metallopolymers **19** exhibited high kinetic stabilities (i.e. no disassembly in polar solvents) attributed to the robust Ru(II) tris-bidentate complexes within the polymer backbone. Detrimentally, the overall structure of the polymers was less defined (compared with similar polymers containing Ru(II) bis-tridentate complexes, Section 4.3.1.3). The phenomenon of metal-centered chirality (i.e. occurrence of Λ- and Δ-enantiomers) did not allow any long-range

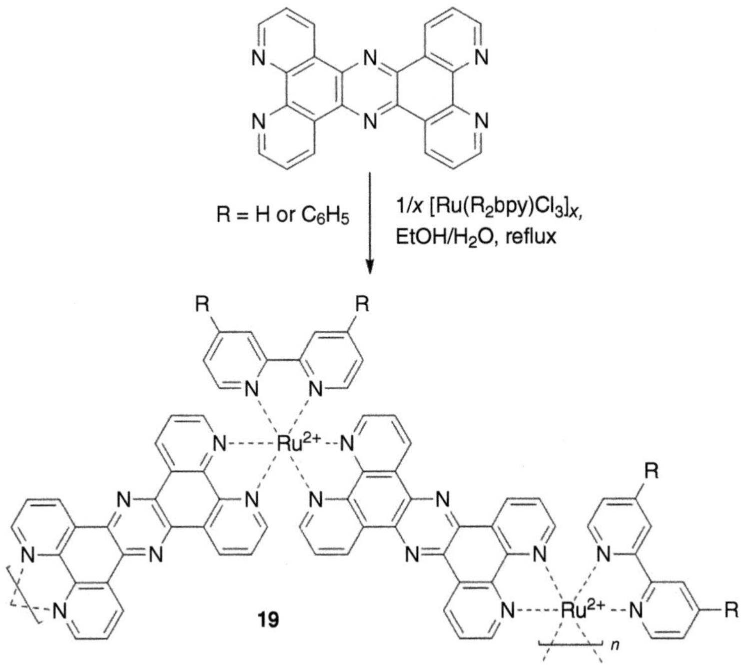

Figure 4.19 Schematic representation of the synthesis of the rigid metallopolymers **19**.

order of the polymer chains, and thus, only randomly coiled chain conformations were adopted.

Cassagneau et al. applied a different synthetic approach for basically the same purpose [169]: these authors first assembled the polymer main chain from a flexible ditopic ligand and $RuCl_3{\cdot}xH_2O$ (2 : 1 ratio of the components); the resultant prepolymer was than treated with bpy to form the Ru(II) tris-bidentate (**20**, Figure 4.20). It has to be noted in this context, that using ditopic ligands in the formation of linear metallo-supramolecular polymers inherently covers the risk that also metallo-macrocyclic assemblies might be formed (at least to some extent) [170]. However, this aspect was not in the focus of the authors interest and even negligible with respect to the intended optoelectronic applications. The beneficial properties of Ru(II) tris-bipyridine complexes, i.e. their stability and rich photophysical properties, make metallopolymers incorporating such photoactive complexes highly attractive materials for photovoltaic or light-emitting devices [171, 172].

The kinetic stability found for the aforementioned (pseudo-)octahedral Ru(II) tris-bidentate complexes cannot be observed for their tetrahedral bis-bidentate counterparts incorporating Cu(I) or Ag(I) ions. Accordingly, the formation of structurally defined metallo-supramolecular polymers, based on this particular binding motive, represents a challenge. The usage of non-coordinating, apolar solvents was identified as pivotal to obtain the stable assemblies **21** (Figure 4.21a); all other solvents (even in traces) acted as competitive ligands leading to a fast dissociation

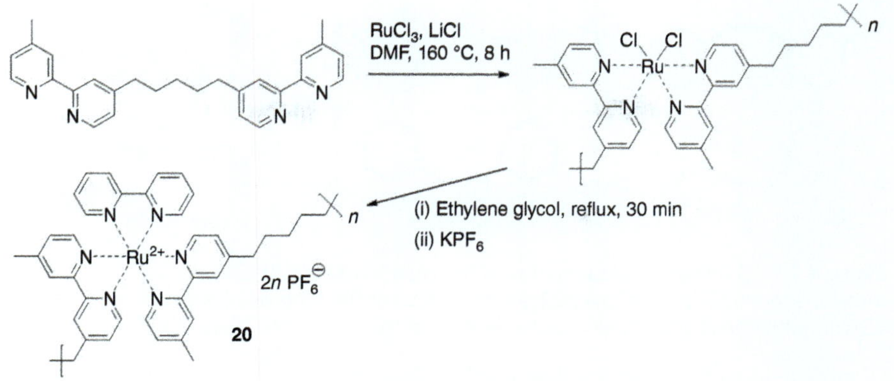

Figure 4.20 Schematic representation of the synthesis of the flexible metallo-supramolecular polymer **20**.

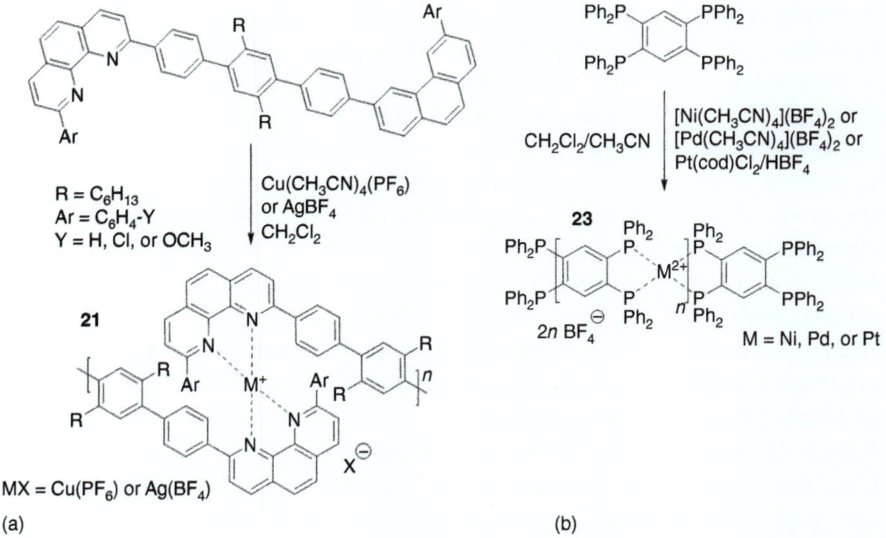

Figure 4.21 Schematic representation of the synthesis of metallopolymers **21** and **23** based on tetrahedral bis-bidentate complexes (a) and a phosphine-based ligand (b), respectively.

of **21** into small-molecule solvato-complexes [173, 174]. The metallopolymer **22** possessing a helical conformation was assembled from an *exo*-ditopic macrocyclic ligand with two bpy binding sites in the presence of Ag(I) ions (Figure 4.22). Due to the achiral nature of the organic ligand, the helical chains were formed as racemic mixture only [175]. Similar metallopolymers were also prepared from appropriate ditopic phosphine derivatives [176]. For example, the linear metallo-supramolecular assemblies **23** were synthesized from 1,2,4,5-tetrakis(diphenylphosphino)benzene and Ni(II), Pd(II), or Pt(II) ions (Figure 4.21b); detailed spectroscopic and electrochemical investigations of these materials indicated strong intrachain metal–metal interactions [177–179].

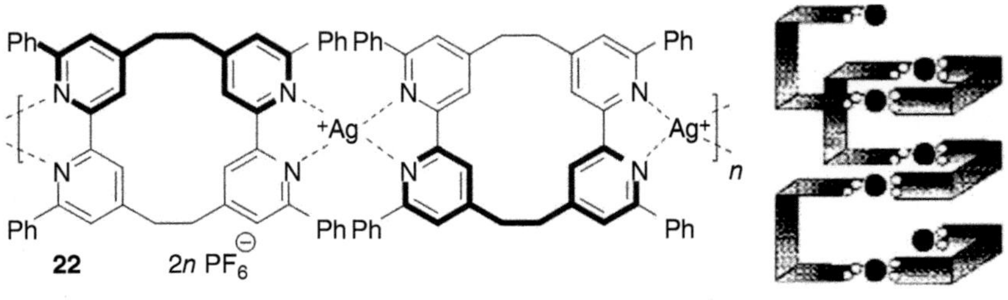

Figure 4.22 Schematic representation of the chemical structure of the metallo-supramolecular polymer **22**, an illustration of the proposed helical 3D-conformation is also shown. Source: Kaes et al. [175]. © 1998 Wiley-VCH.

The phenomenon of electrochromism is related to materials that exhibit a reversible change of their optical properties, when an external electrical field is applied. Electrochromic materials are important for the realization of, e.g. electronic papers or smart windows [180]. Traditionally, metal oxides (e.g. WO_3) or redox-active polymers [e.g. Prussian blue, poly(3,4-ethylenedioxythiophene), or polyaniline] have been tested in this respect. To overcome the limitations inherently related to these materials (i.e. their price, the non-satisfying optical contrast, and switching rate), various types of metallo-supramolecular polymers have been introduced as alternative electrochromic materials. For example, the color of the Cu(II)-containing metallo-polymer **24a** could be reversibly changed from green to colorless by electrochemical switching between the +II and +I oxidation states of the metal centers (Figure 4.23) [181]. The high kinetic stability of **24a** was shown by SEC measurements; remarkably, the molar mass (M_W) of c. 7×10^6 g mol^{-1} ($Đ$ of 1.1) was not affected by the applied reduction/re-oxidation cycles. Moreover, the Ni(II)-containing ion-conductive materials **24b** and **25** were synthesized from the respective bis-phenanthroline ligands (Figure 4.23) [182]. The ionic conductivity of the deposited polymer films could be increased by raising the relative humidity of the environment. Moreover, metallopolymers **25** missing the hydrophobic fluorene spacer exhibited a much higher ionic conductivity (by a factor of 500) compared to **24b**: a maximum conductivity 0.75×10^{-3} S cm^{-1} at a relative humidity of 98% was reported for **25a**. The ionic conductivity could be increased even further to 5×10^{-2} S cm^{-1} under the same conditions by replacing the ClO_4^- anions by Cl^- ones [183].

Sing and Sinha reported on the metallopolymers **26** and **27** containing a bridging $[Ni(mnt)_2]^{2-}$ unit and Pd(II), Ag(I), or Hg(II) ions (mnt: maleonitrile-dithiolate) [184]. The $Na[Ni(mnt)_2]$ precursor complex was synthesized first from $Ni(NO_3)_2$ and $Na_2(mnt)$; subsequent cation–metathesis reactions with appropriate precursor salts [i.e. K_2PdCl_4, $AgNO_3$, $Hg(NO_3)_2$] afforded the thermally stable, though hardly soluble materials for which a semiconductor-like behavior in the temperature regime from 306 to 393 K was documented (Figure 4.24). Related to this, also the partially oxidized monosulfonyl ligand mmnt^{2-} was employed to assemble a coordination polymer with Ag(I) ions [185].

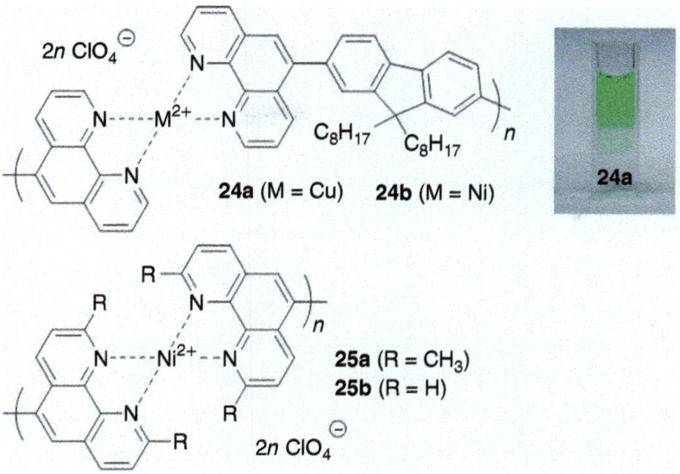

Figure 4.23 Schematic representation of the metallopolymers **24** and **25**. The picture shows the green color of **24a** that could be changed reversibly to colorless upon electrochemical reduction of the Cu(II) centers. Source: Hossain et al. [181]. Figure reproduced with kind permission. © 2013 Wiley-VCH.

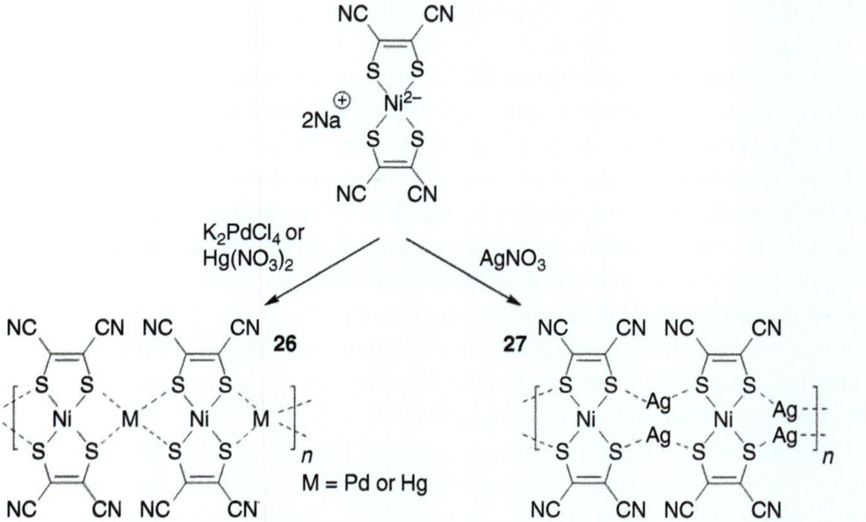

Figure 4.24 Schematic representation of the synthesis of coordination polymers **26** and **27** containing bridging $[Ni(mnt)_2]^{2-}$ units.

4.3.1.3 Coordination Polymers Containing Ditopic Tridentate Ligands

In comparison to the rather scattered examples where metallo-supramolecular polymers have been assembled from transition metal ions and ditopic bis-bidentate ligands, the usage of the corresponding ditopic bis-tridentate ligands represents a highly active field of research. In particular, polypyridyl-type ligands are used to form linear metallo-supramolecular with appropriate transition metal ions in equimolar amounts (Figure 4.25). The various bis-terpyridines are, by far, the most

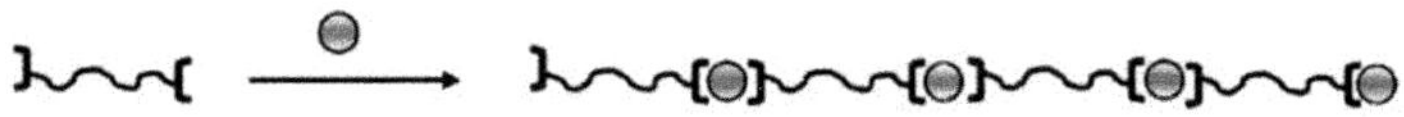

Figure 4.25 Generalized schematic representation of the self-assembly of ditopic bis-tridentate ligands and divalent transition metal ions into linear metallo-supramolecular polymers. Source: Redrawn from Chiper et al. [52]. 2009 John Wiley & Sons.

prominent representatives in this respect [2, 29, 52, 67]. Two types of ligand architectures, flexible and rigid–rigid ones, have been employed to the resultant polymers have been investigated in the context of various potential applications; from these optoelectronics, sensing, and stimuli–responsive or self-healing materials are the most relevant examples [29]. The linking of two tpy units with an organic spacer via their 4′-positions and the pseudo-octahedral geometry of terpyridine bis-complexes enables the formation linear assemblies, thereby avoiding the occurrence of any metal-centered chirality. Nonetheless, metallo-macrocyclic species might still be formed, in particular, in the case of flexible bis-terpyridine ligands [186]. Principally, the ring-chain equilibrium can be driven to favor linear supramolecular polymerization when high-concentration conditions are applied [187].

Storrier et al. were among the first to report on metal-supramolecular polymers, derived from bis-terpyridine ligands [188, 189]. The authors compared two different synthetic routes toward these materials: the polycondensation of the preformed Fe(II)–tpy bis-complexes **29** and the metallo-supramolecular polymerization of the ditopic ligands **28** with Fe(II) ions (Figure 4.26). Similar results in terms of the molar masses of metallopolymers $[Fe(\mathbf{28})]_n(BF_4)_{2n}$ were obtained by the two routes. In the case of Ru(II) ions, which typically require much harsher reaction conditions for the formation of tpy bis-complexes, the metallo-supramolecular polymerization route appeared to be less suited and, therefore, Chan and coworkers synthesized their Ru(II)-containing metallopolymers – having similar structures as the Fe(II)-containing examples reported by Storrier et al. – conveniently via the "complex first" methodology (Figure 4.26) [190].

The self-assembly of bis-terpyridines, featuring flexible OEG-spacers, in the presence of Fe(II) ions was studied independently by the Constable and Schubert groups (OEG: oligo(ethylene glycol)) [191–193]. Mixtures of defined metallo-macrocyclic species and linear metallopolymer chains were found at room temperature. At elevated temperatures, the formation of the thermodynamically less favored linear polymer chains was suppressed, and the defined three- or four-centered macrocycles were observed exclusively. Using Ru(II) instead of Fe(II) ions gave kinetically much more stable bis-complexes, and thus, the equilibration between linear and cyclic assemblies is prevented. The metallo-supramolecular polymerization of an OEG-bridged bis-terpyridine ligand with $RuCl_3{\cdot}xH_2O$ under reducing conditions, i.e. in a mixture of *n*-butanol and *N*-ethylmorpholine, was reported by Meier et al. (Figure 4.27) [194]. It was shown that the formation of metallopolymers of high molar masses was preferred at high monomer concentrations. Remarkably,

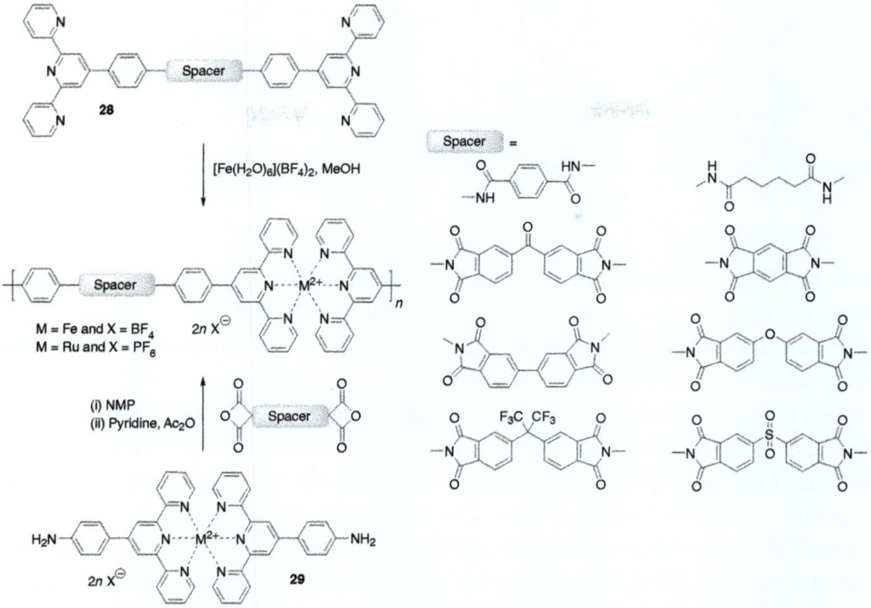

Figure 4.26 Schematic representation of the synthesis of metallopolymers bearing amido- or imido-linkages via two different routes (denoted as methods III and Vb, respectively, see also Figure 4.3).

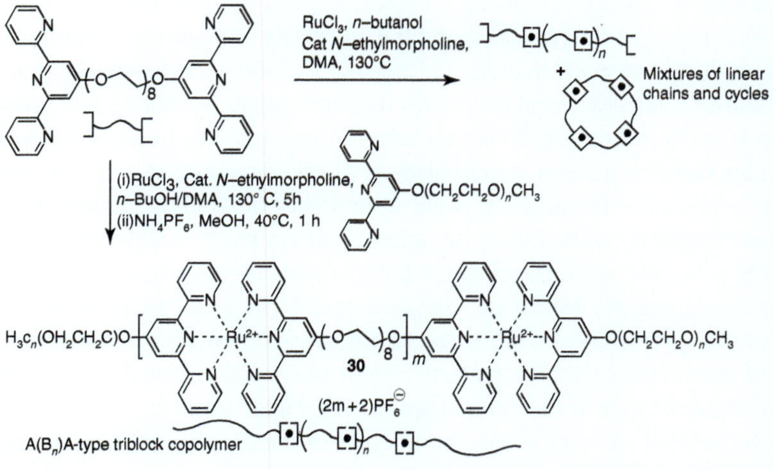

Figure 4.27 Schematic representation of the synthesis of metallo-supramolecular assemblies from a flexible bis-terpyridine ligand and Ru(II) ions.

in the presence of an appropriate chain-stopping agent (10 mol% of MPEG–tpy were added), the metallo-macrocyclization pathway was blocked almost fully and the metallo-supramolecular $A(B_n)A$-type triblock copolymer **30** was isolated. Not only for the Ru(II) terpyridine complexes but also for their Ni(II)-, Co(II)-, and Fe(II)-containing counterparts a certain degree-of-kinetic-stability is given. On this basis, the same self-assembly strategy was later applied by Chiper et al. to obtain

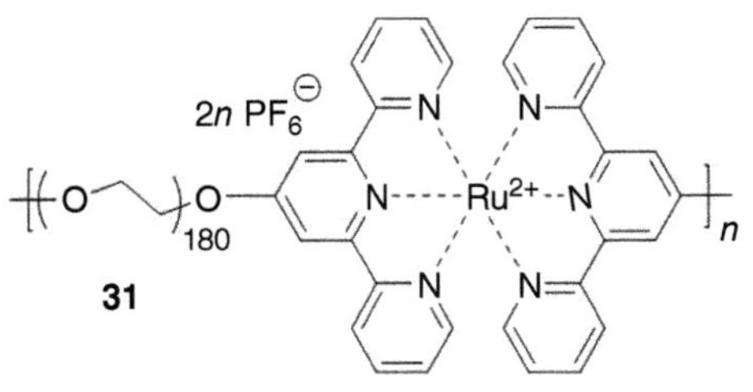

31

Figure 4.28 Schematic representation of the Ru(II)-containing chain-extended polymer **31**.

A(B_n)A triblock copolymers with these metal ions (though cautious control over the reaction and characterization conditions was required) [195].

Besides the former examples where small-molecule ditopic ligands used in the self-assembly into metallo-supramolecular polymers, also ditopic macromolecular ligands ("macroligands") have been used for this purpose [52, 186]. As one representative example, the poly(ethylene glycol)s (PEGs) of various molar masses, functionalized tpy units at both termini, were assembled into chain-extended polymers with Fe(II) or Ru(II) ions (**31**, Figure 4.28) [196–198].

One highly important application of the directional and stable character of some types of metal-to-ligand interactions is the ability to assemble supramolecular block copolymer architectures on the basis of such non-covalent bonding [2, 58, 59, 186]. As also outlined for other types of non-covalent interactions, i.e. H-bonding interactions (Chapter 3) or the various types of host–guest complexation (Chapters 6–11), this concept aims to combine the characteristics of "covalent" block copolymers with those introduced by a transition metal complex. Beyond the potential reversibility of the metal-to-ligand interaction(s) and the possibility to modulate its strength by the proper choice of metal centers and organic ligands, also the special photophysical and/or electrochemical properties that are created by linking two polymer chains might be of interest. This rich behavior prompted us to define such a metal complex within a metallo-supramolecular block copolymer as a "block" on its own [199]. Though this statement is not in-line with the traditional definition from polymer chemistry, which basically relies on chain length and molar mass, a view on the properties of the constituent units of metallo-supramolecular block copolymers does somehow legitimate this description. The term "element-block" was coined by Chujo and coworkers 2015 and defines the "minimum functional unit composed of heteroatoms within a material" [200] and can be adapted to describe the metal-complex linkage in such block copolymer architectures.

It has to be pointed out that by far not all types of metal-to-ligand interactions are suited for assembly of metallo-supramolecular block copolymers – some crucial requirements have to be fulfilled: directionality and heterolepticity. What does this mean? When aiming for the selective assembly of a block copolymer with the general formula A–[M]–B, the two possible homopolymeric species A–[M]–A and B–[M]–B should not be formed in the ideal case. In general, such A–[M]–B diblock copolymers comprising a heteroleptic metal complex are synthesized following a two-step assembly protocol (Figure 4.29). The second important requirement to be met is the directionality of the interaction, i.e. the linear assembly of the components without the occurrence of coordination isomers. From the wide range of possible

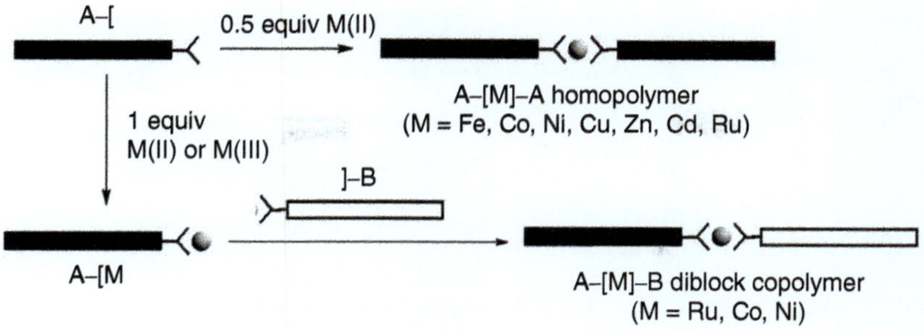

Figure 4.29 Schematic representation of the general structure of a metallo-supramolecular homo as well as a diblock copolymer. Source: Redrawn from Schubert et al. [186].

metal-to-ligand interactions only two types remain, which fulfill these demands and have used to assemble such block copolymer systems: terpyridine bis-complexes and pincer complexes. However, the kinetic lability of many tpy bis-complexes in solution practically limits the range of potential metal ions to a small number: Ru(II), Os(II), Ir(III), Co(III), and Ni(II). From these, Co(III) and Ni(II) have only been used rather recently and only exemplarily in this respect [201, 202]. Up to now, the vast majority of metallo-supramolecular block copolymers has been assembled around Ru(II) ions; whereas, the required harsh reaction conditions basically exclude the usage of Os(II) or Ir(III) ions (attributed to a potential decomposition of the polymer chains) [203–230]. Concerning other transition metal ions, such as Cr(III) or Fe(II), the formation of the required heteroleptic complexes has been reported, but is hampered by the poor stability of these complex in solution [231, 232].

Polymers equipped with a terminal metal-binding site, such as tpy, can be prepared via two basic strategies [233]: either an initiator, already bearing the designated binding site, is used for the polymerization or the binding site is attached to the terminal functionality of a preformed polymer (Figure 4.30). A broad range of mono- and bis-functionalized polymers has been prepared using these approaches. However, their synthesis is beyond the scope of this chapter and has already been summarized comprehensively elsewhere [2, 186, 199]. From the wide range of polymers that has been modified with a terminal metal-binding site, the poly(ethylene glycol) monomethyl ethers (MPEGs) represent, by far, the most commonly used macroligands for complexation by transition metal ions. The reason for this can be found in the special properties of PEGs (e.g. hydrophilicity and water solubility, biocompatibility, softness).

The formation of A–[M]–A homopolymers is based on homoleptic tpy bis-complexes, which are readily formed by broad range of divalent transition metal ions, and thus, many examples for this can be found in the literature [i.e. Ru(II), Fe(II), Co(II), Ni(II), Cu(II), Zn(II), and Cd(II)] [208, 212, 225, 234–238]. Such homopolymers represent commonly used, as reference materials, in investigating more advanced metallo-supramolecular architectures, such as the A–[M]–B diblock copolymers (i.e. by SEC or analytical ultracentrifugation, see Chapter 12) [2, 186]. However, applications for these metallo-supramolecular polymers

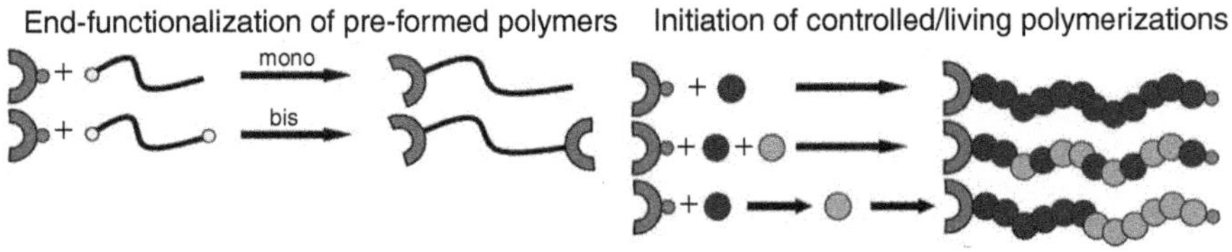

Figure 4.30 Schematic representation of the synthesis of polymers equipped with designated metal-binding sites via the end-functionalization and the initiator route. Source: Hoogenboom and Schubert [233]. © 2006 The Royal Chemical Society.

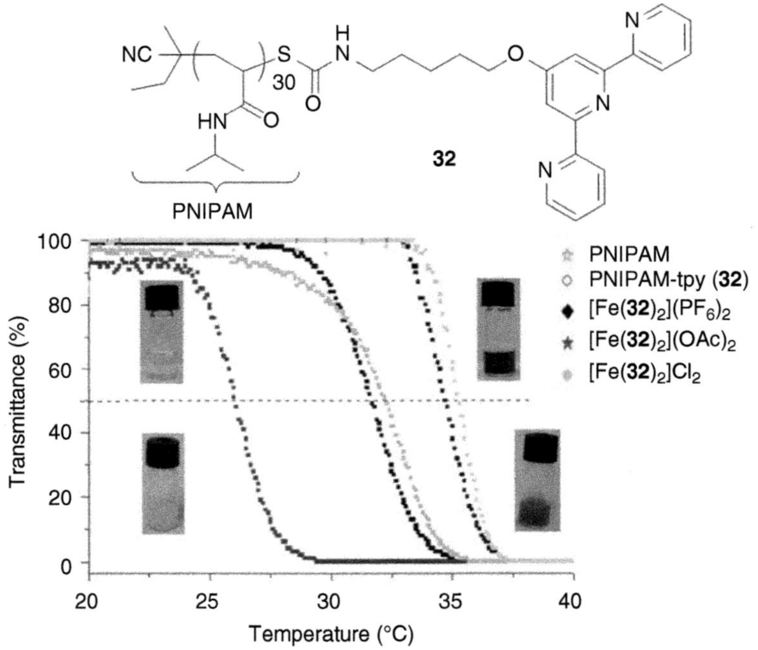

Figure 4.31 LCST behavior of PNIPAM, tpy-PNIPAM (**32**), and $[Fe(\mathbf{32})_2]X_2$ with different counterions X (i.e. Cl^-, AcO^-, and PF_6^-) in water as monitored by transmittance measurements. Images of the vials below (top) and above the LCST (bottom) are also shown for **32** (left) and $[Fe(\mathbf{32})_2]^{2+}$ (right). Source: Chiper et al. [239]. © 2008 Wiley-VCH.

have hardly been discussed. As one of the very few exceptions from this, the homopolymer assembled from tpy-functionalized poly(*N*-isopropylacrylamide) (tpy-PNIPAM, **32**) in the presence of Fe(II) or Zn(II) ions needs to be mentioned (Figure 4.31) [239]. An aqueous solution of PNIPAM, as a typical thermoresponsive polymer, exhibits a transition of solubilized polymer chains to collapse precipitated ones upon heating; the temperature at which precipitation occurs is referred to as the lower critical solution temperature (LCST) [240]. Chiper et al. identified a dependency of the LCST on the size of the metal ion as well as on the hydrophilicity of the charge-balancing anion.

Schubert, Gohy, and their coworkers synthesized a plethora of metallo-supramolecular block copolymers in which the constituent blocks were linked by heteroleptic Ru(II)–tpy bis-complexes. All these assemblies were studied regarding their phase separation in solution (i.e. micellization) or in the solid state (i.e. microphase separation) [58, 59]. In particular, the MPEG–[Ru]–PS diblock copolymers comprising a hydrophilic PEG and a hydrophobic polystyrene (PS) block have been in the focus of interest. In aqueous solution, these materials gave spherical core–corona micelles with the cationic complexes located at the interface between the hydrophobic PS core and the hydrophilic PEG corona (Figure 4.32a/b) [204, 205, 214, 217, 223, 241]. Furthermore, the properties of this material in the solid and molten states were investigated – the aggregation/segregation behavior of the metallo-supramolecular deblock copolymers was found to be strongly dependent on the length of the constituent block chains and on the nature of the counterions (i.e. interacting vs. noninteracting ones) [213, 216, 219, 242]. Following a combinatorial approach, Lohmeijer et al. synthesized a [4×4]-library of PS_x–[Ru]–PEG_y diblock copolymers [216]. These 16 polymers covered a wide range of PS/PEG chain length ratios, and the authors studied the morphology of their thin spin-coated films: a broad range of phase-separated structures with tunable domain sizes could be visualized by atomic-force microscopy (AFM, Figure 4.32c): spherical (samples 3, 4, 8, and 13) as well as cylindrical morphologies (samples 2, 7, 9, 11, and 12) could be identified and rationalized by the theories from solid-state polymer physics. It was shown later that the phase-separation into ordered domains could be improved by applying solvent annealing of the spin-coated films (e.g. with tetrahydrofuran [THF]) [243].

Though Ru(II)-tpy bis-complexes are kinetically highly stable in solution, they can be cleaved oxidatively, e.g. by Ce(IV) under acidic conditions. Fustin et al. utilized this behavior to fabricate nanoporous membranes: first, the MPEG–[Ru]–PS diblock copolymer was synthesized and assembled into cylindrical, phase-separated domains; subsequently, the hydrophilic MPEG–tpy block was cleaved oxidatively and washed out [220]. The authors used AFM imaging (Figure 4.33a/b) and XPS (Figure 4.33c) to confirm the removal of water-soluble block.

Beyond the MPEG–[Ru]–PS deblock copolymers, a broad range of diblock copolymer assemblies has been prepared and analyzed accordingly (Table 4.1). In addition to this, amphiphilic ABC-type triblock copolymers are known to give multicompartment micelles under appropriate conditions. The Schubert and Gohy groups synthesized such metallo-supramolecular triblock copolymers and multicompartment micelles were observed for the assemblies PS-*b*-P2VP–[Ru]–MPEG and PS-*b*-PTFMS–[Ru]–MPEG [PTFMS: poly(*p*-trifluoromethylstyrene)] [207, 246]. In addition to this, also a tetrablock copolymer was reported by the same author consortium [136].

More recently, Chan and coworkers reported on the assembly of a kinetically stable P3HT–[Zn]–MPEG diblock copolymers from the tpy-functionalized macroligands **33** and **34** in the presence of Zn(II) ions (Figure 4.34) [247]. The remarkable formation of a heteroleptic Zn(II)-tpy bis-complex was facilitated by additional π–π stacking interaction of the coordinated ligands (being reminiscent of a molecular tweezers, see Chapter 5). Thereby, the formation of two possible homopolymers, i.e. P3HT–[Zn]–P3HT and MPEG–[Zn]–MPEG, was excluded due to steric

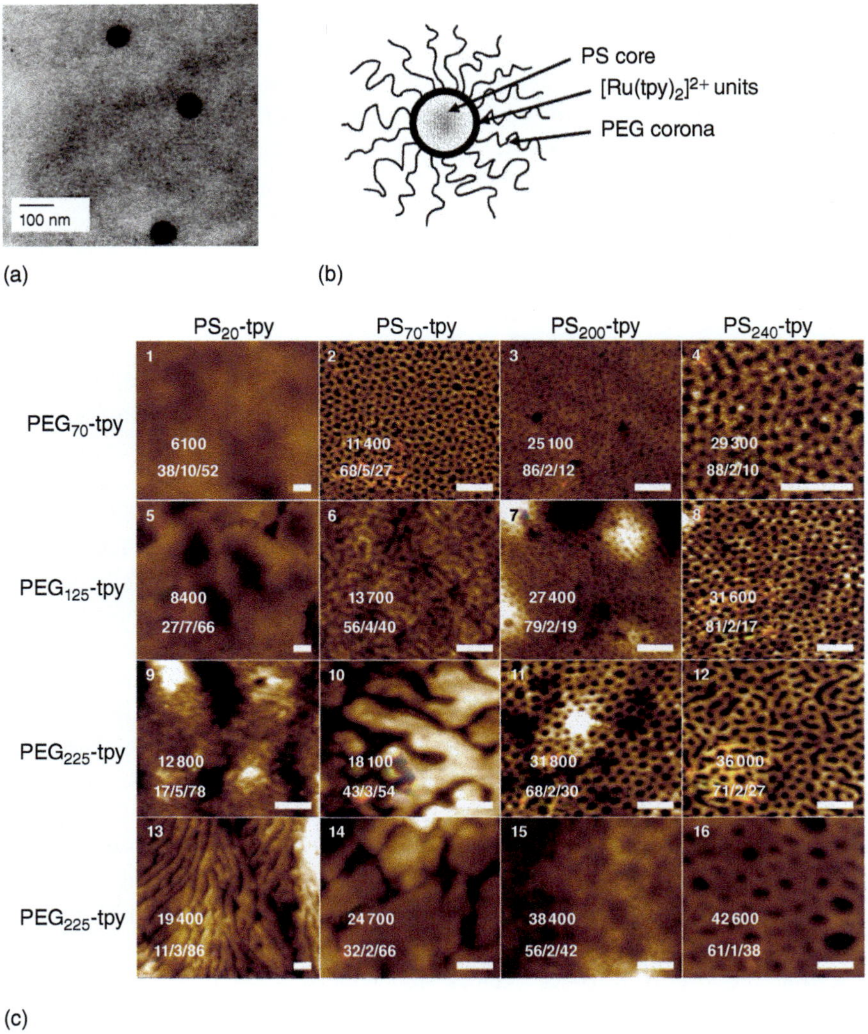

Figure 4.32 (a) Transmission electron microscopy (TEM) image of PEG–[Ru]–PS micelles in water (without staining). Source: Lohmeijer et al. [216]. Figure reproduced with kind permission. © 2004 The Royal Chemical Society. (b) Schematic representation of a typical micelle of PEG–[Ru]–PS diblock copolymer. Source: Gohy et al. [204]. © 2002 American chemical society. (c) AFM phase images of the 16 PS_x–[Ru]–PEG_y diblock copolymers (thin spin-casted films, no annealing). For each sample, the upper number represents the molar mass and the lower number denotes to the volume fractions of the three-component systems. The scale bar is always 100 nm. Source: Lohmeijer et al. [216]. Figure reproduced with kind permission. © 2004 The Royal Chemical Society.

hindrance and kinetic lability, respectively. The authors studied the self-assembly of the metallo-supramolecular deblock copolymers by AFM and transmission electron microscopy (TEM) imaging. As in the aforementioned example, i.e. MPEG–[Ru]–PS, the length fraction of the constituent blocks influenced the morphology of the resultant structures (Figure 4.34): fiber-like assemblies were

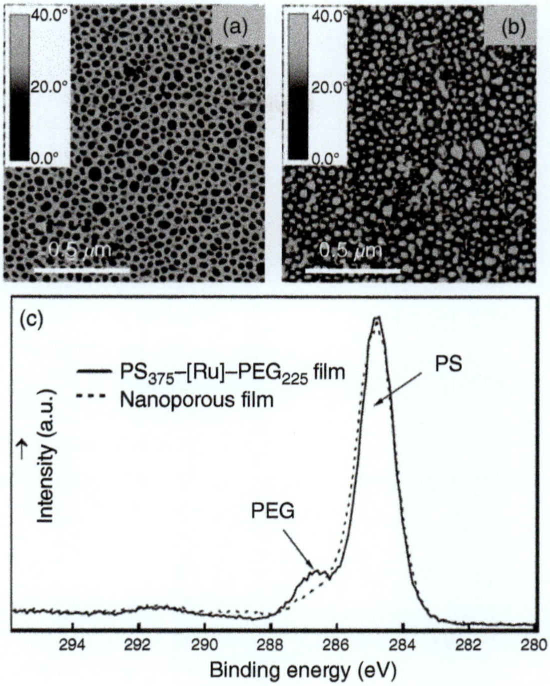

Figure 4.33 Representative AFM images of a spin-coated film (75 nm thickness) of PS_{375}–[Ru]–PEG_{225} on a Si substrate, before (a) and after (b) the removal of the hydrophilic block. (c) XPS signature (C_{1s} regime) of the film before (solid line) and after treatment with the oxidizing agent. Source: Fustin et al. [220]. Figure reproduced with kind permission. © 2005 Wiley-VCH.

preferred when the length of the P3HT block was increased. Moreover, the core of the assemblies, comprising the hard P3HT chains, exhibited a crystallization-driven self-assembly (CDSA) behavior. In the presence of a strong competitive ligand (e.g. ethylenediamine-tetraacetic acid [EDTA]), the Zn(II)–tpy bis-complexes were cleaved and the water-soluble macroligand could be washed out, thus giving a network of P3HT nanofibers. These nanofibers, in turn, could be redispersed by adding a preformed $[Zn(\mathbf{33})]^{2+}$ mono-complex. The concept of CDSA, as introduced by the Manners and Winnik groups, utilizes block copolymer micelles containing a hard and crystallizable core. These initial micelles feature a quasi-living epitaxial growth to eventually generate cylindrical assemblies for which remarkably narrow dispersities have been found [248]. A variety of potential applications, such semiconductors or precursors to ferromagnetic materials, has been suggested to arise from the CDSA of micelles into well-defined nanostructures [249]. Thus, metallo-supramolecular A–[M]–B diblock copolymers, in which the block A is a crystallizable polymer, represent promising building blocks for such applications (the diblock copolymer MPEG–[Ru]–PFDS [PFDS: poly(ferrocenyldimethylsilane)] is one further example for this [215]).

A different type of tpy-containing diblock copolymer was utilized, as a macromolecular building block, for the formation of a metallo-supramolecular polymer [250]. The macroligand bearing a hydrophilic MPEG and a pH-responsive poly[2-(diisopropylamino)ethyl methacrylate] side chain gave, upon complexation with Ru(II) ions, a polymer-brush-like material, which assembled into striped nanosheets in water (Figure 4.35). The size of the nanosheets could be controlled

Table 4.1 Overview over the metallo-supramolecular block copolymers, based on terpyridine bis-complexes.

	Polymer blocks	References
MPEG–[Ru]–PS	MPEG: poly(ethylene glycol) monoethyl ether PS: polystyrene	[204, 205, 213, 214, 216, 217, 219, 223, 241, 242]
MPEG–[Ru]–PDMS	MPEG PDMS: poly(dimethylsiloxane)	[228]
MPEG–[Ru]–PEB	MPEG PEB: poly(ethylene-*co*-butylene)	[203, 210]
MPEG–[Ru]–PFDS	MPEG PFDS: poly(ferrocenyldimethylsilane)	[215]
PNIPAM–[Ru]–PS	PNIPAM: poly(*N*-isopropyl acrylamide) PS	[244]
PS–[Ru]–PAA	PS PAA: poly(acrylic acid)	[245]
PS-*b*-P2VP–[Ru]–MPEG	PS P2VP: poly(2-vinylpyridine)	[207]
PS-*b*-PTFMS–[Ru]–MPEG	PS PTFMS: poly(*p*-trifluoromethylstyrene) MPEG	[246]
PTFMS-*b*-PTBA-*b*-PS–[Ru]–MPEG	PTFMS PTBA: poly(*tert*-butyl acrylate) PS MPEG	[230]
MPEG–[M]–PS M = Ni or Co	MPEG PS	[202]
P3HT–[Zn]–MPEG	P3HT: poly(3-hexyl thiophene) MPEG	[247]

via the DP of the side chains. The metal-free macroligand formed, due to its amphiphilic nature, spherical micelles under the same experimental conditions. The authors concluded that rigid-linear, polycationic supramolecular backbone, the overall brush-shaped architecture as well as the presence of the PF_6^- counterions accounted for the observed hierarchical self-assembly process.

The Gohy group also studied the metallo-supramolecular cross-linking and the postassembly functionalization of micelles, which contained vacant tpy moieties on their outer shell. These core–shell micelles were assembled from

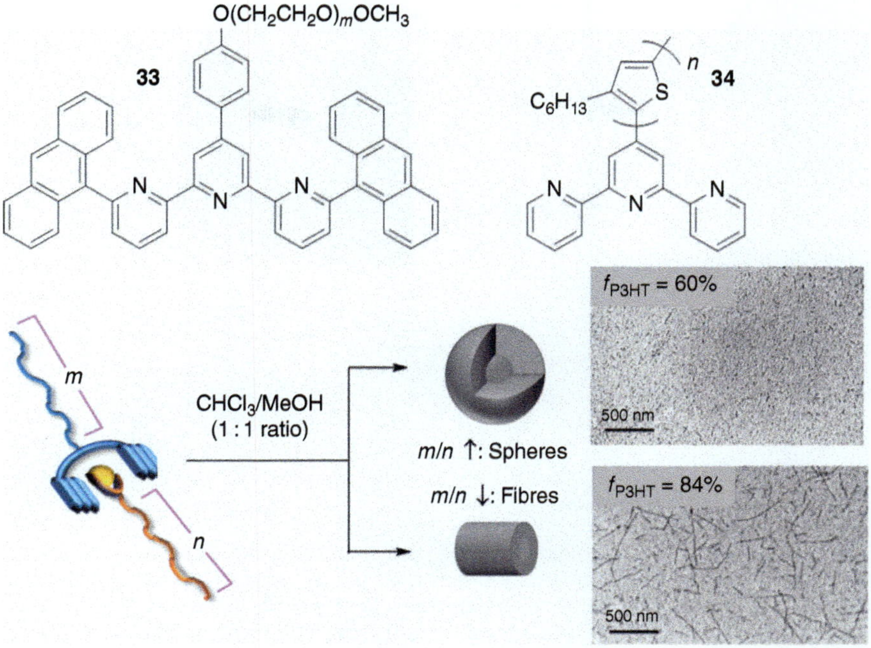

Figure 4.34 Schematic representation of the macroligands **33** and **34**, which gave an A–[M]–B diblock copolymer in the presence of Zn(II) ions. The diblock copolymer assembled further into spheres and fibers as a function of *m*/*n* ratio (*m*: length of the PEG block, *n*: length of the P3HT block). Typical TEM images for both cases (i.e. low and high contents of P3HT, respectively) are also shown. Source: He et al. [247]. Figure reproduced with kind permission. © 2017 American Chemical Society.

a PS-*b*-PTBA-tpy [PTBA: poly(*tert*-butyl acrylate)] diblock copolymer [224]. As depicted in Figure 4.36, a stepwise metallo-supramolecular protocol was applied for the grafting of MPEG chains onto these assemblies, thus generating micelles of core–shell–corona morphologies. In a related study, the same authors also assembled "micellar gels" bridging the initial micelles by metal-to-ligand coordination chemistry (Figure 4.36) [251].

As the last aspect in this context, some oligonuclear metallo-supramolecular block copolymers formed by directed Ru(II)-tpy complexation have to be mentioned. Such architectures might be regarded as an intermediate form between the aforementioned A–[M]–A homopolymers or A–[M]–B diblock copolymers and linear metallo-supramolecular polymers discussed hereafter. In this respect, various A–[M]–B–[M]–A triblock copolymers, comprising monodisperse rigid-rod segments, as the central block B, were synthesized by the Schubert group. The utilized π-conjugated bis-terpyridine ligands represent important building blocks for the formation of metal-supramolecular polymers by the self-assembly with divalent transition metal ions (*vide infra*) [2, 29, 67]. Following the established

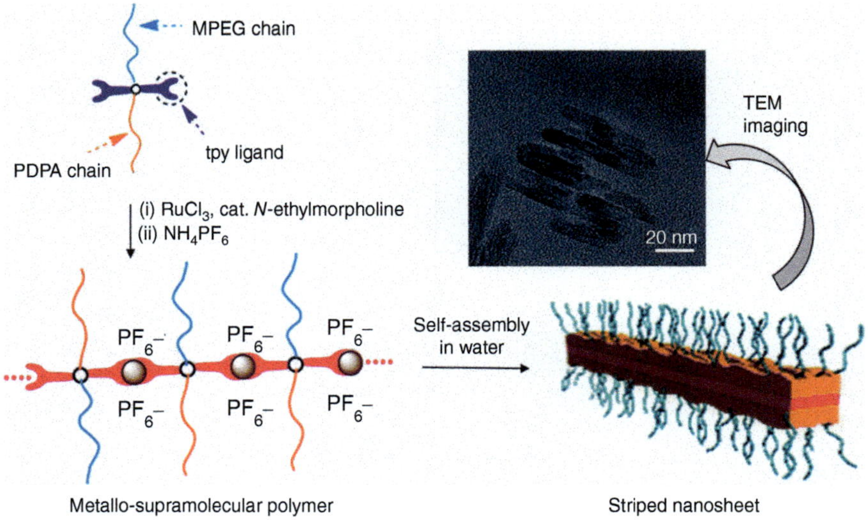

Figure 4.35 Schematic representation of the synthesis of a metallo-supramolecular polymer with pending MPEG and poly(2-diisopropylaminoethyl methacrylate) (PDPA) chains on each repeat unit. This material self-assembled into striped nanosheets in water. A representative TEM image of such a nanosheet is also depicted. Source: Zhang et al. [250]. Figure reproduced with kind permission. © 2020 The Royal Chemical Society.

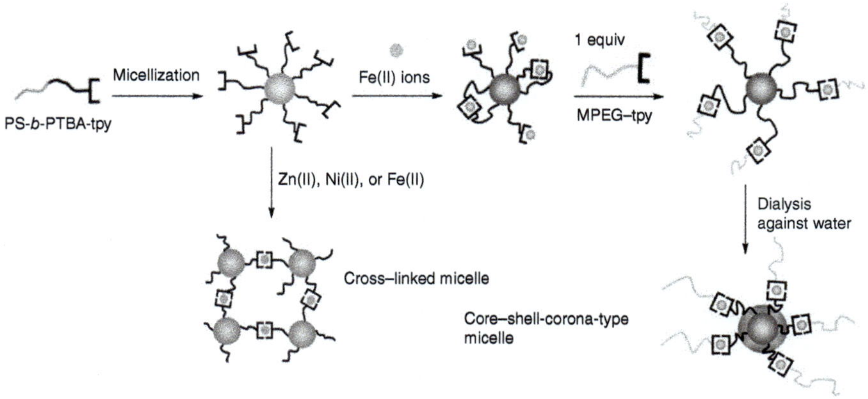

Figure 4.36 Schematic representation of the metallo-supramolecular modification of tpy-functionalized micelles via grafting-onto and cross-linking methodologies. Source: Refs. [224, 251]. © 2008 and 2009 The Royal Chemical Society.

protocols for the synthesis of the A–[M]–B diblock copolymers (Figure 4.29), a two-step assembly sequence was utilized to prepare the coil–rod–coil-type triblock copolymers **35** with hydrophilic outer MPEG blocks (in contrast to earlier work on this field, the more reactive $Ru(DMSO)_4Cl_2$ (DMSO: dimethylsulfoxide) precursor was used, Figure 4.37) [227, 229]. The same authors extended the scope of this concept also to star-shaped copolymers in which three or four such MPEG chains were attached to rigid core molecules via directed Ru(II)-tpy complexation [252].

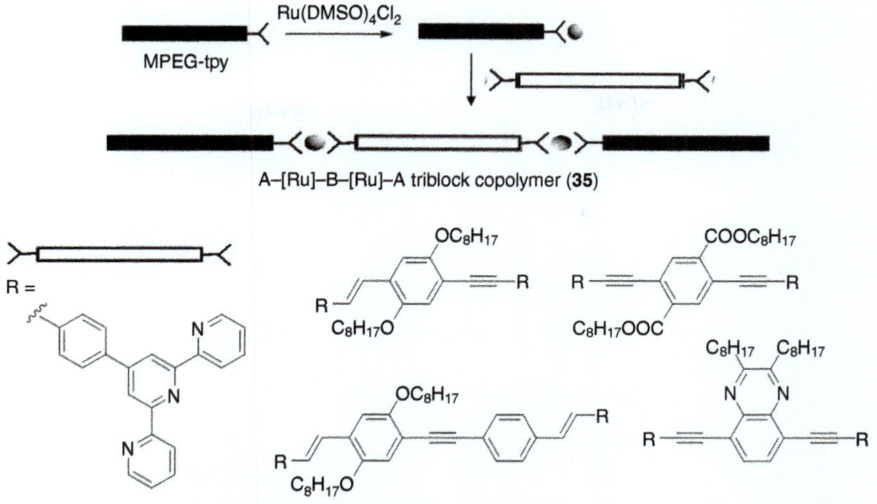

Figure 4.37 Schematic representation of the synthesis of the metallo-supramolecular ABA-type triblock copolymers **35**.

The tpy motive definitively represents the most prominent example for a tridentate ligand used in metallo-supramolecular chemistry; however, it is not the exclusive example with respect to denticity and binding strength. Rowan, and coworkers introduced 2,6-bis(1-methyl-1*H*-benzo[*d*]imidazol-2-yl)pyridine (mebip), as an alternative ligand in this respect. The flexible ditopic ligand **36**, for example, was prepared by attaching mebip units to the termini of a monodisperse OEG segment [253–258]. This particular ligand was combined with transition [e.g. Co(II) or Zn(II)] and rare-earth metal ions [e.g. La(III) or Eu(III)] in a 97 : 3 ratio to afford a metallo-supramolecular polymer network (Figure 4.38). These gel-like materials exhibited a reversible thermoresponsive and a thixotropic behavior, i.e. they turned fluid upon heating or shaking, respectively. Moreover, the bright light emission of the gels [when incorporating Zn(II) and Eu(III) ions] was quenched when the temperature was increased or formic acid, as a strong competitive ligand for the Eu(III) ions, was added. Due to the multiresponsive character of these metallo-supramolecular polymer networks, the authors tested them further in self-healing applications [30, 259]. These authors later also used ditopic ligands with polymeric linkers between the terminal mebip sites, such as poly(ethylene-*co*-butylene)s (PEBs) and fully (or partially) π-conjugated segments [260–273]. The metallo-supramolecular polymers assembled from these ligands in the presence of transition and rare-earth metal ions were investigated regarding their stimuli–responsive properties and their potential applications as self-healing materials [30].

As summarized by Wild et al., π-conjugated segments have been widely used as spacer units in rigid-rod bis-terpyridine ligands (some examples are depicted in Figure 4.37) [67]. Following Ziessel's pioneering work in this field [274, 275], these ligands – covering a broad range of flexibility and π-conjugation – were

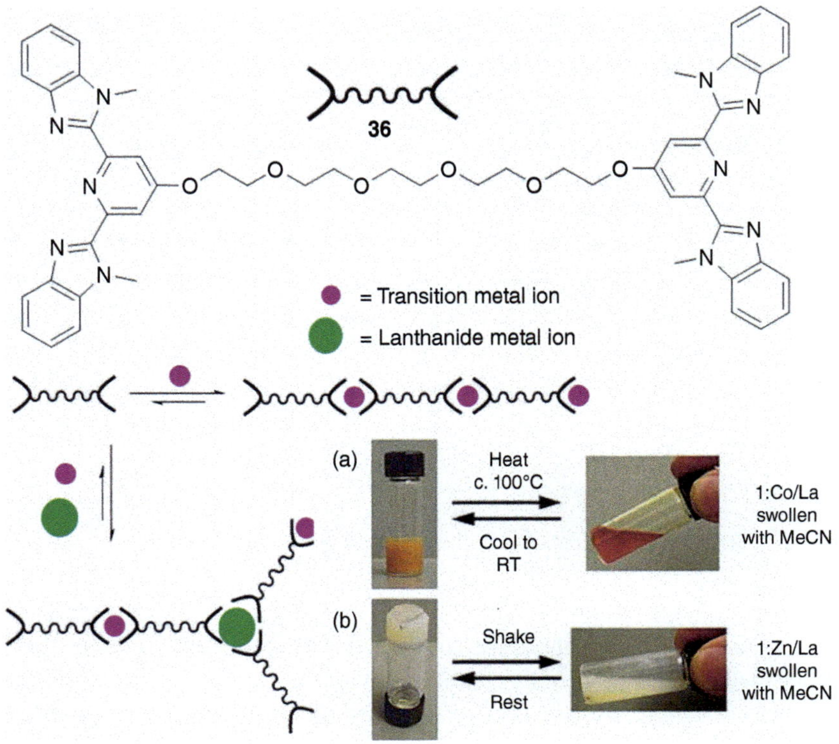

Figure 4.38 Schematic representation of the metallo-supramolecular polymerization of the ditopic ligand **36** in the presence of transition and rare-earth metal ions. The pictures illustrate the thermoresponsive (a) and the thixotropic behavior (b) of representative combinations of metal ions. Source: Beck and Rowan [253]. Figure reproduced with kind permission. © 2003 American Chemical Society.

assembled into linear metallo-supramolecular polymers and the properties of these materials, in solution as well as in the solid state, were studied intensively [2, 3, 67, 186, 276–323]. As for the aforementioned ditopic bis-bidentate ligands (Figure 4.19), the metallo-supramolecular polymerization of bis-tridentate ligands with Ru(II) ions intrinsically hampered by the relatively harsh self-assembly conditions (Figure 4.39) [324]. On the contrary, the divalent first-row transition metal ions easily generate metallo-supramolecular polymers with such ligands, typically at ambient temperature.

The Kurth and Higuchi groups assembled a range of such linear metallo-supramolecular polymers from the ditopic ligands **37** and some divalent redox-active metal ions (Figure 4.40) [317, 318]. Sufficient solubility of these metallopolymers in different solvents, including water, enabled an in-depth investigation of their optical, electrochemical, and electrochromic properties. These studies allowed one to deduce structure–property relationships and enabled the design of metallo-supramolecular materials for electrochromic applications [180]: the external substituents on the tpy fragment were found to strongly affected the switching rates and, as important, the kinetic stability of the entire assemblies.

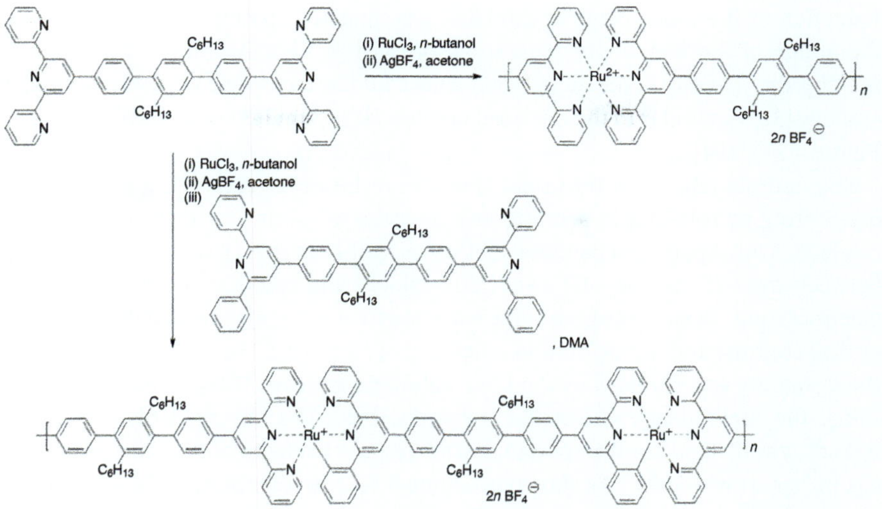

Figure 4.39 Schematic representation of the metallo-supramolecular polymerization of rigid bis-tridentate ligands with Ru(II) ions.

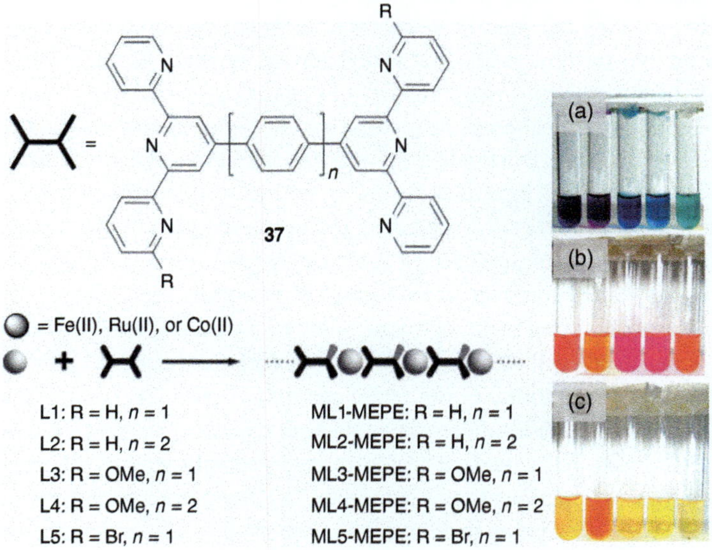

Figure 4.40 Schematic representation of the metallo-supramolecular polymers reported by the Kurth and Higuchi groups. The colors of the metallopolymers in dilute solution are also shown [a: Fe(II), b: Ru(II), and c: Co(II)]. Source: Han et al. [318]. Figure reproduced with kind permission. © 2004 American Chemical Society.

In short, electron-donating groups enabled a faster electrochromic switching, whereas the switching was slowed down by electron-withdrawing moieties (which, in succession decreased the stability). Moreover, steric constraints in proximity to the metal center had a profound influence on the optical properties: bulky substituents, for example, supported the optical memory effect and hampered the reduction of the metal centers; extended π-conjugated spacers, in turn, increased the ability of the metallopolymers to absorb visible light. Moreover, the DP and, thereby, the viscosity of the metallopolymers could be controlled via a chain-stopper strategy [322] – similar to the one used previously by Schubert and coworkwers (see Figure 4.27) [194].

The authors fabricated thin solid films from the Fe(II)-containing polymers by dip-coating or role-to-role printing and investigated their electrochromic properties [325, 326]. Applying a potential of 4.1 V vs. Li/Li^+ enabled a reversible switching between the +II and the +III state (blue-colored and colorless, respectively). The electrochromic device featured, at the wavelength of 590 nm, very high values for the optical contrast and the coloration efficiency of 71% and 525 $cm^2\ C^{-1}$, respectively; the durability was revealed by the high stability, even over 10 000 switching cycles. Thus, the metallo-supramolecular polymers, based on the rigid bis-terpyridine ligands, are of significant importance with respect to electrochromic applications, e.g. in "smart windows." As shown in Figure 4.40, the color of such electrochromic devices depends on the nature of the incorporated metal ion and/or the structure of the organic ligands [317]. When using two types of metal ions with different redox potentials within one metallopolymer, devices with a three-color scheme can be realized. For example, a metallopolymer incorporating Fe(II) and Ru(II) ions offered the ability to switch between three colors at different potentials – ranging from purple via orange to light green [327]. In addition to this, electrochromic devices were also fabricated from the same ditopic ligands and Os(II) or Cr(III) ions [328–331]. These recent examples impressively reflect the modularity of the approach: a wide range of electrochromic materials can readily be assembled from transition metal ions and ditopic ligands. By utilizing sophisticated fabrication techniques, Yoshida and Hamaguchi also produced multipatterned and tube-shaped electrochromic devices incorporating Fe(II)- and Ru(II)-containing polymers, as redox-active materials [332]. Moreover, when using a photoluminescent dye, as the counterion, the Fe(II)-containing metallopolymer showed a dual responsive behavior, i.e. electrochromism as well as a redox-switchable emission [333]. The latter was attributed the quenching ability of the metal centers – in the Fe(II) state, the emission of the sulforhodamine-B anion was quenched, whereas this deactivation pathway was blocked in the Fe(III) state. These authors also combined two different types of metal-to-ligand interactions in one material: Fe(II)–tpy bis-complexes and Pt(II) bis-acetylides in an alternating fashion [334]. Two geometrical isomers were assembled, a linear and a zigzag-shaped supramolecular polymer. However, only the former gave a reversible electrochromic response.

Kurth and coworkers studied the hierarchical self-assembly of their metallo-supramolecular polymers into supramolecular structures of higher orders. In particular, the polymer derived from ligand L1 and mainly Fe(II) ions was used in this context (Figure 4.40). The polycationic metallopolymers can be considered as polyelectrolyte-like materials and, thus, allowed the co-assembly with appropriate polyanions, such as poly(styrene sulfonate) (PSS), in a layer-to-layer-type fashion on PEI-coated substrates (PEI: polyethylenimine; Figure 4.41a) [298]. In another study, the PEI-coated substrates were replaced by latex particles, which acted as

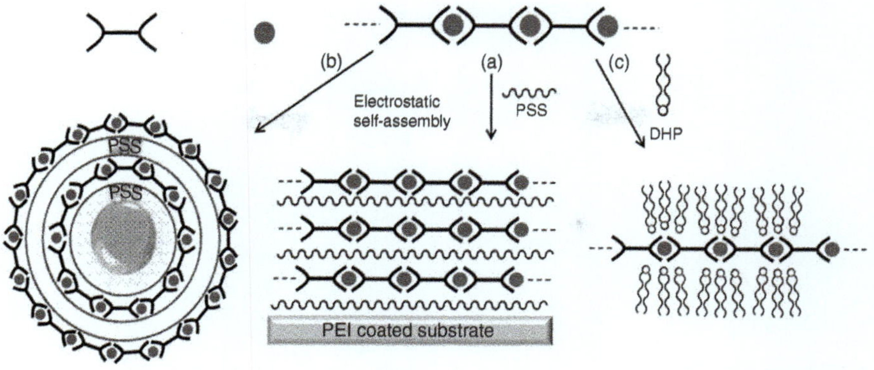

Figure 4.41 Representation of the hierarchical self-assembly of a metallo-supramolecular polymer by electrostatic self-assembly processes. (a) Formation of layer-by-layer assemblies, (b) formation of core-shell or hollow particles and (c) formation of metallopolymer-amphiphile complexes. PSS: poly(styrene sulfonate); PEI: polyethyleneimine; DHP: dihexadecyl phosphonate [298, 299, 308]. Source: Wild et al. [67]. © 2011 The Royal Chemical Society.

templates to generate core–shell architectures. PS and melamine–formaldehyde particles were used in this context; the latter ones gave, after decomposition of the templating polymer, stable hollow-shell particles (Figure 4.41b) [299]. Finally, polyelectrolyte–amphiphile complexes (PACs) could be formed by introducing amphiphilic counterions, such as dihexadecyl phosphonate (DHP, Figure 4.41c) [308]. These organo-soluble PACs also generated Langmuir-type monolayer at the air–water interface, which could be transferred onto solid supports applying the Langmuir–Blodgett (LB) method [311]. The alignment of the PACs on graphite surfaces, when coabsorbed with long-chain alkanes, was also demonstrated by Kurth et al. [310]. A temperature-responsive switching could be observed for the PACs based on Fe(II) ions [302, 306, 307, 312]: the Langmuir–Blodgett films as well as the crystalline samples exhibited a spin-crossover behavior, i.e. a reversible switch between a high-spin and low-spin material.

Very recently, Hossain and coworkers prepared a heterobimetallic metallo-supramolecular material that incorporated Ni(II) and Co(II) centers in an alternating fashion (L1 was used as ligand, Figure 4.40) [335]. When immobilized onto a glassy carbon electrode, this material functioned as an electrochemical nitrite detector. The setup showed a linear respond over a wide regime (from 2.5 μM to 1.73 mM) and offered a low detection limit of 0.45 μM.

The thermodynamics and kinetics for the complexation of Zn(II) ions by tpy ligands, as reported by Würtner and coworkers [86], revealed that Zn(II)–tpy bis-complexes exhibit an interesting combination of properties – reversibility of the complexation in solution (attributed to the low binding strength) and high stability in the solid state. These properties were utilized by the Würtner group to assemble metallo-supramolecular polymers from bis-terpyridine ligands, possessing fluorescent spacer units, and Zn(II) ions (e.g. **38a**, Figure 4.42) [277]. The promising optoelectronic properties of the resultant polymers (e.g. bright emission) suggested

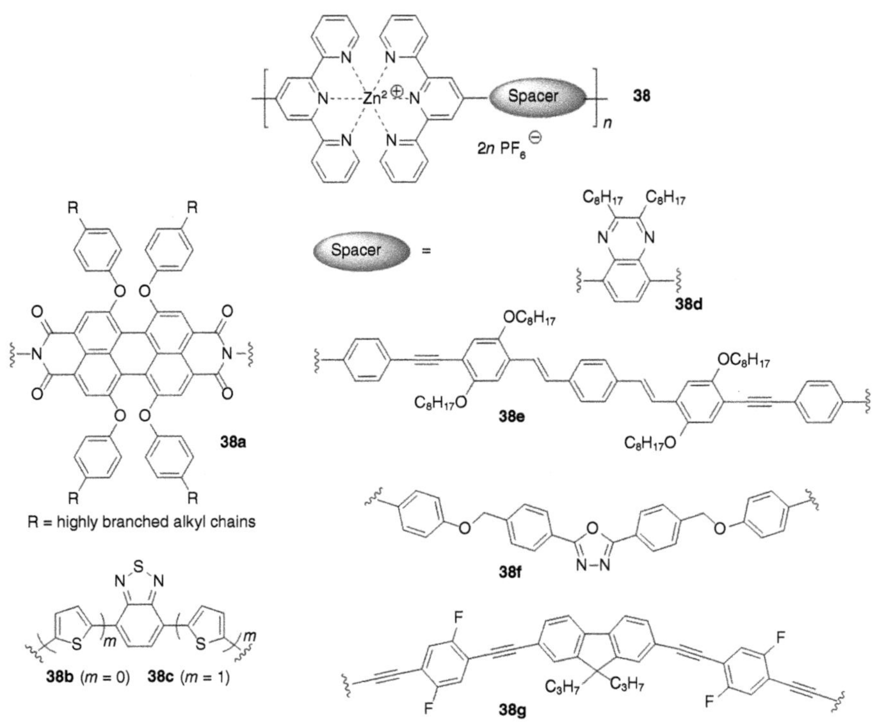

Figure 4.42 Schematic representation of selected metallo-supramolecular polymers assembled from Zn(II) ions and π-conjugated bis-terpyridine ligands.

a potential application as emissive layers in light-emitting diodes [278, 336]. Note that, the role of Zn(II) ions in these assemblies remained a structural one, as the photophysical properties of the organic chromophore were unaffected by the metal ions. In contrast, the fluorescence of the organic chromophores was fully quenched when Fe(II) ions were used for the metallo-supramolecular polymerization.

Inspired by these results many research groups have contributed further to this field, since then: Schubert (e.g. **38b–e**) [280–282], Che (e.g. **38f**) [279], and Lin (e.g. **38g**) [291–293] as well as others [337–345]. In all of these studies, π-conjugated segments, rigid as well as flexible ones, were used as spacers and a plethora of metallopolymers **38** incorporating Zn(II) ions have been prepared (some selected examples are shown in Refs. [2, 67]). As a general characteristic, all of these materials are easy to synthesize and exhibited a bright photoluminescence with the emission maximum simply depending on the nature of the π-conjugated chromophore (thus, the photophysical properties of the metallopolymers are tunable and predictable). Some representatives have been tested as emissive materials in electroluminescent devices. Concerning such applications, the efficient and reproducible processing from solution becomes relevant. It was shown by Friebe et al. that not only spin-coating but also inkjet printing, as much more material-saving technique can be applied for this purpose [284]. Moreover, the kinetic lability of these Zn(II)-containing materials was applied to fabricate thin films of statistical metallo-supramolecular copolymers via inkjet printing [287]. The authors printed three pure-colored inks (i.e. the metallopolymers **38b**, **38c**, and **38d** as red, green, and blue emitters, respectively) separately on top of each other. An additional solvent layer was added to facilitate the formation of statistical copolymers due to disassembly/reassembly processes. The resulting films showed bathochromically shifted emission bands attributed to π–π stacking interactions of the chromophore units. For different ratios of the three metallopolymers, the *x*,*y*-chromaticity coordinates of their emission colors were calculated according to the Commision Internationale l'éclairage (CIE) and enabled a fine-tuning of the emission colors (Figure 4.43). An optimum ratio of 10 : 10 : 80 for **38b**, **38c**, and **38d** was identified where the emission of white light was achieved (CIE coordinates: $x = 0.311$ and $y = 0.307$). This approach is comparable with other supramolecular approaches to assemble white-light-emitting materials [346], for example, the combination of three dyes equipped with functionalized with quadruple hydrogen-bonding entities (see Chapter 3 for details) [347]. A reasonable solubility of a material in an organic solvent represents a prerequisite to allow its reliable processing into thin-solid films. This requirement is not fulfilled for most of the aforementioned rigid linear metallo-supramolecular assemblies – in many cases, only a very low solubility was reported, even when long and/or branched alkyl chains were attached to the chromophores. Winter et al. demonstrated that the attachment of polymeric side chains to the chromophores within the metallopolymers is beneficial for all – the solubility, processability, and thin-solid film formation [288].

Noteworthy, not only 1D supramolecular polymers can be assembled from tpy-containing ligands and transition metal ions. Schlüter and coworkers employed phenyl-centered three- and six-armed terpyridine ligands to assemble 2D CONASHs at the air-to-water interface via a Langmuir–Blodgett technique [348, 349]. In contrast to their soluble 1D counterparts, the 2D assemblies were robust, free-standing films, which could be transferred onto solid supports. Using the differences in the kinetic stability of metal–tpy bis-complexes, a Zn(II)-containing CONASH, deposited onto a quartz slide, was (partially) transmetalated by immersing it into a solution of an Fe(II) salt [350]. More recently, Nishihara and coworkers fabricated a multilayer electrochromic device using a tris-terpyridine ligand and Fe(II) and Co(II) ions [351]. The deposition of the Fe(II)- and Co(II)-containing CONASH on an ITO-coated glass substrate gave a solidified electrochromic device which offered a dual-electrochromic behavior. These authors also employed CONASHs formed from Zn(II) ions and tris-terpyridine ligands – a phenyl- or triphenylamine-centered one – as photoluminescent materials [352]. The emission color could be modulated via the chemical structure of the organic core or the nature of the counterions. This, in turn, suggested that the anion actively contributed to the structure of the Zn(II)–tpy complex: the BF_4^- anion was non-coordinating, whereas for sulfate a $[Zn_2(\mu\text{-}O_2SO_2)(tpy)_2]$ motif was assumed. Park and coworkers fabricated a CONASHs containing a sub-phthalocyanine or phthalocyanine unit, as organic cores [353, 354]. These materials showed electrochromism with high

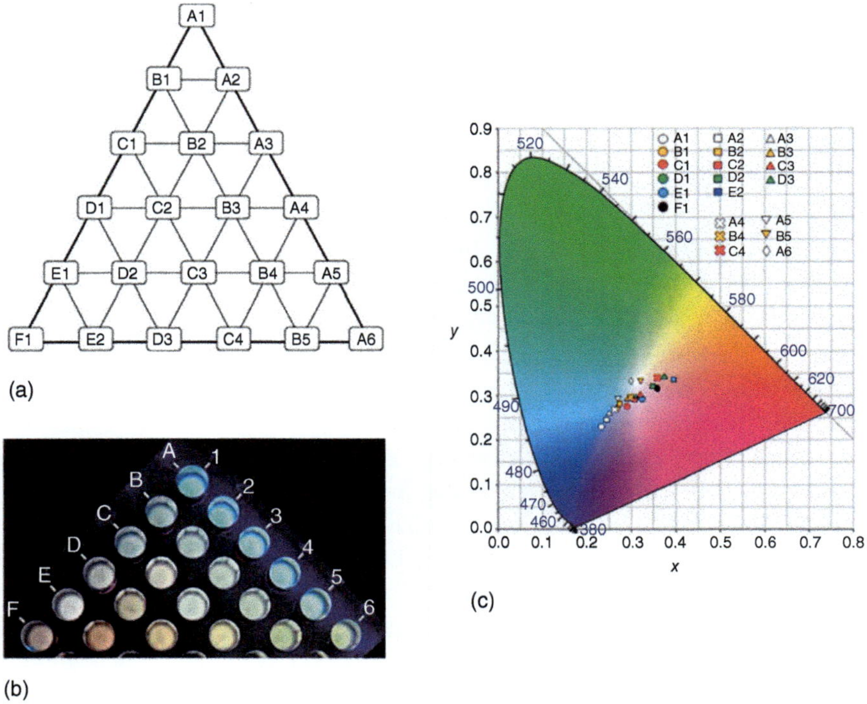

Figure 4.43 (a) Mixing triangle of the metallo-supramolecular polymers **38b**, **38c**, and **38d**. (b) Picture of the corresponding solutions in a quartz microtiter plate (excitation at $\lambda = 365$ nm). (c) Position of the observed emission according to the CIE color scheme. Source: Wild et al. [287]. Figure reproduced with kind permission. © 2013 The Royal Chemical Society.

transmittance and coloration efficiencies (reversible color change from blue to colorless).

Various chiral bis-terpyridine ligands have been used to assemble metallo-supramolecular polymers (the Abruña (**39**/**40**) [355, 356], Kimura (**41**) [357], and Kurth (**42**) [358] groups contributed to this field; Figure 4.44). Enantiomerically pure ditopic ligands were combined with Ru(II) or Fe(II) ions to afford helical superstructures, which were verified by circular dichroism (CD) spectroscopy and microscopy imaging.

Structural modification of the parent tpy motif afforded some new types of rigid ditopic ligands that were employed, as monomers, in metallo-supramolecular polymerizations (Figure 4.45). For example, the ditopic ligand **43** with tridentate 2,6-bis(1*H*-pyrazol-1-yl)pyridine units was reported by Chandrasekhar and coworkers; its metallo-supramolecular polymerization with Fe(II) ions afforded a material, which exhibited a reversible spin-transition at 50 °C as revealed by superconducting quantum interference device (SQUID) and Mössbauer measurements [359–361]. In a similar fashion, metallo-supramolecular polymers were assembled from the ditopic ligands **44** in the presence of Eu(III) or Ru(II) ions [362, 363].

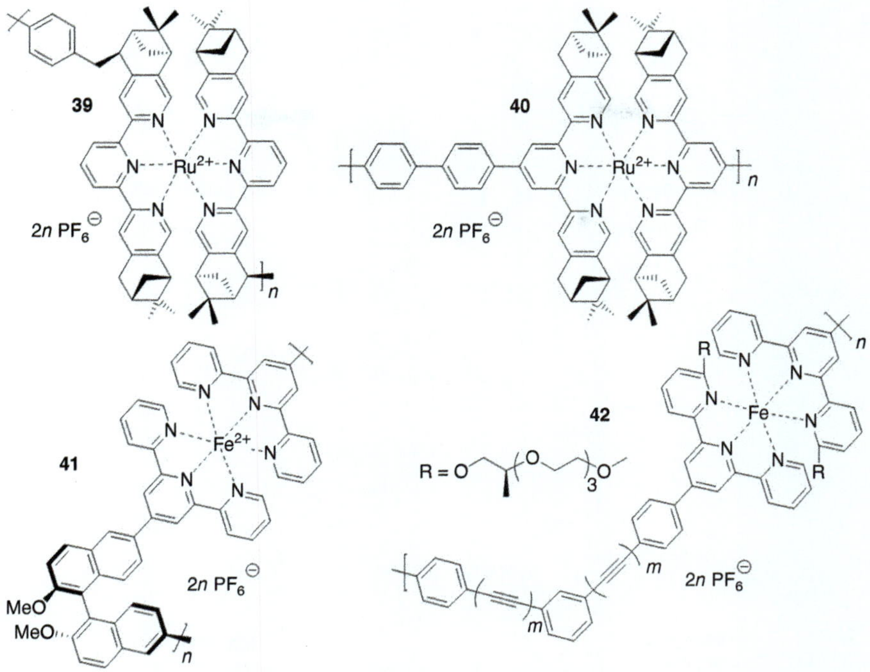

Figure 4.44 Schematic representation of the chiral Ru(II)- and Fe(II)-containing metallopolymers **39–42**.

Zhu and coworkers prepared the ditopic ligands **45** containing tridentate 2,6-bis(oxazol-2-yl)pyridine (pybox) binding sites (Figure 4.45) [364]. These ligands were applied in the metallo-supramolecular polymerization with Zn(II) ions. The polymers exhibited bright photoluminescence with emission colors ranging from blue to yellow (depending on the structure of the ligand). The polymers were tested, as fluorescent chemosensors, for the detection of nitroaromatics (the polymer derived from **45a** offered a remarkably low detection limit of 1.5 ppb for picric acid). The chiral variants of these ligands (**46**, Figure 4.45) were utilized by the same authors [365]. The resulting metallo-supramolecular polymers (containing Fe(II) or Zn(II) ions) were analyzed by temperature-dependent CD spectroscopy; these measurements revealed the dynamic nature of the metal-to-ligand coordination, which more expressed for the polymer containing ligand **46b**.

Finally, a different type of chelating moiety was reported by the Jäkle group [366]: PS was functionalized at both chain ends with tris(pyrazol-1-yl)borate (Tp) moieties. These entities feature a negatively charged boron center and three pyrazol-1-yl rings, acting as a strong tridentate ligand (the so-called "scorpionate" motive). The ditopic macroligand **47** was metallo-polymerized with Fe(II) ions, as shown in Figure 4.46. Complexes of the general type $[Fe(Tp)_2]$, being structurally and electronically similar to ferrocene (Fc), are of particular interest due to their spin-crossover behavior that can be triggered via substituents on the heteroaromatic rings [367]. Remarkably, the metallo-supramolecular polymer

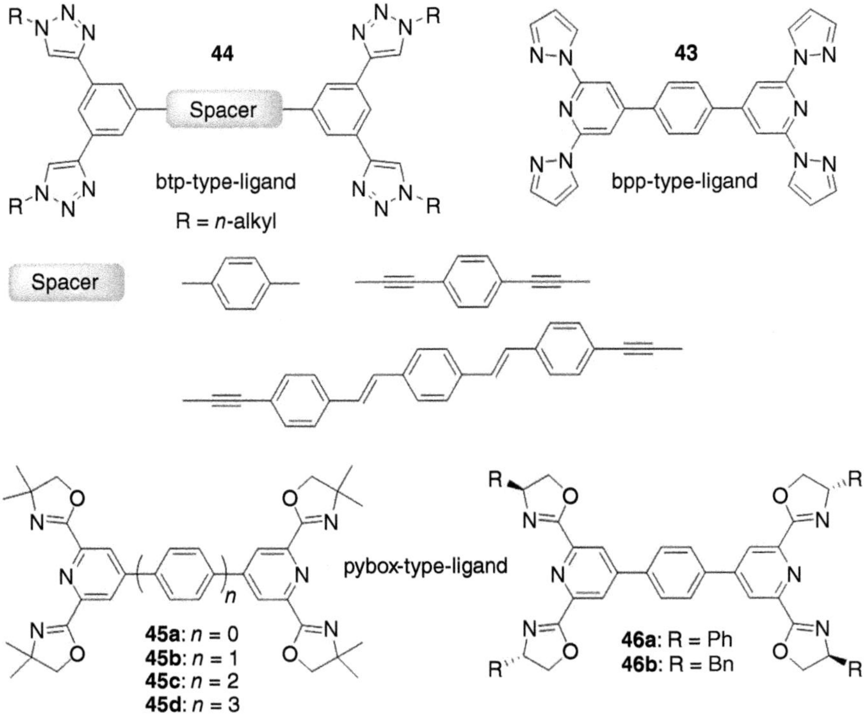

Figure 4.45 Schematic representation of the bpp-, btp-, and pybox-type ligands **43–46** used for the self-assembly with Fe(II), Eu(III), Ru(II), and Zn(II) ions.

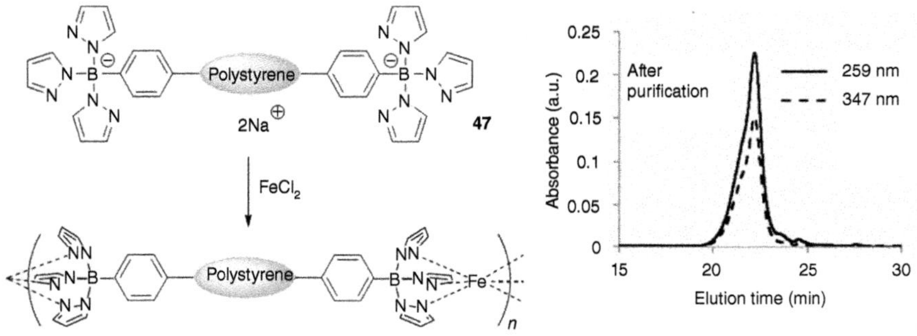

Figure 4.46 Schematic representation of the "scorpionate"-type metallopolymer [Fe(**47**)]$_n$, the SEC trace (UV detector) of the purified polymer is also shown. Source: Qin et al. [366]. Figure reproduced with kind permission. © 2012 Wiley-VCH.

could be investigated by SEC, a molar mass (M_n) of 21 000 g mol^{-1} ($Đ$ = 1.10) could be estimated (Figure 4.46). The high kinetic stability of the non-covalent interactions is revealed by the fact that similar systems, based on Fe(II)-tpy bis-complexes, are prone to dissociation under those measurements conditions (see also Chapter 12).

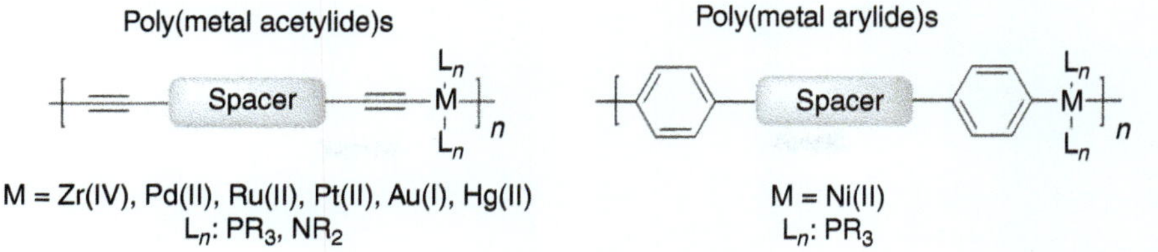

Figure 4.47 Schematic representation of the general structure of poly(metal acetylide)s and poly(metal arylide)s.

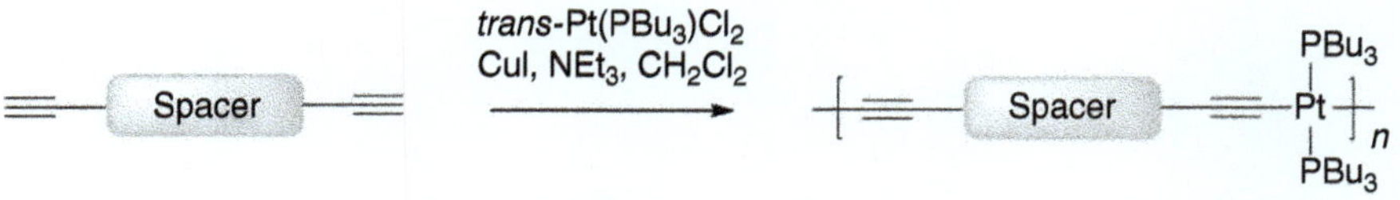

Figure 4.48 Schematic representation of the polyplatinyne synthesis, according to Hagihara's original protocol.

4.3.2 Metallo-supramolecular Polymers, Based on Covalent/Ionic Bonding

Metal-to-carbon bonds are known to exhibit a certain degree-of-covalency, at least in the case of the late and less electropositive transition metals. This fundamental concept from organometallic chemistry has also been adopted by polymer chemists and, accordingly, metallopolymers, based on such a bonding scenario, have been developed. In particular, the poly(metal acetylide)s [40, 43, 47, 368–376], and, though less common, poly(metal arylide)s [377] represent typical examples for such materials to be named in this context (Figure 4.47).

The main benefits of incorporating heavy metal ions, such as Pt(II) centers, into π-conjugated polymers can be found in the photophysical properties of these organometallic polymers: in general, the mixing of singlet and triplet excited states results in an efficient phosphorescence (expressed by high quantum yields and long lifetimes) [40]. The research on such rigid-rod polyplatinynes (**48**) was pioneered by the Hagihara group back in the 1970s. Basically, the synthesis involves a dehydrohalogenation between an ethinyl-functionalized organic chromophore with an appropriate transition metal precursor (Figure 4.48) [378–380]. Still this method represents the most straightforward and frequently used one in this field. Targeting the potential applications of polyplatinynes as active layers in optoelectronic devices or, more recently, also as precursor materials to fabricate metal nanoparticles, the Raithby, Wong, and Harvey groups prepared a broad range of such materials with fine-tuned photophysical properties from which a selection is depicted in Figure 4.49 (for more details see, e.g. Refs. [2, 40, 47, 381] and the literature cited therein).

Though Hagihara's original protocol can be basically employed for the synthesis of other poly(metal acetylide)s, e.g. incorporating Pd(II) or Rh(I) centers [382], some modified methods have been introduced by Lewis [383] and Lo Sterzo [384–386] groups, which are of much wider scope (Figure 4.50). For example, the

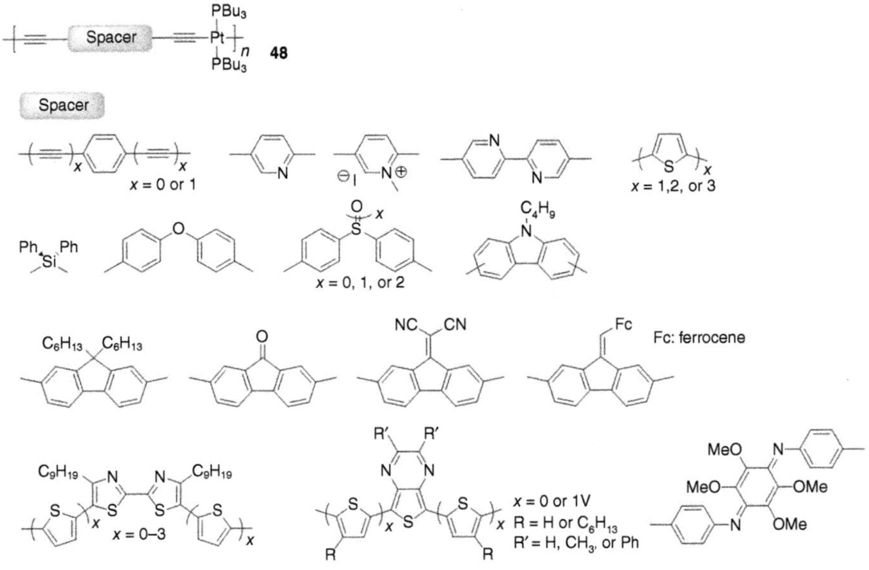

Figure 4.49 Schematic representation of selected examples of polyplatinynes highlighting the broad structural diversity reported in literature.

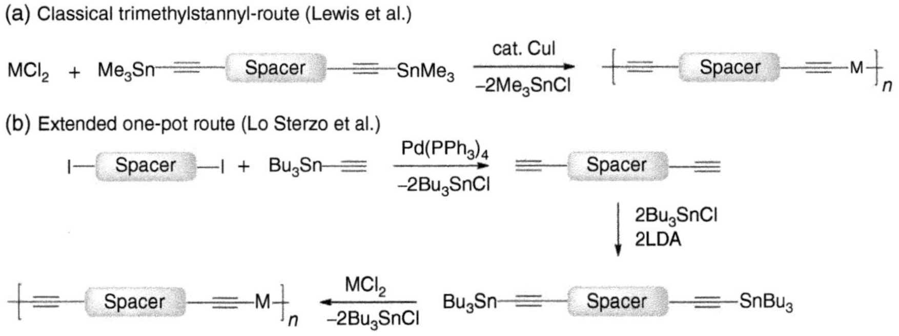

Figure 4.50 Schematic representation of the poly(metal acetylide)s synthesis via metathesis reactions: (a) classical route and (b) extended one-pot route.

original protocol cannot be used to prepare Ni(II)-containing polymers, due to the incompatibility of the involved dihalo complexes in the presence of amines and diethynyl derivatives under oxidizing conditions [387]. Both the modified methods both employ $SnMe_3$-activated aromatic diacetylides, as starting materials and the polyynes are formed by metathesis reactions with appropriate metal salts; poly(metal acetylide) with Fe(II), Ru(II), Os(II), Rh(II), Pd(II), and Ni(II) centers have thusly been obtained [383, 388–396].

Synthesis of poly(metal acetylide)s can be brought in line with the basic theories for (metallo-)supramolecular polymerizations (Chapter 1); however, can these materials be defined as metallo-supramolecular polymers in the strict sense of the meaning? Due to the highly covalent character of the carbon-to-metal bonds, one of the most important criteria to define a bond as supramolecular, i.e. reversibility, is not fulfilled in this case. Thus, the poly(metal acetylides) as well as other related types of organometallic polymers [e.g. poly(metal arylide)s, poly(metallacyclopentadienes), etc.] will not be discussed further within this chapter. Regarding their broad range of applications, the reader is invited to focus on specialized review articles published elsewhere [40, 47, 371, 372, 376, 381].

Basically, the same holds true for the Au(I)-containing metallopolymers featuring almost covalent carbon-to-gold bonds. Gold, in the oxidation state of +1, prefers a coordination number of +2 and, thus, represents an ideal candidate for the assembly into linear rigid-rod metallopolymers with suitable organic ligands. The ability of Au(I) to establish non-covalent Au···Au interactions due to London forces, the so-called aurophilic interactions, represents a remarkable property. The strength of this type of non-covalent interaction is presumably in the same range as that of single hydrogen bonds (5–10 kJ mol^{-1}) [154]. Puddephatt was the pioneer in the field of neutral Au(I)-containing polyynes **49**, which were prepared by reacting the polymeric [AuC≡C—Ar—C≡CAu]$_n$ precursor with *p*-diisocyanoarenes (Figure 4.51) [154]. Due to the supramolecular cross-linking of the metallopolymer chains via aurophilic interactions, the materials were almost insoluble but exhibited an intense solid-state emission behavior [397]. In a related study, Puddephatt and coworkers also employed ditopic phosphine-ligands, which gave the assemblies **50** with a slightly increased solubility and a more pronounced metallo-supramolecular character (Figure 4.51) [398]. A much more complex bonding scenario, involving contributions from p_π–p_π (CC), p_π–d_π (CP), and d_π–d_π (PAu) orbital interactions, resulted in a deviation from the linear molecular structure and, thus, resulted

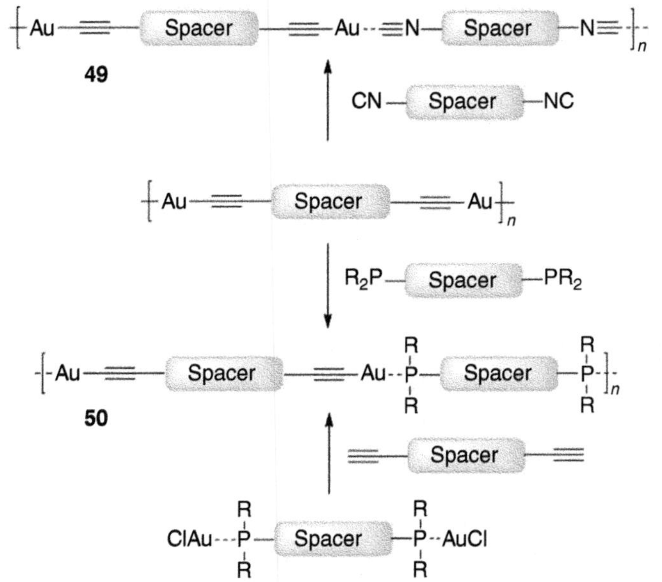

Figure 4.51 Schematic representation of Puddephatt's synthesis of Au(I)-containing polymers **49** and **50**.

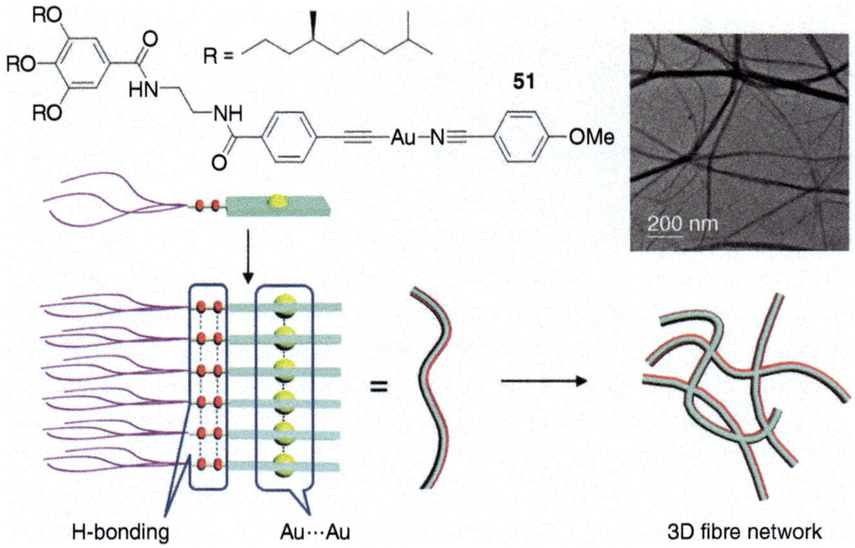

Figure 4.52 Schematic representation of the Au(I)-containing monomer **51** which assembled into supramolecular fibers due to aurophilic and H-bonding interactions. A representative TEM image of the obtained fiber network is also shown. Source: Chen et al. [405]. Figure reproduced with kind permission. © 2017 The Royal Chemical Society.

in a much weaker degree-of-conjugation along the polymer chains. Besides the aforementioned neutral metallopolymers, some polyanionic materials have also been reported, which include homometallic Au(I)-containing [399] as well as heterobimetallic Pt(II)–Au(I)-containing ones [400]. Another type of rigid-linear, neutral metallo-supramolecular polymer was obtained from pyridine-4-thiol, as a heteroditopic small-molecule ligand, and the Au(I) precursor complex under basic conditions [401]; however, the low solubility of the material, presumably of the composition $[Au(S\text{-}C_5H_4N)]_n$, did not allow any further structural characterization (e.g. by nuclear magnetic resonance [NMR] spectroscopy).

Recently, Wang and coworkers disclosed that metallophilic interactions, Pt· · ·Pt as well as Au· · ·Au ones, can be utilized to assemble supramolecular polymers, according to a cooperative nucleation–elongation mechanism (also see: Section 1.3.3) [402–407]. It was demonstrated that monomers, such as **51**, formed supramolecular fibers in nonpolar solvents due to metallophilic interactions, which were supported by H-bonding interactions; the entanglement of the fibers gave the final 3D network (Figure 4.52) [405]. The assemblies, obtained from Au(I)-containing monomers, were more resistant to changes in solvent polarity and indifferent against competing ligands, such as bpy, compared with their Pt(II)-containing counterparts [407]. This increase in stability was attributed to the low steric hindrance of the linear R—C≡C—Au—N≡C—R fragment, which allowed an efficient interaction between the metal centers.

In the former cases, the metallopolymers exclusively relied on single carbon-to-metal bonds. In addition to this, examples are also known where these monodentate

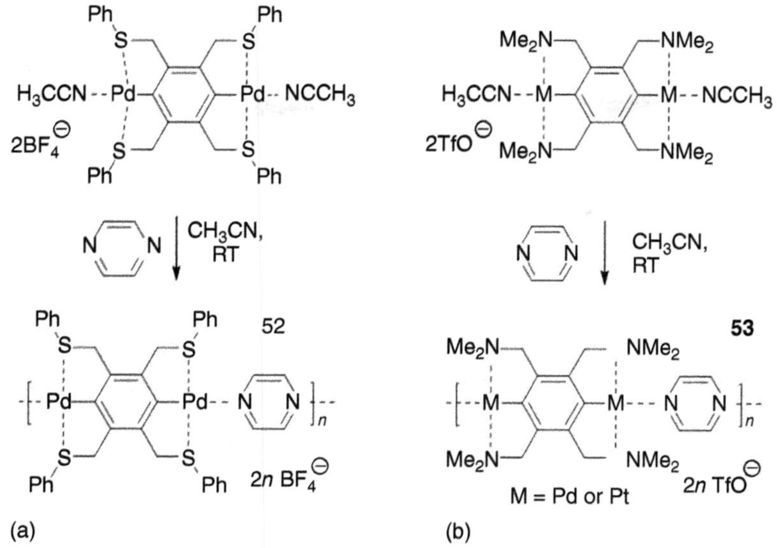

Figure 4.53 Schematic representation of metallo-supramolecular polymers based on pincer-type binding motives according to (a) Loeb and Shimidzu [408] was well as (b) van Koten et al. [409].

interactions (with a certain degree-of-covalency) were supported by additional coordinative ones. In particular, bi- or tridentate ligands, comprising cyclometallating binding sites, have to be named in this context. Such ligands can be seen as polypyridyl-type ligands (e.g. bpy, tpy, etc.) in which one of the pyridine rings has been replaced by benzene. Using these ligands in the coordination sphere of, e.g. Ru(II) or Ir(III) ions enables a strong manipulation of the photophysical properties of the resultant complexes. However, the preparation of metal-supramolecular polymers on the basis of such complexes has hardly been realized due to the required considerably harsher reaction conditions (i.e. high temperatures and long reaction times) and the accompanying (in the best cases) moderate yields [324]. The formation of linear metallopolymers incorporating such photoactive complexes is conveniently performed by (co)polymerizing preformed functionalized complexes (denoted as Method Vb in Figure 4.3) [61, 67].

A dinuclear pincer-type precursor complex was employed by Loeb and Shimidzu for the assembly of Pd(II)-containing metallo-supramolecular polymers in the presence of ditopic monodentate *N*-heteroaromatic ligands, such as 4,4′-bipyridine or pyrazine (**52**, Figure 4.53a). Pd(II) complexes always possess square-planar coordination geometries; thus, the self-assembly process gave strictly linear assemblies [408]. van Koten and coworkers basically applied the same strategy to assemble the metallo-supramolecular polymer **53** (Figure 4.53b) [409].

Yount et al. investigated the reversible self-assembly of polymers **51** (Figure 4.54) [410]. The authors demonstrated that the dynamics of the disassembly depended mainly on the structure of the substituents on the chelating amino groups (i.e. R = Me or Et). According to NMR spectroscopic studies, the K_a values remained

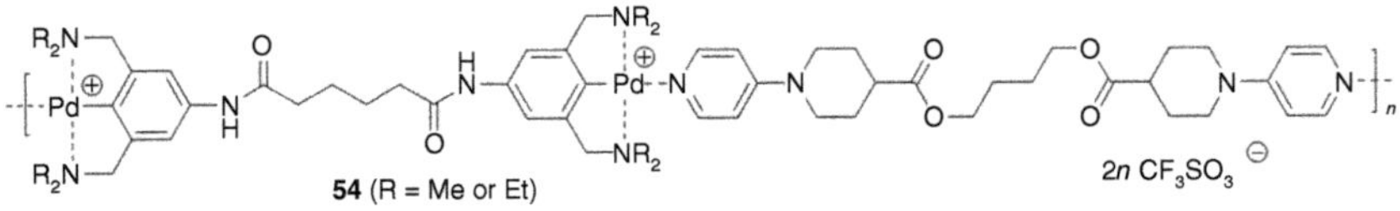

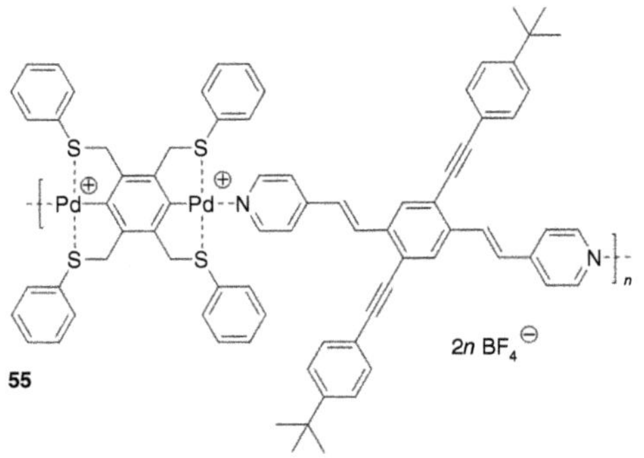

Figure 4.54 Schematic representation of two coordination polymers based on Pd(II)-pincer complexes according to the Craig (**51**) and the Weck groups (**52**), respectively.

almost constant in various solvents (i.e. $10^3\,M^{-1}$ in $CDCl_3$, MeOD, and $CDCl_3$, MeOD, DMSO, and D_2O/DMSO), whereas the rates for the disassembly differed by almost 2 orders of magnitude. These effects were rationalized by the different steric crowding around the Pd(II) centers. Besides this rather fundamental study, Weck and coworkers also targeted the application of metallo-supramolecular polymers in optoelectronic devices (e.g. polymer-based light-emitting diodes [PLEDs]), liquid crystal displays (LCDs), thin-film transistors, or optical storage devices [411, 412]. For example, the "supramolecular cruciform" **52** was assembled from a ditopic pyridyl-functionalized cruciform-shaped ligand and appropriate dinuclear pincer-type precursor complexes (Figure 4.54). These cross-conjugated materials featured high degrees of π-conjugation and were highly emissive. Moreover, the spatially orthogonal orientation of the highest occupied molecular orbitals (HOMO) and the lowest unoccupied molecular orbitals (LUMO) on the cruciform ligand allowed a remarkable tuning of the photophysical and electrochemical properties of the resultant metallopolymers by introducing electron-donating and -withdrawing substituents [413]. The disassembly of **52**, as induced by adding strong competitive ligands (e.g. PPh_3), could be monitored by UV/vis absorption spectroscopy [414]. Furthermore, the dependency of the DP on the stoichiometry of the employed building blocks was studied by viscosity measurements and gave the expected maximum in viscosity at the 1 : 1 ratio of the two components [412].

Back et al. envisioned *para*-ethynyl-substituted Pt(II) pincer complexes to be highly versatile AB-type monomers giving strictly linear metallo-supramolecular polymers, such as **56**, via self-condensation reactions (Figure 4.55a) [415]. This idea was taken up later by Schubert and coworkers who prepared the linear metallopolymer **57** by self-condensation of a protected Pt(II) precursor complex; according to SEC analysis, a DP of 13 was reached, corresponding to a molar mass (M_n) of c. 7100 g mol^{-1} (Figure 4.55b) [416].

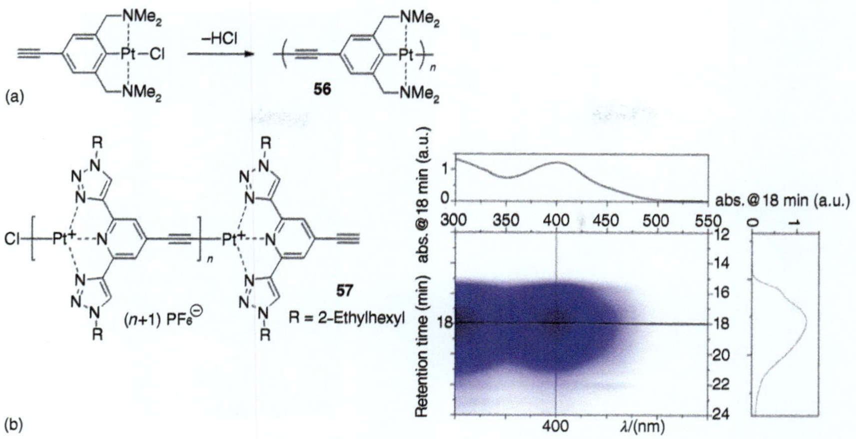

Figure 4.55 (a) Schematic representation of the proposed synthesis of metallopolymer **56** featuring pincer-type coordination and carbon-to-metal bonds. (b) Schematic representation of the metallopolymer **57** featuring bis(triazol-4-yl)pyridine coordination and carbon-to-metal bonds; a representation of the 3D SEC analysis is also depicted. Source: Schulze et al. [416]. Figure reproduced with kind permission. © 2017 American Chemical Society.

The carbon-centered interaction of organic ligands with metal centers represents not the only possibility for an ionic/covalent bonding; as common, heteroatoms (often as a part of multidentate ligands) are known to strongly bind to metal centers in an anionic fashion. The luminescent dyes of the dipyrrin family are currently in the focus of intense research and the boron-containing derivatives of the BODIPY type are the most important systems to be named (BODIPY: boron dipyrromethene) [417]. Besides, also the binding of divalent transition metal ions, such as Zn(II), to such ligands is also known to give complexes with rich photophysical properties [418]. Maeda et al. adopted this binding motive to assemble the charge-neutral, linear coordination polymers **55**, based on ditopic bis-dipyrromethane-type ligands (Figure 4.56) [419, 420]. The authors demonstrated that the size as well as the shape of the assemblies could be controlled via the geometry of the spacer moiety and solvent composition. For example, when using an angular ditopic ligand, spherical assemblies were formed by the self-assembly with Zn(II) ions in THF, whereas bell-shaped structures were identified in a THF/water mixture (2 : 11 ratio). Greenish solid-state emission was observed from the noncrystalline metallopolymer samples, which was consistent with the typical emission of Zn(II)-dipyrromethane dyes (λ_{em} of c. 540 nm).

The complexation of transition metal ions by β-diketonate derivatives is widely employed in the coordination chemistry of, e.g. Ir(III) and Pt(II) ions (ligands of the acac-type represent the most typical examples in this case; acac: acetoacetate) [40, 164]. The metallo-supramolecular cross-linking of polymers, possessing β-diketonate moieties in the side chains, with divalent transition metal ions has also been reported [421]. Besides, also structurally well-defined metallo-supramolecular polymers have been assembled from various ditopic polyesters, equipped with

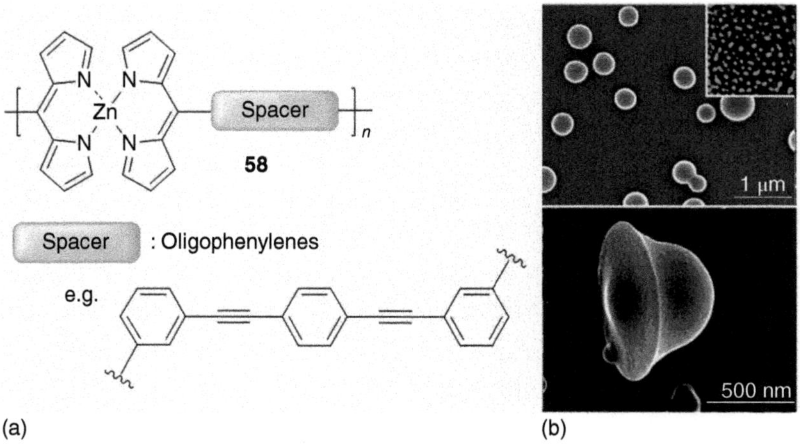

Figure 4.56 (a) Schematic representation of metallo-supramolecular polymers **58**, based on Zn(II)-dpm bis-complexes. (b) Scanning electron microscopy (SEM) images of the assembly **58** containing an angular ligand in THF (top, inset: 25 μm × 25 μm fluorescence micrograph) and from a 2 : 1 THF/water mixture (bottom). Source: Maeda et al. [419]. Figure reproduced with kind permission; © 2006 American Chemical Society.

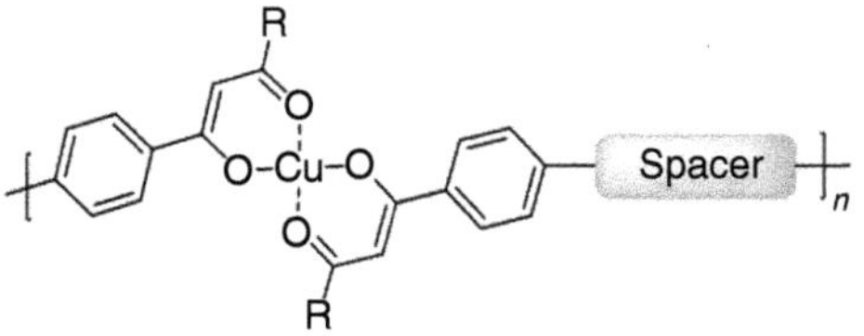

59
R: *n*-alkyl
Spacer: polyester

Figure 4.57 Schematic representation of the Cu(II)-containing chain-extended polymers **59** showing LC behavior.

1,3-diketonate end groups, in the presence of Cu(II) ions; these chain-extended polymers **59** featured thermotropic liquid-crystalline (LC) properties (Figure 4.57) [422]. Moreover, the Fraser group employed Eu(III) ions as well as various divalent transition metal ions in combination with polycaprolactone, bearing a β-diketonate moiety at one chain end, to assemble stable metallo-supramolecular star-shaped or linear polymers, respectively, which could even be analyzed by SEC regarding their molar masses [423].

Tris(8-hydroxyquinolinato)aluminum (Alq_3) plays a prominent role, as emitter material, in optoelectronic applications [424]. In addition, also metallopolymers incorporating this particular complex (or its Zn(II)-containing counterpart Znq_2) have attracted considerable interest. For example, the coordination of the ditopic π-conjugated ligand **60** to Zn(II) or Al(III) ions via its 8-hydroxyquinolinyl binding sites afforded linear and branched metallo-supramolecular polymers, respectively, of good solubility in many organic solvents (Figure 4.58). The polymers exhibited blue/green emission in solution and in the solid state; DPs of 11 and 32, respectively, were estimated by SEC analysis [425, 426].

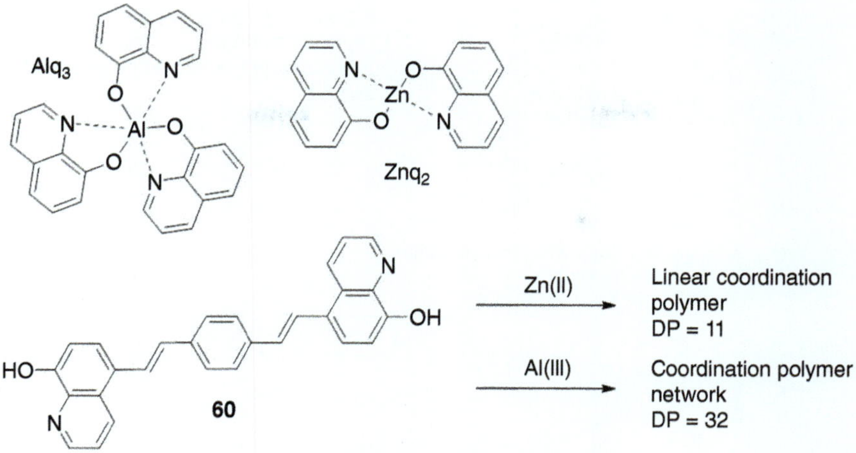

Figure 4.58 Schematic representation of the synthesis of linear and branched coordination polymers, incorporating luminescent Znq_2- or Alq_3-type complexes.

The somehow related homoleptic Zn(II) complexes with two salicylaldimine-type ligands are also of relevance due to their photo- and electroluminescent properties in concert with the high thermal stability. Che and coworkers synthesized a variety of linear metallopolymers **61** comprising such complexes within the main chain; the modularity of this synthesis allowed a high structural freedom (Figure 4.59): the authors employed different ditopic ligands in which the substituents on the N-atoms [427] as well as structure of the (non)conjugated spacer were varied [428]. Thereby, the photoluminescence of the materials could be adjusted; for example, the emission maximum was shifted to lower energies when the length of the π-conjugated spacer was increased. Moreover, PLED devices incorporating these metallopolymers were fabricated and gave, in all cases investigated, strong green electroluminescence. Structurally related to Hanabusa's assemblies (**59**, Figure 4.57) [422], Serrano and coworkers prepared mesogenic metallo-supramolecular polymers from α,ω-bis(salicylaldimine)-functionalized polyesters and Cu(II) ions (**62**, Figure 4.59) [429–431]. The scope of this method to prepare metallopolymers with thermotropic LC properties was extended later by Caruso et al. toward other types of metal centers: Ni(II), Pd(II), and V(IV) [432, 433]. A presumably 2D supramolecular network was obtained from the same ligand with La(III) ions; this material was hardly soluble in many organic solvents and even disassembled solvents, which could act as competitive ligands (e.g. DMSO) [434].

If two salicylaldimine units are linked via their N-atoms, tetradentate ligands of the so-called salen-type are formed [435]. Chen et al. investigated the applicability of ditopic salen-type ligands to assemble linear metallo-supramolecular polymers (some of these materials showed luminescent properties) [436–439]. It has to be noted, that a metal ion requires both a sufficiently high coordination number and size to allow the binding of two tetradentate ligands. Basically, this requirement is only met by the large rare-earth metal ions [i.e. Ln(III)] as well as a very few transition metal ions, such as Y(III) and Zr(IV). Y(III) and the lanthanides prefer a

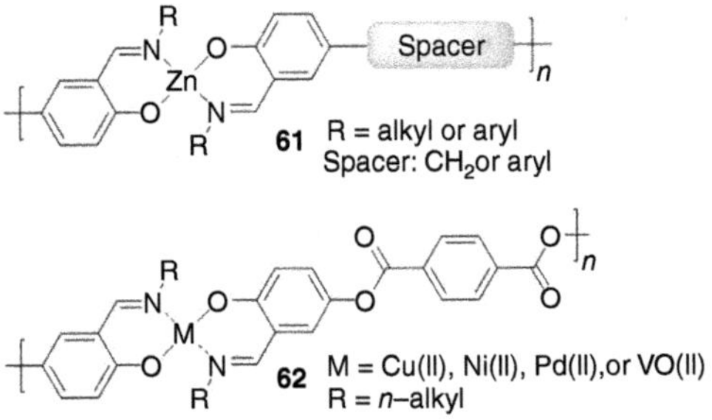

Figure 4.59 Schematic representation of metallo-supramolecular polymers **61** and **62**.

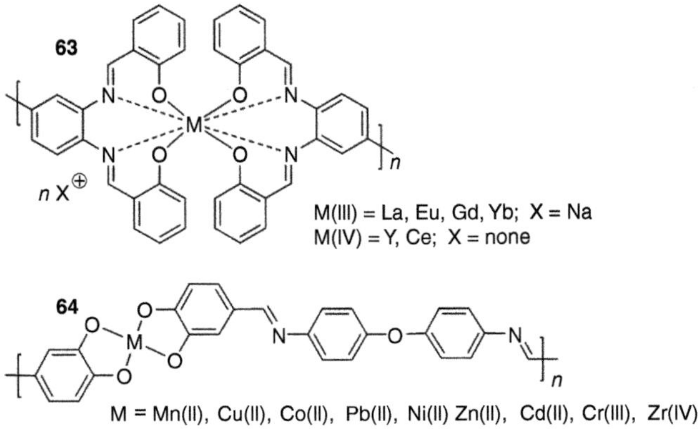

Figure 4.60 Schematic representation of metallo-supramolecular polymers, based on salen (**63**) and catechol bis-complexes (**64**).

square–antiprismatic coordination geometry; whereas, dodecahedral complexes are typically formed with Zr(IV) centers. The highly soluble metallopolymers **63**, e.g. containing Ce(IV) ions (Figure 4.60), were characterized by NMR spectroscopy and viscometry; remarkably, even SEC could be applied to estimate the molar masses (M_n) in the range of 30 000 g mol^{-1}. Moreover, good thermal stability and high T_g values could be evidenced by thermal analyses [438].

A linear ditopic ligand, equipped with two catechol-type binding sites, was employed by Kaya and Yíldírím [440]. The metallo-supramolecular polymers **64** (Figure 4.60) assembled from this ligand and a range of mainly divalent transition metal ions were studied regarding their conductive properties. From these, the Cu(II)- and Mn(II)-containing materials were highly conductive (though only after doping with I_2).

Carboxylates (along with some related functional groups) represent highly versatile building blocks that are widely used to obtain 1D, 2D, and 3D coordination compounds. Thereby, the number of carboxylate moieties on the organic molecule as well as the coordination number of the employed metal ion, serving as connector, dictates the resulting structure [73]. In many cases, such coordination compounds represent highly stable solid-state structures, which readily disassemble into small fragments upon exposure to solvents. However, examples for metallo-supramolecular polymers showing a reasonable kinetic stability in solution

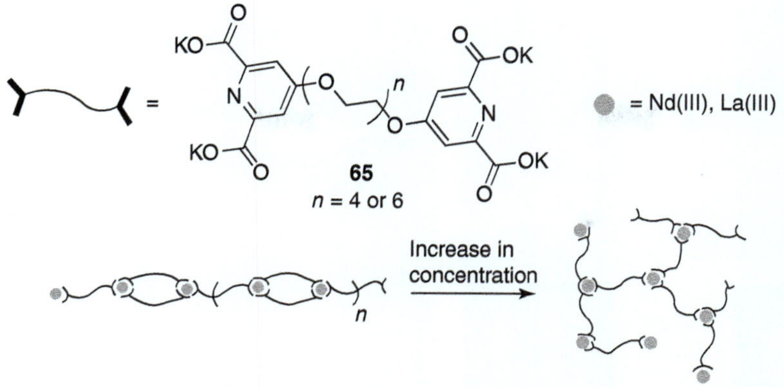

Figure 4.61 Structural characteristics of the coordination assemblies obtained from Ln(III) ions and **65**, as a function of the concentration. Source: Vermonden et al. [443]. Figure reproduced with kind permission; © 2004 Wiley-VCH.

have been reported. For example, the interaction of pyromellitic acid with Cu(II) ions afforded a water-soluble coordination network in which the Cu(II)–carboxylate complexes had to be stabilized by triethanolamine [441, 442]. Vermonden et al. focused on water-soluble coordination polymers assembled from the OEG-bridged (pyridine-2,6-dicarboxylic acid) **65** with Zn(II) and lanthanide ions [e.g. Nd(III) and La(III); Figure 4.61] [443, 444]. The tridentate nature of the binding site offered a much stronger binding of the metal ions and, thus, offered reasonable kinetic stability, even in water. The metallopolymers were analyzed in detail (i.e. viscometry, diffusion-ordered (NMR) spectroscopy [DOSY], and isothermal titration calorimetry [ITC]) to identify the influence of some parameters (i.e. concentration, temperature, and stoichiometry) on the resultant supramolecular structures, i.e. the concurrence between linear chain and macrocycles in the case of Zn(II) ions [or linear chains and 3D networks in the case of the Ln(III) ions, Figure 4.61].

The ditopic ligand **66** containing pentadentate binding sites gave, upon complexation with Eu(III) ions, the linear metallo-supramolecular polymer $K_n[Eu(\mathbf{66})]_n$ which exhibited a reversible vapoluminescent behavior, i.e. the emission of the free-standing polymer film could be "switched off" upon exposure to HCl gas and "switched on" again when treated with NEt_3 vapor (this process could be repeated numerous times without loss of response intensity, Figure 4.62) [445]. Based on the same type of metallo-supramolecular architecture, a proton-conductive material was obtained by Yoshida and coworkers (the ligand contained the same pentadentate binding site, but had a central dialkylfluorene moiety instead of dialkoxybenzene) [446]. The proton conductivity in the solid state was attributed to the presence of the coordinating carboxylic-acid groups, which facilitate the migration of H^+ (in control experiments, related polymers, based on bis-terpyridine ligands, gave a 10^4 times lower conductivity). Remarkably, the proton conductivity increased with increasing relative humidity and temperature – conductivities of 4.45×10^{-2} and 5.6×10^{-2} S cm^{-1} were observed at a relative humidity of 95% and 25 and 75 °C, respectively. The very low activation energy (E_a of 0.04 eV) was indicative

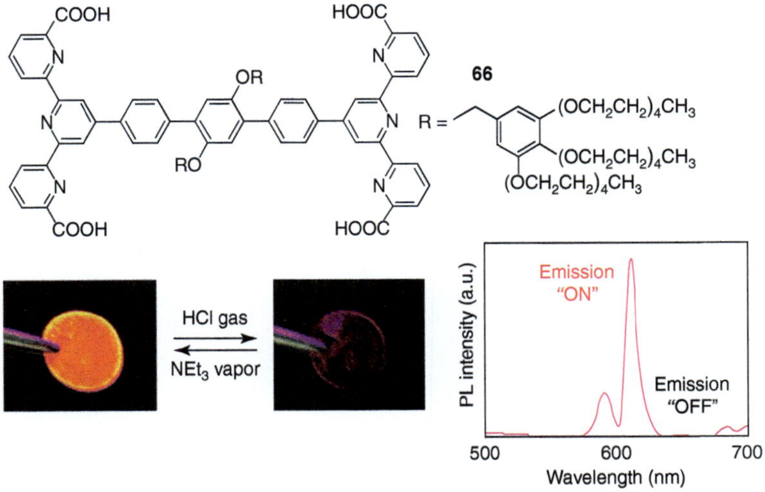

Figure 4.62 Schematic representation of the ditopic ligand, which was used for the complexation of Eu(III) ions. Photographs of the free-standing $K_n[Eu(\mathbf{66})]_n$ polymer film upon excitation with 365-nm-light – under NEt_3 vapor and upon exposure to HCl gas – are also shown. The comparison of the photoluminescence intensity reveals the switchability between the "ON" and "OFF" state. Source: Sato and Higuchi [445]. Figure reproduced with kind permission. © 2012 The Royal Chemical Society.

for a Grotthuss-type mechanism, i.e. proton conductivity resulting from a transport of H^+ through channels in the supramolecular material, which were filled by the moisture. Moreover, the proton conductivity was found to depend on the nature of the counter-cation [447]. The highest water-uptake was found for the lithium salt of the metallopolymer, which, in turn, accelerated the proton conductivity (1.54×10^{-2} S cm^{-1} at 95% relative humidity and 75 °C).

Recently, Sato and Higuchi employed the heteroditopic ligand **67** to assemble a metallo-supramolecular polymer, containing Zn(II) and Eu(III) centers in an alternating fashion [448]. The Eu(III) ions formed homoleptic bis-complexes with the pentadentate binding sites, whereas the Zn(II) ions favored the coordination of the tpy ligands (Figure 4.63). Due to an efficient ligand-to-metal energy-transfer process, the emission of white light (CIE coordinates of $x = 0.31$ and $y = 0.33$) was realized; in ethylene glycol solution, a photoluminescence (PL) quantum yield of 57% was reached.

Chandrasekhar and coworker reported on a metallo-supramolecular polymer with a remarkable molecular structure: **68** possessed ferrocene as well as triorganotin units, as parts of its backbone (Figure 4.64a) [449]. This kinetically highly labile metallopolymer, which readily disassemble in polar solvents, was prepared in a two-step protocol.

Besides all the aforementioned metallo-supramolecular polymers, which are based on the interaction of *N*- and/or *O*-coordinating ligands with (transition) metal ions, some examples have been reported in the literature, where such polymers were assembled from ditopic ligands via their sulfur atoms. Pearson's concept of

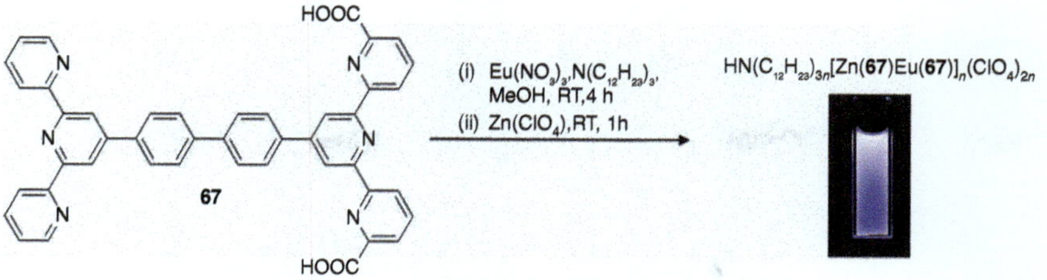

Figure 4.63 Schematic representation of the synthesis of a white-light-emitting heterobimetallic supramolecular polymer. Source: Sato and Higuchi [448]. Figure reproduced with kind permission. © 2019 Elsevier B.V.

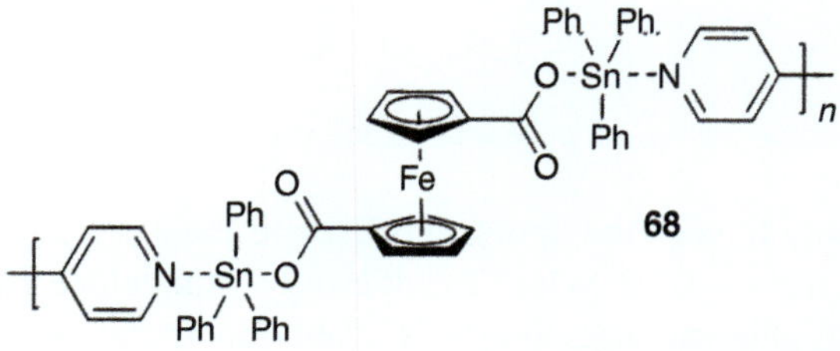

68

(a)

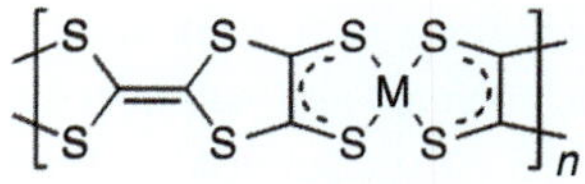

69

M = Fe(II), Pd(II), Pt(II), Cu(II), Ni(II)

(b)

M	σ (S cm^{-1})
Fe(II)	10^{-5}
Pd(II)	10^{-3}
Pt(II)	10^{-2}
Cu(II)	10^{-1}
Ni(II)	30

Figure 4.64 (a) Schematic representation of the heterobimetallic coordination polymer **68**. (b) Schematic representation of the tetrathiolate-based metallopolymers **69** and their electric conductivities.

"hard and soft acids and bases" (HSAB), predicts high oxo- and azo-philicities for the early transition metal ions; whereas, the late transition metal ions exhibit high affinities toward sulfur atoms [43]. Thus, a proper choice of the transition metal ion and the heteroatom-containing ligand is required to assemble stable supramolecular polymers with fine-tuned properties. Already in the late 1970s, Rivera et al. reported on the metallopolymers **69** containing a tetrathiolate-type ligand (Figure 4.64b) [450]. A proper choice of the incorporated transition metal ion enabled the authors to adjust the conductivity over a wide range from 10^{-5} to $30\,\mathrm{S\,cm^{-1}}$ (the systems containing Fe(II) and Ni(II) centers mark these extremes).

Other structurally related thiolate-type ligand systems were also reported [451–455], but the intrinsic low solubility and processability of these materials prevented any further applications. These materials were revisited quite recently in the context of the production of molecular hydrogen from water. In particular, the $[Co(bdt)_2]^-$ (bdt: 1,2-benzenedithiolate) has attracted considerable

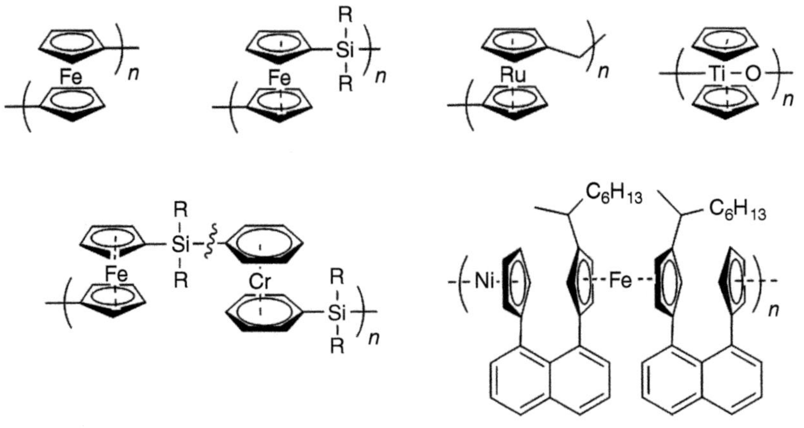

Figure 4.65 Schematic representation of typical polymetallocene structures.

attention due to its applicability in both the photochemical and the electrochemical hydrogen-evolution reactions (HER) [456]. To efficiently immobilize such complexes onto electrode or photocathode materials, their incorporation into stable polymers is required. Downes and Marinescu employed a liquid–liquid interfacial synthesis of a 1D metallopolymer starting from a Co(II) salt and benzene-1,2,4,5-tetrathiol [457]. The material was obtained as a black solid at the air–water interface (after evaporation of the organic solvent) and was successfully tested as active material in the (photo)electrochemical H_2 production. In the same context, the 2D metallopolymers were also formed from ligand molecules equipped with in a total six thiol groups [458, 459].

4.3.3 Metallo-supramolecular Polymers Based on Metal–Arene Interactions

Metallopolymers, which were derived from the arene π-complexation of metal ions, are the last class group of supramolecular materials to be discussed in this chapter. From the wide range organometallic polymers, these polymetallocenes maybe currently the most intensively studied ones: some representative structures are shown in Figure 4.65 (as a common feature, all structures contain arene sandwich-type complexes kinked by an organic [or silane] spacer).

Ferrocene (Fc), as the most prominent metallocene derivative, was first reported in the early 1950s [460, 461] and, only a few years later, Arimoto and Haven accomplished its incorporation into a polymer by the free radical polymerization (FRP) of vinylferrocene [462]. This report marks the date of birth of what is now known as organometallic polymers. The overwhelming majority of such metallopolymers is prepared from preformed sandwich- or half-sandwich-type complexes via the controlled polymerization of appropriate functional groups (i.e. according to the methods denoted as Va or Vb, Figure 4.3) [41, 43, 50, 55, 68]. However, only very few examples have been prepared by a direct "complexation" of metal centers by organic building blocks (i.e. using Method III, Figure 4.3).

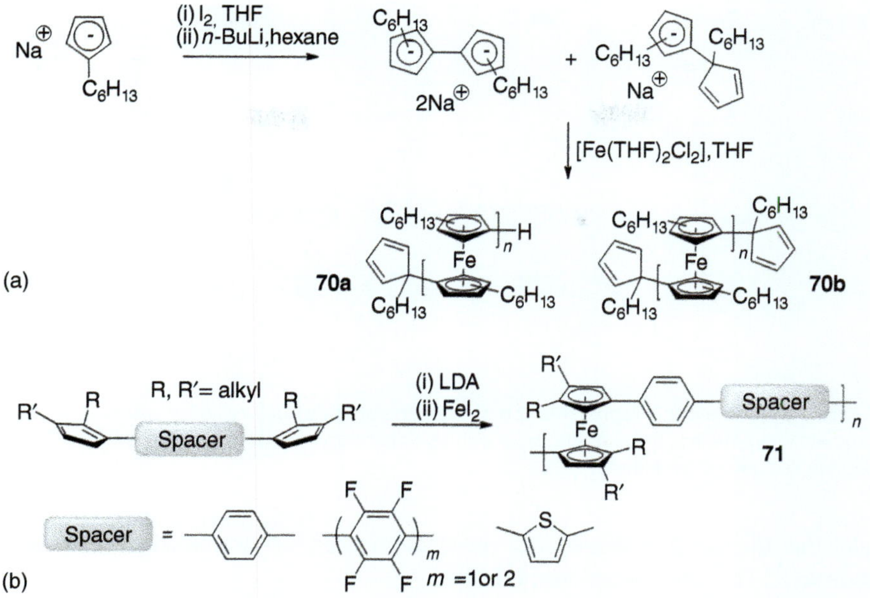

Figure 4.66 Schematic representation of the synthesis of the ferrocene-containing polymers **70** (a) and **71** (b) by a metallo-supramolecular polymerization. Source: Refs [465, 466].

With respect to this method, which features the characteristics of a typical metallo-supramolecular polymerization, appropriate ditopic ligands have to be reacted with suitable metal precursor salts/complexes. The pentafulvalene dianion represents the smallest possible ligand to be used in this context (the molecule comprised two cyclopentadienyl anions linked by a carbon-to-carbon bond). To overcome the poor solubility, reported by Neuse, for the unsubstituted material [463], *n*-hexyl substituents were introduced by Hirao et al. and used to prepare the reasonably soluble metallopolymer **70** (M_n = 4000 g mol^{-1}, $Đ$ = 1.2, Figure 4.66a) [464]. High-performance liquid chromatography (HPLC) was applied to separate single oligomeric species (n = 1–6) from the polymer ($n > 10$); thus, individual species could be investigated independently from each other by photophysical and electrochemical methods.

Southard and Curtis further developed the concept of metallo-supramolecular polymerization by arene π-complexation: these authors assembled the aryl-bridged metallopolymers **71** in a two-step synthesis; high molar masses (M_n values in the range of 42 000–50 000 g mol^{-1}) and very broad dispersities ($Đ$ values of 10–15) were estimated for the soluble materials **71** by SEC analysis (Figure 4.66) [465, 466]. Electrochemical studies revealed moderate Fe$\cdots$Fe interactions (i.e. two reversible oxidation waves with redox splitting ΔE of 0.17 and 0.19 V, respectively). Molecular modeling studies were conducted to verify this assumption [465]: the coupling between two metal centers was increased due to the presence of more polarizable groups, whereas less polarizable groups retarded such interactions. It was shown

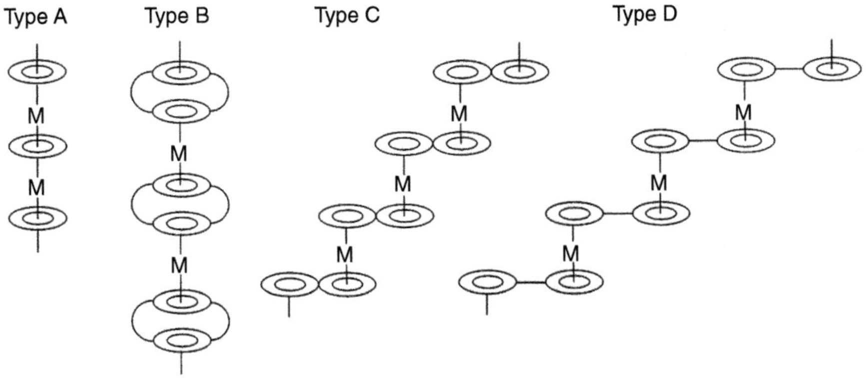

Figure 4.67 Schematic representation of the generalized structures of polydecker sandwich complexes (Types A and B) as well as linear polymetallocenes (Types C and D). Source: Redrawn from Manners [41].

later that the metallopolymer with the electron-poor perfluorinated spacer indeed showed no Fe· · ·Fe interactions [467, 468].

Besides the aforementioned linear metallopolymers, the so-called polydecker-type architectures represent one further class of metallocene-containing polymers that can basically be assembled via a quasi-supramolecular, self-assembly strategy. Their chemical structures, a generalized drawing of the two possible motives (Types A and B) is depicted in Figure 4.67, considerably deviation from the more common linear assemblies of Types C and D.

Siebert synthesized a range of type-A polydecker assemblies [469, 470]. For example, the polycondensation of substituted tris(allyl)dinickel-(μ-2,3-dihydro-1,3-diborolyl) complexes generated the Ni(II)-containing polydecker complexes **72**, as black products (Figure 4.68a); relatively harsh reaction conditions were required to obtain these materials (i.e. heating in vacuo at 200 °C) [471–473]. A Rh(II)-containing triple-decker precursor complex was used in another study. The thermal cleavage of one terminal 2,3-dihydro-1,3-diborole ligand generated a reactive double-stack which, in turn, assembled into the corresponding polydecker **73** (Figure 4.68b) [473]. These polydecker assemblies were found to be thermally stable but of very limited solubility in almost any organic solvent and highly sensitive toward air. Thus, a detailed characterization or even application of these materials remains problematic. One further example for such polydecker assemblies was shown by Grimes, who synthesized the tetradecker sandwich precursor with 2-chloro-4,5-diethyl-C_2B_3 units, as ligands. Subsequent deprotonation and metathesis reactions with an appropriate metal salt (i.e. $NiCl_2$ or $CoCl_2$) afforded the oligometallocenes **74** with up to 17 metal centers (Figure 4.69) [474].

Rosenblum and coworkers focused on the synthesis of type-B polydecker architectures (Figure 4.70), in which naphthalene or biphenyl spacers are used to orientate the metallocenes in a face-to-face-type manner [475–479]. In the initial studies, the Nigishi-type cross-coupling of 1,1′-bis(chlorozinc)-functionalized ferrocene (or ruthenocene) with 1,8-diiodonaphthalene was tested, but this method

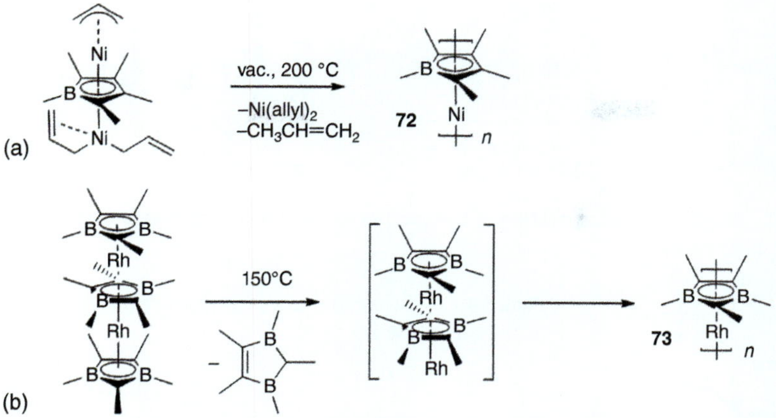

Figure 4.68 Schematic representation of the synthesis of type-A polydecker complexes incorporating (a) Ni(II) (**72**) and (b) Rh(II) centers (**73**).

Figure 4.69 Schematic representation of the oligometallocene **74**.

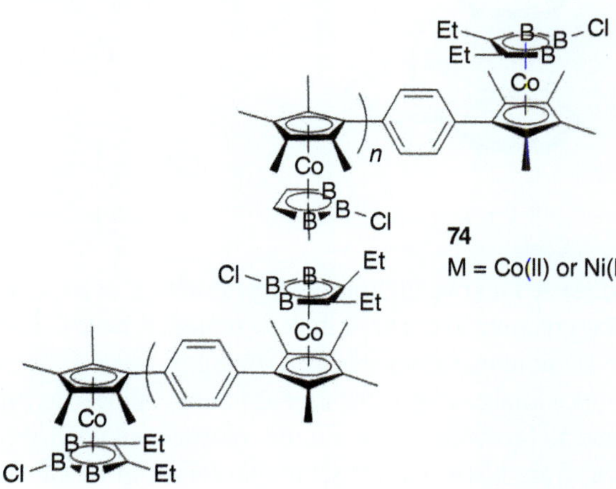

74
M = Co(II) or Ni(II)

gave only oligomers, for which a molar mass (M_n) of below 4000 g mol^{-1} was estimated by viscosity measurements [475, 476]. The authors then introduced solubilizing side chains on the cyclopentadienyl ring and changed the synthetic strategy – the metallo-supramolecular polymerization of the ferrocene-centered monomer with Fe(II) ions in the presence of a strong base occurred at room temperature (Figure 4.70); these modifications enabled the synthesis of polymetallocenes **75** in good yield and with high molar masses ($M_n \leq 18\,000$ g mol^{-1}) [477]. This is by-far the more applicable route that not only allowed the assembly of the Fe(II)- and Ni(II)-containing homometallic polymers, but also their heterobimetallic Fe(II)/Ni(II) or Fe(II)/Co(II) counterparts with a strictly alternating order of the two types of metal centers [479]. Rosenblum et al. investigated the conductive and magnetic properties of these remarkable metallo(co)polymer structures. The iodine-doped samples showed electrical conductivities in the range of 6.7×10^{-3} S cm^{-1} [479]. The solid-state structure of these metallopolymers, as evidenced from the X-ray single-crystal analysis of isolated oligomeric samples,

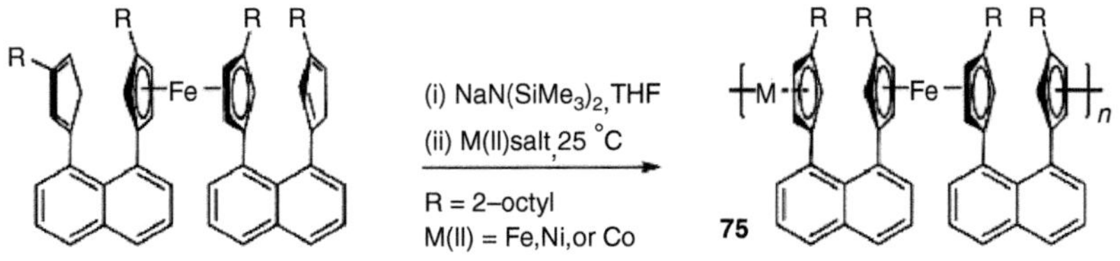

Figure 4.70 Schematic representation of the synthesis of the homo- and heterometallic polydecker assemblies **75**.

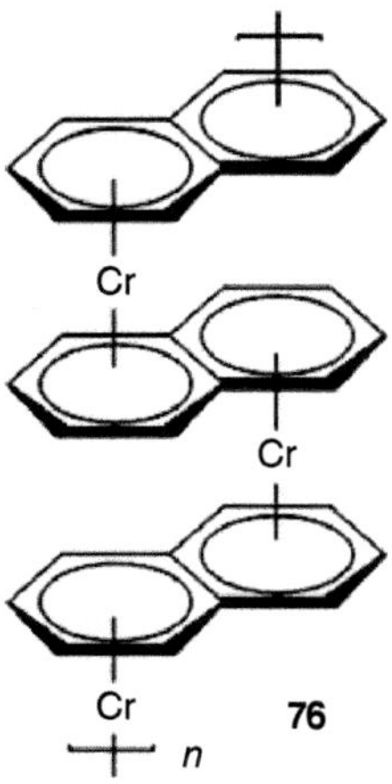

Figure 4.71 Schematic representation of the type-C polymetallocene **76**.

suggested an overall helical superstructure of the chains (i.e. 45° rotation of the Cp rings relative to the naphthalene moieties, helical diameter of c. 68 Å corresponding to 16 monomer units per turn) [476].

As a final example, the type-C polymetallocenes, reported by Bush and Lagowski, have to be noted. These authors synthesized poly[(μ-η^6,η^6-naphthalene)chromium] (**76**) from bis(η^6-naphthalene)chromium in a ligand exchange (Figure 4.71) [480]. The proposed highly regular all-eclipsed conformation of **76** was concluded from NMR spectroscopic studies in liquid ND_3.

4.4 Concluding Remarks

Metallo-supramolecular polymers represent an important group within the broad field of supramolecular polymers due to their unique combination of the special properties arising from metal-to-ligand coordination with the features of conventional polymer science. A diversity of methods is known, which allows the synthesis of such main-chain metallopolymers; herein, only those examples have been included, where the polymer chains have been formed via metallo-supramolecular chemistry, i.e. by the direct interaction of the metal centers with appropriate organic ligand molecules. In this context, coordinative bonding, ionic/covalent bonding, and arene π-complexation, as the three fundamental types of metal-to-ligand interactions, have been summarized. Polymers with metal centers, as integral parts of the backbone, can of course also be prepared via other methods; however, their coverage is beyond the scope of this chapter. The ring-opening polymerization of strained metallocenes and the cross-coupling of preformed metal complexes are two examples to be named in this context [40, 61, 68, 165]. The emerging field of electrochemical polymerization [481], directly depositing thin polymer films onto electrode or semiconductor surfaces represents the third prominent example for the strategy "polymerization of complexes" (referred to as Method V in Figure 4.3).

Not only has the synthetic toolbox to prepare these metallo-supramolecular polymers been tremendously expanded within the past decades, but also significant improvements have been made regarding their structural characterization (see Chapter 12). This, in turn, represents a prerequisite with respect to all potential applications that require well-defined polymers.

Utilizing metallo-supramolecular polymers, i.e. bringing these materials "into action," is today the main driving force for continuous research in this field (going far beyond – just offering new types of structures to the scientific community). The most relevant topic, which is the focus of interest, is the implementation of metallopolymers, as self-healing [30], opto-electronic [40, 61, 165], or electrochromic materials [325, 482, 483]. In addition to this, metal-containing polymers emerge as templating precursor materials for the fabrication of bi- or trimetallic nanoparticles, which show novel conductive or magnetic properties [1]; as well as their role in future biomedical and pharmaceutical applications, i.e. the formulation of therapeutically active metallodrug–polymer conjugates is worth mentioning at this stage [484].

References

1 Whittle, G.R., Hager, M.D., Schubert, U.S., and Manners, I. (2011). *Nat. Mater.* 10: 176–188.

2 Winter, A. and Schubert, U.S. (2016). *Chem. Soc. Rev.* 45: 5311–5357.

3 Dobrawa, R. and Würthner, F. (2005). *J. Polym. Sci., Part A: Polym. Chem.* 43: 4981–4995.

4 Eryazici, I., Moorefield, C.N., and Newkome, G.R. (2008). *Chem. Rev.* 108: 1834–1895.

5 Liu, H.-K. and Sadler, P.J. (2011). *Acc. Chem. Res.* 44: 349–359.

6 Gottschaldt, M. and Schubert, U.S. (2009). *Chem. Eur. J.* 15: 1548–1557.

7 Gill, M.R. and Thomas, J.A. (2012). *Chem. Soc. Rev.* 41: 3179–3192.

8 Bogani, L. and Wernsdorfer, W. (2008). *Nat. Mater.* 7: 179.

9 Sorace, L., Benelli, C., and Gatteschi, D. (2011). *Chem. Soc. Rev.* 40: 3092–3104.

10 Gatteschi, D., Bogani, L., Cornia, A. et al. (2008). *Solid State Sci.* 10: 1701–1709.

11 Cassoux, P., Valade, L., Kobayashi, H. et al. (1991). *Coord. Chem. Rev.* 110: 115–160.

12 Robertson, N. and Cronin, L. (2002). *Coord. Chem. Rev.* 227: 93–127.

13 Di Bella, S. (2001). *Chem. Soc. Rev.* 30: 355–366.

14 Evans, O.R. and Lin, W. (2002). *Acc. Chem. Res.* 35: 511–522.

15 Chen, C.-T. and Suslick, K.S. (1993). *Coord. Chem. Rev.* 128: 293–322.

16 Hang, T., Zhang, W., Ye, H.-Y., and Xiong, R.-G. (2011). *Chem. Soc. Rev.* 40: 3577–3598.

17 Grätzel, M. (2001). *Nature* 414: 338.

18 Nazeeruddin, M.K., Klein, C., Liska, P., and Grätzel, M. (2005). *Coord. Chem. Rev.* 249: 1460–1467.

19 Bailey, D.C. and Langer, S.H. (1981). *Chem. Rev.* 81: 109–148.

20 Buchmeiser, M.R. (2009). *Chem. Rev.* 109: 303–321.

21 Leadbeater, N.E. and Marco, M. (2002). *Chem. Rev.* 102: 3217–3274.

22 Guiglion, P., Butchosa, C., and Zwijnenburg, M.A. (2016). *Macromol. Chem. Phys.* 217: 344–353.

23 Happ, B., Winter, A., Hager, M.D., and Schubert, U.S. (2012). *Chem. Soc. Rev.* 41: 2222–2255.

24 Baytekin, B., Baytekin, H.T., and Schalley, C.A. (2006). *Org. Biomol. Chem.* 4: 2825–2841.

25 Guo, K., Guo, Z., Ludlow, J.M. et al. (2015). *Macromol. Rapid Commun.* 36: 1539–1552.

26 Liu, Y., Wang, Z., and Zhang, X. (2012). *Chem. Soc. Rev.* 41: 5922–5932.

27 Pastor, A. and Martinéz-Viviente, E. (2008). *Coord. Chem. Rev.* 252: 2314–2345.

28 Rasa, M. and Schubert, U.S. (2006). *Soft Matter* 2: 561–572.

29 Yang, L., Tan, X., Wang, Z., and Zhang, X. (2015). *Chem. Rev.* 115: 7196–7239.

30 Hager, M.D., Greil, P., Leyens, C. et al. (2010). *Adv. Mater.* 22: 5424–5430.

31 Eloi, J.-C., Chabanne, L., Whittle, G.R., and Manners, I. (2008). *Mater. Today* 11: 28–36.

32 Abd-El-Aziz, A.S. (2002). *Macromol. Rapid Commun.* 23: 995–1031.

33 Humphrey, M.G. (2004). *Macromol. Symp.* 209: 1–22.

34 Andres, P.R. and Schubert, U.S. (2004). *Adv. Mater.* 16: 1043–1068.

35 Hofmeier, H. and Schubert, U.S. (2004). *Chem. Soc. Rev.* 33: 373–399.

36 Hofmeier, H. and Schubert, U.S. (2005). *Chem. Commun.*: 2423–2432.

37 Holliday, B.J. and Swager, T.M. (2005). *Chem. Commun.*: 23–36.

38 Kingsborough, R.P. and Swager, T.M. (2007). *Prog. Inorg. Chem.* 48: 123–231.

39 Manners, I. (2001). *Science* 294: 1664–1666.

40 Ho, C.-L. and Wong, W.-Y. (2011). *Coord. Chem. Rev.* 255: 2469–2502.

41 Manners, I. (2004). *Synthetic Metal-Containing Polymers*. Weinheim: VCH-Wiley.

42 Newkome, G.R., He, E., and Moorefield, C.N. (1999). *Chem. Rev.* 99: 1689–1746.

43 Nguyen, P., Gómez-Elipe, P., and Manners, I. (1999). *Chem. Rev.* 99: 1515–1548.

44 Schubert, U.S., Newkome, G.R., and Manners, I. (eds.) (2006). *Metal-Containing and Metallosupramolecular Polymers and Materials*. Washington, DC: American Chemical Society.

45 Abd-El-Aziz, A.S. and Manners, I. (2007). *Frontiers in Transition Metal-Containing Polymers*. Hoboken, NJ: Wiley.

46 Williams, K.A., Boydston, A.J., and Bielawski, C.W. (2006). *Chem. Soc. Rev.* 36: 729–744.

47 Wong, W.-Y. (2007). *Dalton Trans.*: 4495–4510.

48 Whittle, G.R. and Manners, I. (2007). *Adv. Mater.* 19: 3439–3468.
49 Stanley, J.M. and Holliday, B.J. (2012). *Coord. Chem. Rev.* 256: 1520–1530.
50 Hardy, C.G., Zhang, J., Yan, Y. et al. (2014). *Prog. Polym. Sci.* 39: 1742–1796.
51 Ramanathan, M. and Darling, S.B. (2013). *Polym. Int.* 62: 1123–1134.
52 Chiper, M., Hoogenboom, R., and Schubert, U.S. (2009). *Macromol. Rapid Commun.* 30: 565–578.
53 Schubert, U.S. and Eschbaumer, C. (2002). *Angew. Chem. Int. Ed.* 41: 2892–2926.
54 Hoogenboom, R., Fournier, D., and Schubert, U.S. (2008). *Chem. Commun.*: 155–162.
55 Rehahn, M. (1998). *Acta Polym.* 49: 201–224.
56 Friese, V.A. and Kurth, D.G. (2008). *Coord. Chem. Rev.* 252: 199–211.
57 Friese, V.A. and Kurth, D.G. (2009). *Curr. Opin. Colloid Sci.* 14: 81–93.
58 Fustin, C.A., Guillet, P., Schubert, U.S., and Gohy, J.F. (2007). *Adv. Mater.* 19: 1665–1673.
59 Gohy, J.-F. (2009). *Coord. Chem. Rev.* 253: 2214–2225.
60 Shunmugam, R., Gabriel, G.J., Aamer, K.A., and Tew, G.N. (2010). *Macromol. Rapid Commun.* 31: 784–793.
61 Ulbricht, C., Beyer, B., Friebe, C. et al. (2009). *Adv. Mater.* 21: 4418–4441.
62 Elemans, J.A.A.W., van Hameren, R., Nolte, R.J.M., and Rowan, A.E. (2006). *Adv. Mater.* 18: 1251–1266.
63 Zhang, X. and Wang, C. (2011). *Chem. Soc. Rev.* 40: 94–101.
64 Li, S.-L., Xiao, T.-X., Lin, C., and Wang, L.-Y. (2012). *Chem. Soc. Rev.* 41: 5950–5968.
65 Yan, X., Wang, F., Zheng, B., and Huang, F. (2012). *Chem. Soc. Rev.* 41: 6042–6065.
66 Ciferri, A. (2010). *Chem. Eur. J.* 16: 10930–10945.
67 Wild, A., Winter, A., Schlütter, F., and Schubert, U.S. (2011). *Chem. Soc. Rev.* 40: 1459–1511.
68 Abd-El-Aziz, A.S., Shipman, P.O., Boden, B.N., and McNeil, W.S. (2010). *Prog. Polym. Sci.* 35: 714–836.
69 Abd-El-Aziz, A.S. and Strohm, E.A. (2012). *Polymer* 53: 4879–4921.
70 Wang, Y., Astruc, D., and Abd-El-Aziz, A.S. (2019). *Chem. Soc. Rev.* 48: 558–636.
71 Hagrman, P.J., Hagrman, D., and Zubieta, J. (1999). *Angew. Chem. Int. Ed.* 38: 2638–2684.
72 Robin, A.Y. and Fromm, K.M. (2006). *Coord. Chem. Rev.* 250: 2127–2157.
73 Kitagawa, S., Kitaura, R., and Noro, S.-i. (2004). *Angew. Chem. Int. Ed.* 43: 2334–2375.
74 Janiak, C. (2003). *Dalton Trans.*: 2781–2804.
75 James, S.L. (2003). *Chem. Soc. Rev.* 32: 276–288.
76 Zhou, H.-C., Long, J.R., and Yaghi, O.M. (2012). *Chem. Rev.* 112: 673–674.
77 Batten, S.R., Champness, N.R., Chen, X.-M. et al. (2012). *CrystEngComm* 14: 3001–3004.
78 Murray, L.J., Dinca, M., and Long, J.R. (2009). *Chem. Soc. Rev.* 38: 1294–1314.

79 Czaja, A.U., Trukhan, N., and Müller, U. (2009). *Chem. Soc. Rev.* 38: 1284–1293.
80 Langmi, H.W., Ren, J., North, B. et al. (2014). *Electrochim. Acta* 128: 368–392.
81 Suh, M.P., Park, H.J., Prasad, T.K., and Lim, D.-W. (2011). *Chem. Rev.* 112: 782–835.
82 Wang, B., Xie, L.-H., Wang, X. et al. (2018). *Green Energy Environ.* 3: 191–228.
83 Furukawa, H., Cordova, K.E., O'Keeffe, M., and Yaghi, O.M. (2013). *Science* 341: 1230444.
84 Winter, A., Hager, M.D., and Schubert, U.S. (2012). Supramolecular polymers. In: *Polymer Science: A Comprehensive Review*, vol. 5 (eds. H.-W. Schmidt and M. Ueda), 269–310. Amsterdam: Elsevier B.V.
85 Ciardelli, F., Tsuchida, E., and Wöhrle, D. (eds.) (1996). *Macromolecule-Metal Complexes*. Berlin: Springer-Verlag.
86 Dobrawa, R., Ballester, P., Saha-Möller, C.R., and Würthner, F. (2006). Thermodynamics of 2,2′:6′,2″-terpyridine-metaö ion complexes. In: *Metal-Containing and Metallosupramolecular Polymers and Materials*, vol. 928 (eds. U.S. Schubert, G.R. Newkome and I. Manners), 43–62. Washington, DC: American Chemical Society.
87 Holyer, R.H., Hubbard, C.D., Kettle, S.F.A., and Wilkins, R.G. (1966). *Inorg. Chem.* 5: 622–625.
88 Irving, H. and Williams, R.J.P. (1953). *J. Chem. Soc.*: 3192–3210.
89 Schmolke, W., Ahmadi, M., and Seiffert, S. (2019). *Phys. Chem. Chem. Phys.* 21: 19623–19638.
90 Khlobystov, A.N., Blake, A.J., Champness, N.R. et al. (2001). *Coord. Chem. Rev.* 222: 155–192.
91 Liu, M.-H., Kira, A., and Nakahara, H. (1997). *Langmuir* 13: 4807–4809.
92 Lang, H. and Leschke, M. (2002). *Heteroat. Chem* 13: 521–533.
93 Paulusse, J.M.J., Huijbers, J.P.J., and Sijbesma, R.P. (2006). *Chem. Eur. J.* 12: 4928–4934.
94 Paulusse, J.M.J. and Sijbesma, R.P. (2004). *Angew. Chem. Int. Ed.* 43: 4460–4462.
95 Johnson, B.F.G., Sanderson, K.M., Shephard, D.S. et al. (2000). *Chem. Commun.*: 1317–1318.
96 Phang, L.-T., Gan, K.-S., Lee, H.K., and Hor, T.S.A. (1993). *J. Chem. Soc., Dalton Trans.*: 2697–2702.
97 Efraty, A., Feinstein, I., and Frolow, F. (1982). *Inorg. Chem.* 21: 485–488.
98 Efraty, A., Feinstein, I., Frolow, F., and Goldman, A. (1980). *J. Chem. Soc., Chem. Commun.*: 864–865.
99 Efraty, A., Feinstein, I., Frolow, F., and Wackerle, L. (1980). *J. Am. Chem. Soc.* 102: 6341–6343.
100 Efraty, A., Feinstein, I., Wackerle, L., and Frolow, F. (1980). *Angew. Chem. Int. Ed. Engl.* 19: 633–634.
101 Feinstein-Jaffe, I. and Barash, C. (1991). *Inorg. Chim. Acta* 185: 3–5.
102 Feinstein-Jaffe, I., Biran, I., Mahalu, D. et al. (1988). *Inorg. Chim. Acta* 154: 129–131.
103 Feinstein-Jaffe, I. and Efraty, A. (1986). *J. Mol. Catal.* 35: 285–302.

104 Feinstein-Jaffe, I., Frolow, F., Wackerle, L. et al. (1988). *J. Chem. Soc., Dalton Trans.*: 469–476.

105 Feinstein-Jaffe, I. and Maisuls, S.E. (1987). *J. Organomet. Chem.* 326: C97–C100.

106 Deger, S. and Hanack, M. (1986). *Synth. Met.* 13: 319–324.

107 Keppeler, U. and Hanack, M. (1986). *Chem. Ber.* 119: 3363–3381.

108 Keppeler, U., Schneider, O., Stöffler, W., and Hanack, M. (1984). *Tetrahedron Lett.* 25: 3679–3682.

109 Dartiguenave, M., Dartiguenave, Y., Mari, A. et al. (1988). *Can. J. Chem.* 66: 2386–2394.

110 Fournier, E., Sicard, S., Decken, A., and Harvey, P.D. (2004). *Inorg. Chem.* 43: 1491–1501.

111 Bradford, A.M., Kristof, E., Rashidi, M. et al. (1994). *Inorg. Chem.* 33: 2355–2363.

112 Bérubé, J.-F., Gagnon, K., Fortin, D. et al. (2006). *Inorg. Chem.* 45: 2812–2823.

113 Irwin, M.J., Jia, G., Vittal, J.J., and Puddephatt, R.J. (1996). *Organometallics* 15: 5321–5329.

114 Sicard, S., Bérubé, J.-F., Samar, D. et al. (2004). *Inorg. Chem.* 43: 5321–5334.

115 Zhang, T., Drouin, M., and Harvey, P.D. (1999). *Inorg. Chem.* 38: 957–963.

116 Clément, S., Aly, S.M., Gagnon, K. et al. (2008). *J. Inorg. Organomet. Polym. Mater.* 18: 104–110.

117 Carraher, C.E., Scott, W.J., Schroeder, J.A., and Giron, D.J. (1981). *J. Macromol. Sci. Part A, Chem.* 15: 625–631.

118 Callari, M., Aldrich-Wright, J.R., de Souza, P.L., and Stenzel, M.H. (2014). *Prog. Polym. Sci.* 39: 1614–1643.

119 Lo, W.-Y., Zhang, N., Cai, Z. et al. (2016). *Acc. Chem. Res.* 49: 1852–1863.

120 Hanack, M., Dürr, K., Lange, A. et al. (1995). *Synth. Met.* 71: 2275–2278.

121 Hanack, M., Deger, S., and Lange, A. (1988). *Coord. Chem. Rev.* 83: 115–136.

122 Collman, J.P., McDevitt, J.T., Yee, G.T. et al. (1986). *Proc. Natl. Acad. Sci. U.S.A.* 83: 4581–4585.

123 Taylor, P.N. and Anderson, H.L. (1999). *J. Am. Chem. Soc.* 121: 11538–11545.

124 Drobizhev, M., Stepanenko, Y., Rebane, A. et al. (2006). *J. Am. Chem. Soc.* 128: 12432–12433.

125 Grozema, F.C., Houarner-Rassin, C., Prins, P. et al. (2007). *J. Am. Chem. Soc.* 129: 13370–13371.

126 Screen, T.E.O., Thorne, J.R.G., Denning, R.G. et al. (2002). *J. Am. Chem. Soc.* 124: 9712–9713.

127 Screen, T.E.O., Thorne, J.R.G., Denning, R.G. et al. (2003). *J. Mater. Chem.* 13: 2796–2808.

128 Winters, M.U., Kärnbratt, J., Eng, M. et al. (2007). *J. Phys. Chem. C* 111: 7192–7199.

129 You, C.-C. and Würthner, F. (2004). *Org. Lett.* 6: 2401–2404.

130 Davidson, G.J.E., Lane, L.A., Raithby, P.R. et al. (2008). *Inorg. Chem.* 47: 8721–8726.

131 Asai, Y., Onishi, K., Miyata, S. et al. (2001). *J. Electroanal. Soc.* 148: A305–A310.

132 Fernández-Rodríguez, O., Fernández-Lázaro, F., Cabezón, B. et al. (1997). *Synth. Met.* 84: 369–370.

133 Hanack, M. (1988). *Mol. Cryst. Liq. Cryst.* 160: 133–137.

134 Hanack, M., Hees, M., and Witke, E. (1998). *New J. Chem.* 22: 169–172.

135 Hanack, M., Hirsch, A., Kamenzin, S. et al. (1991). *Synth. Met.* 42: 2633–2636.

136 Hanack, M., Hirsch, A., and Lehmann, H. (1990). *Angew. Chem. Int. Ed. Engl.* 29: 1467–1468.

137 Hanack, M., Knecht, S., Pollcy, R., and Subramanian, L.R. (1996). *Synth. Met.* 80: 183–189.

138 Hanack, M., Knecht, S., Witke, E., and Haisch, P. (1993). *Synth. Met.* 55: 873–878.

139 Hanack, M., Renz, G., Straehle, J., and Schmid, S. (1991). *J. Org. Chem.* 56: 3501–3509.

140 Knecht, S., Polley, R., and Hanack, M. (1996). *Appl. Organomet. Chem.* 10: 649–660.

141 Metz, J. and Hanack, M. (1987). *Chem. Ber.* 120: 1307–1313.

142 Pawlik, J., Kautz, C., and Baumgarten, M. (1994). *J. Inorg. Organomet. Polym.* 4: 237–250.

143 Ryu, H., Kang, Y.-G., Knecht, S. et al. (1997). *Synth. Met.* 87: 69–73.

144 Stuzhin, P.A., Vagin, S.I., and Hanack, M. (1998). *Inorg. Chem.* 37: 2655–2662.

145 Vagin, S.I., Stuzhin, P.A., and Hanack, M. (1999). *Russ. J. Gen. Chem.* 69: 308–313.

146 Senge, M.O. and Smith, K.M. (1994). *J. Chem. Soc., Chem. Commun.*: 923–924.

147 Burrell, A.K., Officer, D.L., Reid, D.C.W., and Wild, K.Y. (1998). *Angew. Chem. Int. Ed. Engl.* 37: 114–117.

148 Satake, A., Shoji, O., and Kobuke, Y. (2007). *J. Organomet. Chem.* 692: 635–644.

149 Michelsen, U. and Hunter, C.A. (2000). *Angew. Chem. Int. Ed.* 39: 764–767.

150 Brandys, M.-C., Jennings, M.C., and Puddephatt, R.J. (2000). *J. Chem. Soc., Dalton Trans.*: 4601–4606.

151 Brandys, M.-C. and Puddephatt, R.J. (2001). *Chem. Commun.*: 1508–1509.

152 Irwin, M.J., Rendina, L.M., Vittal, J.J., and Puddephatt, R.J. (1996). *Chem. Commun.*: 1281–1282.

153 Irwin, M.J., Vittal, J.J., Yap, G.P.A., and Puddephatt, R.J. (1996). *J. Am. Chem. Soc.* 118: 13101–13102.

154 Puddephatt, R.J. (1998). *Chem. Commun.*: 1055–1062.

155 Visbal, R. and Gimeno, M.C. (2014). *Chem. Soc. Rev.* 43: 3551–3574.

156 Boydston, A.J. and Bielawski, C.W. (2006). *Dalton Trans.*: 4073–4077.

157 Boydston, A.J., Rice, J.D., Sanderson, M.D. et al. (2006). *Organometallics* 25: 6087–6098.

158 Boydston, A.J., Williams, K.A., and Bielawski, C.W. (2005). *J. Am. Chem. Soc.* 127: 12496–12497.

159 Williams, K.A., Boydston, A.J., and Bielawski, C.W. (2007). *J. R. Soc. Interface* 4: 359–362.

160 Peris, E. and Crabtree, R.H. (2005). *Coord. Chem. Rev.* 248: 2239–2246.

161 Boydston, A.J., Khramov, D.M., and Bielawski, C.W. (2006). *Tetrahedron Lett.* 47: 5123–5125.

162 Khramov, D.M., Boydston, A.J., and Bielawski, C.W. (2006). *Angew. Chem. Int. Ed.* 45: 6186–6189.

163 Wu, C.-S., Lee, Y.-T., and Chen, Y. (2012). *Polym. Chem.* 3: 2776–2784.

164 Winter, A. and Schubert, U.S. (2017). Polymeric architectures containing phosphorescent Ir(III) complexes. In: *Iridium(III) in Optoelectronic and Photonics Applications* (ed. E. Zysman-Coleman), 145–203. Weinheim: Wiley-VCH.

165 Zhao, Q., Liu, S.-J., and Huang, W. (2010). *Macromol. Rapid Commun.* 31: 794–807.

166 Bera, M.K., Mori, T., Yoshida, T. et al. (2019). *ACS Appl. Mater. Interfaces* 11: 11893–11903.

167 Kelch, S. and Rehahn, M. (1997). *Macromolecules* 30: 6185–6193.

168 Knapp, R., Schott, A., and Rehahn, M. (1996). *Macromolecules* 29: 478–480.

169 Cassagneau, T., Fendler, J.H., Johnson, S.A., and Mallouk, T.E. (2000). *Adv. Mater.* 12: 1363–1366.

170 Leininger, S., Olenyuk, B., and Stang, P.J. (2000). *Chem. Rev.* 100: 853–908.

171 Holder, E., Langeveld, B.M.W., and Schubert, U.S. (2005). *Adv. Mater.* 17: 1109–1121.

172 Marin, V., Holder, E., Hoogenboom, R., and Schubert, U.S. (2007). *Chem. Soc. Rev.* 36: 618–635.

173 Velten, U., Lahn, B., and Schubert, U.S. (1997). *Macromol. Rapid Commun.* 198: 2789–2816.

174 Velten, U. and Rehahn, M. (1996). *Chem. Commun.*: 2639–2640.

175 Kaes, C., Hosseini, M.W., Rickard, C.E.F. et al. (1998). *Angew. Chem. Int. Ed. Engl.* 37: 920–922.

176 James, S.L. (2009). *Chem. Soc. Rev.* 38: 1744–1758.

177 Fox, M.A. and Chandler, D.A. (1991). *Adv. Mater.* 3: 381–385.

178 Wang, P.-W. and Fox, M.A. (1994). *Inorg. Chem.* 33: 2938–2945.

179 Wang, P.-W. and Fox, M.A. (1994). *Inorg. Chim. Acta* 225: 15–22.

180 Higuchi, M. (2009). *Polym. J.* 41: 511–520.

181 Hossain, M.D., Sato, T., and Higuchi, M. (2013). *Chem. Asian J.* 8: 76–79.

182 Pandey, R.K., Hossain, M.D., Moriyama, S., and Higuchi, M. (2013). *J. Mater. Chem. A* 1: 9016–9018.

183 Pandey, R.K., Hossain, M.D., Moriyama, S., and Higuchi, M. (2014). *J. Mater. Chem. A* 2: 7754–7758.

184 Singh, N. and Sinha, R.K. (2003). *Inorg. Chem. Commun.* 6: 97–101.

185 Sugimoto, K., Kuroda-Sowa, T., Maekawa, M., and Munakata, M. (2001). *Bull. Chem. Soc. Jpn.* 74: 391–394.

186 Schubert, U.S., Winter, A., and Newkome, G.R. (2011). *Terpyridine-based Materials for Catalytic, Optoelectronic and Lifescience Applications*. Weinheim: VCH-Wiley.

187 Knops, P., Sendhoff, N., Mekelburger, H.-B., and Vögtle, F. (1992). *Top. Curr. Chem.* 161: 1–36.

188 Storrier, G.D. and Colbran, S.B. (1996). *J. Chem. Soc., Dalton Trans.*: 2185–2186.
189 Storrier, G.D., Colbran, S.B., and Craig, D.C. (1997). *J. Chem. Soc., Dalton Trans.*: 3011–3028.
190 Ng, W.-Y., Gong, X., and Chan, W.-K. (1999). *Chem. Mater.* 11: 1165–1170.
191 Andres, P.R. and Schubert, U.S. (2004). *Synthesis* 2004: 1229–1238.
192 Andres, P.R. and Schubert, U.S. (2004). *Macromol. Rapid Commun.* 25: 1371–1375.
193 Constable, E.C., Housecroft, C.E., and Smith, C.B. (2003). *Inorg. Chem. Commun.* 6: 1011–1013.
194 Meier, M.A.R., Wouters, D., Ott, C. et al. (2006). *Macromolecules* 39: 1569–1576.
195 Chiper, M., Meier, M.A.R., Wouters, D. et al. (2008). *Macromolecules* 41: 2771–2777.
196 Chiper, M., Hoogenboom, R., and Schubert, U.S. (2008). *e-Polymers* 8: 1881–1819.
197 Hofmeier, H., Schmatloch, S., Wouters, D., and Schubert, U.S. (2003). *Macromol. Chem. Phys.* 204: 2197–2203.
198 Schubert, U.S., Hien, O., and Eschbaumer, C. (2000). *Macromol. Rapid Commun.* 21: 1156–1161.
199 Winter, A. and Schubert, U.S. (2018). Block copolymers with element blocks: the metal-bisterpyridine linkage. In: *Polymers with Element Blocks* (ed. Y. Chujo). Heidelberg: Springer-Verlag.
200 Gon, M., Tanaka, K., and Chujo, Y. (2017). *Polym. J.* 50: 109.
201 Mansfeld, U., Winter, A., Hager, M.D. et al. (2013). *Polym. Chem.* 4: 113–123.
202 Mugemana, C., Guillet, P., Hoeppener, S. et al. (2010). *Chem. Commun.* 46: 1296–1298.
203 Gohy, J.-F., Lohmeijer, B.G.G., and Schubert, U.S. (2002). *Macromol. Rapid Commun.* 23: 555–560.
204 Gohy, J.-F., Lohmeijer, B.G.G., and Schubert, U.S. (2002). *Macromolecules* 35: 4560–4563.
205 Gohy, J.-F., Lohmeijer, B.G.G., Varshney, S.K., and Schubert, U.S. (2002). *Macromolecules* 35: 7427–7435.
206 Lohmeijer, B.G.G. and Schubert, U.S. (2002). *Angew. Chem. Int. Ed.* 41: 3825–3829.
207 Gohy, J.-F., Lohmeijer, B.G.G., Varshney, S.K. et al. (2002). *Macromolecules* 35: 9748–9755.
208 Lohmeijer, B.G.G. and Schubert, U.S. (2003). *Macromol. Chem. Phys.* 204: 1072–1078.
209 Lohmeijer, B.G.G. and Schubert, U.S. (2003). *J. Polym. Sci., Part A: Polym. Chem.* 41: 1413–1427.
210 Gohy, J.-F., Lohmeijer, B.G.G., and Schubert, U.S. (2003). *Chem. Eur. J.* 9: 3472–3479.
211 Lohmeijer, B.G.G., Schlaad, H., and Schubert, U.S. (2003). *Macromol. Symp.* 196: 125–135.

212 Meier, M.A.R., Lohmeijer, B.G.G., and Schubert, U.S. (2003). *Macromol. Rapid Commun.* 24: 852–857.
213 Al-Hussein, M., Lohmeijer, B.G.G., Schubert, U.S., and de Jeu, W.H. (2003). *Macromolecules* 36: 9281–9284.
214 Regev, O., Gohy, J.-F., Lohmeijer, B.G.G. et al. (2004). *Colloid. Polym. Sci.* 282: 407–411.
215 Gohy, J.-F., Lohmeijer, B.G.G., Alexeev, A. et al. (2004). *Chem. Eur. J.* 10: 4315–4323.
216 Lohmeijer, B.G.G., Wouters, D., Yin, Z., and Schubert, U.S. (2004). *Chem. Commun.*: 2886–2887.
217 Mayer, G., Vogel, V., Lohmeijer, B.G.G. et al. (2004). *J. Polym. Sci., Part A: Polym. Chem.* 42: 4458–4465.
218 Guerrero-Sanchez, C., Wouters, D., Fustin, C.-A. et al. (2005). *Macromolecules* 38: 10185–10191.
219 Al-Hussein, M., de Jeu, W.H., Lohmeijer, B.G.G., and Schubert, U.S. (2005). *Macromolecules* 38: 2832–2836.
220 Fustin, C.A., Lohmeijer, B.G.G., Duwez, A.S. et al. (2005). *Adv. Mater.* 17: 1162–1165.
221 Lohmeijer, B.G.G. and Schubert, U.S. (2005). *J. Polym. Sci., Part A: Polym. Chem.* 43: 6331–6344.
222 Ott, C., Lohmeijer, B.G.G., Wouters, D., and Schubert, U.S. (2006). *Macromol. Chem. Phys.* 207: 1439–1449.
223 Guillet, P., Fustin, C.-A., Lohmeijer, B.G.G. et al. (2006). *Macromolecules* 39: 5484–5488.
224 Guillet, P., Fustin, C.-A., Mugemana, C. et al. (2008). *Soft Matter* 4: 2278–2282.
225 Rasa, M., Tziatzios, C., Lohmeijer, B.G.G. et al. (2006). *Prog. Colloid Polym. Sci.* 131: 165–171.
226 Ott, C., Kranenburg, J.M., Guerrero-Sanchez, C. et al. (2009). *Macromolecules* 42: 2177–2183.
227 Chiper, M., Winter, A., Hoogenboom, R. et al. (2008). *Macromolecules* 41: 8823–8831.
228 Landsmann, S., Winter, A., Chiper, M. et al. (2008). *Macromol. Chem. Phys.* 209: 1666–1672.
229 Schlütter, F., Pavlov, G.M., Gohy, J.-F. et al. (2011). *J. Polym. Sci., Part A: Polym. Chem.* 49: 1396–1408.
230 Gohy, J.-F., Ott, C., Hoeppener, S., and Schubert, U.S. (2009). *Chem. Commun.*: 6038–6040.
231 Constable, E.C., Housecroft, C.E., Neuburger, M. et al. (2014). *Dalton Trans.* 43: 7227–7235.
232 Fallahpour, R.-A., Neuburger, M., and Zehnder, M. (1999). *New J. Chem.* 23: 53–61.
233 Hoogenboom, R. and Schubert, U.S. (2006). *Chem. Soc. Rev.* 35: 622–629.
234 Chiper, M., Meier, M.A.R., Kranenburg, J.M., and Schubert, U.S. (2007). *Macromol. Chem. Phys.* 208: 679–689.

235 Meier, M.A.R., Lohmeijer, B.G.G., and Schubert, U.S. (2003). *J. Mass Spectrom.* 38: 510–516.

236 Oudhoff, K.A., Schoenmakers, P.J., and Kok, W.T. (2004). *Chromatographia* 60: 475–480.

237 Tziatzios, C., Schubert, D., Precup, A.A. et al. (2004). *Prog. Colloid Polym. Sci.* 127: 54–60.

238 Zhuge, F., Hawke, L.G.D., Fustin, C.-A. et al. (2017). *J. Rheol.* 61: 1245–1262.

239 Chiper, M., Fournier, D., Hoogenboom, R., and Schubert, U.S. (2008). *Macromol. Rapid Commun.* 29: 1640–1647.

240 Lutz, J.-F., Akdemir, Ö., and Hoth, A. (2006). *J. Am. Chem. Soc.* 128: 13046–13047.

241 Vogel, V., Gohy, J.-F., Lohmeijer, B.G.G. et al. (2003). *J. Polym. Sci., Part A: Polym. Chem.* 41: 3159–3168.

242 Guillet, P., Fustin, C.-A., Wouters, D. et al. (2009). *Soft Matter* 5: 1460–1465.

243 Fustin, C.-A., Guillet, P., Misner, M.J. et al. (2008). *J. Polym. Sci., Part A: Polym. Chem.* 46: 4719–4724.

244 Zhou, G. and Harruna, I.I. (2005). *Macromolecules* 38: 4114–4123.

245 Ievins, A.D., Moughton, A.O., and O'Reilly, R.K. (2008). *Macromolecules* 41: 3571–3578.

246 Ott, C., Hoogenboom, R., Hoeppener, S. et al. (2009). *Soft Matter* 5: 84–91.

247 He, Y.-J., Tu, T.-H., Su, M.-K. et al. (2017). *J. Am. Chem. Soc.* 139: 4218–4224.

248 Gilroy, J.B., Gädt, T., Whittell, G.R. et al. (2010). *Nat. Chem.* 2: 566.

249 Hailes, R.L.N., Oliver, A.M., Gwyther, J. et al. (2016). *Chem. Soc. Rev.* 45: 5358–5407.

250 Zhang, X., Liu, Y., Wang, X. et al. (2020). *RSC Adv.* 10: 9686–9692.

251 Guillet, P., Mugemana, C., Stadler, F.J. et al. (2009). *Soft Matter* 5: 3409–3411.

252 Gohy, J.-F., Chiper, M., Guillet, P. et al. (2009). *Soft Matter* 5: 2954–2961.

253 Beck, B.J. and Rowan, S.J. (2003). *J. Am. Chem. Soc.* 125: 13922–13923.

254 Beck, J.B. and Stuart, J.R. (2006). The preparation of metallosupramolecular polymers and gels by utilizing 2,6-bis-(1′-metylbenzimidazoyl)pyridine-metal ion interactions. In: *Metal-Containing and Metallosupramolecular Polymers and Materials*, vol. 928 (eds. U.S. Schubert, G.R. Newkome and I. Manners), 97–112. Washington, DC: American Chemical Society.

255 Weng, W., Beck, J.B., Jamieson, A.M., and Rowan, S.J. (2006). *J. Am. Chem. Soc.* 128: 11663–11672.

256 Weng, W., Jamieson, A.M., and Rowan, S.J. (2007). *Tetrahedron* 63: 7419–7431.

257 Weng, W., Li, Z., Jamieson, A.M., and Rowan, S.J. (2008). *Macromolecules* 42: 236–246.

258 Weng, W., Li, Z., Jamieson, A.M., and Rowan, S.J. (2009). *Soft Matter* 5: 4647–4657.

259 Wojtecki, R.J., Meador, M.A., and Rowan, S.J. (2011). *Nat. Mater.* 10: 14–27.

260 Balkenende, D.W.R., Coulibaly, S., Balog, S. et al. (2014). *J. Am. Chem. Soc.* 136: 10493–10498.

261 Burnworth, M., Knapton, D., Rowan, S., and Weder, C. (2007). *J. Inorg. Organomet. Polym. Mater.* 17: 91–103.

262 Burnworth, M., Rowan, S.J., and Weder, C. (2011). *Macromolecules* 45: 126–132.

263 Knapton, D., Iyer, P.K., Rowan, S.J., and Weder, C. (2006). *Macromolecules* 39: 4069–4075.

264 Rowan, S.J. and Beck, J.B. (2005). *Faraday Discuss.* 128: 43–53.

265 Miller, A.K., Li, Z., Streletzky, K.A. et al. (2012). *Polym. Chem.* 3: 3132–3138.

266 Coulibaly, S., Roulin, A., Balog, S. et al. (2013). *Macromolecules* 47: 152–160.

267 Heinzmann, C., Coulibaly, S., Roulin, A. et al. (2014). *ACS Appl. Mater. Interfaces* 6: 4713–4719.

268 Kumpfer, J.R., Wie, J.J., Swanson, J.P. et al. (2011). *Macromolecules* 45: 473–480.

269 Burnworth, M., Tang, L., Kumpfer, J.R. et al. (2011). *Nature* 472: 334–337.

270 Lee, S.W., Kumpfer, J.R., Lin, P.A. et al. (2012). *Macromolecules* 45: 8201–8210.

271 Kumpfer, J.R., Jin, J., and Rowan, S.J. (2010). *J. Mater. Chem.* 20: 145–151.

272 Kumpfer, J.R. and Rowan, S.J. (2011). *J. Am. Chem. Soc.* 133: 12866–12874.

273 Terada, K.-i., Nakamura, H., Kanaizuka, K. et al. (2012). *ACS Nano* 6: 1988–1999.

274 Harriman, A. and Ziessel, R. (1996). *Chem. Commun.*: 1707–1716.

275 Romero, F.M., Ziessel, R., Dupont-Gervais, A., and Van Dorsselaer, A. (1996). *Chem. Commun.*: 551–553.

276 Schwarz, G., Bodenthin, Y., Geue, T. et al. (2010). *Macromolecules* 43: 494–500.

277 Dobrawa, R., Lysetska, M., Ballester, P. et al. (2005). *Macromolecules* 38: 1315–1325.

278 Dobrawa, R. and Würthner, F. (2002). *Chem. Commun.*: 1878–1879.

279 Yu, S.-C., Kwok, C.-C., Chan, W.-K., and Che, C.-M. (2003). *Adv. Mater.* 15: 1643–1647.

280 Schlütter, F., Wild, A., Winter, A. et al. (2010). *Macromolecules* 43: 2759–2771.

281 Winter, A., Friebe, C., Chiper, M. et al. (2009). *J. Polym. Sci., Part A: Polym. Cheme.* 47: 4083–4098.

282 Wild, A., Schlütter, F., Pavlov, G.M. et al. (2010). *Macromol. Rapid Commun.* 31: 868–874.

283 Wild, A., Winter, A., Hager, M.D., and Schubert, U.S. (2012). *Chem. Commun.* 48: 964–966.

284 Friebe, C., Wild, A., Perelaer, J., and Schubert, U.S. (2012). *Macromol. Rapid Commun.* 33: 503–509.

285 Wild, A., Winter, A., Hager, M.D., and Schubert, U.S. (2012). *Analyst* 137: 2333–2337.

286 Siebert, R., Tian, Y., Camacho, R. et al. (2012). *J. Mater. Chem.* 22: 16041–16050.

287 Wild, A., Teichler, A., Ho, C.-L. et al. (2013). *J. Mater. Chem. C* 1: 1812–1822.

288 Winter, A., Friebe, C., Hager, M.D., and Schubert, U.S. (2008). *Macromol. Rapid Commun.* 29: 1679–1686.

289 El-ghayoury, A., Schenning, A.P.H.J., and Meijer, E.W. (2002). *J. Polym. Sci., Part A: Polym. Chem.* 40: 4020–4023.

290 Vellis, P.D., Mikroyannidis, J.A., Lo, C.-N., and Hsu, C.-S. (2008). *J. Polym. Sci., Part A: Polym. Chem.* 46: 7702–7712.

291 Chen, Y.-Y. and Lin, H.-C. (2007). *J. Polym. Sci., Part A: Polym. Chem.* 45: 3243–3255.

292 Chen, Y.-Y. and Lin, H.-C. (2007). *Polymer* 48: 5268–5278.

293 Chen, Y.-Y., Tao, Y.-T., and Lin, H.-C. (2006). *Macromolecules* 39: 8559–8566.

294 Bozdemir, Ö.A., Büyükcakir, O., and Akkaya, E.U. (2009). *Chem. Eur. J.* 15: 3830–3838.

295 Chen, X., Ma, L., Cheng, Y. et al. (2007). *Polym. Int.* 56: 648–654.

296 Hinderberger, D., Schmelz, O., Rehahn, M., and Jeschke, G. (2004). *Angew. Chem. Int. Ed.* 43: 4616–4621.

297 Stepanenko, V., Stocker, M., Müller, P. et al. (2009). *J. Mater. Chem.* 19: 6816–6826.

298 Schütte, M., Kurth, D.G., Linford, M.R. et al. (1998). *Angew. Chem. Int. Ed. Engl.* 37: 2819–2893.

299 Caruso, F., Schüler, C., and Kurth, D.G. (1999). *Chem. Mater.* 11: 3394–3399.

300 Kurth, D.G., Caruso, F., and Schuler, C. (1999). *Chem. Commun.*: 1579–1580.

301 Kurth, D.G. and Osterhout, R. (1999). *Langmuir* 15: 4842–4846.

302 Bodenthin, Y., Pietsch, U., Möhwald, H., and Kurth, D.G. (2005). *J. Am. Chem. Soc.* 127: 3110–3114.

303 Kurth, D.G., Schütte, M., and Wen, J. (2002). *Colloids Surf., A* 198–200: 633–643.

304 Akcakayiran, D., Mauder, D., Hess, C. et al. (2008). *J. Phys. Chem. B* 112: 14637–14647.

305 Kolb, U., Büscher, K., Helm, C.A. et al. (2006). *Proc. Natl. Acad. Sci. U.S.A.* 103: 10202–10206.

306 Bothenthin, Y., Pietsch, U., Grenzer, J. et al. (2005). *J. Phys. Chem. B* 109: 12795–12799.

307 Bothenthin, Y., Schwarz, G., Tomkowicz, Z. et al. (2009). *J. Am. Chem. Soc.* 131: 2934–2941.

308 Kurth, D.G., Lehmann, P., and Schütte, M. (2000). *Proc. Natl. Acad. Sci. U.S.A.* 97: 5704–5707.

309 Kurth, D.G., Meister, A., Thünemann, A.F., and Förster, G. (2003). *Langmuir* 19: 4055–4057.

310 Kurth, D.G., Severin, N., and Rabe, J.P. (2002). *Angew. Chem. Int. Ed.* 41: 3681–3683.

311 Lehmann, P., Symietz, C., Brezesinski, G. et al. (2005). *Langmuir* 21: 5901–5906.

312 Bothenthin, Y., Schwarz, G., Tomkowicz, Z. et al. (2009). *Coord. Chem. Rev.* 253: 2414–2422.

313 Zhang, C.-F., Liu, A., Chen, M. et al. (2009). *ACS Appl. Mater. Interfaces* 1: 1250–1258.

314 Kurth, D.G., Pitarch Lopez, J., and Dong, W.-F. (2005). *Chem. Commun.*: 2119–2121.

315 Han, F.S., Higuchi, M., Akasaka, Y. et al. (2008). *Thin Solid Films* 516: 2469–2473.

316 Han, F.S., Higuchi, M., Ikeda, T. et al. (2008). *J. Mater. Chem.* 18: 4555–4560.

317 Han, F.-S., Higuchi, M., and Kurth, D.G. (2007). *Adv. Mater.* 19: 3928–3931.

318 Han, F.-S., Higuchi, M., and Kurth, D.G. (2008). *J. Am. Chem. Soc.* 130: 2073–2081.

319 Li, J.-H. and Higuchi, M. (2010). *J. Inorg. Organomet. Polym. Mater.* 20: 10–18.

320 Mehr, H.S., Romano, N.C., Altamimi, R. et al. (2015). *Dalton Trans.* 44: 3176–3184.

321 Pal, R.R., Higuchi, M., Negishi, Y. et al. (2010). *Polym. J.* 42: 336–341.

322 Munzert, S.M., Schwarz, G., and Kurth, D.G. (2016). *RSC Adv.* 6: 15441–15450.

323 Munzert, S.M., Schwarz, G., and Kurth, D.G. (2016). *Inorg. Chem.* 55: 2565–2573.

324 Schmelz, O. and Rehahn, M. (2002). *e-Polymers* 2: 659–687.

325 Schott, M., Lorrmann, H., Szczerba, W. et al. (2014). *Sol. Energy Mater. Sol. Cells* 126: 68–73.

326 Schott, M., Niklaus, L., Clade, J., and Posset, U. (2019). *Sol. Energy Mater. Sol. Cells* 200: 110001.

327 Hu, C.-W., Sato, T., Zhang, J. et al. (2013). *J. Mater. Chem. C* 1: 3408–3413.

328 Bera, M.K., Chakraborty, C., Rana, U., and Higuchi, M. (2018). *Macromol. Rapid Commun.* 39: 1800415.

329 Yoshida, T., Ninomiya, Y., and Higuchi, M. (2020). *RSC Adv.* 10: 10904–10909.

330 Bera, M.K., Ninomiya, Y., Yoshida, T., and Higuchi, M. (2020). *Macromol. Rapid Commun.* 41: 1900384.

331 Bera, M.K., Ninomiya, Y., and Higuchi, M. (2020). *ACS Appl. Marer. Interfaces* 12: 14376–14385.

332 Yoshida, T. and Higuchi, M. (2018). *J. Photopolym. Sci. Technol.* 31: 343–347.

333 Suzuki, T., Sato, T., Zhang, J. et al. (2016). *J. Mater. Chem. C* 4: 1594–1598.

334 Chakraborty, C., Pandey, R.K., Rana, U. et al. (2016). *J. Mater. Chem. C* 4: 9428–9437.

335 Islam, T., Hasan, M.M., Akter, S.S. et al. (2020). *ACS Appl. Polym. Mater.* 2: 273–284.

336 Würthner, F. (2004). *Chem. Commun.*: 1564–1579.

337 Li, F., Hu, Z., Qiao, H. et al. (2016). *Dyes Pigm.* 132: 142–150.

338 Hrma, M., Šichová, K., Svoboda, J., and Vohlídal, J. (2017). *Polymer* 122: 22–33.

339 Štenclová, P., Šichová, K., Šloufová, I. et al. (2016). *Dalton Trans.* 45: 1208–1224.

340 Štenclová-Bláhová, P., Svoboda, J., Šloufová, I., and Vohlídal, J. (2015). *Phys. Chem. Chem. Phys.* 17: 13743–13756.

341 Vitvarová, T., Svoboda, J., Hissler, M., and Vohlídal, J. (2017). *Organometallics* 36: 777–786.

342 Yuan, M., Wang, F., and Tian, Y.-K. (2018). *RSC Adv.* 8: 40794–40797.

343 Qian, Y., Yang, H., and Wang, Y. (2020). *Dyes Pigm.* 176: 108251.

344 Liu, L., Li, F., Wu, C. et al. (2019). *Macromol. Res.* 27: 534–542.

345 Liu, L., Hu, Z., Wu, C. et al. (2019). *J. Macromol. Sci. A* 56: 918–925.

346 Li, D., Wang, J., and Ma, X. (2018). *Adv. Opt. Mater.* 6: 1800273.
347 Abbel, R., Grenier, C., Pouderoijen, M.J. et al. (2009). *J. Am. Chem. Soc.* 131: 833–843.
348 Bauer, T., Zheng, Z., Renn, A. et al. (2011). *Angew. Chem. Int. Ed.* 50: 7879–7884.
349 Zheng, Z., Ruiz-Vargas, C.S., Bauer, T. et al. (2013). *Macromol. Rapid Commun.* 34: 1670–1680.
350 Zheng, Z., Opilik, L., Schiffmann, F. et al. (2014). *J. Am. Chem. Soc.* 136: 6103–6110.
351 Takada, K., Sakamoto, R., Yi, S.-T. et al. (2015). *J. Am. Chem. Soc.* 137: 4681–4689.
352 Tsukamoto, T., Takada, K., Sakamoto, R. et al. (2017). *J. Am. Chem. Soc.* 139: 5359–5366.
353 Beneto, A.J., Jeong, J.Y., and Park, J.S. (2018). *Dalton Trans.* 47: 16036–16039.
354 Arockiam, J.B., Son, H., Han, S.H. et al. (2019). *ACS Appl. Energy Mater.* 2: 8416–8424.
355 Barron, J.A., Glazier, S., Bernhard, S. et al. (2003). *Inorg. Chem.* 42: 1448–1455.
356 Bernhard, S., Takada, K., Díaz, D.J. et al. (2001). *J. Am. Chem. Soc.* 123: 10265–10271.
357 Kimura, M., Sano, M., Muto, T. et al. (1999). *Macromolecules* 32: 7951–7953.
358 Pal, R.R., Higuchi, M., and Kurth, D.G. (2009). *Org. Lett.* 11: 3562–3565.
359 Basak, S., Hui, P., Boodida, S., and Chandrasekar, R. (2012). *J. Org. Chem.* 77: 3620–3626.
360 Narayana, Y.S.L.V., Basak, S., Baumgarten, M. et al. (2013). *Adv. Funct. Mater.* 23: 5875–5880.
361 Rajadurai, C., Fuhr, O., Kruk, R. et al. (2007). *Chem. Commun.*: 2636–2638.
362 Chandrasekhar, N. and Chandrasekhar, R. (2010). *J. Org. Chem.* 75: 4852–4855.
363 Schulze, B., Friebe, C., Hoeppener, S. et al. (2012). *Macromol. Rapid Commun.* 33: 597–602.
364 Li, H.-Q., Ding, Z.-Y., Pan, Y. et al. (2016). *Inorg. Chem. Front.* 3: 1363–1375.
365 Wang, Y.-Q., Pan, Y., Gao, W.-Q. et al. (2019). *Tetrahedron* 75: 3809–3814.
366 Qin, J., Shipman, P.O., and Jäkle, F. (2012). *Macromol. Rapid Commun.* 33: 562–567.
367 Trofimenko, S. (1993). *Chem. Rev.* 93: 943–980.
368 Puddephatt, R.J. (2008). *Chem. Soc. Rev.* 37: 2012–2027.
369 Wolf, M.O. (2006). *J. Inorg. Organomet. Polym. Mater.* 16: 189–199.
370 Wong, W.-Y. (2007). *Coord. Chem. Rev.* 251: 2400–2427.
371 Wong, W.-Y. (2008). *Macromol. Chem. Phys.* 209: 14–24.
372 Wong, W.-Y. and Ho, C.-L. (2006). *Frontiers in Transition Metal-Containing Polymers*, 247–286. Wiley.
373 Wong, W.-Y. and Ho, C.-L. (2010). *Acc. Chem. Res.* 43: 1246–1256.
374 Wong, W.-Y. (2010). *Macromolecules Containing Metal and Metal-Like Elements*, 289–324. Wiley.
375 Wong, W.-Y. and Ho, C.-L. (2006). *Coord. Chem. Rev.* 250: 2627–2690.
376 Zhou, G.-J. and Wong, W.-Y. (2011). *Chem. Soc. Rev.* 40: 2541–2566.

377 Guo, X.A., Sturge, K.C., Hunter, A.D., and Williams, M.C. (1994). *Macromolecules* 27: 7825–7829.

378 Fujikura, Y., Sonogashira, K., and Hagihara, N. (1975). *Chem. Lett.* 4: 1067–1070.

379 Sonogashira, K., Fujikura, Y., Yatake, T. et al. (1978). *J. Organomet. Chem.* 145: 101–108.

380 Sonogashira, K., Kataoka, S., Takahashi, S., and Hagihara, N. (1978). *J. Organomet. Chem.* 160: 319–327.

381 Ho, C.-L., Yu, Z.-Q., and Wong, W.-Y. (2016). *Chem. Soc. Rev.* 45: 5264–5295.

382 Wild, A., Winter, A., Hager, M.D., and Schubert, U.S. (2010). Synthetic strategies to conjugated Maion-Chain metallopolymers. In: *Conjugated Polymer Synthesis* (ed. Y. Chujo), 251–287. Weinheim: Wiley-VCH.

383 Khan, M.S., Davies, S.J., Kakkar, A.K. et al. (1992). *J. Organomet. Chem.* 424: 87–97.

384 Antonelli, E., Rosi, P., Lo Sterzo, C., and Viola, E. (1999). *J. Organomet. Chem.* 578: 210–222.

385 Lo Sterzo, C. (1999). *Synlett*: 1704–1722.

386 Pizzoferrato, R., Berliocchi, M., Di Carlo, A. et al. (2003). *Macromolecules* 36: 2215–2223.

387 Takahashi, S., Murata, E., Sonogashira, K., and Hagihara, N. (1980). *J. Polym. Sci: Polym. Chem. Ed.* 18: 661–669.

388 Johnson, B.F.G., Kakkar, A.K., Khan, M.S., and Lewis, J. (1991). *J. Organomet. Chem.* 409: C12–C14.

389 Atherton, Z., Faulkner, C.W., Ingham, S.L. et al. (1993). *J. Organomet. Chem.* 462: 265–270.

390 Davies, S.J., Johnson, B.F.G., Lewis, J., and Raithby, P.R. (1991). *J. Organomet. Chem.* 414: C51–C53.

391 Faulkner, C.W., Ingham, S.L., Khan, M.S. et al. (1994). *J. Organomet. Chem.* 482: 139–145.

392 Khan, M.S., Kakkar, A.K., Ingham, S.L. et al. (1994). *J. Organomet. Chem.* 472: 247–255.

393 Green, K.A., Simpson, P.V., Corkery, T.C. et al. (2012). *Macromol. Rapid Commun.* 33: 573–578.

394 Davies, S.J., Johnson, B.F.G., Khan, M.S., and Lewis, J. (1991). *J. Chem. Soc., Chem. Commun.*: 187, 188.

395 Yang, M., Zhang, L., Cai, Z. et al. (1995). *J. Polym. Sci., Part A: Polym. Chem.* 33: 2203–2208.

396 Yang, M., Zhang, L., Lei, Z. et al. (1998). *J. Appl. Polym. Sci.* 70: 1165–1172.

397 Irwin, M.J., Jia, G.-C., Payne, N.C., and Puddephatt, R.J. (1996). *Organometallics* 15: 51–57.

398 Jia, G.-C., Puddephatt, R.J., Scott, J.D., and Vittal, J.J. (1993). *Organometallics* 12: 3565–3574.

399 Vicente, J., Chicote, M.T., Abrisqueta, M.D., and Alvarez-Falcón, M.M. (2002). *J. Organomet. Chem.* 663: 40–45.

400 Vicente, J., Chicote, M.T., Alvarez-Falcón, M.M., and Jones, P.G. (2005). *Organometallics* 24: 2764–2772.

401 Jia, G., Puddephatt, R.J., and Vittal, J.J. (1992). *Polyhedron* 11: 2009–2014.

402 Tian, Y.-J., Meijer, E.W., and Wang, F. (2013). *Chem. Commun.* 49: 9197–9199.

403 Wang, X., Han, Y., Liu, Y. et al. (2017). *Angew. Chem. Int. Ed.* 56: 12466–12470.

404 Gao, Z., Han, Y., and Wang, F. (2018). *Nat. Commun.* 9: 3977.

405 Chen, J., Zhang, Z., Wang, C. et al. (2017). *Chem. Commun.* 53: 11552–11555.

406 Gao, Z., Zhu, J., Han, Y. et al. (2016). *Polym. Chem.* 7: 5763–5767.

407 Wang, C., Chen, Z., Liu, M. et al. (2019). *Polym. Chem.* 10: 3210–3216.

408 Loeb, S.J. and Shimizu, G.K.H. (1993). *J. Chem. Soc., Chem. Commun.*: 1395–1397.

409 Steenwinkel, P., Kooijman, H., Smeets, W.J.J. et al. (1998). *Organometallics* 17: 5411–5426.

410 Yount, W.C., Juwarker, H., and Craig, S.L. (2003). *J. Am. Chem. Soc.* 125: 15302–15303.

411 Gerhardt, W.W., Zucchero, A.J., South, C.R. et al. (2007). *Chem. Eur. J.* 13: 4467–4474.

412 Gerhardt, W.W., Zucchero, A.J., Wilson, J.N. et al. (2006). *Chem. Commun.*: 2141–2143.

413 Mardsen, J.A., Miller, J.J., Shirtcliff, L.D., and Haley, M.W. (2005). *J. Am. Chem. Soc.* 127: 2464–2476.

414 van Manen, H.-J., Nakashima, K., Shinkai, S. et al. (2000). *Eur. J. Inorg. Chem.* 2000: 2533–2540.

415 Back, S., Albrecht, M., Spek, A.L. et al. (2001). *Organometallics* 20: 1024–1027.

416 Schulze, B., Winter, A., Friebe, C. et al. (2017). *ACS Macro Lett.* 6: 181–184.

417 Solomonov, A.V., Marfin, Y.S., and Rumyantsev, E.V. (2019). *Dyes Pigm.* 162: 517–542.

418 Baudron, S.A. (2013). *Dalton Trans.* 42: 7498–7509.

419 Maeda, H., Hasegawa, M., Hashimoto, T. et al. (2006). *J. Am. Chem. Soc.* 128: 10024–10025.

420 Maeda, H., Kobayashi, H., and Akuta, R. (2013). *J. Porphyrins Phthalocyanines* 17: 86–91.

421 Yeh, H.C., Eichinger, B.E., and Andersen, N.H. (1982). *J. Polym. Sci.: Polym. Chem. Ed.* 20: 2575–2602.

422 Hanabusa, K., Isogai, T., Koyama, T., and Shirai, H. (1993). *Makromol. Chem.* 194: 197–210.

423 Bender, J.L., Shen, Q.-D., and Fraser, C.L. (2004). *Tetrahedron* 60: 7277–7285.

424 Chen, C.H. and Shi, J. (1998). *Coord. Chem. Rev.* 171: 161–174.

425 Du, F., Wang, H., Bao, Y. et al. (2011). *J. Mater. Chem.* 21: 10859–10864.

426 Huang, H., Zhong, C., and Zhou, Y. (2008). *Eur. Polym. J.* 44: 2944–2950.

427 Kwok, C.-C., Yu, S.-C., Sham, I.H.T., and Che, C.-M. (2004). *Chem. Commun.*: 2758–2759.

428 Peng, Q., Xie, M.-G., Huang, Y. et al. (2005). *Macromol. Chem. Phys.* 206: 2373–2380.

429 Alonso, P.J., Puertolas, J.A., Davidson, P. et al. (1993). *Macromolecules* 26: 4304–4309.

430 Marcos, M., Oriol, L., and Serrano, J.L. (1992). *Macromolecules* 25: 5362–5368.

431 Oriol, L., Alonso, P.J., Martinez, J.I. et al. (1994). *Macromolecules* 27: 1869–1874.

432 Caruso, U., Panunzi, B., Roviello, A. et al. (2001). *J. Polym. Sci., Part A: Polym. Chem.* 39: 2342–2349.

433 Caruso, U., Roviello, A., Sirigu, A., and Troise, C. (1998). *Macromolecules* 31: 1439–1445.

434 Radecka-Paryzek, W., Pospieszna-Markiewicz, I., and Kubicki, M. (2007). *Inorg. Chim. Acta* 360: 488–496.

435 Whiteoak, C.J., Salassa, G., and Kleij, A.W. (2012). *Chem. Soc. Rev.* 41: 622–631.

436 Chen, C.-T. and Archer, R.D. (1996). *Macromolecules* 29: 1957–1964.

437 Chen, H.-Y. and Archer, R.D. (1995). *Macromolecules* 28: 1609–1617.

438 Chen, H.-Y., Cronin, J.A., and Archer, R.D. (1994). *Macromolecules* 27: 2174–2180.

439 Chen, H.-Y., Cronin, J.A., and Archer, R.D. (1995). *Inorg. Chem.* 34: 2306–2315.

440 Kaya, İ. and Yíldírím, M. (2008). *J. Inorg. Organomet. Polym. Mater.* 18: 325–333.

441 Karabach, Y.Y., Kirillov, A.M., da Silva, M.F.C.G. et al. (2006). *Cryst. Growth Design* 6: 2200–2203.

442 Karabach, Y.Y., Kirillov, A.M., Haukka, M. et al. (2008). *J. Inorg. Biochem.* 102: 1190–1194.

443 Vermonden, T., de Vos, W.M., Marcelis, A.T.M., and Sudhölter, E.J.R. (2004). *Eur. J. Inorg. Chem.* 2004: 2847, 2852.

444 Vermonden, T., van der Gucht, J., de Waard, P. et al. (2003). *Macromolecules* 36: 7035–7044.

445 Sato, T. and Higuchi, M. (2012). *Chem. Commun.* 48: 4947–4949.

446 Narayana, Y.S.L.V., Chakraborty, C., Rana, U. et al. (2018). *RSC Adv.* 8: 37193–37199.

447 Narayana, Y.S.L.V., Yoshida, T., Chakraborty, C. et al. (2020). *ACS Appl. Polym. Mater* 2: 326–334.

448 Sato, T. and Higuchi, M. (2019). *Tetrahedron Lett.* 60: 940–943.

449 Chandrasekhar, V. and Thirumoorthi, R. (2010). *Dalton Trans.* 39: 2684–2691.

450 Rivera, N.M., Engler, E.M., and Schumaker, R.R. (1979). *J. Chem. Soc., Chem. Commun.*: 184–185.

451 Böhm, M.C. (1984). *Phys. Status Solidi B* 121: 255–264.

452 Reynolds, J.R., Chien, J.C.W., and Lillya, C.P. (1987). *Macromolecules* 20: 1184–1191.

453 Ribas, J. and Cassoux, P.C. (1981). *C.R. Acad. Sci., Ser. IIc: Chim.* 293: 665.

454 Teo, B.-K., Wudl, F., Hauser, J.J., and Krüger, A. (1977). *J. Am. Chem. Soc.* 99: 4862–4863.

455 Wang, F. and Reynolds, J.R. (1990). *Macromolecules* 23: 3219–3225.

456 McNamara, W.R., Han, Z., Alperin, P.J. et al. (2011). *J. Am. Chem. Soc.* 133: 15368–15371.

457 Downes, C.A. and Marinescu, S.C. (2015). *J. Am. Chem. Soc.* 137: 13740–13743.

458 Clough, A.J., Yoo, J.W., Mecklenburg, M.H., and Marinescu, S.C. (2015). *J. Am. Chem. Soc.* 137: 118–121.

459 Dong, R., Pfeffermann, M., Liang, H. et al. (2015). *Angew. Chem. Int. Ed.* 54: 12058–12063.

460 Kealy, T.J. and Pauson, P.L. (1951). *Nature* 168: 1039.

461 Miller, S.A., Tebboth, J.A., and Tremaine, J.F. (1952). *J. Chem. Soc.*: 632–635.

462 Arimoto, F.S. and Haven, A.C. Jr., (1955). *J. Am. Chem. Soc.* 77: 6295–6297.

463 Neuse, E.W. (1981). *J. Macromol. Sci. Chem.* A16: 3–72.

464 Hirao, T., Kurashina, M., Aramaki, K., and Nishihara, H. (1996). *J. Chem. Soc., Dalton Trans.*: 2929–2933.

465 Southard, G.E. and Curtis, M.D. (2001). *Organometallics* 20: 508–522.

466 Southard, G.E. and Curtis, M.D. (1997). *Organometallics* 16: 5618–5620.

467 Deck, P.A., Lane, M.L., Montgomery, J.L. et al. (2000). *Organometallics* 19: 1013–1024.

468 Hollandsworth, C.B., Hollis, W.G. Jr.,, Slebodnick, C., and Deck, P.A. (1999). *Organometallics* 18: 3610–3614.

469 Siebert, W. (1985). *Angew. Chem. Int. Ed. Engl.* 24: 943–958.

470 Siebert, W. (1991). *Russ. Chem. Rev.* 60: 784–791.

471 Kuhlmann, T., Roth, S., Rozière, J., and Siebert, W. (1986). *Angew. Chem. Int. Ed. Engl.* 25: 105–107.

472 Kuhlmann, T., Roth, S., Roziere, J. et al. (1987). *Synth. Met.* 19: 757–762.

473 Siebert, W. (1988). *Pure Appl. Chem.* 60: 1345–1348.

474 Grimes, R.N. (1992). *Chem. Rev.* 92: 251–268.

475 Arnold, R., Matchett, S.A., and Rosenblum, M. (1988). *Organometallics* 7: 2261–2266.

476 Foxman, B.M., Gronbeck, D.A., and Rosenblum, M. (1991). *J. Organomet. Chem.* 413: 287–294.

477 Nugent, H.M., Rosenblum, M., and Klemarczyk, P. (1993). *J. Am. Chem. Soc.* 115: 3848–3849.

478 Foxman, B.M., Rosenblum, M., Sokolov, V., and Khrushchova, N. (1993). *Organometallics* 12: 4805–4809.

479 Rosenblum, M., Nugent, H.M., Jang, K.-S. et al. (1995). *Macromolecules* 28: 6330–6342.

480 Bush, B.F. and Lagowski, J.J. (1988). *Organometallics* 7: 1945–1948.

481 Friebe, C., Hager, M.D., Winter, A., and Schubert, U.S. (2012). *Adv. Mater.* 24: 332–345.

482 Hsu, C.-Y., Zhang, J., Sato, T. et al. (2015). *ACS Appl. Mater. Interfaces* 7: 18266–18272.

483 Shankar, S., Lahav, M., and van der Boom, M.E. (2015). *J. Am. Chem. Soc.* 137: 4050–4053.

484 Wani, W.A., Prashar, S., Shreaz, S., and Gómez-Ruiz, S. (2016). *Coord. Chem. Rev.* 312: 67–98.

5

Supramolecular Polymers, Based on π-Electronic Interactions

5.1 General Aspects

The interactions of conjugated systems via their π-electronic systems represent an important motif in supramolecular chemistry [1–3]. Thereby, the interaction strength between two π-conjugated moieties is strongly dependent on the environment, i.e. on the solvent. In aqueous medium, the π–π stacking interaction between two aromatic compounds is mainly the result of the so-called hydrophobic effect. Basically, the aromatic π-electronic surface is solvated by water molecules that have a much higher energy than bulk water; thus, the aromatic molecules will stack onto each other to reduce the total π-electronic surface exposed to the environment. The interactions between solvent molecules will intrinsically be weaker than that for water, and thus, the solvophobic forces will always be less expressed. Hunter and Sanders analyzed the nature of π–π stacking interactions and proposed a practical model to predict the geometry of a favorable interaction (Figure 5.1) [4]. Basically, three-point charges are considered in this merely electrostatic model and, for an aromatic compound, they are located as follows: the first is placed at the site of a selected atom which is typically a positively charged carbon center; the other two are located above and below the plane of the π-conjugated system (these π-point charges each count as −1/2). An important message of this model is that geometries other than the face-to-face-type orientation can hardly be found and employed with respect to the design of supramolecular architectures. Accordingly, a large variety of assemblies featuring a columnar-type stacking of molecules, mainly due to π–π stacking interactions, has already been published and reviewed [3, 5–8]. Thus, the supramolecular polymerization of monomers, small molecules as well as macromolecular ones, aligned into highly ordered solid-state assemblies driven by π–π stacking interactions, represents an established field in modern materials chemistry. However, the supramolecular polymerization of such building blocks *in solution* remains problematic due their rather weak and nondirectional character. This issue epitomizes the major disadvantage, in particular, when compared to the highly directional supramolecular polymerizations discussed in most of the other chapters, e.g. metal-to-ligand coordination (Chapter 4) or the various types of host-guest complexations (Chapters 6–10).

Supramolecular Polymers and Assemblies: From Synthesis to Properties and Applications, First Edition.
Ulrich S. Schubert, George R. Newkome, and Andreas Winter.

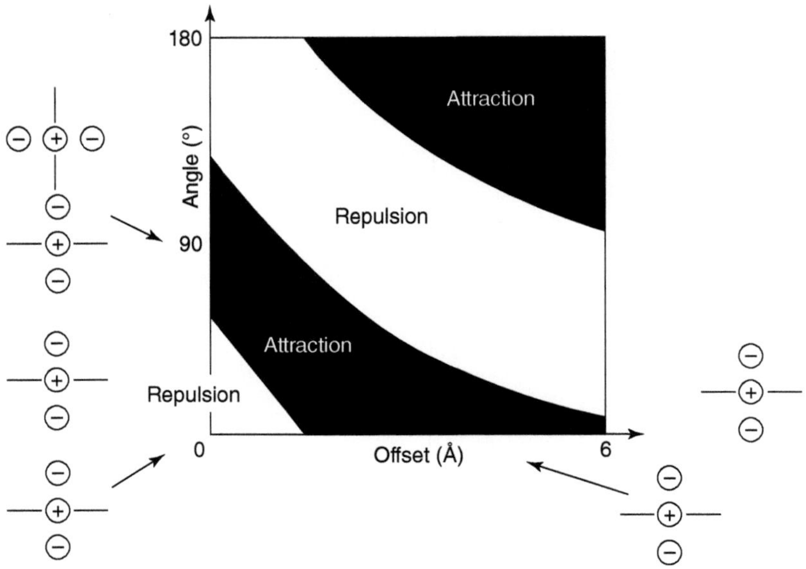

Figure 5.1 Illustration of the interactions between two idealized *p* atoms as a function of their orientation – two "attractive" geometries and the "repulsive" face-to-face-type geometry are shown. Source: Hunter and Sanders [4]; © 1990 American Chemical Society.

Würthner and coworkers have noted that the chemical equilibria, which involve the monomeric unit and the aggregated species, basically describe the self-assembly of π-conjugated systems in solution [8]; however, if the latter is represented by various types of assemblies, multiple equilibria have to be considered instead. A range of mathematical models has been introduced to describe the process of π–π aggregation in a quantitative fashion; these methods have been summarized elsewhere [8, 9] and their discussion is beyond the scope of this chapter.

Diverse factors have an influence on both the resultant structure of the π–π stacked aggregates and the thermodynamics of the underlying self-assembly process. Amongst others, the strength of the electrostatic interactions and the size and geometry of the π-conjugated molecule are known to contribute considerably to this self-assembly process. These aspects have been analyzed by Würthner and coworkers for a broad range of functional π-conjugated systems, including simple aromatics (e.g. benzenes), polycyclic aromatic hydrocarbons (PAHs, e.g. triphenylenes and hexabenzocoronenes [HBCs]), heteroaromatic porphyrins and phthalocyanines, quadrupolar naphthalene bisimides (NBIs) and perylene bisimides (PBIs), dipolar or even zwitterionic merocyanines and squaraines, etc. [8]. Remarkably, from all of these systems only the merocyanines allowed a reliable prediction of their aggregate's structure and of the binding strength from rather simple physical relationships due to their highly dipolar character and the resulting dominant electrostatic contribution within the self-assembly process [10].

In special cases, the dimerization of π-conjugated molecules, either in a homo- or hetero-association mode, can be facilitated by adding an appropriate macrocyclic host. The π-conjugated systems will enter the host's cavity, thereby stabilizing the aggregate via a combination of non-covalent interactions offered by its structural composition. This approach has also been used to assemble supramolecular polymers and will be discussed within the Chapters 6–10 that deal with the different types of host–guest interactions. In particular, the formation of π–π stacked dimers within the cavities of cucurbiturils is worth mentioning in this context (Chapter 7).

5.2 Columnar Supramolecular Polymers, Based on π–π Stacking Interactions

The ability of extended π-conjugated systems to assemble via π–π stacking interactions can be utilized to prepare supramolecular materials in a highly controlled fashion, in particular when making use of the more directional charge–transfer (CT) or cation–anion π-interactions (see Sections 5.3 and 5.4). Accordingly, a broad range of π-conjugated materials – polymers as well as monodisperse "oligomers" (i.e. rod- or disk-shaped π-conjugated small molecules) – has been assembled into conductive or liquid-crystalline (LC) solid-state nanostructures [1–3]. Despite the intrinsic limitations arising from rather low K_a values, the assembly of supramolecular polymers in solution on the basis of π–π stacking interactions has been realized for a variety of unimers.

As a prominent representative in this respect, the supramolecular polymerization of hexa-*peri*-substituted HBCs has to be mentioned, since they represent discotic derivatives of graphene that are solubilized via their side chains. Their LC properties, attributed to their high anisotropic charge-carrier mobility, are well documented in literature [11]. Müllen and coworkers investigated the supramolecular polymerization of the HBC derivatives **1** in dilute solutions (Figure 5.2) [12]. On the basis of concentration- and temperature-dependent nuclear magnetic resonance (NMR) spectroscopy, an isodesmic supramolecular polymerization (IDP) mechanism was concluded for **1a–e** in a range of solvents of different polarities [i.e. benzene, cyclohexane, tetrahydrofuran (THF), and tetrachloroethane]. The van't Hoff analysis of the experimental data suggested that the supramolecular polymerization of **1** to be generally favored for enthalpic reasons, but entropically disfavored (thus, the ability to polymerize is increased at lower temperatures). Due to the unhindered stacking, **1e** exhibited the highest K_a value within the series (K_a of 5500 M^{-1} in THF), whereas lower K_a values were found for the derivatives with branched side chains that decreased the tendency of these molecules to associate. In the same context, the self-assembly of the amphiphilic HBCs **2** in THF or THF/water mixtures was studied by Aida and coworkers [13]. In contrast to the 1D columnar structures obtained for **1**, quasi-2D tubular assemblies were found, in which the alkyl chains intertwined within the bilayer and crystallized (representing the driving force for tubular growth); the oligo(ethylene glycol) (OEG) chains were located at the periphery, thereby allowing solubility in polar media (Figure 5.3). From a mechanistic point of view, an allosteric cooperativity rather than an IDP was proposed [2], since the alkyl chains and phenyl rings in the substituents were found to be crucial to obtain the tubular materials with an enormous number of HBC molecules.

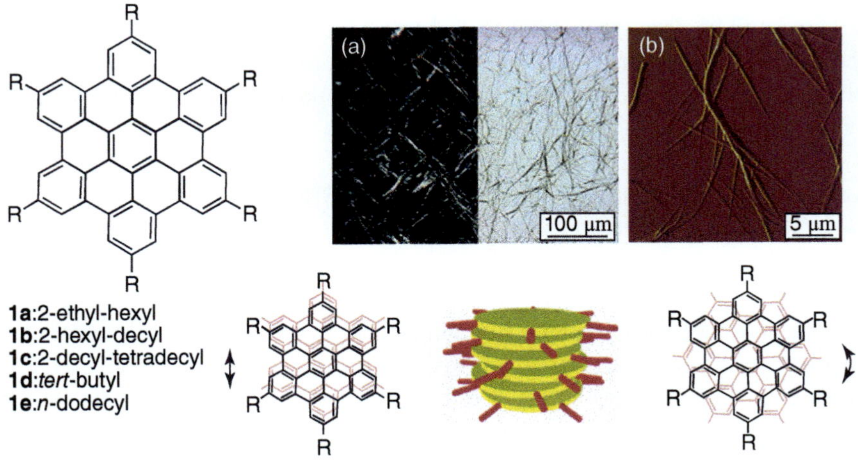

Figure 5.2 Schematic representation of the hexa-*peri*-substituted hexabenzocoronenes **1** along with the proposed stacking mode. The polarized optical microscopy (POM) (a) and atomic-force microscopy (AFM) and (b) images of drop-casted fibers of **1e** are also shown. Source: Kastler et al. [12]. Figure reproduced with kind permission; © 2005 American Chemical Society.

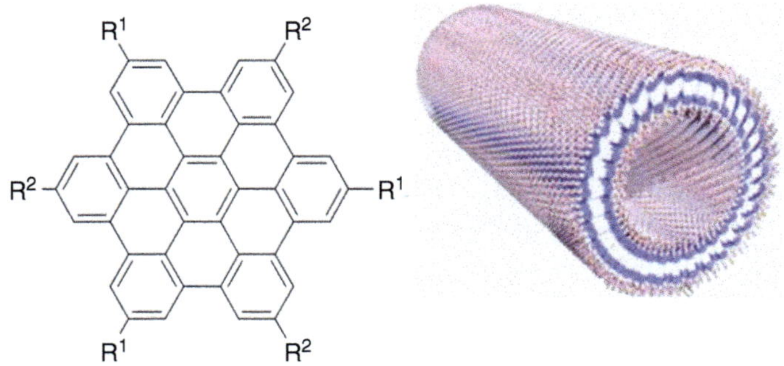

2a: $R^1 = C_{12}H_{25}$ $R^2 = C_6H_4\text{-}O(CH_2CH_2O)_3CH_3$
2b: $R^1 = C_{12}H_{25}$ $R^2 = C_6H_4\text{-}O(CH_2CH_2O)_2CH_3$
2c: $R^1 = C_{12}H_{25}$ $R^2 = C_6H_4\text{-}O(CH_2CH_2O)_2CH_3$
2d: $R^1 = C_{12}H_{25}$ $R^2 = C_6H_4\text{-}OCH_3$
2e: $R^1 = C_{12}H_{25}$ $R^2 = C_6H_5$
2f: $R^1 = C_{16}H_{33}$ $R^2 = C_6H_4\text{-}O(CH_2CH_2O)_3CH_3$
2g: $R^1 = C_{13}H_{27}$ $R^2 = C_6H_4\text{-}O(CH_2CH_2O)_3CH_3$
2h: $R^1 = C_{12}H_{25}$ $R^2 = CCO(CH_2CH_2O)_3CH_3$

Figure 5.3 Schematic representation of the amphiphilic HBCs **2** and their supramolecular polymerization into tubular structures. Source: Modified from Hill et al. [13].

Figure 5.4 Schematic representation of the self-assembly of a discotic amphiphile in aqueous medium. Source: Aida [15], © 2020 John Wiley and Sons.

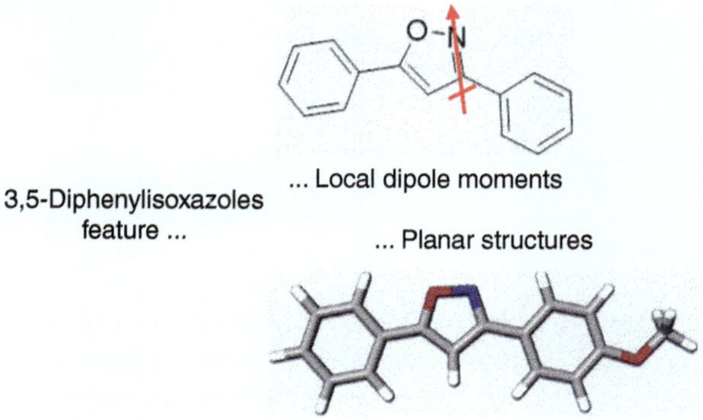

Figure 5.5 Schematic representation of 3,5-diphenylisoxazoles featuring a local dipole moment and a planar structure. Source: Haino et al. [17], © 2008 Royal Society of Chemistry.

For example, a precursor assembly with a width of 22 nm would eventually generate a nanotube of c. 10 μm length – corresponding to c. 2×16 unidirectionally aligned columnar π–π stacks, each having approximately 5×10^4 molecules. It is noteworthy that the ability of such discotic amphiphiles to assemble into ordered structures was already recognized by Aida et al. in 1988 [14]. A porphyrin derivative, equipped with four OEG arms, generated a 1D cofacial supramolecular polymer in aqueous media (Figure 5.4); ultrathin and ultralong fibers were obtained by slow solvent evaporation.

π-Conjugated molecules, possessing 3,5-diphenylisoxazole moieties, represent one further class of compounds that assemble in solution according to the concepts of supramolecular polymerizations [16]. Basically, such fragments possess planar conformations due to the absence of any *peri*-steric hindrance, and thus, intermolecular π–π stacking interactions are highly facilitated; moreover, the localized dipole along the C=N double bond is known to further enhance the stacking ability due to intermolecular dipole–dipole interactions (Figure 5.5). Haino et al. showed by concentration-dependent ^{1}H NMR and UV/vis absorption measurements (in $CHCl_3$ and methylcyclohexane, respectively) that the self-assembly of **3** followed an IDP pathway. An interplay of solvophobic, π–π stacking, and dipole–dipole interactions gave helical fibers, which, in turn, acted as organogelators in various solvents (e.g. methylcyclohexane, Et_2O, acetone, EtOAc, dimethylsulfoxide [DMSO], and dimethyl formamide [DMF]). The critical concentration for this highly thermoreversible gelation process was in the range of 0.8–3.5 wt% (depending on the nature of the solvent). Field-emission scanning electron microscopy (FE-SEM) was used to investigate the morphology of the obtained fibers; for **3b** much thicker fibers were formed, presumably due to a stronger entwinement of its olefinic side chains (Figure 5.6).

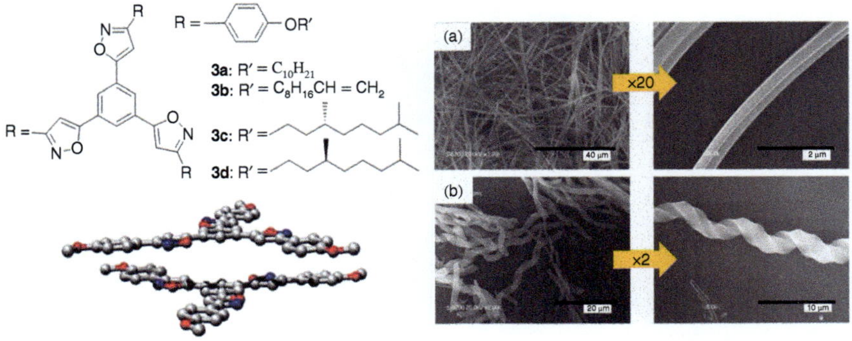

Figure 5.6 Schematic representation of the C_3-symmetric diphenylisoxazoles **3** along with the proposed packing mode. FE-SEM images of the fibers obtained from **3a** (a) and **3b** (b). Source: Haino et al. [17]. Figure reproduced with kind permission; © 2008 The Royal Chemical Society.

Diketopyrrolopyrroles (DPPs) are established industrial pigments, which have attracted scientific attention due to their potential usages as dyes in organic solar cells or field-effect transistors as well as bioimaging. The parent **4a** was first reported in 1974, and the most prominent example is the well-known red "Ferrari" pigment **4b** (Figure 5.7a) [19]. Increasing interest can be rationalized by the intense color of DPP dyes as well as their relative ease of synthesis and functionalization. The latter allows for one to adjust the opto-electronic properties, and thus, a broad range of DPP derivatives has been reported in the last decades [20]. In this context, Ajayaghosh et al. emphasized that the supramolecular self-assembly of DPPs into ordered arrays, i.e. achieving a spatial arrangement and hierarchical ordering on the macroscopic length scale, still represents a challenge. Fréchet and coworkers demonstrated that the supramolecular polymerization of the donor–acceptor–donor triads **5** into tight, aligned arrays occurred via π–π stacking of the terminal chromophores (Figure 5.7b) [18]. As a result, the photovoltaic devices incorporating these dyes gave high power-conversion efficiencies (PCE) and fill factors (FF). The highest values were observed for the pyrene-containing derivative **5c**, since the π–π stacking ability of pyrene is most pronounced. Similar derivatives, equipped with long, branched *N*-alkyl chains, were studied by Beaujuge, Pisula, and coworkers; in this work, the self-assembly from solution gave highly ordered morphologies, which exhibited transistor-like properties [21]. Beyond this, the self-assembly of DPPs can also proceed via double H-bonding interactions of their lactam sites; this approach has been widely applied to organize DPPs into ordered arrays (see Section 3.2.2) [20].

Würthner et al. mainly focused on the supramolecular polymerization of PBIs and their rich supramolecular chemistry as well as the potential applications arising therefrom. This topic has been reviewed mainly by these authors [8, 22]. The colored PBI dyes, used in automotive industry or as *n*-type materials, are chemically robust and possess an inherently flat molecular structure (a requirement for efficient supramolecular π–π stacking interactions). Regarding the supramolecular polymerization of PBI dyes in solution two different types of structures have been employed:

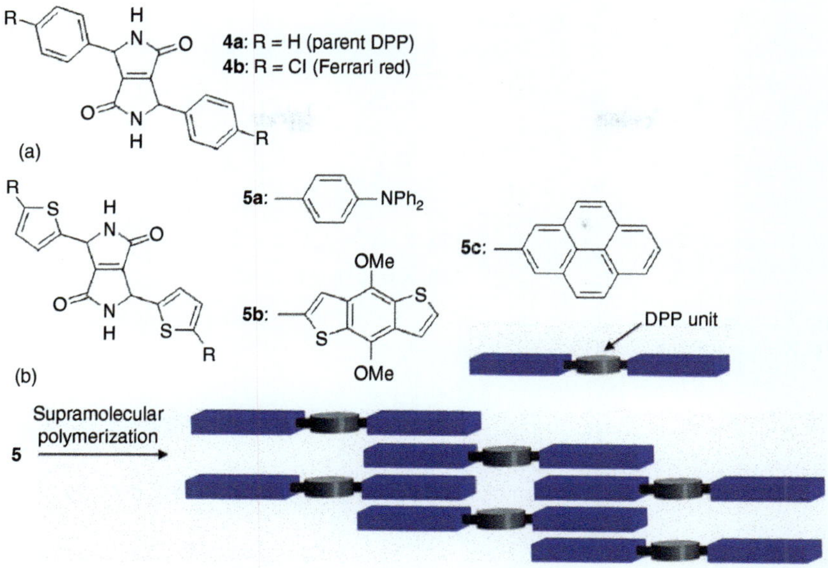

Figure 5.7 (a) Schematic representation of the basic diketopyrrolopyrrole structures **4**. (b) Schematic representation of the donor–acceptor–donor molecules **5**, which assembled due to π–π stacking interactions of their outer electron-rich chromophores. Source: Lee et al. [18], © 2011 John Wiley and Sons.

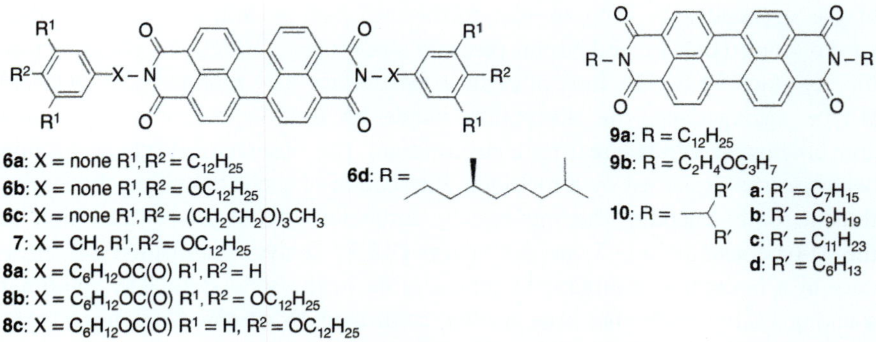

Figure 5.8 Schematic representation of representative core-unsubstituted PBI dyes (**6–10**).

core-unsubstituted (**6–10**, Figure 5.8) and core-substituted derivatives (**11–12**, Figure 5.9). In general, the association of PBI molecules is a typical example for a system in which monomeric and aggregated species are in equilibrium. As detailed in Chapter 1, different models can be employed to analyze such self-assembly process in solution [2, 9, 23, 24]: (i) the monomer–dimer model, (ii) the (modified) isodesmic model, and (iii) the (anti-) cooperative model.

These same authors used concentration-dependent UV/vis absorption and emission spectroscopy to determine the K_a values for the association of their PBI dyes. The derivative **6a** (Figure 5.8), for example, self-assembled in solvents

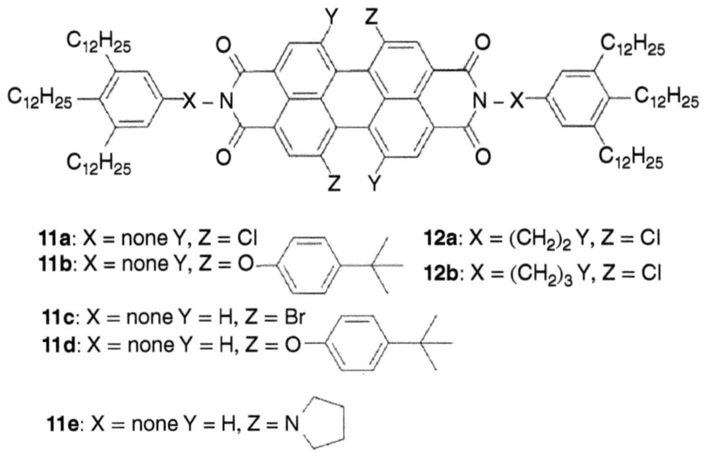

Figure 5.9 Schematic representation of representative bay-substituted PBI dyes (**11** and **12**).

of low polarity (e.g. methylcyclohexane) according to an IDP-type mechanism, i.e. an equal binding constant was observed throughout the entire self-assembly process ($K = 9.7 \times 10^4\ M^{-1}$) [25]. The negative values for the enthalpy and entropy, as extracted from the van't Hoff data analysis, signified the enthalpy-driven nature of the supramolecular association of **6a** in non-polar solvents. The size of the aggregates in solution was further verified by diffusion-ordered NMR spectroscopy (DOSY) and vapor pressure osmometry (VPO). As characteristic for the stacking of PBI molecules in a face-to-face-type manner (the so-called H-type aggregation), the absorption bands became hypochromically shifted and broadened with increasing concentration. The observed UV/vis absorption behavior was explained by a rotational dislocation of the PBIs within the stacks, thus partially enabling the intrinsically forbidden optical J-type transition in the H-type assemblies. X-ray diffraction (XRD) analysis confirmed the presence of a hexagonal columnar LC phase of **6a** in the solid state. The columnar stacking of the molecules was further confirmed by atomic-force microscopy (AFM) imaging; a polarized optical microscopy (POM) image of the LC phase is shown in Figure 5.10b. PBI **6a** exhibited a color-tunable fluorescence behavior – emission colors ranging from green to red were observed (depending on the degree of aggregation), together with an increased emission lifetime (attributed to a retarded $S_1 \rightarrow S_0$ transition in H-type aggregates, Figure 5.10a) and depolarization. Moreover, the aggregates showed a directional charge-carrier mobility of up to $0.42\ cm^2\ V^{-1}\ s^{-1}$.

Extensive theoretical studies suggested a preferred rotational displacement of the stacked dye molecules by c. 30° in the ground state (thereby, neglecting any further influence from the substituents, Figure 5.11a), whereas the angle was close to 0° in the excited state [27]. Thus, the overall supramolecular structure was changed upon optical excitation as reflected by an energetic relaxation and a bathochromic shift of the excimer-type emission band. These assumptions were

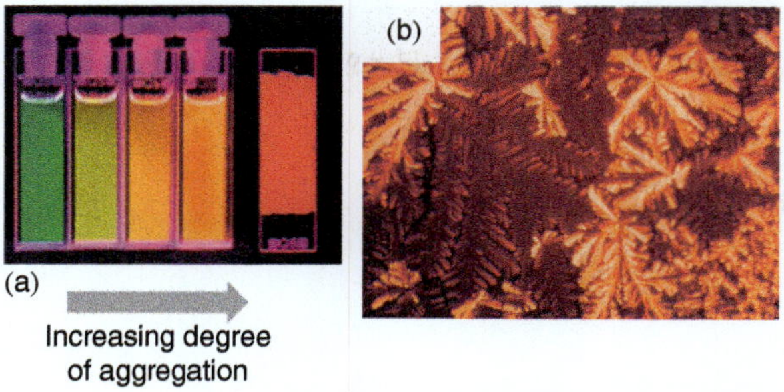

Figure 5.10 (a) Representation of the color tunability of the emission of **6a** in solution, the red emission from a spin-coated film of **6a** is shown on the very right. (b) POM image of the LC phase of **6a**. Source: Chen et al. [25]. Figure reproduced with kind permission; © 2007 Wiley-VCH.

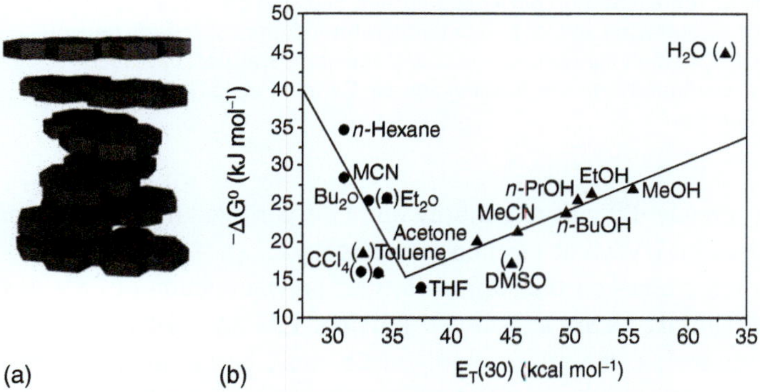

Figure 5.11 (a) Schematic representation of the proposed columnar stacking mode of PBI molecules. (b) Plot of the standard Gibbs free energy (ΔG^0) for an isodesmic aggregation process vs. the solvent polarity parameter [$E_T(30)$] for **6a** (●x025CF;) and **6c** (▲x025B2;). Source: Chen et al. [26], © 2012 The Royal Chemical Society.

later proven by time-resolved spectroscopy [28] and supported by quantum-dynamic simulations [29].

The impact of the solvent polarity on the self-assembly process of PBIs was also studied by the Würthner group [26]. For **6a** and **6c**, a biphasic dependency of the Gibbs free energy (ΔG_0) of the aggregation on the polarity of the solvent was found (Figure 5.11b): the K_a value decreased when aliphatic solvents with low polarity [i.e. (cyclo)alkanes] were replaced by the ones with moderate polarity (e.g. toluene, THF), attributed to the negligible electrostatic interaction of the highly quadrupolar PBI molecules in such solvents. However, the enhancement of the binding strength in highly dipolar (e.g. acetone, MeCN) or even protic (e.g. MeOH) solvents was attributed to dispersion interactions, which are known to increase in strength with the solvent polarity. Halogenated solvents (e.g. CCl_4 and $CHCl_3$) feature high polarizabilities and, thus, offered the best solvation of PBIs as reflected by a reduced binding strength. Finally, the hydrophobic effect of water resulted in the highest

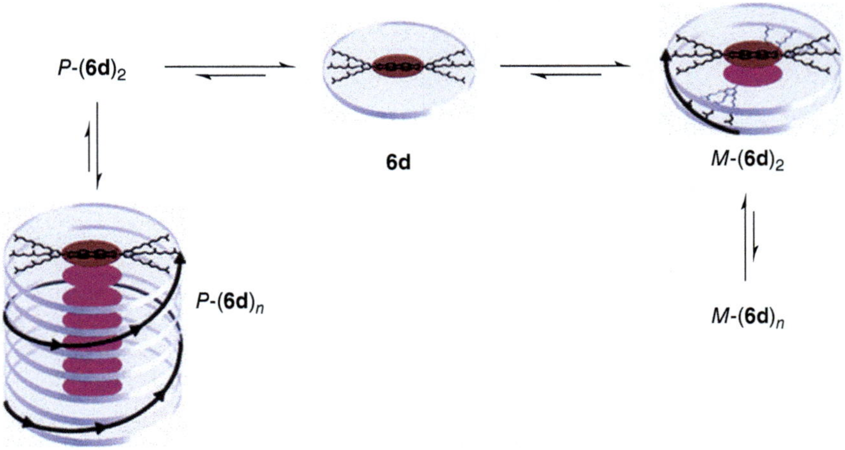

Figure 5.12 Schematic representation of the equilibrium between **6d** and its aggregates in solution. At low a concentration and/or a high temperature, dimers of *M*-type helicity were preferred; whereas, at high concentrations and/or low temperatures, extended polymeric stacks of *P*-configuration were formed. Source: Dehm et al. [31], © 2007 American Chemical Society.

K_a values that are already difficult to quantify due to experimental limitations of concentration-dependent UV/vis absorption spectroscopy.

In one of the initial studies on the supramolecular polymerization of PBIs, the self-assembly of the hexa*kis*alkoxy-substituted PBIs **6b**, **11b**, and **11d** was studied (Figure 5.8) [30]. As for **6a**, the self-assembly of **6b** could be rationalized by an isodesmic model; however, fluorescence of **6b** was absent, independent from the state of aggregation, and assigned to a photoinduced electron-transfer process from the donor-type substituents to the acceptor-type PBI core. The helical growth of a columnar PBI assembly was found for the chiral derivative **6d** [31]. Dehm et al. concluded from the absence of any isosbestic points in the temperature-dependent circular dichroism (CD) spectra that at a low concentration (or at a high temperature) the left-handed dimer $(\mathbf{6d})_2$ was formed. At a low temperature (or at an elevated concentration), this *M*-dimer, in turn, transformed into the right-handed *P*-dimer and subsequent chain growth resulted in an extended overall right-handed helical assembly (Figure 5.12).

Meijer and coworkers investigated, *inter alia*, the self-assembly of the PBI derivative **7** (Figure 5.8), for which a similar aggregation behavior as for **6a** was found. However, the increased flexibility of **7** reduced the π–π stacking ability, and thus, a much lower K_a value as for **6a** was determined (K_a of $1.3 \times 10^4\ M^{-1}$) [32]. These authors also performed the co-assembly of **7**, as an acceptor-type compound, with a donor-type oligo(*p*-phenylene vinylene) (OPV) derivative. The orthogonal self-assembly of these two materials in methylcyclohexane gave separate nanosized stacks; in toluene, a much better solvent for both compounds, only molecularly dissolved molecules were present (Figure 5.13). The thin-solid films, deposited from either of these solutions, featured a photoinduced donor-to-acceptor electron

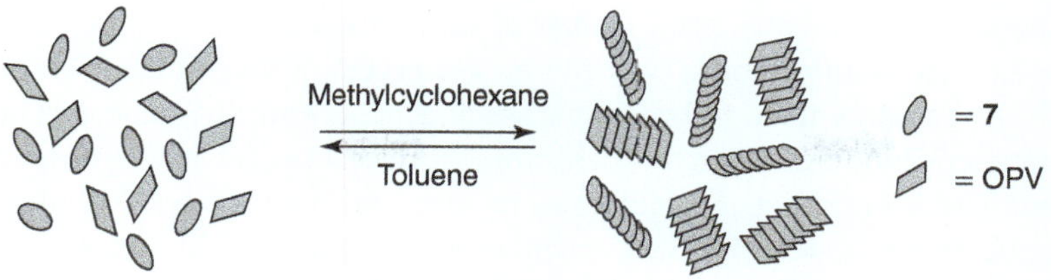

Figure 5.13 Schematic representation of the solvent-dependent self-assembly/disassembly behavior observed for a binary mixture of **7** and an OPV derivative. Source: van Herrikhuyzen et al. [32], © 2004 American Chemical Society.

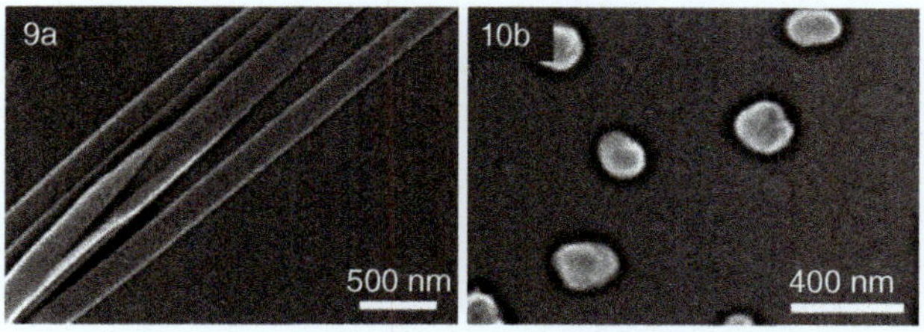

Figure 5.14 Scanning electron microscopy images of the nanobelts and nanoparticles obtained from the self-assembly of **9a** and **10b**, respectively. Source: Balakrishnan et al. [34]. Figure reproduced with kind permission; © 2006 American Chemical Society.

transfer. Microscopy imaging revealed the presence of an ordered fibrous network in the films obtained from methylcyclohexane solutions, whereas the films prepared from toluene solutions exhibited no such ordering.

In a related study, also the effects of long alkoxy substituents on the self-assembly behavior were investigated (PBIs **8** with *in toto* zero, two, or six such substituents, Figure 5.8) [33]. Balakrishnan et al. utilized PBIs, in which alkyl chains – linear (**9**) as well as branched (**10a–c**) – were directly attached to the *N*-atoms of the PBI cores (Figure 5.8) [34, 35]. For the PBIs **7**, 1D belt-type nanocrystals were obtained (Figure 5.14); however, thermodynamic studies of the self-assembly process could not be performed due to the limited solubility of **7**. On the contrary, the PBIs **10a–c** with branched alkyl chains were much more soluble and only low K_a values were determined (for a comparison of K_a values for various PBI derivatives, see Ref. [26]). Moreover, steric hindrance prevented the controlled 1D growth of the aggregates and 0D nanoparticles were obtained instead (Figure 5.14). A different situation was, however, reported for **10d**, from which ultralong semiconducting fibrils, i.e. up to 1.1 mm in length, were grown by drop-casting from dodecane solution [36]. The analogous tetra-*ortho*-alkylated PBI gave only thin fibers due to isotropic crystallization.

Thus, the nature of the N-substituents has a profound influence on the self-assembly behavior of core-unsubstituted PBI derivatives in solution as well as solid state [22]. This behavior can be adjusted further by introducing substituents on the core, either in the headland or, more commonly, in the bay positions. Basically, introducing substituents in these positions of the PBI core results in some steric constraint and, thus, considerable distortion/twisting of the primary planar molecule. Consequently, the K_a values of such PBIs will be lower compared to those with unsubstituted cores, typically by 2–3 orders of magnitude [37]. In addition, if the substituents in bay positions are directed toward the same side of the PBI core, steric hindrance is on the π-electronic surface, thus preventing further aggregation. The backside of the molecule, in turn, will interact with another molecule much easier and stable dimers might even be accessible. This behavior is of relevance in view of the formation of linear supramolecular polymers based on the selective dimerization of distorted PBI molecules (also see Section 5.5); however, examples making use of this concept have – to the best of our knowledge – not yet been reported. One further aspect to be dealt with is the twisting angle that strongly effects the supramolecular assembly of bay-substituted PBIs as well as their electronic and LC properties [22]. For example, the core-disubstituted derivatives **11c–e** feature rather small twist angles (15°–25°); whereas, the tetra-substituted **11a/b** possess twist angles of up to 36°. The former gives closer packed assemblies, thus featuring a stronger excitonic coupling behavior and hypochromically shifted absorption bands.

It has also been shown that the degree-of-distortion affects the aggregation behavior of PBIs in the solid state: for the strongly distorted PBI derivative **11a**, a rectangular columnar mesophase was found; whereas, the less twisted PBIs **11c/d** instead gave hexagonal mesophases (Figure 5.15) [37]. As a result, from the different packing behavior, the assemblies showed distinct differences in their properties: due to the longitudinally slipped stacking of **11a** fluorescent J-aggregates were obtained. It was concluded that the distortion of the PBI core induced a transition from the preferred π–π stacking mode, which is the rotational displacement of molecules within highly ordered columnar mesophases, to a longitudinal one (i.e. formation of a rectangular columnar mesophase). Regarding the charge-carrier mobility, the core-substituted PBI **12a** was compared to the core-unsubstituted derivative **6b** – both compounds featured similarly good charge-carrier mobilities in their LC phases with a significantly longer charge-carrier lifetime for **12a** (by a factor of c. 100); this drastic increase was attributed to the higher degree-of-ordering within the stacks [38]. The Percec group focused on the self-assembly of dendronized PBIs – tetrachlorinated as well as non-chlorinated ones. The molecularly distorted PBIs provided 3D crystalline phases; in contrast to the former thermodynamically controlled process, kinetic control was identified as the key driving force to assemble the core-unsubstituted molecules into 2D periodic arrays and 3D crystalline phases [39].

One further aspect to be mentioned in the context of core-substituted PBIs is chirality: such molecules exist as atropo-enantiomers, and thus, various aggregation modes are basically possible, i.e. a homochiral and a heterochiral one with different geometries and presumably (binding) energies [40, 41]. To prevent the interconversion between the *P*- and *M*-atropo-enantiomers, Würthner and coworkers synthesized conformationally locked PBI derivatives in which the bay positions were bridged with diethylene glycol moieties in a crosswise fashion. The strapped enantiomers were resolved by chiral high-performance liquid chromatography (HPLC) and investigated regarding their self-assembly behavior in solution (notably

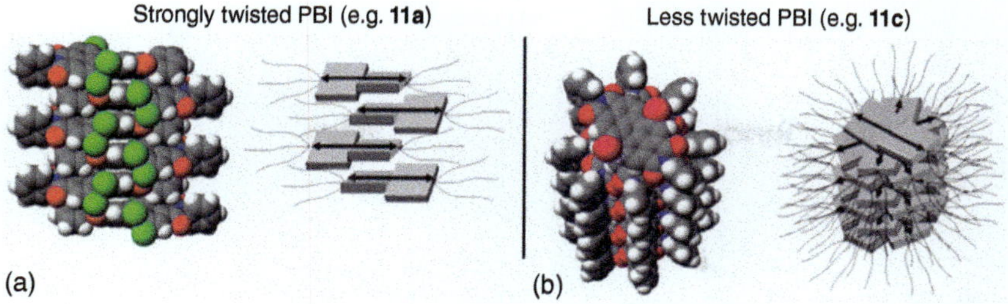

Figure 5.15 Schematic representation of the two different stacking modes: Longitudinally displaced J-type stacking of **11a** (a) and rotationally displaced cofacial stacking of **11c** (b). Source: Chen et al. [37]. © 2007 John Wiley and Sons.

the pure enantiomers as well as various enantiomeric mixtures were studied and compared). Due to a more efficient overlap of the π-electronic surfaces, self-sorting of the molecules preferentially gave the homodimers [42]. This process could be correlated to the rigidity of the molecules – the least rigid PBIs featured the poorest self-sorting behavior [43]. In the solid state, racemic PBIs assembled into soft columnar crystalline phases; whereas, the enantiopure samples formed lamellar LC mesophases.

Finally, amphiphilic PBIs with poly(ethylene glycol) residues in the core positions were employed by the Rybtchinski group to assemble supramolecular structures. For example, the self-assembly of such an amphiphile with an additional poly(phenyl acetylene) (PPA) side chain in aqueous media gave 2D porous networks that were stable in solution as well as in the solid state, as evidenced by cryogenic microscopy imaging [44]. The presence of perfluorinated side chains, instead of the PPA, tremendously influenced the self-assembly behavior – relative to the non-fluorinated counterpart; the K_a value in water was higher by 2 orders of magnitude (attributed to the large hydrophobic surface area of the perfluorocarbon chain) [45]. The addition of THF to the aqueous medium decreased the K_a value and changed the mechanism of aggregation as well as the resulting morphologies. At a low THF content, an isodesmic self-assembly process produced tube-like fibers, whereas a cooperative self-assembly pathway facilitated the formation of helical columnar stacks at high THF ratios. This cooperativity effect was, however, absent for the non-fluorinated PBIs that exhibited a higher conformational rigidity. Cooperativity in the self-assembly of PBIs was achieved by attaching cholesterol moieties to the π-conjugated chromophore via carbonate linkers [46]. The dipole–dipole interactions between these units converged to a macro-dipole character of the supramolecular polymer, as revealed by the bulk-phase polarization of the material. George et al. concluded that cooperativity was due to dipole–dipole interactions, as one type of anisotropic long-range non-covalent interactions, representing an important tool to control the supramolecular self-assembly of discotic molecules (thus, complementary to hydrophilic/hydrophobic or H-bonding interactions). Examples for this can be found in Li and coworkers' recent review on the supramolecular polymerization of PBIs, assisted by functional groups [47]. Beyond the aforementioned structurally rather simple PBI derivatives, also more

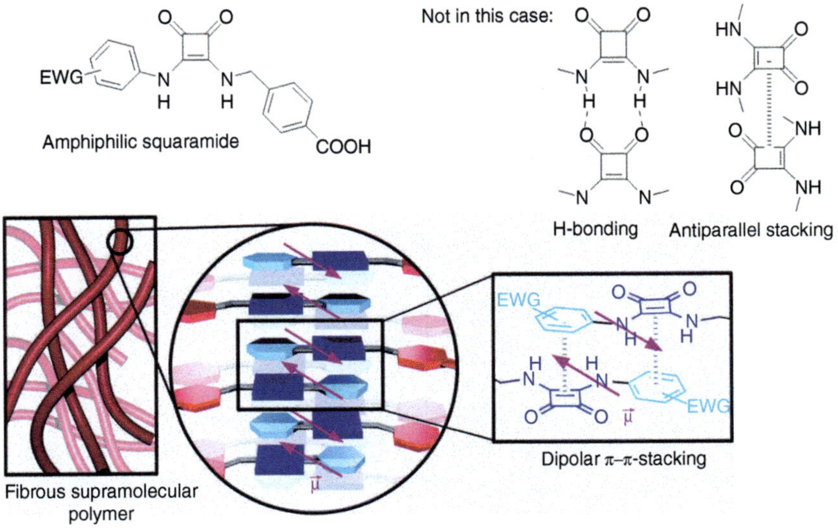

Figure 5.16 Schematic representation of amphiphilic squaramide dyes which gave supramolecular polymers due to dipolar π–π stacking interactions. Source: Bujosa et al. [49], © 2020 The Royal Chemical Society.

complex monomers (e.g. containing multiple PBI units) were studied with respect to their self-assembly in solution or in the solid state [22].

The particular role of dipole-assisted self-assembly of π-conjugated molecules was shown in various examples, such as merocyanine dyes, which gave defined nanoscale objects in low-polar solvents [48]. Thus, dipole–dipole interactions can increase the directionality of π–π stacking interactions, which are intrinsically low. In the same context, Soberats and coworkers assembled amphiphilic aryl-substituted squaramides into fibrous supramolecular polymers and hydrogels [49]. These authors showed that these assemblies were stabilized by dipole–dipole interactions and not by H-bonding or antiparallel stacking, as the common interaction modes of squaramides (Figure 5.16).

The planar members from the heteroatom-containing triangulene family represent the last important example for discotic structures that can be utilized to assemble supramolecular stacks via π–π stacking interactions of their extended π-electronic surfaces [50, 51]. Besides the cationic carbon-centered derivatives that assemble into columnar structures via π-electronic ion-pairing interactions (see Section 5.4) [52, 53], the charge-neutral, heteroatom-centered derivatives might undergo self-assembly into columnar arrays with LC properties [54]. For example, the boron-centered derivative **13** was found to form a hexagonal columnar LC phase in which an interfacial distance of 3.57 Å between the disc-shaped molecules was found (Figure 5.17) [55]. This LC material exhibited ambipolar charge-carrier transport properties even at 25 °C with electron- and hole-mobility values of c. 10^{-3} and 3×10^{-5} cm^2 V^{-1} s^{-1}, respectively.

In contrast to the planar molecules, the convex/concave or bowl-shaped molecules, such as the P-centered triangulenes show a more complex packing

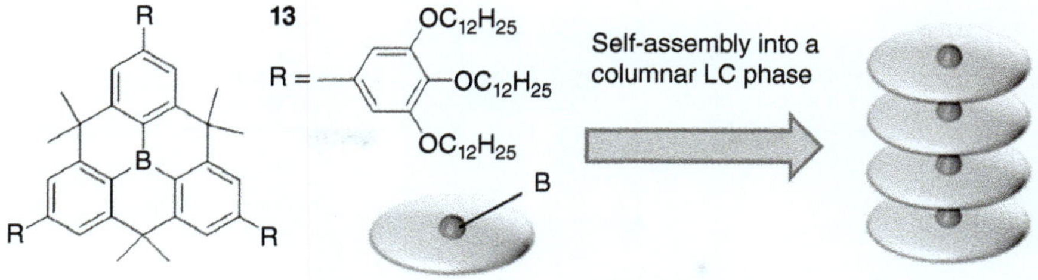

Figure 5.17 Schematic representation of the boron-centered triangulene **13** that self-assembled into a columnar LC phase. Source: Kushida et al. [55]. © 2015 John Wiley and Sons.

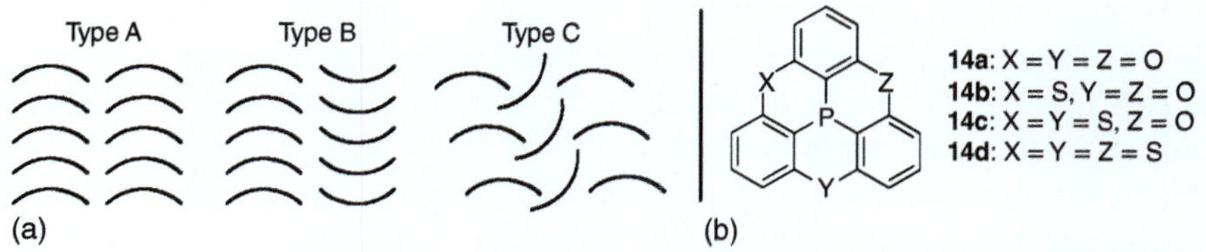

Figure 5.18 (a) Schematic representation of the different packing modes of bowl-shaped molecules (from the non-columnar type-C assemblies only one representative example is shown). (b) Schematic representation of the P-centered triangulenes **14**. Source: Yamamura et al. [56], © 2016 American Chemical Society.

behavior from which the most important modes are summarized in Figure 5.18. Within this context, Yamamura et al. prepared a series of O-/S-bridged P-centered triangulenes (**12**, Figure 5.18). For the C_{3v}-symmetric derivatives **14a** and **14d**, a columnar packing according to types A and B, respectively, was found; however, the C_s-symmetric **14b** and **14c** assembled in a *zig–zag*-type fashion (type C). Thus, the packing mode could be controlled basically by the nature of the bridging heteroatoms.

5.3 From π–π Stacking to Advanced Donor–Acceptor-Type Charge–Transfer Interactions

Aromatic CT interactions represent a special type of non-covalent forces, closely related to the previously discussed π–π stacking interactions. The latter is fundamentally based on the interaction between two similar (aromatic) molecules; whereas, aromatic CT interactions only exist between electron-rich donor-type and electron-poor acceptor-type moieties, the so-called D–A pairs [57, 58]. Such combinations with CT character can offer unique optical, electrical or magnetic properties to be utilized in materials science. In practice, the hetero-association of donor- and acceptor-type molecules, driven solely by CT interactions, will always be in competition with the homo-association pathway driven by π–π stacking interactions (Figure 5.19) [60]. As a result, a precise long-range order of a D–A-type supramolecular polymer is, in essence, difficult to achieve.

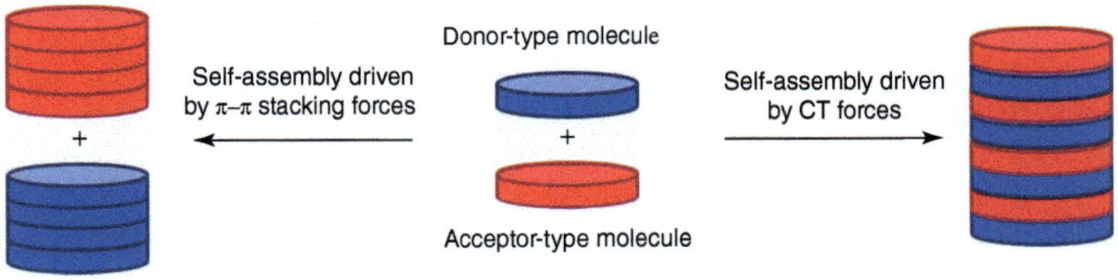

Figure 5.19 Schematic representation of the competition between the homo- and hetero-type stacking of donor–acceptor-type molecules driven by π–π and charge–transfer stacking forces, respectively. Source: Modified from Han et al. [59].

As one further limiting factor, the strength of CT interactions is highly dependent on the solvent polarity. In particular, solvents with a high refractive index, such as $CHCl_3$ or CH_2Cl_2, are hardly suited for this purpose – in such an environment, only rather low K_a values in the range of 10^{-1}–10^2 M^{-1} can be reached [8, 58]. Taking the theory of supramolecular polymerization into account (Chapter 1), assemblies of high molar masses cannot be formed under such conditions. As pointed out above, the encapsulation of π-conjugated systems within the cavity of an appropriate macrocyclic host represents one possibility to overcome this limitation. As shown by Liu et al., the CT interaction between a donor- (e.g. 2,6-dihydroxynaphthalene) and an acceptor-molecule (e.g. methyl viologen or MV^{2+}) can be increased significantly by their co-inclusion into the cavity of a cucurbit[8]uril (**CB[8]**) host [61]. This particular host-stabilized charge–transfer (HSCT) interaction was then utilized for the formation of water-soluble, well-defined supramolecular assemblies, including chain-extended polymers and block copolymers (see also: Chapter 7).

In continuation of their work on amphiphilic CT complexes between organic dyes and their alternate self-assembly into advanced nanostructures [62, 63], the George group suggested a more sophisticated approach to control the self-assembly of D- and A-type moieties via CT interactions [64]: the folded bis-pyrene donor (**15**) was combined with a functionalized NBI derivative (**16**) to afford a non-covalent amphiphilic 1 : 1 "CT foldamer." In the presence of one further equivalent of **16**, the mixed 1D stack with an alternate D–A arrangement was formed (Figure 5.20). In aqueous solution, this type of supramolecular polymerization was driven by a cooperative combination of CT, hydrophobic, and ionic non-covalent interactions. The same combination of CT chromophores was also utilized by Ghosh and coworker to assemble an alternate stack in which H-bonding interactions assisted the ordering of the molecular building blocks [58].

More recently, the same authors reported on an amphiphile **17** that adopted a folded conformation in water due to intramolecular CT interactions [65]. Dynamic light-scattering (DLS) measurements and cryogenic transmission electron microscopy (cryo-TEM) imaging revealed the presence of defined patterns (average diameter of c. 70 nm); these vesicles were formed due to the self-assembly of the folded conformer with a packing factor (f) of 0.68 (Figure 5.21). For an extended conformer (f = 1), a packing into a sheet-like nanostructure was predicted on the basis of molecular modeling studies. The transition from a vesicular to sheet-like

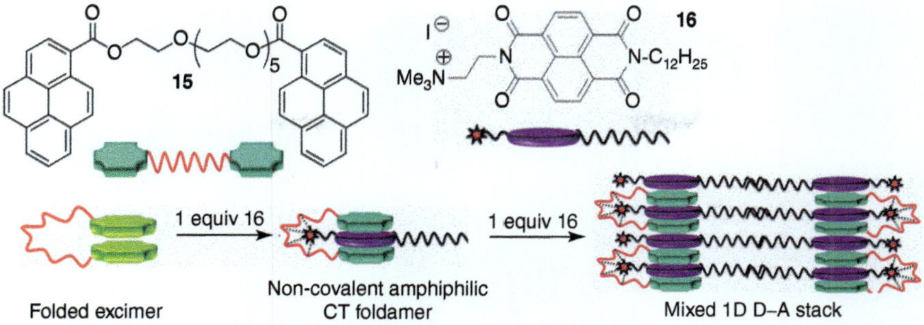

Figure 5.20 Schematic representation of the synthesis of an amphiphilic D–A-type foldamer and its self-assembly into a 1D nanostack. Source: Jalani et al. [64], © 2013 Royal Society of Chemistry.

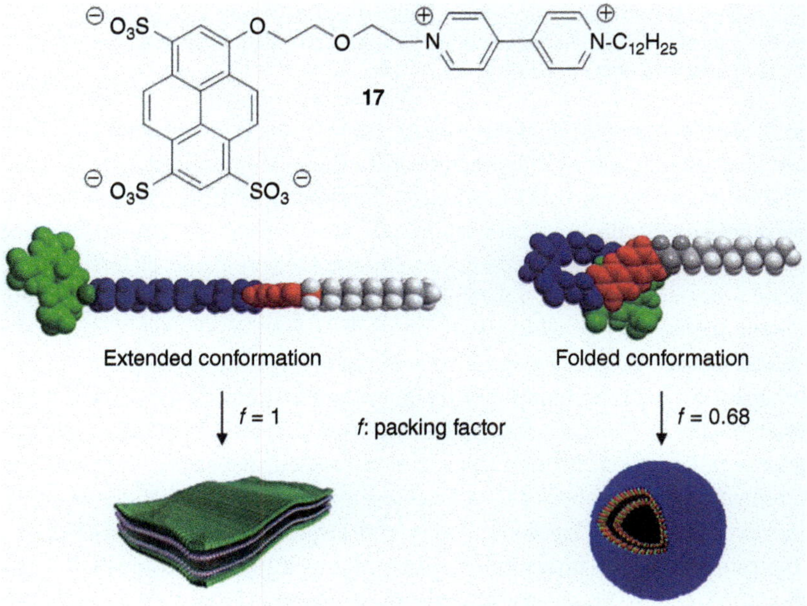

Figure 5.21 Schematic representation of the amphiphilic donor–acceptor molecule **17** that could assemble into nanosheets or vesicles depending on its conformation. Source: Jalani et al. [65]. Licenced under CC BY 3.0.

morphology could be achieved by threading the electron-rich MV^{2+} part into the cavity of a curcurbit[7]uril (CB[7]) host, thus rigidifying the molecular structure of **17** (the process could be reversed by adding adamantane, as a strong competitive guest for CB[7]). Moreover, the MV^{2+} moiety could be reduced by sodium dithionite to afford a radical cation species that could not establish a CT interaction with the pyrene unit (re-oxidation was achieved enzymatically by using glucose oxidase).

Besides this "foldamer approach," the incorporation of donor–acceptor moieties into a more confined structure of a "molecular tweezers," which can complex an appropriate guest, via its pincers, represents a highly versatile strategy to assemble

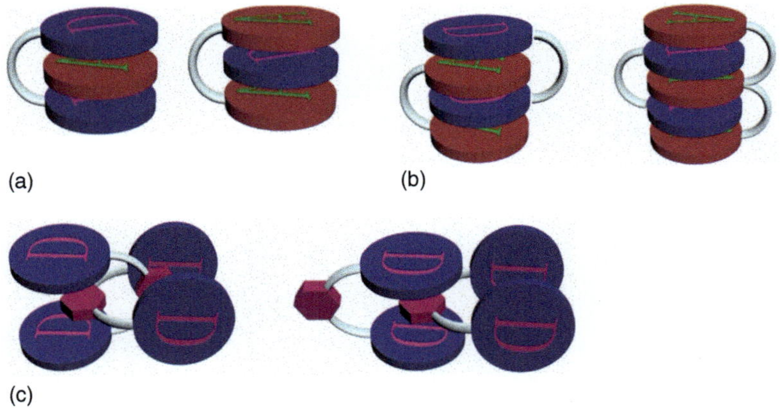

Figure 5.22 Schematic representation of the various types of molecular tweezers/guest recognition motifs. (a) DAD- and ADA-type assemblies, (b) DADA- and ADADA-type assemblies and (c) self-complementary and head-to-tail dimerization. Source: Han et al. [59], © 2018 The Royal Chemical Society.

supramolecular polymers using highly directional complexation of guests with high K_a values, in particular when compared to conventional D–A pairing.

In 1978, Whitlock and coworker introduced the concept of molecular tweezers in which two complexing chromophores of either donor- or acceptor-type, the so-called pincers were linked by a rigid spacer; thereby, the ideal centroid distance of the syn-cofacial pincers should be c. 7 Å to allow for an efficient host–guest interaction [66]. The complementary guest can than enter the open slot-shaped cavity of the preorganized supramolecular receptor. In such an array, twofold D–A CT interactions account for the strong binding of the guest and two basic types of structures are accessible – the DAD- and ADA-modes (Figure 5.22a) [59]. In addition to these two fundamental systems, further types of non-covalent tweezering systems can be obtained when the number of D–A stacking units is increased. As reported by Yam and coworkers DADA- or even ADADA-type structures can be assembled and, due to the multivalent CT interactions, much higher K_a values (compared to the conventional ADA-/DAD-type assemblies) can be realized (Figure 5.22b) [67]. Disadvantageously, the synthetic effort, required to prepare such multivalent molecular tweezers, will be much higher and, for more practical reasons, research with respect to supramacromolecular assemblies has mainly focused on the structurally simpler molecular tweezers. Beyond the aforementioned motifs, also other types of architecture, i.e. self-complementary or head-to-tail ones, might, however, be formed if the rigid spacer of a molecular tweezers is planar and, at the same time, electronically complementary to the pincers (Figure 5.22c) [68].

As a general rule, preorganized binding sites offer a higher directionality and, for thermodynamic reasons, also a higher strength in binding compared to more flexible ones. In one of the first contributions in this field, Zimmerman and VanZyl prepared a set of molecular tweezers with two complexing acridine-type pincers and studied the binding of 2,4,7-trinitrofluorenone (TNF), as a model guest; K_a values in the range of 150 M^{-1} were found (in comparison, the interaction of simple acridines with

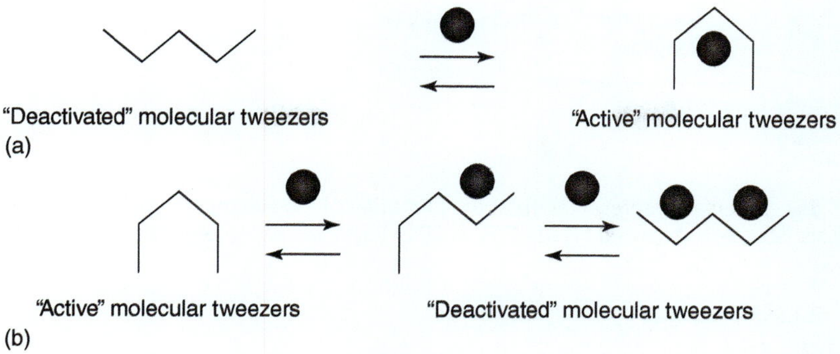

Figure 5.23 Schematic representation of the metal-induced activation (a) and deactivation (b) of molecular tweezers according to Petitjean et al. [71].

TNF gave very low K_a values of <5 M^{-1}) [69]. In the following, also other arene-based molecular tweezers were reported by these authors and showed basically the same behavior [70].

Furthermore, Lehn and coworkers showed that stimuli responsiveness can also be implemented in the spacer's architecture to control the highly reversible binding and enable the release of a guest from a molecular tweezers (Figure 5.23). For example, molecular tweezers were designed and only functioned when a transition metal ion was coordinated to the binding sites present in the spacer, thus generating the proper molecular geometry to bind a guest in a DAD- or ADA-fashion only when in the complexed state [71]. Upon abstraction of the metal ion, a conformational change of the spacer occurred, thereby disfavoring the binding of planar organic dyes. The orthogonal approach, i.e. deactivation of a molecular tweezers as a result from metal-to-ligand coordination can also be envisioned (given that the metal-to-ligand coordination is stronger than the CT interaction offered by the molecular tweezers).

5.4 From Charge-Transfer to π-Electronic Ion-Pairing Interactions

Recently, Haketa and Maeda summarized how the ion-pairing of π-electronic charged molecules can be employed in the context of controlled supramolecular self-assembly [72, 73]. The ionic self-assembly, arising from electrostatic interactions between oppositely (or identically) charged species, has been widely used in the field of supramolecular chemistry (see Chapter 2). Basically, the electron transfer from a donor-type molecule to an appropriate acceptor results in a charge-separated state between them. Thus, replacing the π-electronic D-/A-molecules undergoing self-assembly via CT interactions (Figure 5.19) by a pair of planar π-electronic species, i.e. a cation and an anion, will also allow the controlled self-assembly facilitated by concerted ionic and π–π stacking interactions. Such ion pairs represent a genuine type of π-electronic ion pairs (for the other possible types of

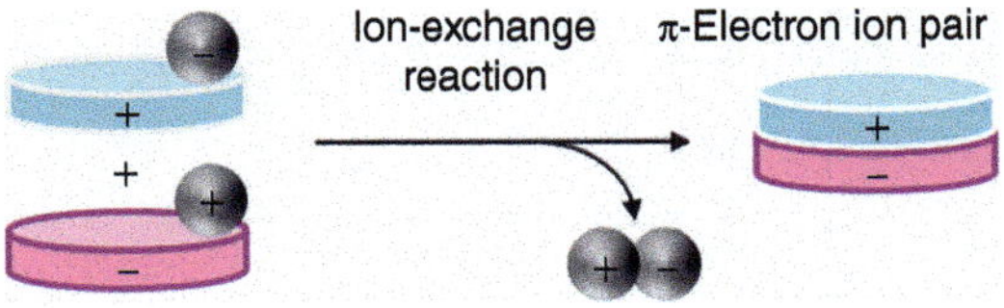

Figure 5.24 Schematic representation of the formation of a π-electron ion pair via ion metathesis. Source: Haketa and Maeda [72], © 2017 Royal Society of Chemistry.

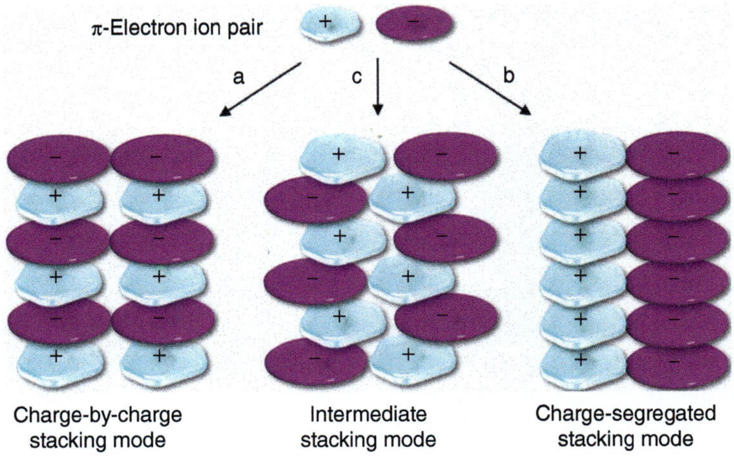

Figure 5.25 Schematic representation of the three different self-assembly modes of π-electron ion pairs: charge-by-charge (a), charge segregated (b), and intermediate stacking (c). Source: Dong and Maeda [74].

structures, see Ref. [72]). The synthesis of π-electronic ion pairs is based on a metathesis reaction, in which the accompanying counterions are replaced and separated (Figure 5.24). Thereby, the separation of the resultant π-electronic ion pairs from the counterions is of utmost importance to guarantee purity and precise stoichiometry – prerequisites for a controlled supramolecular self-assembly.

With respect to a controlled arrangement of planar π-electron ion pairs into columnar assemblies, different stacking modes can be obtained [52, 72, 74]: first of all, the charge-by-charge assembly represents a packing mode in which positively and negatively charged species are stacked on each other in a strictly alternate fashion (Figure 5.25a; this mode corresponds to the hetero-stacking of a D–A pair, see also Figure 5.19). If identically charged species are stacked, thus overcoming the electrostatic repulsion, a charge-segregated assembly is formed (Figure 5.25b). From these, the former is more suited for the preparation of columnar assemblies due to the attractive forces between the oppositely charged species, supported by π–π stacking interactions. However, the formation of an intermediate stacking mode, showing contributions of the charge-by-charge and charge-segregated stacking, represents the most common one (Figure 5.25c).

To fabricate such assemblies from π-electron ion pairs, the appropriate cations and anions need to be synthesized. Considering only those structures where

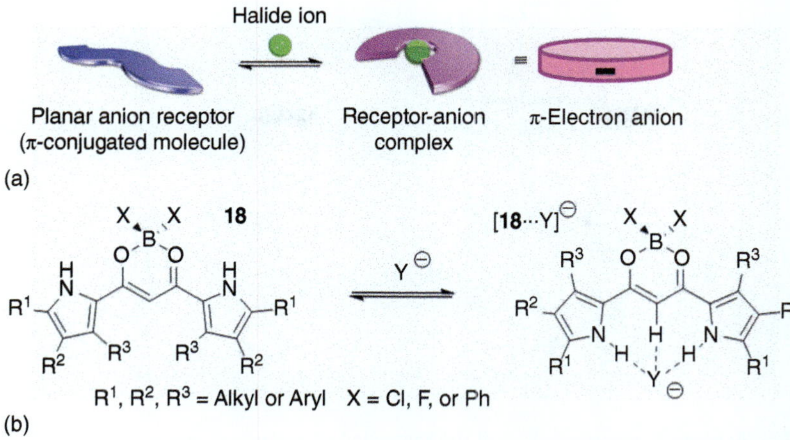

Figure 5.26 Schematic representation of the formation of a π-electron anion via the supramolecular complexation of a (halide) ion by a planar anion receptor. (a) cartoon representation and (b) at the example of **18**. Source: Dong and Maeda [74], © 2013 Royal Society of Chemistry.

the charges are delocalized over the (almost) planar π-conjugated system, some restriction arise. Cationic species featuring a certain degree-of-delocalization of the positive charge are already known, for early examples, see Refs. [75–77], and a broad range of planar π-conjugated cations has been prepared [74] – such as, carbocations, N-based cations, cationic porphyrins, and metal complexes (*vide infra*). On the contrary, planar π-conjugated anions are much more difficult to achieve (a substantial degree-of-delocalization is required to compensate the reactivity, i.e. shield the electrons from electrophilic attack). It has been shown that this issue can be circumvented by forming planar anionic supramolecular complexes between appropriate π-conjugated anion receptors and anions (Figure 5.26) [72, 74]. In particular, the 1,3-bis(pyrrol-2-yl)-1,3-diketone difluoroboron complexes (**18**), as introduced by Maeda and Kusunose in 2005 [78], have been widely used in this context (the K_a values of **18** for the binding of, e.g. F^- in CH_2Cl_2 are in the range of 10^4–10^5 M^{-1}) [74].

Furthermore, two other types of π-electronic anions have been utilized, though to only a minor extent. They both rely on the deprotonation in O–H and C–H acidic positions, respectively. 2,6-Bis(pyrrol-2-yl)-4-nitrophenols, such as **19** [79, 80], as well as the related 1,3-bis(pyrrol-2-yl)-1,3-diketone derivative **20** feature a conformational switch (Figure 5.27) [81]; these reported anions were recently used in the context of π-electron ion pairing [72]. More recently, also deprotonated 2,3-bis(pyrrol-2-yl)quinoxaline and 1,3-bis(dicyanomethylidene)indane were reported, as planar π-electron anions [82, 83]. The latter gave charge-by-charge and charge-segregated assemblies with various π-electron cations (according to density functional theory (DFT) calculations, dipole–dipole interactions helped to overcome the electrostatic cation–cation, and anion–anion repulsions). The penta-substituted cyclopentadienyl anions **21**⁻ represents one further π-electron

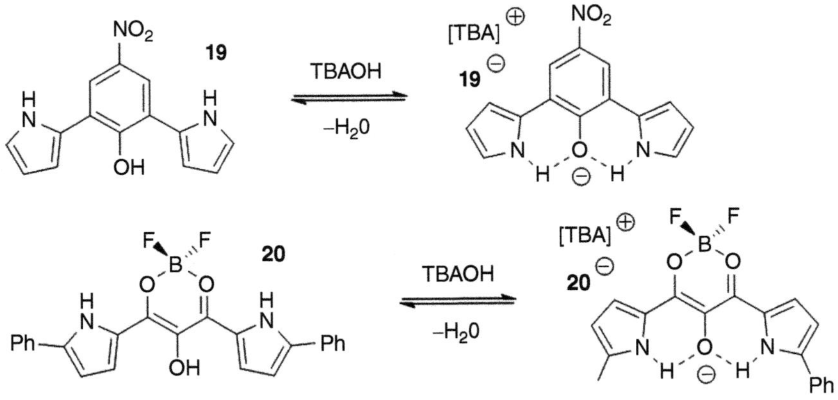

Figure 5.27 Schematic representation of the formation of π-electron anions by deprotonation of OH-groups.

anion in which aromaticity and electron-withdrawing nature of the substituents account for their stability. From these, genuine π-electron ion pairs were formed with the aromatic cycloheptatrienyl (Ch^+) cation [84, 85]. The solid-state structures of [Ch][**21**], as revealed by X-ray single crystal analysis, featured an almost perfect charge-by-charge stacking mode, in which the columnar structures were stabilized by a combination of electrostatic and π–π stacking interactions (Figure 5.28). In contrast, the more complex crystal violet and 4,8,12-tripropyl-4,8,12-triazatriagulenium ($TATA^+$) cations generated less defined stacks with **21b** in the solid state, showing contributions of the charge-by-charge and charge-segregated modes, according to Figure 5.25 [85].

$TATA^+$ (**22**) and its relatives from the triangulenium family – the parent compound, 4,8,12-trioxa-4,8,12,12c-tetrahydrodibenzo[*cd*,*mn*]pyrene ($TOTA^+$), was already synthesized in 1964 [75] – represent prominent π-electronic cations whose photophysical and electrochemical properties can be tuned over a wide range by structural modifications of the central framework and/or on the periphery [53]. Due to their extended planar structure, their usage as cationic components in ion-pairing assemblies is very common [51, 72, 74]. For example, LC materials were formed from the ion pairs; thereby, the LC properties could be adjusted via the nature of the employed cations and anions. For the combination of $TATA^+$ with [**23**···Cl]$^-$, POM imaging indicated a mesophase with a focal conic texture [with transition temperatures of 94, 88, and 42 °C upon cooling from the isotropic liquid state, Figure 5.29(a)] [86]. The synchrotron XRD patterns featured peaks that were assigned to the Col_h phase with dimension of $a = 4.64$ nm and $c = 0.73$ nm [a tetrameric assembly with $Z = 4$ for $\rho = 1.12$, see Figure 5.29(b)]. The c value, representing the stacking distance between $TATA^+$ and [**23**···Cl]$^-$, suggested a charge-by-charge-type of assembly. For comparison, also the ion-pair assembly of the bulky tetra(*n*-butyl)ammonium (TBA) cation and [**23**···Cl]$^-$ was studied: also in this case, the Col_h mesophase resulted from a charge-by-charge-type assembly mode [Figure 5.29(b)] [87]; however, the XRD data indicated a much weaker stacking as in the case of the planar $TATA^+$ cation.

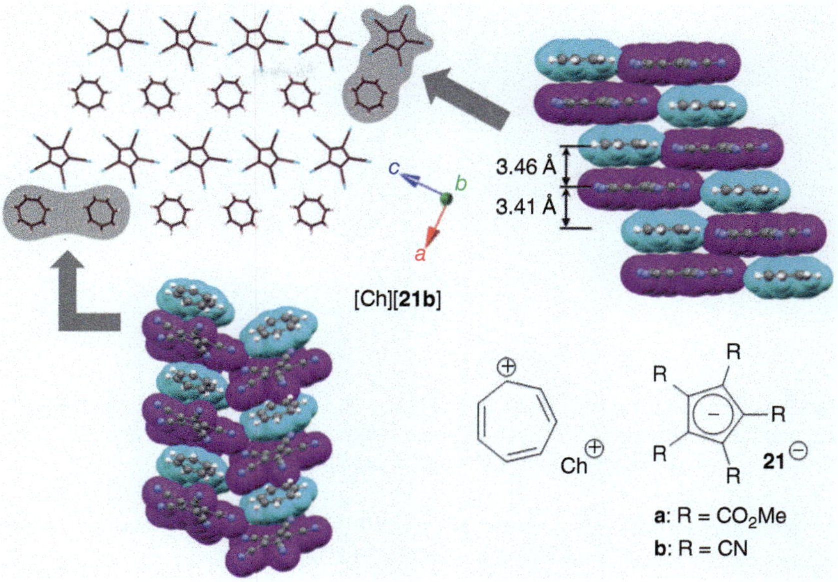

Figure 5.28 Schematic representation of the π-electron ion pairs, based on the cycloheptatrienyl (Ch^+) cation and cyclopentadienyl anions (21^-). The representations show the packing mode of [Ch][**21b**] in the solid state (three different views are depicted). Source: Bando et al. [85], © 2016 John Wiley and Sons.

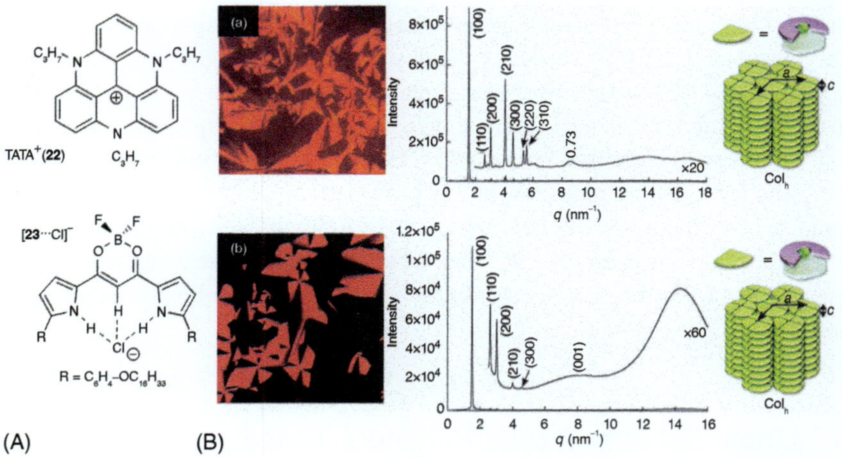

Figure 5.29 (a) Schematic representation of the $TATA^+$ cation **22** and of the receptor–anion complex [**23**···Cl]$^-$. (b) POM images, XRD patterns (at 101 °C) and proposed packing modes (from left to right) of the ion pairs [**22**][**23**···Cl] (a). Source: Haketa et al. [86]. Figure reproduced with kind permission; © 2010 Wiley-VCH and [TBA][**23**···Cl] (b). Source: Dong et al. [87]. Figure reproduced with kind permission; © 2010 Wiley-VCH.

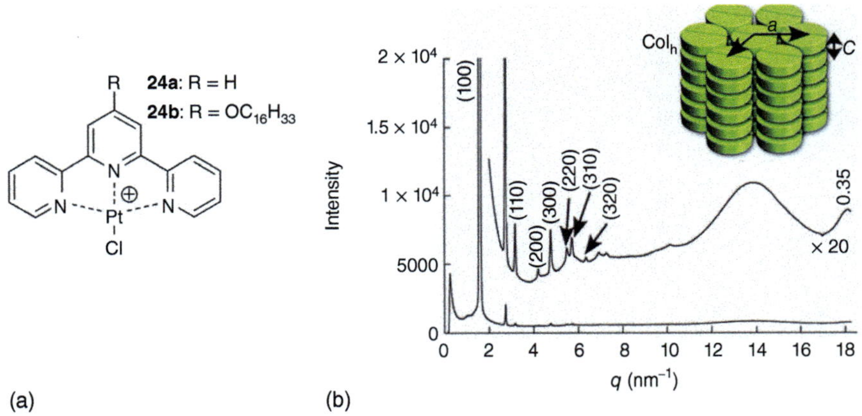

Figure 5.30 Schematic representation of the square-planar Pt(II) complexes **24**. The XRD pattern and the proposed packing mode for the ion-pairing assembly of **24b** and [**23**···Cl]$^-$ is also shown. Source: Sekiya et al. [89], © 2014 Royal Society of Chemistry.

As noted above, not only organic π-electronic cations but also cationic metal complexes can be utilized for the self-assembly, based on ion-pairing interactions. This, however, applies only for those of square-planar coordination geometry, such as the Pt(II) complexes with 2,2′:6′,2″-terpyridine (tpy) ligands [88]. For example, the cationic Pt(II) complexes **24** were combined with anion [**23**···Cl]$^-$ [89]. The self-assembly of these ion pairs from 1,4-dioxane solutions gave, in the case of [**24b**][**23**···Cl], again a $\mathrm{Col_h}$ mesophase with $a = 4.57$ nm and $c = 0.35$ nm based on a dimeric assembly ($Z = 2$ for $\rho = 1.36$, Figure 5.30). As concluded from the prominent XRD diffraction peak at 0.35 nm, the stacking of the components occurred rather in a charge-segregated mode. UV/vis absorption spectroscopy of the mesophase at 100 °C revealed an absorption band at 667 nm, which was indicative of a dσ*(Pt_2) → π*(tpy) metal-metal-to-ligand charge–transfer (MMLCT) process involving Pt(II)–Pt(II) interactions. Thus, the resultant charge-separated stacking mode was stabilized by strong metallophilic interactions (thereby, overcoming the electrostatic repulsion of the piled cationic units). Sekiya et al. also investigated the charge-carrier mobility of the thin film of [**24b**][**23**···Cl]: hole- and electron-transport mobilities of c. 0.7 and 0.6 $\mathrm{cm^2\ V^{-1}\ s^{-1}}$, respectively, were found suggesting that the ion-paired assemblies were effective ambipolar conductive materials [89].

5.5 Linear Supramolecular Polymers, Based on π-Electronic Interactions

According to the general considerations outlined in Chapter 1, linear supramolecular polymers can be formed in solution via the self-association of AB-type heteroditopic monomers. Alternatively, stoichiometric mixtures of AA- and BB-type homoditopic monomers can be used for the very same purpose. Hyperbranched supramolecular architectures are formed if AB_2-type monomers or a combination of A_2- and B_3-type monomers are used. More complex architectures will be obtained from polytopic A_n- and B_n-type monomers; these include cross-linked or dendronized supramolecular polymers. To prepare such materials on the basis of π-electronic interactions, one crucial aspect needs to be mentioned that is directionality. In contrast to conventional π–π stacking interactions as well as the related CT and π-electronic ion pairing interactions, where the formation of columnar assemblies is highly favored, only the utilization of molecular tweezers fulfills this criterion. Thus, supramolecular polymers can be obtained from appropriate monomers equipped with molecular tweezers sites (A) and proper guest units (B). The topology of a supramolecular polymer can then be controlled by the number of these functional units within the monomers. Recently, Han et al. summarized the donor–acceptor-type polymers resulting from this particular approach [59].

Based on Zimmerman's pioneering work on CT complexes of preorganized molecular tweezers (see Section 5.3) [70], sophisticated derivatives representing AB-type monomers for the self-assembly into supramolecular polymers have been designed in recent years. The Sessler group, for example, investigated the self-assembly behavior of the pyrene-based molecular tweezers **25**, representing strong hosts for the binding of electron-poor guests, such as nitroaromatics (e.g. 1,3,5-trinitrobenzene [TNB] or 2,4,6-trinitrotoluene [TNT]) [90]. In the case of **25a**, UV/vis absorption and emission spectroscopy suggested excimer formation due to the close proximity of the pyrene entities, even in dilute solutions. According to NMR spectroscopy and XRD analysis, the self-assembly of **25a** in a head-to-tail fashion occurred selectively at high concentration as well as in the solid state. The kinetically labile supramolecular polymer $(\mathbf{25a})_n$ could readily be disassembled in the presence of a strong competitive guest, thus affording a deep-red, non-fluorescent CT complex of 1 : 1 stoichiometry (Figure 5.31). Due to the more electron-rich nature of the spacer, **25b** did, on the contrary, not give any supramolecular polymer. The supramolecular polymerization of **25a** was, thus, rationalized by the slight electron-deficient nature of the spacer in concert with intramolecular

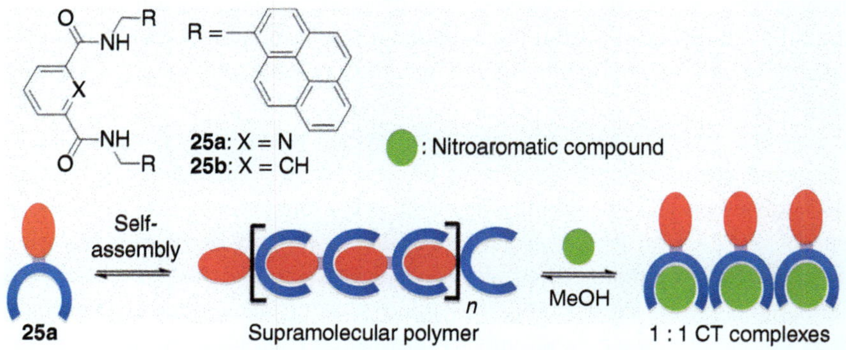

Figure 5.31 Schematic representation of the molecular tweezers **25** and their reversible self-assembly of **25a** into a linear supramolecular polymer. Source: Kim et al. [90], © 2012 American Chemical Society.

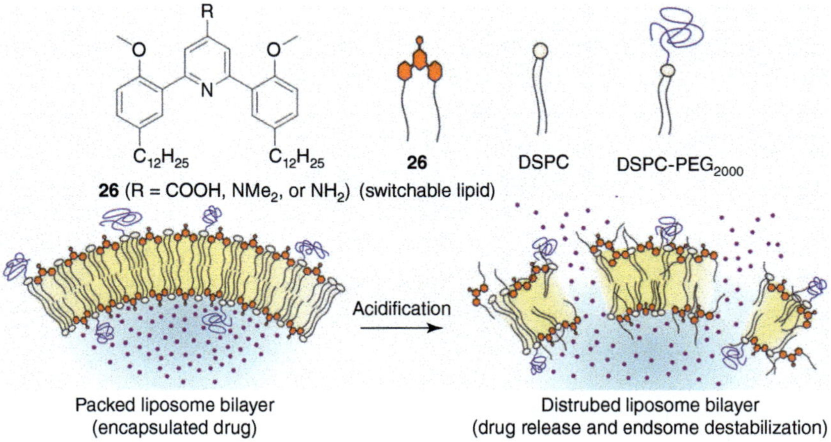

Figure 5.32 Schematic representation of the pH-switchable molecular tweezers **26** that were used for the assembly of a packed liposome bilayer with 1,2-dioctadecanoyl-*sn*-glycero-3-phosphocholine (DSPC) and *N*-(carbonyl-methoxypolyethyleneglycol-2000)-1,2-distearoyl-*sn*-glycero-3-phosphoethanolamine (DSPE-PEG_{2000}). The packed liposome bilayer collapsed upon acidification. Source: Viricel et al. [94].© 2015 John Wiley and Sons.

N—H···N hydrogen-bonding interactions within the planar pyridine dicarboxamide fragment (additionally, the outward facing carbonyl moieties were identified as crucial for an intermolecular interaction). The same type of molecular tweezers unit was also employed for the formation of other types of supramolecular polymer architectures; for instance, small-molecule molecular tweezers were complexed by polymers featuring electron-poor chromophores within the backbone [91, 92]. Alternatively, polymers with molecular tweezers moieties in their side chains were reported to reversibly bind small-molecule, electron-deficient guests [93].

Reversibility of binding can also be achieved by (de)protonation of a molecular tweezers, which can undergo a conformational change. This concept was applied by Viricel et al. to assemble a packed liposome bilayer using the amphiphilic molecular tweezers **26** [94]. Organic dyes as drug mimics, such as sulforhodamine B, could be encapsulated within the supramolecular structure. It was shown that the assembly collapsed upon protonation and, thus, released the encapsulated drug (Figure 5.32).

Besides the PAHs as pincers units in molecular tweezers, also other planar π-conjugated chromophores, have been utilized for this purpose. In particular, tetrathiafulvalenes (TTFs), porphyrins, and some organometallics have to be named in this context. With respect to the latter ones, only derivatives featuring a sufficient π-surface might be employed as pincers in molecular tweezers. This requirement is met for the d^8-configured square-planar transition metal complexes, such as the cationic alkynylplatinum(II) terpyridine complexes as introduced by Yam and coworkers [95]. It was shown that the molecular tweezers **27** with a rigid 2,6-diphenylpyridine spacer and two such planar pincers were capable to sandwich electron-rich PAH guests with moderate K_a values (Figure 5.33).

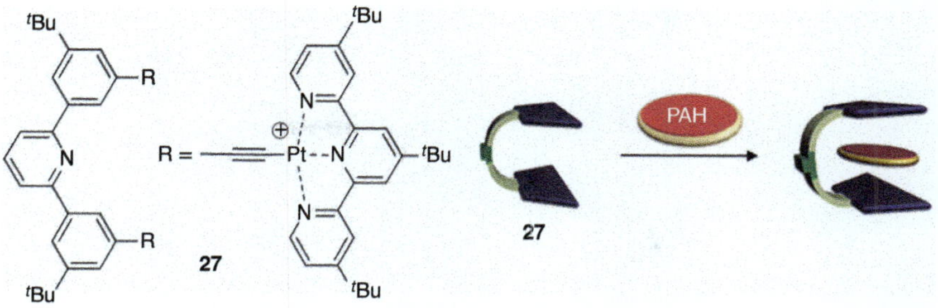

Figure 5.33 Schematic representation of the molecular tweezers **27** containing two positively-charged organoplatinum(II) pincers. Source: Tanaka et al. [95], © 2013 John Wiley and Sons.

This basic approach was extended by Wang and coworkers to assemble linear supramolecular polymers. For this purpose, the heteroditopic monomer **28** – comprising a host end based on an organoplatinum(II)-containing molecular tweezers and a pyrene guest moiety at the other end of the flexible spacer were synthesized (Figure 5.34) [96]. The sufficiently high binding affinity in $CHCl_3$ (K_a of 2270 M^{-1} for the interaction of the molecular tweezers with pyrene) allowed the self-assembly of **28** into a linear supramolecular polymer, though only at high concentrations >20 mM. However, the K_a value was too low to obtain a supramolecular polymer of high molar mass (a degree-of-polymerization (DP) of only 13 was estimated at a concentration of 70 mM). Due to the even stronger binding of 9-methylanthracene to the molecular tweezers (K_a of 3330 M^{-1}), the polymer could be fully disassembled in the presence of 1 equiv of this competitive guest (Figure 5.34). The depolymerization process was monitored by DOSY: an increase of the diffusion coefficient from 2.45×10^{-10} $m^2\ s^{-1}$ for $(\mathbf{28})_n$ to 8.20×10^{-10} $m^2\ s^{-1}$ confirmed the formation of the 1 : 1 complex of **28** and 9-methylanthracene. Upon addition of a strong dienophile, the complexed anthracene derivative quantitatively underwent a Diels–Alder reaction even at room temperature and, as a result, the initial supramolecular polymer $(\mathbf{28})_n$ was restored.

The limitation in reaching high DPs due to a too low K_a value could be overcome by using instead the AA- and BB-type monomers **29** and **30**. These homoditopic components assembled in a DADA-type fashion (see Figure 5.22b) and showed a significantly increased binding affinity (i.e. K_a values in the range of 10^6 M^{-1}). Thus, an equimolar mixture of **29** and **30** afforded a supramolecular polymer of high molar mass as concluded from DOSY (i.e. diffusion coefficients of 2.24×10^{-9} and 1.15×10^{-11} $m^2\ s^{-1}$ at 2.00 and 21.0 mM, respectively) and viscosity measurements (Figure 5.35). A DP of 395 was calculated for a monomer concentration of 17.5 mM. Thus, the cooperativity of CT and H-bonding interactions appeared to be beneficial when aiming for supramolecular polymers of high molar masses.

Porphyrin molecules are known for their large π-surfaces and rich photophysical properties; thus, such compounds have been recognized as important building blocks in supramolecular chemistry and, regarding molecular tweezers, a range of porphyrin-based molecular tweezers has already been documented [98]. Thereby,

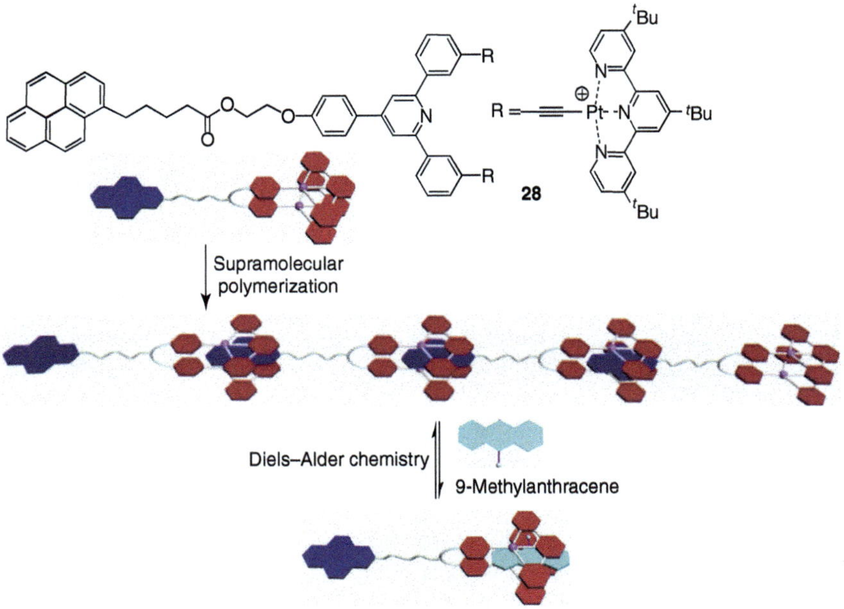

Figure 5.34 Schematic representation of the reversible supramolecular polymerization of **28**. Source: Tian et al. [96], © 2014 John Wiley and Sons.

the flexibility and length of their spacer have been varied and planar (e.g. TNF, 7,7,8,8-tetracyanoquinodimethane [TCNQ], or MV^{2+}) or even concave-shaped electron-poor guest molecules (e.g. C_{60}) have been complexed by the pincers of such preorganized receptors.

In line with their arene-based counterparts **25** (Figure 5.31), the self-assembly behavior of **31** in solution was studied by Haino et al. [68]. In a similar fashion, **31a** gave a dimer in a non-polar environment due to a self-complementary interaction (K_a = 310 and 5500 M^{-1} in $CHCl_3$ and in toluene, respectively; Figure 5.36). The ability of **31a** to dimerize was attributed to the planar structure of the pyridine dicarboxamide spacer; consequently, the non-planar phenyl-centered analog **31b** did not allow this type of association under the same conditions. The dimer **(31a)$_2$** could be disassembled by adding a strong competitive guest, such as TNF that formed a stable 1 : 1 complex with **31a** ($K_a = 4.1 \times 10^5$ M^{-1} in $CHCl_3$). Based on these observations, the S-shaped homoditopic molecular tweezers **32** was designed to enable supramolecular polymerization via a self-association process (Figure 5.37) [99]. AFM imaging was applied to analyze the morphology of the resultant polymer **(32)**$_n$. Coiled fibers with a length of c. 1 mm were observed on a mica substrate. The authors proposed that individual polymer chains adopted a bundled helical structure that eventually gave the observed coiled superstructure.

Besides the supramolecular polymerization due to self-complementary interactions, Haino et al. also employed the heteroditopic monomers **33** to obtain linear supramolecular assemblies (Figure 5.38) [100]. For the metal-free representative **33a**, a concentration-dependent supramolecular polymerization behavior by way of

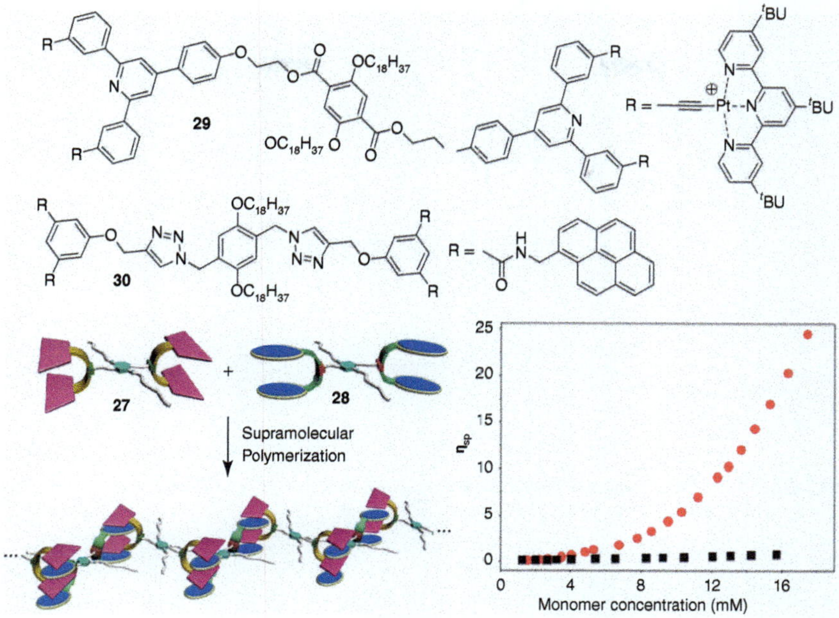

Figure 5.35 Schematic representation of the complementary homoditopic monomers **29** and **30** and their supramolecular polymerization in a DADA-type fashion. The evolution of the specific viscosity as a function of the monomer concentration is also shown. Source: Tian et al. [97], © 2016 American Chemical Society.

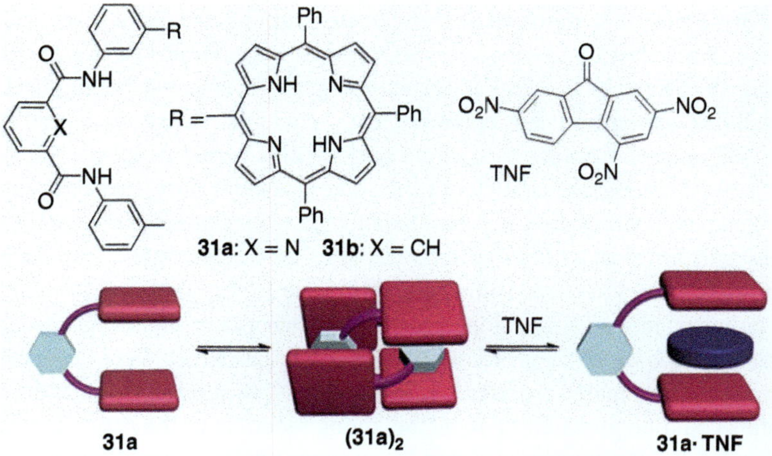

Figure 5.36 Schematic representation of the porphyrin-based molecular tweezers **31** and of the self-complementary dimerization of **31a**. Source: Haino et al. [68], © 2006 American Chemical Society.

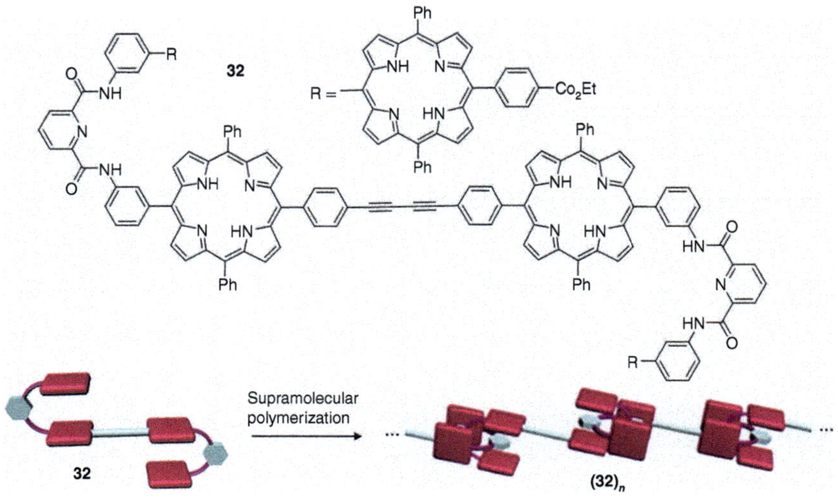

Figure 5.37 Schematic representation of the supramolecular polymerization of **32**. Source: Modified from Haino et al. [99].

a ring-chain mechanism was found (for more details on this particular mechanism, see Chapter 1): the double-logarithmic plot of specific viscosity vs. concentration exhibited a pronounced change of the slope (Figure 5.38). In the low concentration regime, a slope value of 0.95 suggested the presence of cyclic oligomers of rather low molar mass, whereas the increased slope values at higher concentrations were indicative of linear supramolecular polymer chains of high molar masses. For **33a**, a critical polymerization concentration (cpc) of c. 24 mM was estimated from the viscosity measurements in CH_3Cl or toluene. For the metal-containing monomer **33b**, however, a much more complicated supramolecular polymerization behavior was found that was mainly ascribed to the presence of the Zn(II) centers being able to establish additional metal-to-ligand coordinative bonds [101]. The initial linear supramolecular polymer $(\mathbf{33b})_n$ could be transformed into a supramolecular polymer network by adding 1 equiv of a bis-pyridine ligand, such as **34** (Figure 5.39). A strong cooperativity of the supramolecular host–guest binding and metal-to-ligand coordination was concluded from viscosity measurements. In the double-logarithmic plot of the specific viscosity vs. concentration, the slope changed from 1.5 to 4.6 at a cpc value of 9.0 mM; due to the presence of highly entangled polymer networks, the slope became even steeper at higher monomer concentrations (slope of 8.7). Furthermore, viscoelastic properties were observed for the supramolecular material derived from **33b** to **34** – elastic and free-standing polymer films could be prepared thereof (the Young modulus of the material was c. 1 GPa, Figure 5.39). It was concluded that the supramolecular yet macroscopically robust material could be a promising candidate for implementation in self-healing applications [102].

The molecular tweezers, containing a chiral spacer and two Zn(II)–porphyrin arms, was recently reported by Rath and coworkers [103]. The combination with

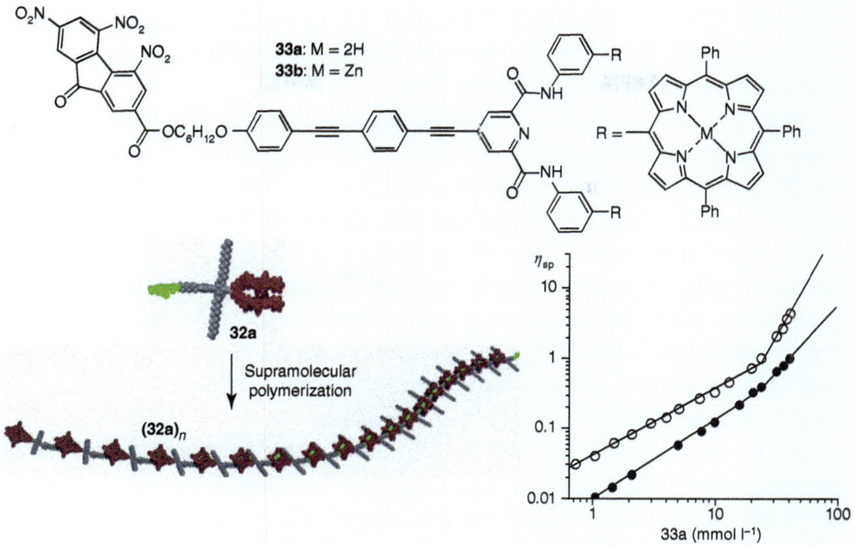

Figure 5.38 Schematic representation of the heteroditopic monomers **33** and of the supramolecular polymerization of **33a**. The specific viscosity of **33a** in $CHCl_3$ (white circles) and toluene (black circles) at 293 K is also depicted. Source: Haino et al. [100], © 2012 John Wiley and Sons.

Figure 5.39 Schematic representation of the bis-pyridine derivative **34** that was used for the formation of a supramolecular polymer network in combination with **33b**. A picture of the free-standing polymer film of this material is also shown. Source: Hager et al. [102]. Figure reproduced with kind permission; © 2015 Wiley-VCH.

chiral and achiral guests afforded 1 : 1 sandwich-type complexes and indicated an ability of chiral discrimination of the host. Though not yet shown, this strategy might be useful to assemble supramolecular helices with high precision via π-electronic host–guest complexation of chiral monomers.

Due to its planar π-electronic surface and highly electron-rich character, TTFs represent highly versatile donor components for the preparation of CT complexes with appropriate π-conjugated acceptor molecules. In this respect, the CT complex of parent TTF with TCNQ is by far the most popular example for which a metal-like electronic behavior was already reported in the early 1970s [104, 105]. In the following, many organic CT complexes incorporating TTF (or its derivatives) have been reported and studied regarding electronic and/or magnetic properties [106]. However, all these materials resemble supramolecular assemblies in the solid state (see Section 5.3) and supramolecular polymers in solution, formed by CT complexes of structurally simple TTFs, have not yet been reported.

Aiming for the employment of TTF's CT complexes for the construction of supramolecular polymers, Sessler and coworkers developed the calix[4]pyrrole receptors **35** (Figure 5.40) for the binding of neutral electron-poor guest molecules in a 1 : 2 fashion [107]. Thereby, the macrocycle of the receptors favored an 1,3-alternate conformation, thus **35** represented a back-to-back-type ditopic molecular tweezers. The conformation of **35** could be reversibly converted into a cone-shaped assembly upon addition of chloride ions. Due to the strong H-bonding interactions between the anion and the N—H groups of the pyrrole rings, coordinated guest molecules could be released from the CT complex of 1 : 2 stoichiometry (the binding constant toward Cl^- in CH_2Cl_2 was about $10^5\ M^{-1}$). Moreover, **35a** featured a cooperative allosteric recognition behavior as concluded from the high K_{a2} and K_{a1} values: TNB was bound to the receptor with K_{a1} and K_{a2} values in CH_2Cl_2 of 20 and 900 M^{-1}, respectively, tetrafluoro-*p*-benzoquinone even gave a K_{a2}

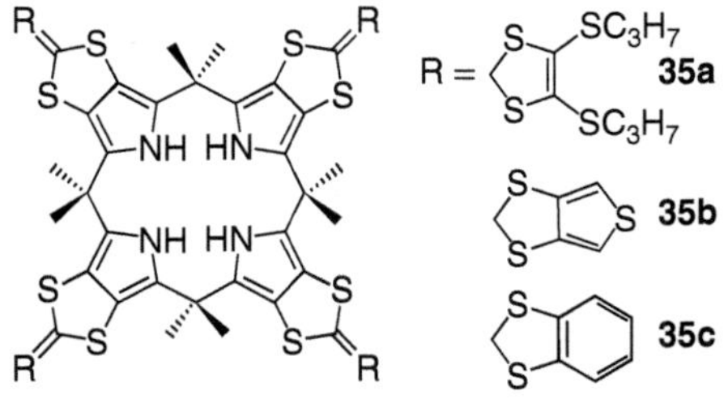

K_a values for the binding of TNB

	$K_{a1}(M^{-1})$	$K_{a2}(M^{-1})$	K_{a2}/K_{a1}
35a	390	1 400	3.6
35a	1 300	31 000	24
35c	2 800	17 000	6.2

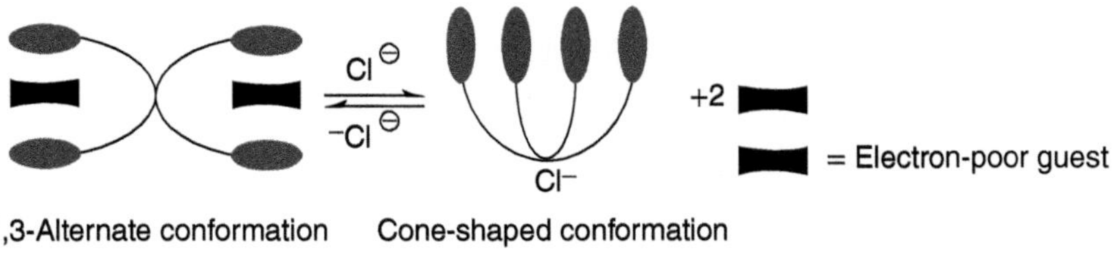

Figure 5.40 Schematic representation of the calix[4]pyrroles **35** that adopted a 1,3-alternate conformation, which could accommodate two guest molecules via CT interactions. The conformation of **35** could be switched reversibly by adding chloride anions. Source: Nielsen et al. [107], © 2004 American Chemical Society.

value of 3500 M^{-1} (isothermal titration calorimetry [ITC] was used to determine these data). When the π-conjugated surface of the TTF sidewalls was enlarged, the binding strength could be increased significantly (**35b** and **35c**, Figure 5.40) [108]. According to Adair's two-site binding model, K_{a2}/K_{a1} ratios of >20 were found (Figure 5.40). These high values supported the assumption of a positive homotropic allosteric-binding process, since for a non-cooperative system, where the binding sites are identical and independent from each other, the theoretical K_{a2}/K_{a1} ratio is 0.25 [109].

To eventually utilize **35**, as monomer, in supramolecular polymerizations, the same authors also prepared the corresponding calix[4]pyrrole monomers *cis*-**36** and *trans*-**36**, both equipped with two dinitrophenyl guest moieties [110]. Mixing either of these with a back-to-back receptor **35**, in equimolar amounts, gave supramolecular polymers that assembled into long needle-like fibers, as shown by scanning electron microscopy (SEM) imaging (Figure 5.41). K_a values in the range of 5.9×10^4 M^{-1} (**35c**/*trans*-**36**) to 7.9×10^5 M^{-1} (**35b**/*cis*-**36**) were determined. XRD analysis of the structure of model CT complexes revealed a close distance (c. 3.26–3.60 Å) of the TTF pincers and the embedded dinitrobenzene guests; moreover, strong H-bonding interactions between the N—H groups of the pyrrole rings and the *O*-atoms of the nitro groups accounted to the stability of the assemblies. Based on the model for an IDP (see Chapter 1), DPs of 21 and 72 were calculated for **35c**/*trans*-**36** and **35b**/*cis*-**36**, respectively (14 mM in $CHCl_3$). It was further shown that the DP could be modulated via the polarity of the solvent and the rigidity of the linker. For example, when using the ditopic guest **37** with a short ester-type linkage instead of the rigid calix[4]pyrrole one in combination with **35a**, a DP of 16 was observed in methylcyclohexane; whereas, a DP of only 1.7 was found in $CHCl_3$ at the same concentration [111]. The supramolecular polymers could be disassembled either by adding triethylammonium chloride (to achieve a conformational switch to the non-coordinating cone-shaped form) or in the presence of TNB as a highly competitive guest [110]. As depicted in Figure 5.41, the stimuli responsiveness of the supramolecular polymers could be distinguished on a molecular level due to different pathways leading to defined small-molecule assemblies. This behavior represents a prerequisite to establish a NAND logic gate. As a proof of concept, Sessler and coworkers realized such a system on the basis of a supramolecular polymer (assembled from receptor **35** and guest **36**) showing a highly reversible stimuli-responsive behavior toward the tetraethylammonium (TEA^+) cation and iodide (I^-) anion [112].

Besides the binding of planar guest molecules, as previously shown, spherical molecules – such as C_{60} – can also be coordinated to molecular tweezers if their pocket matches the dimensions of such guests. The high potential of C_{60}'s supramolecular chemistry was recognized in the late 1990s and diverse architectures ranging from rotaxanes and catenanes to DNA complexes, dyads and triads for photo-induced energy-/electron–transfer processes as well as highly ordered thin films were reported since then [113]. For example, γ-cyclodextrin (γ-CD) gave a 2 : 1 complex with C_{60}; a capsule-like structure was suggested for this supramolecular assembly (Figure 5.42) [114].

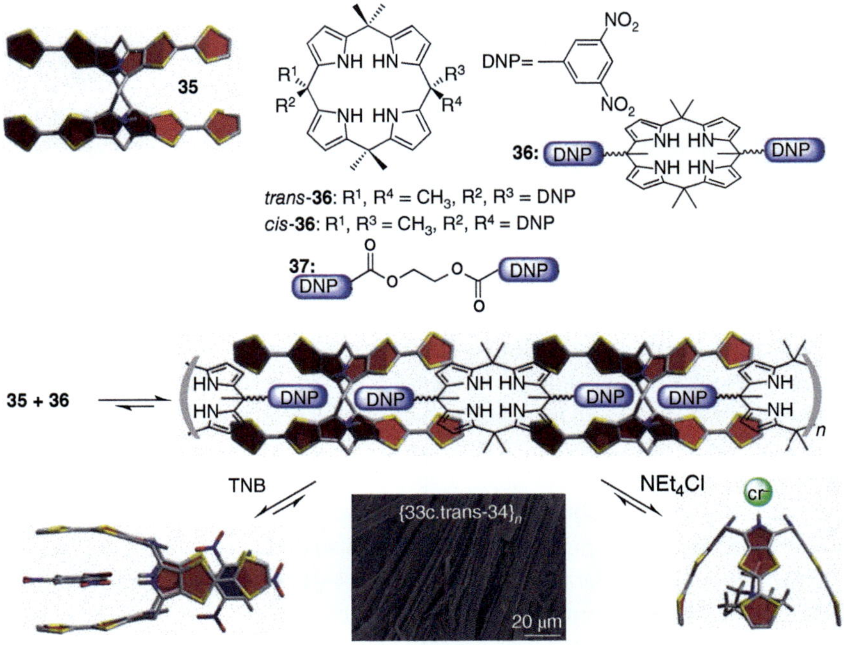

Figure 5.41 Schematic representation of the supramolecular polymers derived from **35** and **36**. A representative SEM image of the fiber-like assemblies also shown. The polymers could be disassembled in the presence of competitive guest molecules or anions. Source: Park et al. [110]. Figure reproduced by kind permission; © 2011 National Academy of Science.

Figure 5.42 Schematic representation of the supramolecular 2 : 1 complex formed by γ-CD and C_{60} in water. Source: Based on Andersson et al. [114].

In a similar fashion, the interaction between fullerenes and calix[*n*]arenes has been studied. Fukuzawa and coworkers reported on 1 : 1 complexes between substituted calix[5]arenes and C_{60} in solution; whereas, in the solid state, a 2 : 1 complex could be identified by single-crystal X-ray analysis [115]. The covalent linkage of two such calix[5]arenes afforded shape-selective molecular tweezers with defined cavity sizes for the efficient encapsulation of fullerenes (i.e. C_{60}, C_{70}, and higher analogs) in organic solvents [116]. Thereby, the least appropriate

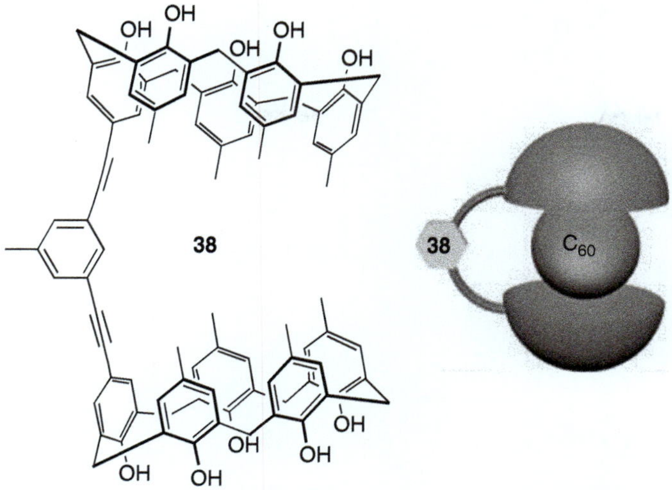

Figure 5.43 Schematic representation of a molecular tweezers on the basis of a bis-calix[5]arene (**38**) and its encapsulation of a C_{60} guest in the cavity. Source: Based on Haino et al. [116].

solvents for fullerenes (i.e. arenes) allowed the strongest binding in an 1 : 1 fashion (Figure 5.43): For the complexation of C_{60} by, e.g. **38** very high K_a values (76 000 M^{-1}) were found in toluene; for C_{70} the binding in the same environment was even stronger (K_a = 163 000 M^{-1}). The application of this supramolecular binding with respect to the formation of supramolecular polymers is detailed in Chapter 8 in the context of calixarenes' supramacromolecular chemistry.

In contrast to the flat π-electronic surface of TTFs, their π-extended analogs (exTTFs) inherently possess a butterfly-shaped geometry that, in combination with the pronounced electron-rich character, makes them ideal candidates for the binding of electron-poor fullerenes and carbon nanotubes (CNTs) in a concave–convex fashion (CT complexes of TTFs with C_{60} have only been reported for the solid state [117]). The Martín group pioneered this field of research by using the bis-exTTF molecular tweezers **39**, reminiscent of Fukuzawa's bis-calix[5]arenes, for the binding of C_{60} [118]. In comparison to parent exTTF, **39** exhibited a significantly increased binding strength (K_a of 3.0×10^3 M^{-1} in C_6H_5Cl). Remarkably, the formation of the CT complexes was found to depend on the solvent composition (Figure 5.44): in C_6H_5Cl, the CT complex of 1 : 1 stoichiometry was observed; whereas, in a $CHCl_3/CS_2$ mixture, a positive homotropic cooperative effect supported the two-step assembly toward a CT complex of a 2 : 2 composition.

To utilize the interaction between bis-exTTF molecular tweezers and C_{60} for the formation of supramolecular polymers, Fernández et al. prepared the heteroditopic monomer **40** that self-assembled in a head-to-tail fashion into a linear donor–acceptor-type polymer (Figure 5.45) [119]. The supramolecular polymerization of **40** was investigated by UV/vis absorption spectroscopy in which the appearance of an absorption band at 487 nm was attributed to the formation of CT interactions between the bis-exTTF molecular tweezers and the phenyl C_{61}

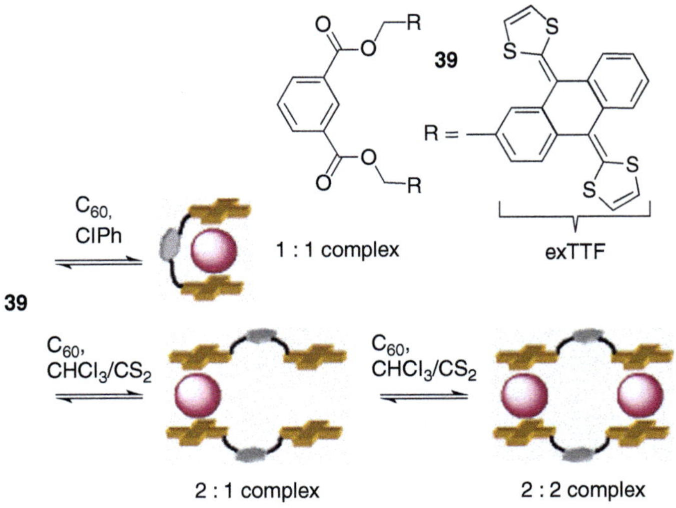

Figure 5.44 Schematic representation of the bis-exTTF-based molecular tweezers **39** and its solvent-dependent interaction with C_{60}. Source: Pérez et al. [118], © American Chemical Society.

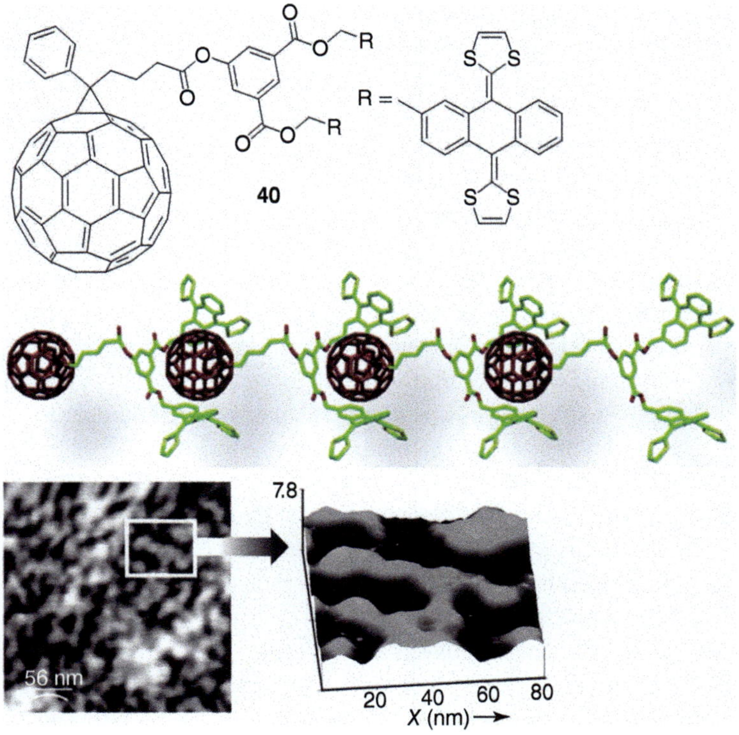

Figure 5.45 Schematic representation of the heteroditopic monomer **40** and of the corresponding supramolecular CT polymer $(\mathbf{40})_n$. The AFM images of $(\mathbf{40})_n$ deposited onto a mica surface are also shown (left: 2D image, 277 nm × 277 nm; right: 3D image, 80 × 65 nm). Source: Fernández et al. [119], Figure reproduced with kind permission; © 2008 Wiley-VCH.

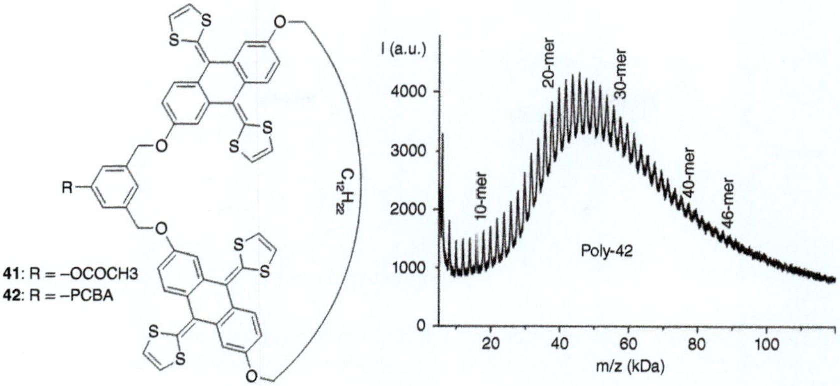

Figure 5.46 Schematic representation of the preorganized exTTF-based receptor **41** and the analogous AB-type monomer **42**. The MALDI-TOF mass spectrum of (**42**)$_n$ is also depicted. Source: Isla et al. [120], © 2014 John Wiley and Sons.

butyric acid (PCBA) moieties. Moreover, cyclic voltammetry (CV) measurements were conducted and indicated a strong electronic D–A-type interaction already in the ground state. The supramolecular polymer (**40**)$_n$ was also investigated by AFM imaging – long (15–300 nm) and winding structures with heights in the range of 1.7–2.5 nm were observed (Figure 5.45).

Martín and coworkers modified their initial molecular tweezers **39** by introducing an alkyl bridge between the two exTTF pincers, thus forming the less flexible macrocyclic receptor **41** (Figure 5.46) [120]. Due to its preorganized molecular structure, **41** exhibited an increased binding affinity toward C_{60} in C_6H_5Cl when compared to **39** (K_a values of $6.3 \times 10^4\ M^{-1}$ vs. $3.0 \times 10^3\ M^{-1}$) and even C_{70} could be complexed efficiently in the same solvent (K_a of c. $10^6\ M^{-1}$). This high binding affinity was further used to assemble a linear supramolecular polymer of high molar mass from the heteroditopic building block **42**. Remarkably, matrix-assisted laser desorption/ionization (MALDI) time-of-flight (TOF) mass spectrometry could be applied to analyze (**40**)$_n$ – signals arising from the 10-mer up to the 46-mer could be assigned (the molar mass of the 46-mer was 90 000 g mol^{-1}, Figure 5.46). For (**40**)$_n$ on the contrary, only short oligomers up to the 5-mer could be identified using the same analytical method. AFM imaging revealed the presence of very long, straight fibers (up to 2 μm in length).

Besides the aforementioned heteroditopic AB-type monomer **38**, the same authors also employed an AB_2-type monomer, i.e. a PCBA unit equipped with two bis-exTTF molecular tweezers, to obtain a hyperbranched supramolecular polymer (Figure 5.47) [121]. This material exhibited triangular patterns, as shown by AFM imaging. Furthermore, the Martín group extended their concept of fullerene fixation by molecular tweezers also to dendritic architectures. Dendrimers with 3, 6, and 12 such molecular tweezers sites were prepared and the binding of C_{60} was studied [122]. The non-complexed dendrimers were prone to aggregate in which on average 10, 12, and 20 molecules of the respective dendrimers assembled. The apparent binding constants (K_{app}) were 2238, 3085, and 2137 M^{-1} for these systems and

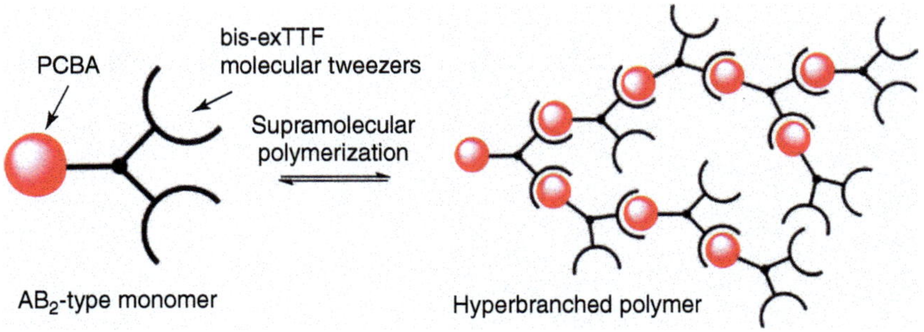

Figure 5.47 Schematic representation of the formation of a supramolecular hyperbranched polymer starting from an AB_2-type monomer. Source: Fernández et al. [121], © 2008 American Chemical Society.

Hill coefficients in the range of 2.6–3.4 were estimated, thus suggesting a positive cooperativity in the course of the multiple binding of C_{60} molecules to the dendritic architectures. The cooperativity of binding was attributed to the dissociation of the dendrimer aggregates prior to the complexation of C_{60}.

5.6 Conclusion and Outlook

π–π Stacking interactions can be used to assemble a broad range of supramolecular structures in which the stacking of the π-electronic surfaces of the building blocks represents the main driving force. Such stacks cannot only be obtained from the simple π–π stacking of molecules, also donor–acceptor CT interactions as well as π-electronic ion pairing have been employed in this context. The resultant supramolecular materials are of interest for opto-electronic applications and often feature LC behavior. However, the poor directionalities and rather weak binding strength represent considerable drawbacks, in particular when aiming for the assembly of defined supramolecular polymers of high molar mass in solution. This issue can, to some extent, be overcome by utilizing preorganized building blocks, the so-called molecular tweezers that offer a high directionality of binding and gives, in general, highly stable sandwich-type complexes. In recent years, a variety of potent molecular tweezers has been developed and employed to assemble stable linear supramolecular polymers in solution either from AB-type monomers or by combining AA- and BB-type monomers, thus making this interaction able to be competitive to the other types of non-covalent interaction with respect to supramolecular polymer science. It has to be expected that supramolecular block copolymers, similar to those based on, e.g. metal-to-ligand coordination [123] or H-bonding interactions [124], can also be prepared on the basis of molecular tweezers and appropriate guest molecules.

Though the stimuli-responsive character of supramolecular polymers based on the complexation of guests by molecular tweezers has already been shown, the development of applications, e.g. as self-healing materials or in drug delivery systems, sensing, or opto-electronics, still needs to be advanced.

References

1 Brunsveld, L., Folmer, B.J.B., Meijer, E.W., and Sijbesma, R.P. (2001). *Chem. Rev.* 101: 4071–4098.
2 de Greef, T.F.A., Smulders, M.M.J., Wolffs, M. et al. (2009). *Chem. Rev.* 109: 5687–5754.
3 Hoeben, F.J.M., Jonkheijm, P., Meijer, E.W., and Schnenning, A.P.H.J. (2005). *Chem. Rev.* 105: 1491–1546.
4 Hunter, C.A. and Sanders, J.K.M. (1990). *J. Am. Chem. Soc.* 112: 5525–5534.
5 Lee, C.C., Grenier, C., Meijer, E.W., and Schenning, A.P.H.J. (2009). *Chem. Soc. Rev.* 38: 671–683.
6 Jonkheijm, P., van der Schoot, P., Schenning, A.P.H.J., and Meijer, E.W. (2006). *Science* 313: 80–83.
7 Keizer, H.M. and Sijbesma, R.P. (2005). *Chem. Soc. Rev.* 34: 226–234.
8 Chen, Z., Lohr, A., Saha-Möller, C.R., and Würthner, F. (2009). *Chem. Soc. Rev.* 38: 564–584.
9 Martin, R.B. (1996). *Chem. Rev.* 96: 3043–3064.
10 Würthner, F., Yao, S., Debaerdemaeker, T., and Wortmann, R. (2002). *J. Am. Chem. Soc.* 124: 9431–9447.
11 Wu, J., Pisula, W., and Müllen, K. (2007). *Chem. Rev.* 107: 718–747.
12 Kastler, M., Pisula, W., Wasserfallen, D. et al. (2005). *J. Am. Chem. Soc.* 127: 4286–4296.
13 Hill, J.P., Jin, W., Kosaka, A. et al. (2004). *Science* 304: 1481–1483.
14 Aida, T., Takemura, A., Fuse, M., and Inoue, S. (1988). *J. Chem. Soc., Chem. Commun.*: 391–393.
15 Aida, T. (2020). *Adv. Mater.* 32: 1905445.
16 Ikeda, T. and Haino, T. (2017). *Polymer* 128: 243–256.
17 Haino, T., Tanaka, M., and Fukazawa, Y. (2008). *Chem. Commun.*: 468–470.
18 Lee, O.P., Yiu, A.T., Beaujuge, P.M. et al. (2011). *Adv. Mater.* 23: 5359–5363.
19 Farnum, D.G., Mehta, G., Moore, G.G.I., and Siegal, F.P. (1974). *Tetrahedron Lett.* 15: 2549–2552.
20 Ghosh, S., Shankar, S., Philips, D.S., and Ajayaghosh, A. (2020). *Mater. Today Chem.* 16: 100242.
21 Zhang, K., Wucher, P., Marszalek, T. et al. (2018). *Chem. Mater.* 30: 5032–5040.
22 Würthner, F., Saha-Möller, C.R., Fimmel, B. et al. (2016). *Chem. Rev.* 116: 962–1052.
23 Winter, A., Hager, M.D., and Schubert, U.S. (2012). Supramolecular polymers. In: *Polymer Science: A Comprehensive Review*, vol. 5 (eds. H.-W. Schmidt and M. Ueda), 269–310. Amsterdam: Elsevier BV.
24 Zhao, D.-H. and Moore, J.S. (2003). *Org. Biomol. Chem.* 1: 3471–3491.
25 Chen, Z., Stepanenko, V., Dehm, V. et al. (2007). *Chem. Eur. J.* 13: 436–449.
26 Chen, Z., Fimmel, B., and Würthner, F. (2012). *Org. Biomol. Chem.* 10: 5845–5855.
27 Fink, R.F., Seibt, J., Engel, V. et al. (2008). *J. Am. Chem. Soc.* 130: 12858–12859.
28 Lim, J.M., Kim, P., Yoon, M.-C. et al. (2013). *Chem. Sci.* 4: 388–397.

29 Schubert, A., Settels, V., Liu, W. et al. (2013). *J. Phys. Chem. Lett.* 4: 792–796.
30 Würthner, F., Thalacker, C., Diele, S., and Tschierske, C. (2001). *Chem. Eur. J.* 7: 2245–2253.
31 Dehm, V., Chen, Z., Baumeister, U. et al. (2007). *Org. Lett.* 9: 1085–1088.
32 van Herrikhuyzen, J., Syamakumari, A., Schenning, A.P.H.J., and Meijer, E.W. (2004). *J. Am. Chem. Soc.* 126: 10021–10027.
33 Jancy, B. and Asha, S.K. (2008). *Chem. Mater.* 20: 169–181.
34 Balakrishnan, K., Datar, A., Naddo, T. et al. (2006). *J. Am. Chem. Soc.* 128: 7390–7398.
35 Balakrishnan, K., Datar, A., Oitker, R. et al. (2005). *J. Am. Chem. Soc.* 127: 10496–10497.
36 Dasgupta, D., Kendhale, A.M., Debije, M.G. et al. (2014). *Chem. Open* 3: 138–141.
37 Chen, Z., Baumeister, U., Tschierske, C., and Würthner, F. (2007). *Chem. Eur. J.* 13: 450–465.
38 Debije, M.G., Chen, Z., Piris, J. et al. (2005). *J. Mater. Chem.* 15: 1270–1276.
39 Partridge, B.E., Leowanawat, P., Aqad, E. et al. (2015). *J. Am. Chem. Soc.* 137: 5210–5224.
40 Osswald, P. and Würthner, F. (2007). *J. Am. Chem. Soc.* 129: 14319–14326.
41 Osswald, P., Reichert, M., Bringmann, G., and Würthner, F. (2007). *J. Org. Chem.* 72: 3403–3411.
42 Safont-Sempere, M.M., Osswald, P., Radacki, K., and Würthner, F. (2010). *Chem. Eur. J.* 16: 7380–7384.
43 Safont-Sempere, M.M., Osswald, P., Stolte, M. et al. (2011). *J. Am. Chem. Soc.* 133: 9580–9591.
44 Weissman, H., Ustinov, A., Shimoni, E. et al. (2011). *Polym. Adv. Technol.* 22: 133–138.
45 Krieg, E., Weissman, H., Shimoni, E. et al. (2014). *J. Am. Chem. Soc.* 136: 9443–9452.
46 Kulkarni, C., Bejagam, K.K., Senanayak, S.P. et al. (2015). *J. Am. Chem. Soc.* 137: 3924–3932.
47 Guo, Z., Zhang, X., Wang, Y., and Li, Z. (2019). *Langmuir* 35: 342–358.
48 Würthner, F. (2016). *Acc. Chem. Res.* 49: 868–876.
49 Bujosa, S., Castellanos, E., Frontera, A. et al. (2020). *Org. Biomol. Chem.* 18: 888–894.
50 Nakatsuka, S., Gotoh, H., Kinoshita, K. et al. (2017). *Angew. Chem. Int. Ed.* 56: 5087–5090.
51 Winter, A. and Schubert, U.S. (2019). *Mater. Chem. Front.* 3: 2308–2325.
52 Haketa, Y., Yamakado, R., and Maeda, H. (2017). Supramolecular assemblies of π-electronic charged species. In: *Conjugated Objects: Developments, Synthesis, and Application* (eds. A. Nagai and K. Takagi), 349–380. Singapore: Pan Sanford.
53 Bosson, J., Gouin, J., and Lacour, J. (2014). *Chem. Soc. Rev.* 43: 2824–2840.
54 Stępień, M., Gońka, E., Żyła, M., and Sprutta, N. (2017). *Chem. Rev.* 117: 3479–3716.

55 Kushida, T., Shuto, A., Yoshio, M. et al. (2015). *Angew. Chem. Int. Ed.* 54: 6922–6925.

56 Yamamura, M., Hasegawa, T., and Nabeshima, T. (2016). *Org. Lett.* 18: 816–819.

57 Kumar, M., Venkata Rao, K., and George, S.J. (2014). *Phys. Chem. Chem. Phys.* 16: 1300–1313.

58 Das, A. and Ghosh, S. (2014). *Angew. Chem. Int. Ed.* 53: 2038–2054.

59 Han, Y., Tian, Y., Li, Z., and Wang, F. (2018). *Chem. Soc. Rev.* 47: 5165–5176.

60 Kaeser, A., Wolffs, M., and Schenning, A.P.H.J. (2012). Multicomponent self-assembled polymers based on π-conjugated systems. In: *Supramolecular Chemistry*, vol. 7 (eds. P.A. Gale and J.W. Steed), 3183–3204. Hoboken, NJ: Wiley.

61 Liu, Y., Yu, Y., Gao, J. et al. (2010). *Angew. Chem. Int. Ed.* 49: 6576–6579.

62 Rao, K.V., Jayaramulu, K., Maji, T.K., and George, S.J. (2010). *Angew. Chem. Int. Ed.* 49: 4218–4222.

63 Rao, K.V. and George, S.J. (2012). *Chem. Eur. J.* 18: 14286–14291.

64 Jalani, K., Kumar, M., and George, S.J. (2013). *Chem. Commun.* 49: 5174–5176.

65 Jalani, K., Dhiman, S., Jain, A., and George, S.J. (2017). *Chem. Sci.* 8: 6030–6036.

66 Chen, C.W. and Whitlock, H.W. Jr., (1978). *J. Am. Chem. Soc.* 100: 4921–4922.

67 Kong, F.K.-W., Chan, A.K.-W., Ng, M. et al. (2017). *Angew. Chem. Int. Ed.* 56: 15103–15107.

68 Haino, T., Fujii, T., and Fukazawa, Y. (2006). *J. Org. Chem.* 71: 2572–2580.

69 Zimmerman, S.C. and VanZyl, C.M. (1987). *J. Am. Chem. Soc.* 109: 7894–7896.

70 Zimmerman, S.C. (1993). *Top. Curr. Chem.* 165: 71–102.

71 Petitjean, A., Khoury, R.G., Kyritsakas, N., and Lehn, J.-M. (2004). *J. Am. Chem. Soc.* 126: 6637–6647.

72 Haketa, Y. and Maeda, H. (2017). *Chem. Commun.* 53: 2894–2909.

73 Haketa, Y., Urakawa, K., and Maeda, H. (2020). *Mol. Syst. Des. Eng.* 5: 757–771.

74 Dong, B. and Maeda, H. (2013). *Chem. Commun.* 49: 4085–4099.

75 Martin, J.C. and Smith, R.G. (1964). *J. Am. Chem. Soc.* 86: 2252–2256.

76 Katritzky, A.R., Zakaria, Z., Lunt, E. et al. (1979). *J. Chem. Soc., Chem. Commun.*: 268–269.

77 Katritzky, A.R., Zakaria, Z., and Lunt, E. (1980). *J. Chem. Soc., Perkin Trans. 1*: 1879–1887.

78 Maeda, H. and Kusunose, Y. (2005). *Chem. Eur. J.* 11: 5661–5666.

79 Maeda, H., Fukui, A., Yamakado, R., and Yasuda, N. (2015). *Chem. Commun.* 51: 17572–17575.

80 Maeda, H., Takeda, Y., Haketa, Y. et al. (2018). *Chem. Eur. J.* 24: 8910–8916.

81 Sasano, Y., Yasuda, N., and Maeda, H. (2016). *Chem. Asian J.* 11: 3423–3429.

82 Yamakado, R., Ishibashi, H., Motoyoshi, Y. et al. (2019). *Chem. Commun.* 55: 326–329.

83 Tanaka, Y., Ichijo, K., Kodama, S. et al. (2019). *Cryst. Growth Des.* 19: 5811–5818.

84 Bruce, M., Humphrey, P., Skelton, B., and White, A. (1986). *Aust. J. Chem.* 39: 165–169.

85 Bando, Y., Haketa, Y., Sakurai, T. et al. (2016). *Chem. Eur. J.* 22: 7843–7850.
86 Haketa, Y., Sasaki, S., Ohta, N. et al. (2010). *Angew. Chem. Int. Ed.* 49: 10079–10083.
87 Dong, B., Terashima, Y., Haketa, Y., and Maeda, H. (2012). *Chem. Eur. J.* 18: 3460–3463.
88 Eryazici, I., Moorefield, C.N., and Newkome, G.R. (2008). *Chem. Rev.* 108: 1834–1895.
89 Sekiya, R., Tsutsui, Y., Choi, W. et al. (2014). *Chem. Commun.* 50: 10615–10618.
90 Kim, S.K., Lim, J.M., Pradhan, T. et al. (2014). *J. Am. Chem. Soc.* 136: 495–505.
91 Zhu, Z., Cardin, C.J., Gan, Y., and Colquhoun, H.M. (2010). *Nat. Chem.* 2: 653.
92 Burattini, S., Greenland, B.W., Hayes, W. et al. (2011). *Chem. Mater.* 23: 6–8.
93 Leblond, J., Gao, H., Petitjean, A., and Leroux, J.-C. (2010). *J. Am. Chem. Soc.* 132: 8544–8545.
94 Viricel, W., Mbarek, A., and Leblond, J. (2015). *Angew. Chem. Int. Ed.* 54: 12743–12747.
95 Tanaka, Y., Wong, K.M.-C., and Yam, V.W.-W. (2013). *Chem. Eur. J.* 19: 390–399.
96 Tian, Y.-K., Shi, Y.-G., Yang, Z.-S., and Wang, F. (2014). *Angew. Chem. Int. Ed.* 53: 6090–6094.
97 Tian, Y.-K., Han, Y.-F., Yang, Z.-S., and Wang, F. (2016). *Macromolecules* 49: 6455–6461.
98 Valderrey, V., Aragay, G., and Ballester, P. (2014). *Coord. Chem. Rev.* 258–259: 137–156.
99 Haino, T., Fujii, T., Watanabe, A., and Takayanagi, U. (2009). *Proc. Natl. Acad. Sci. U.S.A.* 106: 10477–10481.
100 Haino, T., Watanabe, A., Hirao, T., and Ikeda, T. (2012). *Angew. Chem. Int. Ed.* 51: 1473–1476.
101 Kinjo, K., Hirao, T., Kihara, S.-i. et al. (2015). *Angew. Chem. Int. Ed.* 54: 14830–14834.
102 Hager, M.D., Greil, P., Leyens, C. et al. (2010). *Adv. Mater.* 22: 5424–5430.
103 Saha, B., Ikbal, S.A., and Rath, S.P. (2020). *Inorg. Chem.* 59: 7795–7809.
104 Ferraris, J., Cowan, D.O., Walatka, V., and Perlstein, J.H. (1973). *J. Am. Chem. Soc.* 95: 948–949.
105 Coleman, L.B., Cohen, M.J., Sandman, D.J. et al. (1973). *Solid State Commun.* 12: 1125–1132.
106 Goetz, K.P., Vermeulen, D., Payne, M.E. et al. (2014). *J. Mater. Chem. C* 2: 3065–3076.
107 Nielsen, K.A., Cho, W.-S., Jeppesen, J.O. et al. (2004). *J. Am. Chem. Soc.* 126: 16296–16297.
108 Park, J.S., Le Derf, F., Bejger, C.M. et al. (2010). *Chem. Eur. J.* 16: 848–854.
109 Koshland, D.E. Jr.,, Némethy, G., and Filmer, D. (1966). *Biochemistry* 5: 365–385.
110 Park, J.S., Yoon, K.Y., Kim, D.S. et al. (2011). *Proc. Natl. Acad. Sci. U.S.A.* 108: 20913–20917.
111 Bähring, S., Kim, D.S., Duedal, T. et al. (2014). *Chem. Commun.* 50: 5497–5499.

112 Kim, D.S., Lynch, V.M., Park, J.S., and Sessler, J.L. (2013). *J. Am. Chem. Soc.* 135: 14889–14894.

113 Diederich, F. and Gómez-López, M. (1999). *Chem. Soc. Rev.* 28: 263–277.

114 Andersson, T., Nilsson, K., Sundahl, M. et al. (1992). *J. Chem. Soc., Chem. Commun.*: 604–606.

115 Haino, T., Yanase, M., and Fukazawa, Y. (1997). *Angew. Chem. Int. Ed. Engl.* 36: 259–260.

116 Haino, T., Yanase, M., and Fukazawa, Y. (1998). *Angew. Chem. Int. Ed. Engl.* 37: 997–998.

117 Izuoka, A., Tachikawa, T., Sugawara, T. et al. (1992). *J. Chem. Soc., Chem. Commun.*: 1472–1473.

118 Pérez, E.M., Sánchez, L., Fernández, G., and Martín, N. (2006). *J. Am. Chem. Soc.* 128: 7172–7173.

119 Fernández, G., Pérez, E.M., Sánchez, L., and Nazario, M. (2008). *Angew. Chem. Int. Ed.* 47: 1094–1097.

120 Isla, H., Pérez, E.M., and Martín, N. (2014). *Angew. Chem. Int. Ed.* 53: 5629–5633.

121 Fernández, G., Pérez, E.M., Sánchez, L., and Martín, N. (2008). *J. Am. Chem. Soc.* 130: 2410–2411.

122 Fernández, G., Sánchez, L., Pérez, E.M., and Martín, N. (2008). *J. Am. Chem. Soc.* 130: 10674–10683.

123 Gohy, J.-F. (2009). *Coord. Chem. Rev.* 253: 2214–2225.

124 Binder, W.H. and Zirbs, R. (2007). *Adv. Polym. Sci.* 207: 1–78.

6

Supramolecular Polymers, Based on Crown Ether Recognition

6.1 General Aspects

Crown ethers and cryptands, as their *N*-containing 3D analogs, were discovered in the 1960s [1, 2] and their ability to complex a broad range of (transition) metal ions marks the birth of what is today known as supramolecular chemistry [3]. Justifiably, the Nobel Prize for Chemistry in 1987 was awarded to D.J. Cram, J.-M. Lehn, and C.J. Pedersen honoring their early contributions to this field [4–6].

In pioneering work, Stoddard and coworkers showed that organic cations, such as paraquat (i.e. *N*,*N*′-dimethyl-4,4′-bipyridinium or methylviologen/paraquat [MV^{2+}]), can also form host–guest complexes with crown ethers in appropriate solvents [7]. Charge–transfer (CT) interactions in combination with C–H···O or N–H···O hydrogen-bonding (H-bonding)interactions in the case of paraquat derivatives and secondary ammonium salts, respectively, have been identified as the main driving force for such non-covalent binding [8]. Inspired by this, many examples have been reported making use of the association of paraquat derivatives as well as secondary ammonium salts to crown ethers and related macrocyclic hosts. In particular, bis(*p*-phenylene)-34-crown-10 (**1**, BPP34C10), bis(*m*-phenylene)-32-crown-10 (**2**, BMP32C10), and dibenzo-24-crown-8 (**3**, DB24C8) represent the fundamental macrocycles (Figure 6.1) that are most commonly utilized for the complexation of appropriate guest molecules [8, 9].

Concerning the self-assembly process itself (i.e. the reversible threading of an organic molecule into a crown ether cavity), two basic cases have to be distinguished: dethreading is disabled due to the presence of bulky stoppers [10] and dethreading is enabled. These two extremes are generally referred to as rotaxanes and pseudorotaxanes, respectively (Figure 6.2). Based on this simple concept, a broad range of supramolecular "machines" has been developed; the importance of these findings has recently been acknowledged by awarding the 2016 Nobel Prize in Chemistry to J.-P. Sauvage, J.F. Stoddard, and B.L. Feringa [11]. Beyond crown ethers, utilizing other cavity-containing macromolecules may also function as hosts for (pseudo)rotaxane formation: cyclodextrins (CDs) [12], cucurbit[8]urils [13], calixarenes [14], and pillarenes [15]. The supra(macro)molecular chemistry of these is detailed in the subsequent Chapters 7–10. In this part, the focus is on the supramolecular chemistry of crown ethers, in particular, in view of supramolecular

Supramolecular Polymers and Assemblies: From Synthesis to Properties and Applications, First Edition.
Ulrich S. Schubert, George R. Newkome, and Andreas Winter.

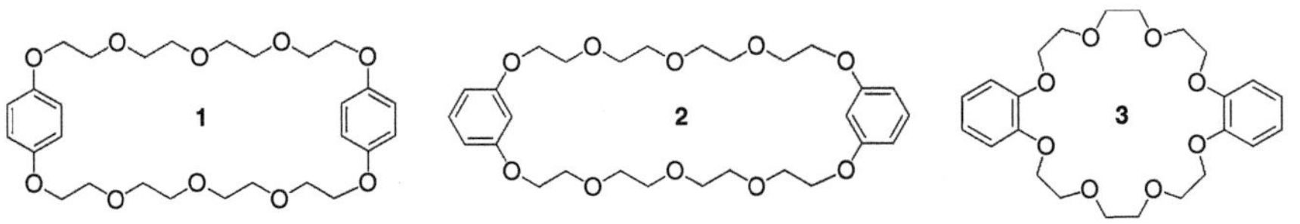

Figure 6.1 Schematic representation of important crown ether hosts.

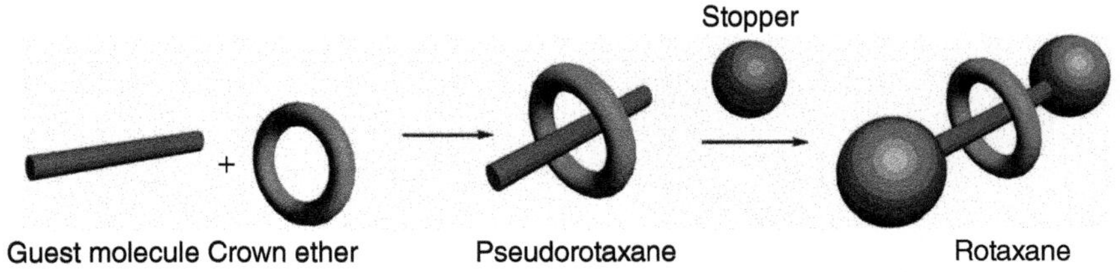

Figure 6.2 Schematic representation of the (pseudo)rotaxane formation. Source: Zheng et al. [9], 2012 Royal Society of Chemistry.

polymers, i.e. poly(pseudo)rotaxanes. The main rule for the nomenclature of (pseudo)rotaxanes is, in simple words, as follows: a prefix in the form of a bracketed integer indicates the total number of molecular components in the supramolecular assembly [16]. Thus, the general structure depicted in Figure 6.2 has to be referred to as a [2]rotaxane.

Stable, small-molecule, pseudorotaxanes from secondary ammonium salts and **3** are well known in the literature [17]. For the threading of the dibenzylammonium cation into **3**, association constants (K_a) of 360 and 420 M^{-1} in acetone and MeCN, respectively, were determined. In contrast, the interaction of **3** with paraquat derivatives is relatively weak [18]. The formation of "taco complexes" as well as pseudorotaxanes was observed for **2** with paraquat derivatives [19, 20]. Stable pseudorotaxane-like complexes of 1 : 1-stoichiometry can be obtained utilizing **1** and MV^{2+} itself; the K_a value for this system was c. 730 M^{-1} in acetone [21].

Many reports on the interaction of crown ethers derivatives with inorganic as well as organic salts have been published [22–25], and the various types of (potential) applications of such host–guest complexes are well documented [26–30]. However, the utilization of this type of non-covalent interaction for the assembly of smart supramolecular polymers represents a notable application that will be detailed in this chapter [9, 31].

6.2 From Crown Ether Molecular Recognition Toward Supramolecular Polymers

The highly reversible character of host–guest interactions, here involving the molecular encapsulation within crown ether derivatives, is notably suited for the creation of advanced materials with self-healing or adaptive properties [9, 32]. For the majority of building blocks, designed as monomers, for the self-assembly into supramacromolecular architectures, a classification scheme can be outlined which is simply based on the numbers of host- and/or guest-type units within the building blocks:

1. To assemble 1D linear supramolecular polymers from appropriate host and guest monomers in a polyassociative process, one has to distinguish between two cases: (i) the self-assembly of heteroditopic AB-type monomers into homopolymers and (ii) the two-component association of a precise equimolar mixture of heterocomplementary, homoditopic AA- and BB-type monomers into regularly alternating heteropolymers. In either case, a competition in the formation of the anticipated linear polymers and cyclic oligomers will occur; however, the regulation of the process by triggering the external conditions (e.g. concentration, temperature) is often possible.
2. 3D Hyperbranched (i.e. dendritic) supramolecular polymers will result from self-assembly when an imbalance of the complementary host and guest units is confirmed. For this purpose, either heterotritopic AB_2-type building blocks or mixtures of complementary components may be used (e.g. a combination of an A_2- and a B_3-type unit). Intrinsically, such materials will always be rather ill defined from a structural point-of-view. Nonetheless, they are of considerable interest, since they offer unique chemical and physical properties (e.g. ease of synthesis, low melting, and solution viscosity as well as a high degree-of-functionalization).
3. Starting from multitopic hosts and guests, a range of supramolecular polymer topologies, including star-shaped, dendronized, or cross-linked ones, can be derived. These materials exhibit an increased viscosity, when compared to their linear analogs.

From this classification scheme, the 1D linear supramolecular polymers will be discussed in this chapter, only a short section will be devoted to the structurally more complex architectures [32, 33]. A range of applications in which poly(pseudo)rotaxanes have already been implemented includes, for example, supramolecular (adaptive) gels or machines, materials with stimuli-responsive or self-healing behavior as well as supramolecular sensors.

6.2.1 Linear Polypseudorotaxanes, Derived from Small-Molecule Building Blocks

A versatile synthetic strategy to obtain the most simple type of structure, i.e. 1D chain-like supramolecular polymers, was developed by the Stoddard group [34]. Based on the paradigm that stiff monomers preferentially lead to daisy chains, instead of forming intramolecular host–guest complexes, various building blocks comprising rigid linkers between the terminal functionalities have been designed. However, even the AB-type monomer **4** with the complementary binding sites being directly adjacent to each other was found to preferentially dimerize into a [*c*2]daisy chain (Figure 6.3) [34, 35]. In solution, this $(\mathbf{4})_2$ dimer as well as its deuterated or fluorinated analogs were confirmed to be the most favorable structures. However,

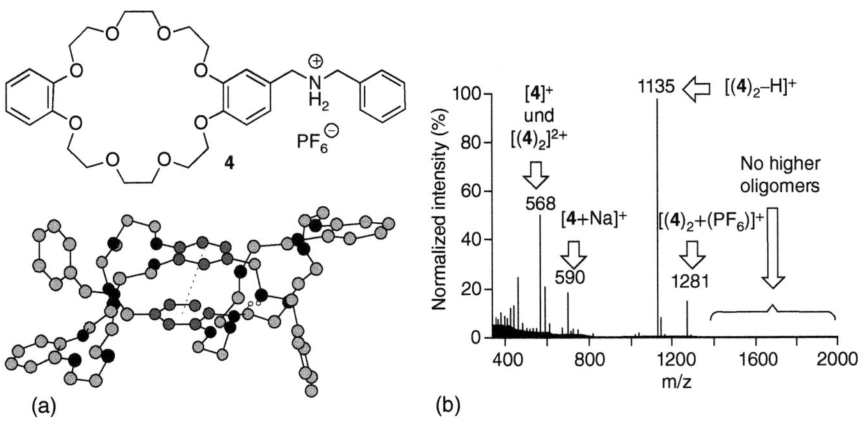

Figure 6.3 (a) Schematic representation of the AB-type monomer **4** and the ball-stick representation of the solid structure of the dimer $(\mathbf{4})_2$ (counter ions were omitted for clarity). (b) MALDI time-of-flight mass spectrum of **4**. Source: Cantrill et al. [35], © 2001 American Chemical Society.

matrix-assisted laser desorption/ionization (MALDI) mass spectrometry, using a competitive H-bonding matrix, permitted the detection of larger aggregates, up to pentamer $(\mathbf{4})_5$ (though with very low signal intensities) [36].

Daisy chains are defined as "a string of daisies with stems linked to form a chain" [37]. Thereby, daisy chains may be acyclic [*a*], i.e. linear or cyclic [*c*], i.e. having its ends linked together to afford a cycle; thus, a [*c*]daisy chain resembles a necklace. In view of molecular structures, an [*a*]daisy chain features a free pair of potentially matching recognition sites at the termini; whereas, a [*c*]daisy chain follows the rule-of-maximal-occupancy of recognition sites [3], as a result from the closure of an [*a*]daisy chain using its last pair of complementary recognition sites located at the termini of the growing chain.

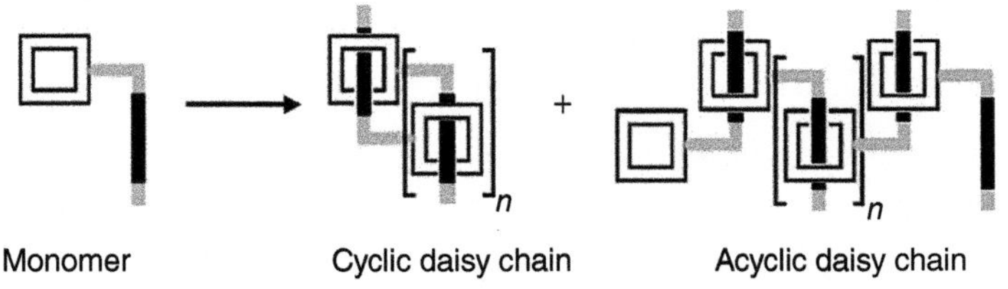

Related to this, the same group prepared the AB-type monomer **5** featuring a larger cavity [38]. Though laser-spray ionization (LSI) mass spectrometric analysis revealed the existence of larger assemblies up to the pentamer, the [*c*2]daisy chain $(\mathbf{5})_2$ was identified as the dominate species in the solid state (Figure 6.4). The stability of this dimer was attributed to an efficient π–π stacking interaction of the acceptor-type MV^{2+} arm of a molecule with the donor-type π-electronic system of a second one that

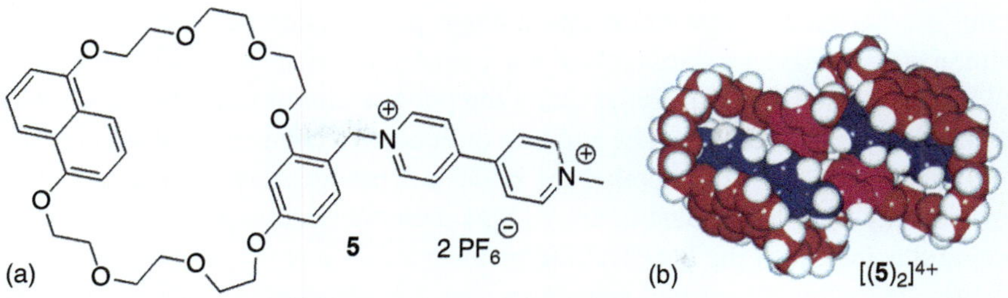

Figure 6.4 (a) Schematic representation of monomer **5**. (b): Space-filling representation of the [*c*2]daisy chain $(\mathbf{5})_2$ (counterions were omitted for clarity). Source: Ashton et al. [38], © 1998 John Wiley and Sons.

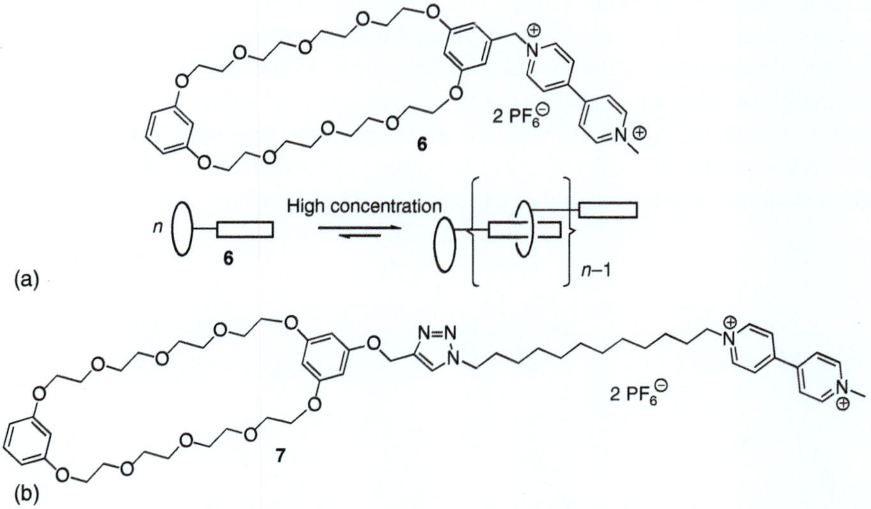

Figure 6.5 (a) Schematic representation of the self-assembly of **6** into a polypseudorotaxane. (b) Schematic representation of the multifunctional monomer **7**.

was further enhanced by C–H···O H-bonding interactions with macrocyclic framework.

Despite these limitations, the ring-chain equilibrium can be well-driven toward the formation of acyclic daisy chains at high monomer concentrations. For example, monomer **6** comprising a BMP32C10-type host moiety and a rod-like MV^{2+} tail, as guest, revealed an efficient intermolecular complexation behavior (Figure 6.5a). According to ^{1}H NMR (nuclear magnetic resonance) spectroscopy, a degree-of-polymerization (DP) of c. 50 was estimated (at a concentration of 2.0 M), which corresponded to a molar mass of 51 000 g mol^{-1} [39]. Freeze-drying the concentrated samples of **6** ($c > 0.1$ M) gave amorphous materials with increased glass-transition temperatures (T_g). The higher T_g values in comparison with that of recrystallized **6** were attributed to the existence of supramolecular polymers of high molar mass.

Using a different monomer design afforded, at high concentration and low temperature, a supramolecular polymer. The monomer **7**, containing a 1,2,3-triazole unit within the flexible tail, assembled into a linear polypseudorotaxane which, upon addition of Pd(II) ions, was cross-linked into a supramolecular network (Figure 6.5b) [40]. Thereby, the triazole moieties along the polypseudorotaxane chains acted as ligands for the Pd(II) centers. The addition of PPh_3, as a competitive ligand for the complexation of Pd(II), regenerated the initial polypseudorotaxane structure. The same strategy was also applied to an AB-type monomer containing a cryptand, as binding site, instead of the BMP32C10-type unit.

It was shown by Huang and coworkers that the self-assembly of heteroditopic monomers can be used to generate materials with interesting bulk-phase properties [41]. The initially formed 1D supramolecular polymer self-organized into a 3D fiber network with gel-like properties. Thereby, the aliphatic spacer between the binding sites physically linked the polymer chains to afford the supramolecular polymer gel. Both a thermo- and a pH-responsive behaviors were observed with a high degree-of-reversibility of the sol–gel phase transition, which could be initiated either thermally (i.e. heating or cooling) or by changing the pH value (i.e. adding either a base or an acid, Figure 6.6). These authors proposed that such behavior might be of relevance for the realization of new types of drug-delivery systems and other biomedical applications. A similar supramolecular 1D polymer could be fabricated

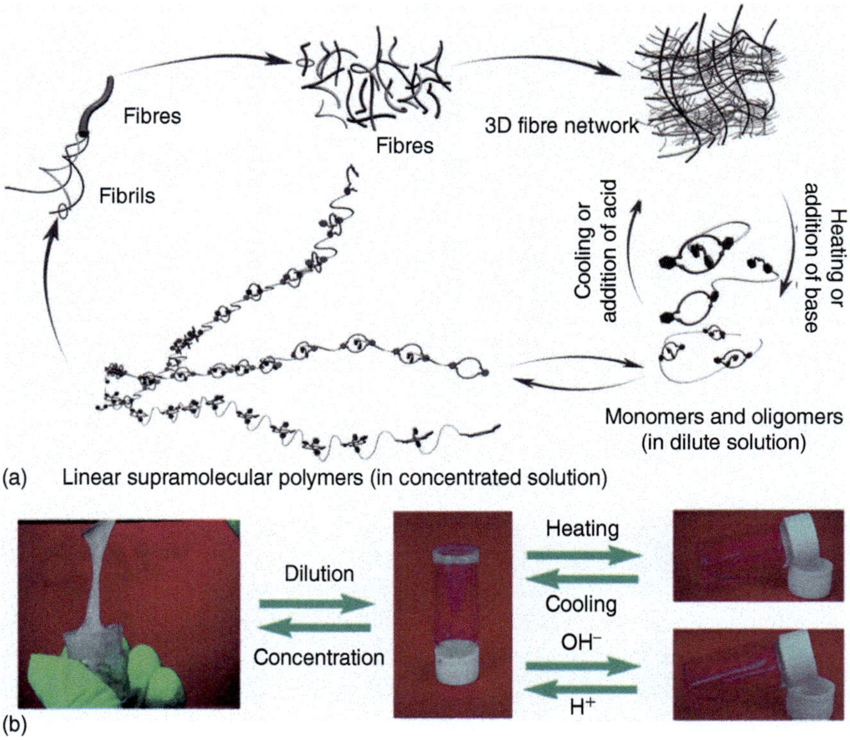

Figure 6.6 (a) Schematic representation of the formation of a supramolecular polymer gel via the self-assembly of a flexible AB-type monomer. (b) Pictures of the supramolecular gel and its stimuli-triggered gel–sol transitions (by changing either the temperature or pH value). Source: Dong et al. [41]. Figure reproduced with kind permission; © 2011 Wiley-VCH.

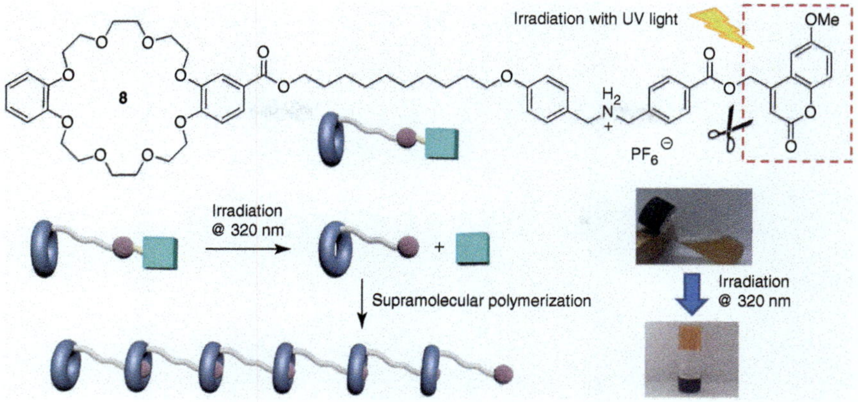

Figure 6.7 Schematic representation of the photoinduced supramolecular polymerization of **8** which was accompanied by sol–gel transition of the solution at high monomer concentrations. Source: Zheng et al. [43]. Figure reproduced with kind permission; © 2018 Elsevier B.V.

into nanofibers by electro-spinning; this report highlights the potential of polypseudorotaxanes for nanoscale scientific applications [42].

The AB-type monomer **8**, containing a DB24C8-type crown ether site and flexible dibenzalammonium-containing tail, was synthesized by Qu and coworkers [43]. To trigger the supramolecular polymerization of this monomer photochemically, a photo-cleavable coumarin moiety was attached, as a protecting group. Due to a mismatch of the size of the coumarin dye with the diameter of the crown ether, the direct supramolecular polymerization of **8** was disabled. Irradiation with UV light, however, cleaved the protecting group and induced the self-assembly process (Figure 6.7). Following a ring-chain-mediated mechanism, monomer and cyclic oligomers predominated at low concentrations; whereas, linear chains of high molecular weight were formed at high monomer counter ions (a critical polymerization concentration [cpc] of 40 mM was derived from viscosity measurements). At very high monomer concentrations, the self-assembly process gave an organogel composed of interwoven fibers.

Not only AB-type monomers, as in the previous examples but also combinations of AA- and BB-type monomers, in equimolar amounts, have been utilized for the formation of (poly)pseudorotaxanes via crown ether recognition. The self-assembly of the homoditopic monomers **9** and **10**, at low concentrations, gave cyclic species, e.g. of 1 : 1 or 2 : 2 stoichiometry (Figure 6.8a) [44]. At elevated concentrations, supramolecular polymerization could be observed. Thus, neglecting the formation of other species but linear chains, the Carothers' equation (see Section 1.3) was applied to estimate a DP of c. 36 (i.e. a molar mass of 83 800 g mol^{-1}). In a similar approach, monomers **11** and **12** with various spacer lengths were investigated with respect to their association behavior (Figure 6.8b) [45–47]. Thereby, an equal spacer length (i.e. l, $m = 6$) was found to be crucial for the formation of linear polypseudorotaxanes in concentrated solutions via reversible chain extension. At a concentration of 2.0 M, a molar mass of c. 18 000 g mol^{-1} was reached (DP of c. 9). Upon

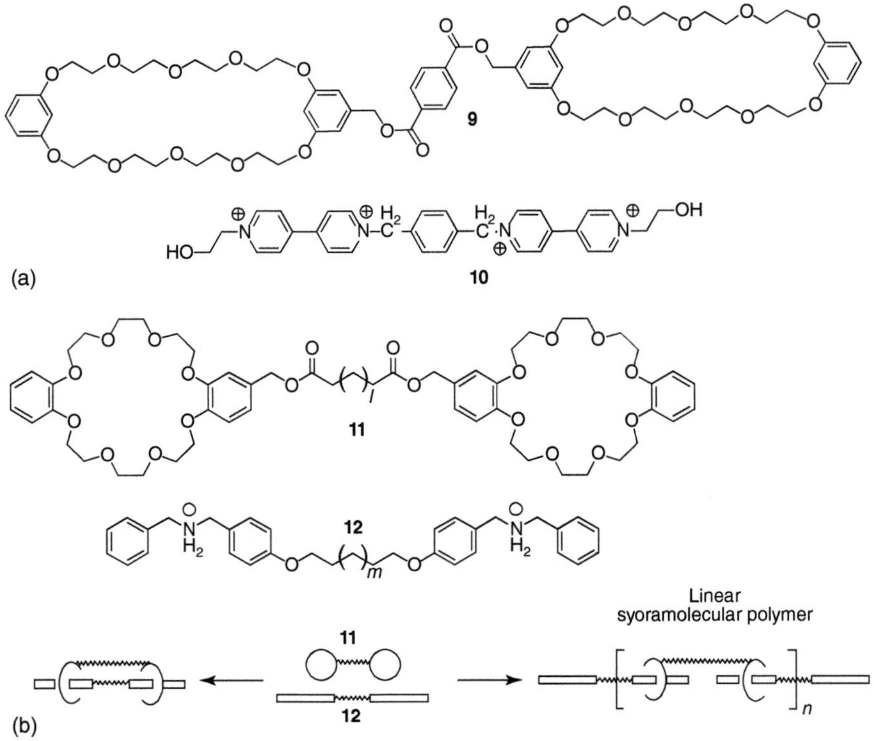

Figure 6.8 Schematic representation of the self-assembly of the complementary homoditopic monomers **9** and **10** (a) as well as **11** and **12** (b); the counterions were omitted for clarity [47]. Source: Figure reproduced with kind permission; © 2003 American Chemical Society.

decreasing the concentration, polypseudorotaxane formation was less favored; thus, a sandwich-type 1 : 1-adduct became the predominant species. In particular, the formation of such cyclic dimers could be controlled by modulating the overall distance between the terminal binding sites: the fraction of cyclic dimers could be decreased significantly by increasing the length of the aliphatic linker in either of the components (**11** or **12**). The effect of mismatching with respect to the twofold threading of **12** into **11** enhanced the linear chain extension toward polypseudorotaxane formation.

The Huang group investigated the supramolecular self-sorting behavior of the AB-type components (**13** and **14**) [48]. Basically, the self-sorting principle relies on the molecule's ability to selectively interact only with the matching complementary binding sites in the complex mixture; thereby, the affinity and fidelity of the binding process represent the requirements, which are fulfilled in many natural systems [9]. The selectivity of DB24C8 (**3**) and BPP34C10 (**1**) to bind dibenzyl ammonium salts and MV^{2+} derivatives, respectively, represents an exquisite example of applying orthogonal host–guest chemistry for the preparation of supramolecular polymers [48]. This particular case is discussed in detail in the context of supramolecular polymerization, based on orthogonal non-covalent interactions in Chapter 11.

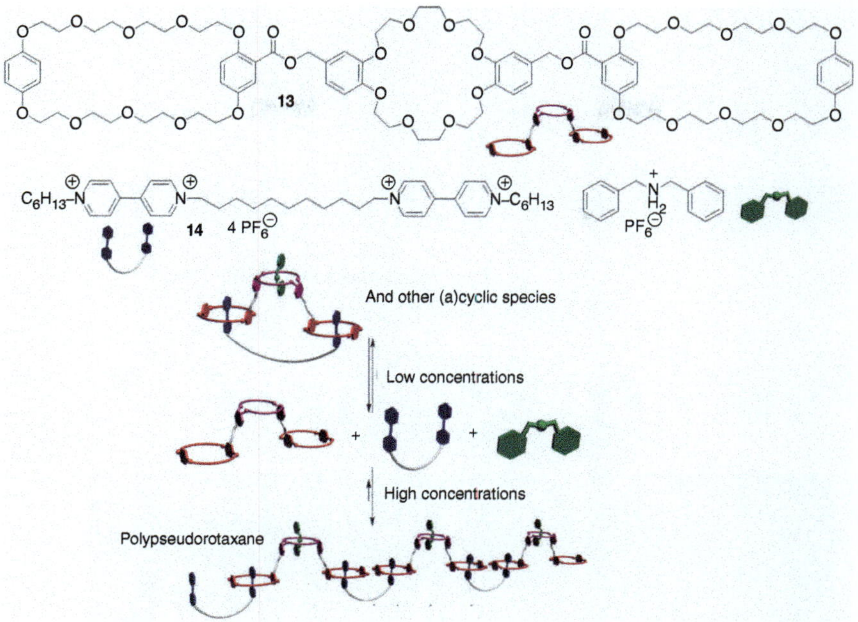

Figure 6.9 Schematic representation of polypseudorotaxane formation via selective crown ether recognition. Source: Wang et al. [49], © 2009 Royal Society of Chemistry.

Monomer **13** – comprising two BPP34C10 units flanking a central DB24C8 – was used for polypseudorotaxane formation by self-assembly with the bis-paraquat derivative **14** (Figure 6.9) [49]. Thereby, the MV^{2+} units were threaded selectively into the BPP34C10 macrocycles, leaving the DB24C8 cavities empty. Upon the addition of a dibenzyl ammonium salt, the latter ones were also filled, thus increasing the rigidity of the polymer. As a result, the material's physical properties drastically changed as revealed by viscometry and rheology measurements.

The bis-(crown ether) derivative **15** featuring a central tetraphenylethylene (TPE) moiety was combined with homoditopic guest **12** ($m = 10$) into a supramolecular polymer (Figure 6.10) [50]. Due to the more rigid environment when incorporated into the polymer, aggregation-induced emission (AIE) arising from the TPE dye was observed from the supramolecular polymer in solution (the fluorescence was by a factor of six higher than that of **15** alone). The polymer could be fabricated into nanofibers with an average diameter of 1 μm by electro-spinning; these fibers also exhibited AIE-based fluorescence (Figure 6.10). Moreover, the fluorescence in aqueous solution could be strongly quenched upon the addition of Pd(II) ions due to their binding to the 1,2,3-triazolyl moieties. Thus, the potential application of supramolecular polymers, featuring this AIE behavior as optical sensor for heavy metal ions in water, such as Pd(II) ions, could be envisioned.

Very recently, Yan and coworkers demonstrated how covalent and supramolecular polymerizations can be combined to afford mechanically robust, yet dynamic materials [51]. The AB-type monomer **16** gave the corresponding supramolecular polymer poly-**16** due to crown ether recognition; whereas, the ring-opening metathesis

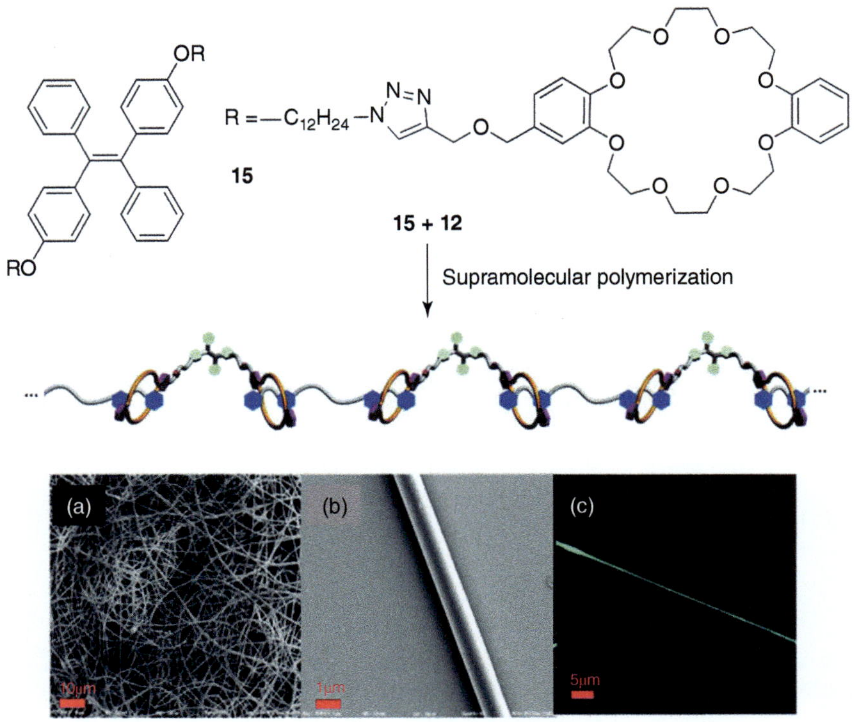

Figure 6.10 Schematic representation of the supramolecular self-assembly of **15** and **12** ($m = 10$). The supramolecular polymer exhibited AIE-based fluorescence in solution as well as in the solid state, i.e. in the electrospun nanofibers. Two representative SEM images (a/b) an a fluorescence microscopy image (c) of the fibers are shown. Source: Chen et al. [50], © 2015 The Royal Chemical Society.

polymerization (ROMP) gave the covalent polymers **17** with pending crown ether and alkylammonium side chains (dibutyl norbornene-2,3-dicarboxylate was used as a comonomer in the ROMPs Figure 6.11). A covalent-supramolecular polymer (CSP) network was then obtained from the self-assembly of these side chains. Due to synergistic effects, this network featured remarkable mechanical properties compared to poly-**16** and **17** (Figure 6.11). Moreover, the mechanical properties of **16** and the CSP could be tuned over a wide range via the fraction of the incorporated comonomer (e.g. the Young's modulus decreased from 285 to 92 MPa with increasing content of the comonomer).

As an extension of the concept to obtain supramolecular polymers from AB-type monomers, Yu and coworkers prepared the tetratopic AA-BB-type monomer **18**, which afforded a supramolecular polymer network due to a self-cross-linking process (Figure 6.12) [52]. This network showed a stimuli-responsive behavior – in solution as well as in the gel state: reversible disassembly could be induced by changing either the temperature or pH. The viscoelasticity of the material was analyzed, and an increased flexibility of the bulk polymer at elevated temperatures was observed, thus poly-**17** behaved as a thermoplastic.

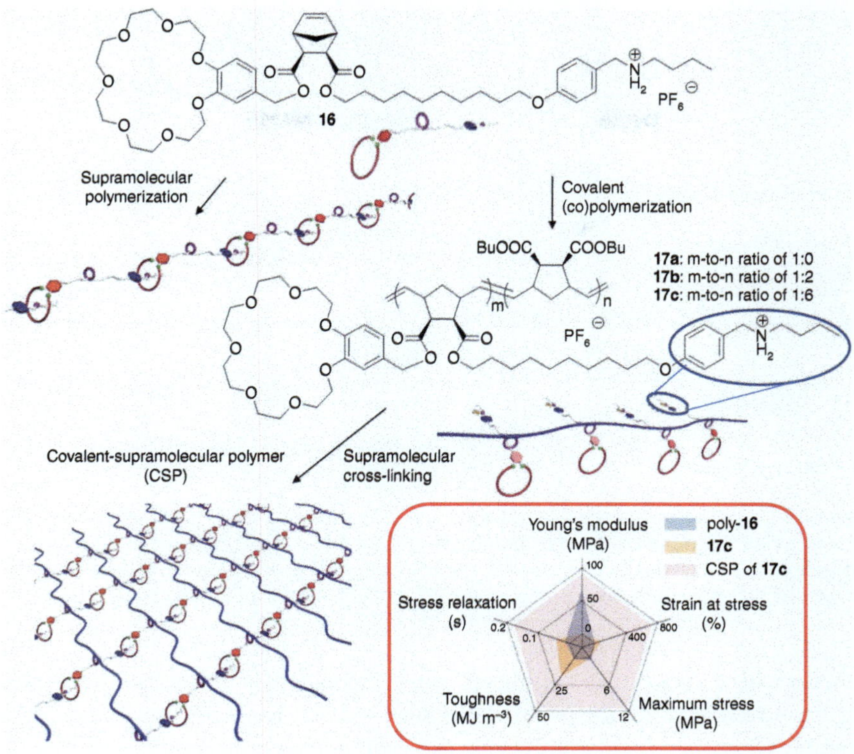

Figure 6.11 Schematic representation of the multifunctional monomer **16** and its transformation into a linear supramolecular polymer (poly-**16**), covalent (co)polymers (**17**) and, eventually, covalent-supramolecular polymers. A summary of the mechanical properties of these materials is also depicted. Source: Zhang et al. [51], © 2020 John Wiley and Sons.

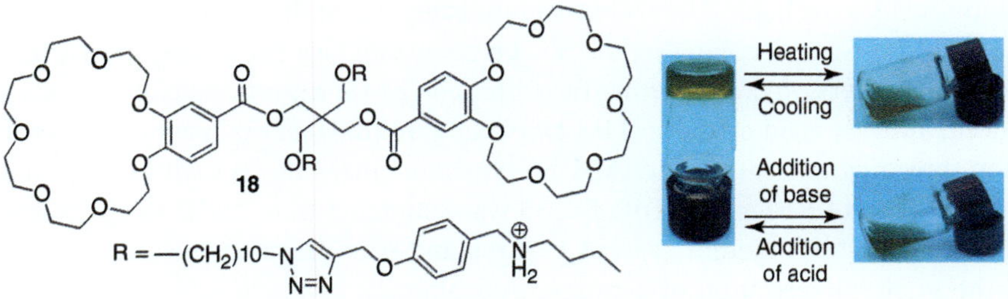

Figure 6.12 Schematic representation of the hetero-tetratopic monomer **18,** which gave a supramolecular polymer network due to a self-cross-linking process. Source: Wang et al. [52]. Figure reproduced with kind permission; © 2020 American Chemical Society.

Cryptands, the 3D relatives of crown ethers, have attracted considerable interest in supramolecular chemistry due to their unique binding properties [53]. For example, in comparison to the simple crown ether BMP32C10 (**2**), the analogous cryptands bind MV^{2+} guests more strongly [54–56]. For instance, for the binding of MV^{2+} to the cryptand **19**, a K_a value of $5.0 \times 10^6\ M^{-1}$ was determined and

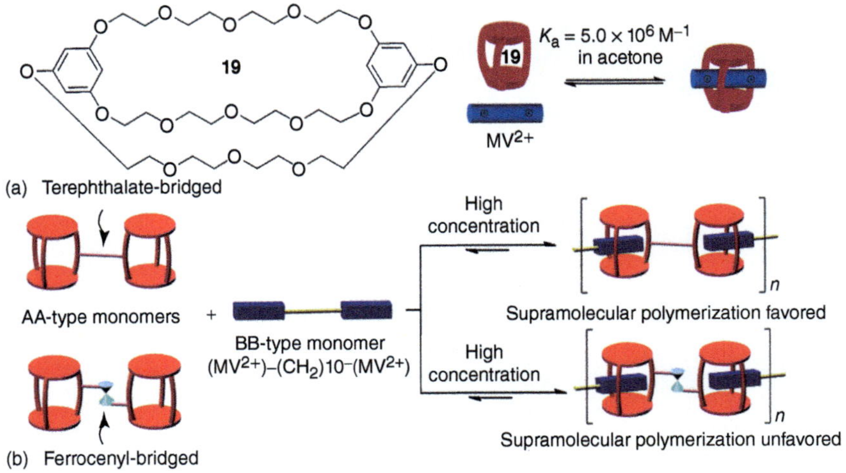

Figure 6.13 (a) Schematic representation of the formation of a host–guest assembly via threading MV^{2+} into the cavity of cryptand **19**. (b) Schematic representation of the self-assembly of AA-type monomers with the BB-type $(MV^{2+})_2$-monomer into linear supramolecular polymers. Source: Niu et al. [57], © 2011 American Chemical Society.

shown to be a factor of 9000 higher when compared to related crown ethers (Figure 6.13a) [56]. Consequently, this host system has been utilized for the assembly of supramolecular polymers, based on cryptand recognition [9]. Due to the high K_a values for the self-assembly of homotelechelic bis-cryptands linked by either terephthalate or ferrocenyl esters, high DPs were expected [57]. However, the more flexible ferrocenyl ester linkage favored the formation of cyclic oligomers; whereas, the rigid terephthalate-based system afforded the linear supramolecular polymers (Figure 6.13b). Accordingly, a larger critical polymerization cpc was determined for the latter. These observations suggest that the loss of entropy in the course of host–guest complexation can be circumvent by increasing the rigidity of the monomers, thus supporting the formation of 1D supramolecular polymers. Indeed, a highly rigid-cylindrical bis-cryptand gave the linear polypseudorotaxane upon threading of a rather stiff $(MV^{2+})_2$-type monomer into its cavities [58]. At a concentration of 6.67 mM, a DP of c. 35 was estimated from NMR spectroscopic measurements. A different type of bis-cryptand was prepared by Cheng et al. by the photo-dimerization of a coumarin-containing cryptand [59]. This ditopic macrocyclic host gave supramolecular polymers when combined with BB-type guests, similar to those depicted in Figure 6.13. Yao et al. showed that bis-cryptands can also be assembled from pyridyl-functionalized cryptands and transition metal complexes [60]. These bis-cryptands and the aforementioned BB-type guests were utilized to assemble linear supramolecular polymers. The two steps of the self-assembly process could also be inverted, i.e. the host–guest complexes were formed and subsequent metallo-supramolecular polymerization afforded the type of supramolecular architecture (the entire self-assembly process also occurred in one-pot in a self-sorting supramolecular polymerization, Figure 6.14).

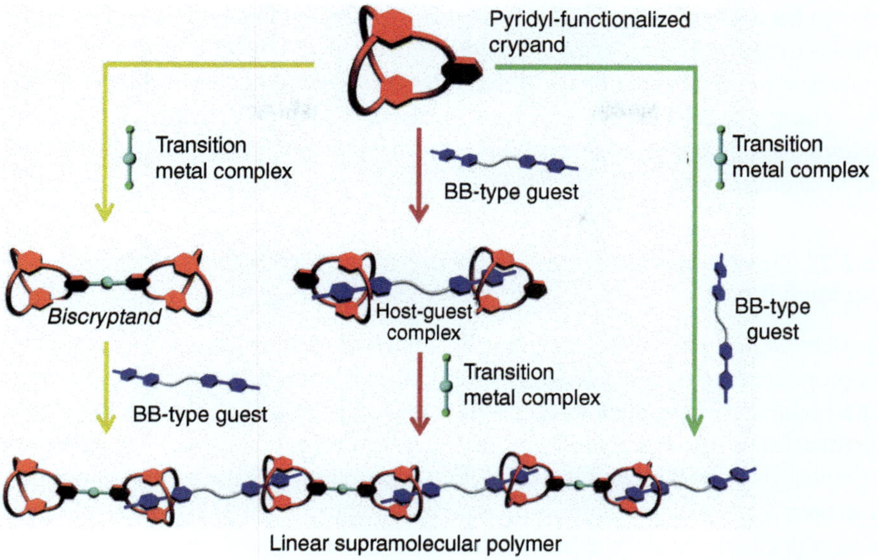

Figure 6.14 Schematic representation of the formation of a cryptand-based linear supramolecular polymer via three different reaction pathways. Source: Wei et al. [60].

The unsymmetrical cryptand **20**, merging the two classical crown ether moieties BPP34C10 and DB24C8, was found to simultaneously accommodate the MV^{2+} and the dibenzylammonium cations in his cavity – despite their electrostatic repulsion (Figure 6.15a/b) [61]. When combined with the BB-type guests **12** and **14**, a supramolecular polymer of high molecular weight was formed accordingly and analyzed by 1D/2D NMR spectroscopy (Figure 6.15c). Uniform rod-like fibers were

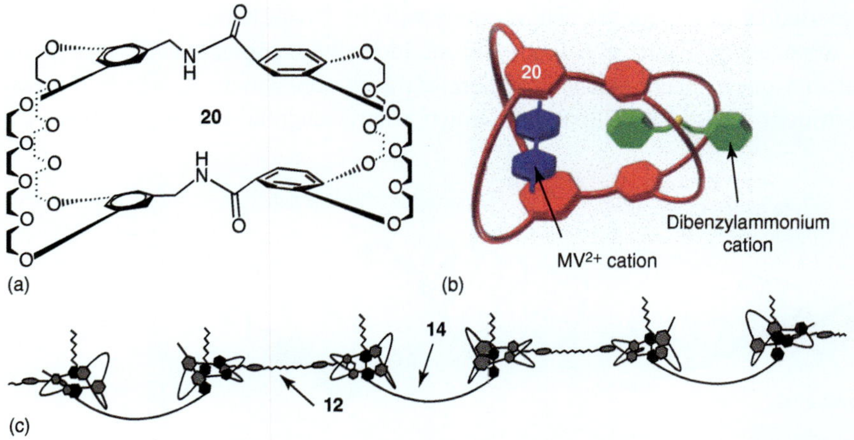

Figure 6.15 (a) Schematic representation of the unsymmetrical cryptand **20**. (b) Schematic representation of the host–guest complex of **20** with the MV^{2+} and dibenzylammonium cation. (c) Schematic representation of the linear supramolecular polymer assembled from **20** and the ditopic guests **12** and **14**. Source: Yao et al. [61], © 2018 John Wiley and Sons.

drawn from a very concentrated monomer mixture, scanning electron microscopy (SEM) imaging revealed a regular diameter of c. 6.6 μm. As for the supramolecular polymers formed by crown ether recognition, the assembly could be depolymerized by adding a base (thus, deprotonating the dibenzylammonium sites). This process featured a high degree-of-reversibility – the supramolecular polymer was reformed upon acidification.

6.2.2 Linear Supramolecular Polymers Derived from Macromolecular Building Blocks

Besides using small-molecule monomers, macromolecular systems, equipped with appropriate binding units, have also been utilized for the formation of supramacromolecular materials, such as (i) chain-extended polymer or (ii) block copolymer architectures. Analogous to macromolecules being able to undergo H-bonding (see Chapter 3) or metal-to-ligand coordination (see Chapter 4), crown ether recognition has been applied for that same purpose. Thus, polymers functionalized at one chain end with either a host, i.e. a crown ether moiety, or a guest system, i.e. a paraquat or dibenzyl ammonium derivative, have been prepared (Figure 6.16).

The nitroxide-mediated polymerization (NMP) was used to synthesize well-defined polystyrene-based, crown ether derivatives. Also, the bulk polymerization of styrene proceeded in a controlled fashion and gave similar host-functionalized polystyrenes (PSs) of low dispersity ($Đ < 1.30$). These macromolecular hosts exhibited the expected preference of recognition: secondary ammonium salts and MV^{2+} derivatives were selectively bound by the PS-DB24C8 and PS-BMP32C10 systems, respectively [62]. It was further shown that linear chain-extended supramolecular polymers can be formed via pseudorotaxane formation – despite the intrinsically moderate association constants for the underlying host–guest interactions. Homoditopic small building blocks, such as **12** or **9**, were used as linkers for the self-assembly with PS-DB24C8 and MV^{2+}-terminated PS, respectively (Figure 6.17a) [63]. In addition, the combination of appropriate host and guest macromolecules generated block copolymers in which the two constituent blocks were linked via crown ether recognition. The assembly of a

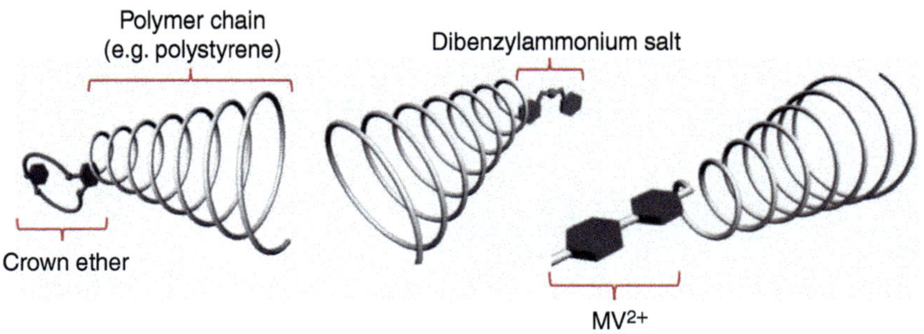

Figure 6.16 Schematic representation of macromolecular building blocks for host–guest complexation, based on crown ether recognition. Source: Zheng et al. [9], © 2012 Royal Society of Chemistry.

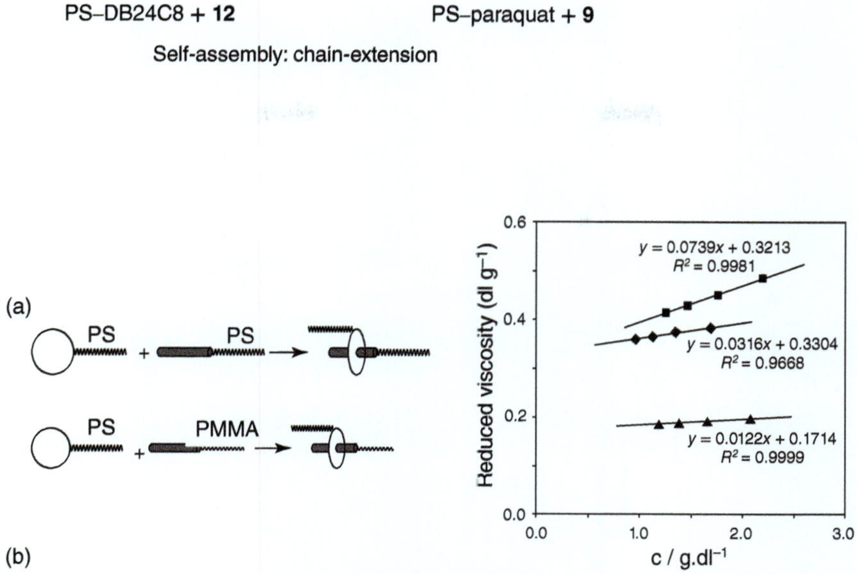

Figure 6.17 (a) Schematic representation of the (a) polymer chain extension via crown ether recognition. Source: Based on Gibson et al. [63], and (b) formation of the supramolecular diblock copolymers utilizing complementary macromolecular building blocks; the viscosimetric analysis (in $CHCl_3$, 25 °C) of the PS (triangles) and PMMA building blocks (squares) as well as of the supramolecular PS-*b*-PMMA (diamonds). Source: Lee et al. [64], © 2009 American Chemical Society.

supramolecular diblock copolymer via the threading of a MV^{2+}-functionalized poly(methyl methacrylate) (PMMA) into the crown ether cavity of a PS derivative was evidenced by viscometry measurements (Figure 6.17b) [64].

The poly(ε-caprolactone)-based heteroditopic macromonomers **21** with a B21C7-type host (B21C7: benzo-21-crown-7) and a secondary ammonium-salt guest tail were assembled into chain-extended supramolecular polymers. In solution, the polymers exhibited a multiple stimuli-responsive behavior (i.e. various cations and anions, as well as changes in pH value or the temperature, affected the supramolecular structure). Moreover, the nanofibers, fabricated by electrospinning from solution, presented an enhanced degradation behavior in the presence of, e.g. K^+, Cl^-, or Et_3N or upon heating (Figure 6.18).

6.2.3 Dendritic, Hyperbranched, Star-shaped, and Grafted Assemblies

Linear supramolecular polymers are obtained in a polycoordination reaction when monomers of the AB-type or, alternatively, stoichiometric combination of AA- and BB-type monomers are employed. Any deviation from this general rule will consequently lead to the formation of supramolecular assemblies of other topologies. For instance, monomers of the AB_2-type, such as **22**, will self-assemble into 3D hyperbranched polymers, with rather irregular properties (Figure 6.19) [65]. Upon increasing the concentration, the size of the aggregates will also increase,

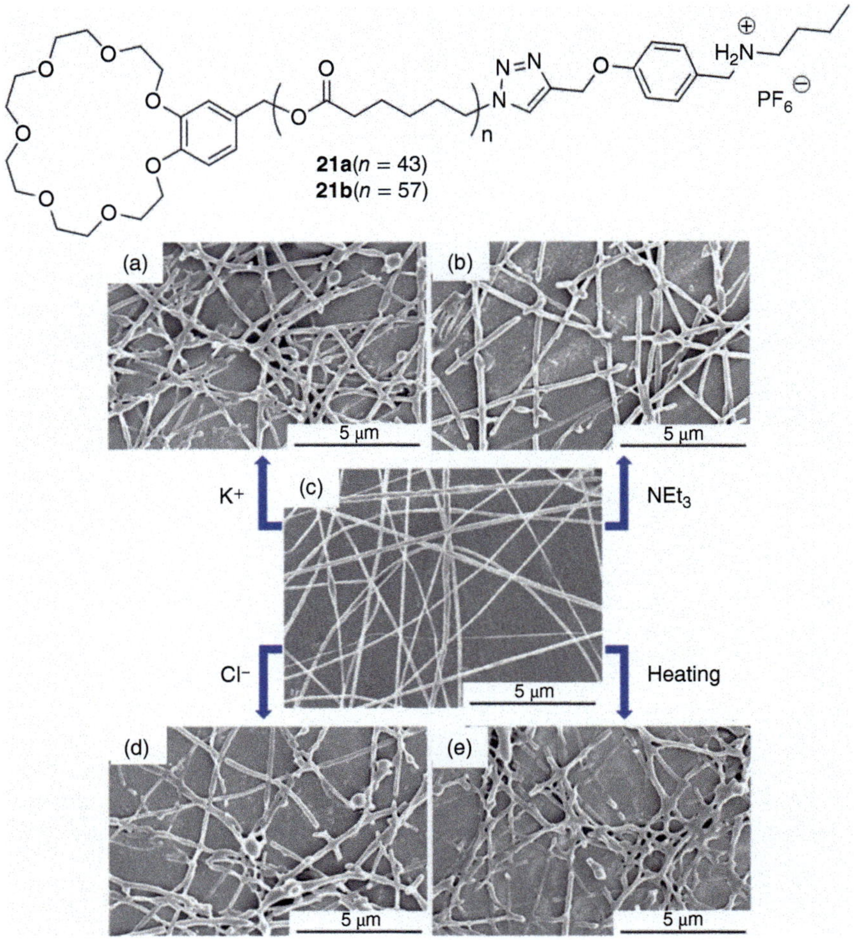

Figure 6.18 Schematic representation of the heteroditopic macromonomers **21**. The nanofibers, fabricated from the chain-extended polymer $(\mathbf{21})_n$ by electrospinning, showed a multiple degradation behavior. The representative SEM images show the fibers before (c) and after the respective treatment with K^+ (a), NEt_3 (b), Cl^- (d), and heat (e). Source: Chen et al. [123]. Figure reproduced with kind permission; © 2016 The Royal Chemical Society.

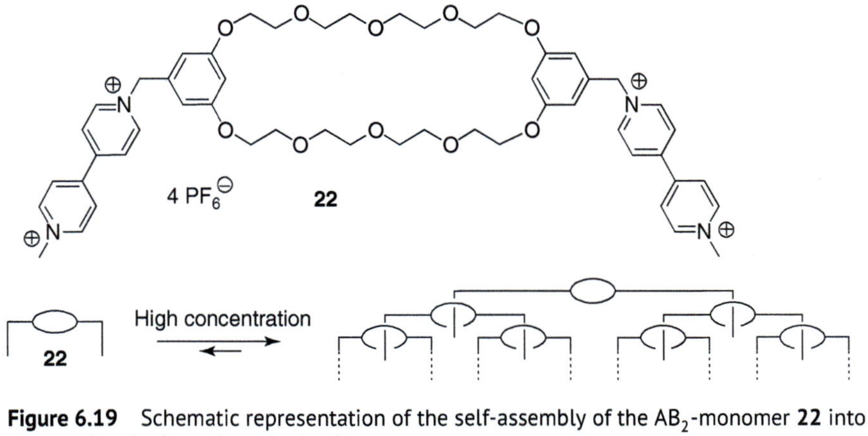

Figure 6.19 Schematic representation of the self-assembly of the AB_2-monomer **22** into a supramolecular hyperbranched polymer.

amounting to a DP of c. 40 at a concentration of 100 mM – this corresponds to a molar mass of 59 400 g mol^{-1}. ^{1}H NMR titration experiments, MALDI-TOF (time-of-flight) mass spectrometry, and viscometric analyses were conducted to investigate the hyperbranched polypseudorotaxane.

In a different approach, a pseudorotaxane, formed by threading a MV^{2+}-derivative into the cavity of a DB24C8 system, was polymerized via esterification of the lateral functional groups (e.g. the paraquat derivative was bearing two OH groups; whereas, the DB24C8 unit was bearing one acid chloride residue) [66]. This particular method refers to the principle of using supramonomers for the preparation of supramolecular polymers (see Section 1.4). Furthermore, the hyperbranched polymer was stabilized by reacting the free OH groups with a "stopper" molecule. Due to high kinetic stability of the material, a characterization by size-exclusion chromatography (SEC) became possible – a M_n value of 19 000 g mol^{-1} was estimated.

When considering their chemical structures, dendrimers and dendrons possess more defined architectures than hyperbranched polymers due to their tree-like monodisperse structures in solution [67]. Such molecules are composed of individual wedges (the so-called dendrons) and the synthetic strategy toward such macromolecules may be divergent, convergent, or a combination thereof. The typically demanding synthetic effort can be significantly bypassed by applying supramolecular chemistry. As an additional benefit, supramolecular dendrimers, e.g. based on crown ether recognition and related non-covalent interactions, may offer properties, which are not accessible from the non-dendritic examples [68, 69].

Poly(propylene imine) (PPI) dendrimers of first and third generation (G1 and G3, respectively) were modified at their peripheries with BMP32C10 moieties [70]. Binding of MV^{2+} was reported, whereby the G1-dendrimer showed a stronger binding. In contrast, the larger G3-dendrimer showed a pH-dependence of the non-covalent interaction; in which under acidic conditions, the structure became more rigid, thereby facilitating the MV^{2+} binding (in an anti-cooperative manner). In contrast, repulsion between the charged guest molecules was dominate for the nonprotonated dendrimer.

The Gibson group also followed a convergent strategy by utilizing a core molecule comprising three dibenzylammonium arms (**23**) [71, 72]. This particular core-bound dendronized crown ether hosts of the DB24C8-type in an 1 : 3 manner, different generations of Fréchet-type dendrons were utilized (e.g. crown ether **24**, Figure 6.20). In general, pseudorotaxane formation was found to be rather slow; this observation was attributed to the low solubility of the charged guest core in concert with steric hindrance of the bulky dendritic hosts; for the less bulky dendrons, host–guest complexation was accelerated. In polar-aprotic solvents, such as $CHCl_3$ or acetone, the interaction of the host with guest had a positive cooperative effect [72]. In less polar solvents, however, almost 90% of the dissolved core molecules were incorporated into the three-armed pseudorotaxane; thereby, encapsulation of the insoluble cationic core was identified as the main driving force for this high degree-of-conversion. With increasing generations, larger K_a values were obtained.

With respect to the formation of star-shaped molecules, in particular the supramolecular ones, the host–guest interaction between small-molecule cores

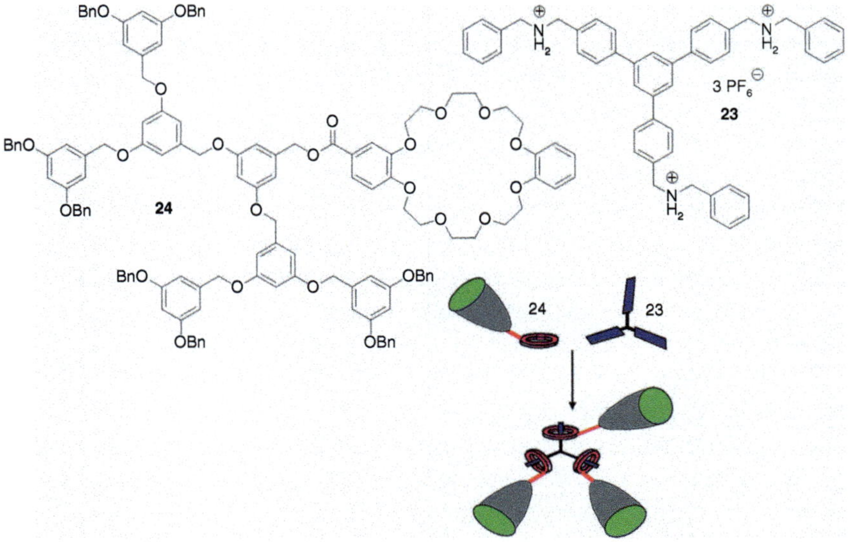

Figure 6.20 Schematic representation of the formation of dendritic pseudorotaxanes via a convergent synthesis. Source: Gibson et al. [72], © 2002 American Chemical Society.

and complementary macroligands, as the designated arms, has been employed (Figure 6.21a). This approach represents an extension of the aforementioned chain-extension process (Figure 6.17a). For example, three-armed star-shaped polymers were obtained by threading DB24C8-PS onto a tricationic dibenzylammonium salt or a tetracationic paraquat derivative [63]. In the former case, the

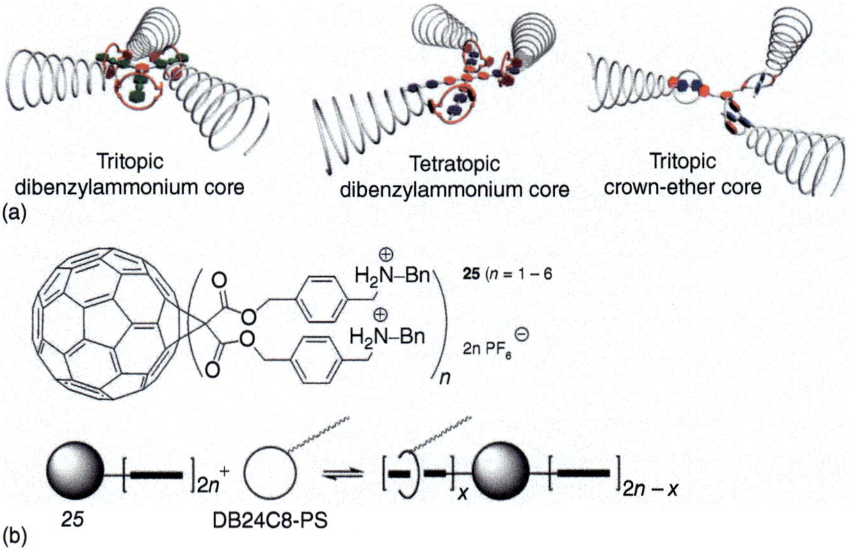

Figure 6.21 Schematic representation of the formation of star-shaped and graft polymers (a and b, respectively), based on crown ether recognition. Source: (a) Zheng et al. [9], © 2012 Royal Society of Chemistry; (b) Gibson et al. [75], © 2009 John Wiley and Sons.

expected three-armed assembly was obtained by threefold crown ether recognition. In contrast, the tetracationic core could not be fully loaded – due to steric hindrance only three of the MV^{2+} units could enter the cavities of the PS-macroligands. In the complementary case, a tris(crown ether) host was used for accommodating three MV^{2+}-terminated polystyrene molecules [73]. A CT interaction between the donor-type crown ethers and the acceptor-type MV^{2+} moieties gave rise to a product, which showed a yellow coloration. Related to this, a BMP32C10 bearing two polystyrene chains was utilized as a host for the inclusion of a MV^{2+}-terminated PS or PMMA guest [64]. In both cases, a three-armed, star-shaped polymer was obtained; however, the latter gave a turbid yellow film upon processing into the solid state and a macrophase separation due to immiscibility of the PS and PMMA blocks in the solid state was observed by optical microscopy. The formation of other supramolecular topologies was also investigated [33]. For instance, a polymer composed of crown ether entities within the polymer chain was used, as a polytopic host, for the formation of comb-like graft copolymers via threading of MV^{2+}-terminated PS units into the macrocycles [74]. In a more sophisticated approach, functionalized C_{60} (**25**) was employed as multitopic guest core for the decoration with up to 12 PS chains (Figure 6.21b) [75].

Diverse strategies for the synthesis of stable supramolecular dendrimers have also followed: the so-called "threading-followed-by-stoppering" [17] and (subsequent) Wittig-exchange approaches [76, 77] as well as the sliding methodology [78] have to be mentioned in this context (see also Section 6.3). Elizarov et al. synthesized a dendritic architecture containing a combination of mechanical (1st generation) and covalent bonds (second and subsequent generations) at the branching points (Figure 6.22a) [79]. Initially, a bis-[2]rotaxane was formed by threading a dibenzylammonium salt, functionalized with two benzylic triphenylphosphonium stoppers, into the cavities of a bis-DB24C8 crown ether derivative; subsequently, a high-yielding Wittig-type reaction with aldehyde-functionalized

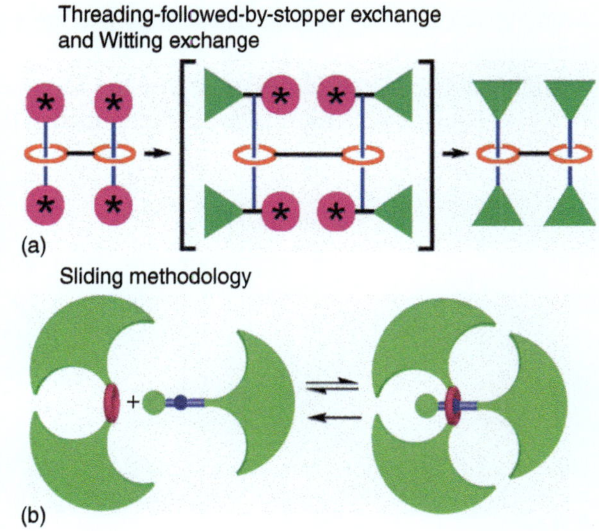

Figure 6.22 Schematic representation of the formation of dendrimers, based on crown ether recognition: (a) threading-followed-by-stoppering and subsequent Wittig exchange. and (b) sliding methodology. Source: Elizarov et al. [80], © 2002 American Chemical Society.

Fréchet-type wedges was conducted to give an interlocked dendrimer. Remarkably, no disassembly of the labile bis-[2]rotaxane was observed during the second synthesis step. The sliding methodology was utilized to construct kinetically labile assemblies (Figure 6.22b) [80]. Due to a high degree-of-steric-loading, the [2]pseudorotaxane exhibited properties similar to those typical for mechanically interlocked species. Disadvantageously, the threading reaction was remarkably slow, taking about 90 days. Due to the enormous time required, the assembly of supramolecular dendrimers via this approach is hardly applicable.

Moreover, a defined cross-linked supramolecular network was formed when appropriate poly(ε-caprolactone) (PCl) monomers were employed. The Liu group self-assembled a DB24C8-terminated four-armed star-shaped PCl and a dibenzylammonium-terminated two-armed PCl into a stimuli-responsive supramolecular gel [81]. The high degree-of-responsiveness of the material was shown by the reversible sol–gel transitions when external stimuli, e.g. changes in temperature or pH, were applied. Further examples for cross-linked or graft-type supramolecular polymers featuring crown ether recognition can be found in a recent review by Arunachalam and Gibson [33].

6.3 Mechanical Interlocking: From Pseudorotaxanes to Rotaxanes

In general, a rotaxane is formed from the corresponding pseudorotaxane by introducing bulky stoppers, thus preventing dethreading of the once complexed guest (Figure 6.2). The Stoddard group reported various approaches for the preparation of such mechanically interlocked supramolecular assemblies. For instance, the so-called click reaction (i.e. formation of a triazole unit via an azide-alkyne cycloaddition, Figure 6.23) [82] and *N*-alkylation (i.e. formation of pyridinium salts) [83] as well as urea, carbamate or imine formation [84] have successfully been applied for this purpose. These strategies are commonly referred to as the "threading-followed-by-stoppering" methods.

Additionally, the transformation of the slim benzyl bromide functionalities into bulky triphenylphosphonium salts has been reported [85] and proved its versatility

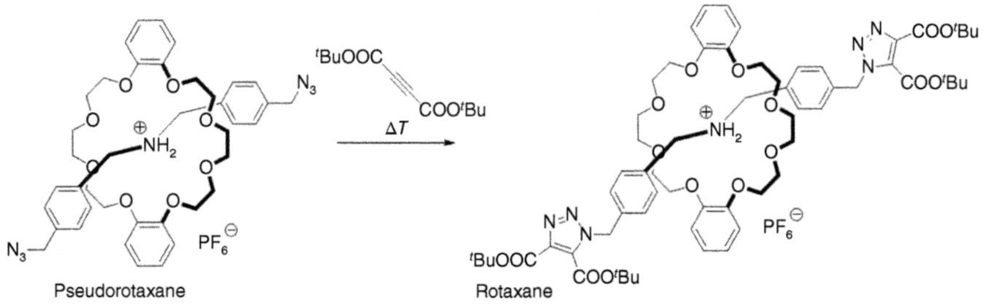

Figure 6.23 Schematic representation of the formation of a [2]rotaxane via a thermal azide to alkyne cycloaddition "click" reaction. Source: Ashton et al. [82], © 1996 John Wiley and Sons.

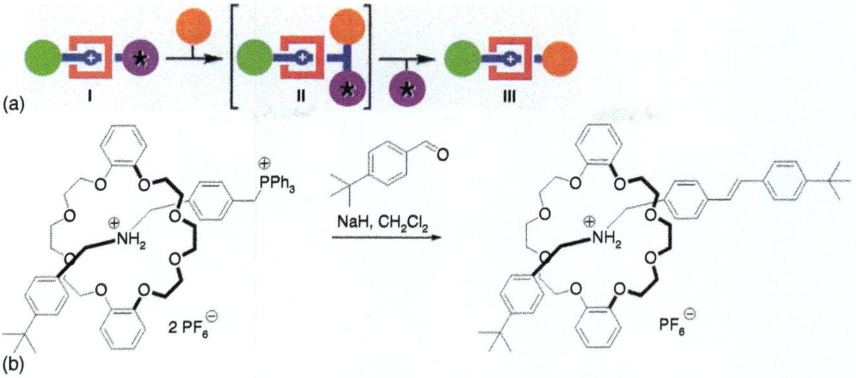

Figure 6.24 (a) Schematic representation of the so-called Wittig exchange – transformation of the "reactive" rotaxane **I** into the "inert" rotaxane **III** via intermediate **II**. Source: Rowan et al. [76], © 2000 American Chemical Society. (b) Schematic representation of the synthesis of a [2]rotaxane via Wittig exchange. Source: Rowan and Stoddard [86].

for the so-called Wittig exchange – using the pending benzylic triphenylphosphonium moieties, as substrate, for Wittig reactions with bulky aldehydes or ketones (Figure 6.24) [86]. It has to be pointed out again, that dethreading in the course of the Wittig condensation is typically not observed. This also holds true for the other aforementioned examples prepared by the "threading-followed-by-stoppering" approach.

Rowan et al. synthesized daisy-chain-type polymers starting from the hetero-bis-functionalized rotaxane **26** (Figure 6.25a); for this particular system, a [*c*2]daisy chain will be formed when the aldehyde functionalities of two rotaxanes react with each other's phosphonium sites [76]. Consequently, larger macrocycles will be generated when additional monomers are incorporated. However, if the termini of such an oligomeric assembly does not "bite back," a supramolecular acyclic daisy chain will be formed [77]. Whereas the [*c*2]daisy chain $(\mathbf{20})_2$ was formed preferentially as confirmed by fast-atom bombardment (FAB) mass spectrometry. More recently, a novel strategy, referred to the "threading-followed-by-swelling" approach, was introduced [87]: a "swellable" dialkenylcyclopropane group was attached to the tail of the heteroditopic monomer **27**, which, in solution, self-assembled into a [*c*2]daisy chain that could be trapped by heating the dialkenylcyclopropane unit, which isomerized quantitatively into a sterically more demanding cycloheptadiene system (Cope rearrangement, Figure 6.25b).

It was shown by Coutrot et al. that a properly designed [*c*2]daisy chain, such as $(\mathbf{20})_2$, can function as a supramolecular muscle or shuttle (Figure 6.26) [88]. The "guest tail" of the AB-type monomer composed of two different recognition sites (i.e. a secondary ammonium salt and a 1,2,3-triazolium unit), which have different affinities to the crown ether host: high for the ammonium salt, moderate for the triazolium cation, and negligible for the secondary amine (i.e. the deprotonated ammonium salt). The interwoven dimer could be contracted by increasing the pH value (thus, deprotonating the ammonium salt). Acidification led to expansion

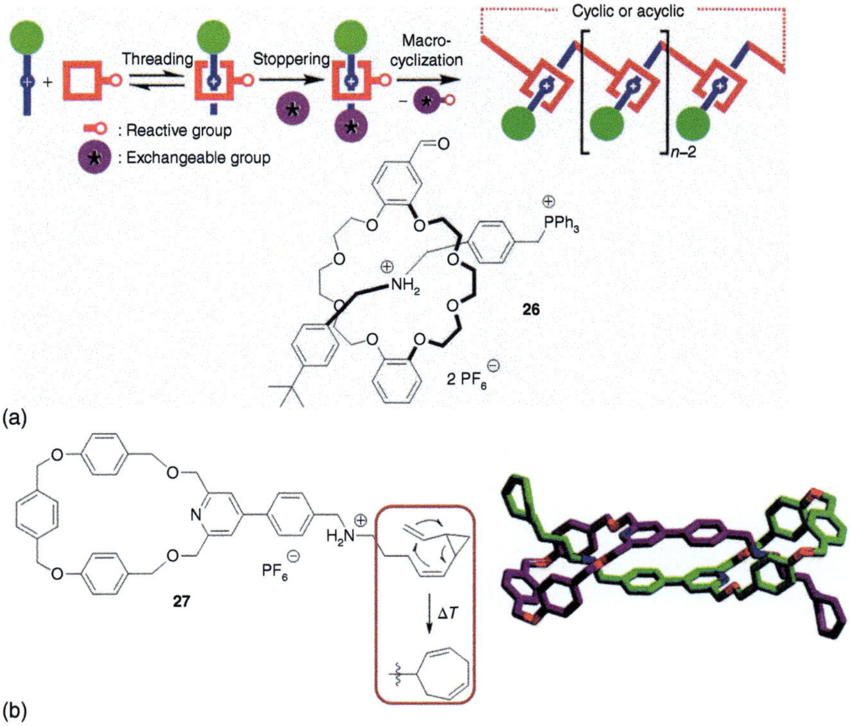

Figure 6.25 Schematic representation of the formation of a mechanically interlocked daisy chain using the Wittig exchange (a) and the "threading-followed-by-swelling" methodology. Source: Rowan et al. [76], © 2000 American Chemical Society. (b); the solid-state structure of the [c2]daisy chain $(\mathbf{27})_2$ is also shown (counterions were omitted for clarity). Source: Ueng et al. [87]. © 2008 Royal Society of Chemistry.

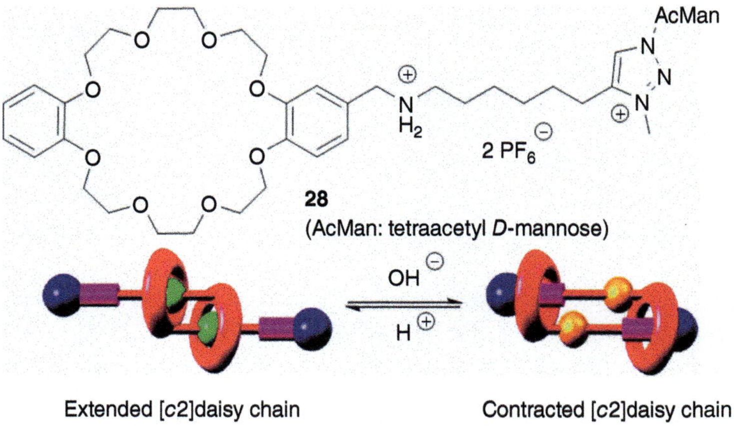

Figure 6.26 Schematic representation of a supramolecular muscle/shuttle, based on the pH-switchable [c2]daisy chain $(\mathbf{28})_2$. Source: Coutrot et al. [88], © 2008 American Chemical Society.

of the molecule, as the regenerated secondary ammonium salt was complexed by the DB24C8 rather than the triazolium cation. The concept of reversible contraction/expansion of mechanically interlocked [*c*2]daisy chains was developed further by the Stoddard group toward polymeric materials, thus aiming for the transfer of chemical stimuli into mechanical motion [89, 90]. For this purpose, the stimuli-active subunits, based on a similar motive as in **28** (a MV^{2+} unit and a dibenzylammonium salt were used as recognition sites within the tail [91]), were polymerized by a Cu(I)-catalyzed azide-alkyne cycloaddition (CuCAAC) reaction. This approach, once again, represents the polymerization of supramonomers, as outlined in Section 1.4. The resulting polymer was investigated by SEC revealing a M_n value of c. 33 000 g mol^{-1}, corresponding to a DP of about 11. The pH-stimulated extraction and contraction movements of the material were studied by ^1H NMR and UV/vis absorption spectroscopy; all repeating units within the supramolecular polymer contracted simultaneously upon addition of base, thereby reducing the length of the macromolecule by 9.9 nm (however, other effects possibly leading to changes in the overall molecular structure were not considered).

The Grubbs group prepared similar mechanically interlocked supramolecular polymers by polymerizing [*c*2]daisy-chain monomers [92, 93]. These monomers, in turn, were formed by a ring-closing metathesis of the precursors **29**; subsequently, the terminal OH-groups of the resulting supramonomer were converted into azide moieties that were subsequently used for polymerization by CuCAAC reaction (Figure 6.27) [93]. According to SEC measurements, the M_n value of the resultant polymer was c. 48 000 g mol^{-1} (DP ≈ 22). However, efficient π–π stacking interactions of the biaryl entities (**29b**) hindered expansion of the molecules upon deprotonation. If the size of the amine units was increased by *N*-acylation, the crown ether moieties were pushed along the rigid arm generating the extended analogue; thus, the length of the polymer chain was increased by 70% to 21.4 nm.

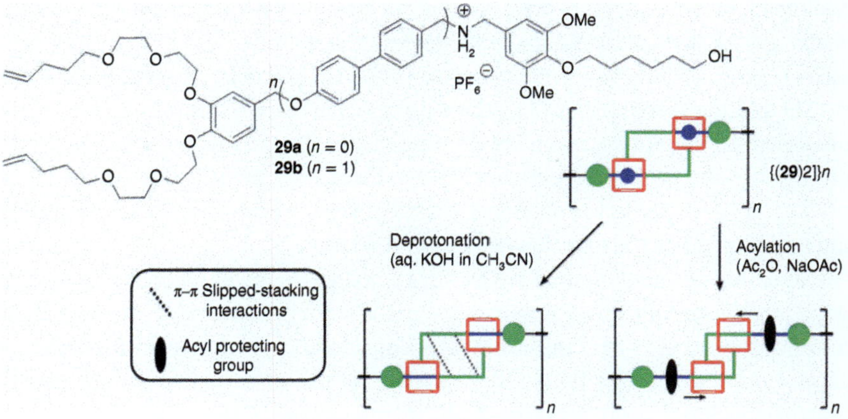

Figure 6.27 Schematic representation of building block **23** and the derived poly([*c*2]daisy chain) that was utilized for expansion/contraction studies. Source: Clark et al. [93], © 2009 American Chemical Society.

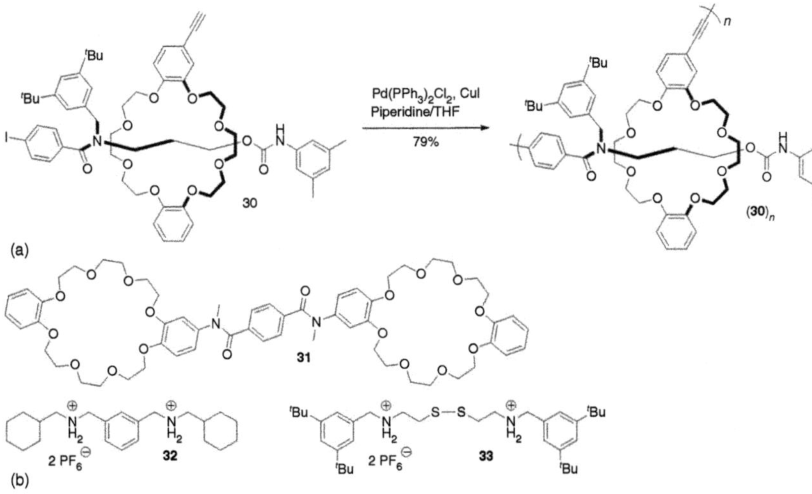

Figure 6.28 (a) Schematic representation of the synthesis of a poly[2]rotaxane via Sonogashira cross-coupling polymerization. Source: Sasabe et al. [94]; (b) schematic representation of the homoditopic building blocks used for the formation of poly[3] rotaxanes via a "polyslipping" methodology. Source: Sohgawa et al. [96] and Oku et al. [96].

A Sonogashira-type cross-coupling polymerization to prepare a poly[2]rotaxane was reported by Sasabe et al. [94]. The [2]rotaxane monomer **30**, composed of an ethinyl moiety on the crown ether macrocycle as well as a pending iodobenzene substituent on the guest molecule, was coupled in good yield (79%) under Pd(0)-catalysis (M_n = 6800 g mol^{-1}, Figure 6.28a). In contrast to the poly(phenylene-ethynylene) polymers (PPEs) with crown ether residues in the side chains, in which (pseudo)rotaxane formation gave supramolecular comb- or cross-linked polymers [95], (**30**)$_n$ features the host–guest complexes, as an integral part of the polymer backbone. Due the uncharged character of the material, good solubility in organic solvents with various polarities was guaranteed. The same research group also prepared polyrotaxane-like assemblies via a "polyslipping" strategy (Figure 6.28b) [96]; thus, upon heating a 1 : 1 mixture of the homoditopic host (**31**) and guest molecules (**32**), the supramolecular polymer was formed. The entire assembly could be "frozen" by cooling to afford a kinetically stable polymer. Similarly, a poly[3]rotaxane was formed by threading **33** into the crown ether cavities of **31** [97]. Due to the chemically labile disulfide bridge, the assembly could easily be degraded by the addition of PBu_3 and water.

The dynamic covalent chemistry (DCC), i.e. using chemical reactions reversibly under equilibrium control, represents one further strategy concerning the formation of advanced molecular structures and is of considerable relevance with respect to supramolecular chemistry [98]. The reversible nature of certain reactions allows for "error-checking" and "proof-reading" in the course of the syntheses that are performed under DCC. Many examples for reactions under kinetic (e.g. formation of acetal, disulfide, ester, or imine bonds) as well as thermodynamic control (e.g. formation of C—C bonds or metal-to-ligand coordination) have been demonstrated

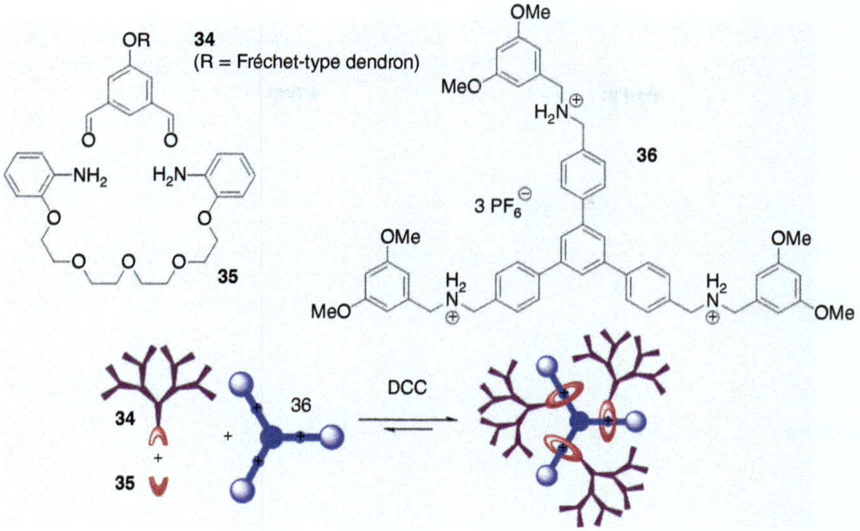

Figure 6.29 Schematic representation of the synthesis of dendritic [4]rotaxanes by "dynamic covalent chemistry" (DCC). Source: Leung et al. [99], © 2005 American Chemical Society.

in DCC. It has been shown that DCC – assisted by supramolecular chemistry – can also be a versatile tool for the formation of interlocked molecules. For instance, the isophthaldehyde derivatives **34**, which were equipped with Fréchet-type dendritic residues, and diamine **35** self-assembled around the arms of **36**, via a template effect, by forming diimine-containing macrocycles (Figure 6.29) [99]. Due to the presence of bulky stoppers on the termini of the cores, dethreading was prevented; the mechanically interlocked dendrimers of the [4]rotaxane-type were obtained in remarkably high yields (>90%). Further stabilization of the product could be achieved by post-assembly reduction of the imine bonds. The versatility of DCC with respect to the formation of rotaxane-based dendrimers was highlighted by its application in combinatorial chemistry [100]: using building blocks of various dendritic generations allowed for the preparation of unsymmetrically functionalized, interlocked dendrimers – in essence, materials that are hardly accessible via traditional dendrimer chemistry.

6.4 Poly(pseudo)rotaxanes, Derived from Preformed Polymers

The formation of poly(pseudo)rotaxanes in which macrocycles are threaded onto the main chain or side chain of polymers represents an important application in polymer science that has, in particular, been applied to CDs [12, 33, 101]. Besides these, crown ethers have also been employed in the same context [102]. Some general aspects regarding poly(pseudo)rotaxane formation, including thermodynamic

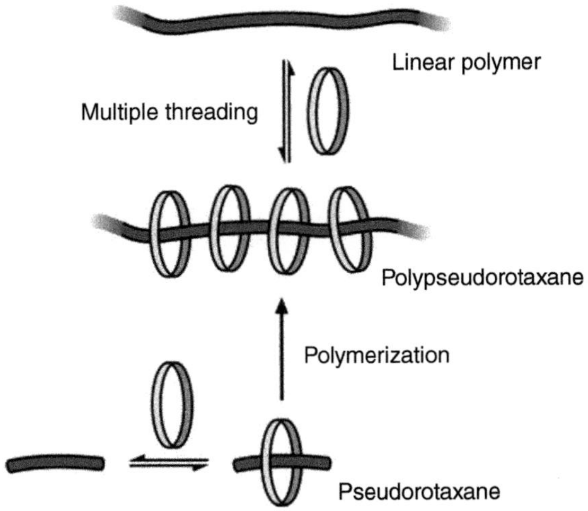

Figure 6.30 Schematic representation of main-chain pseudorotaxane formation via two different synthetic strategies: Multiple threading of macrocycles onto a linear polymer chain (top) and polymerization of preassembled pseudorotaxanes, as supramonomers (bottom). Source: Wenz et al. [101], © 2006 American Chemical Society.

and kinetic considerations, are summarized in Chapter 9. Briefly, two basic synthetic strategies can be employed to prepare linear poly(pseudo)rotaxanes: multiple threading of appropriate macrocycles onto the linear polymer chain will afford these supramolecular polymers; alternatively, the polymerization (either a covalent or a supramolecular one) of preassembled pseudorotaxanes ("supramonomers" according to Section 1.4) can be utilized for the same purpose (Figure 6.30).

One of the earliest examples in this respect was reported by Gibson et al., who synthesized a small library of polyester-type poly(pseudo)rotaxanes by transesterification polymerization or step-growth polycondensation in the presence of various crown ether macrocycles (i.e. 30-crown-10, 42-crown-14, 60-crown-20, or **1**). [103, 104]. After purification by multiple precipitation steps, the topology of the resultant poly(pseudo)rotaxanes was studied by an extensive hydrolysis experiment, supported by NMR spectroscopy, SEC, and vapor pressure osmometry (VPO) measurements. Due to the threading of the crown ether moieties onto the polymer chains, an increased solubility in polar organic solvents and higher intrinsic viscosities, compared with the analogous pristine polyesters, was observed. Remarkably, the poly(pseudo)rotaxanes with a sebacate-type backbone were able to extract metal ions from water (thus showing that the key characteristic of crown ethers was retained despite the threading onto a polymer chain) [105]. Differential scanning calorimetry (DSC) measurements evidenced the presence of two distinct crystalline phases, though only for those poly(pseudo)rotaxanes with high loading densities: the first was attributed to the covalent polyester-type backbone itself and the second one to the threaded crown ether rings. Moreover, a glass-transition temperatures (T_g) could be determined for the crown ether units of the materials. In a similar fashion, polyurethane-based pseudorotaxanes were prepared from tetra(ethylene glycol) and various crown ethers, affording an intermediate pseudorotaxane and subsequent polymerization with a bis-isocyanate [106, 107]. The remarkable stability of these supramolecular assemblies was a result from multiple

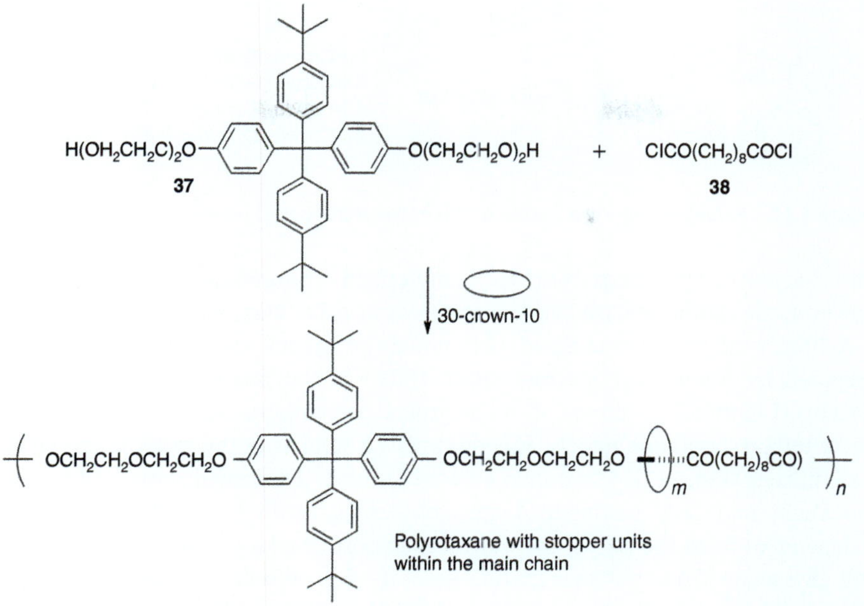

Figure 6.31 Schematic representation of the synthesis of polyrotaxanes with stopper moieties within the polymers main chain.

H-bonding interactions involving the polyurethane backbone and the threaded crown ether moieties [108].

To prevent the crown ether rings from leaving the polymer chains, bulky stoppers were introduced at both chain ends, thus transforming the poly(pseudo)rotaxane into a polyrotaxane [109–113]. Main-chain polyrotaxanes were also obtained when blocking groups were introduced within the main chain [109, 110]. Using the polycondensation of monomers **37** and **38** in the presence of 30-crown-10 afforded main-chain polyrotaxanes with threading densities that were c. five times higher (i.e. m/n ratios of up to 0.061) as for the analogous polymers missing the stopper moieties within the chain [114]. The m/n value could be increased further to 0.172 when both monomers were equipped with central tetraarylmethane units [111]. It was shown further that the m/n value increased with decreasing temperature, since dethreading was prevented for thermodynamic reasons (Figure 6.31).

More recently, Gibson and coworkers synthesized main-chain polyaramides from oxydianiline and isophthalic acid [115]. When the polycondensation was performed in the case of crown ether derivatives, an *in situ* threading of the rings onto the polymer chain was observed. In the case of 30-crown-10, a very low threading density was reached (as concluded from the only marginal increase in mass); whereas, the larger 60-crown-20 entity was threaded more efficiently (c. 17% mass increase). DCS measurements showed an irreversible high-temperature transition of exothermic character, which was assigned to a solid-state reorganization of the intermolecularly H-bonded amide pairs. Moreover, these authors also prepared related polymers with

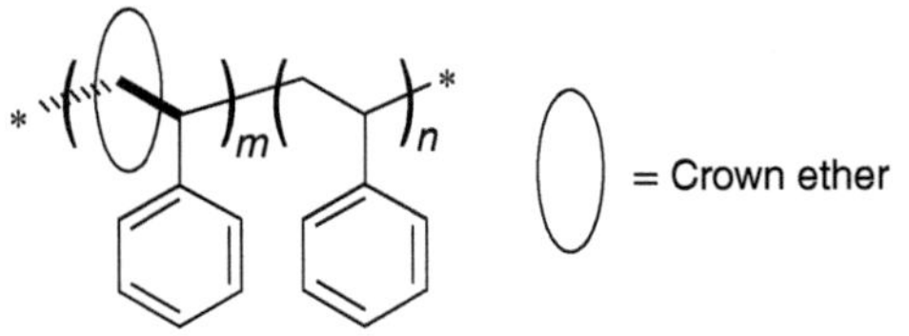

Figure 6.32 Schematic representation of the polystyrene-based poly(pseudo)rotaxanes.

threaded crown ether rings from bis-(*p*-aminophenyl)fluorene and isophthalic acid; however, the exothermic transition was absent in either case.

A library of polystyrene-based poly(pseudo)rotaxanes and polyrotaxanes was prepared by free-radical polymerization (FRP) in the presence of crown ether systems (Figure 6.32). In case of the polyrotaxane synthesis, an initiator bearing a bulky tetraarylmethane moiety, functioning as a stopper in the resultant polymer, was utilized [116, 117]. The crown ether rings were distributed statistically along the chain and only moderate degrees-of-loading could be achieved – almost independent from the size of the macrocyclic host (the larger crown ethers could still dissociate from polymer chain, even in the presence of bulky tetraaryl-methane stoppers). The thermal analysis of the poly(pseudo)rotaxanes revealed two glass-transition temperatures (T_g) that could be assigned to the incorporated polystyrene and crown ethers, which are intrinsically incompatible. Gibson et al. extended the scope of this method to other types of polymers: poly(acrylonitrile) [118], poly(methyl acrylate) [119], and poly(methyl methacrylate) [119]. In the case of the latter two, a much higher loading efficiency as for polystyrene was observed, presumably due to a better compatibility of the more polar backbone with the hydrophilic host.

These fundamental studies on the formation of crown ether-based poly(pseudo) rotaxanes have shown that the threading efficiency depends on a variety of parameters [102]: (1) the presence of appropriate stopper moieties, (2) the interaction strength between the crown ether moieties and the polymer chain that they are threaded onto, (3) the diameter of the macrocycle, (4) the macrocycle-to-monomer feed ratio (only applicable when polymerizing *in situ* formed pseudorotaxanes) and, finally, (5) the threading temperature. When compared to the related poly(pseudo)rotaxanes, based on the threading of CD rings, a lower rigidity and, as a consequence, a lower crystallinity will generally be observed. Though distinct differences between poly(pseudo)rotaxanes and the related “ring free” polymers regarding solubility, viscosity, and thermal properties (i.e. T_g and melting temperature [T_m]) have been reported; applications in, e.g. polymer processing, have not been realized for these particular supramacromolecular materials.

6.5 Supramolecular Amphiphiles

The combination of supramolecular chemistry with that of conventional amphiphiles opened a new field of research – the so-called “supramolecular amphiphiles." Utilizing non-covalent interactions to assemble amphiphiles allows one to generate nanostructures with remarkable degrees-of-structural-complexity and, of utmost relevance for potential applications, an excellent stimuli responsiveness [120]. A broad range of such assemblies that involve host–guest interactions has already been realized and investigated regarding their usage in materials and biomedical science. In particular, the biological and pharmaceutical relevance has to be mentioned in this context (i.e. drug delivery, controlled release, and bioimaging). For all important types of macrocyclic hosts, many examples for supramolecular amphiphiles prepared thereof can be found – with one exception: supramolecular amphiphiles, based on crown ether recognition. This restriction can mainly be attributed to the intrinsically labile nature of crown ether-based supramolecular assemblies in water, the environment that is commonly used to prepare such amphiphiles.

In one example, Huang and coworkers utilized the macromolecular host **39** in which a poly(ethylene glycol) (PEG) chain was conjugated to a BMP32C10-type recognition site, possessing two carboxylate groups [121]. When combined with the paraquat derivative **40**, a supramolecular amphiphile was formed that showed a reversible pH-responsive behavior (Figure 6.33). The critical micellization concentration (cmc) of the host–guest complex between **39** and **40** was estimated to be c. 7.5×10^{-5} M; the high K_a value of this complex (K_a of 1.5×10^3 M^{-1}) was rationalized by additional electrostatic interactions involving the guest's cationic MV^{2+} unit and the host's anionic $-CO_2^-$ moieties. The latter ones became protonated under acidic conditions, thereby weakening the host–guest interaction between **39** and **40** – eventually resulting in the disassembly of the supramolecular micelles (instead, surfactant **40** assembled into rod-like aggregates). It was shown further that hydrophobic molecules, e.g. Nile red, could be encapsulated within the core of the assemblies and released in a controlled fashion by changing the pH

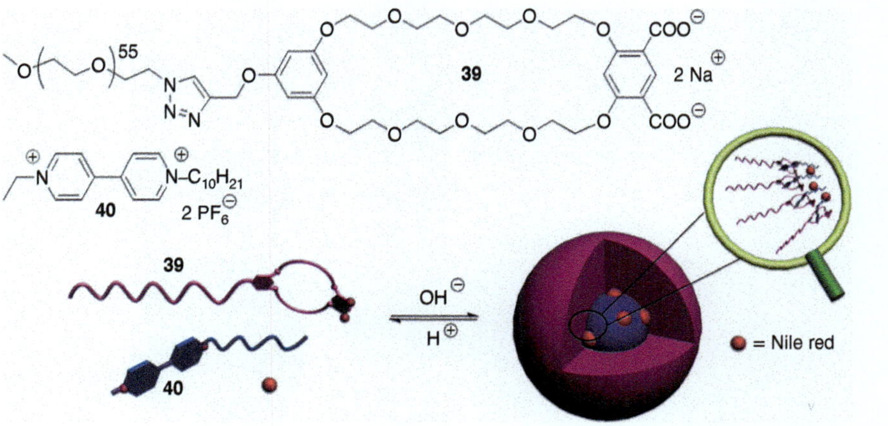

Figure 6.33 Schematic representation of the hydrophilic macromolecular host **39** and the hydrophilic guest **40** that self-assembled into a pH-responsive supramolecular amphiphile; the amphiphile could be used as a nanocontainer for the encapsulation and controlled release of hydrophobic molecules, such as Nile red. Source: Ji et al. [121], © 2012 American Chemical Society.

(Figure 6.33). Thus, the basic criteria of a supramolecular amphiphile to be used in the context of drug-delivery applications were fulfilled.

In extension of the former work, the same authors also assembled an amphiphilic supramolecular diblock copolymer in which a supramolecular and a covalent polymer block comprised the hydrophobic and the hydrophilic segment, respectively. The self-assembly of **41** in *N,N*-dimethylformadide (DMF) afforded the hydrophobic supramolecular block; its terminal MV^{2+} unit was then complexed by the BMP32C10-type binding site of **39** (given that c(**41**) was higher than c(**39**), Figure 6.34) [122]. Thus, the DP of the supramolecular segment and, thereby, the proportion between the hydrophobic and the hydrophilic block could be adjusted via the applied **41** : **39** molar ratio. As a result from this, various types of morphologies could be observed for the supramolecular amphiphiles by transmission electron microscopy (TEM) imaging (Figure 6.34): spherical micelles with an average diameter of c. 90 nm were obtained when the length of the hydrophobic segment was c. 24 nm, increasing the length to c. 30 nm supported the occurrence of flat, disk-like micelles. Finally, vesicles with an average diameter and wall thickness of c. 1 μm and 100 nm, respectively, were assembled. In all cases, the length of

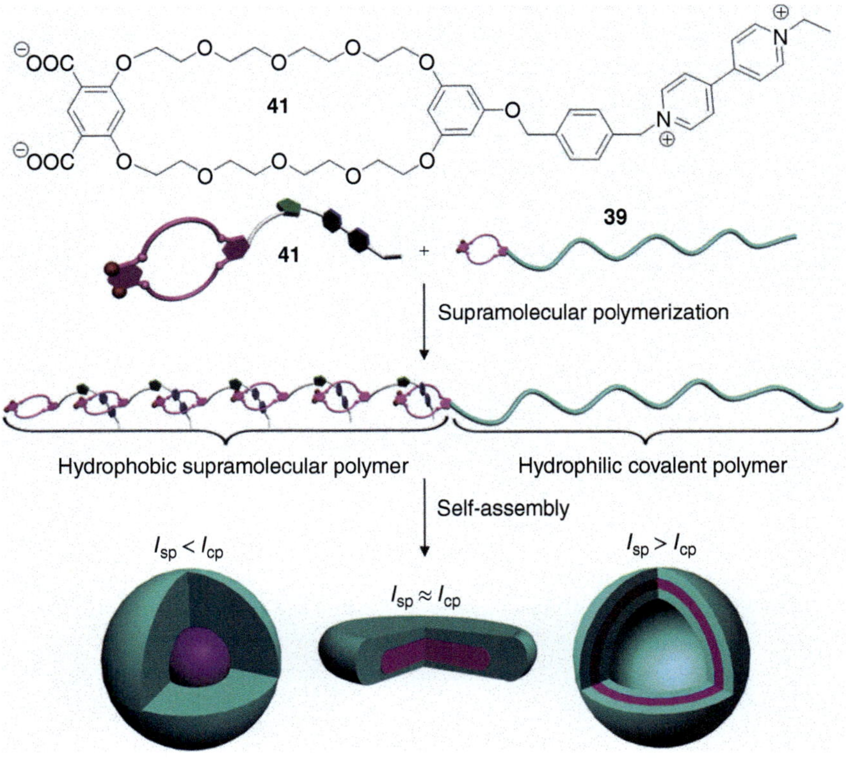

Figure 6.34 Schematic representation of the formation of an amphiphilic supramolecular diblock copolymer from **29** and **31** that assembled further into nanostructures depending on the length ratio of the constituent blocks (l_{sp}: length of the supramolecular block, l_{cp}: length of the covalent block). Source: Ji et al. [122], © 2013 John Wiley and Sons.

the hydrophilic block was kept constant (l = 28.1 nm). As for the supramolecular amphiphile, based on **39** and **40**, also here the pH-triggered controlled release of Nile red, as model substrate, was studied.

6.6 Concluding Remarks

The field of crown ether recognition and poly(pseudo)rotaxane chemistry represents one of the most famous areas in supramolecular chemistry with two Nobel Prizes in Chemistry (1987 and 2016) being given to the pioneers and figureheads, who coined today's supramolecular chemistry. The associated "molecular machines" or "molecular muscles" are fascinating examples, but more importantly are still topics of fundamental research. Nonetheless, crown ether recognition is not only a model systems to study host–guest interactions as well as to investigate (pseudo)rotaxane formation but has become an important tool in tomorrow's material's science arena where "smart," adaptive materials are of particular relevance: stimuli-responsive and self-healing materials, organic–inorganic hybrid materials, chemical sensors, and biomedical applications have to be named in this context [8, 9]. Furthermore, crown ether-modified polymers are widely used as stationary phases in the chromatographic separation of metal ions or organic amines [8]. More recently, supramolecular amphiphiles have attracted considerable interest regarding potential applications in the fields of biology and pharmaceuticals [120]. However, such materials, based on crown ether recognition, have so far only been investigated to a minor content.

References

1. Pedersen, C.J. (1967). *J. Am. Chem. Soc.* 89: 7017–7036.
2. Dietrich, B., Lehn, J.-M., and Sauvage, J.-P. (1969). *Tetrahedron Lett.* 10: 2885–2888.
3. Lehn, J.-M. (1995). *Supramolecular Chemistry – Concepts and Perspectives*. Weinheim: VCH-Wiley.
4. Cram, D.J. (1988). *Angew. Chem. Int. Ed. Engl.* 27: 1009–1020.
5. Lehn, J.-M. (1988). *Angew. Chem. Int. Ed. Engl.* 27: 89–112.
6. Pedersen, C.J. (1988). *Angew. Chem. Int. Ed. Engl.* 27: 1021–1027.
7. Allwood, B.L., Shahriari-Zavareh, H., Stoddard, J.F., and Williams, D.J. (1987). *J. Chem. Soc., Chem. Commun.*: 1058–1061.
8. Gokel, G.W., Leevy, W.M., and Weber, M.E. (2004). *Chem. Rev.* 104: 2723–2750.
9. Zheng, B., Wang, F., Dong, S.-Y., and Huang, F.-H. (2012). *Chem. Soc. Rev.* 41: 1621–1636.
10. Raymo, F.M. and Stoddard, J.F. (1999). *Chem. Rev.* 99: 1643–1664.
11. Leigh, D.A. (2016). *Angew. Chem. Int. Ed.* 55: 14506–14508.
12. Harada, A., Takashima, Y., and Yamaguchi, H. (2009). *Chem. Soc. Rev.* 38: 875–882.

13 Liu, Y., Yu, Y., Gao, J. et al. (2010). *Angew. Chem. Int. Ed.* 49: 6576–6579.
14 Guo, D.-S., Chen, S., Qian, H. et al. (2010). *Chem. Commun.* 46: 2620–2622.
15 Zhang, Z.-B., Luo, Y., Chen, J.-Z. et al. (2011). *Angew. Chem. Int. Ed.* 50: 1397–1401.
16 Yerin, A., Wilks, E.S., Moss, G.P., and Harada, A. (2008). *Pure Appl. Chem.* 80: 2041–2068.
17 Ashton, P.R., Chrystal, E.J.T., Glink, P.T. et al. (1996). *Chem. Eur. J.* 2: 709–728.
18 Huang, F.-H., Jones, J.W., Slebodnick, C., and Gibson, H.W. (2003). *J. Am. Chem. Soc.* 125: 14458–14464.
19 Bryant, W.S., Jones, J.W., Mason, P.E. et al. (1999). *Org. Lett.* 1: 1001–1004.
20 Li, S.-J., Zhu, K.-L., Wen, X.-H. et al. (2009). *Eur. J. Org. Chem.* 2009: 1053–1057.
21 Allwood, B.L., Spencer, N., Shahriari-Zavareh, H. et al. (1987). *J. Chem. Soc., Chem. Commun.*: 1064–1066.
22 Zhang, M.-M., Zhu, K.-L., and Huang, F.-H. (2010). *Chem. Commun.* 46: 8131–8141.
23 Yan, X.-H., Wei, P.-F., Xia, B.-Y. et al. (2012). *Chem. Commun.* 48: 4968–4970.
24 Inokuchi, Y., Boyarkin, O.V., Kusaka, R. et al. (2012). *J. Phys. Chem. A* 116: 4057–4068.
25 Jiang, W. and Schalley, C.A. (2010). *J. Mass Spectrom.* 45: 788–798.
26 Gokel, G.W. and Durst, H.D. (1976). *Synthesis* 1976: 168–184.
27 Song, Y.-Y., Jing, H.-W., Li, B., and Bai, D.-S. (2011). *Chem. Eur. J.* 17: 8731–8737.
28 Tsakunov, A.V., Dubonosov, A.D., Bren, V.A., and Minkin, V.I. (2008). *Chem. Heterocycl. Compd.* 44: 889–923.
29 Zhang, M.-M., Xu, D.-H., Yan, X.-H. et al. (2012). *Angew. Chem. Int. Ed.* 51: 7011–7015.
30 Xu, W., Xiao, J., Wang, D. et al. (2010). *Electrochem. Solid-State Lett.* 13: A48–A51.
31 Yang, L., Tan, X., Wang, Z., and Zhang, X. (2015). *Chem. Rev.* 115: 7196–7239.
32 Yang, X., Yu, H., Wang, L. et al. (2015). *Soft Matter* 11: 1242–1252.
33 Arunachalam, M. and Gibson, H.W. (2014). *Prog. Polym. Sci.* 39: 1043–1073.
34 Ashton, P.R., Baxter, I., Cantrill, S.J. et al. (1998). *Angew. Chem. Int. Ed.* 37: 1294–1297.
35 Cantrill, S.J., Youn, G.J., Stoddard, J.F., and Williams, D.J. (2001). *J. Org. Chem.* 66: 6857–6872.
36 Gibson, H.W., Yamaguchi, N., Niu, Z.-B. et al. (2010). *J. Polym. Sci., Part A: Polym. Chem.* 48: 975–985.
37 Babcock-Gove, P. (ed.) (1993). *Webster's Third New International Dictionary*. Springfield: Merrian-Webster.
38 Ashton, P.R., Parsons, I.W., Raymo, F.M. et al. (1998). *Angew. Chem. Int. Ed.* 37: 1913–1916.
39 Yamaguchi, N., Nagvekar, D.S., and Gibson, H.W. (1998). *Angew. Chem. Int. Ed.* 37: 2361–2364.

40 Wang, F., Zhang, J., Ding, X. et al. (2010). *Angew. Chem. Int. Ed.* 49: 1090–1094.
41 Dong, S.-Y., Luo, Y., Yan, X.-H. et al. (2011). *Angew. Chem. Int. Ed.* 50: 1905–1909.
42 Yan, X.-H., Zhou, M., Chen, J.-H. et al. (2011). *Chem. Commun.* 47: 7086–7088.
43 Zheng, X., Miao, Q., Wang, W., and Qu, D.-H. (2018). *Chin. Chem. Lett.* 29: 1621–1624.
44 Huang, F.-H., Nagvekar, D.S., Zhou, X.-C., and Gibson, H.W. (2007). *Macromolecules* 40: 3561–3567.
45 Yamaguchi, N. and Gibson, H.W. (1999). *Angew. Chem. Int. Ed.* 38: 143–147.
46 Yamaguchi, N. and Gibson, H.W. (1999). *Chem. Commun.*: 789–790.
47 Gibson, H.W., Yamaguchi, N., and Jones, J.W. (2003). *J. Am. Chem. Soc.* 125: 3522–3533.
48 Wang, F., Han, C.-Y., He, Q.-Z. et al. (2008). *J. Am. Chem. Soc.* 130: 11254–11255.
49 Wang, F., Zheng, B., Zhu, K.-L. et al. (2009). *Chem. Commun.*: 4375–4377.
50 Chen, D., Zhan, J., Zhang, M. et al. (2015). *Polym. Chem.* 6: 25–29.
51 Zhang, Z., Cheng, L., Zhao, J. et al. (2020). *Angew. Chem. Int. Ed.* 59: 12139–12146. https://doi.org/10.1002/anie.202004152.
52 Wang, L., Cheng, L., Li, G. et al. (2020). *J. Am. Chem. Soc.* 142: 2051–2058.
53 Ren, X., Zhang, J., Cheng, M. et al. (2018). *Tetrahedron Lett.* 59: 2197–2204.
54 Huang, F.-H., Gibson, H.W., Bryant, W.S. et al. (2003). *J. Am. Chem. Soc.* 125: 9367–9371.
55 Huang, F.-H., Fronczek, F.R., and Gibson, H.W. (2003). *J. Am. Chem. Soc.* 125: 9272–9273.
56 Huang, F.-H., Switek, K.A., Fronczek, F.R. et al. (2005). *J. Org. Chem.* 70: 3231–3241.
57 Niu, Z.-B., Huang, F.-H., and Gibson, H.W. (2011). *J. Am. Chem. Soc.* 133: 2836–2839.
58 Huang, F.-H. and Gibson, H.W. (2005). *Chem. Commun.*: 1696–1698.
59 Cheng, M., Yao, C., Cao, Y. et al. (2016). *Chem. Commun.* 52: 8715–8718.
60 Wei, P., Xia, B., Zhang, Y. et al. (2014). *Chem. Commun.* 50: 3973–3975.
61 Yao, C., Zhang, J., Cheng, M. et al. (2018). *Macromol. Rapid Commun.* 39: 1700218.
62 Gibson, H.W., Ge, Z.-X., Huang, F.-H. et al. (2005). *Macromolecules* 38: 2626–2637.
63 Gibson, H.W., Farcas, A., Jones, J.W. et al. (2009). *J. Poly. Sci., Part A: Polym. Chem.* 47: 3518–5543.
64 Lee, M.-J., Schoonover, D.V., Gies, A.P. et al. (2009). *Macromolecules* 42: 6483–6494.
65 Huang, F.-H. and Gibson, H.W. (2004). *J. Am. Chem. Soc.* 126: 14738–14739.
66 Li, S.-J., Zheng, B., Chen, J.-H. et al. (2010). *J. Polym. Sci., Part A: Polym. Chem.* 48: 4067–4073.
67 Fréchet, J.M.J. and Tomalia, D.A. (eds.) (2001). *Dendrimers and other Dendritic Polymers.* Chichester: Wiley.

68 Lee, J.-W. and Kim, K. (2003). *Top. Curr. Chem.* 228: 111–140.

69 Leung, K.C.-F. and Lau, K.-N. (2010). *Polym. Chem.* 1: 998–1000.

70 Jones, J.W., Bryant, W.S., Bosman, A.W. et al. (2003). *J. Org. Chem.* 68: 2385–2389.

71 Yamaguchi, N., Hamilton, L.M., and Gibson, H.W. (1998). *Angew. Chem. Int. Ed.* 37: 3275–3279.

72 Gibson, H.W., Yamaguchi, N., Hamilton, L.M., and Jones, J.W. (2002). *J. Am. Chem. Soc.* 124: 4653–4665.

73 Huang, F.-H., Nagvekar, D.S., Slebodnick, C., and Gibson, H.W. (2005). *J. Am. Chem. Soc.* 127: 484–485.

74 Lee, M.-J., Moore, R.B., and Gibson, H.W. (2011). *Macromolecules* 44: 5987–5993.

75 Gibson, H.W., Ge, Z.-X., Jones, J.W. et al. (2009). *J. Polym. Sci., Part A: Polym. Chem.* 47: 6472–6495.

76 Rowan, S.J., Cantrill, S.J., Stoddard, J.F. et al. (2000). *Org. Lett.* 2: 759–762.

77 Chiu, S.-H., Rowan, S.J., Cantrill, S.J. et al. (2002). *Chem. Commun.*: 2948–2949.

78 Harrison, I.T. (1972). *J. Chem. Soc., Chem. Commun.*: 231–232.

79 Elizarov, A.M., Chiu, S.-H., Glink, P.T., and Stoddard, J.F. (2002). *Org. Lett.* 4: 679–682.

80 Elizarov, A.M., Chang, T., Chiu, S.-H., and Stoddard, J.F. (2002). *Org. Lett.* 4: 3565–3568.

81 Ge, Z.-S., Hu, J.-M., Huang, F.-H., and Liu, S.-Y. (2009). *Angew. Chem. Int. Ed.* 48: 1798–1802.

82 Ashton, P.R., Glink, P.T., Stoddard, J.F. et al. (1996). *Chem. Eur. J.* 2: 729–736.

83 Ashton, P.R., Ballardini, R., Balzani, V. et al. (1998). *J. Am. Chem. Soc.* 120: 11932–11942.

84 Cantrill, S.J., Fulton, D.A., Fyfe, M.C.T. et al. (1999). *Tetrahedron Lett.* 40: 3669–3672.

85 Rowan, S.J., Cantrill, S.J., and Stoddard, J.F. (1999). *Org. Lett.* 1: 129–132.

86 Rowan, S.J. and Stoddard, J.F. (2000). *J. Am. Chem. Soc.* 122: 164–165.

87 Ueng, S.-H., Hsueh, S.-Y., Lai, C.-C. et al. (2008). *Chem. Commun.*: 817–819.

88 Coutrot, F., Romuald, C., and Busseron, E. (2008). *Org. Lett.* 10: 3741–3744.

89 Fang, L., Hmadeh, M., Wu, J. et al. (2009). *J. Am. Chem. Soc.* 131: 7126–7134.

90 Hmadeh, M., Fang, L., Trabolsi, A. et al. (2010). *J. Mater. Chem.* 20: 3422–3430.

91 Wu, J.-S., Leung, K.C.-F., Benítez, D. et al. (2008). *Angew. Chem. Int. Ed.* 47: 7470–7474.

92 Guidry, E.N., Li, J., Stoddard, J.F., and Grubbs, R.H. (2007). *J. Am. Chem. Soc.* 129: 8944–8945.

93 Clark, P.G., Day, M.W., and Grubbs, R.H. (2009). *J. Am. Chem. Soc.* 131: 13631–13633.

94 Sasabe, H., Inomoto, N., Kihara, N. et al. (2007). *J. Polym. Sci., Part A: Polym. Chem.* 45: 4154–4160.

95 Ji, X., Yao, Y., Li, J. et al. (2013). *J. Am. Chem. Soc.* 135: 74–77.

96 Sohgawa, Y., Fujimori, H., Shoji, J. et al. (2001). *Chem. Lett.* 30: 774–775.

97 Oku, T., Furusho, Y., and Takata, T. (2003). *J. Polym. Sci., Part A: Polym. Chem.* 41: 119–123.

98 Rowan, S.J., Cantrill, S.J., Cousins, G.R.L. et al. (2002). *Angew. Chem. Int. Ed.* 41: 898–952.

99 Leung, K.C.-F., Aricó, F., Cantrill, S.J., and Stoddard, J.F. (2005). *J. Am. Chem. Soc.* 127: 5808–5810.

100 Leung, K.C.-F., Aricó, F., Cantrill, S.J., and Stoddard, J.F. (2007). *Macromolecules* 40: 3951–3959.

101 Wenz, G., Han, B.-H., and Müller, A. (2006). *Chem. Rev.* 106: 782–817.

102 Huang, F.-H. and Gibson, H.W. (2005). *Prog. Polym. Sci.* 30: 982–1018.

103 Wu, C., Bheda, M.C., Lim, C. et al. (1991). *Polym. Commun.* 32: 204–207.

104 Gibson, H.W., Liu, S., Lecavalier, P. et al. (1995). *J. Am. Chem. Soc.* 117: 852–874.

105 Gibson, H.W., Liu, S., Gong, C. et al. (1997). *Macromolecules* 30: 3711–3727.

106 Shen, Y.X. and Gibson, H.W. (1992). *Macromolecules* 25: 2058–2059.

107 Shen, Y.X., Xie, D., and Gibson, H.W. (1994). *J. Am. Chem. Soc.* 116: 537–548.

108 Marand, E., Hu, Q., Gibson, H.W., and Veytsman, B. (1996). *Macromolecules* 29: 2555–2562.

109 Gong, C. and Gibson, H.W. (1997). *Macromol. Chem. Phys.* 198: 2321–2332.

110 Gong, C. and Gibson, H.W. (1996). *Macromolecules* 29: 7029–7033.

111 Gong, C., Ji, Q., Glass, T.E., and Gibson, H.W. (1997). *Macromolecules* 30: 4807–4813.

112 Gong, C. and Gibson, H.W. (1997). *Macromolecules* 30: 8524–8525.

113 Gong, C., Glass, T.E., and Gibson, H.W. (1998). *Macromolecules* 31: 308–313.

114 Gibson, H.W. and Liu, S. (1996). *Macromol. Symp.* 102: 55–61.

115 Lefebvre, H., Bheda, M., and Gibson, H.W. (2016). *Polymer* 90: 317–330.

116 Gibson, H.W., Engen, P.T., and Lee, S.-H. (1999). *Polymer* 40: 1823–1832.

117 Lee, S.-H., Engen, P.T., and Gibson, H.W. (1997). *Macromolecules* 30: 337–343.

118 Nagapudi, K., Leisen, J., Beckham, H.W., and Gibson, H.W. (1999). *Macromolecules* 32: 3025–3033.

119 Gibson, H.W., Bryant, W.S., and Lee, S.-H. (2001). *J. Polym. Sci., Part A: Polym. Chem.* 39: 1978–1993.

120 Yu, G., Jie, K., and Huang, F. (2015). *Chem. Rev.* 115: 7240–7303.

121 Ji, X., Li, J., Chen, J. et al. (2012). *Macromolecules* 45: 6457–6463.

122 Ji, X., Dong, S., Wei, P. et al. (2013). *Adv. Mater.* 25: 5725–5729.

123 Chen, J., Zhang, S., Sun, F. et al. (2016). *Polym. Chem.* 7: 2947–2954.

7

Supramolecular Polymers, Based on Cucurbiturils

7.1 General Aspects

The discovery of what is known today as cucurbiturils actually dates back to 1905 when Behrend et al. investigated the condensation of glycoluril (**1**) with formaldehyde in a highly acidic regime [1]: the "white, amorphous compounds, which were weakly soluble in dilute acid and base, were able to absorb large quantities of water without losing their dusty powdery character." Though a component of the molecular formula $C_{18}H_{18}N_{12}O_6$ could be identified, still little was known about that supposedly mix of differently sized cucurbit[*n*]urils (CB[*n*]). About 75 years later, the Mock group repeated Behrend's initial experiments [2]: upon complexation with $CaSO_4$, they isolated a hydrated macrocycle comprising 6 glycoluril units linked via, in total, 12 CH_2 bridges. The interaction with the Ca(II) ions occurred via the macrocycle's carbonyl-containing rims. The name "cucurbituril" was coined due to the structure being reminiscent of "a gourd or pumpkin" and a cucurbit (in former times, alchemists used such vessel-type distillation apparatus). Today, this particular derivative is referred to as cucurbit[6]uril or, in short, CB[6] (thereby, the number "6" refers to the number of glycoluril moieties within the macrocycle). The ability of CB[6] to encapsulate alkylammonium cations was also reported in the same publication. Presumably, a broad range of other CB[*n*]-type systems was formed in Mock's experiments [3]; however, the homologs CB[5], CB[7], and CB[8] were isolated and identified by X-ray single-crystal analysis in the early 2000s (Figure 7.1) [4]; a remarkable interlocked complex of the type CB[5]@CB[10] was published in that same timeframe [3].

Since these pioneering contributions, the chemistry of cucurbiturils has experienced a rapid development, in particular with respect to their supramolecular properties – CB[*n*]s represent a family of versatile host materials for recognition studies – as well as their potential applications, e.g. as molecular switches, organo-catalysts, carriers in drug delivery, and advanced (nano)materials [5–16]. Taking a closer look on the cavity sizes of CB[*n*]s ($n = 6–8$), a similar environment in terms of the dimensions is offered as for the related α-, β-, and γ-cyclodextrins, respectively (Figure 7.1). However, the various cyclodextrin (CD) derivatives still represent the receptors of choice for industrial applications, since they are commercially available and inexpensive (for more details on CDs, see Chapter 9).

Supramolecular Polymers and Assemblies: From Synthesis to Properties and Applications, First Edition.
Ulrich S. Schubert, George R. Newkome, and Andreas Winter.

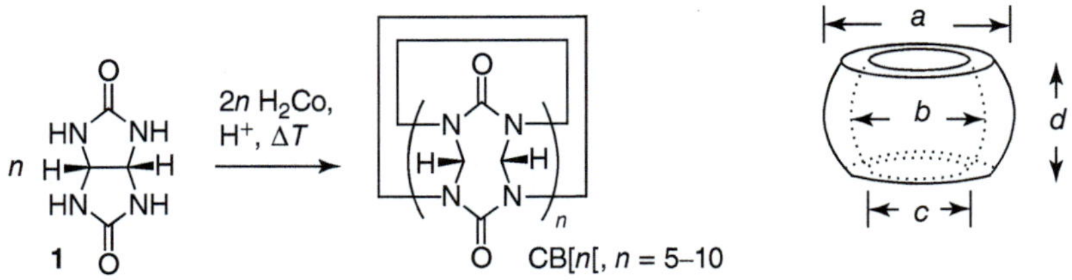

CB[8]

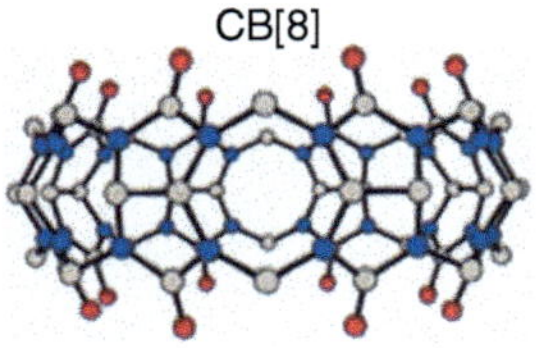

		CB[5]	CB[6]	CB[7]	CB[8]
Outer diameter (Å)	*a*	13.1	14.4	16.0	17.5
Cavity (Å)	*b*	4.4	5.8	7.3	8.8
	c	2.4	3.9	5.4	6.9
Height (Å)	*d*	9.1	9.1	9.1	9.1
Cavity volume (Å$^{3)}$		82	164	279	479

Figure 7.1 Schematic representation of the generalized CB[*n*] synthesis via co-condensation of **1** with formaldehyde under acidic conditions. The dimensions of different CB[*n*] representatives are listed [5]. The ball-stick representation of the solid-state structure of CB[8] is also shown. Source: Kim et al. [4], © 2000 American Chemical Society.

Disadvantageously, CDs suffer from a considerable low solubility in organic solvents, and their selective functionalization still remains, despite all improvements, a synthetic challenge. In contrast, CB[*n*]s feature good solubility in both aqueous and organic media, while their availability and host–guest chemistry can now almost compete with that of CDs [11, 16].

In the following, the utilization of CB[*n*]s as host materials for the formation of supramolecular assemblies – discrete as well as macromolecular architectures – will be evaluated. Here, the interactions of CB[*n*]s with organic compounds are in the focus of interest – small molecules and polymers. The fascinating host–guest chemistry of CB[*n*] with inorganic species has been the subject of review articles published elsewhere [11, 15].

7.2 Interactions of CB[*n*]s with Small Organic Guest Molecules

The ability of CB[*n*]s to encapsulate small organic molecule guests is well documented. The basic binding modes are depicted in Figure 7.2 and include various types of binary 1 : 1 complexes (mainly for CB[7]) and, in the case of CB[8], hetero-ternary 1 : 1 : 1 assemblies. The thermodynamics and kinetics governing these supramolecular interactions have been investigated extensively to determine binding parameters (e.g. binding constants [K_a], enthalpy and entropy) as well as to elucidate the mechanisms of (de)complexation. An extensive summary over such values has been given by Barrow et al. some years ago [16]. Titration experiments, such as isothermal calibration titration (ICT), and UV/vis as well as ^{1}H NMR (nuclear magnetic resonance) spectroscopy turned out to be well suited for deriving

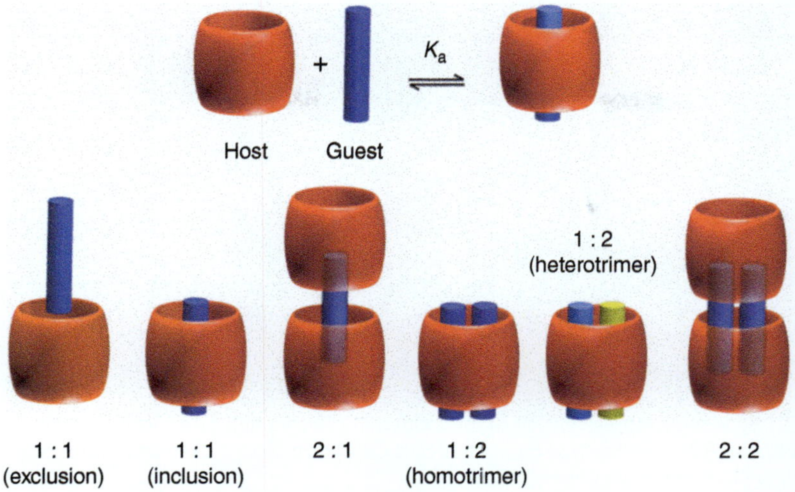

Figure 7.2 Schematic representation of the various exclusion and inclusion complexes of CB[*n*]s. Source: Barrow et al. [16].

the data. In particular, the latter technique is highly sensitive, since the H-atoms located close to the center of the macrocycles' cavity experience a significant upfield shift of c. 1.6 ppm upon guest inclusion (the more remote protons are less sensitive) [17]; whereas, the signals assigned to protons of the cavity's shell are shifted downfield by c. 0.7 ppm [18].

In general, a cationic amphiphilic guest is complexed by a CB[*n*] host via ion–dipole interactions with the carbonylated rims; thereby, the hydrophobic part of the guest is housed in the host's cavity. In early experiments, a K_a value of c. $10^7\,M^{-1}$ was measured for the interaction of spermine with CB[6] in a 50 vol% formic acid solution [19]. Indeed, very high K_a values represent a characteristic feature of CB[*n*]-based host–guest complexation. For instance, remarkably high K_a values of c. $10^{15}\,M^{-1}$ were reported for the interaction of the 1,1′-bis(trimethyl-methanaminium)ferrocene or N^1-(adamantyl)ethane-1,2-diaminium cation with CB[7] [20, 21] – these K_a values reach the strength of the avidin–biotin interaction [22, 23], as benchmark, and represent the strongest non-covalent interactions measured (excluding systems relying, e.g. on polyvalency [24], interactions of enzymes and transition states [25]). The smallest member of the cucurbituril family, CB[5], is highly selective for the recognition of inorganic cations and xenon [15]. The cavity is, however, too small to encapsulate organic species except for some gaseous hydrocarbons: methane, ethylene, and ethane (K_a values from 10^3 to $10^1\,M^{-1}$) [26]. Recently, CB[10] has attracted attention due to its selectivity for encapsulating large molecules, such as calix[4]arene and octahedral polypyridyl complexes of Ru(II), Ir(III), Pt(II), or Au(III) ions [27–29]. Moreover, CB[10] is able to flatten its structure to accommodate extended planar molecules, such as (metallo)porphyrins [30].

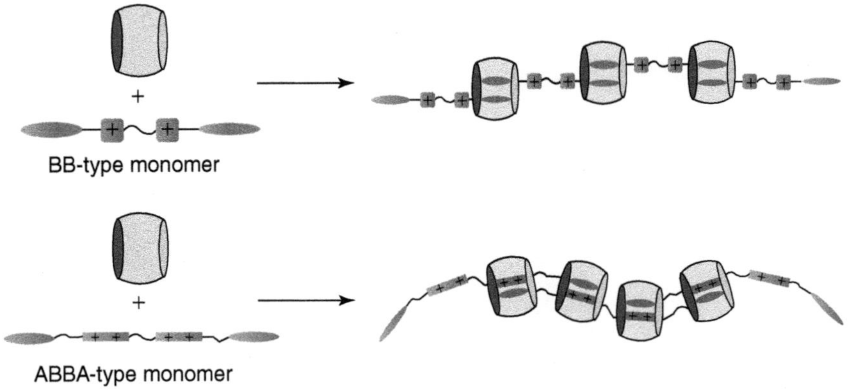

Figure 7.3 Schematic representation of the formation of supramolecular polymers via complexation of BB- or ABBA-type monomers by a CB[*n*] host. Source: Redrawn from Yang et al. [33], © 2015 American Chemical Society.

7.3 Supramolecular Polymers Incorporating CB[*n*] Units

As for all systems involving host–guest interactions, the formation of linear supramolecular polymers is always in competition with various oligomerization pathways, such as intramolecular complexation, dimerization, or (macro)cyclization. According to the theories of supramolecular polymerization (see Chapter 1), oligomerization will always be favored in dilute solutions [31, 32]. However, increasing the concentration is often not possible for CB[*n*]-based systems due to their intrinsically low solubility. On the basis of the general complexation modes (Figure 7.2), two combinations have been established to assemble supramolecular polymers via CB[*n*] complexation: interaction of the host with either BB- or ABBA-type monomers (Figure 7.3) [33]. The latter was introduced to overcome the poor water solubility of such polymers. Concerning CB[*n*]-based polymers, one can distinguish between two different classes: polymers assembled from small organic molecules and those involving the interaction of CB[*n*] with macromolecular species [16, 33–40].

7.3.1 Supramolecular Polymers, Based on Small Organic Molecules

In contrast to the CB[*n*]s with small cavities (n = 5–7), CB[8] represents the most versatile representative for the formation of supramolecular polymers. This has to be attributed mainly to the ability of CB[8] to accommodate two guest molecules (Figures 7.2 and 7.3) [34]. To fully benefit from the special properties of CB[*n*]s and, eventually, to obtain supramolecular polymers of high molar mass, one has to consider the theory of supramolecular polymerizations (see Section 1.3): notably, both a high binding strength and concentration are required to reach a high degree-of-polymerization (DP). In particular, the solubility of the CB[*n*]s, which is influenced by a number of parameters (i.e. polarity of the solvent, temperature, ionic strength of the solution, and the nature of the guest molecule), represents a major issue regarding their usage, as building blocks, in supramolecular polymer chemistry. In general, CB[5] and CB[7] are much more soluble in water as CB[6] and CB[8] (i.e. below 1 mM for CB[8]) [11]. However, the high binding strengths, in particular for CB[8], level out the drawbacks resulting from the limited water solubility [16]. Second, following the typical mechanism for a ring-chain-mediated supramolecular polymerization (see Section 1.3.2), the formation of macrocyclic assemblies at the cost of polymeric ones needs to be mentioned [41–43]. Thereby, the obtained ring-chain ratio will depend on various factors, such as the monomer structure (i.e. shape, length, and flexibility), binding strength, concentration, and temperature. The third aspect that needs to be considered, before detailing the various examples for supramolecular polymers, is the binding mode between the repeating units. The typical terms used in this context are "complementarity" and "self-complementarity" of the building blocks (this is of utmost importance also for other types of supramolecular polymers, such as those involving H-bonding interactions [44]; see Chapter 3) [45]. Complementary systems might be composed of two or three components; whereas, in the latter case, a third moiety is required to enable the association of the other two. CB[8] features the ability to bind two organic molecules in its relatively large cavity and is, therefore, a representative for a three-component binding (Figure 7.2) [46–49].

To prepare supramolecular polymers, incorporating CB[8], three synthetic routes have been proposed [in either case, (hetero)dimerization will occur within the cavity of the host]:

1. Self-assembly of CB[8] host with AB-type monomers,
2. Self-assembly of CB[8] with AA-type monomers, and, finally,
3. Self-assembly of CB[8] with a stoichiometric mixture of AA- and BB-type monomers.

Credit has to be given to Kim et al., who first reported the formation of charge–transfer (CT) complexes within the cavity of CB[8]. As discussed in Chapter 5, the intrinsically weak CT interactions between small organic molecules in solution can be stabilized by offering a rigid environment, e.g. a cavity-containing host, in which the CT complex is included [50]. It was shown that the self-assembly of AB-type monomers featuring an electron-donating and an electron-accepting substituent with CB[8] is strongly affected by the length and flexibility of the spacer between the terminal groups of the monomer. Thus, the flexible and semi-rigid linkers selectively gave 1 : 1 complexes; whereas, for more rigid spacers, the formation of 2 : 2 complexes was favored [51, 52]. The equimolar combination of **2** and CB[8] gave a 2 : 2 complex (along with some oligomeric species with a 3 : 2 ratio of the components) [53]. The hydrodynamic volume of these presumably macrocyclic assemblies was calculated on the basis of diffusion-ordered NMR spectroscopy (DOSY), and from this, a DP of 4 was concluded. Moreover, the growth of linear supramolecular polymers from a gold substrate bearing a monolayer of **3**@CB[8] was shown. The polymer was formed by simply soaking this functionalized substrate into a solution of **2** and CB[8] in water (Figure 7.4). Atomic force microscopy (AFM) and surface plasmon resonance spectroscopy were applied to

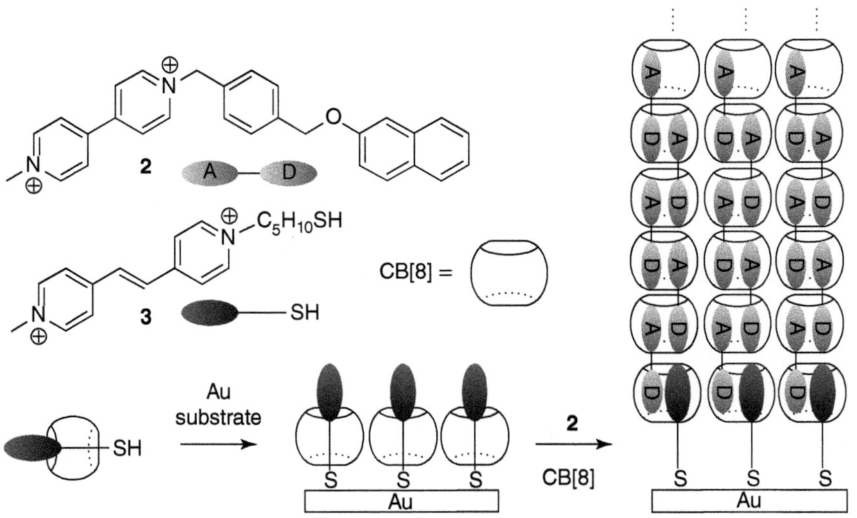

Figure 7.4 Schematic representation of the formation of a surface-anchored supramolecular polymer (counterions omitted for clarity). Source: Kim et al. [53], © 2004 American Chemical Society.

analyze the surface-anchored material; both techniques suggested that a DP of 4 was reached.

A defined supramolecular macrocyclic pentamer was obtained from the self-assembly of an AB-type monomer with a short alkyl spacer between the terminal sites, thus preventing intramolecular complexation within the CB[8] host [54]. It was shown by the Zhang group that macrocyclization can be efficiently suppressed by utilizing an ABBA-type monomer **4** with four designated binding sites for CB[8] [55]. The monomeric structure featured a central butyl spacer, thus preventing intramolecular complexation, and two viologen moieties that suppressed the formation of 2 : 2 complexes due to electronic repulsion of the cationic units (Figure 7.5). At an 1 : 1 ratio of the components, a water-soluble polymer with CT characteristics was formed from the barely soluble molecular components. Applying AFM-based single-molecule force spectroscopy (SMFS) confirmed the presence of long supramolecular polymer chains: the force-extension curves were similar to those of covalent polymers and the measured Kuhn lengths of c. 2.2 nm agreed with rather rigid polymer chains [for comparison, DNA in water and poly(*tert*-butyl methacrylate) in toluene have Kuhn lengths of 100 and 2 nm, respectively]. Furthermore, dynamic light scattering (DLS) measurements supported this assumption – the R_h values of the polymer chains were in the order of 30 nm (the force-extension curves gave a chain length of c. 60 nm). As detailed in Chapter 11, the heteroleptic Fe(II) complex of terpyridyl ligand **5** also gave a supramolecular polymer due to the formation of a 1 : 2 homotrimer complex when CB[8] was added (Figure 7.6) [56]. This example represents a supramolecular polymer assembled from two orthogonal types of non-covalent interactions – metal-to-ligand coordination and CB[8]'s host–guest chemistry.

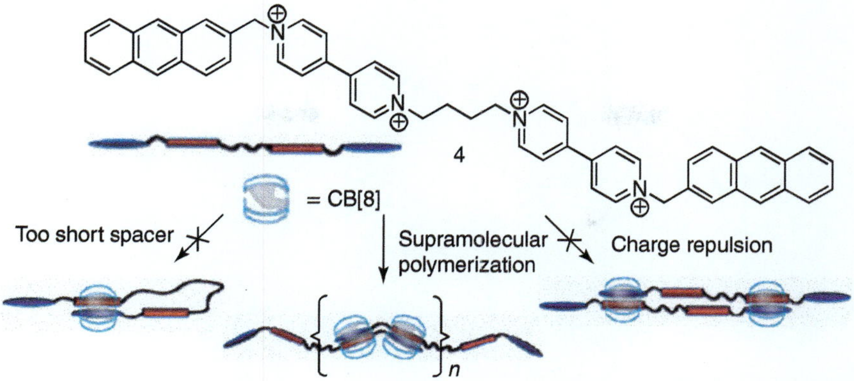

Figure 7.5 Schematic representation of the formation of the supramolecular polymer $\{4@CB[8]\}_n$; the assembly into 1 : 1 or 2 : 2 species is prevented due to steric and electronic constraints, respectively. Source: Liu et al. [55], © 2010 John Wiley and Sons.

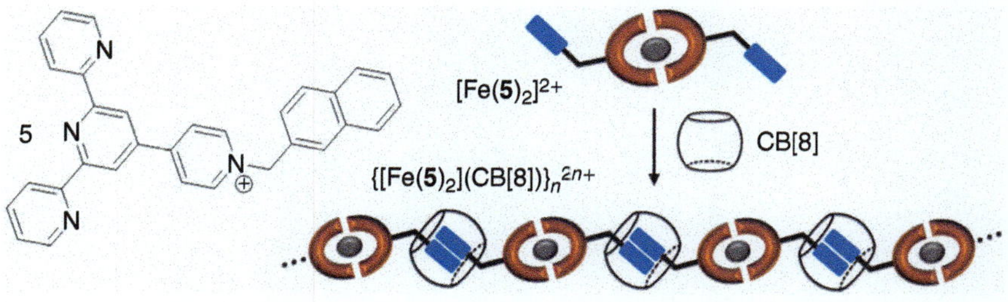

Figure 7.6 Schematic representation of a supramolecular polymer based on metal-to-ligand coordination and host–guest complexation. Source: Liu et al. [56], © 2013 The Royal Chemical Society.

Basically, the same monomer design, but possessing diazobenzene moieties instead of the anthracene, was employed by the Scherman group in a supramolecular polymerization [57]; also the short diazobenzene–viologen conjugate **6** was used in the same context. The formed polymers were investigated in solution and solid state by light scattering as well as synchrotron small-angle X-ray scattering (SAXS) and small-angle neutron scattering (SANS) measurements. For the polymer, based on guest monomer **5**, single crystals suited for X-ray structure analysis were formed (Figure 7.7). The supramolecular polymers exhibited a stimuli-responsive behavior: reversible polymerization/depolymerization could be achieved by photoisomerization of the azobenzene moieties (the Z-form of the guest molecules fully occupied the CB[8] cavities with high binding constants).

Besides the former examples, where host-centered CT interactions enabled the formation of supramolecular polymers, numerous examples can also be found involving host-centered π–π stacking interactions (i.e. assembly of CB[8] with AA-type monomers in a 1 : 2 fashion; see Figure 7.2). Zhang et al. studied a wide range of such divalent monomers with respect to their supramolecular polymerization by dimerization of their self-complementary sites within the CB[8] cavities

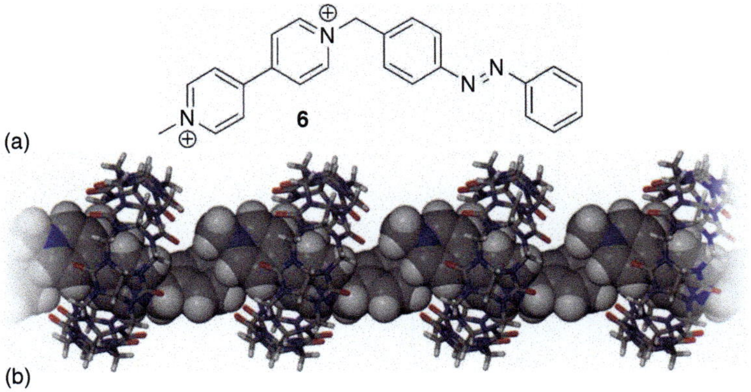

Figure 7.7 (a) Schematic representation of the bifunctional monomer **6** (counterions were omitted for clarity). (b) Side view of the X-ray single crystal structure of {**6**@CB[8]}$_n$. Source: del Barrio et al. [57], © 2013 American Chemical Society.

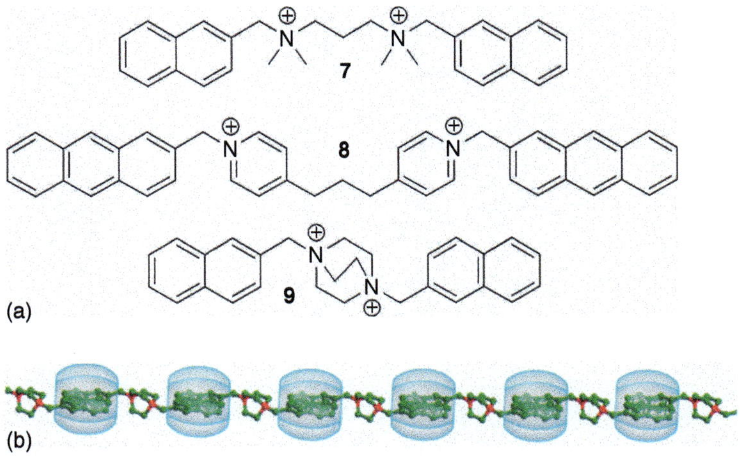

Figure 7.8 (a) Schematic representation of the ditopic monomers **7** to **9**. (b) Schematic representation of the linear supramolecular polymer {**9**@CB[8]}$_n$. Source: Liu et al. [59], © 2013 John Wiley and Sons.

(some selected monomers are depicted in Figure 7.8a) [58–61]. In all cases, the monomers have flat aromatic groups to facilitate π–π stacking interactions within the host's cavities as well as cationic moieties to enhance the solubility of the resulting polymers in water and to simultaneously offer additional stabilization via ion–dipole interactions with the CB[8] host's carbonyl groups. For example, the homoditopic **7** only gave cyclic oligomers [59]; whereas, linear supramolecular polymers were obtained from **8** and CB[8] (the monomeric structure was designed to prevent cyclization and dimerization) [58]. It was further shown that monomer **9** with its rigid, shape-persistent spacer efficiently inhibited cyclization (Figure 7.8b). According to DOSY measurements, a DP of 64 was reached for {**9**@CB[8]}$_n$; whereas, the cyclic oligomers {**7**@CB[8]}$_n$ had a DP of about 7.

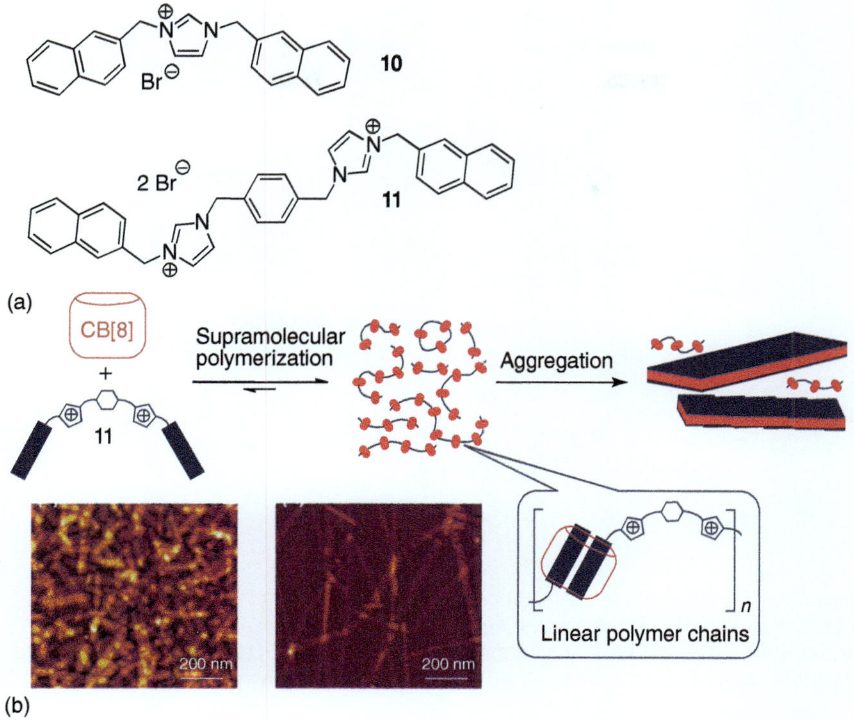

Figure 7.9 Schematic representation of the imidazolium-containing guests **10** and **11**. (a) Schematic representation of the step-wise self-assembly of **11** and CB[8] into linear polymer chains and, subsequently, fibrous aggregates. (b) Representative AFM images of the crystalline nanostructures obtained from **10**/CB[8] (left) and **11**/CB[8] (right). Source: Barrio et al. [62]. Figure reproduced with kind permission; © 2019 American Chemical Society. Licensed under CC BY 4.0.

Del Barrio, Scherman, et al. synthesized the imidazolium-containing guest molecules **10** and **11**, which were assembled into supramolecular polymers upon the addition of CB[8] [62]. As concluded from SANS measurements as well as AFM and transmission electron microscopy (TEM) imaging, these semiflexible polymer chains were in coexistence with crystalline nanostructures (Figure 7.9a). The authors hypothesized that the aggregation of polymer chains into the nanostructures was driven by the crystallization of the host–guest complexes. Depending on the structure of the guest, platelet-like or fibrous aggregates were identified, as the final morphologies (Figure 7.9b). Moreover, the fibers exhibited a viscoelastic and lyotropic behavior.

The rigid yet water-soluble thiazole-orange dye (**12**) was used by Xu et al. for the synthesis of supramolecular polymers [63]. UV/vis absorption and emission spectroscopy were used to monitor the self-assembly process. The formation of H-aggregates within the CB[8] cavities, as revealed by a blue-shifted absorption and strong fluorescence enhancement, was concluded from these measurements; a binding constant of $4 \times 10^{10}\ M^{-3}$ was calculated for the proposed 2 : 2 dimer (Figure 7.10a). When performing a UV/vis titration experiment, also the formation

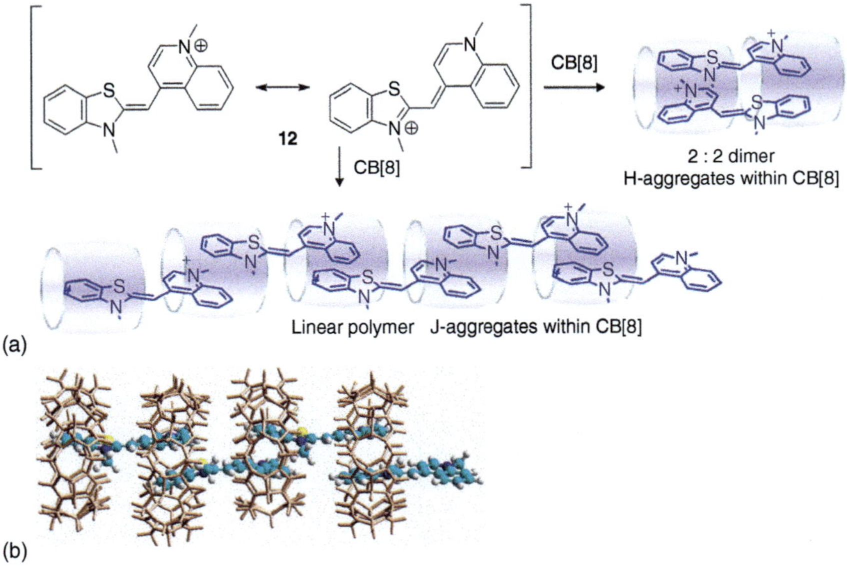

Figure 7.10 (a) Schematic representation of the self-assembly of **12** and CB[8] into 2 : 2 dimers and linear polymers due to the formation of H- or J-type aggregates within the hosts' cavities. (b) Representation of the structure of oligomeric $\{\mathbf{12}@\mathrm{CB}[8]\}_n$ according to molecular modeling. Source: Xu et al. [63], © 2011 The Royal Chemical Society.

of J-aggregates – assigned to the assembly into linear polymer chains – could be evidenced (Figure 7.10a/b). Such assemblies could also be identified by applying multidimensional mass spectrometry (MS), i.e. electro-spray ionization (ESI) MS coupled to a traveling-wave ion-mobility (TWIM) set-up. This particular technique allows for distinguishing between supramolecular ions of the same m/z values due to different sizes/shapes and, therefore, mobilities in the TWIM chamber (for further applications of the method, see Refs. [64, 65] and Chapter 12).

Zhang et al. showed that not only rigid monomers but also those with longer and flexible linkers can form supramolecular polymers upon interaction with CB[8] [60]. The monomer **13** featuring two tripeptide moieties of the sequence PheGlyGly (Figure 7.11a) gave a polymer that was investigated by matrix-assisted laser desorption/ionization time-of-flight (MALDI-TOF) MS and AFM–SMFS; the Gaussian fitting of the SMFS curves suggested a DP of 50–60 (Figure 7.11c). The employed tripeptide sequence is well known to generate stable complexes possessing 2 : 1 stoichiometry with CB[8] (K_a value of 10^{11} M^{-2}) due to a cooperativity effect involving multiple H-bonding, ion–dipole, and hydrophobic interactions (Figure 7.11b) [66].

The interaction of a structurally similar monomer, i.e. having two PheGlyGlyGly tetrapeptides linked via a flexible bridge, with CB[8] was investigated by Ramaekers et al.; these authors employed isothermal titration caloriometry (ITC), DOSY as well as quadrupole-time-of-flight (QTOF), and MALDI-TOF MS to characterize the assembly [67]. In contrast to the results published by Tan et al. [60], these cyclic 1 : 1 complexes were obtained and the species of high molar masses observed in the MS

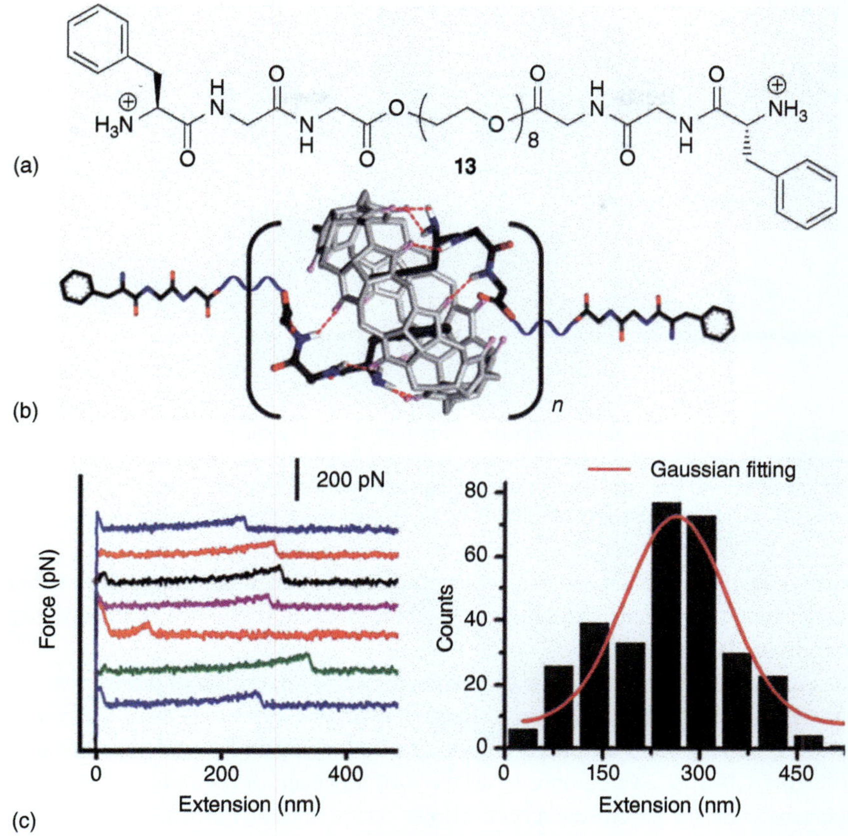

Figure 7.11 (a) Schematic representation of the homoditopic monomer **13**. (b) Schematic representation of the proposed 2 : 1 binding mode. (c) AFM–SMFS curves of the supramolecular polymer $\{\mathbf{10}@CB[8]\}_n$ and the Gaussian fitting thereof. Source: Tan et al. [60], © 2013 Royal Society of Chemistry.

measurements were ascribed to aggregate artifacts rather than to supramolecular polymers. This contradiction points out the necessity that a careful mass spectrometric analysis, ideally via multidimensional techniques, is required for a reliable structure elucidation (see Chapter 12). Ramaekers et al. proposed a ring-chain equilibrium strongly favoring the cyclic 1 : 1 complex with a binding constant (K_{intra}) of $9.0 \times 10^6\ M^{-1}$ and an effective concentration of 750 μM (Figure 7.12a) [67]. It was further shown that the highly stable cyclic species represented an efficient inhibitor for the complexation of proteins [68], even when carrying the same terminal PhGlyGlyGly sequence (Figure 7.12b).

Noteworthy, the concept of homotrimer formation has been extended from linear supramolecular polymers to more complicated topologies, such as hyperbranched polymers [69–71] and 2D polymer networks [72, 73]. To obtain such supramolecular assemblies, monomers bearing at least three binding sites have to be utilized for the binding in CB[8] hosts at appropriate stoichiometries (e.g. in a 2 : 3 ratio in the case of tritopic monomers). Thereby, the flexibility of the monomer had a profound

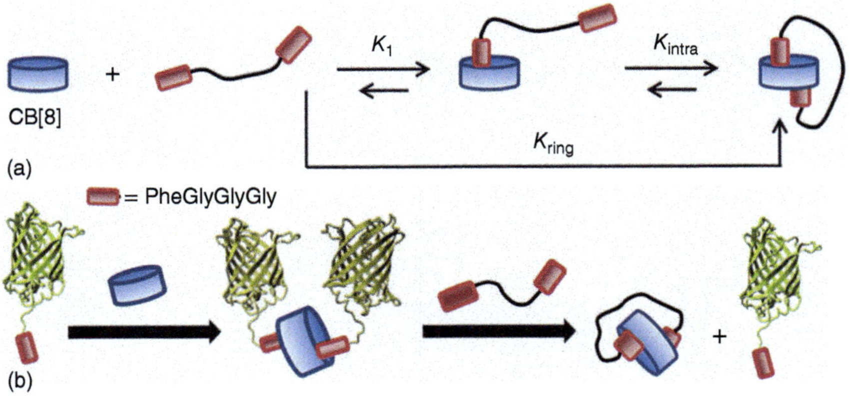

Figure 7.12 (a) Schematic representation of the formation of a cyclic 1 : 1 complex via a ring-chain equilibrium pathway. (b) Schematic representation for the inhabitation of protein complexation due to the formation of the highly stable cyclic complex. Source: Ramaekers et al. [67], © 2013 Royal Society of Chemistry.

impact on the overall macromolecular structure: in general, monomers with flexible linkers gave hyperbranched architectures; whereas, more rigid ones resulted in the formation of 2D networks.

The enhanced dimerization of radicals in a confined environment represents one further driving force that has been utilized to assemble supramolecular polymers. The Kim group reported on the formation of a 1 : 1 complex from methyl viologen (MV^{2+}) and CB[8] (K_a value of c. $10^5\ M^{-1}$) [74]; however, in the reduced state, the radical cation ($MV^{\cdot+}$) preferred a homotrimer complex of 2 : 1 stoichiometry (overall binding strength of $10^7\ M^{-1}$). The high stability of this assembly was explained by strong ion–dipole interactions between the host and the two accommodated radical cations (the cavity had the proper size to efficiently house two such radical $MV^{\cdot+}$ species). The reversibility of the process was revealed by the reoccurrence of the 1 : 1 complex after re-oxidation of the $MV^{\cdot+}$. This initial concept was taken up in 2004 by the Kaifer group, who reversibly dimerized the Newkome-type dendrons with MV^{2+}-moieties at the focal points with CB[8] [75]. More recently, the same type of interaction was employed to assemble a 2D supramolecular framework [72]: Zhang et al. utilized shape-persistent, three-armed MV^{2+} monomers that could be reduced in aqueous solution by sodium dithionite, thereby inducing the self-assembly process. The less bulky triply reduced monomer strongly stacked in a face-to-face-type manner and gave discrete complexes of 2 : 3 stoichiometries in the presence of CB[8] (Figure 7.13a). The monomer with bulky substituents, however, could not undergo any face-to-face stacking in its reduced state. Instead, an extended single-layer 2D supramolecular structure was established (the strong binding of the $MV^{\cdot+}$ dimers within the CB[8] units contributed to the stabilization of this material; Figure 7.13b). The supramolecular network was studied by UV/vis absorption and electron paramagnetic resonance (EPR) spectroscopy, DLS, as well as small-angle X-ray diffraction (XRD) measurements (in solution as well as the solid state). The single-layered nature of the assembly was further verified by AFM imaging; these

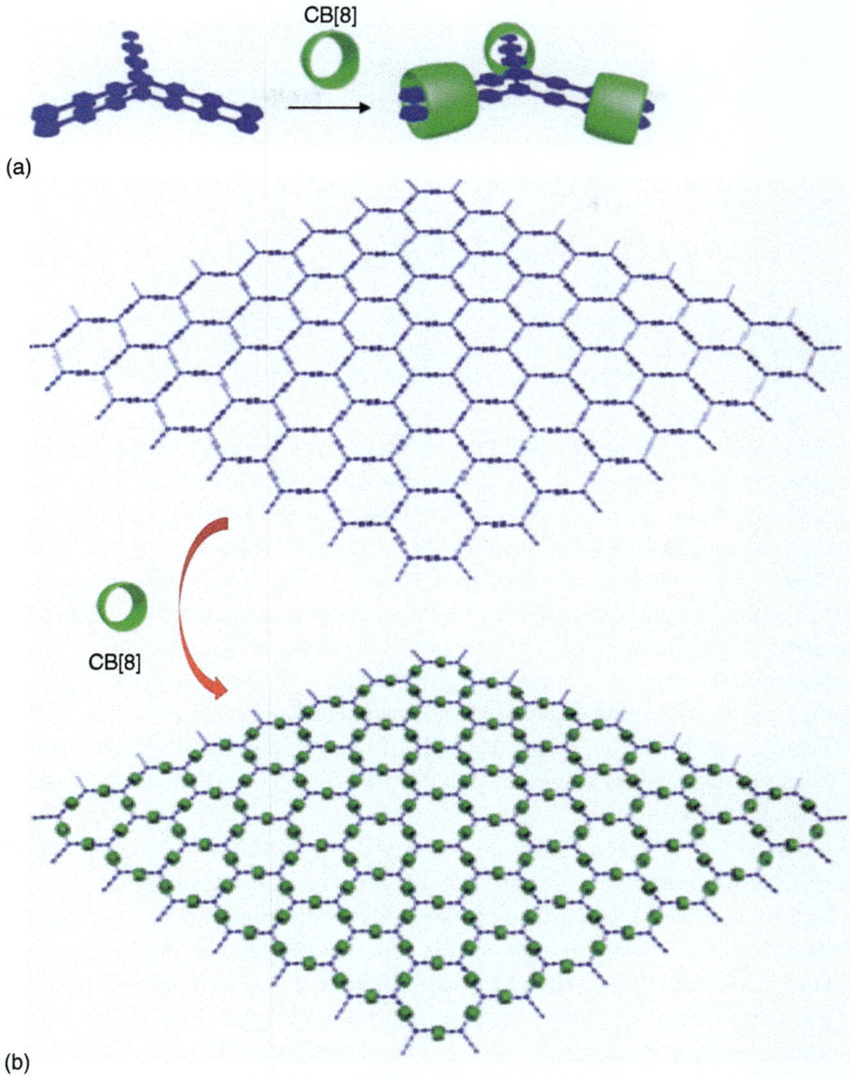

Figure 7.13 (a) Schematic representation of the formation of a discrete 2 : 3 complex and (b) a 2D supramolecular organic framework due to radical dimerization stabilized by CB[8]. Source: Zhang et al. [72], © 2014 Royal Society of Chemistry.

experiments gave a good agreement of the measured average thickness with the theoretical diameter of the CB[8] (1.75 nm). Related to this, monomers also featuring three 4,4′-bipyridin-1-ium units self-assembled in their reduced state with CB[8] into supramolecular polymer networks [73].

Recently, Xu and coworkers reported on a light-driven dissipative supramolecular polymerization in which the polymerization process proceeded in a far-from-equilibrium state [76]. Examples for such materials, which are driven by chemical reactions, can widely be found in Nature; typically, adenosine or

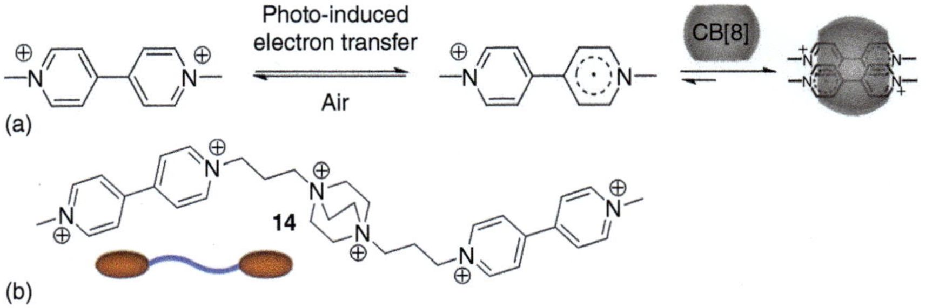

Figure 7.14 (a) Schematic representation of the self-assembly of the $MV^{\cdot+}$ radical cation and CB[8] into a 2 : 1 complex. (b) Schematic representation of the hexacationic ditopic guest **14**. Source: Redrawn from ref. Yin et al. [76], © 2013 Chinese Chemical Society.

guanosine triphosphate (ATP and GTP, respectively) represents the chemical "fuel" for such systems. Inspired by this, research on artificial far-from-equilibrium assemblies has emerged quite recently [77]. As pointed out before, MV^{2+} gives 1 : 1 complexes with CB[8]; whereas, the $MV^{\cdot+}$ radical cation prefers homotrimer complexes of 2 : 1 stoichiometry (Figure 7.14a). Accordingly, mixing the ditopic guest **14** with CB[8] did not generate a supramolecular polymer (Figure 7.14b) [76]. However, photoreduction of the MV^{2+} sites enabled the supramolecular polymerization due to the formation of the aforementioned 2 : 1 complexes. Upon exposure to air, the MV^{2+} units were regenerated and, instantaneously, depolymerization into the molecular components occurred. The light-induced polymerization (as well as the oxidative depolymerization) was monitored by DOSY and UV/vis absorption spectroscopy as well as AFM imaging.

Besides the former examples of radical-cation dimerization, other types of radical cations have also been found to stabilize dimers within the cavity of CB[8] (i.e. forming 1 : 2 homotrimers). In particular, the highly reactive tetrathiafulvalene (TTF) radical cation is worth mentioning in this context [78]. Though the dimerization of $TTF^{\cdot+}$ has already been applied in the formation of various polymeric systems [79, 80] or mechanically interlocked architectures [81–83], examples for the assembly of supramolecular polymers, based on the interaction of $TTF^{\cdot+}$ and CB[8], are not yet in hand.

A similar binding behavior as for CB[8] can also be observed for CB[10], and consequently, attempts have been made to utilize this particular host in aqueous supramolecular polymerizations [84, 85]. For example, Nally et al. employed a bis-CB[10] derivative in combination with bis-adamantyl guests to assemble linear polymers (Figure 7.15). As concluded from extensive NMR spectroscopic measurements, 1 : 1 or 2 : 2 complexes were formed preferentially; whereas, supramolecular polymers could not (yet) be identified.

Recently, Wang et al. introduced CB[n]-analogue materials in which cyclopentano-moieties were placed on the equator of the cucurbituril-like macrocycles. The synthesis of these host compounds, named $CyP_nTD[n]$ (n = 4 or 5), involved the condensation of the glycoluril-like cyclopentanopropanediurea with paraformaldehyde (Figure 7.16); for the $CyP_5TD[5]$ derivative, a good solubility in organic

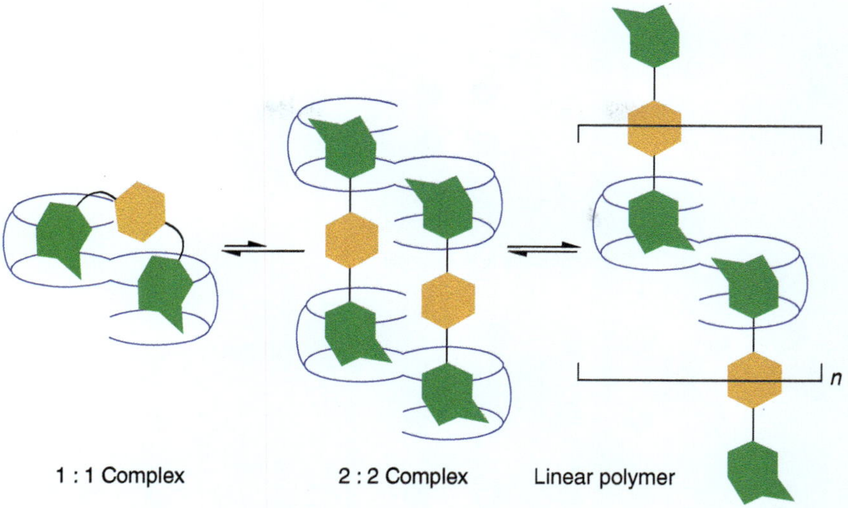

Figure 7.15 Schematic representation of the proposed equilibrium for the self-assembly of a bis-CB[10] derivative with a bis-adamantyl guest into discrete complexes and linear supramolecular polymers. Source: Nally and Isaacs [85], © 2009 Elsevier.

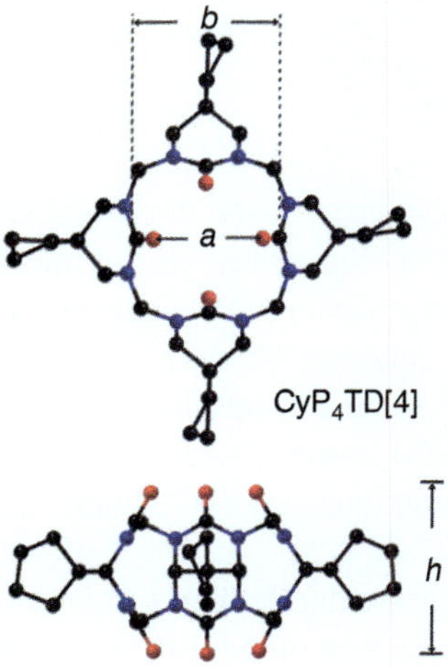

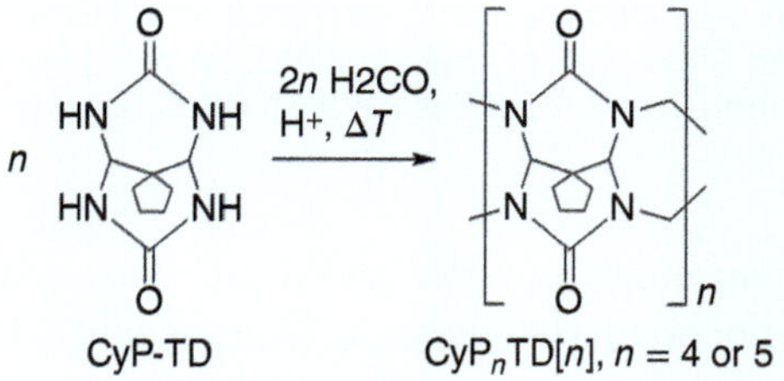

		CyP$_4$TD[4]	CyP$_4$TD[5]
Portal diameter (Å)	a	1.2	2.2
Cavity width (Å)	b	3.8	4.8
Height (Å)	h	8.6	8.2

Figure 7.16 Schematic representation of the synthesis of CyP$_n$TD[n] macrocycles. The solid-state structure of CyP$_4$TD[4], as determined by single crystal X-ray analysis, as well as the cavity dimensions of the CyP$_n$TD[n] derivatives are also shown. Source: Wu et al. [86], © 2017 Royal Society of Chemistry.

solvents, compared to CB[n]s, was reported [86]. For the smaller representative of the CyP$_n$TD[n] family ($n = 4$), a high affinity toward Ag^+ ions was found [87]; whereas, the larger CyP$_5$TD[5] macrocycles exhibited an efficient binding of two primary ammonium cations to the exterior of the cavity [88]. Taking advantage of this property, CyP$_5$TD[5] was combined with a tetraphenylethylene (TPE) dye equipped with four appended primary ammonium arms. The self-assembly of the

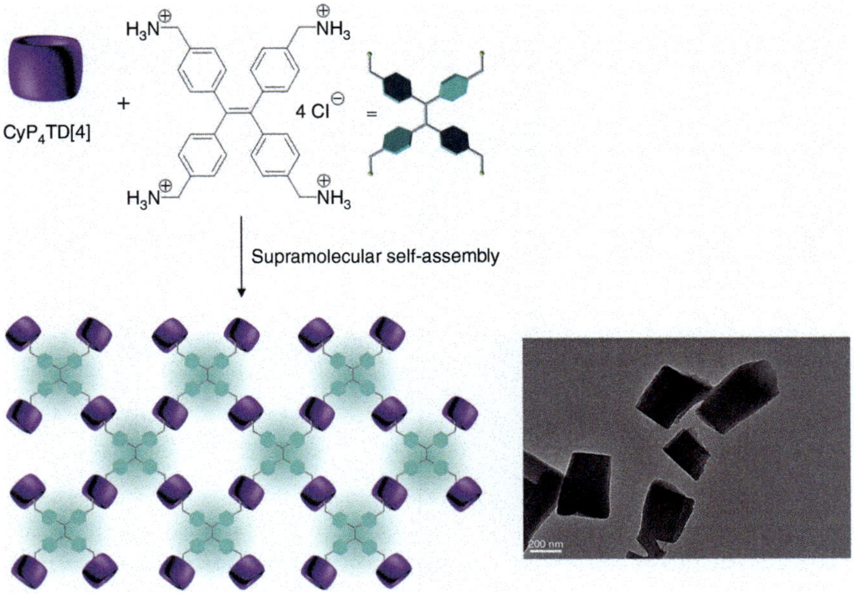

Figure 7.17 Schematic representation of the formation of a supramolecular polymer network with AIE behavior. A representative TEM image of the polymer particles is also depicted. Source: Wu et al. [88], © 2018 Royal Society of Chemistry. Figure reproduced with kind permission; © 2018 The Royal Chemical Society.

two components in water gave a supramolecular polymer network that featured aggregation-induced emission (AIE; Figure 7.17). TPE represents one of most prominent examples for dyes that show an increase of emission upon aggregation [89]. Due to the rigid environment offered by the polymer network, the TPE molecules became planarized and the emission was enhanced significantly by a factor of c. 60. NMR spectroscopy, ITC, as well as TEM imaging were applied to investigate this supramolecular polymer; the latter technique revealed square-like particles with a length of 200–400 nm (Figure 7.17).

The TPE-type dyes **15** and **16** assembled into supramolecular networks upon the addition of CB[8] [90]. Depending on the molecular structure of the dye, planar or curved supramolecular sheets were obtained; these layered assemblies aggregated further to give cubic or spherical nanostructures, respectively (Figure 7.18). Due to the AIE properties of the TPE dyes, the nanomaterials were fluorescent. The fluorescence could be switched off when a strong competitive guest for CB[8] was offered (e.g. 3,5-dimethyl-1-adamantylamine hydrochloride).

A fundamentally different strategy to assemble a 2D supramolecular network was followed by Zhou et al. [91]. One CB[7] and two CB[6] units were threaded onto a linear guest, thus forming a "hetero-wheel" pseudorotaxane. This assembly, in turn, gave a 2D pseudorotaxane network in which water molecules and H-bonding interactions contributed to the ordered structure. The proposed binding was verified, *inter alia*, by XRD analysis of solid, crystalline samples.

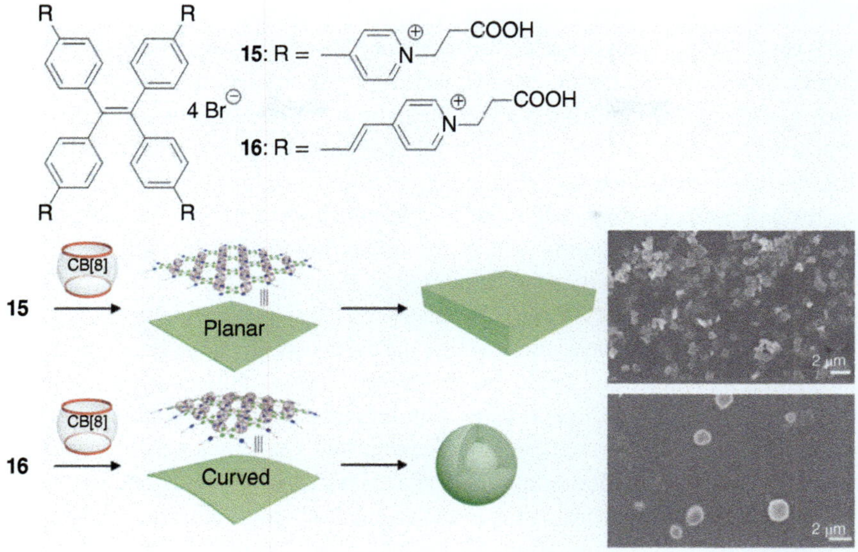

Figure 7.18 Schematic representation of the two-step self-assembly of the TPE dyes **15** and **16** into cubic or spherical nanostructures, respectively. Representative TEM images visualizing these nanostructures are also depicted. Source: Li et al. [90]. Figure reproduced with kind permission; © 2018 Wuiley-VCH.

7.3.2 Supramolecular Polymers Incorporating (Bio)macromolecular Building Blocks

Besides the interaction of CB[8] with small-molecule organic guests, eventually leading to supramolecular polymers, also the tethering of molecules (e.g. small organic molecules or (bio)polymers) to polymers by CB[8]-based host–guest interactions is in the focus of the current research. Binding may occur either at the polymers termini or in their side chains, and thus, a wide range of supramacromolecular architectures can be derived [16, 40], such as chain-extended polymers, block copolymers, polymer networks, hydrogels, biopolymer conjugates, etc.

The chain extension of polymers via complexation of their end-group functionalities by CB[8] was reported by Ji et al. [92]. For this purpose, reversible addition–fragmentation chain-transfer (RAFT) polymerization was applied to prepare poly(*N*,*N*-dimethylacrylamide) (PDMA) and poly(*N*-isopropylacrylamide) (PNIPAM) of low molar masses featuring naphthyl moieties at both termini. These polymers gave chain-extended supramolecular polymers when combined with an alkyl-bridged bis-MV^{2+} derivative and CB[8] due to the formation of host-stabilized CT complexes. Viscometry measurements were performed to prove the existence of assemblies of high molar masses. Due to the supramolecular nature of these assemblies, the PNIPAM-based polymer exhibited an LCST behavior (LCST: lower critical solution temperature) that was strongly dependent on the concentration (i.e. low LCST values were obtained at high concentrations); in contrast, "covalent" PNIPAM shows a less pronounced concentration dependence of the LCST [93].

The Kaifer group reported on the reversible, non-covalent binding of two different, centrally functionalized dendrimers, one bearing an electron-poor MV^{2+} group and the other having an electron-rich 1,4-dialkoxybenzene unit [94]. In the presence of CB[8], the dendrimers assembled into a 2 : 1 heterotrimer, forming a CT complex within the host's cavity (the efficiency of complexation could be correlated to the size of the respective dendrimers). Upon one-electron reduction of the MV^{2+} unit, the heterodimers dissociated and 2 : 1 homotrimers were obtained, based on the favored $MV^{\cdot+}$ dimerization within the CB[8] host's cavity (*vide supra*). The reversibility of this process was shown by the re-formation of the 2 : 1 heterotrimers upon re-oxidation of the $MV^{\cdot+}$ radical cations.

Supramolecular block copolymers, where the different constituent blocks are connected via (reversible) non-covalent bonds, can be obtained if the two major requirements are fulfilled [95]: (i) directionality of the employed supramolecular interaction and (ii) a sufficient kinetic stability thereof in solution. This concept has successfully been applied to assemble these supramolecular materials based on H-bonding interactions, metal-to-ligand coordination, and various host–guest inclusion complexes. With respect to CB[8] complexation, the formation of CT complexes, which are stabilized in these cavities, i.e. the assembly into 2 : 1 heterotrimers, represents a facile option. The Scherman group investigated the CB[8]-assisted hetero-dimerization of MV^{2+}-terminated polymers with Np-terminated ones (Np: naphthalene) [96]. The formation of linear AA-type homopolymers as well as AB-type diblock copolymers, as depicted in Figure 7.19, was shown by NMR spectroscopy and DLS measurements. The supramolecular amphiphilic diblock copolymers with a water-soluble poly(ethylene glycol) (PEG) block and a hydrophobic polyisoprene (PI) or octadecane block were found to assemble into large vesicular structures due to hydrophobic interactions. In a related study, a supramolecular diblock copolymer, composed of a linear hyperbranched polyphosphate and a poly(D,L-lactide) (PLA) block, complexed via their terminal MV^{2+} and indole units within the CB[8] host, was investigated by the Ji group [97].

A five-component, ABA-type triblock copolymer was assembled from the bis-viologen derivative **17** and Np-terminated PEGs (**18a**, M_n values ranging from 1000 to 8000 g mol^{-1}) or a dibenzofuran-terminated poly(*N*-isopropyl acrylamide) (**18b**, PNIPAM, M_n = 19 000 g mol^{-1}) in the presence of CB[8] (Figure 7.20) [98]. As in the previous examples, CT interactions within the CB[8]'s cavity (i.e. 2 : 1 heterotrimers) facilitated the formation of the supramolecular triblock copolymer in water. Viscometry and DOSY measurements were applied to verify the formation of the supramolecular assembly (see also Chapter 12).

The concept of supramolecular block copolymer micelle formation was developed further by the Scherman and Ji groups aiming for drug-delivery applications [99, 100]. As a proof-of-concept, the delivery of doxorubicin (DOX) – one of the most common chemotherapeutics – from the self-assembled supramolecular micelles was studied in this context. In the contribution by Scherman and coworkers, the three-component mixture of CB[8], Np-functionalized poly(dimethylaminoethyl methacrylate) (PDMAEMA), and MV^{2+}-functionalized PNIPAM in water gave multiple stimuli-responsive micelles that served as nanocontainers for the

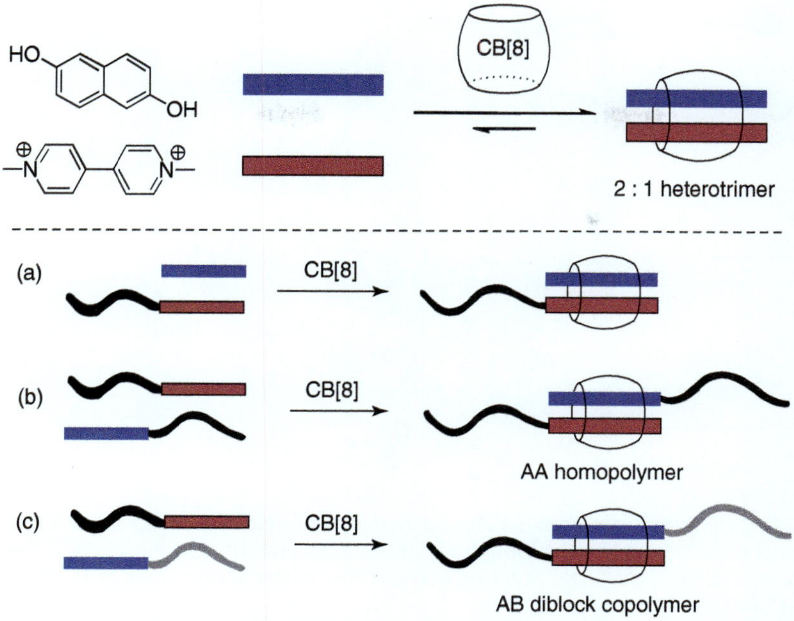

Figure 7.19 Schematic representation of the self-assembly of 2 : 1 heterotrimers due the formation of a CT complex within the CB[8] cavity; (a–c) three applications of this particular interaction using polymeric building blocks are also shown. Source: Rauwald and Scherman [96], © 2008 John Wiley and Sons.

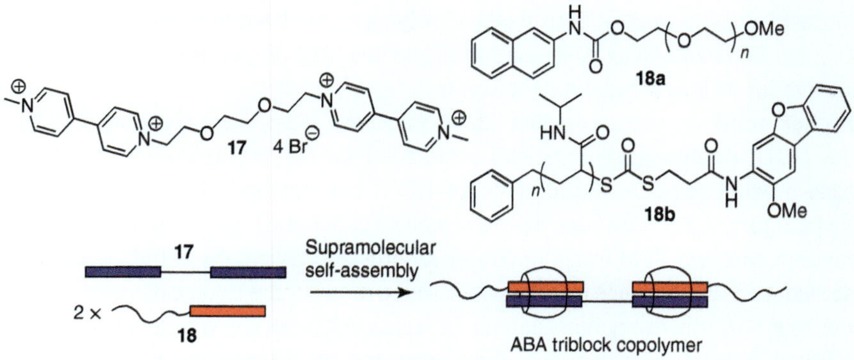

Figure 7.20 Schematic representation for the formation of a supramolecular ABA-type triblock copolymer. Source: Zayed et al. [98], © 2014 Royal Society of Chemistry.

encapsulation of DOX [100]. Changing the pH value or temperature as well as adding competitive guest molecules (i.e. adamantylamine) resulted in the disintegration of these nanocontainers, and thus, the drug was release. The triggered release was found to reduce significantly the viability of HeLa cells. A different type of supramolecular 2 : 1-heterotrimer linkage was reported by the Ji group [99] in which indole-terminated PLA was utilized as the hydrophobic block and MV^{2+}-functionalized PEG served as the hydrophilic part. As in the previous studies

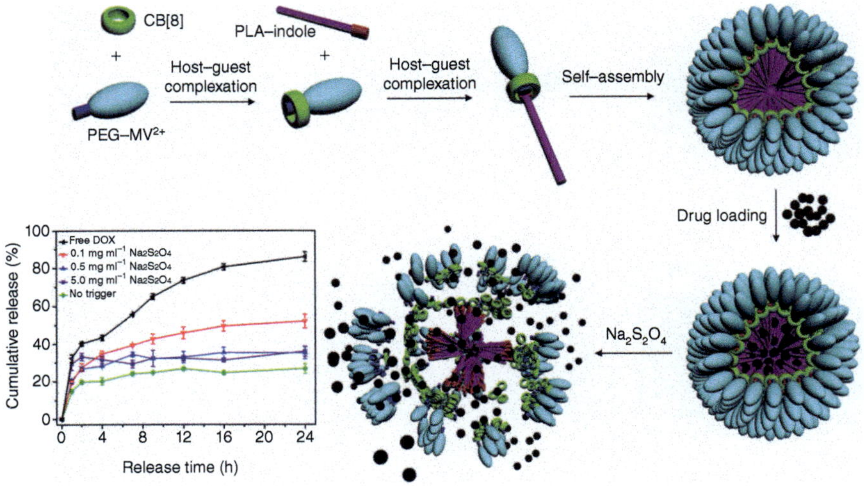

Figure 7.21 Schematic representation of the step-wise, self-assembly of supramolecular micelles that could be used for the controlled release of doxorubicin, as a model drug. The release, as a function of the added reducing agent, is also shown. Source: Zhao et al. [99], © 2014 Royal Society of Chemistry.

by the same authors (*vide supra*), the end-group functionalities established a CT complex within the CB[8] cavity. Similar to Scherman's work, DOX could be encapsulated in the hydrophobic micellar core and the drug release could efficiently be induced by reducing the MV^{2+} moieties by simply adding sodium dithionite (Figure 7.21). Noteworthy, both types of polymer have been approved by the FDA (FDA: US Food and Drug Administration) for medical use, and in particular, PLA is known for its biodegradability behavior.

Higher order structures, other than micelles, were also observed by Jiao et al. [101]: double-layered vesicles resulted from the self-assembly of a short, pyrene-bearing peptide strand with an MV^{2+}-bearing lipid in the presence of CB[8] (Figure 7.22). The size of the vesicles (c. 200 nm) was estimated by DLS measurements and TEM imaging. Moreover, the cellular uptake of the vesicles was observed, and fluorescence imaging was used to monitor the uptake of a competitive guest (e.g. 1,6-dihydroxynaphthalene or adamantylamine), which disassembled the initial heterotrimer complex, thus liberating the fluorescent pyrene–peptide conjugate. When using MV^{2+}-modified PNIPAM instead of a short lipid, vesicles were also formed. These were capable to even accommodate large biomolecules, such as the basic fibroblast growth factor (bFGF) [102]. The assemblies could be used as a protective shield for bFGF (i.e. against various freeze–thaw cycles) and to enable the delivery into cells. The latter one is of relevance when considering an application as an injectable therapy.

Finally, the rearrangement of micelles into vesicles – driven by host–guest interactions with CB[8] – was shown by Das and coworkers [103]. In aqueous medium, the combination of MV^{2+} amphiphiles and tetraalkylammonium salts gave mixed micelles that reorganized into vesicles in the presence of a CB[8] host.

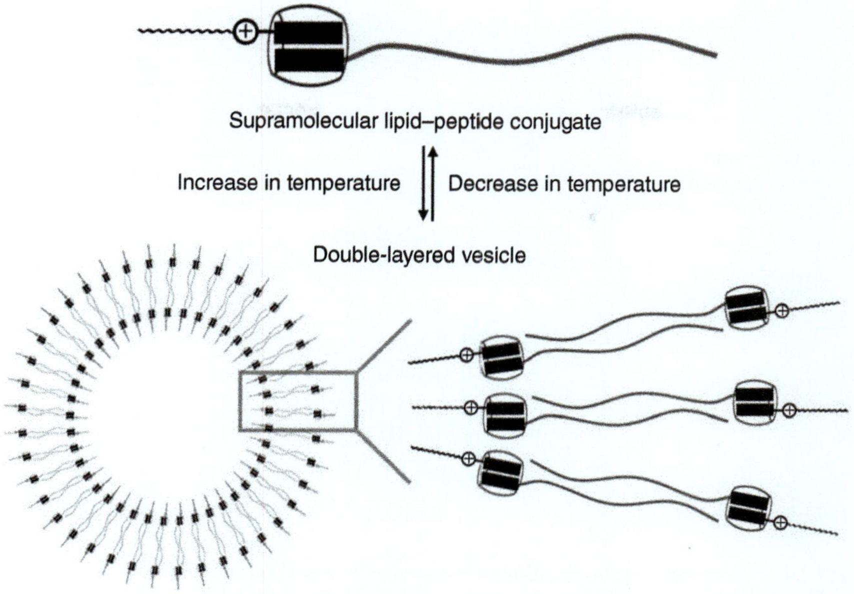

Figure 7.22 Schematic representation of the temperature-triggered formation of double-layered vesicles from supramolecular lipid–peptide conjugates. Source: Loh et al. [102], © 2014 Royal Society of Chemistry.

In a related study, mechanistic studies were conducted to establish a mechanism for the micelle-to-vesicle transformation [104]. For this purpose, a plethora of analytical techniques including NMR, MS, ITC, as well as microscopy imaging were utilized. The authors identified a stable intermediate nonspherical state, which was obtained by adding CB[8] to a solution of the mixed micelles. These aggregates, in turn, experienced a transition into vesicles when appropriate guest molecules, such as water-soluble naphthalene derivatives, were added (Figure 7.23).

7.4 Concluding Remarks

Though the chemistry of cucurbit[*n*]urils represents a rather young field of research – the first CB[*n*]s were first isolated and characterized just 30 years ago – a broad range of applications has been derived from this type of supramolecular host–guest complexation [16]: materials and medical chemistry, catalysis, as well as polymer science. Concerning the latter one, different types of polymer architectures, including block copolymers, chain-extended polymers, and even biomacromolecule–polymer conjugates, have been reported. Thus, CB[*n*]-based host–guest complexation is a versatile tool in supramacromolecular chemistry being able to compete with the more “traditional” types of interactions (e.g. metal-to-ligand complexation or H-bonding interactions) regarding directionality and (kinetic) stability of supramolecular system formed. However, the poor solubility of several CB[*n*] derivatives remains to be a limiting factor. This is expected to be

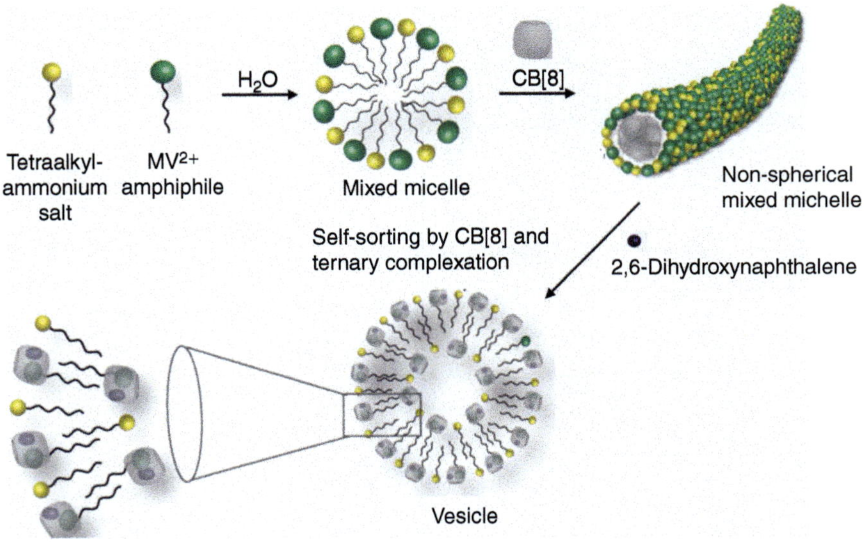

Figure 7.23 Schematic representation of the stepwise formation of vesicles from an initial mixed micellar system by CB[8]-triggered self-sorting. Source: Mondal et al. [104], © 2014 American Chemical Society.

overcome in the near future by utilizing functionalized host molecules of different cavity sizes. Thus, the employment of other types of supramolecular binding within CB[n]s will widen the scope of water-based, supramacromolecular chemistry.

References

1 Behrend, R., Meyer, E., and Rusche, F. (1905). *Justus Liebigs Ann. Chem.* 339: 1–37.

2 Freeman, W.A., Mock, W.L., and Shih, N.Y. (1981). *J. Am. Chem. Soc.* 103: 7367–7368.

3 Day, A., Arnold, A.P., Blanch, R.J., and Snushall, B. (2001). *J. Org. Chem.* 66: 8094–8100.

4 Kim, J.-H., Jung, I.-S., Kim, S.-Y. et al. (2000). *J. Am. Chem. Soc.* 122: 540–541.

5 Lee, J.W., Samal, S., Selvapalam, N. et al. (2003). *Acc. Chem. Res.* 36: 621–630.

6 Kim, K. (2002). *Chem. Commun.* 31: 96–107.

7 Geras'ko, O.A., Samsonenko, D.G., and Fedin, V.P. (2002). *Russ. Chem. Rev.* 71: 741–760.

8 Geras'ko, O.A., Sokolov, M.V., and Fedin, V.P. (2004). *Pure Appl. Chem.* 76: 1633–1646.

9 Kim, K., Selvapalam, N., and Oh, D. (2004). *J. Inclusion Phenom. Macrocyclic Chem.* 50: 31–36.

10 Krasia, T.C., Khodabakhsh, S., Tuncel, D., and Steinke, J.H.G. (2004). Cucurbituril: a versatile "bead"for polyrotaxane-synthesis. In: *Macromolecular*

Nanostructured Materials, vol. 78 (eds. N. Ueyama and A. Harada), 41–59. Berlin Heidelberg: Springer.

11 Lagona, J., Mukhopadhyay, P., Chakrabati, S., and Isaacs, L. (2005). *Angew. Chem. Int. Ed.* 44: 4844–4870.

12 Kim, K., Selvapalam, N., Ko, Y.-H. et al. (2007). *Chem. Soc. Rev.* 36: 267–279.

13 Huang, W.-H., Liu, S.-M., and Isaacs, L. (2008). Cucurbit[n]urils. In: *Modern Supramolecular Chemistry: Strategies for Macrocycle Synthesis* (eds. F. Diederich, P.J. Stang and R.R. Tykwinski), 113–142. Weinheim: Wiley-VCH.

14 Isaacs, L. (2009). *Chem. Commun.*: 619–629.

15 Masson, E., Ling, X.-X., Joseph, R. et al. (2012). *RSC Adv.* 2: 1213–1247.

16 Barrow, S.J., Kasera, S., Rowland, M.J. et al. (2015). *Chem. Rev.* 115: 12320–12406.

17 Moon, K. and Kaifer, A.E. (2004). *Org. Lett.* 6: 185–188.

18 Buschmann, H.-J., Wego, A., Zielesny, A., and Schollmeyer, E. (2006). *J. Inclusion Phenom. Macrocyclic Chem.* 54: 241–246.

19 Mock, W.L. and Shih, N.Y. (1986). *J. Org. Chem.* 51: 4440–4446.

20 Rekharsky, M.V., Mori, T., Yang, C. et al. (2007). *Proc. Natl. Acad. Sci. U.S.A.* 104: 20737–20742.

21 Moghaddam, S., Yang, C., Rekharsky, M.V. et al. (2011). *J. Am. Chem. Soc.* 133: 3570–3581.

22 Pazy, Y., Kulik, T., Bayer, E.A. et al. (2002). *J. Biol. Chem.* 277: 30892–30900.

23 Green, N.M. (1990). *Methods Enzymol.* 184: 51–67.

24 Rao, J.-H., Lahiri, J.-D., Weis, R.M., and Whitesides, G.M. (2000). *J. Am. Chem. Soc.* 122: 2696–2710.

25 Houk, K.N., Leach, A.G., Kim, S.P., and Zhang, X.-Y. (2003). *Angew. Chem. Int. Ed.* 42: 4872–4897.

26 Florea, M. and Nau, W.M. (2011). *Angew. Chem. Int. Ed.* 50: 9338–9342.

27 Yang, X., Liu, F., Zhao, Z. et al. (2018). *Chin. Chem. Lett.* 29: 1560–1566.

28 Yang, X., Zhao, Z., Zhang, X., and Liu, S. (2018). *Sci. China Chem.* 61: 787–791.

29 Kuang, S., Hu, Z., Zhang, H. et al. (2018). *Chem. Commun.* 54: 2169–2172.

30 Liu, S., Shukla, A.D., Gadde, S. et al. (2008). *Angew. Chem. Int. Ed.* 47: 2657–2660.

31 de Greef, T.F.A., Smulders, M.M.J., Wolffs, M. et al. (2009). *Chem. Rev.* 109: 5687–5754.

32 Winter, A., Hager, M.D., and Schubert, U.S. (2012). Supramolecular polymers. In: *Polymer Science: A Comprehensive Review*, vol. 5 (eds. H.-W. Schmidt and M. Ueda), 269–310. Amsterdam: Elsevier BV.

33 Yang, L., Tan, X., Wang, Z., and Zhang, X. (2015). *Chem. Rev.* 115: 7196–7239.

34 Liu, Y., Yang, H., Wang, Z., and Zhang, X. (2013). *Chem. Asian J.* 8: 1626–1632.

35 Dong, S., Zheng, B., Wang, F., and Huang, F. (2014). *Acc. Chem. Res.* 47: 1982–1994.

36 Hu, J. and Liu, S. (2014). *Acc. Chem. Res.* 47: 2084–2095.

37 Li, H. and Wu, L. (2014). *Soft Matter* 10: 9038–9053.

38 Yan, X., Wang, F., Zheng, B., and Huang, F. (2012). *Chem. Soc. Rev.* 41: 6042–6065.

39 Guo, D.-S. and Liu, Y. (2012). *Chem. Soc. Rev.* 41: 5907–5921.
40 Zou, H., Liu, J., Li, Y. et al. *Small* 14, Article ID: 1802234.
41 Wittmer, J.P., van der Schoot, P., Milchev, A., and Barrat, J.L. (2000). *J. Chem. Phys* 113: 6992–7005.
42 Vermonden, T., van der Gucht, J., de Waard, P. et al. (2003). *Macromolecules* 36: 7035–7044.
43 Kricheldorf, H.R. (2003). *Macromol. Symp.* 199: 15–22.
44 Binder, W.H. and Zirbs, R. (2007). *Adv. Polym. Sci.* 207: 1–78.
45 Brunsveld, L., Folmer, B.J.B., Meijer, E.W., and Sijbesma, R.P. (2001). *Chem. Rev.* 101: 4071–4098.
46 Kim, H.-J., Heo, J., Jeon, W.S. et al. (2001). *Angew. Chem. Int. Ed.* 40: 1526–1529.
47 Wang, R., Bardelang, D., Waite, M. et al. (2009). *Org. Biomol. Chem.* 7: 2435–2439.
48 Yang, C., Mori, T., Origane, Y. et al. (2008). *J. Am. Chem. Soc.* 130: 8574–8575.
49 Jiao, D., Biedermann, F., Tian, F., and Scherman, O.A. (2010). *J. Am. Chem. Soc.* 132: 15734–15743.
50 Han, Y., Tian, Y., Li, Z., and Wang, F. (2018). *Chem. Soc. Rev.* 47: 5165–5176.
51 Lee, J.W., Kim, K., Choi, S. et al. (2002). *Chem. Commun.*: 2692–2693.
52 Ko, Y.H., Kim, K., Kim, E., and Kim, K. (2007). *Supramol. Chem.* 19: 287–293.
53 Kim, K., Kim, D., Lee, J.W. et al. (2004). *Chem. Commun.*: 848–849.
54 Ko, Y.H., Kim, K., Kang, J.-K. et al. (2004). *J. Am. Chem. Soc.* 126: 1932–1933.
55 Liu, Y., Yu, Y., Gao, J. et al. (2010). *Angew. Chem. Int. Ed.* 49: 6576–6579.
56 Liu, Y., Huang, Z., Tan, X. et al. (2013). *Chem. Commun.* 49: 5766–5768.
57 del Barrio, J., Horton, P.N., Lairez, D. et al. (2013). *J. Am. Chem. Soc.* 135: 11760–11763.
58 Liu, Y., Liu, K., Wang, Z., and Zhang, X. (2011). *Chem. Eur. J.* 17: 9930–9935.
59 Liu, Y., Fang, R., Tan, X. et al. (2012). *Chem. Eur. J.* 18: 15650–15654.
60 Tan, X., Yang, L., Liu, Y. et al. (2013). *Polym. Chem.* 4: 5378–5381.
61 Yang, H., Liu, Y., Liu, K. et al. (2013). *Langmuir* 29: 12909–12914.
62 Del Barrio, J., Liu, J., Brady, R.A. et al. (2019). *J. Am. Chem. Soc.* 141: 14021–14025.
63 Xu, Y., Guo, M., Li, X. et al. (2011). *Chem. Commun.* 47: 8883–8885.
64 Crotty, S., Gerişlioğlu, S., Endres, K.J. et al. (2016). *Anal. Chim. Acta* 932: 1–21.
65 Guo, K., Guo, Z., Ludlow, J.M. et al. (2015). *Macromol. Rapid Commun.* 36: 1539–1552.
66 Heitmann, L.M., Taylor, A.B., Hart, P.J., and Urbach, A.R. (2006). *J. Am. Chem. Soc.* 128: 12574–12581.
67 Ramaekers, M., Wijnands, S.P.W., van Dongen, J.L.J. et al. (2015). *Chem. Commun.* 51: 3147–3150.
68 Nguyen, H.D., Dang, D.T., van Dongen, J.L.J., and Brunsveld, L. (2010). *Angew. Chem. Int. Ed.* 49: 895–898.
69 Fang, R., Liu, Y., Wang, Z., and Zhang, X. (2013). *Polym. Chem.* 4: 900–903.
70 Yang, H., Ma, Z., Wang, Z., and Zhang, X. (2014). *Polym. Chem.* 5: 1471–1476.
71 Liu, Y., Huang, Z., Liu, K. et al. (2014). *Polym. Chem.* 5: 53–56.

72 Zhang, L., Zhou, T.-Y., Tian, J. et al. (2014). *Polym. Chem.* 5: 4715–4721.

73 Zhang, K.-D., Tian, J., Hanifi, D. et al. (2013). *J. Am. Chem. Soc.* 135: 17913–17918.

74 Jeon, W.S., Kim, H.-J., Lee, C., and Kim, K. (2002). *Chem. Commun.*: 1828–1829.

75 Moon, K., Grindstaff, J., Sobransingh, D., and Kaifer, A.E. (2004). *Angew. Chem. Int. Ed.* 43: 5496–5499.

76 Yin, Z., Song, G., Jiao, Y. et al. (2019). *CCS Chem.* 1: 335–342.

77 Boekhoven, J., Hendriksen, W.E., Koper, G.J.M. et al. (2015). *Science* 349: 1075–1079.

78 Ziganshina, A.Y., Ko, Y.H., Jeon, W.S., and Kim, K. (2004). *Chem. Commun.*: 806–807.

79 Tian, J., Ding, Y.-D., Zhou, T.-Y. et al. (2014). *Chem. Eur. J.* 20: 575–584.

80 Chen, L., Zhang, S.-C., Wang, H. et al. (2014). *Tetrahedron* 70: 4778–4783.

81 Spanggaard, H., Prehn, J., Nielsen, M.B. et al. (2000). *J. Am. Chem. Soc.* 122: 9486–9494.

82 Coskun, A., Spruell, J.M., Barin, G. et al. (2011). *J. Am. Chem. Soc.* 133: 4538–4547.

83 Frasconi, M., Kikuchi, T., Cao, D. et al. (2014). *J. Am. Chem. Soc.* 136: 11011–11026.

84 Liu, S., Zavalij, P.Y., and Isaacs, L. (2005). *J. Am. Chem. Soc.* 127: 16798–16799.

85 Nally, R. and Isaacs, L. (2009). *Tetrahedron* 65: 7249–7258.

86 Wu, Y., Xu, L., Shen, Y. et al. (2017). *New J. Chem.* 41: 6991–6994.

87 Wu, Y., Xu, L., Shen, Y. et al. (2017). *Chem. Commun.* 53: 4070–4072.

88 Wu, Y., Hua, H., and Wang, Q. (2018). *New J. Chem.* 42: 8320–8324.

89 Mei, J., Leung, N.L.C., Kwok, R.T.K. et al. (2015). *Chem. Rev.* 115: 11718–11940.

90 Li, Y., Dong, Y., Miao, X. et al. (2018). *Angew. Chem. Int. Ed.* 57: 729–733.

91 Zhou, Q., Li, Y., Han, Z. et al. (2017). *Supramol. Chem.* 29: 176–182.

92 Ji, Z., Liu, J., Chen, G., and Jiang, M. (2014). *Polym. Chem.* 5: 2709–2714.

93 Lutz, J.-F., Akdemir, Ö., and Hoth, A. (2006). *J. Am. Chem. Soc.* 128: 13046–13047.

94 Wang, W. and Kaifer, A.E. (2006). *Angew. Chem. Int. Ed.* 45: 7042–7046.

95 Yang, S.K., Ambade, A.V., and Weck, M. (2011). *Chem. Soc. Rev.* 40: 129–137.

96 Rauwald, U. and Scherman, O.A. (2008). *Angew. Chem. Int. Ed.* 47: 3950–3953.

97 Chen, C.-J., Li, D.-D., Wang, H.-B. et al. (2013). *Polym. Chem.* 4: 242–245.

98 Zayed, J.M., Biedermann, F., Rauwald, U., and Scherman, O.A. (2010). *Polym. Chem.* 1: 1434–1436.

99 Zhao, J., Chen, C., Li, D. et al. (2014). *Polym. Chem.* 5: 1843–1847.

100 Loh, X.J., del Barrio, J., Toh, P.P.C. et al. (2012). *Biomacromolecules* 13: 84–91.

101 Jiao, D., Geng, J., Loh, X.J. et al. (2012). *Angew. Chem. Int. Ed.* 51: 9633–9637.

102 Loh, X.J., del Barrio, J., Lee, T.-C., and Scherman, O.A. (2014). *Chem. Commun.* 50: 3033–3035.

103 Mondal, J.H., Ahmed, S., and Das, D. (2014). *Langmuir* 30: 8290–8299.

104 Mondal, J.H., Ghosh, T., Ahmed, S., and Das, D. (2014). *Langmuir* 30: 11528–11534.

8

Supramolecular Polymers, Based on the Host–Guest Chemistry of Calixarenes

8.1 General Aspects

Host–guest complexation represents, along with ionic, π–π stacking, and hydrogen-bonding (H-bonding) interactions as well as metal-to-ligand coordination, one of the most important types of non-covalent interaction in the field of supra(macro)molecular chemistry [1]. In general, macrocyclic species can serve as hosts for the complexation (or inclusion) of appropriate molecular guests. A variety of such systems has successfully been utilized in the field of supramolecular chemistry and, in particular, for the preparation of supramolecular polymers via self-assembly strategies: crown ethers [2], cyclodextrins [3, 4], cucurbiturils [5], and more recently, pillarenes [6]. The cyclic oligomers, derived from the hydroxyalkylation of a phenolic arene with an aldehyde, today also play a central role in supramolecular chemistry [7]: calix[*n*]arenes, resorcin[*n*]arenes and, much less common, pyrogallo[*n*]arenes (Figure 8.1).

Basically, the chemistry of calixarenes dates back to the 1870s, when von Baeyer studied the reaction of various aldehydes, including formaldehyde, with phenols under strong acidic conditions. The insoluble products that could not be characterized at that time represented typical products of a phenol-aldehyde polycondensation – today referred to as phenolic resins. Baekeland recognized that these tars could be cured into a brittle material, which he named "Bakelite" after himself. This polymer was the first commercial synthetic plastic with a patent being issued in 1909 [8]. Later, Zinke and Ziegler obtained crystalline compounds from the reaction of 4-alkylphenols and formaldehyde by utilizing freezing-point depression measurements in which their molar mass could be determined and revealed the presence of cyclic tetramers [9, 10]. In the same time frame, Niederl and Vogel isolated similar cyclic tetramers by an acid-catalyzed condensation of resorcinol with diverse aldehydes [11]. Cornforth was, however, the first to spectroscopically confirm the formation of cyclic tetramers (with some other cyclomers as by-products) [12]; the first X-ray single crystal structure of a calix[4]arene was published in 1988 [13]. Gutsche coined the name "calixarene" for the cyclic tetramers due to their calix-like shape that has been formed by the reaction of arenes ("calix" is the Greek name for a chalice) [14]. Accordingly, the name "resorcinarenes" was established for the cyclic species derived from resorcinol.

Supramolecular Polymers and Assemblies: From Synthesis to Properties and Applications, First Edition.
Ulrich S. Schubert, George R. Newkome, and Andreas Winter.

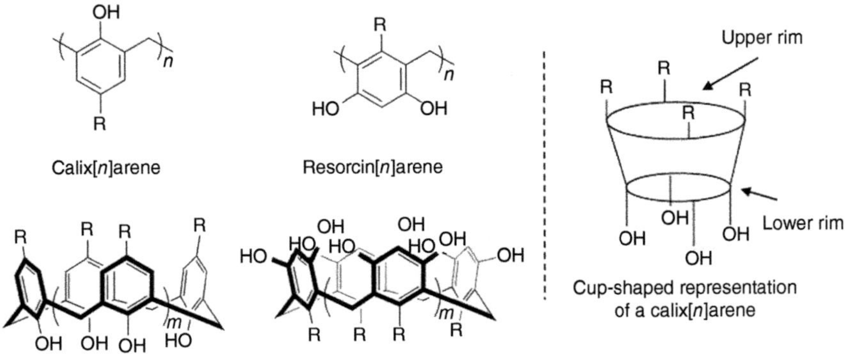

Figure 8.1 Schematic representation of the macrocyclic hosts calix[*n*]arene and resorcin[*n*]arene (R denotes any substituent; whereas, *n* and $m = n - 3$ refer to the number of phenyl moieties within the macrocycle).

Calixarenes have been employed in polymer science in a highly diverse fashion: catalysts in metal-based polymerizations [15, 16], synthesis of linear, star-shaped, or graft polymers with calixarenes within the backbone, as the core or in the pending side chain, respectively [16, 17]. Besides these usages, calixarenes have been employed in the field of supramolecular polymers, i.e. using the calixarenes' host–guest chemistry to form such assemblies. In this respect, three different types of supramolecular architectures can be distinguished [18, 19]:

1) Calixarene-based supramolecular polymeric capsules, the so-called "supercaps," which rely on multiple H-bonding in concert with guest complexation;
2) Supramolecular polymers in which the cavities of the calixarene scaffolds remain empty; and, finally,
3) Supramolecular polymers where the interaction between calixarene-type hosts and molecular guests represents the driving force in linking (macro)molecular units in supramolecular fashion.

These three basic types of calix[*n*]arene-based polymers will be highlighted in the following. A broad range of coordination polymers based, on the interaction of calix[*n*]arenes with transition metal ions is known from the literature [20]; however, these solid-state materials of 1D–3D structures, which often exhibit rich photophysical and/or magnetochemical properties, are not included herein.

8.2 Calixarene-Based Supramolecular Polycaps

Rebek and coworkers pioneered the topic of "polycaps" and were the first to report on a supramolecular polymer, which resulted from the dimerization of calixarene moieties [21]. Structurally, polycaps resemble "beads-on-a-string" and their formation depends, at least to some extent, on the nature of the guests based on their size, shape, and chemical surface. In general, polycaps are highly stimuli-responsive materials that can be addressed, e.g. thermally or chemically. Guo and Liu searched

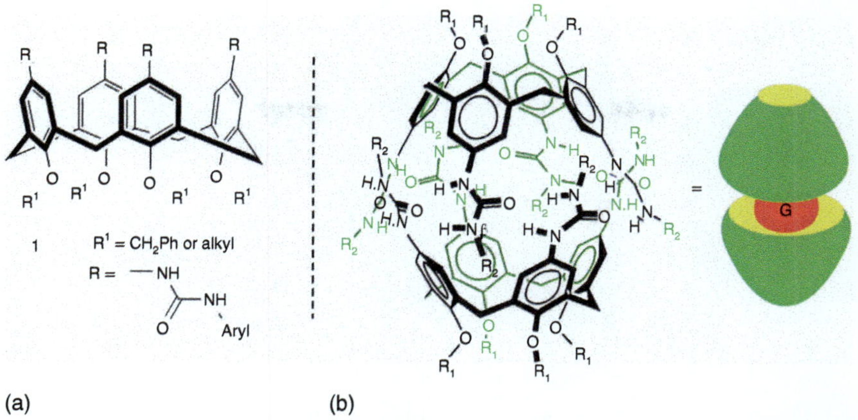

Figure 8.2 Schematic representation of the tetraurea-functionalized calix[4]arene (a) and of its dimer with an encapsulated small guest molecule (b) [21]. Source: Redrawn from Dalcanale and Pinalli [19], © 2015 Springer Nature.

their fascinating properties to realize that by fine tuning the chemical structure of these supramolecular materials [18]: informational encoding, viscoelasticity, liquid crystallinity, mechanical integrity, or gelation were possible. While supramolecular polymers, based on the interaction of a macrocyclic host and an organic guest, can be derived from the entire range of macrocyclic hosts (also see Chapters 6, 7, 9, and 10), supercaps represent an unique supramacromolecular structure that can be exclusively obtained from the calix[*n*]arenes.

Independently, the Rebek [22] and Böhmer groups [23] reported on the dimerization of calix[4]arenes (**1**), functionalized with four urea moieties on their upper rims, in apolar solvents, thus affording a capsular host with a rigid cavity of c. 200 Å^3 in which solvent (e.g. benzene or $CHCl_3$) or other appropriate guest molecules can be encapsulated (Figure 8.2). Due to the efficient multiple H-bonding interaction between the urea units of the calixarene half capsules in concert with a templating effect of the guest, high association constants (K_a) ranging from 10^6 to 10^8 M^{-1} were observed. The covalent linkage of two such calixarenes via their lower rims gave the homoditopic bis-calixarenes **2** that self-assembled in $CHCl_3$ to generate polycaps, according to Figure 8.3. It was further shown that depolymerization of the assembly can be induced by solvents that can compete for the H-bonding (e.g. DMSO). At a certain concentration of added DMSO, the "melting point" was reached, where a 1 : 1 ratio of **2** and polycaps was present in solution. In general, dimers **2** were favored at high DMSO concentration; whereas, polycap formation was dominate at low DMSO concentrations. Also the addition of a simple dimeric capsule, as shown in Figure 8.2, was found to induce disassembly of the supramolecular material, eventually leading to dumbbell-shaped complexes possessing a 2 : 1 stoichiometry (Figure 8.4).

Castellano and Rebek prepared various structurally related homoditopic and heteroditopic bis-calixarenes as well as homotritopic tris-calixarenes, as monomers for supramolecular polycap formation [24]. Remarkably, heterodimerization of two calixarene units occurred when two different types of urea were present (e.g. aryl

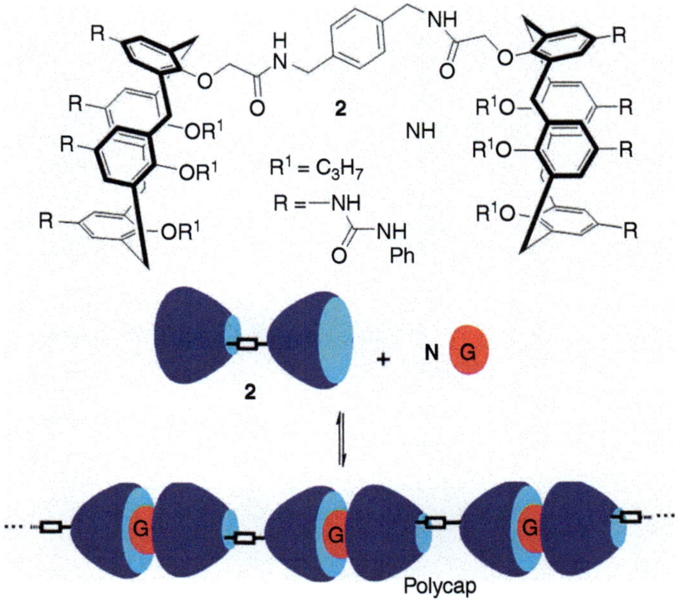

Figure 8.3 Schematic representation of the self-assembly of calix[4]arene dimers **2** into polycaps [21]. Source: Redrawn from Dalcanale and Pinalli [19], © 2015 Springer Nature.

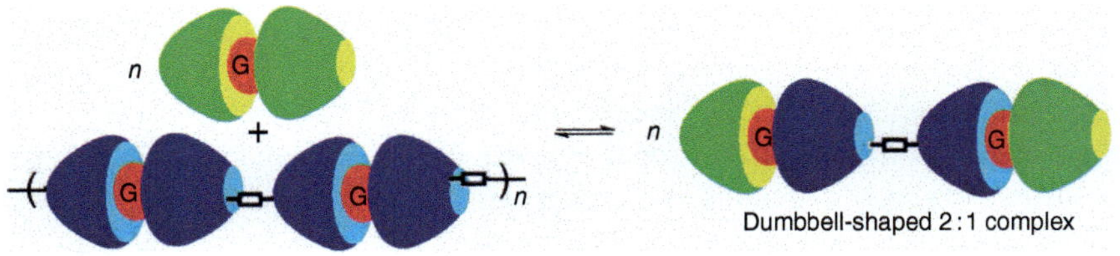

Figure 8.4 Schematic representation of the depolymerization of a polycap in the presence of a dimeric capsule into a dumbbell-shaped 2 : 1 complex [21]. Source: Redrawn from Dalcanale and Pinalli [19]. © 2015 Springer Nature.

and sulfonyl). It was concluded that this selectivity of binding translated memory to the molecular level, which could be used to design smart polymer-based memory devices. Accordingly, the self-assembly of two homoditopic bis-calixarenes, one equipped with aryl urea moieties and the other with sulfonyl urea ones, will afford polycaps with a strictly alternating order of the subunits (Figure 8.5b). Consequently, the supramolecular polymerization of either will generate the corresponding polycaps (Figure 8.5a). Moreover, self-complementary heteroditopic derivatives allow for the assembly of supercaps in which the capsules are ordered in a “head-to-tail” fashion (Figure 8.5c). Furthermore, the self-assembly of a homotritopic tris-calixarene with $CHCl_3$ into a supercap-network was investigated; however, the resultant supramolecular gel was obtained which could not be further characterized.

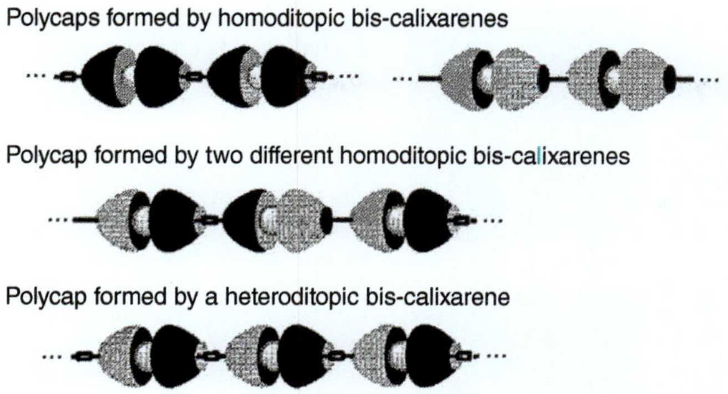

Figure 8.5 Schematic representation of different polycaps derived from homo- or heteroditopic bis-calixarenes (black calixarenes are equipped with aryl ureas, gray calixarenes carry sulfonyl ureas). Source: Castellano et al. [24], © 1998 American Chemical Society.

The Rebek group also reported on polycaps with liquid crystalline (LC) behavior [25]. In contrast to the polycaps obtained from the self-assembly of **2**, the derivative with *n*-decyloxy moieties on the lower ring gave micrometer-scaled fibrous assemblies with LC properties. In detail, the material featured the typical Schlieren-type textures of lyotropic, nematic LC phases (Figure 8.6a). Thus, the long alkyl chains were defined as crucial for the orientation of the polycap chains in a liquid-like environment (the short residues, as in **2**, facilitated instead a [partial] crystallization of the polycaps); furthermore, the rigid linkage of the calixarene units was identified to be as important: an *n*-dodecyl-functionalized bis-calixarenes with a flexible spacer also failed to give the LC-type phases. By applying external forces, the LC formulations could be ordered: if the textures were sheared, oriented fibrous assemblies were obtained; however, when pulled with a needle from the LC sample, fibers were formed (Figure 8.6b). The viscosity of the LC polycap in *o*-dichlorobenzene slowly increased with increasing concentration until a critical value of 0.6 wt% was reached; however, increasing the concentration further led to a sharp increase in viscosity [26]. Moving from the diluted to the semidiluted regime enabled the polycap chains to overlap and entangle giving rise to a viscoelastic behavior. The mechanical properties of the polycap were investigated by rheology measurements: the normal stress in solution was measured and indicated a high stability of the supramolecular assembly, which was high enough to resist forcing disruptive forces. Gelation of the linear assembly was compared with that derived from a related tetrakiscalixarene derivative. Due to the formation of a highly cross-linked network, the latter required a much lower concentration to allow gelation to occur.

The Rudkevich group published a multiple-responsive supercap in which both the capsules and the linker contributed to this behavior. In particular, addressing the spacer allowed the authors to utilize "on–off" properties without disassembling the supramolecular polymer. The urea-functionalized bis-calixarene **3**, for example, where the binding sites were linked by a short dipeptide unit gave the corresponding polycaps in benzene [27]. Upon protonation of the NH_2 groups of the linkers, the

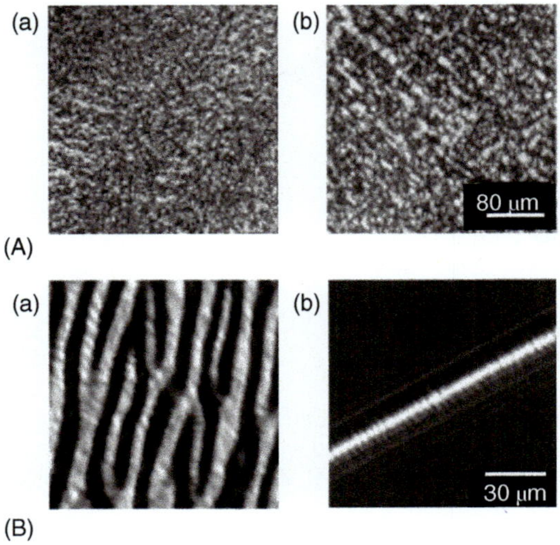

Figure 8.6 (A) Photomicrographs of a typical Schlieren texture observed from a LC polycap in $CHCl_3$ (a) and *p*-difluorobenzene (b) as viewed between crossed polarizers. (B) Laser confocal microscopy images of fibers assembled from the LC phases of the polycap in $CHCl_3$ either by sample shearing (a) or fiber pulling from the sample (b). Source: Castellano et al. [25]. Figure reproduced with kind permission; © 1999 Wiley-VCH.

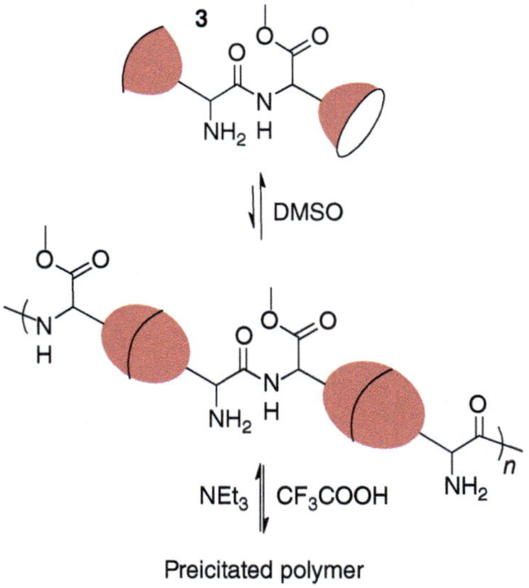

Figure 8.7 Schematic representation of the bis-calixarene **3** comprising a pH-sensitive dipeptide linker that could be used to switch between a soluble and insoluble form of the resulting polycap. Source: Redrawn from Xu et al. [27], © 2003 American Chemical Society.

material became insoluble and precipitated; the soluble form could be regenerated by cautious neutralization with NEt_3 without inducing any dissociation of the supramolecular polymer chain.

The dynamic and reversible chemistry of CO_2 represents a versatile method for preparing reversible polymers or organogels [28]. Basically, CO_2 readily reacts with amines to generate carbamates, even at ambient temperature and pressure; however, carbamates are thermally unstable and will release CO_2 upon heating. Rudkevich and coworkers utilized this special behavior in combination with the supramolecular chemistry of calixarenes to prepare two-parameter switchable polycaps and organogels [29, 30]. These responsive materials could then be utilized for entrapment, storage, and controlled-release applications. The supramolecular polymer was prepared in a two-step assembly approach (Figure 8.8): first, the

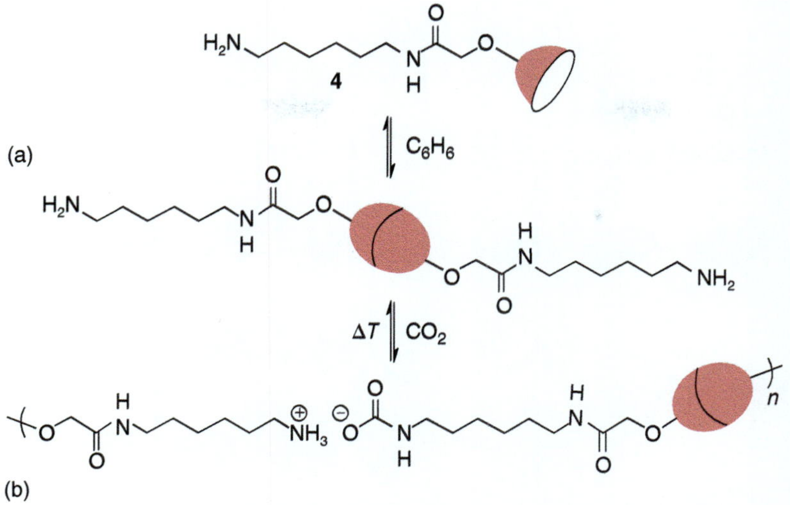

Figure 8.8 Schematic representation of the two-step assembly of a supercap by (a) calixarene dimerization in an apolar solvent and (b) binding of CO_2. Source: Redrawn from Xu et al. [29].

tetraurea-functionalized calix[4]arene **4**, featuring a primary NH_2 group on the periphery of the lower rim, was assembled into a dimeric capsule encapsulating one solvent molecule (i.e. $CHCl_3$, benzene). Subsequently, CO_2 was bubbled through the solution and precipitation of the so-formed polycap occurred. Due to the presence of two different types of binding – capsules based on H-bonding and linkage of the capsules via carbamate salt bridges – the polycap could be disassembled in two different ways by changing the solvent polarity (by adding a competitive solvent, such as DMSO) or heating to c. 100 °C.

It was further shown that a transition from a linear supramolecular polycap to a 3D polymer network could be achieved, based on reversible CO_2 fixation [30]. The bis-calixarene derivative **5** self-assembled into a supercap with the "CO_2-philic" amino-groups oriented in a perpendicular fashion to the polymer chain (Figure 8.9). Viscometry measurements confirmed the linear nature of the assembly and cyclic oligomers could, thus, be excluded. The 3D organogel was obtained by purging a solution of the supercap in either benzene or $CHCl_3$ with CO_2; the resultant supramolecular network also featured a dual-responsive behavior (i.e. solvent polarity and temperature; Figure 8.9). The voids in the organogel had sizes ranging from 15 to 20 Å, enabling the material to be employed further for the entrapping of dyes (e.g. coumarin-314 or tetraphenylporphyrin); a controlled release of these dyes could be achieved via changes of the solvent polarity, temperature, pH value, or concentration [31]. Xu and Rudkevich also utilized the bis-calixarene **6** with a pending pyrene moiety for the analogous step-wise self-assembly of an organogel (Figure 8.10) [32]. The organogel, assembled in benzene, was fluorescent; whereas, the material obtained from a benzene–nitrobenzene solvent mixture (in a 95 : 5 ratio) was not. The emission quenching was rationalized by a photoinduced

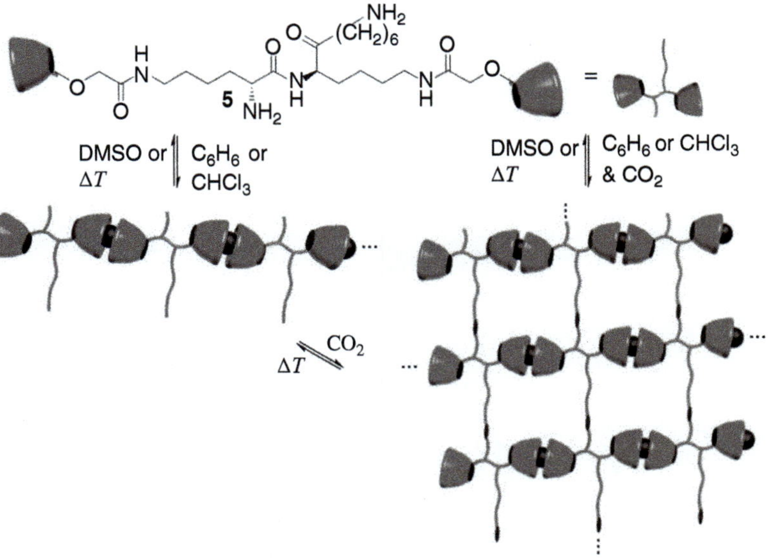

Figure 8.9 Schematic representation of the bis-calixarene **5**, as a building block for the two-step assembly of a stimuli-responsive organogel [30]. Source: Redrawn from Guo et al. [18], © 2012 Royal Society of Chemistry.

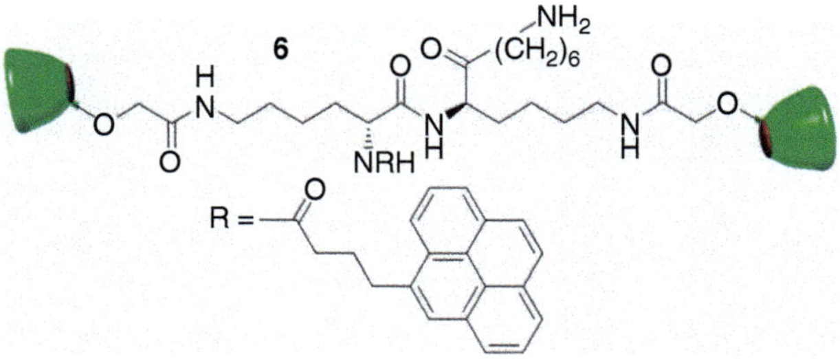

Figure 8.10 Schematic representation of the bis-calixarene **6** equipped with a pending pyrene moiety. Source: Modified from Xu and Rudkevich [32].

energy-transfer process from the chromophores to the solvent; thus, such a responsive organogel could be employed further in supramolecular light harvesting or related photophysical applications.

In all aforementioned examples, the bis-calixarenes were formed by linking them via their lower rims. Alternatively, their connection can also be made via the upper rims, i.e. covalently connecting one of their four urea-functionalized arms (**7**, Figure 8.11). Depending on the chemical nature of the linker – having either a rigid-linear or a flexible structure – intermolecular and/or intramolecular dimerization is enabled. Unimolecular capsules were observed by Brody et al. in the self-assembly of bis-calixarenes **7a** in which flexible alkyl chains were used to connect the calix[4]arene entities [33]. It was further shown that the addition of a double amount of **1** induced a reorganization of the assembly into a bis-heterodimer. Thus, for bis-calixarenes of the **7a**-type both types of interaction are

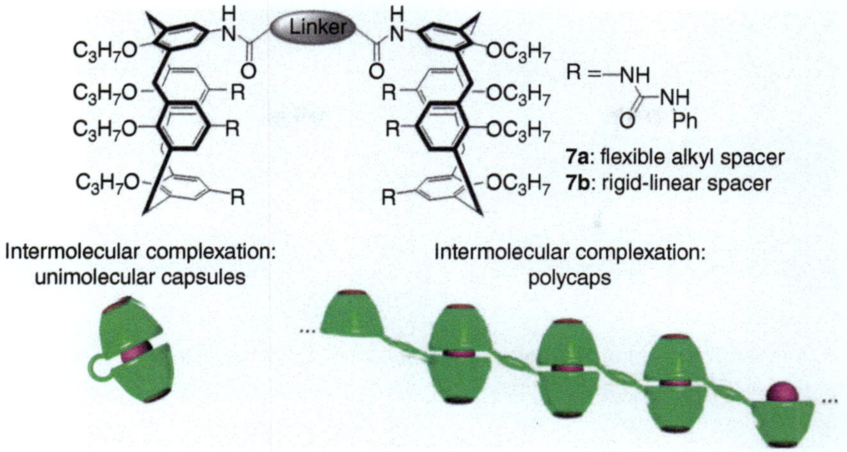

Figure 8.11 Schematic representation of bis-calixarenes, connected via their upper rims. Depending on the structure of linker inter- and/or intramolecular association is enabled. Source: Guo and Liu [18], © 2012 Royal Society of Chemistry.

possible and can be adjusted by the absence/presence of competitive calixarenes. Böhmer and coworkers focused on bis-calixarenes with π-conjugated spacers (**7b**) where intramolecular complexation was prevented due to the rigid-linear structure of the linker [34]. Accordingly, linear polymer chains were obtained from their self-assembly in appropriate solvents (Figure 8.11). Due to their more rigid structure, compared with the polycaps with lower-rim linkages, these polycaps were much more restricted in their conformational freedom.

8.3 Supramolecular Polymers Featuring Vacant Calixarene Scaffolds

As for the covalent polymers with attached calixarene moieties, e.g. serving as receptor units in fluorescence quenching applications [17], supramolecular polymers have also been assembled in which the binding sites of the appended calixarene moieties remained vacant. As a prominent approach, the utilization of calixarene–cyclodextrin conjugates needs to be mentioned, since it can be assembled into supramolecular polymers via the selective formation of host–guest-type inclusion complexes exclusively involving cyclodextrin (CD) moieties [35–39]. A representative example for such a supramolecular polymerization is depicted in Figure 8.12. The supra(macro)molecular chemistry of CDs is discussed further in Chapter 9. Basically, calixarenes only form inclusion complexes after special chemical modification(s) of their scaffold, and thus, their use in the formation of supramolecular polymers is less prominent, which is in sharp contrast to the other macrocyclic hosts (i.e. CDs, crown ethers, cucurbiturils, or pillarenes).

One further example to be highlighted in this context is the formation of columnar nanostructures, based on the calix[4]arenes **8** and **9**, which feature two melamine

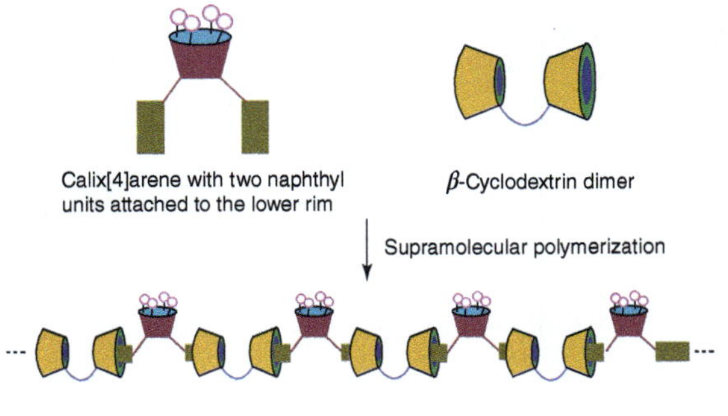

Figure 8.12 Schematic representation of the formation of a supramolecular polymer with vacant calixarene moieties. Source: Liu et al. [37], © 2002 Taylor & Francis.

and two cyanurate units, respectively, on their upper rims (Figure 8.13). These particular substituents represent a prominent pair for complementary H-bonding entities, which are known to generate rosette-like architectures and are widely utilized to assemble supramolecular materials via H-bonding interactions (see Chapter 3) [40]. The Timmerman and Reinhoudt groups investigated various types of calix[4]arenes equipped with these moieties – in particular, addressing the supramolecular chirality of the resulting assemblies and studying how regio-, diastereo-, and enantio-selection can be achieved [41]. Initially, the self-organization of the assembly formed from **8a** with diethyl barbiturate (DEB), i.e. [(**8a**)$_3$·(DEB)$_6$], on graphite was investigated by scanning-force microscopy (SFM) [42]. Rigid-rod assemblies were only obtained on the substrate surface; whereas, in solution

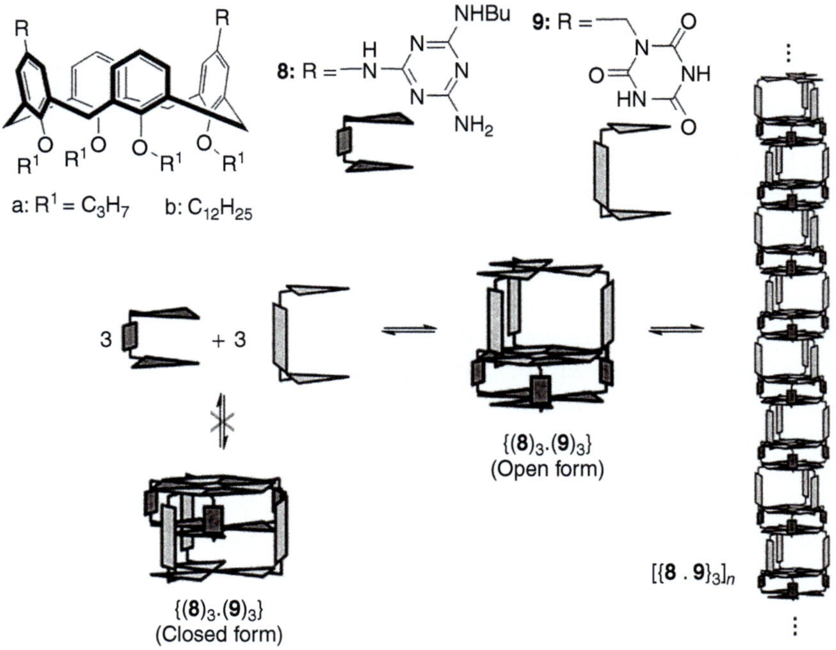

Figure 8.13 Schematic representation of the calix[4]arenes **8** and **9** equipped with H-bonding moieties. These were used for the self-assembly of supramolecular polymeric nanostructures. Source: Klok et al. [42], © 1999 American Chemical Society.

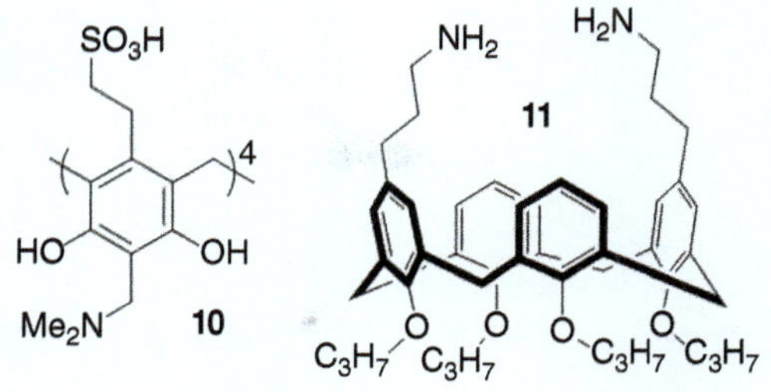

Figure 8.14 Schematic representation of resorcin[4]arene **10** and calix[4]arene **11**. Source: Baldini et al. [45], © 2011 American Chemical Society.

the formation of such structures was absent, even at high concentrations. The occurrence of these structures was attributed to the aggregation of disk-shaped [43] assemblies in a face-to-face-type fashion; solvophobic interactions were suggested as the driving force for this process. Equimolar mixtures of **8** and **9** gave, due to the nonmatching of the components, selectively an assembly of the "open" form {[(**8**)$_3$·(**9**)$_{39}$], Figure 8.13}. In contrast to the "closed" isomer [(**8a**)$_3$·(**9**)$_3$], the top side was complementary to the bottom one, thus facilitating its self-assembly into a rod-like polymeric nanostructure comprising a hydrophobic outer shell and a polar core {[(**8a**)$_3$·(**9**)$_3$]$_n$, Figure 8.13}.

Only a few examples are known in the literature, where supramolecular architectures have been formed from resorcinarenes. Korshin et al. employed the resorcin[4]arene **10** equipped with four sulfonato and four tertiary amino units on the lower and the upper rim, respectively (Figure 8.14) [44]. Due to electrostatic interactions, **10** assembled in water in a head-to-tail-type manner to generate a supramolecular polymer. The polymer was able to bind 1,5-diaminonaphthalene and could be reversibly disassembled by changing the pH. Finally, a supramolecular assembly was reported by Baldini et al. which was formed due to the binding of CO_2 to the diamino-functionalized calix[4]arene **11** (Figure 8.14); the supramacromolecular polymerization was again attributed to electrostatic interactions involving ammonium-carbamate salt bridges [45].

8.4 Supramolecular Polymers, Formed by Host–Guest Interactions

In Section 8.2, polymeric architectures were discussed, where capsule formation in the presence of appropriate solvent molecules represented the basis for supramolecular polymerization. In contrast to this, supramolecular polymers that are directly formed by molecular recognition of calixarene cavities by proper guest molecules will be highlighted in the following. This type of supramolecular host–guest polymers is closely related to those materials obtained from the other macrocyclic hosts and their appropriate guests. Basically, two types of architectures can be utilized for the preparation of linear supramolecular polymers: first, AB-type monomers where a guest tail is attached to the host and, second, host dimers – linked via a spacer moiety – are combined with a ditopic guest (i.e. using a combination of AA- and BB-type

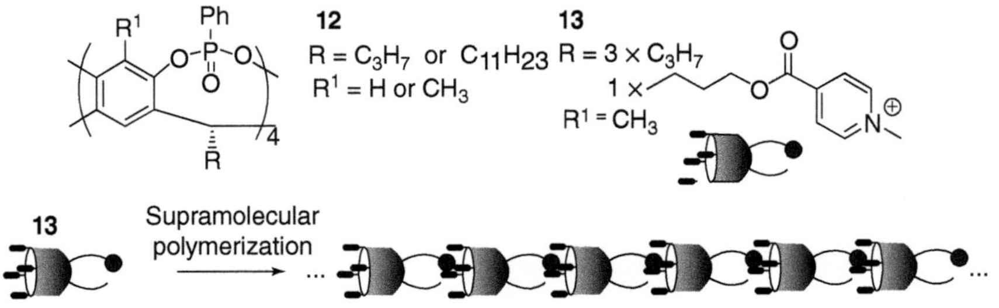

Figure 8.15 Schematic representation of cavitands **12** and **13**. The latter one was used as AB-type monomer in a supramolecular polymerization. Source: Dalcanale and Pinalli [19], © 2015 Springer Nature.

monomers). In particular, cavitands, i.e. the rigid resorcinarene-based analogs of calixarenes, have mainly been employed in this respect [46].

8.4.1 Supramolecular Polymers, Derived from AB-Type Monomers

The synthetic toolbox enables a selective functionalization of calixarenes and resorcinarenes of their upper or lower rims. Thus, heteroditopic host–guest conjugates (i.e. AB-type monomers) can be readily prepared and subsequently employed in the context of a supramolecular polymerization. The first indication for the self-assembly of such monomers into supramolecular assemblies was observed for the structurally simple calix[4]arene bearing a butylpyridinium moiety on the lower rim [47]. Due to the weak cation–π interactions of this system, Ishihara and Takeoka only obtained oligomers instead of polymers; the 1D structure of these oligomers was demonstrated by X-ray diffraction studies.

Cavitands **12**, where the OH groups of resorcin[4]arenes have been bridged by phosphonate moieties, have attracted considerable interest regarding their use in supra(macro)molecular chemistry (Figure 8.15). Such cavitands are known for their remarkable molecular recognition behavior toward cationic *N*-methylated guests – not only in solution [48] but also in both the solid state [49] and gas phase [50]. The high K_a values, which often exceed 10^7 M^{-1} in CH_2Cl_2, result from the synergy of up to three interaction modes involving the inward-facing P=O fragments [51]: cation–dipole ($N^+\cdots O=P$), Me–π (acidic ^+NMe with the π-basic cavity), and H-bonding interactions between two neighboring P=O bridges and two nitrogen protons; though, the latter applies only for protonated secondary ammonium cations.

Dalcanale and coworkers employed the cavitand **13** with one pending *N*-methylpyridinium arm, as an AB-type monomer, for the self-assembly into a linear supramolecular polymer (Figure 8.15) [52]. The polymer was investigated in the solid state by X-ray single-crystal analysis (Figure 8.16a) and in solution by NMR spectroscopy and dynamic light-scattering (DLS) measurements. The latter method suggested a maximum molar mass (M_w) of 26 300 g mol^{-1}, corresponding to a DP of c. 18. The DP could be controlled by adding an appropriate cavitand as chain-stopping agent, such as the tetrathiophosphonate analog to **12** (the exchange

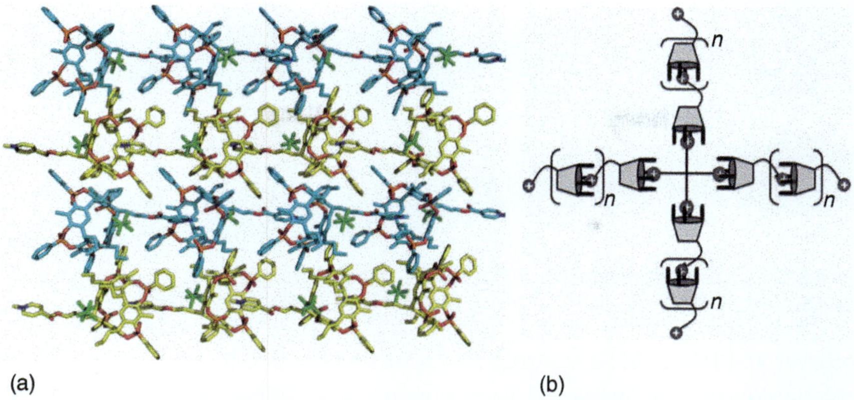

Figure 8.16 (a) Representation of the solid-state structure of poly-**13** (perpendicular view to the polymer chains) [52]. (b) Schematic representation of the star-shaped supramolecular polymer, derived from a template-driven self-assembly using a planar tetratopic guest, as core. Source: Yebeutchou et al. [52], © 2008 John Wiley and Sons.

from P=O to P=S reduced the driving force for complexation by weakening the cation–dipole interactions). Moreover, the addition of a tetratopic guest molecule, i.e. a porphyrin derivative equipped with four *N*-methylpyridinium arms, increased the molar mass of the polymer significantly: the initial linear supramolecular polymer chains interacted with the tetratopic guest, serving as a template, to give a star-shaped supramolecular polymer (M_w values of >40 000 g mol^{-1} were determined by DLS analysis). It was also shown that depolymerization could be achieved by adding an efficient competitive guest molecule, such as *N*-methyl butylammonium iodide. Due to its ability to establish additional H-bonding interactions within the cavity, a significantly higher binding strength was reached (K_a of 10^8 M^{-1}) and the *N*-methylpyridinium units were substituted. The depolymerization could be reversed by adding 1,8-diazabicyclo[5.4.0]undec-7-ene (DBU), as a bulky base, that could not be included into the cavity but deprotonate the secondary ammonium guest – as a result, the initial supramolecular polymer was restored.

Parisi and coworkers showed that *p-tert*-butylcalix[5]arenes **14** (Figure 8.17), having a fixed cone-like arrangement of C_{5V} symmetry, were able to form stable inclusion complexes of 1 : 1 stoichiometry with linear alkylammonium cations [53, 54]; when using *n*-alkyldiammonium salts of sufficient chain length, analogous, capsule-like complexes of 1 : 2 stoichiometry were obtained with **14** [55]. The stability of these inclusion complexes was rationalized by cation–π (as primary recognition motif) and CH–π interactions, supported by H-bonding interactions between the tripodal ammonium cation with ethereal *O*-atoms buried deep within the cavity. Consequently, linking a linear alkylamino group to a rigid *p-tert*-butylcalix[5]arene afforded the AB-type monomers **15** that could be employed in supramolecular polymerizations (Figure 8.17) [56]. Since neutral, primary amines are not bound by calix[5]arenes, protonation of **15** was required to enable the self-assembly in a selective head-to-tail-type fashion. The supramolecular polymerization could again be reversed under basic conditions due to deprotonation of the ammonium cations;

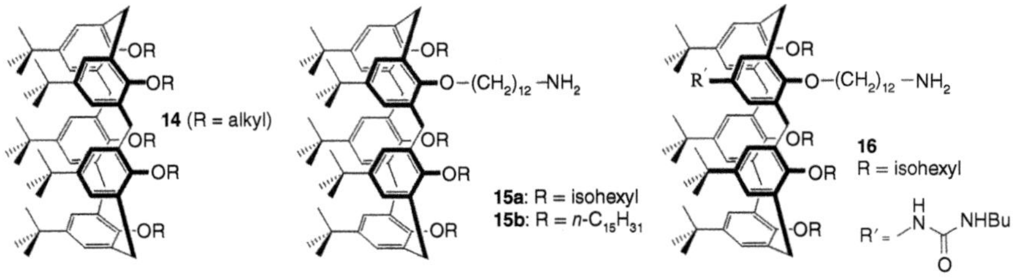

Figure 8.17 Schematic representation of the *p-tert*-butylcalix[5]arenes **14–16**. Source: Pappalardo et al. [56].

moreover, the DP could be controlled by adjusting the monomer concentration or the counterion. Utilizing various acids to induce polymerization, a dependency on the strength of the ion-pairing interaction between the ammonium cation and the accompanying anion was found. Accordingly, the highest DPs were observed for picric and hexafluorophosphoric acid; whereas, HCl or HBr only gave oligomers, even at elevated concentrations [56, 57]. In the case of the more lipophilic derivative **15b**, the DP could be regulated by employing orthogonal chain stoppers that were able to interact with either of the moieties of the AB-type monomer (or its polymer) [57]: both types of "polymerization-inert" competitors, a calix[5]arene “cap” or a *n*-butylammonium “plug” could be used to regulate the molar mass of the resulting polymer.

To suppress the issue of ion-pairing interactions affecting the DP, a modified calix[5]arene (**16**, Figure 8.17) was synthesized and offered a further binding site for anions, as a pending substituent on the upper rim (i.e. an urea moiety) [58]. The structural feature was introduced to facilitate salt dissociation and, thus, making the supramolecular polymerization more efficient. NMR spectroscopy and DLS measurements revealed a threefold increase of the DP for **16** compared with **15**, even in the presence of Cl^- as the strongest anion for ion-pairing. The authors also showed by circular dichroism (CD) spectroscopy that, when using chiral mandelic acid for protonation, chiral supramolecular polymers could be obtained; due to the "chiral memory" effect, the CD signature was retained if the chiral anion was replaced by a nonchiral one.

8.4.2 Supramolecular Polymers, Derived from AA- and BB-Type Monomers

In Section 8.4.1, heteroditopic monomers were employed for the formation of linear supramolecular polymers. Utilizing two complementary homoditopic monomers (AA- and BB-type) represents the alternative approach that also affords such architectures.

The Parisi group employed the bis-calix[5]arenes **17**, connected with rigid xylene units via their lower rims in a centrosymmetric divergent fashion, for the supramolecular polymerization with long-chain α,ω-alkanediammonium dipicrates (**18**, Figure 8.18) [59, 60]. Thus, the authors retained the type of supramolecular interaction (*vide infra*) but changed from a heteroditopic [56] to two homoditopic

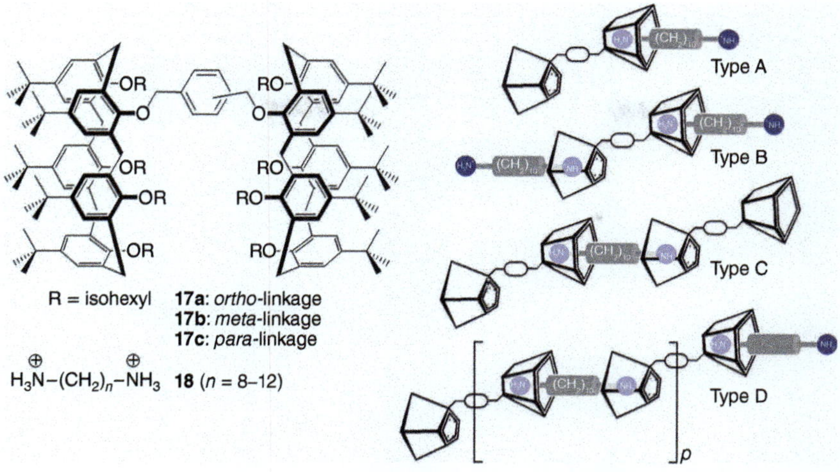

Figure 8.18 Schematic representation of the bis-calix[5]arenes **17** and the alkyl-diammonium salts **18** used for the self-assembly into the supramolecular architectures A-D. Source: Gattuso et al. [60], © 2008 American Chemical Society.

monomers. Extensive MS and NMR studies were conducted to identify the different types of supramolecular species being formed. Both their composition and their dynamics were dependent a variety of parameters: the spacer length of **18**, the substitution pattern of the xylyl bridge of **17** and the concentration (and/or molar ratios) of the two components. Four different types of assemblies could be identified (Figure 8.18). Molecular-modeling studies, supported by CPK models, excluded the presence of cyclic assemblies due to the rigidity of **17** and the short chains of **18**. When using the short-chain homoditopic guests (i.e. **18** with $n = 8$ or 9), the length of the alkyl chain was too short to allow the guests to enter the cavities of two host sites; thus, type-C and type-D assemblies could not be formed in this case. For the other dimeric guests, however, all four types could be assembled reversibly and the reorganization from one to the other could be triggered by various basic external input (e.g. dilution, addition of additional dimeric guests or hosts).

The Fukuzawa and Haino groups prepared the AA-type monomer **19** comprising two covalently linked double-calix[5]arenes units on each side (Figure 8.19). Such double-calix[5]arenes represent highly versatile nanocontainers (the so-called "molecular tweezers") for the binding of fullerenes [61, 62], and this particular interaction was used by these authors for the assembly of supramolecular polymers (in combination with the dumbbell-shaped bis-fullerene **20**) [63] and polymer networks (in combination with copolymer **21** with C_{60} moieties in the side chains) [64]. According to the concept of molecular tweezers (see Chapter 5), the twofold charge–transfer (CT) interaction between the two-armed electron-rich host and the electron-poor guest is strongly enhanced (much weaker binding is observed between a single calix[5]arene and C_{60}) [65]. In both cases, the supramolecular polymerization was found to be solvent dependent and offered the highest K_a values in toluene. The association constants for binding two C_{60} spheres according to a 2 : 1

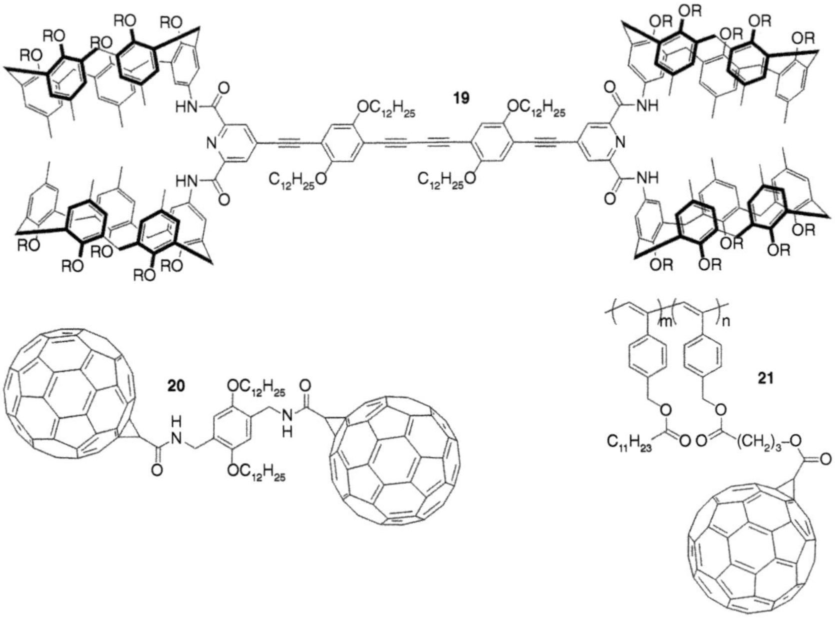

Figure 8.19 Schematic representation of the homoditopic "bis-container" **19** for the complexation of fullerene derivatives, such as the dumbbell-shaped bis-fullerene **20** and the polyacetylene **21** with fullerene side chains.

model (i.e. K_1 and K_2), determined by nonlinear regression analysis of fluorescence titration experiments, were as high as $10^4\ M^{-1}$. Thus, highly stable supramolecular polymers, such as the cross-linked material obtained from the self-assembly of **19** with **21**, could be obtained. The authors employed diffusion-ordered NMR spectroscopy (DOSY), size-exclusion chromatography (SEC), and microscopy imaging to conclude the molar mass (distribution) as well as on the morphology of the materials. For example, **21** tended to agglomerate into nanoparticle-like aggregates; whereas, the cross-linked supramolecular polymer revealed a high degree-of-ordering due to the formation of well-oriented fibrous nanostructures. The supramolecular polymerization of **19** and **20** gave, attributed to the poor solubility of **20** in $CHCl_3$, only short oligomers (DP of c. 3). This restriction could be overcome by introducing a wedge-shaped substituent with long alkyl chains onto the backbone of the bis-fullerene [66]. In this case, the DP could be raised to >80 highlighting the necessity of the solubilizing groups to support supramolecular polymerization and to enable the assembly of polymers of high molar masses. SEM imaging was applied to investigate the morphology of the supramolecular polymer; fibrillar nanostructures attributed to the linear chain-like structure of the polymer were observed. Very recently, the same authors combined building block **19** with a fullerene, equipped with a poly(methyl methacrylate) (PMMA) chain, to obtain a stimuli-responsive, supramolecular triblock copolymer of coil–rod–coil structure (Figure 8.20) [67]. Accordingly, a tritopic derivative, as rigid core, afforded a three-arm star-shaped assembly.

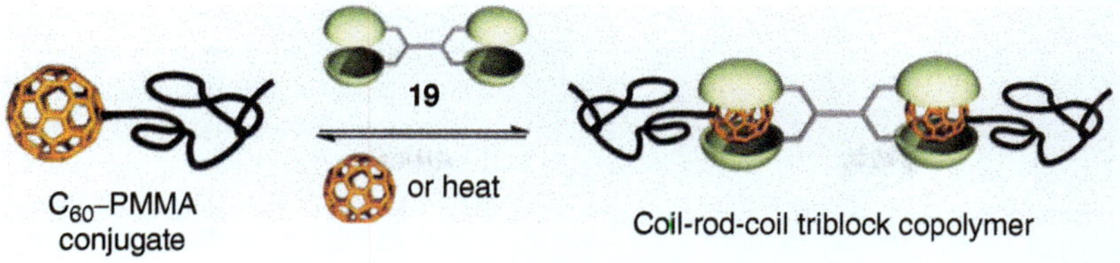

Figure 8.20 Schematic representation of the formation of a supramolecular coil–rod–coil triblock copolymer. Source: Hirao et al. [67], © 2020 American Chemical Society.

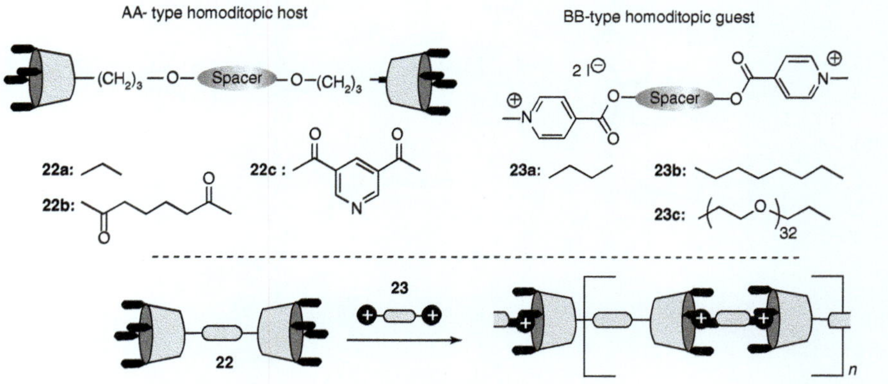

Figure 8.21 Schematic representation of the homoditopic AA-type bis-cavitand hosts (**22**) and BB-type guests (**23**) used for supramolecular polymerization. Source: Tancini et al. [69], © 2010 John Wiley and Sons.

Related to this, Pappalardo and coworkers prepared a poly(*p*-phenylene ethynylene) with two pending calix[5]arene moieties on every second phenyl unit [68]. The self-assembly of this homopolytopic host with a C_{60} fulleropyrrolidine, as appropriate guest, gave a bicontinuous supramolecular network. AFM imaging revealed an uniform distribution of structural features, which were assigned to the C_{60}-calix[5]arene bis-complexes within the network.

Further, Dalcanale and coworkers synthesized a set of AA-type monomers **22** by linking two tetraphosphonate cavitands (e.g. **12**) via their lower rims. These building blocks were employed in the supramolecular polymerization with the BB-type bis-*N*-methylpyridinium guests **23** (Figure 8.21) [69]. In particular, combining **22c** with commercial *N*,*N*′-dimethyl-4,4′-bipyridinium iodide (i.e. paraquat or methylviologen, MV^{2+}) generated a crystalline material whose structure was solved by X-ray diffraction analysis – thus giving a strong physical evidence for the existence of linear supramolecular chains (Figure 8.22). For all other combinations of **22** and **23**, viscometry and static light-scattering (SLS) measurement were carried out to estimate the molar mass (M_w) of the assemblies. In the case of **22b/23c** possessing the most flexible system within this series, a ring-chain equilibrium (see also Chapter 1) was found: at low concentrations, cyclic oligomers were observed [i.e. (**22b·23c**) and (**22b·23c**)$_2$]; whereas, with increasing concentration, a gradual transition into open oligomers was concluded from the measurements. Finally, at high concentration,

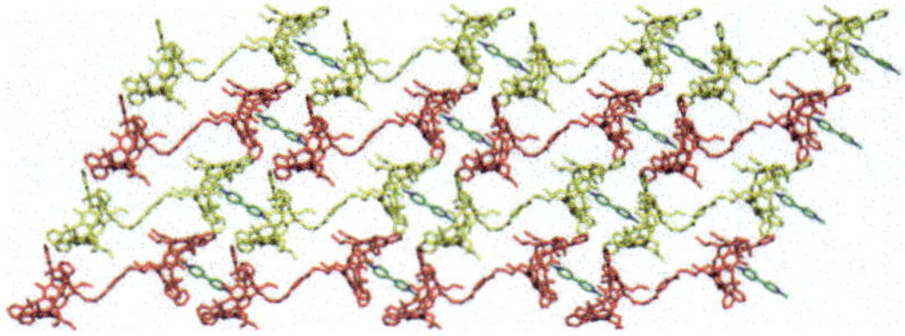

Figure 8.22 Representation of the crystal structure of the linear supramolecular polymer obtained from the self-assembly of 22c and MV^{2+} (the counterions are omitted for clarity). Source: Tancini et al. [69], © 2010 John Wiley and Sons.

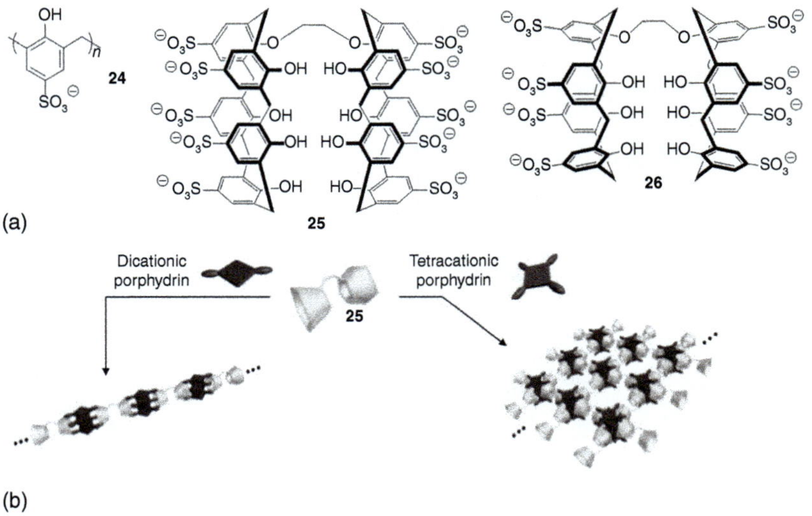

Figure 8.23 (a) Schematic representation of the *p*-sulfonatocalix[*n*]arene **24** and the corresponding homoditopic hosts **25** and **26**. (b) Schematic representation of the self-assembly of **25** into a linear or net-like polymer upon complexation of a dicationic or tetracationic guest. Source: Guo et al. [71], © 2009 John Wiley and Sons.

the presence of linear polymer chains was deduced from the viscometry results. Temperature was identified as a second important parameter contributing to the ring-chain equilibrium where the thermodynamics of the self-assembly process also had to be considered. Due to the presence of a potential binding site for transition metal ions (i.e. the pyridyl moiety), the homoditopic host **22c** allowed for a cross-linking of the linear supramolecular polymer chains via metal-to-ligand coordination. The neutral precursor complex $Pd(MeCN)_2Cl_2$ was used for this purpose: the labile MeCN-ligands were replaced by the pyridine sites of the polymer without interference with the P=O units of the cavitands.

para-Sulfonatocalix[*n*]arenes (**24**, Figure 8.23a) represent robust, yet water-soluble scaffolds, which show remarkable binding affinities and selectivity for the binding of small-molecule organic cations. This behavior has been attributed to the presence of additional binding sites, offered by the sulfonate moieties, in combination with those intrinsically present in the cavities [70]. To prepare linear supramolecular polymers in water, the Liu's group used the homoditopic hosts **25** and **26** featuring two fully *p*-sulfonated calix[5]arene or calix[4]arene host moieties, respectively, linked via their lower rims by an ethylene bridge (Figure 8.23a). The ditopic host **25** was combined with a dicationic or a tetracationic porphyrin derivative to generate a linear or a net-like supramolecular polymer, respectively (Figure 8.23) [71]. However, the determination of the binding stoichiometries, according to Job's method, revealed a significant deviation from the expected molar ratio of 1 : 1 (i.e. one guest site per cavity): stoichiometries of 1 : 5 and 2 : 5 were obtained for the di- and tetracationic guest, respectively. This finding was attributed to a charge balancing of the anionic sulfonate groups by the cationic guest rather than the formation of the anticipated host–guest inclusion complexes. Porphyrins are well-known antenna-like molecules, which are widely utilized in light-harvesting applications, and calixarenes, featuring electron-rich cavities, might be employed as electron-donating components. Thus, the combination of these two photochemically active arrays into supramolecular architectures represents a versatile approach to construct copolymers of the donor–acceptor (D–A) type. The light-induced electron transfer from the calixarene sites to the porphyrins was investigated by means of (time-resolved) fluorescence and transient absorption spectroscopy as well as (spectro)-electrochemical methods.

In a related work, D'Urso et al. studied the self-assembly of an octacationic calix[4]arene dimer in the presence of a tetraanionic porphyrin derivative.[72]. The 2D or 3D supramolecular structures were formed in a stepwise fashion, thus making use of the particular host–guest molecular ratio in each assembly step. Thereby, a cruciform-shaped host–guest complex (4 : 5 ratio of the components) represented the key focal point.

The supramolecular host–guest polymerization of bis-calix[4]arene **26** with the homoditopic bis-viologen guests **27** was reported by the Liu group [73, 74]. In contrast to the linear polymer that was obtained from **26** and **27a**, **26**, and **27b** gave – due to the longer and more flexible spacer – only cyclic oligomers (Figure 8.24). The issue of macrocyclization could, however, be suppressed by adding a cucurbit[*n*]uril derivative ($n = 7$ or 8) to the binary mixture of **26** and **27b**. As already outlined in Chapter 7, cucurbit[*n*]urils (CB[*n*]s) represent macrocyclic hosts for the inclusion of, in particular, aliphatic guests (thus, forming [2] pseudorotaxane-type assemblies). Due to the threading of the alkyl bridge of **27b** into the cavity of the CB[8], the molecule became rigidified and, thereby, the ability to assemble into linear polymers at the cost of cyclic oligomers was greatly enhanced (Figure 8.25).

It has been shown that the association of viologen to *p*-sulfonatocalix[*n*]arenes can be modulated via the oxidation state of the former [75]. Basically, viologens exhibit a three-step redox process that is characterized by its high reversibility – MV^{2+} can be reduced from its dicationic state via the radical cation to the neutral form and oxidized back to the initial MV^{2+} species. The K_a value for the interaction with a *p*-sulfonatocalix[*n*]arene readily decreases from the dication to the neutral molecule. This property was employed by the Liu group for the reversible association–dissociation of the linear binary and ternary polymers, based on **26** and **27**, due to an electrochemical stimulus. However, in the case of **26** and **27a**, disassembly only occurred after the two-electron reduction of the MV^{2+} units to the neutral state, which is insoluble in water, and reversibility of the process was difficult to achieve [73]. In contrast, the ternary polymer of **26**, **27b**, and CB[8] exhibited a fully reversible dissociation upon one-electron reduction of the MV^{2+}

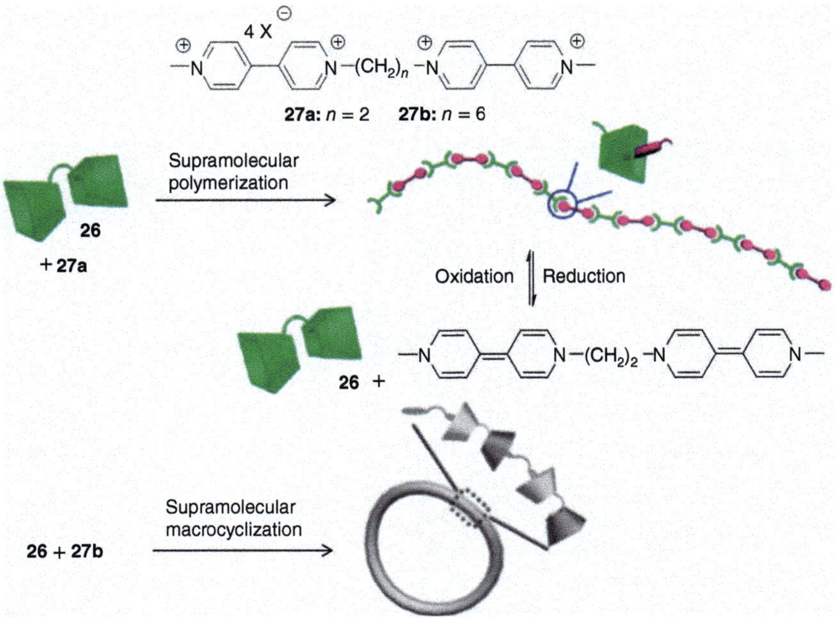

Figure 8.24 Schematic representation of the tetracationic bis-viologen guests **27** and their use in the self-assembly with bis-calixarene **26** to afford linear polymer or cyclic oligomers. Source:(a) Redrawn from Guo et al. [73, 74] © 2010 Royal Society of Chemistry; (b) Redrawn from Qian et al. © 2012 John Wiley and Sons.

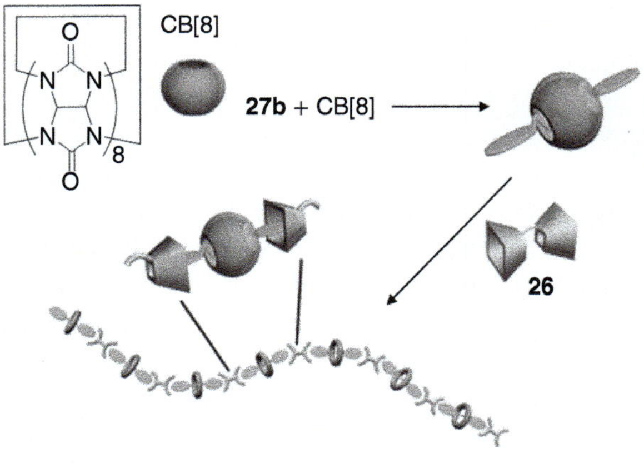

Figure 8.25 Schematic representation of the synthesis of a ternary supramolecular polymer via a two-step self-assembly approach. Source: Redrawn from Qian et al. [74], © 2012 John Wiley and Sons.

moieties to their radical cations [74]. Thereby, the initial [2] pseudorotaxane of **27b** and CB[*n*] reorganized into a so-called "molecular loop" [76]. The CB[8] moiety not only controlled the topology of the resultant supramolecular assembly (i.e. cycles vs. chains) but also facilitated the reversibility of the electrochemically driven assembly and disassembly.

The concept of stimuli-responsive supramolecular polymers was further developed by Tian and coworkers, who prepared a heteroditopic guest (**28**) for complexation by **26** [77]. Guest **28** featured a redox-active MV^{2+} and a pH-sensitive

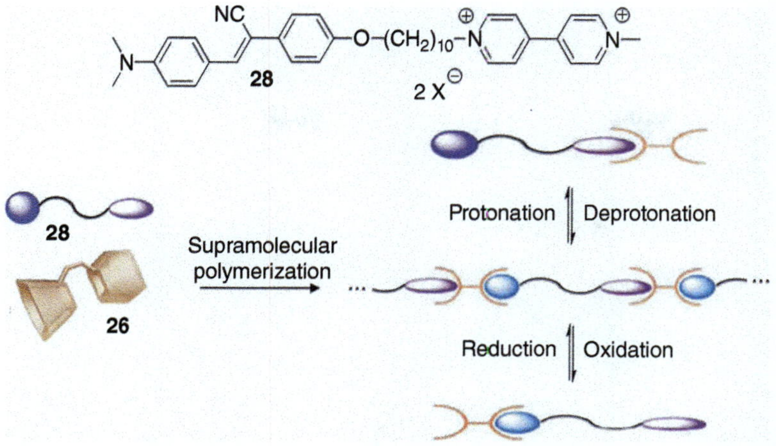

Figure 8.26 Schematic representation of the pH- and redox-sensitive supramolecular polymer assembled from bis-calixarene **26** and the heteroditopic guest **28**. Source: Redrawn from Ma et al. [77], © 2011 Royal Society of Chemistry.

dimethylamino moiety at the chain ends, thus offering a dual responsiveness (Figure 8.26). Supramolecular polymerization only occurred upon protonation of the dimethylamino group; whereas, in the non-protonated state, a 1 : 1 inclusion complex was formed in which the MV^{2+} site was bound in the cavity of one calixarene unit of **26**. The polymer could also be disassembled when the MV^{2+} part was reduced to the neutral state, which is not recognized by the calixarene. Thus, a dual switching between two different types of 1 : 1 assemblies and a linear supramolecular polymer was enabled by proper design of the heteroditopic guest **28** (Figure 8.26).

An example for a photoresponsive supramolecular polymer was published by Sun et al. [78]: the multifunctional guest **29** contained a 1,1′-binaphthyl unit of axial-chiral, photoresponsive azobenzene moieties within the arms as well as binding sites for CD- and calixarene-recognition (Figure 8.27). First, **29** gave a [3] pseudorotaxane by threading the two arms through the cavities of two α-CD rings (thereby, the azobenzene moieties were located within the cavities of the α-CDs). Due to pseudorotaxane formation, its solubility of the hydrophobic guest in water was enabled. Moreover, the position of the α-CD rings on the arms could be modulated by light irradiation: irradiation with UV light induced the *trans–cis* isomerization of the azobenzene moieties and pushed the threaded α-CD rings away from the azobenzene units. Due to the chiral nature of the core fragment, the stereochemistry was translated into the polymer chain, thus allowing the ultimate assembly of a helical supramolecular structure. The authors used NMR and DLS measurements to confirm the supramolecular polymerization. CD spectroscopy was further used to show the chirality of the polymeric material; UV/vis absorption spectroscopy on the supramolecular polymer and its molecular components (i.e. **26**, α-CD, and **29**) also confirmed the success of the assembly process in aqueous solution via the interplay of non-covalent interactions. Microscopy imaging of the

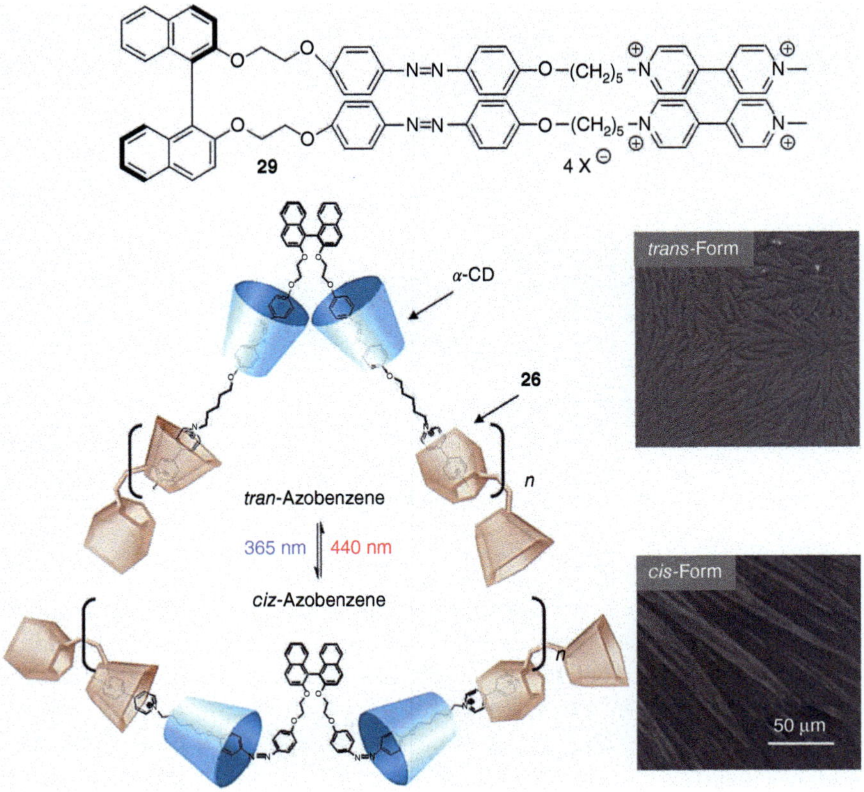

Figure 8.27 Schematic representation of the multifunctional guest **29** and the chiral supramolecular polymer derived by binding of α-CD and **26**. The polymer showed a light-driven isomerization that could be monitored by SEM imaging of the dried cast films on glass slides. Source: Sun et al. [78]. Figure reproduced with kind permission; © 2013 American Chemical Society.

supramolecular polymer revealed significant changes in the morphology upon irradiation with light. As deduced from the SEM images, the supramolecular assembly in its native *trans*-form showed a star-like morphology with multiple branches; whereas, linear belt-like nanostructures were present after exposure to UV light. The change in morphology was attributed to the straighter structure of the *cis*-form compared with that of the *trans*-isomer (Figure 8.27). For the irradiated polymer, a length of c. 2 μm was estimated by atomic-force microscopy (AFM); finally, cryogenic transmission electron microscopy (cryo-TEM) imaging of the polymer in its native state confirmed the helical structure of the assembly.

8.5 Beyond Classical Calix[*n*]arenes: Calix[4]pyrroles

Calix[4]pyrroles, as synthetic nonplanar, nonaromatic tetrapyrrolic macrocycles, have emerged recently as an alternative, to the well-known calix[*n*]arene-type

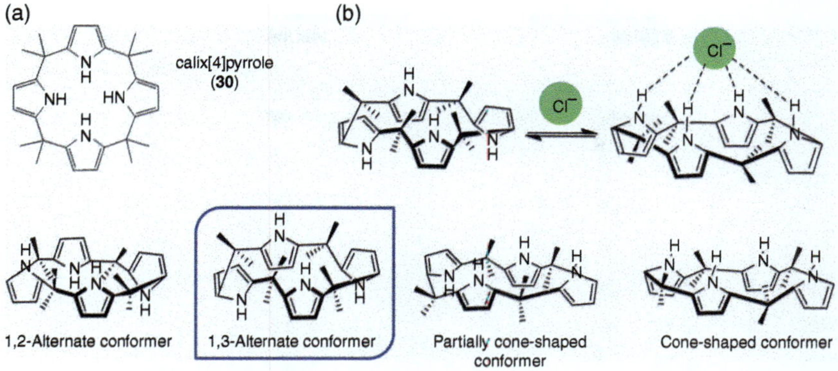

Figure 8.28 Schematic representation of calix[4]pyrrole (**30**). The four most relevant conformations (a) as well as the anion-induced transformation from the 1,3-alterante to the cone-shaped conformation (b) are also shown. Source: Wu et al. [80], © 2001 The Royal Chemical Society.

receptors [79]. Parent calix[4]pyrrole (**30**) and its functionalized derivatives can be readily prepared by an acid-catalyzed co-condensation of the corresponding pyrrole and carbonyl derivatives. Similar to their calix[4]arene counterparts, calix[4]pyrroles can adopt four different stable conformations. Sessler and coworkers reported that, at least for structurally simple derivatives, the 1,3-alternate conformer is thermodynamically favored in the absence of anionic guests (according to molecular modeling studies, the torsional interaction energy is c. 1–2 kcal mol^{-1} lower compared with the other conformers; Figure 8.28) [80].

The presumably first synthesis of **30** dates back to 1886, when Baeyer reported the HCl-catalyzed reaction between pyrrole and acetone [81]; the porphyrin-like macrocyclic structure was suggested 30 years later by Chelintzev and Tronov [82]. Remaining a little-noticed molecule for almost a century, until Sessler and coworkers discovered the anion-binding properties of **30** by N–H hydrogen-bonding (H-bonding) interactions [83]. This finding boosted research in this area and their applications, e.g. in the fields of sensing and organo-catalysis, have been established [84–86].

With respect to supramolecular architectures, the dimerization of calix[4]pyrroles into capsules is worth mentioning [87]. Such assemblies are also known from calix[4]arenes (Figure 8.2), but, in contrast to these, polymeric assemblies, the supercaps, have not yet been derived from calix[4]pyrroles (see Section 8.2).

On the other hand, a variety of supramolecular polymers has been reported in which the preferred 1,3-alternate conformation of calix[4]pyrroles has been exploited. Sessler and coworkers synthesized the tetrathiafulvalene-functionalized calix[4]pyrroles **31**, which were able to form 1 : 2 complexes with electron-deficient guest molecules, e.g. 1,3,5-trinitrobenzene (TNB, Figure 8.29) [88, 89]. In the 1,3-alternate conformation, each two TTF moieties offered a binding site – a "molecular tweezers" – where appropriate guest molecules can be accommodated via charge-transfer (CT) interactions (TTF: tetrathiafulvalene). As already detailed in Chapter 5, this approach was utilized in the following to assemble

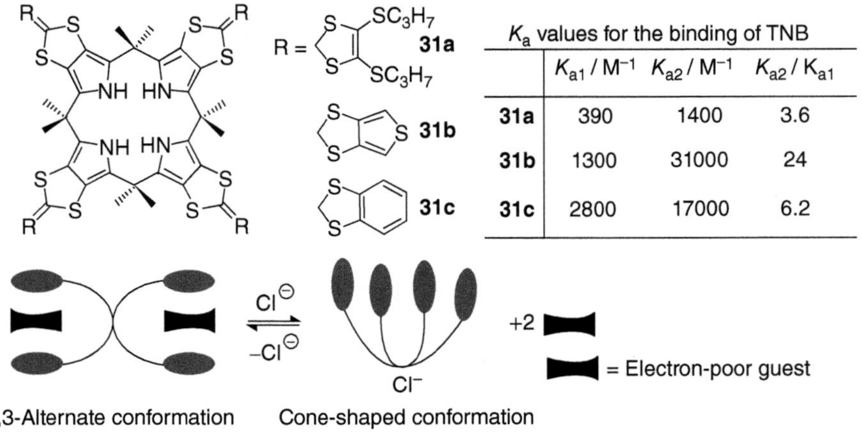

K_a values for the binding of TNB

	K_{a1} / M^{-1}	K_{a2} / M^{-1}	K_{a2} / K_{a1}
31a	390	1400	3.6
31b	1300	31000	24
31c	2800	17000	6.2

Figure 8.29 Schematic representation of the TTF-functionalized calix[4]pyrroles **31**, which, in their 1,3-alternate conformation, accommodated two electron-poor guests. The conformation of **33** could be switched reversibly by adding chloride anions. Source: Nielsen et al. [88], © 2004 American Chemical Society.

a supramolecular polymer; a bis(dinitrobenzene)-calix[4]pyrrole was used, as a ditopic guest, in this study [90].

In related work, Sessler and coworkers combined the calix[4]pyrrole-based host **31c** with the dicationic *trans-meso*-substituted calix[4]pyrroles **32** [91]. In the case of **32a**, where a noncoordinating anion was present, a linear supramolecular chain was obtained in $CHCl_3$ at a 1 : 1 ratio of the monomers. According to 1D/2D NMR and UV/vis absorption spectroscopy, a DP of c. 23 was estimated for (**31c**···**32a**)$_n$; in the solid state, extended, needle-shaped microstructures were visualized by SEM imaging. However, in the case of I^-, **31c** was forced into the cone-shaped conformation, and thus, supramolecular polymerization did not occur for the combination of **31c** and **32b**. Moreover, the supramolecular polymer (**31c**···**32a**)$_n$ was found to be susceptible to a structural transformation triggered by added salts. When tetraalkylammonium iodide was added, depolymerization occurred [TBAI: tetra(*n*-butyl)ammonium iodide, TEAI: tetraethylammonium iodide]. In the case of TBAI, a 2 : 1 capsule was formed; whereas, when using TEAI, an ion-pair complex [**31c**@TEAI] was obtained. The different behavior was attributed to the smaller size of the TEA cation, which fit into the cavity of the **31c** in its cone-shaped conformation, stabilized by the coordinated I^- (a similar ion-pair complex was, for steric reasons, not possible for TBAI).

Sessler and coworkers also studied the weakly associated supramolecular polymers, which resulted from the self-assembly of **31a** with the flexible ditopic guests **33** (Figure 8.31) [92]. In the solid state, all three monomer combinations gave linear supramolecular polymers; whereas, the situation was more complicated in solution. The observed K_a values for the interaction between **31a** and its designated hosts suggested a weak binding; steric effects were identified as the main reason (the guest with the longest alkyl spacer [**33c**] gave the highest K_a value of 3.8×10^3 l mol^{-1} in $CHCl_3$). Moreover, the nature of solvent had a profound impact on the self-assembly

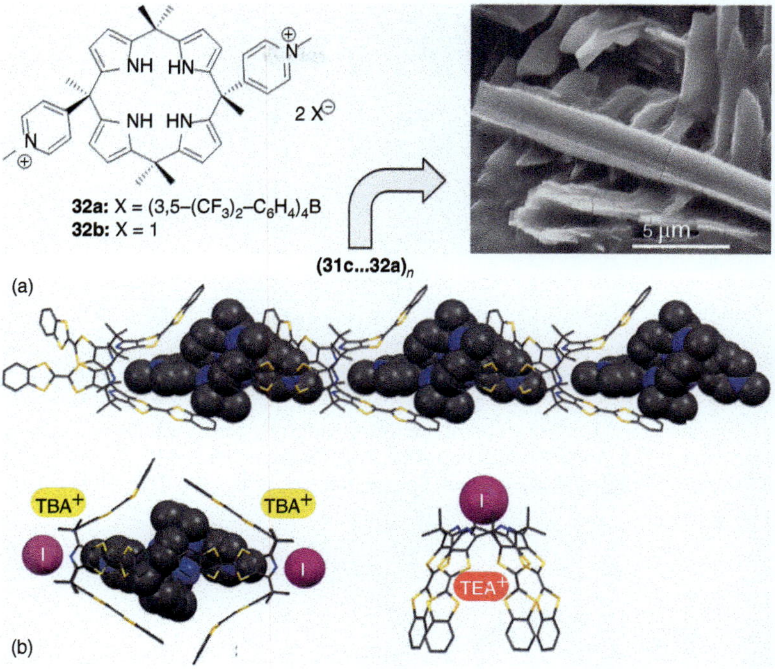

Figure 8.30 Schematic representation of the dicationic calix[4]pyrrole-based guests **32**. (a) Schematic representation of the structure of the supramolecular polymer assembled from **31c** and **32a**; a representative SEM image visualizing the solid-state morphology of the material is also shown. (b) Schematic representation of the supramolecular assemblies formed by the depolymerization of $(\mathbf{31c}\cdots\mathbf{32a})_n$ in the presence of TBAI (left) and TEAI (right). Source: Kim et al. [91], © 2013 American Chemical Society. Figure reproduced with kind permission; © 2013 American Chemical Society.

process – in apolar solvents, e.g. methylcyclohexane, significantly higher K_a values were determined by UV/vis/NIR titration studies. Thus, supramolecular polymers were obtained from **31a** and **33** in this solvent (e.g. a DP of c. 16 was calculated for the combination of **31a** and **33a**, Figure 8.31).

A different type of supramolecular polymer was recently reported by Yuvayapan and Aydogan [93]. These authors utilized the host–guest interactions between AA- and BB-type monomers in equimolar amounts, i.e. the dimeric calix[4]pyrrole **34** and TBA suberate **35** (Figure 8.32a). In this case, the calix[4]pyrrole moieties of **34** acted as anion receptor for the binding of the carboxylate sites of **35**. Following the pathway of a ring-chain-mediated supramolecular polymerization (see Section 1.3.2), cyclic species were found to dominate at low concentrations; above the critical polymerization concentration (cpc), linear polymer chains became favored. Robust, rod-like fibers were drawn from the highly concentrated $CHCl_3$ solutions and corroborated the high DP of the material. Remarkably, the cpc value of 30 mM was very low and even outcompeted the cpc values reported by the same authors for the supramolecular polymerization of analogous AB-type monomers. These calix[4]pyrroles, each equipped with a pending carboxylate side chain, had cpc

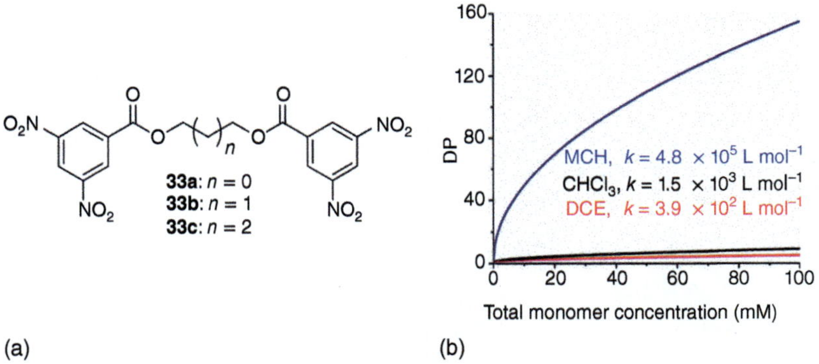

Figure 8.31 (a) Schematic representation of the flexible, ditopic guests **33**. (b) Evolution of the DP as a function of the monomer concentration in different solvents (MCH: methylcyclohexane; DCE: 1,2-dichloroethane). Source: Bähring et al. [92], © 2014 The Royal Chemical Society.

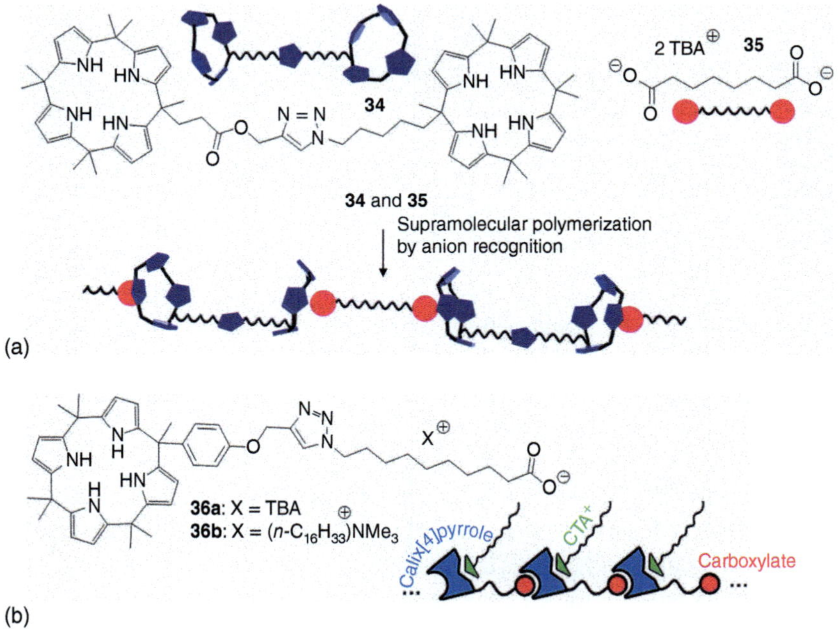

Figure 8.32 (a) Schematic representation of the complementary ditopic monomers **34** and **35**, which gave a linear supramolecular polymer due to calix[4]pyrrole–carboxylate anion recognition. Source: Yuvayapan et al. [93], © 2019 The Royal Chemical Society. (b) Schematic representation of the AB-type monomers **36**; a representation of the assumed supramolecular structure of poly-**36b** is also shown. Source: Yuvayapan et al. [95], © 2019 John Wiley and Sons.

values in the range of 70–118 mM [94]. For the AB-type monomer **36b**, an active participation of the surfactant-type cation in the supramolecular structure was supported by 2D NMR and viscosity measurements [95]. Compared with the supramolecular polymer derived from **36a**, the poly-**36b** showed a higher rigidity, which was attributed to a binding of the cetyltrimethylammonium cation to backside of the calix[4]pyrrole via cation–π interactions (Figure 8.32b).

8.6 Miscellaneous Supramacromolecular Assemblies, Based on Calixarenes

Up to this point, supramolecular polymers, derived from host–guest-type interactions between calixarenes and appropriate guests, were discussed. These polymers were either formed via the homo-association of AB-type monomers or from the hetero-association of AA- and BB-type monomers in a strictly alternating fashion.

A distinctively different supramacromolecular architecture was reported by Guo et al., who utilized calixarenes to induce the self-assembly of perylene bisimides (PBIs) [96]. As mentioned above, *p*-sulfonatocalix[*n*]arenes (**24**) can efficiently support the assembly of certain organic dyes, thereby significantly outreaching the 1 : 1 stoichiometry for the basic host–guest interaction (Figure 8.23) [71]. In water, the PBI dye **37** assembled into oligomers via π–π stacking interactions between their discotic polycyclic aromatic backbones. In the presence of **24** (n = 4 or 5), the host–guest interactions were supported by the electrostatic ones, thus strongly enhancing the ability of **37** to assemble into fiber-like nanostructures (Figure 8.33). The synergy of these interactions caused the hierarchical supramolecular polymerization to proceed in a well-defined fashion. The obtained nanostructures were analyzed by microscopy techniques: for example, TEM imaging revealed the formation of nanorods, assigned to bundles of fibers, with an average length of 220 nm – a value that was in fair agreement with the one obtained from DLS measurements.

As an important example, the formation of "supramolecular amphiphiles" has to be mentioned. Such non-covalent materials have attracted considerable attention as smart materials bridging the gap between molecular architectures and functional assemblies [97, 98]. In contrast to a conventional amphiphile, where the hydrophilic and hydrophobic components are linked covalently, supramolecular amphiphiles are assembled via non-covalent interactions. As a consequence, these materials exhibit high degrees-of-adaptability and -reversibility in the conformational transformation. Mainly cyclodextrins have been employed in this context, but also for the other macrocyclic host materials examples for such materials can be found in the literature [98–100]. Regarding calix[*n*]arenes, the rigid macrocyclic structure is predestined to stabilize supramolecular amphiphiles according to Figure 8.34 [101].

Moreover, their intrinsic cone-like architectures represent the key requirement for the aggregations of supramolecular amphiphiles into highly ordered nanosized structures. García-Río and coworkers were the first to observe the self-assembly

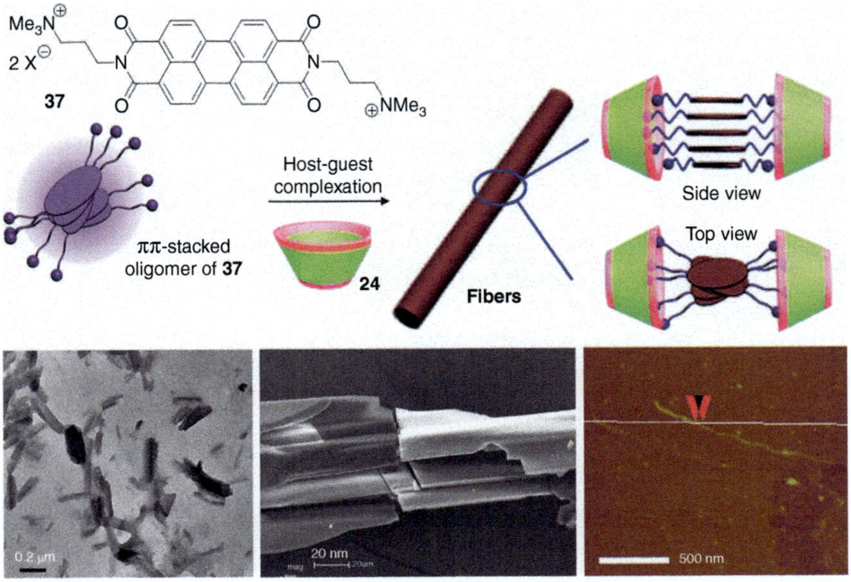

Figure 8.33 Schematic representation of the PBI dye **30** that was assembled into fibers upon addition of a *p*-sulfonatocalix[*n*]arene (**24**). The TEM, SEM, and AFM images (from the left to the right) of the nanostructures are also depicted. Source: Guo et al. [96]. Figure reproduced with kind permission; © 2012 The Royal Chemical Society.

Figure 8.34 Schematic representation of the reversible self-assembly of an amphiphilic molecule in the presence of a macrocyclic host. Source: García-Rio and Basílio [101], © 2019 Elsevier.

of sulfonated calixarenes with small-molecule surfactants into micelles [102] and vesicles [103]. The freeze-dried vesicles could be stored long-term and redispersed in water taking up their initial supramolecular structure in terms of size and morphology.

Wang et al. studied the self-assembly of **24** ($n = 5$) and the pyren-1-yl-methyl ammonium cation (**38**), at a molar ratio of 1 : 4, into vesicular species [104]. These assemblies exhibited a narrow size distribution and an average diameter of c. 90 nm was determined by TEM imaging and DLS measurements (Figure 8.35). If an excess of **24** was added, the vesicles disassembled to generate host–guest complexes of 1 : 1 stoichiometry in which the polar head of **38** was bound in the cavity of **24**. Furthermore, the vesicles exhibited a thermal reversibility in the temperature range of

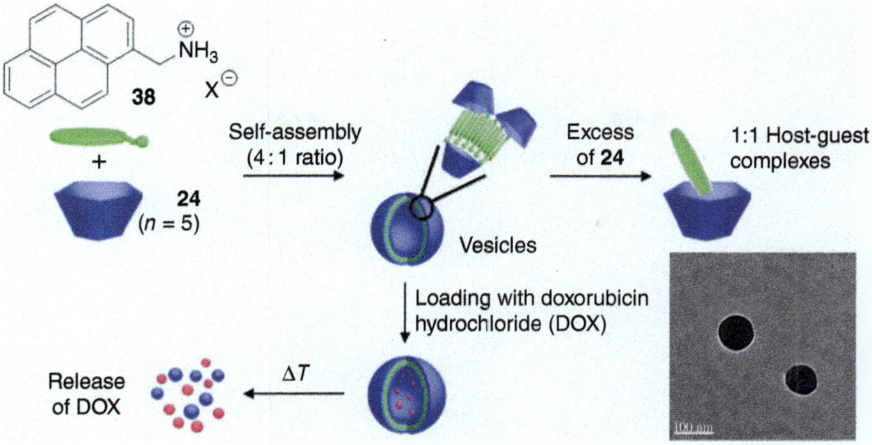

Figure 8.35 Schematic representation of the formation of a supramolecular amphiphile, as a thermoresponsive carrier for doxorubicin hydrochloride. A representative TEM image of the vesicles is also depicted. Source: Wang et al. [104]. Figure reproduced with kind permission; © 2010 Wiley-VCH.

35–40 °C. The authors showed that the vesicles could be loaded with doxorubicin hydrochloride (DOX), as a common model substrate for drug delivery studies. The drug could be released upon heating the vesicles to their point-of-dissociation.

An even more sophisticated supramolecular amphiphile was published by the same authors making use of the asymmetric viologen derivative **39**, as a guest for the interaction with **24** ($n = 4$) [105]. The vesicles formed by the self-assembly of these building blocks were analyzed by microscopy imaging and DLS measurements. A remarkable multiple stimuli-responsive behavior was found to arise from the chemical structure of **39** (Figure 8.36): in detail, the reduction of the MV^{2+} unit to the radical cation induced a shrinkage of the spheres – a process that could be fully reversed. However, the reduction to the neutral MV state, an increase in temperature or the threading of the pending alkyl chain into the cavity of added α-CD led to the disassembly of the vesicles. As in the previous case, these vesicles were also tested in cell experiments regarding the applicability in drug delivery. Loading of the vesicles with DOX did not influence its therapeutic effect toward cancer cells, but interestingly reduced the harm DOX caused to healthy cells.

Zhang et al. utilized the selectivity of host–guest interactions to prepare dynamic supramolecular amphiphiles, which showed a complex self-assembly behavior in solution (Figure 8.37) [106]. The heteroditopic guest **40** alone formed spherical micelles in water (due to the electrostatic repulsion of the highly cationic corona, the micelles were, however, rather large and loose). Addition of bis-calixarene **26** to these micelles induced a transformation into an amorphous worm-like network, thereby **26** functioned as a cross-linking agent establishing host–guest complexes with the cationic MV^{2+} sites in the micelles' coronas. At a stoichiometric ratio of 2 : 1, the micelles were destroyed, however, instead a ternary host–guest complex [**40**·**26**·**40**] was obtained. This assembly, in turn, could be polymerized in the

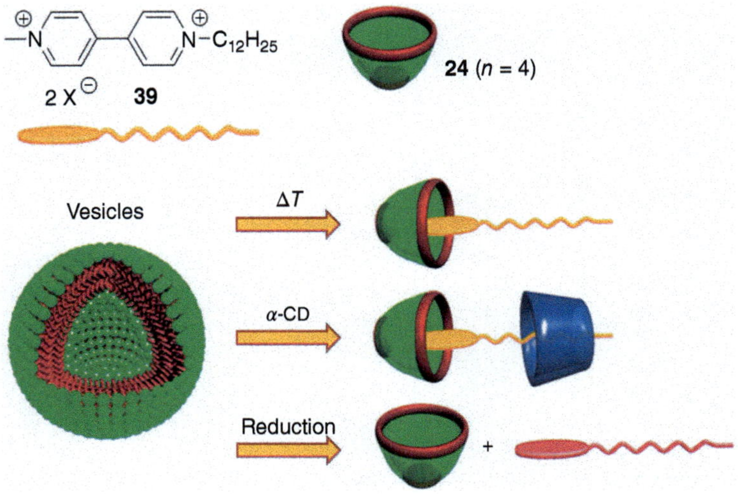

Figure 8.36 Schematic representation of the multiple stimuli-responsive vesicles obtained from the self-assembly of **24** ($n = 4$) and **39**. Source: Wang et al. [105], © 2011 American Chemical Society.

presence of γ-CD, an efficient host to accommodate two coumarin units in its cavity (see also Chapter 9). This linear supramolecular polymer had a size distribution from 100 to 600 nm according to DLS measurements, and the worm-like species had hydrodynamic radii in the range of 1000 nm; whereas, the average diameter of the initial micelles was 16 nm.

The calix[6]arene **41** with three imidazol-2-yl arms and three sulfonate units in an alternating fashion on the upper and lower rims, respectively, was found to assemble into multilamellar vesicles at a pH value of c. 8 (Figure 8.38). According to microscopy imaging, these vesicles showed a broad size distribution ranging from 50 to 250 nm. The morphology as well as the shape of these assemblies altered over time (due to the fusion of smaller into giant vesicles) and by adjusting the pH value (i.e. changing the protonation state of the basic imidazole moieties). The three imidazol-2-yl sites of **41** were able to bind an Ag(I) ion to the upper rim in a tripodal fashion, thus generating a hybrid-type supramolecular amphiphile. These species, in turn, organized themselves into almost monodisperse micelles of spherical shape. The same type of micellar structure could also be derived from the metal-free vesicles by adding Ag(I) ions, emphasizing the control over this two-step self-assembly process.

The phenomenon of "calixarene-induced aggregation" (CIA), as initially reported for **24** (and its derivatives) and some amphiphilic guest molecules, is based on the decrease of the critical aggregation concentration (cac) of a system in concert with a stabilization of the resultant aggregates [18]. Making use of CIA and combining it with aggregation-induced emission (AIE) allowed the Liu group to fabricate water-soluble fluorescent nanoparticles, based on the self-assembly of the tetraphenylethylene-type (TPE) dye **42** and **24** ($n = 4$) [107]. The bolaamphiphile **42** aggregated into spherical nanoparticles that were too loose to allow AIE to occur

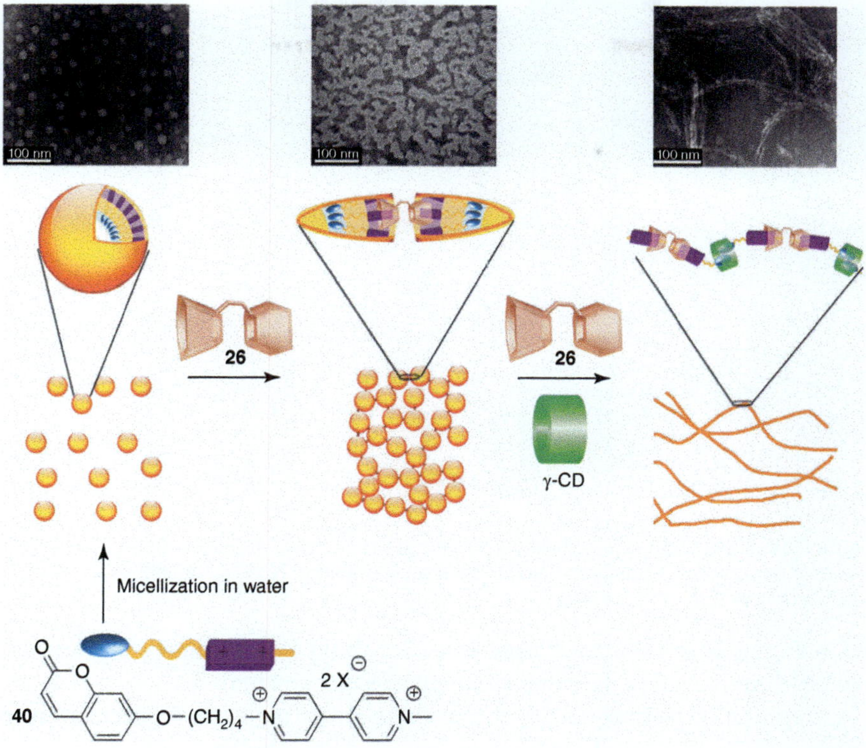

Figure 8.37 Schematic representation of the multistep assembly of a linear supramolecular polymer from the heterodifunctional guest **40** and two different types of hosts. The TEM images of the three types of assemblies observed in the study are also depicted. Source: Zhang et al. [106]. Figure reproduced with kind permission; © 2014 The Royal Chemical Society.

(attributed to electrostatic repulsion between the terminal quaternary ammonium sites). In the presence of **24**, however, AIE could be observed – due to host–guest and electrostatic interactions between **24** and **42**, a supramolecular host–guest amphiphile was formed that assembled further into compact multilamellar nanoparticles (Figure 8.39a). Within these species, for a hydrodynamic diameter of c. 60 nm was determined, a cooperativity of various non-covalent interactions (i.e. host–guest, electrostatic, π–π stacking, and hydrophobic ones) established a highly rigid environment from which AIE could occur. In the case of bis-calixarene **26**, a different morphology was formed with an overall host–guest molar ratio of 1 : 4. The occurrence of spherical nanoparticles with a size of c. 120 nm was attributed to the coiling of the initially linear supramolecular polymer, with π–π stacking interactions stabilizing the coil (Figure 8.39b). Also from this nanostructure, AIE could be observed. It is well known that TPE derivatives can be cyclized upon UV irradiation into the corresponding diphenylphenanthrene (DPP) species, that exhibit aggregation-caused quenching (ACQ) in a condensed or rigid environment [108]. Accordingly, the parent nonfluorescent TPE dye **42** turned into a fluorescent DPP, whose fluorescence was quenched upon self-assembly with **24** or **26**. Also

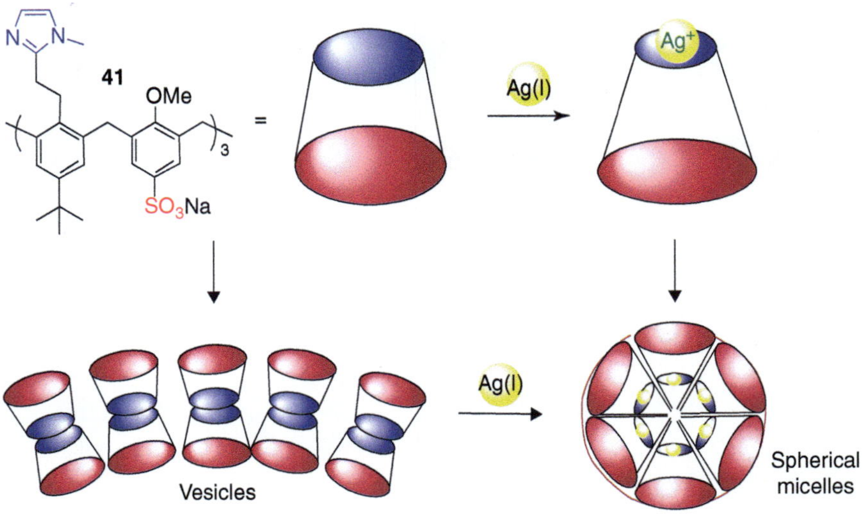

Figure 8.38 Schematic representation of the self-assembly of **41** into vesicles and, in the presence of Ag(I) ions, spherical micelles. Source: Redrawn from ref. Houmadi et al. [122], © 2007 American Chemical Society.

the AIE of the preassembled nanostructures could be quenched due to a fast photocyclization within the compact supramolecular environment.

Aggregation of hydrophobic dyes represents a limiting factor in utilizing such compounds in biomedical applications, e.g. as photosensitizers (PS) in the photodynamic therapy (PDT). This restriction can (at least partially) be overcome by using certain types of nanocarriers for the controlled delivery of such photosensitizers [109]. Among others (e.g. mesoporous Si nanoparticles, liposomes, or polymeric micelles) [110], supramolecular assemblies can offer a homogeneous dispersion of the PSs, and thus, suppress the loss of activity due to self-aggregation. Beyond this, the tumor-targeting selectivity of delivery systems can be increased by introducing special targeting units. Against this background, Zhu and coworkers prepared the water-soluble calix[4]arene **43** with four hydrophilic PEG arms on its lower rim (Figure 8.40) [111]. Due to its large π-conjugated cavity, the hydrophobic PS chlorin-e6 (Ce6) could be bound in a host–guest-type fashion. These supramolecular amphiphiles assembled further into micelles of core–shell morphologies with a **43**-to-Ce6 ratio of 1 : 1 (average diameter of 139 nm, polydispersity of 0.23; Figure 8.40). These supramolecular micelles were tested regarding their applicability in the PDT and a preferential delivery of these into tumor cells was observed. The negative charges of the PS were efficiently shielded due to micelle formation, thus significantly enhancing the cellular uptake. The PDT efficiency of Ce6 was improved significantly by its incorporation into the supramolecular assembly: for the micelles, an IC_{50} value, representing the inhibition of the cellular growth by 50%, of 1.6 $\mu mol\, l^{-1}$ was found; a much higher value was observed when Ce6 was administered alone.

In a similar fashion, an amphiphilic calix[4]arene was employed by Di Bari et al. for the formation of micellar nanocarriers for the encapsulation of a hydrophilic

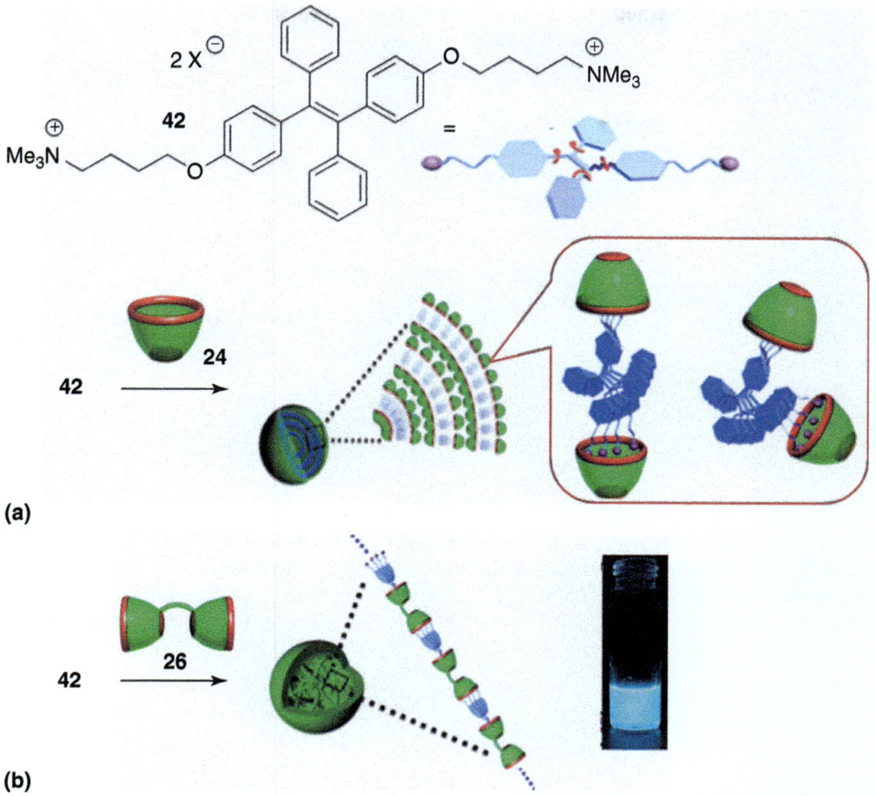

Figure 8.39 Schematic representation of the homoditopic guest **42** that was assembled into spherical nanostructures in the presence of the sulfonated calix[4]arene **24** (a) or bis-calix[4]arene **26** (b). Both types of nanostructures exhibited aggregation-induced emission. Source: Redrawn from ref. Jiang et al. [107], © 2014 American Chemical Society.

phthalocyanine or a hydrophobic porphyrin PS [112]. The PDT performance of these assemblies was proven by their inactivation of *Staphylococcus aureus* and *Pseudomonas aeruginosa*, as examples for Gram-positive and Gram-negative bacteria, respectively, upon irradiation with visible light.

The Liu group prepared a supramolecular amphiphile with enzyme-responsive properties from **24** ($n = 4$), as a macrocyclic host, and myristoylcholine, a naturally occurring and enzyme-cleavable surfactant (Figure 8.41) [113]. Due to the CIA effect, self-assembly of the components into spherical vesicles occurred at concentrations far below the cac of the surfactant (0.02 vs. 2.5 mM). According to DLS measurements, an average diameter of 194 nm and a narrow size distribution were found. High-resolution and cryo-TEM were applied to analyze the structure of the hollow spheres: the vesicles had an average membrane thickness of c. 4 nm, matching twice the length of a supramolecular amphiphile. In the proposed bilayer-type membrane, the alkyl chains of the host–guest assembly were densely packed via van-der-Waals interactions to give the hydrophobic phase and, consequently, the hydrophilic moieties of the amphiphile were located on the inside and outside

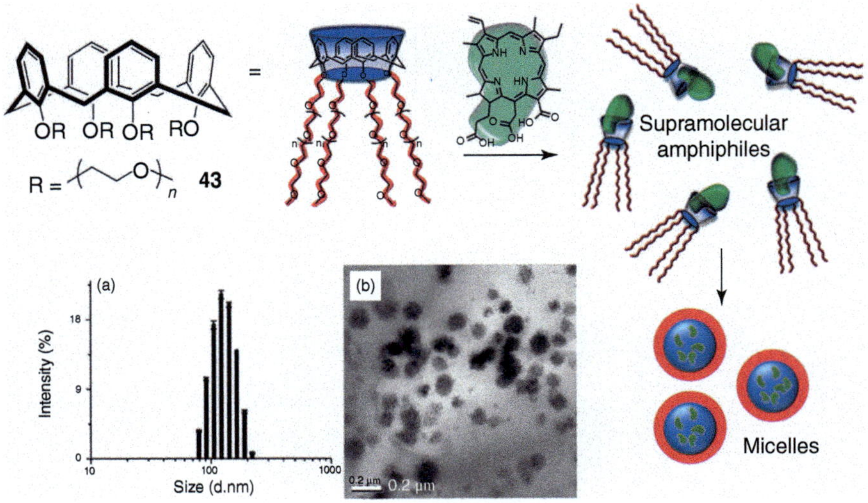

Figure 8.40 Schematic representation of calix[4]arene **43** as a host for the self-assembly into supramolecular micelles with chlorin-e6 (Ce6). The size distribution as determined by DLS measurements (a) and a TEM image of the micelles (b) are also depicted. Source: Tu et al. [111]. Figure reproduced with kind permission; © 2011 The Royal Chemical Society.

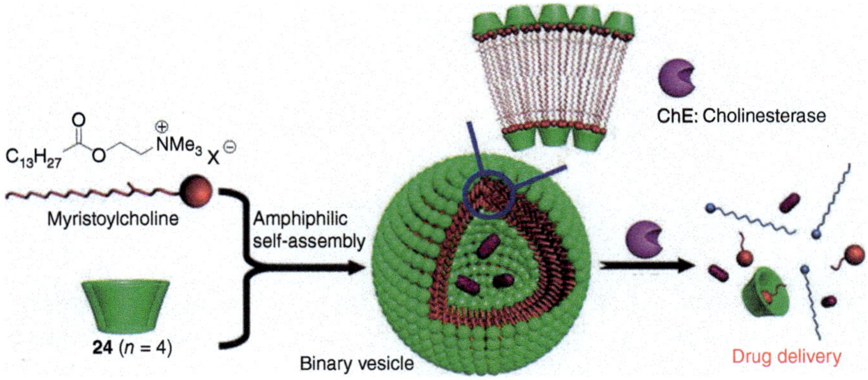

Figure 8.41 Schematic representation of an enzyme-responsive supramolecular amphiphile for drug-delivery applications. Source: Guo et al. [113], © 2012 American Chemical Society.

of this membrane (Figure 8.41). The enzyme responsiveness of the surfactant allowed a cleavage by a cholinesterase (ChE), thus leading to the full collapse of the vesicular structure, as monitored by DLS measurements and microscopy imaging. Moreover, the vesicles served as nanocarriers for tacrine, a prominent water-soluble ChE inhibitor. The enzyme-induced hydrolysis of myristoylcholine destroyed the hydrophilic–hydrophobic balance of the binary vesicles and released the drug molecules from the collapsing nanocontainers. A remarkable sustained-release effect was observed, since the drug-loaded vesicles were only partially broken in the presence of ChE; whereas, the drug-free vesicles were fully hydrolyzed under the

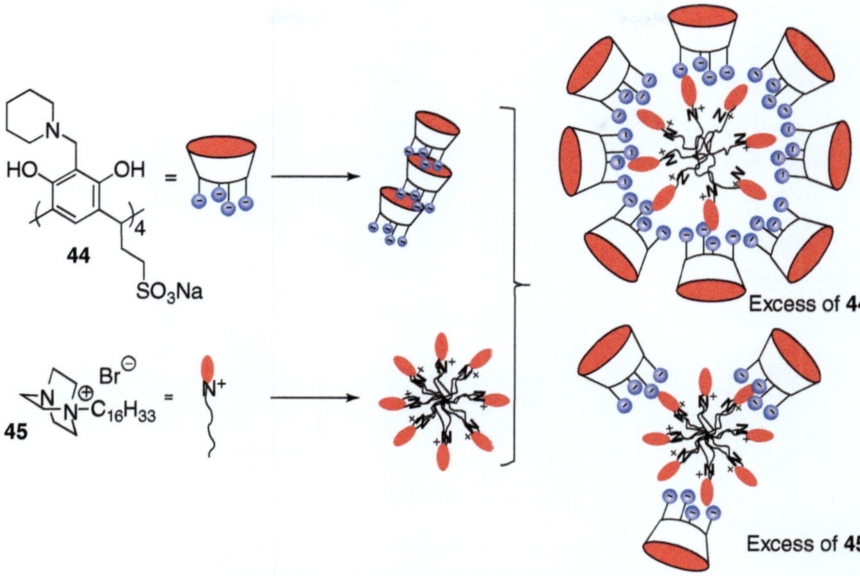

Figure 8.42 Schematic representation of the self-assembly of **44** and **45** into supramolecular amphiphiles. Source: Redrawn from Kharlamov et al. [115], © 2013 American Chemical Society.

same conditions. This finding might be of relevance to improve the drug activity and, simultaneously, suppress the undesired side effects, which result from an inflated tacrine release [114].

Not only calix[*n*]arenes but also some other, structurally related macrocycles have been utilized for the formation of supramolecular amphiphiles, e.g. resorcin[4]arene [115] and an azacalix[6]pyridine [116]. Kharlamov et al. synthesized supramolecular amphiphiles, based on the resorcin[4]arene host **44** and the DABCO-derived cationic guest surfactant **45** (DABCO: 1,4-diazabicyclo[2.2.2]octane, Figure 8.42) [115]. For **44**, its self-assembly in water occurred leading to a supramolecular polymer in which the monomers were ordered in a head-to-tail-type manner; the cationic surfactant afforded spherical micelles at concentrations above the critical micellization concentration (cmc) of 0.8 mM. CIA decreased this value down to 0.1 mM when **44** was added to a solution of **45**. Due to electrostatic repulsion between the negatively charged upper rim of **44** and the cationic head of the surfactant, the formation of an inclusion complexes was prevented. Instead, electrostatic interactions between the surfactant and anionic residues on the lower rim afforded a supramolecular amphiphile. These assemblies, in turn, amassed further into spherical micelle-like structures, whose morphologies could be adjusted via the molar ratio of the underlying small-molecule components.

Yi et al. used the Se-containing amphiphilic guest **46** and combined it with the *N*-rich azacalix[6]pyridine **47** to obtain a supramolecular amphiphile [116]. In contrast to the latter one, which only gave ill-defined precipitates from water due to its low solubility, **46** gave spherical micelles featuring a narrow size distribution

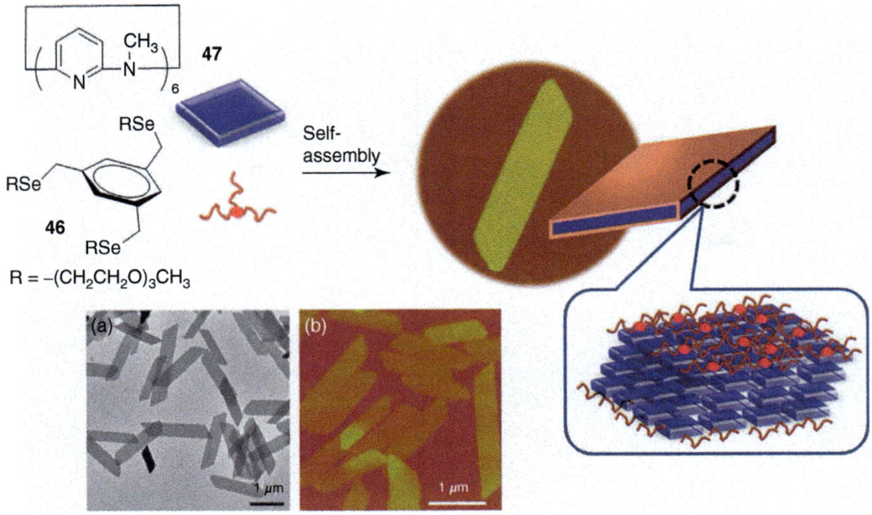

Figure 8.43 Schematic representation of the self-assembly of **46** and **47** into supramolecular nanosheets. The TEM (a) and AFM (b) images of these nanosheets are also depicted. Well-defined nanosheets, which are 1–2 mm long and 300–500 nm wide, are observed. Source: Yi et al. [116]. Figure reproduced with kind permission; © 2012 The Royal Chemical Society

(average diameter of 140 nm). The supramolecular amphiphile comprising **46** and **47**, however, gave highly ordered nanosheets in water (Figure 8.43). Non-covalent, intermolecular Se···N interactions in concert with hydrophobic interactions between **46** and **47** were identified as driving forces for this self-assembly process. TEM imaging as well as electron and X-ray diffraction (XRD) measurements revealed a triclinic single-crystalline structure of the nanosheets, with an interlayer distance of 0.93 nm; the multilayered composition was concluded from AFM studies that indicated an overall thickness of 24–35 nm (Figure 8.43). The nanosheets were 1–2 mm in length and had a width of 300–500 nm. The authors suggested the following mechanism for the self-assembly process: first, the solubility of **47** was enhanced by the surfactant **46**, thus facilitating the initial crystal growth. The host–guest inclusion complexes of **46** and **47** were subsequently added to the nanocrystals' surface, thereby not only suppressing the growth of the face packing dimensions, but also supporting the nanosheet formation. Finally, the adsorption of surfactant molecules fully terminated the crystal growth, eventually affording the well-defined nanosheets. To clarify the structural role of the Se-atoms, an analogous guest was employed in which the three Se centers were replaced by oxygen. This guest gave only irregular aggregates from the self-assembly with **47**. The authors also showed that the supramolecular amphiphile of **46** and **47** exhibited a dual-responsive character. Upon treatment with HCl, the nanosheets could be fully decomposed due to protonation of the *N*-rich host **47**; alternatively, the nanosheets could be destroyed (partially) by oxidizing the Se atoms with H_2O_2.

To the end of this section, the formation of polypseudorotaxanes – covalent polymer chains onto which macrocyclic species have been threaded – needs to be

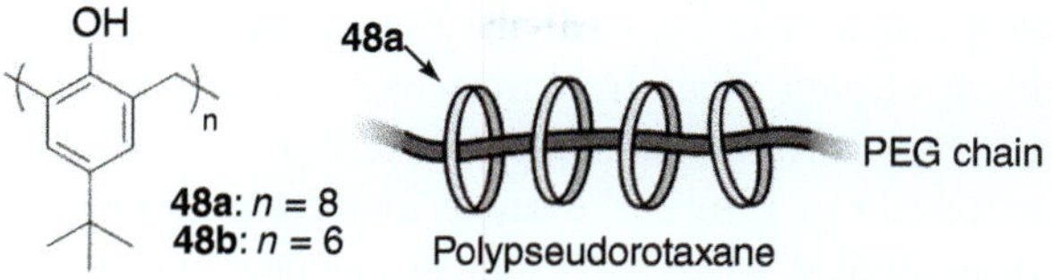

Figure 8.44 Schematic representation of a calix[8]arene-based polypseudorotaxane. Source: Yamagishi et al. [119], © 2001 American Chemical Society.

mentioned [117, 118]. Such supramolecular polymers have attracted considerable attention regarding applications in material's science and the threading of CDs [3, 4] and, though to a minor extent, crown ethers [117] have been reported (for examples in this respect see Chapters 9 and 6). As outlined there, the formation of main-chain polypseudorotaxanes can be realized via two basic strategies [4]: polymerization of preassembled pseudorotaxanes and threading of macrocycles onto preformed polymers. Regarding calixarene-based systems neither of these has yet been realized. However, the *in situ* calixarene formation from 4-*tert*-butylphenol and paraformaldehyde in the presence of PEGs afforded polypseudorotaxanes comprised of calixarene moieties threaded onto the PEG chains [119]. Thereby, the molar mass of the employed PEG was found to be crucial to form such assemblies: products were only obtained in the molar mass regime of 1000 to 10 000 g mol^{-1}. The resultant materials were investigated by FT-IR and ^1H NMR spectroscopy as well as DSC measurements. The FT-IR spectra and DSC curves were significantly different from those of binary blends of linear 4-*tert*-butylphenolic resins or various 4-*tert*-butylcalix[*n*]arenes with pure PEGs. Dethreading of the calixarene rings from the PEG backbone could be achieved by thermal treatment of the polypseudorotaxane (ethylene glycol at 180 °C). NMR spectroscopic analysis of the dethreaded macrocycles revealed the presence of 4-*tert*-butylcalix[8]arene (**48**) as the almost exclusive derivative. Thus, the polypseudorotaxane prepared by the *in situ* polycondensation approach featured calixarenes **48** rings threaded onto the linear PEG chains. It has to be assumed that the PEG chains offered a template effect that favored the formation of the calix[8]arene on the cost of smaller or larger analogous. In a similar fashion, Sokolova et al. synthesized a polyrotaxane in which **48b** was threaded onto a poly(dimethylsiloxane) (PDMS) backbone; due to the presence of the bulky triphenylmethyl end groups, dethreading was disabled [120].

8.7 Concluding Remarks

Calixarenes and their structural relatives exhibit a rich supramolecular chemistry ranging from the so-called supercaps to supramolecular polymers based on host–guest inclusion complexes to, finally, supramolecular amphiphiles. However, considering the number of publications, calixarenes still remain a rather less used type of macrocycle employed in the field of supramolecular polymerization, in particular, when compared to the much more common crown ethers or CDs [18]. Böhmer proclaimed calixarenes to be "macrocycles with (almost) unlimited possibilities" [121] and, indeed, calixarenes can easily be functionalized on their rims to fine-tune their binding affinity. The structural characteristics of their cavities offer a remarkable advantage over, e.g. crown ethers or CDs: for the latter ones, there are two pathways a guest may enter the cavity, and thus, competitive self-assembly arises as an issue to deal with. For example, AB-type monomers can establish an equilibrium between intra- and intermolecular association (i.e. self-inclusion complexes vs. linear oligomers/polymers). Similar considerations can also be made when using AA- and BB-type monomers. These restrictions, however, do not apply for calixarenes, since a guest will, for steric reasons, always enter the cavity via the wider upper rim.

Many calixarene-based supramolecular assemblies exhibit a (multiple) stimuli-responsive character that can be utilized for a wide range of applications that have also been addressed within the scope of this chapter [121], such as self-healing, entrapment, storage, controlled release, sensing, photochemistry, informational encoding, and chiral memory. Thus, supramolecular polymers based on calixarenes can be expected to remain an emerging field of research in the forthcoming years.

References

1 Yang, L., Tan, X., Wang, Z., and Zhang, X. (2015). *Chem. Rev.* 115: 7196–7239.
2 Zheng, B., Wang, F., Dong, S.-Y., and Huang, F.-H. (2012). *Chem. Soc. Rev.* 41: 1621–1636.
3 Harada, A., Takashima, Y., and Yamaguchi, H. (2009). *Chem. Soc. Rev.* 38: 875–882.
4 Wenz, G., Han, B.-H., and Müller, A. (2006). *Chem. Rev.* 106: 782–817.
5 Barrow, S.J., Kasera, S., Rowland, M.J. et al. (2015). *Chem. Rev.* 115: 12320–12406.
6 Ogoshi, T., Yamagishi, T.-a., and Nakamoto, Y. (2016). *Chem. Rev.* 116: 7937–8002.
7 Gutsche, C.R. (1998). *Calixarenes Revisited*. Cambridge: The Royal Chemical Society.
8 L. Bakeland (1909). Method of making insoluble products of phenol and formaldehyde. US Patent 942699, USA. 07 December 1909.
9 Kappe, T. (1994). *J. Inclusion Phenom. Mol. Recogn. Chem.* 19: 3–15.
10 Zinke, A. and Ziegler, E. (1944). *Ber. Dtsch. Chem. Ges.* 77: 264–272.
11 Niederl, J.B. and Vogel, H.J. (1940). *J. Am. Chem. Soc.* 62: 2512–2514.
12 Cornforth, J.W., Morgan, E.D., Potts, K.T., and Rees, R.J.W. (1973). *Tetrahedron* 29: 1659–1667.
13 Andreetti, G.D., Ori, O., Ugozzoli, F. et al. (1988). *J. Inclusion Phenom.* 6: 523–536.
14 Gutsche, C.D. and Muthukrishnan, R. (1978). *J. Org. Chem.* 43: 4905–4906.
15 Homden, D.M. and Redshaw, C. (2008). *Chem. Rev.* 108: 5086–5130.
16 Zhu, W., Gou, P., and Shen, Z. (2008). *Macromol. Symp.* 261: 74–84.

17 Wosnick, J.H. and Swager, T.M. (2004). *Chem. Commun.*: 2744–2745.
18 Guo, D.-S. and Liu, Y. (2012). *Chem. Soc. Rev.* 41: 5907–5921.
19 Dalcanale, E. and Pinalli, R. (2015). *Encyclopedia of Polymeric Nanomaterials* (eds. S. Kobayashi and K. Müllen), 1–17. Heidelberg: Springer.
20 Ovsyannikov, A., Solovieva, S., Antipin, I., and Ferlay, S. (2017). *Coord. Chem. Rev.* 352: 151–186.
21 Castellano, R.K., Rudkevich, D.M., and Rebek, J. Jr., (1997). *Proc. Natl. Acad. Sci. U.S.A.* 94: 7132–7137.
22 Shimizu, K.D. and Rebek, J. Jr., (1995). *Proc. Natl. Acad. Sci. U.S.A.* 92: 12403–12407.
23 Mogck, O., Böhmer, V., and Vogt, W. (1996). *Tetrahedron* 52: 8489–8496.
24 Castellano, R.K. and Rebek, J. Jr., (1998). *Macromolecules* 120: 3657–3663.
25 Castellano, R.K., Nuckolls, C., Eichhorn, S.H. et al. (1999). *Angew. Chem. Int. Ed.* 38: 2603–2606.
26 Castellano, R.K., Clark, R., Craig, S.L. et al. (2000). *Proc. Natl. Acad. Sci. U.S.A.* 97: 12418–12421.
27 Xu, H., Stampp, S.P., and Rudkevich, D.M. (2003). *Org. Lett.* 5: 4583–4586.
28 George, M. and Weiss, R.G. (2002). *Langmuir* 18: 7124–7135.
29 Xu, H., Hampe, E.M., and Rudkevich, D.M. (2003). *Chem. Commun.*: 2828–2829.
30 Xu, H. and Rudkevich, D.M. (2004). *Chem. Eur. J.* 10: 5432–5442.
31 Xu, H. and Rudkevich, D.M. (2005). *Org. Lett.* 7: 3223–3226.
32 Xu, H. and Rudkevich, D.M. (2004). *J. Org. Chem.* 69: 8609–8617.
33 Brody, M.S., Schalley, C.A., Rudkevich, D.M., and Rebek, J. Jr., (1999). *Angew. Chem. Int. Ed.* 38: 1640–1644.
34 Podoprygorina, G., Janke, M., Janshoff, A., and Böhmer, V. (2008). *Supramol. Chem.* 20: 59–69.
35 Bügler, J., Sommerdijk, N.A.J.M., Visser, A.J.W.G. et al. (1999). *J. Am. Chem. Soc.* 121: 28–33.
36 Bügler, J., Engbersen, J.F.J., and Reinhoudt, D.N. (1998). *J. Org. Chem.* 63: 5339–5344.
37 Liu, Y., Li, Z.F., Zhang, H.-Y. et al. (2002). *Nano Lett.* 2: 257–261.
38 Liu, Y., Wang, H., Zhang, H.-Y. et al. (2003). *Chem. Lett.* 32: 884–885.
39 Liu, Y., Wang, H., Zhang, H.-Y., and Liang, P. (2004). *Chem. Commun.*: 2266–2267.
40 Whitesides, G.M., Mathias, J.P., and Seto, C.T. (1991). *Science* 254: 1312–1319.
41 Timmerman, P. and Prins, L.J. (2001). *Eur. J. Org. Chem.* 2001: 3191–3205.
42 Klok, H.-A., Jolliffe, K.A., Schauer, C.L. et al. (1999). *J. Am. Chem. Soc.* 121: 7154–7155.
43 Choi, I.S., Li, X., Simanek, E.E. et al. (1999). *Chem. Mater.* 11: 684–690.
44 Korshin, D.E., Kashapov, R.R., Murtazina, L.I. et al. (2009). *New J. Chem.* 33: 2397–2401.
45 Baldini, L., Melegari, M., Bagnacani, V. et al. (2011). *J. Org. Chem.* 76: 3720–3732.

46 Sliwa, W. and Kozlowski, C. (2009). *Calixarenes and Resorcinarenes*. Weinheim: Wiley-VCH.

47 Ishihara, S. and Takeoka, S. (2006). *Tetrahedron Lett.* 47: 181–184.

48 Dutasta, J.-P. (2004). *Top. Curr. Chem.* 232: 55–91.

49 De Zorzi, R., Dubessy, B., Mulatier, J.-C. et al. (2007). *J. Org. Chem.* 72: 4528–4531.

50 Kalenius, E., Moiani, D., Dalcanale, E., and Vainiotalo, P. (2007). *Chem. Commun.*: 3865–3867.

51 Biavardi, E., Battistini, G., Montalti, M. et al. (2008). *Chem. Commun.*: 1638–1640.

52 Yebeutchou, R.M., Tancini, F., Demitri, N. et al. (2008). *Angew. Chem. Int. Ed.* 47: 4504–4508.

53 Francoise, A.-N., Saowarux, F., Anna, N. et al. (1998). *Angew. Chem. Int. Ed.* 37: 112–114.

54 Ferguson, G., Notti, A., Pappalardo, S. et al. (1998). *Tetrahedron Lett.* 39: 1965–1968.

55 Garozzo, D., Gattuso, G., Kohnke, F.H. et al. (2002). *Tetrahedron Lett.* 43: 7663–7667.

56 Pappalardo, S., Villari, V., Slovak, S. et al. (2007). *Chem. Eur. J.* 13: 8164–8173.

57 Gargiulli, C., Gattuso, G., Notti, A. et al. (2011). *Tetrahedron Lett.* 52: 6460–6464.

58 Capici, C., Cohen, Y., D'Urso, A. et al. (2011). *Angew. Chem. Int. Ed.* 50: 11956–11961.

59 Garozzo, D., Gattuso, G., Kohnke, F.H. et al. (2003). *Org. Lett.* 5: 4025–4028.

60 Gattuso, G., Notti, A., Pappalardo, A. et al. (2008). *J. Org. Chem.* 73: 7280–7289.

61 Haino, T., Yanase, M., and Fukazawa, Y. (1998). *Angew. Chem. Int. Ed.* 37: 997–998.

62 Yanase, M., Haino, T., and Fukazawa, Y. (1999). *Tetrahedron Lett.* 40: 2781–2784.

63 Haino, T., Matsumoto, Y., and Fukazawa, Y. (2005). *J. Am. Chem. Soc.* 127: 8936–8937.

64 Haino, T., Hirai, E., Fujiwara, Y., and Kashihara, K. (2010). *Angew. Chem. Int. Ed.* 49: 7899–7903.

65 Han, Y., Tian, Y., Li, Z., and Wang, F. (2018). *Chem. Soc. Rev.* 47: 5165–5176.

66 Takehiro, H., Masatoshi, T., Shigeru, Y., and Takeharu, H. (2014). *Chem. Eur. J.* 20: 16138–16146.

67 Hirao, T., Fukuta, K., and Haino, T. (2020). *Macromolecules* 53: 3563–3570.

68 Fraix, A., Torrisi, V., Marletta, G. et al. (2016). *Supramol. Chem.* 28: 485–492.

69 Tancini, F., Yebeutchou, R.M., Pirondini, L. et al. (2010). *Chem. Eur. J.* 16: 14313–14321.

70 Guo, D.-S., Wang, K., and Liu, Y. (2008). *J. Inclusion Phenom.Macrocyclic Chem.* 62: 1–21.

71 Guo, D.-S., Chen, K., Zhang, H.-Q., and Liu, Y. (2009). *Chem. Asian J.* 4: 436–445.

72 D'Urso, A., Cristaldi, D.A., Fragalà, M.E. et al. (2010). *Chem. Eur. J.* 16: 10439–10446.

73 Guo, D.-S., Chen, S., Qian, H. et al. (2010). *Chem. Commun.* 46: 2620–2622.

74 Qian, H., Guo, D.-S., and Liu, Y. (2012). *Chem. Eur. J.* 18: 5087–5095.

75 Guo, D.-S., Wang, L.-H., and Liu, Y. (2007). *J. Org. Chem.* 72: 7775–7778.

76 Jeon, W.S., Ziganshina, A.Y., Lee, J.W. et al. (2003). *Angew. Chem. Int. Ed.* 42: 4097–4100.

77 Ma, X., Sun, R., Li, W., and Tian, H. (2011). *Polym. Chem.* 2: 1068–1070.

78 Sun, R., Xue, C., Ma, X. et al. (2013). *J. Am. Chem. Soc.* 135: 5990–5993.

79 Rather, I.A., Wagay, S.A., Hasnain, M.S., and Ali, R. (2019). *RSC Adv.* 9: 38309–38344.

80 Wu, Y.-D., Wang, D.-F., and Sessler, J.L. (2001). *J. Org. Chem.* 66: 3739–3746.

81 Baeyer, A. (1886). *Ber. Dtsch. Chem. Ges.* 19: 2184–2185.

82 Chelintzev, V.V., Tronov, B.V., and Russ, J. (1916). *Phys. Chem. Soc.* 48: 105–155.

83 Gale, P.A., Sessler, J.L., Král, V., and Lynch, V. (1996). *J. Am. Chem. Soc.* 118: 5140–5141.

84 Saha, I., Lee, J.T., and Lee, C.-H. (2015). *Eur. J. Inorg. Chem.* 2015: 3859–3885.

85 Kim, D.S. and Sessler, J.L. (2015). *Chem. Soc. Rev.* 44: 532–546.

86 Ballester, P. (2011). *Isr. J. Chem.* 51: 710–724.

87 Adriaenssens, L. and Ballester, P. (2013). *Chem. Soc. Rev.* 42: 3261–3277.

88 Nielsen, K.A., Cho, W.-S., Jeppesen, J.O. et al. (2004). *J. Am. Chem. Soc.* 126: 16296–16297.

89 Park, J.S., Le Derf, F., Bejger, C.M. et al. (2010). *Chem. Eur. J.* 16: 848–854.

90 Park, J.S., Yoon, K.Y., Kim, D.S. et al. (2011). *Proc. Natl. Acad. Sci. U.S.A.* 108: 20913–20917.

91 Kim, D.S., Lynch, V.M., Park, J.S., and Sessler, J.L. (2013). *J. Am. Chem. Soc.* 135: 14889–14894.

92 Bähring, S., Kim, D.S., Duedal, T. et al. (2014). *Chem. Commun.* 50: 5497–5499.

93 Yuvayapan, S. and Aydogan, A. (2019). *Chem. Commun.* 55: 8800–8803.

94 Amharar, S., Yuvayapan, S., and Aydogan, A. (2018). *Chem. Commun.* 54: 829–832.

95 Yuvayapan, S. and Aydogan, A. (2019). *Eur. J. Org. Chem.* 2019: 633–639.

96 Guo, D.-S., Jiang, B.-P., Wang, X., and Liu, Y. (2012). *Org. Biomol. Chem.* 10: 720–723.

97 Zhang, X. (ed.) (2017). *Supramolecular Amphiphiles*. Croydon: The Royal Chemical Society.

98 Yu, G., Jie, K., and Huang, F. (2015). *Chem. Rev.* 115: 7240–7303.

99 Zhang, X. and Wang, C. (2011). *Chem. Soc. Rev.* 40: 94–101.

100 Zhang, H., Liu, Z., and Zhao, Y. (2018). *Chem. Soc. Rev.* 47: 5491–5528.

101 García-Rio, L. and Basílio, N. (2019). *Curr. Opin. Colloid Interface Sci.* 44: 225–237.

102 Basilio, N. and García-Río, L. (2009). *Chem. Eur. J.* 15: 9315–9319.

103 Francisco, V., Basilio, N., Garcia-Rio, L. et al. (2010). *Chem. Commun.* 46: 6551–6553.

104 Wang, K., Guo, D.-S., and Liu, Y. (2010). *Chem. Eur. J.* 16: 8006–8011.
105 Wang, K., Guo, D.-S., Wang, X., and Liu, Y. (2011). *ACS Nano* 5: 2880–2894.
106 Zhang, Q., Yao, X., Qu, D.-H., and Ma, X. (2014). *Chem. Commun.* 50: 1567–1569.
107 Jiang, B.-P., Guo, D.-S., Liu, Y.-C. et al. (2014). *ACS Nano* 8: 1609–1618.
108 Aldred, M.P., Li, C., and Zhu, M.-Q. (2012). *Chem. Eur. J.* 18: 16037–16045.
109 Yi, G., Hong, S.H., Son, J. et al. (2018). *Quant. Imaging Med. Surg.* 8: 433–443.
110 Lucky, S.S., Soo, K.C., and Zhang, Y. (2015). *Chem. Rev.* 115: 1990–2042.
111 Tu, C., Zhu, L., Li, P. et al. (2011). *Chem. Commun.* 47: 6063–6065.
112 Di Bari, I., Fraix, A., Picciotto, R. et al. (2016). *RSC Adv.* 6: 105573–105577.
113 Guo, D.-S., Wang, K., Wang, Y.-X., and Liu, Y. (2012). *J. Am. Chem. Soc.* 134: 10244–10250.
114 Sameem, B., Saeedi, M., Mahdavi, M., and Shafiee, A. (2017). *Eur. J. Med. Chem.* 128: 332–345.
115 Kharlamov, S.V., Kashapov, R.R., Pashirova, T.N. et al. (2013). *J. Phys. Chem. C* 117: 20280–20288.
116 Yi, Y., Fa, S., Cao, W. et al. (2012). *Chem. Commun.* 48: 7495–7497.
117 Huang, F.-H. and Gibson, H.W. (2005). *Prog. Polym. Sci.* 30: 982–1018.
118 Arunachalam, M. and Gibson, H.W. (2014). *Prog. Polym. Sci.* 39: 1043–1073.
119 Yamagishi, T.-A., Kawahara, A., Kita, J. et al. (2001). *Macromolecules* 34: 6565–6570.
120 Sokolova, M.P., Bronnikov, S.V., Sukhanova, T.E. et al. (2010). *Russ. J. Appl. Chem.* 83: 109–114.
121 Böhmer, V. (1995). *Angew. Chem. Int. Ed.* 34: 713–745.
122 Houmadi, S., Coquière, D., Legrand, L. et al. (2007). *Langmuir* 23: 4849–4855.

9

Cyclodextrins in the Field of Supramolecular Polymers

9.1 General Aspects

Cyclodextrins (CDs) are the cyclic degradation products of starch – comprising glucopyranose units, which are connected selectively in an α-(1-4)-glycosidic manner – and were first isolated in 1891 [1]. Their oligosaccharide structure was identified by Schardinger just a few years later [2]. Initially only representing a laboratory object of oddity, they have emerged as an important family of compounds with a broad range of applications [3, 4]: biotechnology [5], food [6], industry [7], pharmaceutics [8, 9], and/or textile applications [10]. Their ease of functionalization, solubility in water, and biocompatibility are their most important properties that are responsible for their rapid evolution into diverse market places [11]. For a detailed view on the remarkable history of cyclodextrins, the reader is referred to Crini's review on this topic [12].

The most important CD structures are composed of 6–8 glucose moieties and are named the α-, β-, and γ-CD, respectively (Figure 9.1). Each representative features a hydrophilic shell and a hydrophobic cavity with a characteristic diameter capable of forming inclusion complexes in water due to host–guest interactions with a variety of organic molecules (cavity sizes of 4.7–5.3 Å and 6.0–6.5 Å for **1** and **2**, respectively) [4].

Similar to their crown ether relatives, the CDs also readily form pseudorotaxanes and rotaxanes simply via their threading onto an axis-type guest molecule through the cavity of the CD host. Dethreading, in turn, is enabled when one caps the termini of the guest with typical bulky stopper groups [13]. In the case of pseudorotaxanes, where dethreading is basically enabled, the high free activation energy ($\Delta G^{\ddagger}{}_{\text{diss}}$) required for dissociation offers sufficient stability to support these supramolecular assemblies (Figure 9.2) [11].

In this chapter, supramolecular polymers comprising CD-based, host–guest complexes will be evaluated. Concerning the binding of small molecules and the resulting potential applications, the reader is referred to the abovementioned review articles. As already presented in Chapter 6, the covalent linkage of (pseudo)rotaxanes gives oligomeric or even polymeric species, the so-called poly(pseudo)rotaxanes. Thereby, linkage of the subunits may occur via the CD entities, the axis molecules, or a combination of both. Concerning the nomenclature

Supramolecular Polymers and Assemblies: From Synthesis to Properties and Applications, First Edition.
Ulrich S. Schubert, George R. Newkome, and Andreas Winter.

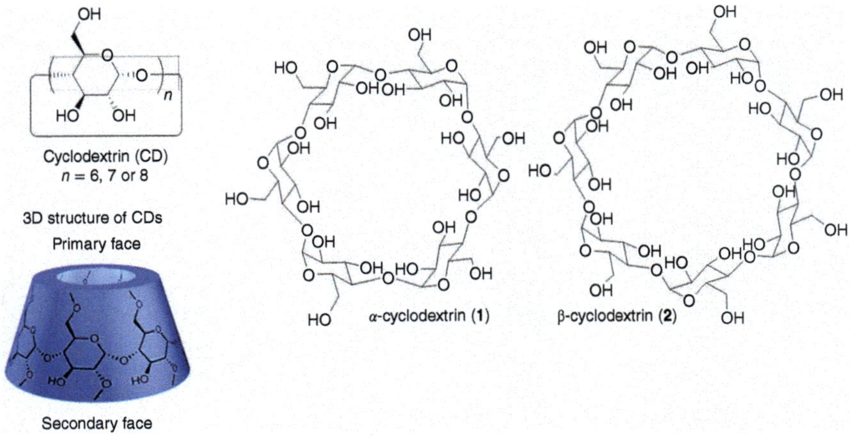

Figure 9.1 Schematic representation of basic cyclodextrin (CD) structures. Source: Szejtli [4]. © 1997 American Chemical Society.

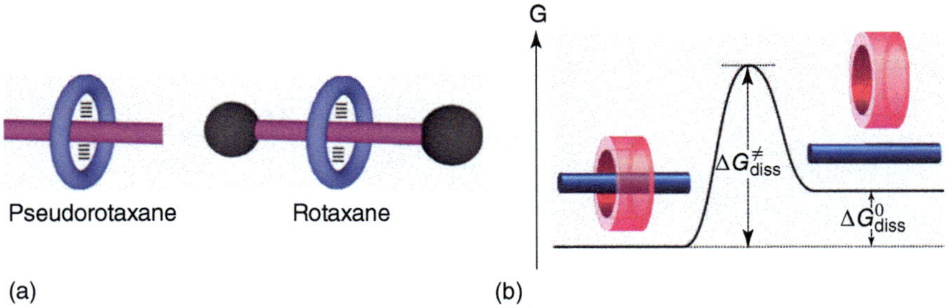

Figure 9.2 (a) Schematic representations of a pseudorotaxane and a rotaxane. (b) Qualitative representation of the energy scheme for the disassembly of a [2] rotaxane into its molecular components. Source: Wenz et al. [11]. © 2006 American Chemical Society.

of rotaxanes, the reader is reminded that a [*n*]rotaxane is a system formed by *n* molecular components, i.e. ring and axis molecules (the same definition holds true for pseudorotaxanes) [14].

9.2 Cyclodextrins and Supramolecular Polymers

Various types of polymer topologies have been realized during the basic research on the guest inclusion within the cavities of CDs. Besides the linear, hyperbranched, or cross-linked species that are obtained from the supramolecular polymerization of appropriate unimers, supramolecular polymers can also be obtained from the threading of CD molecules onto the main chain (or the side chain) of the preformed polymers [15]. Thus, linear, graft/comb, or star-shaped architectures of poly(pseudo)rotaxane type can be observed [16]. The two fundamentally different approaches to assemble linear polypseudorotaxanes are depicted in Figure 9.3 [11].

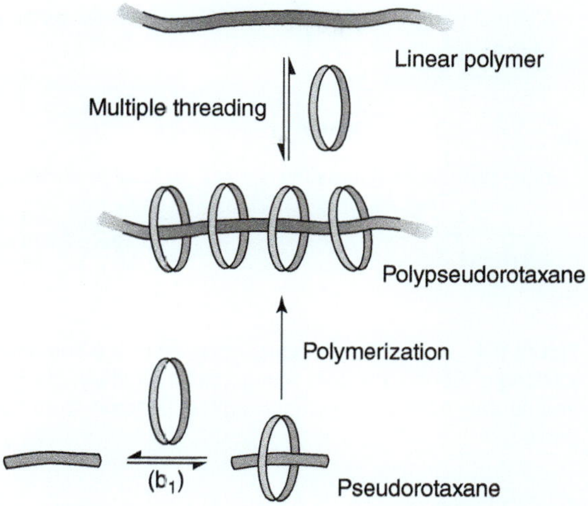

Figure 9.3 Schematic representation of the synthesis of polypseudorotaxanes via two different approaches: multiple threading onto a preformed polymer and polymerization of a preformed pseudorotaxane monomer (a so-called supramonomer). Source: Redrawn from Wenz [11]. © 2006 American Chemical Society.

9.2.1 Linear Polypseudorotaxanes, Derived from Preformed Linear Polymers

In 1990, Harada and Kamachi were the first to report the multiple threading of CD moieties onto a linear polymer chain, as shown in Figure 9.3 [17]. This approach was recognized to be highly versatile to generate linear CD-based polypseudorotaxanes and was applied in subsequent years to a plethora of (commercially available) polymers [11, 18, 19]. The ability of CDs to be threaded onto preformed polymer chains resembles a unique feature that can seldom be observed for a range of related macrocyclic hosts. Basically, the length of the polymer chain dictates the length of the resulting polypseudorotaxane, thus offering simple and straightforward control over the size of the entire supramacromolecular assembly. To predict if such a polypseudorotaxane can be formed via the multiple-threading method, some general considerations still have to be made [11]: a sufficiently high binding enthalpy is required for the threading of a polymer chain through a CD ring – the entropy involved in this process is typically small and may even favor the reverse process. The binding itself depends on an interplay of various non-covalent forces: a combination of hydrophobic and van der Waals interactions accounts for the binding of threaded CD rings onto a linear polymer chain; further, hydrogen bonds between neighboring CD moieties, threaded onto the polymer, aid to increase the strength of the interaction(s). The latter one was found to be the most efficient when the CD rings are arranged in strictly alternating orientation along the chain, thereby maximizing the overall number of H-bonding interactions and stabilizing the resultant polypseudorotaxane (i.e. the interaction of two primary faces and two secondary faces with each other is much stronger than the interaction between a primary and secondary face) [20]. Such highly ordered polypseudorotaxanes can crystallize into channel-type structures and are almost insoluble in water; yet, they are soluble in polar solvents (e.g. dimethylsulfoxide [DMSO]), thus also prone to dissociation in these environments. The alternate packing mode of the threaded CD rings could be confirmed by X-ray diffraction (XRD) studies on model

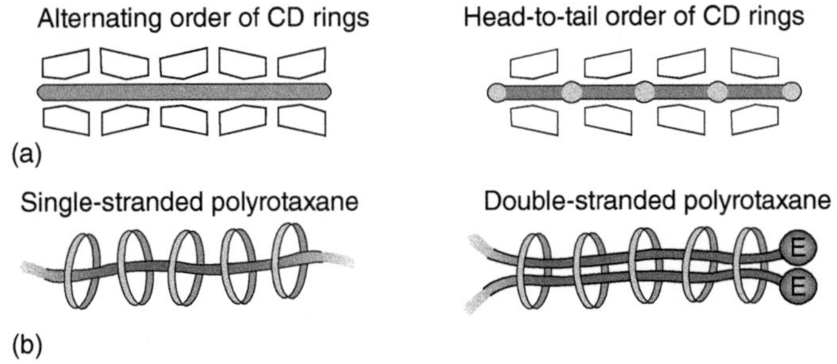

Figure 9.4 (a) Schematic representation of the two possible highly ordered modes for the packing of CD macrocycles along a polymer chain. (b) Schematic representation of a single- and double-stranded assembly. Source: Redrawn from Ref. [11]. © 2006 American Chemical Society.

assemblies formed by α-CD and monodisperse oligo(ethylene glycol) chains [21]. The alternative region-regular head-to-tail orientation can only be observed for those polymers having charged moieties along their main chain, thus preventing a dense packing of the threaded CD rings (e.g. polyamines [22, 23], polycarboxylates [24] or polysulfonates [25]; Figure 9.4a). As a rule of thumb, the interactions between a polymer and the CDs will be strong, if the polymer efficiently fills the CD cavity (otherwise, the distance-dependent van der Waals forces will be too weak to offer any structural stabilization). For example, poly(ethylene glycol) (PEG) will efficiently interact with α-CD to give a polypseudorotaxane but not with the wider β-homolog [17]; the even wider γ-CD was found to accommodate two polymer chains, thus forming a double-stranded assembly (Figure 9.4b) [26]. On the other hand, if the cavity of the host is too narrow to allow inclusion of a polymer, polypseudorotaxane formation will not occur at all (e.g. polyisobutylene [PIB] is, however, complexed selectively by γ-CD [27]). A more detailed overview for the matching/mismatching of polymers with the differently sized CDs is presented elsewhere [11, 18, 28].

Introducing substituents, on either the CD ring or polymer, will disfavor the polypseudorotaxane formation. Substituted CDs, in particular the per-methylated, basically allow for the synthesis of polypseudorotaxane with enhanced solubility in a range of solvents, including water [29–31]. Disadvantageously, such substitution excludes high threading densities for steric reasons, and, even more importantly, the formation of stabilizing hydrogen bonds between the CD rings is restricted. As a result, polypseudorotaxanes derived from linear polymers and substituted CDs are typically less stable. The chemical nature of the polymer is also of utmost importance for polypseudorotaxane formation [11]: the presence of hydrophilic groups along the polymer backbone make polyrotaxane formation in water basically becomes a thermodynamically disfavored process (due to the high desolvation energy that needs to be overcome to enable the threading of CD rings). The kinetics for polypseudorotaxane formation is, however, accelerated for highly water-soluble polymers. Consequently, polymers with hydrophilic moieties within the main chain readily assemble into polypseudorotaxane (under the conditions that the polymer chain is not likely to allow efficient threading, such as PEG). On the other hand, polymers with hydrophilic side chains do not give polypseudorotaxanes in water (e.g. poly(ethylene imine) (PEI) [32], poly(vinyl alcohol), polyacrylamide, or poly(*N*-vinylpyrrolidone) [33]). For solubility reasons, polypseudorotaxane formation with poly(tetrahydrofuran) (PTHF) in water is slow [34] and even impossible with highly hydrophobic polymers, such as polyethylene (PE).

9.2.1.1 Polypseudorotaxanes from of Poly(ethylene Glycol)

As mentioned above, Harada's polypseudorotaxane formed from PEG and α-CD represents the seminal example in this context [17, 33]. The combination of the water-soluble components gave the insoluble polypseudorotaxane, thus making the synthesis highly convenient (polypseudorotaxane formation was even observed under solvent-free conditions in liquid PEGs [35, 36]). It was shown that a minimal degree-of-polymerization (DP) is required to allow polypseudorotaxane formation (however, for $DP < 5$, a polypseudorotaxane was not formed due to disturbance with the terminal OH-groups). The authors used ^{2}H NMR spectroscopy to elucidate the conformation of the PEG chains on which the CD rings were threaded (for this purpose, per-deuterated PEG was used for polypseudorotaxane formation) [37]: the measurements revealed a preferred trans-conformation with a small fraction of confined cis-configured segments along the chain. With increasing molar mass, a full loading of the polymer chain with CD rings was, however, difficult to achieve [38, 39]. For kinetic reasons, the formation of polypseudorotaxane with polymers possessing very high molar masses (i.e. PEG with a DP of >1000) is greatly retarded [17]. To suppress H-bonding between the PEG's terminal OH-groups with those of the CD moieties that retard the threading process, inclusion of the corresponding methyl ethers readily occurred, even for PEGs with high molar masses [40].

Besides NMR spectroscopy, XRD analysis of crystalline polypseudorotaxanes – derived from monodisperse oligo(ethylene glycol)s (Figure 9.5a) [21] – and microscopy imaging (Figure 9.5b) are used to analyze the resultant polypseudorotaxanes. Turbidity measurements were also conducted to investigate the threading kinetics (Figure 9.5c) [41, 42]. After a certain induction period, attributed to the threading of CDs onto the chain, aggregation of the polypseudorotaxane occurred – leading to a steep increase in turbidity [43]. With increasing concentration of either component (i.e. PEG and/or α-CD) or the temperature, the induction period also increased – these findings were symbolic of supramolecular polypseudorotaxane formation representing an entropically disfavored process.

One unique example of a double-stranded polypseudorotaxane, which is reminiscent of the binding of the DNA double helix by DNA polymerases, was presented by Harada et al. (Figure 9.4b) [26]. In contrast to the α-CD, where single-stranded polyrotaxanes are formed in the presence of PEG, γ-CD does not give polypseudorotaxanes with PEG itself; however, with appropriate end groups, such as terminal naphthalene moieties, a double-stranded assembly was obtained. As verified by NMR and UV/vis spectroscopic studies, intermolecular binding of two naphthyl units within a huge γ-CD cavity (facilitated by π–π-stacking interactions between the

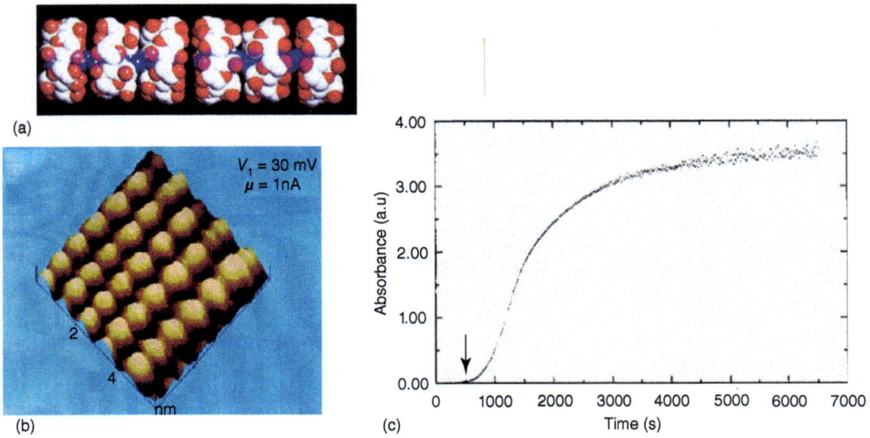

Figure 9.5 (a) Solid-state structure of an oligo(ethylene glycol) with threaded α-CD rings and (b) scanning-microscopy image of PEG with threaded α-CD rings. (c) Plot of the threading kinetics of PEG and α-CD (turbidity was measured by UV/vis absorption spectroscopy; the arrow indicates the end of the proposed induction period). Source: (a) Harada et al. [18]. © 2007 The Royal Chemical Society, (b) Arunachalam and Gibson [41]. © 2009 The Royal Chemical Society and © 1997 American Chemical Society, (c) Ceccato et al. [43]. © 1997 American Chemical Society.

aromatic systems) induced a template effect that forced the two polymer chains into the double-stranded assembly.

To better understand the threading kinetics, Inoue et al. employed a PEG–CD conjugate, in which the PEG was attached to a α-CD ring via a hydrocinnamoyl moiety [44]. The self-threading of this conjugate (to afford a [1]rotaxane) as well as the dethreading process in the presence of a competitive guest was monitored by ^{1}H NMR spectroscopy. The threading rate depended strongly on the PEG chain length was observed (PEG600: 382 s^{-1}; PEG3000: 177 s^{-1}). Placing a photoactive azobenzene moiety at the ends of the PEG chain enabled a thermal and a photochemical switching of the [1]rotaxane structure with high reversibility (Figure 9.6) [45].

9.2.1.2 Polypseudorotaxanes from Other Polyethers

The homologs to PEG, i.e. poly(oxotrimethylene) ($[CH_2O]_3$, P3OM) [46] and PTHF [34], were also investigated in view of their ability to form polypseudorotaxanes with CDs. In both cases, the decreased water solubility (compared to that of PEGs with similar molar masses) slowed the polypseudorotaxane formation that notably could still be accelerated by sonication. For P3OM, the self-assembly with β-CD was observed, and single crystals, suitable for X-ray structure analysis, were obtained [47]. As for the combination of PEG and α-CD, selective packing of the rings along the chain in a strictly alternate order was confirmed; due to the strong hydrogen-bonding interaction of adjacent CDs, the polypseudorotaxane featured a high degree-of-(kinetic)-stability. Methylated α- or β-CD could be threaded onto the PTHF strands. At low loading of the polymer with CD rings, reasonable solubility was achieved; whereas, crystalline precipitates were, however, formed at high CD concentrations. The methylation of CD generally reduces its ability to form

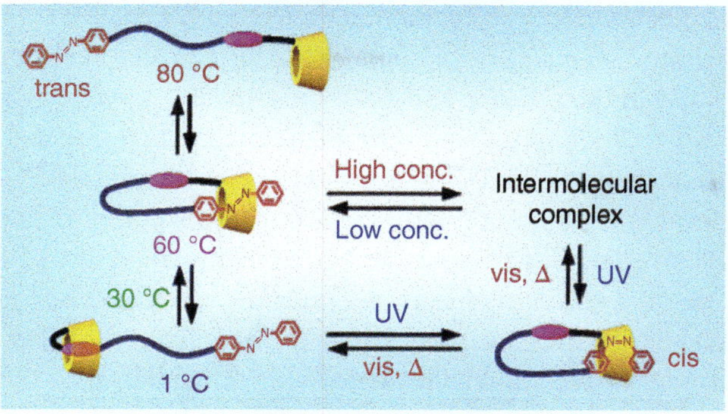

Figure 9.6 Schematic representation of the stimuli-responsive self-assembly of an azobenzene-PEG-CD conjugate. Source: Inoue et al. [45]. © 2007 The Royal Chemical Society.

stabilizing intermolecular H-bonds when threaded onto a polymer – thus, the threading density remained rather low [30].

The pending methyl groups in poly(propylene oxide) (PPO) only allowed the self-assembled into polypseudorotaxane-type structures when using the wider β-CD host [48, 49]. This combination gave a highly space-filling quotient; thus, polypseudorotaxane formation proceeded smoothly and in high yields. Moreover, the authors also investigated per-methylated and partially methylated CD host; from these, the former ones did not form any polypseudorotaxane (attributed to the absence of stabilization by H-bonding). In the case of partially methylated β-CD, a very slow threading process was observed due to solubility behavior [30]: at the onset of the polyrotaxane formation, the polymer remained solubilized; however, with increasing threading density (over days), the chains stiffened eventually leading to precipitation of the final channel-like assemblies.

The pending bulky methoxy groups of poly(methylvinyl ether) (PMVE) resulted in a high cross-sectional size, and this polymer could, therefore, only be threaded into γ-CDs. Harada et al. showed that high threading densities could be reached, even for polymers of high molar masses [20]; these authors proposed a high thermodynamic driving force to be responsible for the formation of such polypseudorotaxanes, which are basically water insoluble.

Besides the "conventional" poly(alkyl ether)s, also their perfluorinated counterparts have been studied regarding their ability to form inclusion complexes with CDs. Due to a significantly higher cross-sectional area, polypseudorotaxane formation only occurred with β- and γ-CDs. It was shown, for example, that oligo(hexafluoropropylene oxide) selectively assembled into a polypseudorotaxane in the presence of γ-CD [50].

9.2.1.3 Polypseudorotaxanes from Polyesters

Various polyester-type polymers, including poly(ε-caprolactone) (PCl) [51] and some aliphatic polyadipinates [52], were also tested by Harada and coworkers for their ability to form polypseudorotaxane with CDs. Remarkably, the inclusion complex obtained from PCl and α-CD in water turned out to be highly unstable [51]: hydrolyses of the polyester into oligomers and monomers occurred (note that PCl is stable in a pH-neutral water in the absence of α-CD); thus, the CD served as an efficient catalyst for PCl degradation. Applying a different preparative protocol, however, gave stable inclusion complexes of PCl with α-CD [53, 54]: treating the PCl melt with sonication at elevated temperature with a concentrated solution of the CD in water generated the corresponding polypseudorotaxane with an almost 1 : 1 stoichiometry of CD moieties and ε-hydroxy hexanoate units. However, increasing the molar mass of the used PCl led to a distinct drop in yield. As observed by differential scanning calorimetry (DSC) and wide-angle X-ray diffraction (WAXD) measurements as well as infrared spectroscopy, crystallization of the PCl chains was efficiently suppressed [55]. The same behavior was observed for the polypseudorotaxanes, derived from the related poly(3-hydroxypropionate) (P3HP) and poly(4-hydroxybutyrate) (P4HB).

The synthesis of polypseudorotaxane from the aliphatic polyadipinates, using basically the same methodology, based on PCl, was applied; under these forcing conditions, polypseudorotaxane was formed as an insoluble material [52]. Moreover, poly(ethylene terephthalate) (PET) and polycarbonate (PaC) were loaded with γ-CD rings, though special reaction conditions had to be applied [56]; the polypseudorotaxane of PET and γ-CD was assembled by adding a solution of PET in phenol/$(Cl_2CH_2)_2$ to a concentrated solution of the CD in water.

The polypseudorotaxane of α- and γ-CDs with poly(butylene succinate) (PBS) of high molar mass was prepared and analyzed by DSC and WAXD measurements as well as Fourier-transform infrared (FT-IR) and NMR spectroscopy [57]. In the solid state, the polypseudorotaxanes featured channel-like structures with stabilizing H-bonding interactions between the PBS chains and threaded CDs. Furthermore, the FT-IR spectra suggested a kinked conformation of the segments in the polymer chain that were included in the CD cavities.

9.2.1.4 Polypseudorotaxanes from Polyamides and Polyurethanes

Polyamides represent a class of polymer that is characterized by its intrinsic low solubility and strong intermolecular association via H-bonding interactions. Consequently, their applicability for the threading of CDs appears (at this point) limited. The first example for CD moieties threaded onto a polyamide chain, i.e. nylon-11, was reported by Wenz et al. [58]; however, the authors applied the "polymerization of pseudorotaxanes" route instead of a self-assembly approach (i.e. the supramonomer, assembled *in situ* from 11-aminoundecanoic acid and two α-CD units, was polymerized; see also Figure 9.3). This 2 : 1 assembly of the molecular building blocks was retained during the polycondensation at high temperatures. The thus-obtained polypseudorotaxane was water soluble until dethreading of the CD rings finally led to the precipitation of nylon-11. A similar approach, i.e. the polyaddition of an *in situ* formed supramonomer, was used to obtain a polypseudorotaxane with a polyurethane backbone; in detail, various diols gave the [2]pseudorotaxanes when combined with permethylated β-CD; these

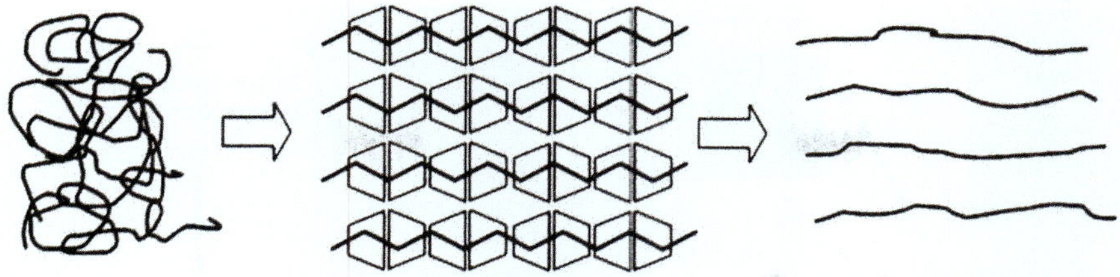

Figure 9.7 Schematic representation of the transformation of a randomly coiled polymer into a highly ordered crystalline state. Source: Tonelli [62]. © 2009 Springer Nature.

assemblies, in turn, reacted with methylenediphenyl-4,4′-diisocyanate (MDI) to afford the corresponding polypseudorotaxanes [59].

Within the context of this section, commercial nylon-6 was employed as the main-chain polyamide for polypseudorotaxane formation. For this purpose, the polymer was dissolved in formic acid and treated with a solution of α-CD in either water or DMSO. A macromolecular inclusion complex precipitated therefrom [60]. The reversibility of polypseudorotaxane formation was employed by Wei et al. to manipulate the crystal structure of nylon-6 [61]. For this purpose, the polymer was released from the polypseudorotaxane by treatment with acid and then coalesced (Figure 9.7). The initial nylon-6 was compared with the recovered material regarding its crystallinity: According to XRD analysis, the crystalline α-phase was preferred; whereas, the γ-phase, which was dominate in the native material, was absent in the coalesced one. Furthermore, the entire assembly/disassembly process gave a highly ordered array of polymer chains with an increased (c. 80%) crystallinity as well as, an elevated melting, crystallization, and degradation temperatures.

Today, this methodology of processing polymers via their intermediate CD-based polypseudorotaxane into ordered nanostructured or crystalline materials represents a versatile method in the modern polymer science toolbox and has been adapted to a broad range of homopolymers as well as binary and ternary polymer blends [62, 63]. A general workflow for this process is summarized in Figure 9.8. In general, the reorganization of amorphous polymers typically gives materials with elevated glass-transition temperatures (T_g); whereas, for semi-crystalline polymers, both higher melting (T_m) and crystallization temperatures (T_c) can be found (often accompanied by a change in the crystalline polymorph). Moreover, when applying the polyrotaxane/coalescence approach to a mixture of different homopolymers or block copolymers, one can often observe an almost perfect mixing between the different homopolymer chains or the different constituent blocks of the copolymers (i.e. absence of phase separation) [63]. Tonelli and coworkers further pointed out that such reorganized samples generally exhibit remarkable stability: the orientation of the polymer chains as well as the resulting properties are often retained even if the samples are temperature annealed extensively above their T_g or T_m values.

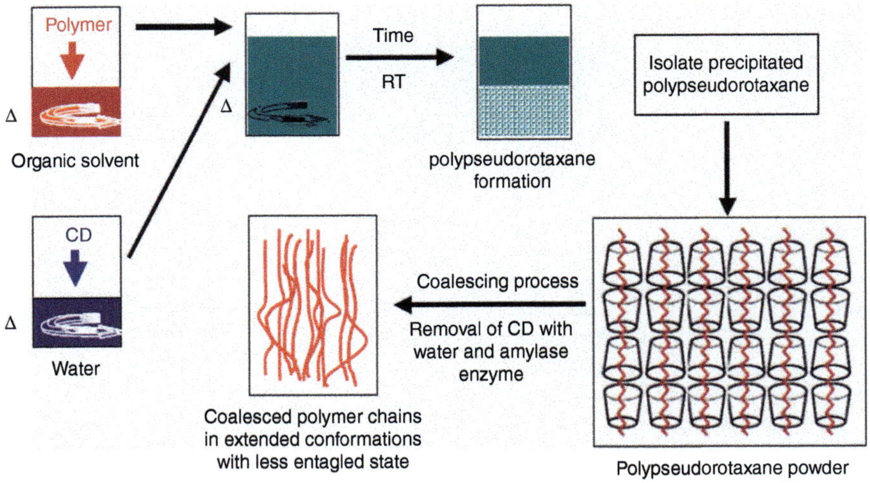

Figure 9.8 Schematic representation of the polypseudorotaxane formation and isolation, the coalescence process, and the coalesced polymer. Source: Tonelli [62]. © 2009 Springer Nature.

9.2.1.5 Polypseudorotaxanes from Polyolefins

Even for various polyolefins, despite their pronounced hydrophobicity, polypseudorotaxane-type assemblies can be formed in the presence of CD derivatives. As for the polyethers, a good correlation between the diameters of the polymer chains and the cavity size of the CD was found: α-CD generated complexes with oligoethylene, β-CD with polypropylene (PP) [64], and γ-CD with PIB [64, 65]. The same also holds true for polybutadiene (PBD) [66] and polyisoprene (PI) [67], both allowing the selective threading of CD rings, according to their relative size.

Due to the poor solubility of PE, only oligomers could be used for polypseudorotaxane formation. A stoichiometry quotient of three was reached, indicating a rather dense packing of the threaded α-CD moieties into a channel-like architecture [31, 68]. According to molecular modeling studies, PE loaded with α-CDs is expected to adopt an extended structure with a preferred trans-conformation of the chain [69]; the same type of structure was also predicted for the polypseudorotaxane, derived from *trans*-polyacetylene, though a reduced stability arising from the smaller cross-sectional area compared to PE had to be expected. Substituted polyolefins, such as PP, gave polypseudorotaxanes with CD derivatives more readily due to their increased solubility and decreased inter-chain van der Waals interactions [11]. Two isotactic polyolefins of high molar mass, PP and poly(1-butene), generated polypseudorotaxanes with γ-CD [70]. WAXD and DSC measurements as well as solid-state NMR and FT-IR spectroscopy were used to investigate the resultant assemblies with channel-like structures. The main aim of forming polypseudorotaxane from polyolefins is to utilize the rigid solid-state structure to extend and reorganize their conformations as well as to retain this after coalescence – similar to the approach discussed in Section 9.2.1.4.. In the case of PP, neither the crystalline form nor the T_m could be modified, even though an increase of 27% in crystallinity was observed.

In a textile application, knitted PP fabrics were finished with β-CD and, subsequently, cross-linked with citric acid [71]. These modified fabrics showed a weight increase of up to 30 wt%. Dyeing the material using disperse, acidic or reactive dyes was performed, and standard tests for wash fastness revealed a good color durability indicating the fixation of the employed dyes either by inclusion complexation or by hydrogen-bonding and/or ionic interactions.

9.2.1.6 Polyrotaxanes from Inorganic Polymers

Not only organic but also inorganic polymers have been studied regarding their ability to generate inclusion complexes via host–guest interactions with CDs. Poly(dimethylsiloxane)s (PDMS) [72, 73] and polysilanes [74, 75], representing homochain- and heterochain-type materials, respectively, were found to give polypseudorotaxanes in the presence of appropriate CD derivatives. For steric reasons, α-CD cannot be threaded onto PDMS, and only short oligomers (DP up to 4) formed inclusion complexes with β-CD. In contrast, however, liquid PDMS forms polypseudorotaxanes with γ-CD upon sonication in water. Solid-state NMR spectroscopy indicated the formation of a tight and rigid channel-type structure. Polydimethylsilane also threaded CD rings, depending on the DP of polymer: short oligomers ($DP > 5$) were able to thread into β-CD; whereas, longer oligomers ($DP > 16$) showed a preference to be included into γ-CDs. In agreement with the polypseudorotaxanes formed from PDMS, a dense packing of the polymer chain was observed. Within the channel-like environment, the polydimethylsilane chains adopted a single-handed helical conformation, and thus, even induced optical activity could be observed by circular dichroism measurements (Figure 9.9) [76].

9.2.1.7 Polypseudorotaxanes from Conductive/Conjugated Polymers

The formation of polypseudorotaxanes from polymers featuring extended π-conjugated systems is demanding, mainly due to two reasons: typically, such polymers are barely water soluble and they might exhibit sensitivity against water and/or oxygen. Nonetheless, examples are known in the literature where the threading onto such preformed polymers has been achieved. For example, polyaniline (PANI) was reacted with β-CD in aqueous *N*-methylpyrrolidone (NMP) to give a blue precipitate at temperatures <2 °C [77]. The blue color was attributed

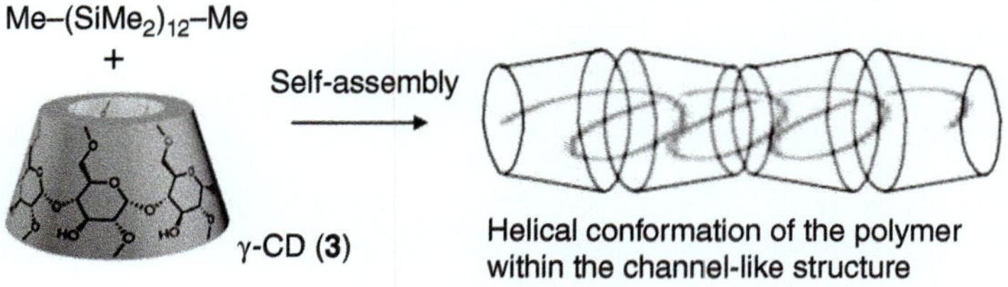

Figure 9.9 Schematic representation of the polydimethylsilane-based polypseudorotaxane featuring a helix-type conformation within the supramolecular assembly. Source: Redrawn from Ref. [76]. © 2003 The Royal Chemical Society.

to the presence of the so-called emeraldine base of PANI (see also Section 10.2). The polypseudorotaxane was studied by means of frequency-domain electric birefringence (FEB) spectroscopy and scanning tunneling microscopy (STM) imaging. According to the FEB measurements, PANI experienced a conformational change from a coil- to a rod-like structure upon polypseudorotaxane formation. It was further shown that doping of complexed PANI with iodine at 273 K was inhibited due to the encapsulation of the polymer by the threaded CD rings, thus making the material an "insulated molecular wire." Such materials are considered to be one of the next generation's material to be applied in organic electronics [78]. At higher temperatures, oxidation with I_2 proceeded to what was attributed to a dissociation of the polypseudorotaxane into plain PANI and β-CD [79]. Remarkably, a similar assembly was obtained when using a cyclotriphosphazene with aromatic substituents as cyclic host for the threading of PANI; in contrast to the material obtained by threading of CD rings, doping with iodine was possible to give a conductivity of $5 \times 10^{-4}\,\mathrm{S\,cm^{-1}}$ (the conductivity of I_2-doped plain PANI is only $10^{-3}\,\mathrm{S\,cm^{-1}}$) [80].

In a similar fashion, polythiophene (PT) and its 3-methylated counterpart (P3MT) could thread β-CD rings in water [81]. Initially, polypseudorotaxane formation improved the solubility of the conjugated polymers in water; however, when a high degree-of-threading was reached, precipitation of the insoluble polypseudorotaxanes occurred. Though not representing a conjugated polymer, the polypseudorotaxane formation of single-walled carbon nanotubes (SWCNTs) has also been mentioned in this context. Due to their outer diameter of 1.2 nm, significantly larger CD derivatives, than the common γ-CD, were utilized for the threading onto SWCNTs. In particular, η-CD (comprising 12 glucose units and a cavity size of 1.8 nm) was found to generate a soluble material comprising η-CD rings loosely threaded onto SWCNTs (Figure 9.10) [82].

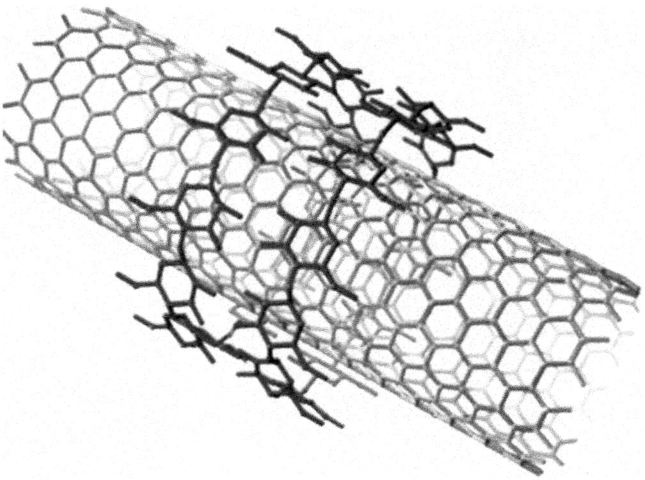

Figure 9.10 Schematic representation of a polypseudorotaxane formed by SWCNTs and η-CD. Source: Dodziuk et al. [82]. © 2003 The Royal Chemical Society.

Various other types of conjugated polymers have also been utilized for the formation of polypseudorotaxane with CDs; however, their synthesis involved the polymerization of an *in situ* formed or preformed pseudorotaxane-type supramonomers and will be discussed later in Section 9.4.

9.2.1.8 Polypseudorotaxanes from Polyamines

Polyamines represent another polymer class that is, at least at first glance, less suited for the formation of polypseudorotaxane via the direct threading of CDs due to their pronounced hydrophilicity, particularly in the protonated state [11]. Consequently, polypseudorotaxane formation will thus be pH-dependent. For linear PEI, as a prominent example, backbone protonation prevents the threading of α- or γ-CD rings onto the polymer at pH values below eight. In the intermediate regime (8 < pH value < 11), polypseudorotaxane formation is possible; whereas, deprotonation of the polymer in an even more basic medium again disfavors its interaction with CDs for electrostatic reasons. As for PEG, also α- or γ-CD represents appropriate hosts for PEI complexation, and accordingly similar stoichiometric ratios of two and four for α-CD and γ-CD, respectively, were found, thus indicating a compact, channel-like packing [32].

In contrast to PEI, polyamines comprising long alkyl units in the backbone allowed the threading of CD rings even at low pH values – the hydrophobic blocks were of sufficient lengths to house the CD rings, thereby preventing any interference with the protonated amino sites. Two types of such polyamines were investigated regarding polypseudorotaxane formation: poly(imino-oligomethylene)s **4** with secondary amino groups capable of protonation within the polymer chain [23, 83] and polyamines with cationic ammonium sites along the backbone (**5** and **6**; Figure 9.11) [22]. These materials are commonly referred to as poly(bolaamphiphile)s, and both types assembled with α-CD into polypseudorotaxane under homogeneous conditions. The threading kinetics were investigated by NMR and revealed a decrease-of-threading-rate with an increasing pH value for **4** and a pronounced temperature dependency for **5**; notably, two days of assembly time at 80 °C were required to load 50% of the alkyl blocks with CD rings (at 25 °C, it took c. two years to reach the same values) [22].

Based on the methylviologen-containing polymer **6**, Stoddard and coworkers developed a biological lactose-containing receptor for galectin-1 (Gal-1) [84]. Threading of the lactoside-functionalized α-CD derivative **7** gave a flexible polypseudorotaxane with one CD ring per aliphatic segment (Figure 9.12a). This assembly was found to efficiently bind the ditopic Gal-1 with an affinity that was up to 30-fold higher as observed for native lactose in a T-cell agglutination assay

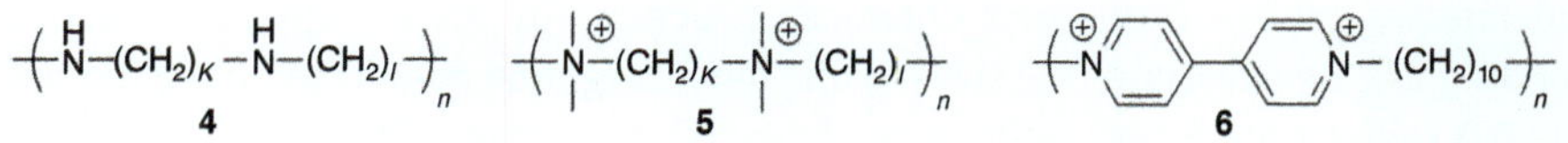

Figure 9.11 Schematic representation of poly(bolaamphiphile)s **4–6** used as macromolecular guests for polypseudorotaxane formation (the counterions are omitted for clarity). Source: Herrmann et al. [22]. © 1997 American Chemical Society.

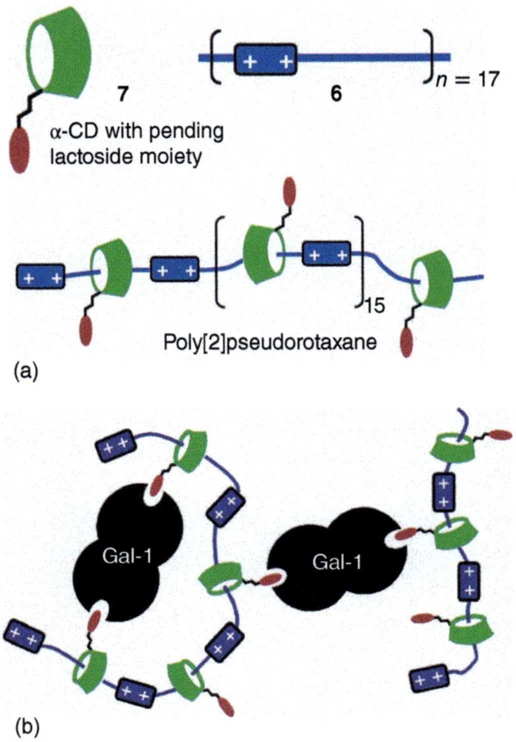

Figure 9.12 (a) Schematic representation of the self-assembly of components **6** and **7** into a flexible poly[2]rotaxane; (b) possible modes of binding of Gal-1 to this supramolecular receptor. Source: Nelson et al. [84]. © 2004 American Chemical Society.

(Figure 9.12b) [85]. This study shows that supramolecular architectures can be used as tools for investigating protein–carbohydrate interactions and might be useful for biomedical applications.

9.2.2 Polypseudorotaxanes, Derived from Other Polymeric Architectures

To widen the scope of polypseudorotaxane formation, polymer architectures other than the previously discussed linear ones have also been used for the binding of CD moieties via host–guest complexation (Figure 9.13) [11, 18]: such as block copolymers and star-shaped or graft/comb-like polymers. Thereby, large differences in the ability to bind CDs were found that could be correlated to the chemical structure of the polymers.

9.2.2.1 Site-Selective Complexation of Linear Block Copolymers

In block copolymer architectures, the constituent blocks might exhibit different thicknesses arising from their chemical structure. In such a system, selective binding of CDs of certain ring sizes to one block might be possible, while the other block(s) offer only a poor affinity to these CD derivatives. Thus, only the block offering the strongest binding of a particular CD derivative will preferentially be loaded with CD rings. This approach is commonly referred to as the "site-selective complexation" and can, for example, be used to solubilize rigid, channel-like

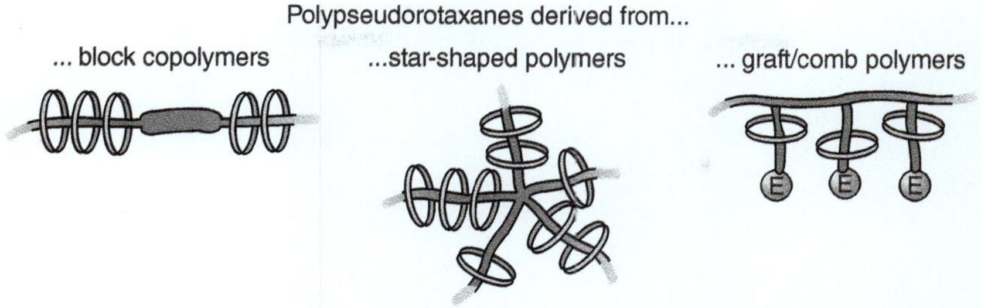

Figure 9.13 Schematic representation of polypseudorotaxanes, based on block copolymers, star-shaped polymers, and graft/comb polymers. Source: Redrawn from Wenz et al. [11]. © 2006 American Chemical Society.

polypseudorotaxanes in which the non-complexed polymer block can still offer sufficient solubility.

Wenz et al. observed that even a controlled complexation of block copolymers with differently sized CDs is possible [11]. However, one has to be aware of the intrinsic disadvantages of such strategy: if once a "wrong" CD ring has been threaded my mistake onto a polymer block, all CD rings threaded thereafter will prevent this ring from leaving the chain. Thus, error correction is only enabled if all rings dissociate from the chain until the "wrong" one is also free to leave. Thus, a very long time could be required, if, e.g. the two constituent blocks of a diblock copolymer shall be loaded selectively with CDs of different cavity sizes in a one-pot self-assembly approach.

The polyester of octanedicarbocyclic acid and PEG900 represents one of the earliest examples for a polymer, where the site-selective complexation by CDs has been studied [86]. Due to the water solubility of the polyester and the resulting polypseudorotaxane, the threading processes could be monitored by NMR spectroscopy or isothermal titration calorimetry (ITC). It was shown that either α- or β-CD was selectively threaded onto the hydrophobic segments that offered a much stronger binding of the CD rings than the hydrophilic PEG segments.

Site-selective polypseudorotaxane formation from Triton X-405 was reported as the first example regarding a deblock-type copolymer architecture [87]. This amphiphilic polymer generated an insoluble polypseudorotaxane in which a dense packing of α-CD moieties along the hydrophilic PEG block established a rigid, channel-like structure. Whereas the threading of β-CDs ring occurred selectively at the bulkier and hydrophobic 4-isooctyl-benzene fragment and solubility in water was still ensured by the uncovered PEG chains.

Pluronics, copolymers comprising PEG and PPO blocks, were found to preferentially bind α-CDs to the slim PEG block; whereas, the bulkier PPO block could be instead loaded with β-CDs [88, 89]. For example, partially methylated β-CD recognized the hydrophobic segments of the triblock copolymers PEG-*b*-PPO-*b*-PEG and PPO-*b*-PEG-*b*-PPO. These polypseudorotaxanes remained water-soluble due to their "free" PEG blocks and could, therefore, be analyzed by light-scattering methods. Likewise, α-CDs could be threaded by the triblock copolymers generating water-insoluble products (the PPO blocks free of CD rings offered negligible

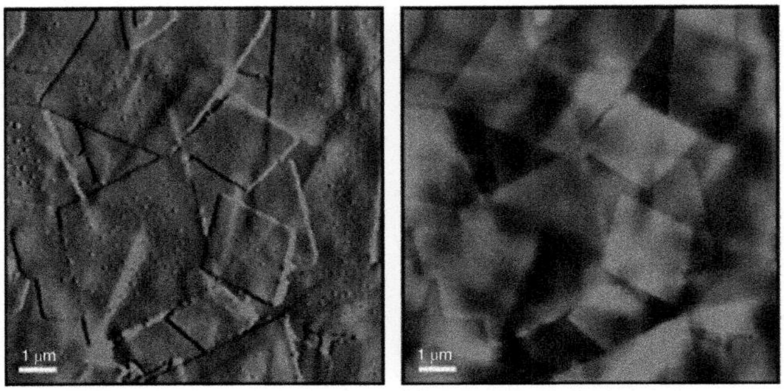

Figure 9.14 Atomic-force microscopy (AFM) images of the platelets formed by the aggregation of pluronics-based polypseudorotaxane [(a) phase image, (b) topography image, scale bar: 1 μm]. Source: Perry et al. [91]. Figure reproduced with kind permission; © 2011 The Royal Chemical Society.

solubility in this medium). Furthermore, a random copolymer of PEG and PPO (molar mass ratio of 4 : 1) could be threaded into α-CDs – the threading density was only c. 80% of that observed for PEG of similar molar mass, thus indicating that only the PEG-rich domains were able to bind α-CD rings [90]. The Schlatter group investigated the aggregation of PEG-*b*-PPO-*b*-PEG, complexed with β-CD, in water by light-scattering measurements and microscopy imaging [91]. Initially, β-CD-swollen micelles were formed at high temperatures; from these, polypseudorotaxane formation occurred upon cooling. With time, bundles of polypseudorotaxane strands aggregated into thin platelets having large planar dimensions in the mm regime (Figure 9.14).

In recent years, polypseudorotaxanes, *inter alia* formed by pluronics and CDs, have experienced considerable interest due to their ability to form hydrogels that, in turn, can be employed for drug-delivery applications [92]. The addition of α-CD to PEG-*b*-PPO-*b*-PEG, which changed the hydrophobicity of the system, allowed hydrogel formation to occur at much lower concentrations then for the initial triblock copolymer [93]. However, careful control of the composition, i.e. loading of the polymer with CDs as well as chemical composition of the pluronic itself, was found to be crucial parameters to obtain stable supramolecular hydrogels. The importance of this was shown as an example of the controlled release of molecules from such supramolecular hydrogels, where only stable hydrogels offered a sustained delivery [the authors used the bovine serum albumin (BSA) protein as test substrate protein in their *in vitro* study] [93]. Hybridizing such a hydrogel with carbon nanotubes (CNTs) reduced its density and, thereby, enabled a better diffusion of embedded drugs [94]; the authors particularly addressed the antimicrobial activity, the skin-adhesion and the water-retention properties of their material with respect to a potential wound-dressing application. Simões et al. developed a viscoelastic, yet syringeable hydrogel formulation for the sustained delivery of vancomycin [95]. The rate for the controlled drug release was found to be slow when the threading density of the employed pluronic was high. The drug-loaded hydrogels also featured an activity against *Staphylococcus aureus*. He et al. utilized hydroxypropyl-functionalized α-CD for the preparation of hydrogels with pluronics; from these, stable formulation with doxycycline for ophthalmic delivery was realized [96]. Recently, Jansook et al. prepared eye-drop formulations containing similar hydrogels, as carriers for dexamethasone and amphotericin B; this observation demonstrates the high potential of these supramolecular materials in pharmaceutical applications, going beyond all *in vitro* studies [97].

Besides pluronics as the most popular block copolymer investigated regarding site-selective complexation, some other examples also need to be mentioned here: symmetrical PPO-*b*-PCl-*b*-PPO triblock copolymers were recognized regioselectively by α-CD, thus allowing for a site-selective complexation of their central PCl blocks (in contrast, γ-CD exhibited no selectivity since both block types were complexed equally) [98]. For the slim PCl-*b*-PEG-*b*-PCl triblock copolymer also no site-selectivity could be observed, not even when threading α-CDs [99]. A similar behavior was reported for the systems PEG-*b*-PHB-*b*-PEG {PHB: poly[(*R*)-3-hydroxybutyric acid]} [100, 101] and PLLA-*b*-PEG-*b*-PLLA {PLLA: poly[L-lactic acid]} [102].

Qi et al. prepared the triblock copolymer **8** in which the central polyurethane-type segment, prepared from 2,2-dimethylolbutyric acid and isophorone diisocyanate, was flanked by two PEG arms [103]. Due to the differences in the binding affinity toward α-CD, a site-selective threading of CD rings onto the PEG arms was achieved. The resultant rod-coil-rod polymer self-assembled into highly ordered nanostructures; in particular, the crystallization of the rigid-rod polypseudorotaxane segments was identified as driving force in these processes (Figure 9.15). The morphology of these assemblies could be switched by adjusting the concentration and/or pH value: for example, 1D nanostructures with different curvatures were observed at different pH values – ring-shaped micelles (pH value of 7.5) and linear rod-like micelles (pH value of 9.2).

9.2.2.2 Polypseudorotaxanes from Star-Shaped Polymers

Compared to linear polymers, their star-shaped counter parts offer some advantages that are also of relevance for polypseudorotaxane formation [11]: higher molar masses can be reached without changing the threading length, the solubility is increased, the viscosity is decreased (comparing polymers of the molar mass), the formed polypseudorotaxanes are more soluble, and, finally, cooperativity during the threading process can be better exploited.

As an example, a series of star-shaped polymers – featuring either three, four or six PEG arms – was utilized for the threading of CD rings [104]. In the case of α-CD, a high threading density, comparable to that observed for linear PEG, was found (see also Section 9.2.1.1). Remarkably, double-stranded polypseudorotaxanes were formed in the presence of γ-CD – due to the close proximity of the PEG arms, the threading of two of these into one CD host was facilitated. Also, multi-arm PEGs were studied in the same context; these architectures comprising 13 or 15 arms were complexed by both α-CD and γ-CD. In contrast to PEGs of high molar masses that gave only insoluble materials, gelation was observed for the star-shaped systems [105]. In more advanced topologies where, for example, hyperbranched

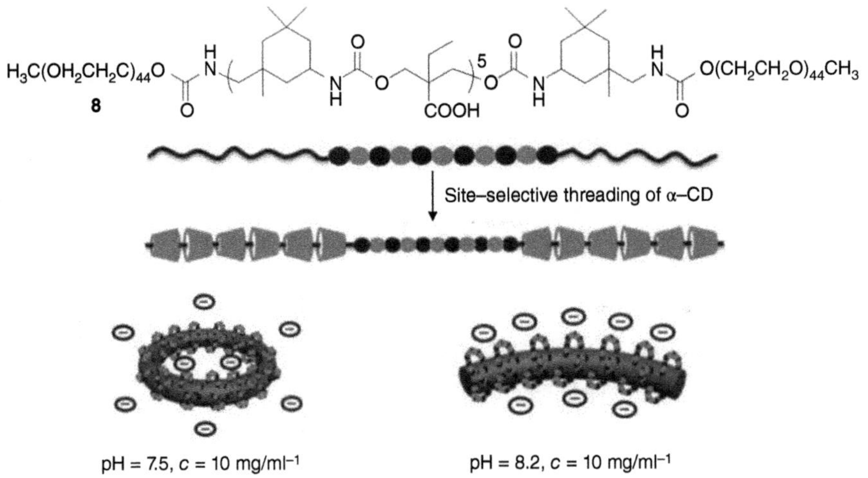

Figure 9.15 Schematic representation of the triblock copolymer **8** used for the site-selective complexation by α-CD. The resultant rod-coil-rod polymer featured a crystallization-driven self-assembly that could be modulated via the concentration and/or the pH value (two different morphologies, i.e. nanorings and nanorods, observed at different pH values are also shown). Source: Qi et al. [103]. © 2015 John Wiley and Sons.

polyether alcohols [106] or polymethacrylates with pending PEG chains were used [107, 108], supramolecular gels or lamellar phases were formed from the interaction of the polymers with α-CDs.

9.2.2.3 Polypseudorotaxanes from Side-Chain Polymers

There are different synthetic approaches available to prepare polypseudorotaxanes in which the inclusion complexes are located within the side chains of the polymers (Figure 9.13). From these, the (co)polymerization of preformed pseudorotaxanes or the grafting of pseudorotaxanes to a preformed polymer chain will not be addressed within the scope of this chapter (examples for these methods can be found elsewhere [11, 15]). However, if the CDs are complexed by the side chains of a preformed polymer, the polypseudorotaxane formation represents a supramolecular process and will, therefore, be discussed here briefly. In such a self-assembly process, the side chains will react almost independently from each other and, consequently, polypseudorotaxane formation is faster when compared to the multiple threading onto a linear chain. On the other hand, the stability of the assembly is decreased since dissociation is also faster (attributed to the absence of stabilizing non-covalent interactions between neighboring CD rings). Concerning the selectivity of the various CD moieties to thread onto a side chain, their bulkiness is decisive.

Further, various poly(alkyl acrylate)s [109] and copolymers of alkyl acrylates and acrylamide [110] were studied regarding the binding of α-CD to the pending linear side chains. NMR spectroscopy revealed an increase of the binding strength with increasing side-chain length (the lauryl moiety gave the highest K_a value). Due to hydrophobic interactions, the lauryl-containing polyacrylate exhibited hydrogel formation in aqueous medium that could be switched-off by adding α-CD. The larger cavity of β-CD shows a preference to include *tert*-butyl groups. This preference was exploited by Harada et al., who observed selective binding of β-CD to copolymers with pending *tert*-butyl moieties [111]. Likewise, poly(maleic acid-*alt*-isobutene) comprising *t*-butylanilide units in the side chains could be complexed by β-CD (a high binding constant of $K_a = 25\,900\,M^{-1}$ was observed for this combination) [112].

Chitosan with grafted PEG side chains was complexed with α-CD [113]. The thus-obtained polymer formed nanoparticles with micellar or hollow-sphere morphology, which could be controlled by the threading density, i.e. the rigidity of the side chains. The nanoparticles were investigated by microscopy imaging and XRD measurements.

The site-selective complexation in a diblock copolymer was reported by the Allcock group (Figure 9.16) [114]. They prepared a pre-polymer featuring a polystyrene (PS) block to which a polyphosphazene chain was grafted. The reactive sites of the polyphosphazene block were subsequently substituted by alkyladamantane units, thus forming a hydrophobic diblock copolymer. The adamantyl group is known for its efficient complexation by β-CD, and thus, the diblock copolymer could selectively be loaded with CDs in the side chains of the polyphosphazene segment. As a result, this polypseudorotaxane block became hydrophilic and the supramolecular amphiphilic copolymer assembled into micelles at a very low critical micelle concentration (cmc = $0.93\,mg\,l^{-1}$). These micelles were analyzed by dynamic light-scattering (DLS) measurements and microscopy imaging

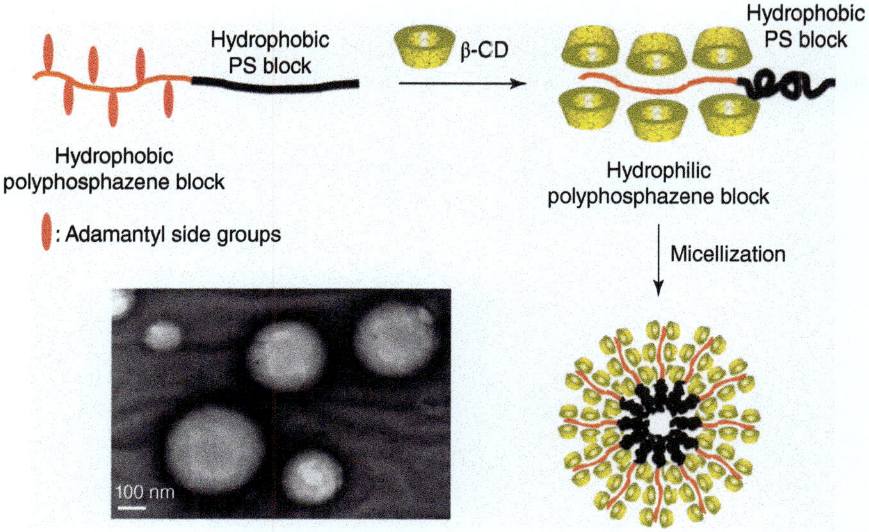

Figure 9.16 Schematic representation of the site-selective complexation of a diblock copolymer by β-CD moieties. A representative TEM image of the micelles assembled from the resultant amphiphilic pseudorotaxane-PS diblock copolymer in water is also shown. Source: Cho and Allcock [114]. Figure reproduced with kind permission; © 2009 American Chemical Society.

(Figure 9.16), the former method suggested an effective diameter of 193 nm and a narrow size distribution.

9.3 End-Capping: From Polypseudorotaxanes to Polyrotaxanes

In contrast to the kinetically labile polypseudorotaxanes where the leaving of CD macrocycles from the polymer chain is basically enabled, polyrotaxanes with bulky substituents do not show this behavior. Basically, such stopper moieties can be introduced not only at the termini of a polypseudorotaxane chain but also within or along the assembly (Figure 9.17). However, the function as a stopper is only given if specific criteria are met [11]: sufficient size to block the CDs from threading off and hydrophilicity to prevent precipitation/aggregation of the polyrotaxanes. Moreover, the coupling reaction should fulfill the criteria of a click-type reaction according to Sharpless, i.e. compatibility with water (as the typical solvent for polypseudorotaxane formation), high yielding, as well as the avoidance of additional reagents and/or by-products [115]. Concerning the polyrotaxanes with stopper moieties within the chain, the yield is less important owing to the fact that a high number of stoppers will be readily available. The different possibilities for polyrotaxane formation by introducing stopper moieties have been summarized in various review articles; a general overview of the various synthetic pathways toward polyrotaxane formation is given in Figure 9.17 [11, 116].

From these methods, the chain-end modifications of a preformed polypseudorotaxane represent the most common method, in particular when compared to the one where the stopper units have to be attached along a polymer chain that is already covered with CDs. Disubstituted phenyl moieties or the bulkier polycyclic

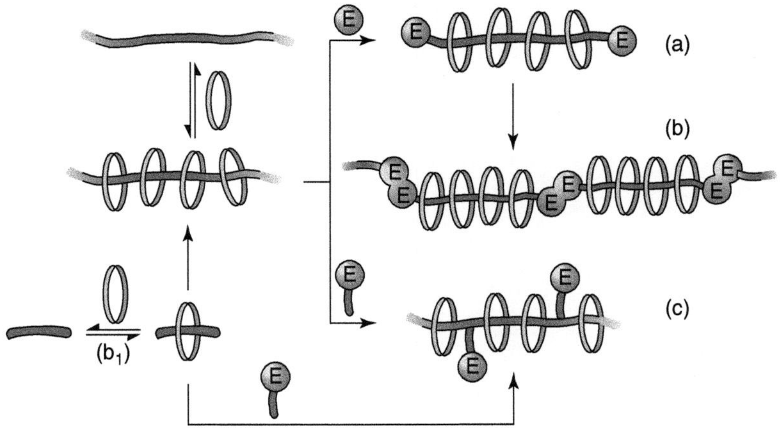

Figure 9.17 Schematic representation of the synthetic routes toward (a) polyrotaxanes with terminal stopper moieties, (b) polyrotaxanes with stopper moieties within the main chain, and (c) polyrotaxanes with stopper moieties attached to the main chain. Source: Wenz et al. [11]. © 2006 American Chemical Society.

aromatics represent appropriate stopper units in the case of α-CD and β-CD, respectively. Such moieties have to be attached to the reactive functionalities at both ends of the pseudorotaxane via high-yielding procedures. The NH_2-groups at the termini of a polymer chain do not interfere with the threading of CDs but feature a sufficient reactivity; therefore, they represent versatile moieties for introducing bulky stoppers at the chain's termini. Sanger's reagent, i.e. 2,4-dinitrofluorobenzene, has often been employed to transform a polypseudorotaxane into a polyrotaxane via a simple nucleophilic aromatic substitution reaction. For example, H_2N-PEG-NH_2, covered with α-CD rings, was reacted with this particular reactive reagent to give a polyrotaxane – one of the earliest and simplest examples reported [117, 118]. The thusly-obtained material was found to be almost water insoluble, though it could be dissolved under basic conditions or in DMSO, allowing its purification by gel chromatography and characterization by 1D and 2D NMR spectroscopy. Other examples for polyrotaxane formation employed an amidation with activated esters (either on the polypseudorotaxane or on the stopper) [119, 120], addition of isothiocyanates to the terminal NH_2-groups [121], or nucleophilic substitution of the *O*-tosylated end groups [122].

Though the introduction of such stoppers at the termini of a polypseudorotaxane represents a common and simple approach to obtain polyrotaxanes, the main disadvantage, however, remains: the end-group functionalization needs to be quantitative to fully prevent the dethreading pathway. Alternatively, stoppers might be attached along the polymer chain, where a modification of each single repeat unit of the chain in a polymer-analogous reaction is not essential (only a moderate degree-of-functionalization will guarantee a sufficient dethreading prohibition). As a prerequisite, the included polymer has to bear reactive side chains and needs to be loosely covered with CDs to allow access [11]. The aforementioned polypseudorotaxanes, derived from poly(bolaamphiphile)s (Section 9.2.1.8), are versatile representatives in this respect. It was shown that the extended hydrophobic aliphatic blocks of such polyamines (e.g. poly[imino-undecamethylene]) could be loaded with α-CDs, thereby leaving their hydrophilic secondary amino groups exposed to the environment. These moieties, in turn, were functionalized with bulky acyl chlorides. The removing of dethreaded CDs by dialysis was monitored by optical spectroscopy: With an increasing degree-of-functionalization, i.e. stopper density, the more CD rings remained threaded onto the polymer [23].

In a different approach, supramolecular chemistry was utilized to facilitate the transformation of a polypseudorotaxane into a polyrotaxane. Wenz and coworkers threaded a mixture of β- and γ-CDs onto the poly(bolaamphiphile) **9**, which was composed of stilbene and decamethylene segments [123]. Due to their cavity size, the threaded γ-CDs were able to accommodate an additional guest molecule (**10**). The formation of these co-inclusion complexes further stabilized the polypseudorotaxane against dissociation and, furthermore, enabled its light-induced transformation into a polyrotaxane: photochemical [2+2]-cycloaddition between the stilbene derivatives within the γ-CDs gave tetraphenyl-substituted cyclobutane units, which functioned as efficient stoppers along the polymer chain (Figure 9.18). Alupei et al. also showed that other cycloaddition reactions, such as the Diels–Alder-type

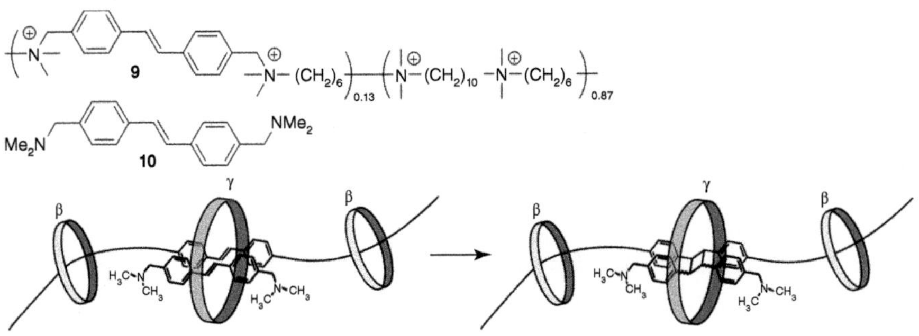

Figure 9.18 Schematic representation of a polyrotaxane formation facilitated by supramolecular host–guest chemistry. Source: Herrmann et al. [123]. © 1997 John Wiley and Sons.

[4+2]-cycloaddition, can basically be employed as polymer analogs [124]. The inclusion complex of cyclopentadiene and per-methylated β-CD was found to thread onto the polyester composed of PEG and maleic acid. The double bonds along the polyester chain served as dienophile for the subsequent Diels–Alder reaction; however, the Diels–Alder adduct was too small to efficiently prevent the sliding of the CD rings down from the polymer chain.

9.4 Polymerization of Pre-assembled Pseudorotaxanes

In Sections 9.2 and 9.3, mainly preformed polymers have been discussed in which they served as macromolecular axis for polypseudorotaxane formation via the threading of CD moieties; however, this method is often hampered by the poor solubility of the polymers in aqueous media. As an alternative approach, the "polymerization of supramonomers," in this case pseudorotaxanes, has been introduced and applied to synthesize polypseudorotaxanes featuring rigid, water-insoluble backbones. The general concept of using supramonomers in the synthesis of supramolecular polymers has been outlined in Section 1.4.

Wenz et al. reported one of the first such examples: a polypseudorotaxane with a nylon-11 backbone was prepared in a polycondensation from the [3]pseudorotaxane assembled from 11-aminoundecanoic acid and two α-CD rings [58]. The 2 : 1 stoichiometry of molecular components in the assembly was retained during the polymerization at high temperatures. The thusly-obtained polypseudorotaxane was soluble in water until dethreading of the CD rings finally led to the precipitation of nylon-11. The basically same approach was followed by Yamaguchi et al. to prepare polypseudorotaxane, based on polyurethane backbones [59]. This polypseudorotaxane formation was conducted in DMF solution using linear diols, MDI, and per-methylated CDs, as building blocks. The resulting supramolecular materials were investigated by size-exclusion chromatography (SEC), NMR, and DSC measurements. The material exhibited lower glass-transition temperatures (T_g) than analogous neat polyurethanes, presumably due to the suppressed intra-chain H-bonding interactions. Depending on the cavity size of the employed

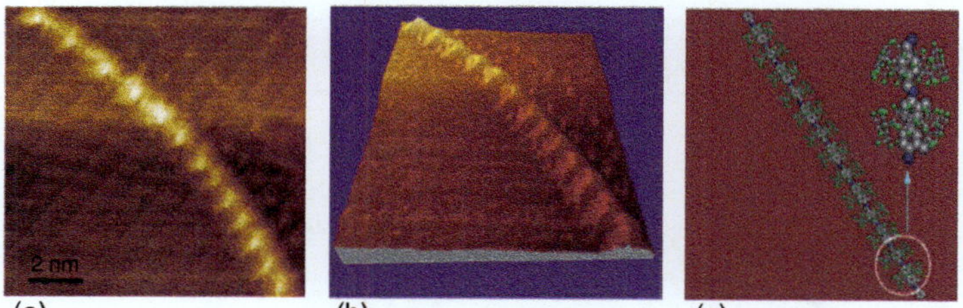

Figure 9.19 Representative scanning-tunneling microscopy (STM) images of the polyazamethine-based polyrotaxane on a highly ordered pyrolytic graphite (HOPG) surface (a) 2D representation, (b) 3D representation, (c) Molecular model of the structure. Source: (a, b) Liu et al. [126]. Figure reproduced with kind permission. © 2004 American Chemical Society.

CD derivative, different polypseudorotaxane architectures were formed: in the case of β-CD, the rings were coordinated to the aromatic rings; whereas, α-CD was found to be located on the aliphatic chains. Supercritical CO_2 was employed as the reaction medium by Wang et al. for the one-step preparation of a polypseudorotaxane by the oxidative polymerization of aniline by benzoyl peroxide in the presence of β-CD [125]. This method represents a versatile alternative to the direct threading of CDs onto PANI, as described in Section 9.2.1.7.

In general, the synthesis of polypseudorotaxanes by performing a polycondensation in the presence of CD was found to be problematic for numerous reasons. For example, the polycondensation of bis(*o*-toluidine) and terephthaldehyde in the presence of β-CD afforded a polypseudorotaxane with a polyazomethine backbone [126]. To stabilize the assembly against dethreading, the terminal amino groups were end-capped with Sanger's reagent. A quantitative threading efficiency was reached; however, a poor DP of only three was concluded from NMR spectroscopic measurements. Microscopy imaging confirmed the linear microstructure of the polyrotaxane (Figure 9.19). Though a higher DP could be reached when using terephthaldehyde and 1,4-diaminobenzene in a similar polycondensation, the efficiency of loading with β-CD rings remained low (c. 30% only) [127]. Fullerene was utilized for end-capping the intermediate polypseudorotaxane, and a molar mass in the range 80 000 g mol^{-1} was observed for the final product ($Đ$ value of 2.14).

Various other polymerization techniques, in the presence of CD derivatives, have also been employed to obtain polypseudorotaxanes possessing π-conjugated polymer backbones: chemical [128, 129] or electrochemical oxidative polymerization [130, 131] of appropriate monomers, such as thiophene derivatives, pyrrole or 4-aminobiphenyl; however, these methods suffered from a poor control over polypseudorotaxane formation and the inability to characterize the final products.

Anderson and coworkers employed the Pd(0)-catalyzed Suzuki cross-coupling polymerization between β-CD-based pseudorotaxanes, equipped with two boronic-ester moieties, and water-soluble diiodo compounds (Figure 9.20) [25]. This work shows how metal-catalyzed polymerizations can be used to obtain well-defined, soluble polypseudorotaxanes featuring a π-conjugated polymer

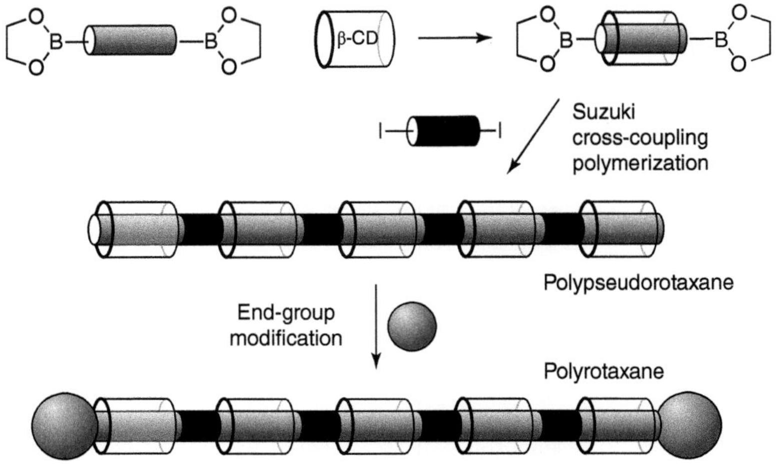

Figure 9.20 Schematic representation of the three-step synthesis of polyrotaxanes, featuring π-conjugated backbones. Source: Michels et al. [25]. © 2003 John Wiley and Sons.

backbone covered with CD rings – the so-called insulated molecular wires. To prevent slipping of the CD rings from the polymer, bulky end groups were introduced. Extensive ultrafiltration was employed as a convenient purification technique. This method was found to be highly versatile to differentiate whether an assembly is a polyrotaxane or a polypseudorotaxane: in general, extensive ultrafiltration will enforce the dethreading of the CDs in the absence of appropriate stopper moieties on the polymer chain. Thus, the content of threaded CD rings will decrease to zero with time in the case of a polypseudorotaxane; whereas, it will remain constant for polyrotaxanes. The authors applied this method to their polyrotaxanes and analyzed the materials by NMR, mass spectrometry, SEC as well as analytical ultracentrifugation (AUC) regarding their DPs. On the basis of the latter, molar masses (M_w) in the range of 25 800–58 500 g mol^{-1} were obtained (corresponding to DPs of 7–10). The threaded CD rings offered a confined environment for the embedded π-conjugated polymer chains whose flexibility became strongly reduced and, as a result, the emission maxima appeared blue-shifted, and the photoluminescence quantum yield was increased (in solution as well as in the solid state).

The β-CD moiety was utilized by van den Boogaard et al. to form a host–guest complex with 5,5′-dibromobithiophene [132]. The Ni(0)-catalyzed Yamamoto coupling reaction of this [2]pseudorotaxane, followed by end-group modification using a bulky anthracene stopper, afforded a polyrotaxane, which was well soluble in water and DMSO (a loading of the chains with an average of 7 CD rings and a DP of 12 was estimated from ^{1}H NMR). The material was also studied by small-angle neutron scattering (SANS) measurements and STM imaging.

In a similar fashion, Terao applied a Glaser-type coupling of the alkyne-functionalized π-conjugated pseudorotaxane segments [133]. To avoid dethreading of the CD rings from the prepared π-conjugated polymer chain, permethylated α-CD with a pending guest arm was prepared and the intramolecular [2]pseudorotaxane assembled thereof, i.e. the self-inclusion complex, was end-capped with a stopper

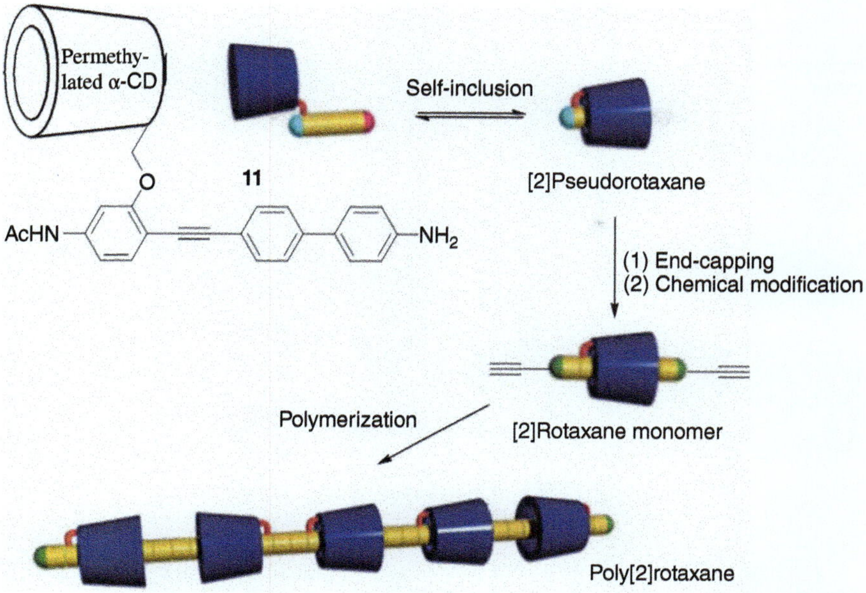

Figure 9.21 Schematic representation of the synthesis of a poly[2]rotaxane with a π-conjugated polymer backbone. Source: Redrawn from Terao [133]. © 2011 The Royal Chemical Society..

unit to afford the corresponding [2]rotaxane. After further chemical modification, the obtained monomer **11** was polymerized into a π-conjugated, still organo-soluble polyrotaxane (Figure 9.21). NMR spectroscopy, mass spectrometry, and SEC were applied to its characterization – a molar mass (M_w) of $3.88 \times 10^4\,g\,mol^{-1}$ (corresponding to a DP of 25) and a $Đ$ value of 2.01 were estimated by the latter technique. The material exhibited a high photoluminescence efficiency and good charge mobility. Moreover, the highly rigid polyrotaxanes showed a liquid crystalline (LC) behavior – the existence of a cholesteric LC phase was attributed to the chirality of the material, which was introduced by the inclusion of the conjugated polymer chain into the CD cavities [134]. In a similar fashion, the Sonogashira-type cross-coupling of the same [2]rotaxane monomer with 1,4-diiodobenzene was reported; the resulting polyrotaxane, containing a poly(phenylene ethynylene) backbone, featured a remarkably high hole mobility, which almost reached the value known for amorphous Si [135].

As a final example, the polymerization of polypseudorotaxanes by the coupling of their end groups has to be mentioned. If the moieties that result from the coupling reaction are sufficiently large, they can function as stoppers in the thusly-obtained chain-extended polyrotaxane. The 2-anthryl group represents one example in this respect, since it is slim enough to allow threading through a β-CD ring; in contrast; however, to this, its photodimer is bulky enough to block the β-CDs from threading. A PPO chain, functionalized with a 2-anthryl and a bulky triphenylmethyl at either chain end, was threaded into β-CDs; subsequently, the terminal anthryl moieties were dimerized photochemically by visible light (Figure 9.22a). The same approach

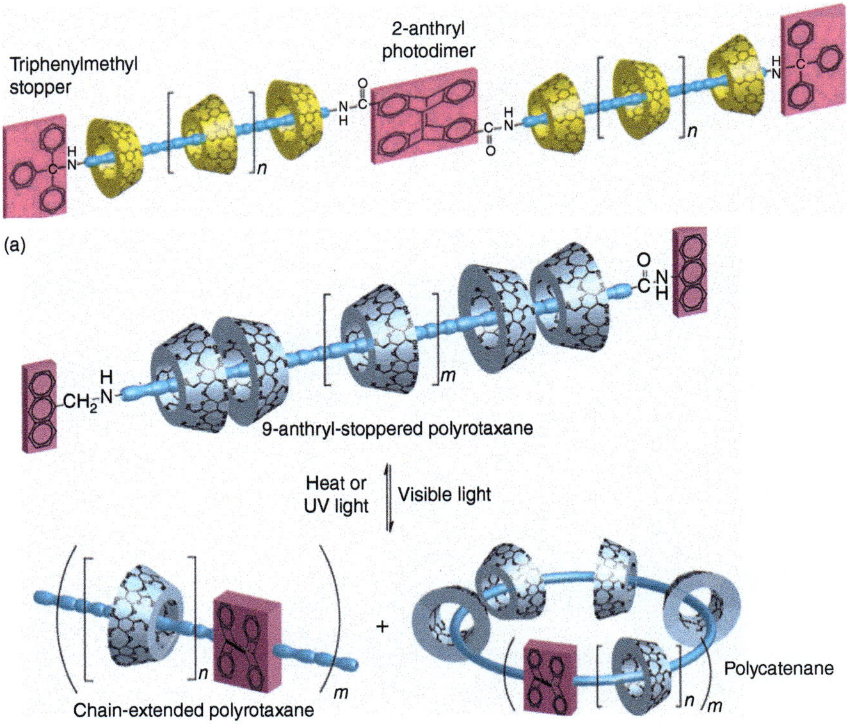

Figure 9.22 (a) Schematic representation of a chain-extended polyrotaxane formed by a photochemical dimerization of the attached 2-anthryl end groups. (b) Schematic representation of the reversible formation of linear chain-extended polyrotaxanes and polycatenanes from a polyrotaxane. Source: (a) Okada et al. [136]. © 2004 American Chemical Society, (b) Okada et al. [137]. © 2003 American Chemical Society.

was also followed by functionalizing PPO at both termini with 9-anthryl entities (these slim moieties still allowed the threading of the polymer chain through the cavities of γ-CDs) [136]. The resultant polypseudorotaxane was irradiated with visible light to photodimerize the anthryl entities (the photodimer is too bulky to allow a γ-CD ring to pass). The 9-anthryl-terminated polyrotaxane of PEG with α-CD was treated in the same manner to give a mixture of linear chain-extended polyrotaxanes along with cyclic polycatenanes (Figure 9.22b) [137]. The initial polyrotaxane could be regenerated by exposing the material to UV light or by thermal cleavage of the photodimers at 130 °C.

9.5 Supramolecular Polymerization, Based on CD Recognition

The supramolecular polymers summarized in the previous sections all featured CD rings threaded onto a covalent polymer chain. A completely different type of material is obtained if the interaction of CD hosts and organic guests directly

(a)

(b)

(c)

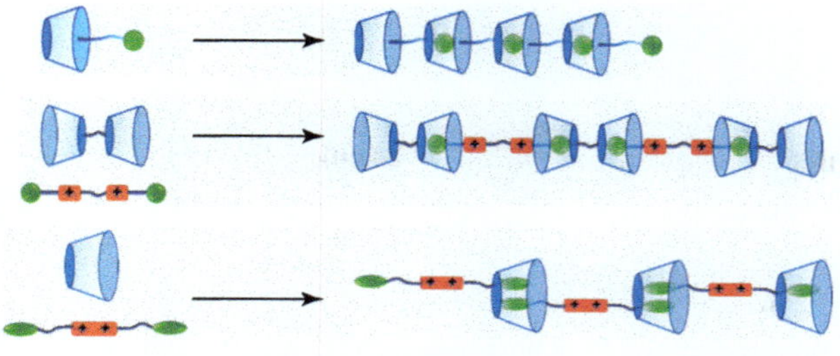

Figure 9.23 Schematic representation of the supramolecular polymerization using (a) a heteroditopic AB-type CD–guest conjugate, (b) an equimolar mixture of homoditopic CD–CD dimer and guest dimer (AA- and BB-type monomers, respectively), and (c) a CD of sufficient cavity size and a BB-type guest dimer. Source: Redrawn Yang et al. [138]. 2015 American Chemical Society.

contributes to polymeric structure itself. In line with the considerations already made in Section 6.2, molecular host–guest conjugates, so-called AB-type monomers, or combinations of homoditopic host–host and guest–guest conjugates in equimolar ratios (i.e. AA- and BB-type monomers) have at times the ability to assemble into linear supramolecular polymers (Figure 9.23) [138]. Furthermore, the combination of CDs of sufficient cavity size to accommodate two guest molecules and appropriate BB-type guest–guest units can alternatively be utilized.

All three combinations, however, inherently share an intrinsic problem, that is, the rivalry between supramolecular polymerization (thus affording linear polymer chains) and oligomerization (i.e. self-inclusion, dimerization, or [macro]cyclization). Increasing the monomer concentration represents one general possibility to circumvent the latter, since supramolecular polymerization is favored over oligomerization at high concentrations (see also Chapter 1). Alternatively, rigid linkers between the host and the guest in these AB-type monomers or between the terminal units in the AA- and BB-type monomers might help to suppress intramolecular complexation.

9.5.1 Supramolecular Polymerization of AB-Type Monomers

A broad range of CDs equipped with one pending side chain, thus representing AB-type monomeric units, is known and has been studied regarding their self-assembly behavior; however, the vast majority of these systems only exhibited the formation of self-inclusion complexes or dimers [139].

As one of the earliest reports for an intermolecular self-assembly process, Harada et al. investigated CDs equipped in one 6-position with a hydrocinnamoyl group (**11**; Figure 9.24). Intramolecular complexation was concluded for **12b**, based on NMR studies; whereas, **12a** revealed a distinct concentration dependency in vapor pressure osmometry (VPO) – indicative of the formation of larger assemblies via intermolecular complexation (a DP of three was estimated from these experiments; Figure 9.24) [140]. Inclusion of the guest occurred from the narrower side of the CD moieties, thus favoring dimerization or cyclization of oligomers due to the more

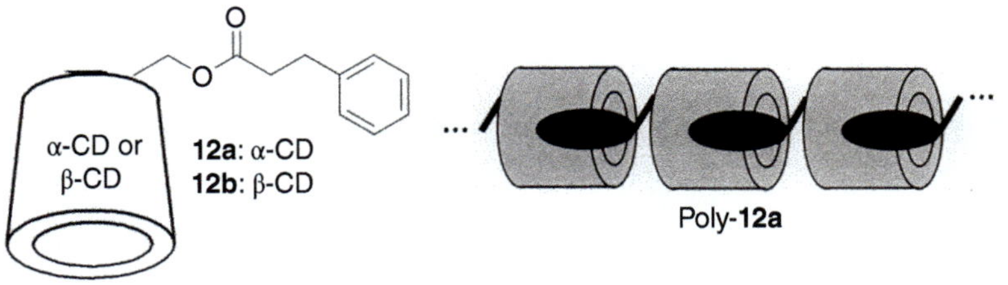

Figure 9.24 Schematic representation of the host–guest conjugates **12** and the supramolecular polymer poly-**12a**. Source: Harada et al. [140]. © 2001 Springer Nature.

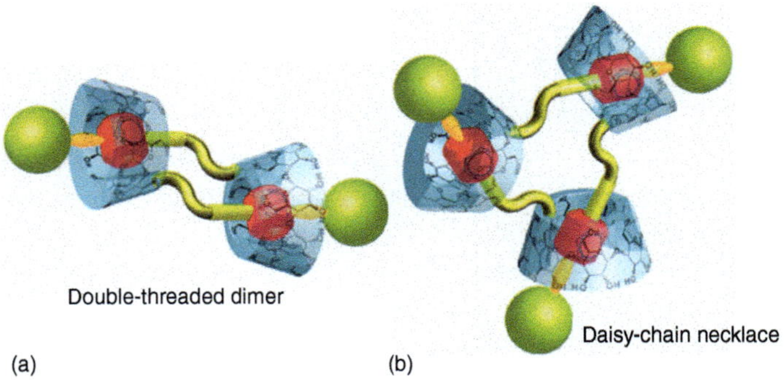

Figure 9.25 Schematic representation of the (a) double-threaded dimer and (b) daisy-chain trimer, obtained from the self-assembly of CD–C=;C–cinnamoyl conjugates. Source: Hoshino et al. [142]. © 2000 The Royal Chemical Society.

inclined structure. Replacing the hydrocinnamoyl moiety by an adamantylamido moiety gave, though only in the case of the conjugate with β-CD, a linear assembly as confirmed by XRD analysis of crystallized material [141]. The formation of a linear supramolecular polymer in favor of dimers or cyclic oligomers was attributed to steric effects arising from the bulky guest unit.

To further increase the rigidity of the host-guest conjugate, C=;C double bonds were introduced between the β-CD ring and the hydrocinnamoyl fragment. The resulting insoluble compounds were analyzed by solid-state NMR as well as XRD analysis of the powdered and single-crystalline samples. The latter suggested formation of a cyclic dimer (Figure 9.25a). The analogous conjugate comprising a α-CD host again showed a concentration dependency in VPO measurements; the presence of cyclic trimers was concluded from NMR and mass spectrometric analyses of the trinitrophenyl-stoppered derivatives (Figure 9.25b) [142].

To prevent inclusion of the guest moiety from the narrower side of the CDs, Miyauchi et al. modified the structure of the host–guest conjugate [143]: the linkage was established via one of the secondary OH-groups of the CDs (in contrast to the previous example where the primary OH-group in a 6-position was employed for attaching a cinnamamide tail), thus forcing guest inclusion to occur from the wider side. The β-CD-containing derivative gave a supramolecular polymer in the solid state; whereas, the α-CD–cinnamamide conjugate assembled into water-soluble linear polymers that were studied by VPO measurements and pulse field gradient

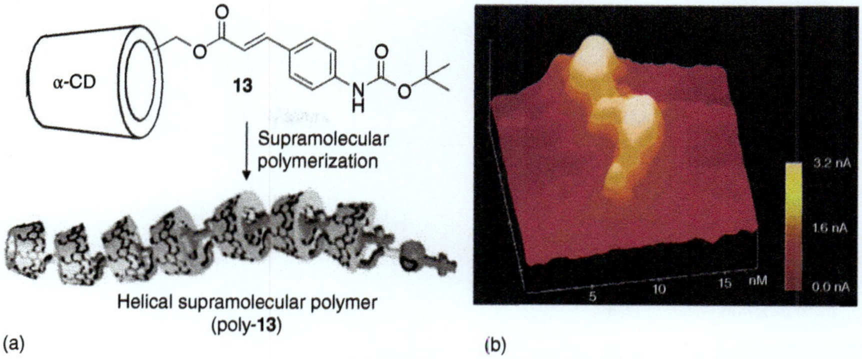

Figure 9.26 (a) Schematic representation of the host–guest conjugate **13** and its self-assembly into a helical supramolecular polymer. (b) STM image of poly-**13** deposited onto a MoS_2 substrate. Source: (a) Miyauchi et al. [145]. © 2005 American Chemical Society.

spin-echo (PFGSE) NMR spectroscopy, a solution-based method that is of high value to conclude on the actual size of supramolecular assemblies [144]. According to these analyses, a molar mass of c. 16 000 $g\,mol^{-1}$, corresponding to a DP of 12, was assumed.

The host–guest conjugate **13** was also employed, as an AB-type monomer, in a supramolecular polymerization (Figure 9.26a). Thereby, the terminal *tert*-butyl group was introduced to further facilitate the formation of linear polymer chains [145]. It was shown that the guest tail of **13** was included within the CD cavities in an inclined fashion giving rise to an overall left-handed anti-configuration in the resulting supramolecular polymer; for poly-**13**, a DP of 15 was found. Circular dichroism spectroscopy and STM imaging confirmed the assumed helical shape of the polymer strands (Figure 9.26b).

The issue to circumvent the competing pathways of AB-type monomers to undergo supramolecular polymerization, the formation of self-inclusion complexes, the head-to-head association, and the so-called sugar tumbling was recently addressed by the groups of Bouteiller and Sollogoub [146]. For this purpose, these authors designed the CD derivative **14** in which the hydrophobic tail was placed on a bridge, which diametrically linked two opposed sugar units of the CD ring (Figure 9.27). This monomer assembled into a supramolecular polymer for which a DP of c. 7 was concluded from DOSY and SANS measurements. Using ITC to describe quantitatively the supramolecular polymerization, a K_a value of $1.1 \times 10^4\,l\,mol^{-1}$ and a binding energy of $-2.9\,kcal\,mol^{-1}$ were obtained (assuming an isodesmic supramolecular polymerization pathway). Remarkably, **14** showed, presumably due to its cationic nature, a pronounced cooperative interaction with DNA and SiRNA; the latter is of particular interest for a potential application in gene transfection.

A strictly alternating supramolecular copolymer was designed by the Harada group making use of the selectivity of adamantyl and cinnamoyl moieties to be included into the cavities of β-CD and α-CD, respectively (complexes of adamantyl derivatives with β-CDs have binding strengths in the order of $10^4\,M^{-1}$) [147].

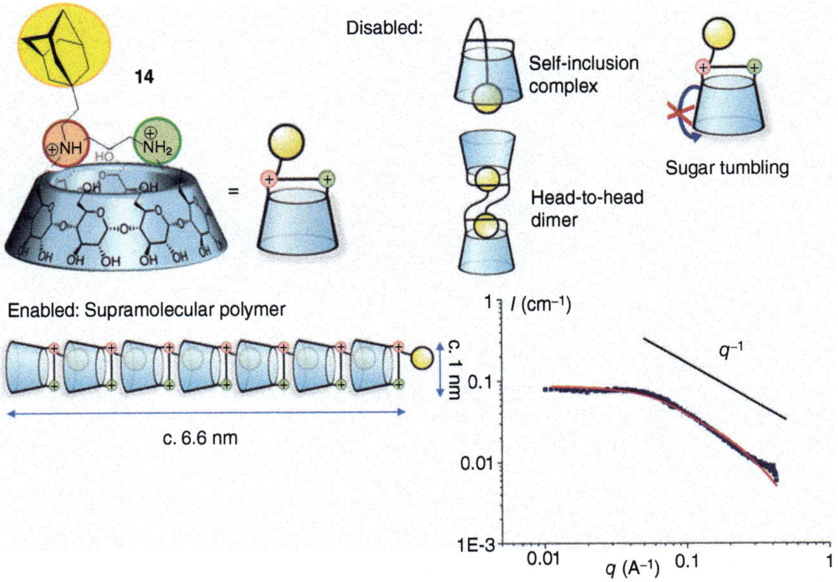

Figure 9.27 Schematic representation of the AB-type monomer **14** for which supramolecular polymerization was enabled; whereas, the competing pathways were disabled. A representation of the SANS data, which corroborated a cylindrical structure of poly-**14** with a length and diameter of c. 6.6 and 1 nm, respectively, is also shown. Source: Evenou et al. [146]. © 2018 John Wiley and Sons.

Utilizing a *tert*-butyl-cinnamoyl–β-CD (**15**) and an adamantyl–α-CD monomer (**16**) in a 1 : 1 ratio, the supramolecular polymerization in water gave the self-sorted supramolecular polymer with alternating α-CD–cinnamoyl and β-CD–adamantyl complexes along the polymer chain (poly-[**15**-*alt*-**16**]; Figure 9.28). Neither of the monomers generated a supramolecular polymer on its own (only self-inclusion complexes were observed). This example highlights how the self-sorting supramolecular polymerization can be employed to assemble a complex macromolecular structure in a controlled fashion from designer small-molecule components (see also Section 1.4).

Finally, a supramolecular polymer, obtained in water from the self-assembly of preformed [2]rotaxane **17**, was published by Harada and coworkers [148]. **17** was prepared from 6-aminocinnamoyl β-CD; the initial self-inclusion complex of this AB-type monomer was transformed into a [2]pseudorotaxane by adding 1-adamantane carboxylic acid, as a competitive guest. Next, the pending 6-aminocinnamoyl moiety was complexed by α-CD, and, subsequently, this pseudo-[2]rotaxane was end-functionalized at the NH_2-group of the cinnamoyl residue with a trinitrophenyl stopper to afford – after rinsing off the adamantyl guest – a stable [2]rotaxane with pending β-CD and trinitrobenzene arms (**15**). It was shown by NMR spectroscopy that the AB-type monomer **17** generated a supramolecular polymer in water (Figure 9.29); whereas, **17** existed only as a monomer in organic solvents.

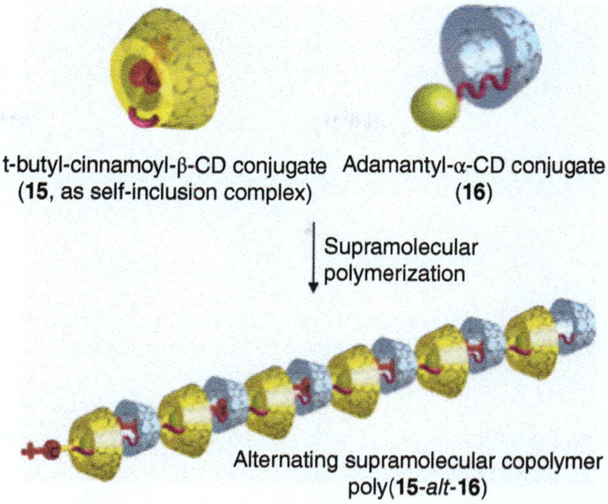

Figure 9.28 Schematic representation of the self-assembly of **15** and **16** into an alternating supramolecular copolymer. Source: Miyauchi and Harada [147]. © 2004 American Chemical Society.

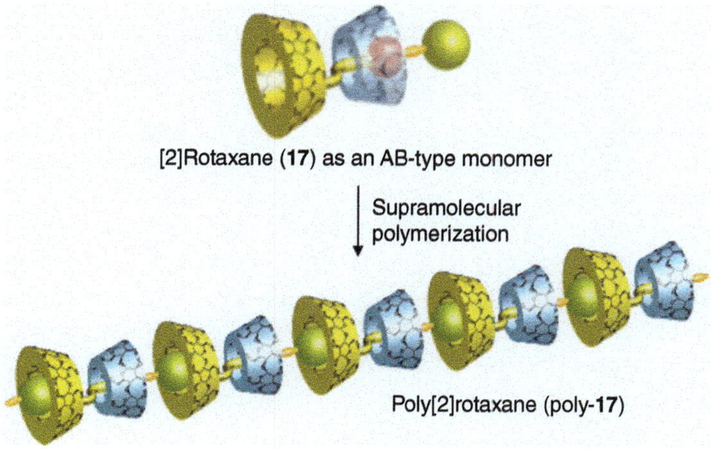

Figure 9.29 Schematic representation of the synthesis of a poly[2]rotaxane from the self-assembly of the AB-type [2]rotaxane monomer **17**. Source: Miyauchi et al. [148]. © 2005 American Chemical Society.

9.5.2 Supramolecular Polymers, Derived from AA- and BB-Type Monomers

The combination of homoditopic monomers of the AA and BB types in precise equimolar amounts also enables the assembly of supramolecular polymers, although in competition with cyclic oligomers (Figure 9.23b). It was shown that the formation of one or the other could be favored by adjusting the rigidity and length of the linkers in these monomers. For example, various ditopic adamantane dimers (**18**) were utilized in the self-assembly with the β-CD dimer **19** (Figure 9.30) [149]. The homoditopic guest **18b/c** having a flexible alkyl unit, as central part, afforded cyclic oligomers (DP of 3–5); whereas, the rigid linear derivative **18a** gave a supramolecular polymer of high molar mass (according to VPO measurements:

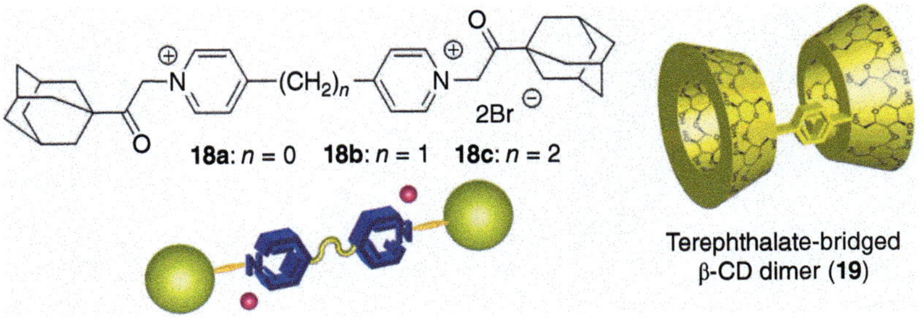

Figure 9.30 Schematic representation of the homoditopic AA- and BB-type components **18** and **19**, respectively, utilized for the supramolecular polymerization. Source: Ohga et al. [149]. © 2005 American Chemical Society.

$M_n > 90\,000\,\text{g mol}^{-1}$, DP > 30). A supramolecular polymer of high molar mass (i.e. M_n of c. $100\,000\,\text{g mol}^{-1}$) was also obtained by combining a homoditopic bis-adamantyl guest with a homoditopic β-CD dimer, both featuring a long PEG spacer ($M_n = 600\,\text{g mol}^{-1}$) [150]. PFGSE–NMR spectroscopy and VPO measurements were applied to estimate the molar mass and overall size of the assembly. Based on SAXS and light-scattering measurements, supported by the study of molecular dynamics, Leggio et al. rationalized the high DP by a solvation effect of the hydrophilic linker that suppressed intra-chain coiling (ideally a linear, well-solvated linker) [151]. Thus, the quality of the solvent was identified as crucial parameter for accessing supramolecular polymers of high molar mass.

Moreover, the self-assembly of two heteroditopic monomers afforded a linear supramolecular polymer, as reported by Harada and coworkers [152]. A dimer composed of two differently sized CD moieties (α-CD and β-CD one) was combined with a heteroditopic guest molecule (featuring a β-CD-selective adamantane and an α-CD-selective phenyl group) to afford a supramolecular polymer, due to a supramolecular self-sorting polymerization (i.e. selectivity of the host–guest complexation).

9.5.3 Supramolecular Polymers, Derived from AA-Type Monomers and CDs

The self-assembly of homoditopic BB-type guests and CD hosts represents the third basic method to prepare supramolecular polymers as shown in Figure 9.23. In contrast to other macrocyclic hosts (i.e. cucurbiturils and pillarenes, seen in Chapters 7 and 10), CDs have hardly been utilized in this respect. Tian and coworkers employed the homoditopic guest **20** with coumarin moieties at both ends [153]. Due to its large cavity size, the γ-CD is able to accommodate two such dye units and, thus, allowed the fabrication of a supramolecular structure by mixing γ-CD and **20** in a 1 : 1 ratio (Figure 9.31). The reversibility of the photochemical dimerization of the complexed chromophores enabled a switching between a covalent polymer (upon irradiation at 365 nm) and a supramolecular polymer (upon irradiation at 254 nm). The morphology of the polymers was investigated by DLS measurements and microscopy imaging: for both types of polymer (i.e. covalent vs. supramolecular), rod-like fibers

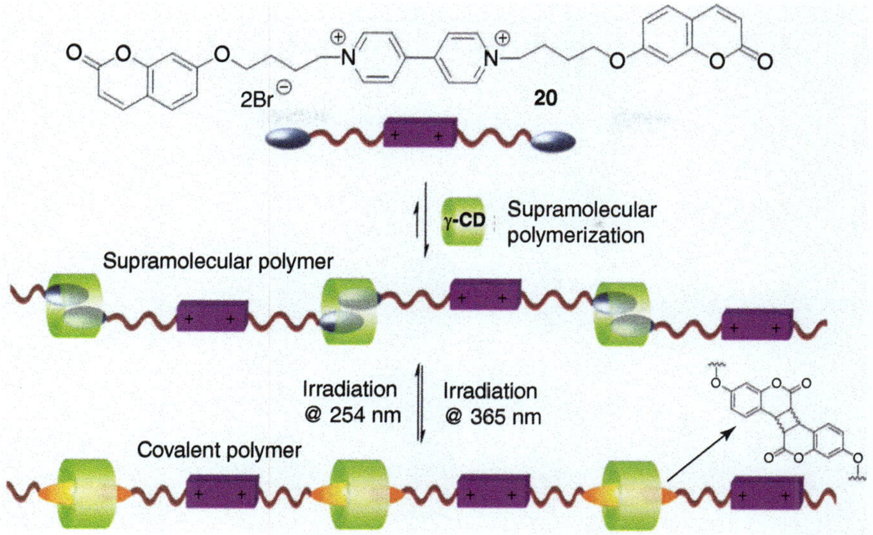

Figure 9.31 Schematic representation of the formation of a supramolecular polymer using the monomers γ-CD and **20**. The supramolecular polymer could reversibly be converted into a covalent polymer, upon light irradiation. Source: From Zhang et al. [153]. © 2013 American Chemical Society.

with a length of over 1000 nm were observed and attributed to the aggregation of individual polymer chains.

In the same context, the Anderson-type polyoxometalate (POM) **21**, equipped with two anthracene moieties, was assembled into a linear supramolecular polymer with γ-CD in a 1 : 1 ratio (the large γ-CD is able to accommodate two anthracene units in its cavity) [154]. Anderson-type POMs represent versatile heteropolyanions, which can easily be organo-functionalized by tris-alkoxylation of both faces of the planar {Mo_6Mn} oxo-cluster [155]. As in the previous example, the photo-induced dimerization of the organic chromophores within the CD cavities gave a stable polyrotaxane, which was analyzed by NMR, UV/vis absorption spectroscopy, and TEM imaging (Figure 9.32). Fibrous bundles as well as single polymer chains with a maximum length of 630 nm (corresponding to a DP of >175) were visualized; static light-scattering measurements revealed a molecular weight (M_w) of up to 3.24×10^5 g mol^{-1}.

9.6 Amphiphilic Supramolecular Diblock Copolymers

The assembly of supramolecular amphiphiles represents an emerging field of research targeting the preparation of smart, responsive materials for potential applications, mainly in the biological and pharmaceutical fields [156, 157]. As for the other cavity-containing macromolecules, CD recognition has also been employed in this context. Besides the inclusion complexes of small molecules,

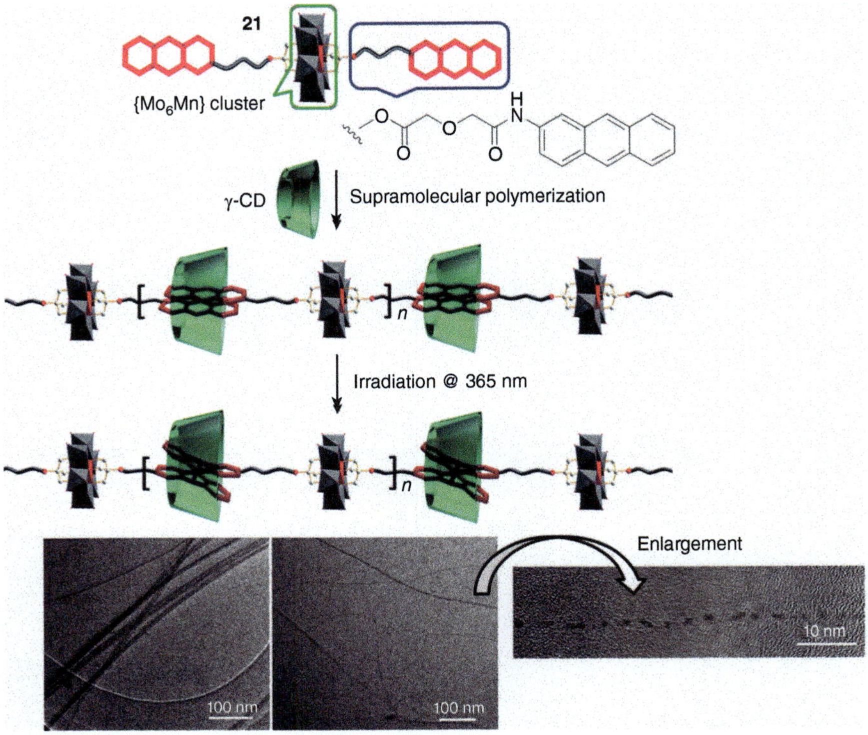

Figure 9.32 Schematic representation of the formation of a POM-containing polyrotaxane by a supramolecular polymerization and subsequent light-driven anthracene dimerization. Representative TEM images showing fibrous bundles and isolated polymer chains are also depicted. Source: Guan et al. [154]. Figure reproduced with kind permission; © 2019 The Royal Chemical Society.

various types of polymer architectures have been applied to prepare supramolecular amphiphiles: diblock copolymers (Figure 9.16) [114, 158], graft copolymers [159, 160], hyperbranched polymers [161–163], dendrimers [164], and biopolymers [165]. In an alternative approach, amphiphilic supramolecular diblock copolymers have been assembled via CD–guest interaction between to appropriately functionalized polymer chains – a hydrophilic and a hydrophobic one. The concept was introduced by Stadermann et al., who combined two water-soluble polymers into a supramolecular polymer via their CD and adamantyl moieties (Figure 9.33). The initial, double-hydrophilic copolymer was reversibly converted into an amphiphilic one due to the temperature-responsive nature of the poly(*N*-isopropyl acrylamide) (PNIPAM) block. PNIPAM is well known for its lower critical solution temperature in water (LCST, at c. 32 °C), at which a phase transition from a swollen hydrated, i.e. hydrophilic, state to a shrunken dehydrated, i.e. hydrophobic, state occurs. Thus, the supramolecular diblock copolymer formed spherical core-shell micelles upon heating above the LCST. For the synthesis of the involved CD-end-functionalized polymers, the living polymerization using CD-containing initiators has recently emerged as the method of choice [166]. The alternative

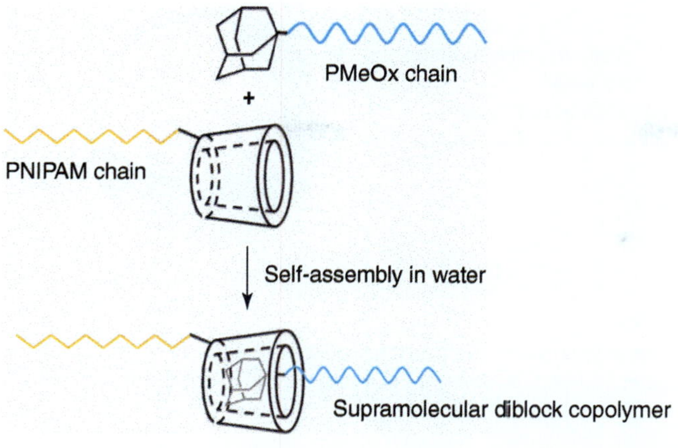

Figure 9.33 Schematic representation of the formation of a supramolecular diblock copolymer, based on a CD–adamantyl host–guest complex [PNIPAM: poly(*N*-isopropyl acrylamide), PMeOx: poly(2-methyl-2-oxazoline)]. Source: Redrawn from Ref. [172]. © 2011 American Chemical Society.

method, i.e. the functionalization of preformed polymers with CD moieties, is known to be more demanding, due to solubility issues.

The formation of amphiphilic supramolecular diblock copolymers by the non-covalent interaction of two appropriately functionalized polymers has been realized for various types of such interactions. In particular, H-bonding interactions, metal-to-ligand coordination, and crown ether recognition have been widely employed in this context (examples for this can be found in the respective chapters). However, utilizing the host–guest chemistry of CDs for this purpose is demanding mainly for solubility reasons (most hydrophilic polymers, such as PS, are barely soluble in water). It was shown that PS, end-functionalized with a β-CD moiety, already represents an amphiphile, which formed reverse-type micelles in less polar solvents (e.g. toluene or chloroform) [167]. The micellization into core-corona assemblies was achieved from a solution of PS-β-CD in DMF, as good solvent, by extensive dialysis against water [168]; thereby, the PS chains formed the hydrophobic core environed by the corona of β-CDs (Figure 9.34).

Yan et al. prepared an amphiphilic, redox-responsive diblock copolymer by assembling PS-β-CD and PEG-Fc (Fc: ferrocene) [169]. Due to strong hydrophobic interactions, stable associates of β-CD with Fc with a 1 : 1 stoichiometry were formed; whereas, the oxidized Fc^+ moiety readily dissociated from the host – a process that could readily be reversed electrochemically [170]. Above the critical aggregation concentration (cac) of 0.28 mg ml^{-1}, vesicles with an average diameter of 102 nm (according to DLS measurements) were formed. Making use of the redox-responsive character of the CD-Fc inclusion complex, a voltage stimulus of +1.5 V was applied to entirely disassemble the supramolecular structure within five hours (as shown by TEM imaging, the process was highly reversible; Figure 9.35). Moreover, rhodamine-B, as model substrate, was encapsulated by the vesicles and released therefrom in a highly controlled fashion; thus, the release

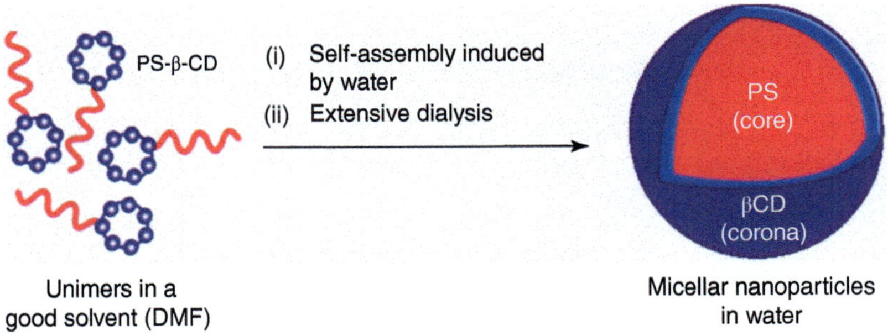

Figure 9.34 Schematic representation of the assembly of core–corona micelles from the polystyrene–β-CD conjugate. Source: Giacomelli et al. [168]. © 2009 American Chemical Society.

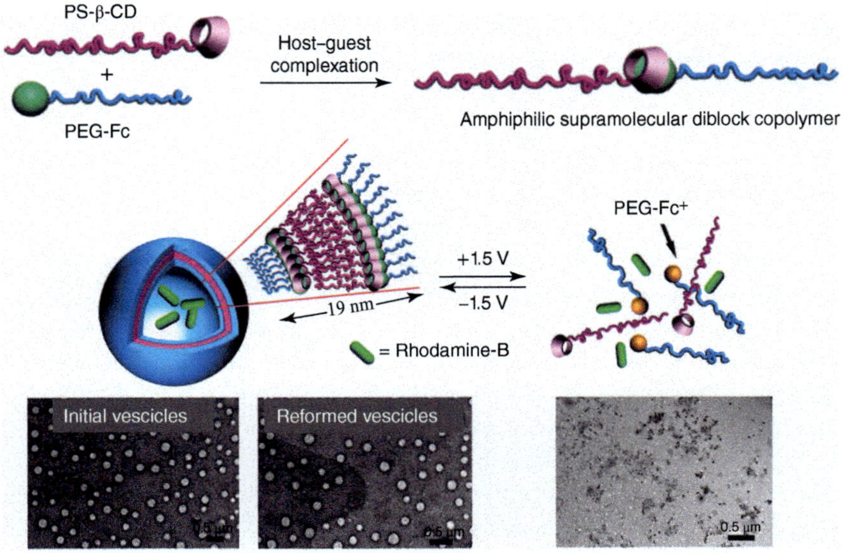

Figure 9.35 Schematic representation of the amphiphilic supramolecular diblock copolymer, which exhibited a redox-responsive behavior due to the selectivity of β-CD to bind Fc and to release Fc^+. The TEM images show the reversibility of the voltage-driven disassembly/reassembly process. Source: Yan et al. [169]. © 2010 The Royal Chemical Society.

rate could be adjusted via the applied voltage from 450 minutes (+1.0 V) through 120 minutes (+2.0 V) to 32 minutes (+4.0 V).

Not only have redox-responsive supramolecular diblock copolymers been reported, but also light-responsive analogs have been considered regarding the controlled release of encapsulated (drug) molecules. For example, poly(acrylic acid) (PAA) and PCl were equipped with an α-CD and a *trans*-azobenzene moiety, respectively [171]. The amphiphilic diblock copolymer was assembled on the basis of a 1 : 1 inclusion complex between the two end groups (K_a of $1.32\times10^4\ M^{-1}$ in water). Above the cac of 0.56 mg ml^{-1}, tubular aggregates with a length of 230 nm

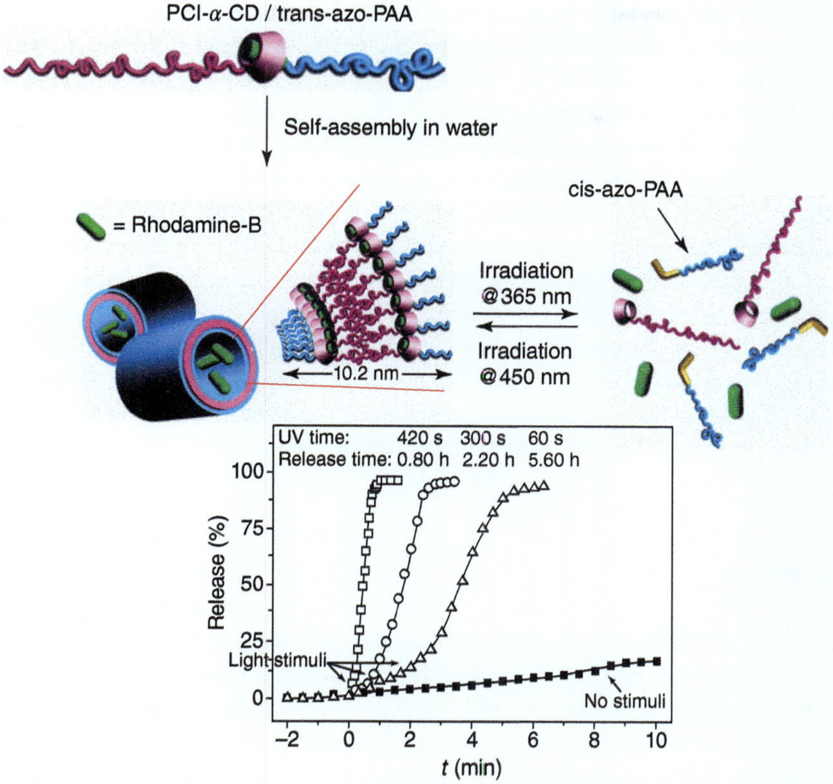

Figure 9.36 Schematic representation of the formation of light-responsive nanotubes, assembled from an amphiphilic supramolecular diblock copolymer, for the controlled release of rhodamine-B. The release rate, as a function of the irradiation time, is also shown. Source: Yan et al. [171]. Figure reproduced with kind permission; © 2010 American Chemical Society.

and a diameter of 90 nm were formed (according to TEM imaging). The aggregates had a wall thickness of c. 10 nm, thus suggesting a bilayer-type packing due to an orthogonal assembly (Figure 9.36). As in the previous example, the nanotubes could be disassembled upon irradiation with UV light for 300 seconds. The nonmatching of the light-generated cis-azobenzene with the cavity size of α-CD was the reason for this behavior that could be fully reversed by photochemically restoring the initial trans-configuration. The release of nanotube-encapsulated rhodamine-B was studied as a function of the irradiation time.

9.7 Concluding Remarks

From all of the macrocyclic species that can be employed as hosts for the formation of inclusion complexes, cyclodextrins (CDs) have up to now attracted the greatest attention – this holds true for small-molecule complexes as well as for polymeric assemblies. Their special reputation results from the natural abundance in starch and their pronounced water solubility. Whereas for supramolecular polymers, based on, e.g. crown ethers, fascinating systems of academic interest, such as "molecular machines" or "molecular muscles," have been realized, the combination of CDs with polymers already allows direct applications in the fields of polymer science (i.e. polymer processing and structuring) and organic electronics (i.e. insulated molecular wires). Moreover, supramolecular hydrogels, formed by the interaction of CDs with biocompatible polymers, can even open the opportunities to utilize such materials in biomedical applications (i.e. drug delivery, controlled drug release). Further, the stimuli-responsive supramolecular polymers might be of interest to realize new types of self-healing materials; however, the reversibility of polymerization and durability still need to be improved. Utilizing supramolecular amphiphiles in the context of biomedical applications represents an emerging field of research, and stimuli-responsive CD-based systems have already shown their versatility in this respect.

References

1 Villiers, A. (1891). *Compt. Rend. Fr. Acad. Sci.*: 435–438.
2 Schardinger, F. (1903). *Z. Untersuch. Nahr. u. Genussm* 6: 865–880.
3 Fourmentin, S., Crini, G., and Lichtfouse, E. (eds.) (2018). *Cyclodextrin Fundamentals, Reactivity and Analysis*. Springer Nature.
4 Szejtli, J. (1998). *Chem. Rev.* 98: 1743–1754.
5 Singh, M., Sharma, R., and Banerjee, U.C. (2002). *Biotechnol. Adv.* 20: 341–359.
6 Astray, G., Gonzalez-Barreiro, C., Mejuto, J.C. et al. (2009). *Food Hydrocolloids* 23: 1631–1640.
7 Hedges, A.R. (1998). *Chem. Rev.* 98: 2035–2044.
8 Loftsson, T. and Duchêne, D. (2007). *Int. J. Pharm.* 329: 1–11.
9 Tiwari, G., Tiwari, R., and Rai, A. (2010). *J. Pharm. Bioallied. Sci.* 2: 72–79.
10 Szejtli, J. (2003). *Starch* 55: 191–196.
11 Wenz, G., Han, B.-H., and Müller, A. (2006). *Chem. Rev.* 106: 782–817.
12 Crini, G. (2014). *Chem. Rev.* 114: 10940–10975.
13 Raymo, F.M. and Stoddard, J.F. (1999). *Chem. Rev.* 99: 1643–1664.
14 Safarowsky, O., Windisch, B., Mohry, A., and Vögtle, F. (2000). *J. Prakt. Chem.* 342: 437–444.
15 Hashidzume, A., Yamaguchi, H., and Harada, A. (2019). *Eur. J. Org. Chem.* 2019: 3344–3357.
16 Schmidt, B.V.K.J., Hetzer, M., Ritter, H., and Barner-Kowollik, C. (2014). *Prog. Polym. Sci.* 39: 235–249.
17 Harada, A. and Kamachi, M. (1990). *Macromolecules* 23: 2821–2823.
18 Harada, A., Takashima, Y., and Yamaguchi, H. (2009). *Chem. Soc. Rev.* 38: 875–882.
19 Harada, A. (1996). *Coord. Chem. Rev.* 148: 115–133.
20 Harada, A., Li, J., and Kamachi, M. (1993). *Chem. Lett.* 22: 237–240.

21 Harada, A., Li, J., Kamachi, M. et al. (1997). *Carbohydr. Res.* 305: 127–129.

22 Herrmann, W., Keller, B., and Wenz, G. (1997). *Macromolecules* 30: 4966–4972.

23 Wenz, G. and Keller, B. (1992). *Angew. Chem. Int. Ed.* 31: 197–199.

24 Taylor, P.N., O'Connell, M.J., McNeill, L.A. et al. (2000). *Angew. Chem. Int. Ed.* 39: 3456–3460.

25 Michels, J.J., O'Connell, M.J., Taylor, P.N. et al. (2003). *Chem. Eur. J.* 9: 6167–6176.

26 Harada, A., Li, J., and Kamachi, M. (1994). *Nature* 370: 126–128.

27 Harada, A., Okada, M., Kawaguchi, Y., and Kamachi, M. (1999). *Polym. Adv. Technol.* 10: 3–12.

28 Wang, J., Qiu, Z., Wang, Y. et al. (2016). *Beilstein J. Org. Chem.* 12: 50–72.

29 Okada, M., Kamachi, M., and Harada, A. (1999). *Macromolecules* 32: 7202–7207.

30 Okada, M., Kamachi, M., and Harada, A. (1999). *J. Phys. Chem. B* 103: 2607–2613.

31 Harada, A. and Okada, M. (1999). *Polym. J.* 31: 1095.

32 Choi, H.S., Ooya, T., Lee, S.C. et al. (2004). *Macromolecules* 37: 6705–6710.

33 Harada, A., Li, J., and Kamachi, M. (1993). *Macromolecules* 26: 5698–5703.

34 Harada, A. (1997). *Carbohyr. Polym.* 34: 183–188.

35 Harada, A., Okada, M., and Kawaguchi, Y. (2005). *Chem. Lett.* 34: 542–543.

36 Peet, J., Rusa, C.C., Hunt, M.A. et al. (2005). *Macromolecules* 38: 537–541.

37 Girardeau, T.E., Leisen, J., and Beckham, H.W. (2005). *Macromol. Chem. Phys.* 206: 998–1005.

38 Harada, A. (1996). *Supramol. Sci.* 3: 19–23.

39 Girardeau, T.E., Zhao, T., Leisen, J. et al. (2005). *Macromolecules* 38: 2261–2270.

40 Tait, C.N. and Davies, D.M. (2002). *Langmuir* 18: 2453–2454.

41 Arunachalam, M. and Gibson, H.W. (2014). *Prog. Polym. Sci.* 39: 1043–1073.

42 Guidry, E.N., Li, J., Stoddard, J.F., and Grubbs, R.H. (2007). *J. Am. Chem. Soc.* 129: 8944–8945.

43 Ceccato, M., Lo Nostro, P., and Baglioni, P. (1997). *Langmuir* 13: 2436–2439.

44 Inoue, Y., Miyauchi, M., Nakajima, H. et al. (2006). *J. Am. Chem. Soc.* 128: 8994–8995.

45 Inoue, Y., Kuad, P., Okumura, Y. et al. (2007). *J. Am. Chem. Soc.* 129: 6396–6397.

46 Harada, A., Okada, M., and Kamachi, M. (1995). *Acta Polym.* 46: 453–457.

47 Kamitori, S., Matsuzaka, O., Kondo, S. et al. (2000). *Macromolecules* 33: 1500–1502.

48 Harada, A. and Kamachi, M. (1990). *J. Chem. Soc., Chem. Commun.*: 1322–1323.

49 Harada, A., Okada, M., Li, J., and Kamachi, M. (1995). *Macromolecules* 28: 8406–8411.

50 Jiao, H., Goh, S.H., Valiyaveettil, S., and Zheng, J. (2003). *Macromolecules* 36: 4241–4243.

51 Kawaguchi, Y., Nishiyama, T., Okada, M. et al. (2000). *Macromolecules* 33: 4472–4477.

52 Harada, A., Nishiyama, T., Kawaguchi, Y. et al. (1997). *Macromolecules* 30: 7115–7118.

53 Harada, A., Kawaguchi, Y., Nishiyama, T., and Kamachi, M. (1997). *Macromol. Rapid Commun.* 18: 535–539.

54 Huang, L., Allen, E., and Tonelli, A.E. (1998). *Polymer* 39: 4857–4865.

55 Kyung-moo, S., Tungalag, D., Yong, H. et al. (2004). *Macromol. Biosci.* 4: 1075–1083.

56 Bullions, T.A., Wei, M., Porbeni, F.E. et al. (2002). *J. Polym. Sci., Part B: Polym. Phys.* 40: 992–1012.

57 Dong, T., He, Y., Shin, K.m., and Inoue, Y. (2004). *Macromol. Biosci.* 4: 1084–1091.

58 Wenz, G., Steinbrunn, M.B., and Landfester, K. (1997). *Tetrahedron* 53: 15575–15592.

59 Yamaguchi, I., Takenaka, Y., Osakada, K., and Yamamoto, T. (1999). *Macromolecules* 32: 2051–2054.

60 Huang, L., Allen, E., and Tonelli, A.E. (1999). *Polymer* 40: 3211–3221.

61 Wei, M., Davis, W., Urban, B. et al. (2002). *Macromolecules* 35: 8039–8044.

62 Tonelli, A.E. (2009). *Adv. Polym. Sci.* 222: 55–77.

63 Gurarslan, A., Joijode, A., Shen, J. et al. (2017). *Polymers* 9: 673.

64 Harada, A., Suzuki, S., Okada, M., and Kamachi, M. (1996). *Macromolecules* 29: 5611–5614.

65 Harada, A., Li, J., Suzuki, S., and Kamachi, M. (1993). *Macromolecules* 26: 5267–5268.

66 Michishita, T., Okada, M., and Harada, A. (2001). *Macromol. Rapid Commun.* 22: 763–767.

67 Michishita, T., Takashima, Y., and Harada, A. (2004). *Macromol. Rapid Commun.* 25: 1159–1162.

68 Jun, L., Akira, H., and Mikiharu, K. (1994). *Bull. Chem. Soc. Jpn.* 67: 2808–2818.

69 Pozuelo, J., Mendicuti, F., and Saiz, E. (2002). *Polymer* 43: 523–531.

70 Rusa, C.C., Rusa, M., Gomez, M. et al. (2004). *Macromolecules* 37: 7992–7999.

71 Ghoul, Y.E., Martel, B., Achari, A.E. et al. (2010). *Polym. J.* 42: 804.

72 Okumura, H., Okada, M., Kawaguchi, Y., and Harada, A. (2000). *Macromolecules* 33: 4297–4298.

73 Okumura, H., Kawaguchi, Y., and Harada, A. (2001). *Macromolecules* 34: 6338–6343.

74 Okumura, H., Kawaguchi, Y., and Harada, A. (2003). *Macromolecules* 36: 6422–6429.

75 Hiromichi, O., Yoshinori, K., and Akira, H. (2002). *Macromol. Rapid Commun.* 23: 781–785.

76 Sanji, T., Yoshiwara, A., Sakurai, H., and Tanaka, M. (2003). *J. Chem. Soc., Chem. Commun.*: 1506–1507.

77 Yoshida, K.-i., Shimomura, T., Ito, K., and Hayakawa, R. (1999). *Langmuir* 15: 910–913.
78 Frampton, M.J. and Anderson, H.L. (2007). *Angew. Chem. Int. Ed.* 46: 1028–1064.
79 Takeshi, S., Ken-ichi, Y., Kohzo, I., and Reinosuke, H. (2000). *Polym. Adv. Technol.* 11: 837–839.
80 Yamaguchi, I. and Yamamoto, T. (2006). *Polym. J.* 38: 495.
81 Isao, Y., Kohki, K., and Takakazu, Y. (2004). *Macromol. Rapid Commun.* 25: 1163–1166.
82 Dodziuk, H., Ejchart, A., Anczewski, W. et al. (2003). *J. Chem. Soc., Chem. Commun.*: 986–987.
83 Wenz, G. and Keller, B. (1994). *Macromol. Symp.* 87: 11–16.
84 Nelson, A., Belitsky, J.M., Vidal, S. et al. (2004). *J. Am. Chem. Soc.* 126: 11914–11922.
85 Belitsky, J.M., Nelson, A., Hernandez, J.D. et al. (2007). *Chem. Biol.* 14: 1140–1151.
86 Meik, W. and Gerhard, W. (1997). *Macromol. Rapid Commun.* 18: 1109–1115.
87 Harada, A., Okumura, H., Okada, M. et al. (2000). *Chem. Lett.* 29: 548–549.
88 Polyakov, V.A., Kolominikova, E.L., Topchieva, I.N., and Kabanov, V.A. (1993). *Vysokomol. Soedin., Ser. B* 35: 285.
89 Mayer, B., Klein, C.T., Topchieva, I.N., and Köhler, G. (1999). *J. Comput.-Aided Mol. Des.* 13: 373–383.
90 Li, J., Li, X., Toh, K.C. et al. (2001). *Macromolecules* 34: 8829–8831.
91 Perry, C., Hebraud, P., Gernigon, V. et al. (2011). *Soft Matter* 7: 3502–3512.
92 Li, J. (2010). *NPG Asia Mater.* 2: 112.
93 Ni, X., Cheng, A., and Li, J. (2009). *J. Biomed. Mater. Res. Part A* 88A: 1031–1036.
94 Hui, Z., Zhang, X., Yu, J. et al. (2010). *J. Appl. Polym. Sci.* 116: 1894–1901.
95 Simões, S.M.N., Veiga, F., Torres-Labandeira, J.J. et al. (2012). *Eur. J. Pharm. Biopharm.* 80: 103–112.
96 He, Z.-x., Wang, Z.-h., Zhang, H.-h. et al. (2011). *Acta Pharma. Sin. B* 1: 254–260.
97 Jansook, P., Pichayakorn, W., Muankaew, C., and Loftsson, T. (2016). *Drug Dev. Ind. Pharm.* 42: 1446–1454.
98 Chan, S.-C., Kuo, S.-W., and Chang, F.-C. (2005). *Macromolecules* 38: 3099–3107.
99 Lu, J., Shin, I.D., Nojima, S., and Tonelli, A.E. (2000). *Polymer* 41: 5871–5883.
100 Li, J., Li, X., Ni, X., and Leong, K.W. (2003). *Macromolecules* 36: 2661–2667.
101 Li, X., Li, J., and Leong, K.W. (2003). *Macromolecules* 36: 1209–1214.
102 Choi, H.S., Ooya, T., Sasaki, S. et al. (2003). *Macromolecules* 36: 9313–9318.
103 Qi, R., Jin, Y., Cheng, X. et al. (2015). *Macromol. Rapid Commun.* 36: 1402–1408.
104 Jiao, H., Goh, S.H., and Valiyaveettil, S. (2002). *Macromolecules* 35: 1980–1983.
105 Sabadini, E. and Cosgrove, T. (2003). *Langmuir* 19: 9680–9683.
106 Zhu, X., Chen, L., Yan, D. et al. (2004). *Langmuir* 20: 484–490.

107 He, L., Huang, J., Chen, Y., and Liu, L. (2005). *Macromolecules* 38: 3351–3355.

108 He, L., Huang, J., Chen, Y. et al. (2005). *Macromolecules* 38: 3845–3851.

109 Tomatsu, I., Hashidzume, A., and Harada, A. (2005). *Macromolecules* 38: 5223–5227.

110 Tomatsu, I., Hashidzume, A., and Harada, A. (2005). *Macromol. Rapid Commun.* 26: 825–829.

111 Harada, A., Adachi, H., Kawaguchi, Y., and Kamachi, M. (1997). *Macromolecules* 30: 5181–5182.

112 Weickenmeier, M., Wenz, G., and Huff, J. (1997). *Macromol. Rapid Commun.* 18: 1117–1123.

113 Huang, L., Chen, J., He, M. et al. (2017). *Chin. Chem. Lett.* 30: 163–166.

114 Cho, S.Y. and Allcock, H.R. (2009). *Macromolecules* 42: 4484–4490.

115 Kolb, H.C., Finn, M.G., and Sharpless, K.B. (2001). *Angew. Chem. Int. Ed.* 40: 2004–2021.

116 Nepogodiev, S.A. and Stoddart, J.F. (1998). *Chem. Rev.* 98: 1959–1976.

117 Harada, A., Li, J., and Kamachi, M. (1992). *Nature* 356: 325.

118 Harada, A., Li, J., and Kamachi, M. (1994). *J. Am. Chem. Soc.* 116: 3192–3196.

119 Ooya, T., Mori, H., Terano, M., and Yui, N. (1995). *Macromol. Rapid Commun.* 16: 259–263.

120 Fujita, H., Ooya, T., Kurisawa, M. et al. (1996). *Macromol. Rapid Commun.* 17: 509–515.

121 Fujita, H., Ooya, T., and Yui, N. (1999). *Macromol. Chem. Phys.* 200: 706–713.

122 Zhao, T. and Beckham, H.W. (2003). *Macromolecules* 36: 9859–9865.

123 Herrmann, W., Schneider, M., and Wenz, G. (1997). *Angew. Chem. Int. Ed.* 36: 2511–2514.

124 Alupei, V., Alupei, I.C., and Ritter, H. (2005). *Macromol. Rapid Commun.* 26: 40–45.

125 Wang, B., He, J., Sun, D. et al. (2005). *Eur. Polym. J.* 41: 2483–2487.

126 Liu, Y., Zhao, Y.-L., Zhang, H.-Y. et al. (2004). *Macromolecules* 37: 6362–6369.

127 Nepal, D., Samal, S., and Geckeler, K.E. (2003). *Macromolecules* 36: 3800–3802.

128 Storsberg, J., Ritter, H., Pielartzik, H., and Groenendaal, L. (2000). *Adv. Mater.* 12: 567–569.

129 Takashima, Y., Oizumi, Y., Sakamoto, K. et al. (2004). *Macromolecules* 37: 3962–3964.

130 Bergamini, J.F., Lagrost, C., Chane Ching, K.I. et al. (1999). *Synth. Met.* 102: 1538–1539.

131 Bergamini, J.F., Jouini, M., Aeiyach, S. et al. (2005). *J. Electroanal. Chem.* 579: 125–131.

132 van den Boogaard, M., Bonnet, G., van't Hof, P. et al. (2004). *Chem. Mater.* 16: 4383–4385.

133 Terao, J. (2011). *Polym. Chem.* 2: 2444–2452.

134 Terao, J., Tsuda, S., Tanaka, Y. et al. (2009). *J. Am. Chem. Soc.* 131: 16004–16005.

135 Terao, J., Tanaka, Y., Tsuda, S. et al. (2009). *J. Am. Chem. Soc.* 131: 18046–18047.

136 Okada, M., Takashima, Y., and Harada, A. (2004). *Macromolecules* 37: 7075–7077.

137 Okada, M. and Harada, A. (2003). *Macromolecules* 36: 9701–9703.

138 Yang, L., Tan, X., Wang, Z., and Zhang, X. (2015). *Chem. Rev.* 115: 7196–7239.

139 Harada, A. and Takashima, Y. (2012). Cyclodextrin-based supramolecular polymers. In: *Supramolecular Polymer Chemistry* (ed. A. Harada), 29–50. Weinheim: Wiley-VCH.

140 Harada, A., Kawaguchi, Y., and Hoshino, T. (2001). *J. Inclusion Phenom.Macrocyclic Chem.* 41: 115–121.

141 Tellini, V.H., Jover, A., Galantini, L. et al. (2004). *Acta Cryst., Sect. B: Struct. Sci.* 60: 204–210.

142 Hoshino, T., Miyauchi, M., Kawaguchi, Y. et al. (2000). *J. Am. Chem. Soc.* 122: 9876–9877.

143 Miyauchi, M., Kawaguchi, Y., and Harada, A. (2004). *J. Inclusion Phenom.Macrocyclic Chem.* 50: 57–62.

144 Cameron, K.S. and Fielding, L. (2001). *J. Org. Chem.* 66: 6891–6895.

145 Miyauchi, M., Takashima, Y., Yamaguchi, H., and Harada, A. (2005). *J. Am. Chem. Soc.* 127: 2984–2989.

146 Evenou, P., Rossignol, J., Pembouong, G. et al. (2018). *Angew. Chem. Int. Ed.* 57: 7753–7758.

147 Miyauchi, M. and Harada, A. (2004). *J. Am. Chem. Soc.* 126: 11418–11419.

148 Miyauchi, M., Hoshino, T., Yamaguchi, H. et al. (2005). *J. Am. Chem. Soc.* 127: 2034–2035.

149 Ohga, K., Takashima, Y., Takahashi, H. et al. (2005). *Macromolecules* 38: 5897–5904.

150 Hasegawa, Y., Miyauchi, M., Takashima, Y. et al. (2005). *Macromolecules* 38: 3724–3730.

151 Leggio, C., Anselmi, M., Di Nola, A. et al. (2007). *Macromolecules* 40: 5899–5906.

152 Takahashi, H., Takashima, Y., Yamaguchi, H., and Harada, A. (2006). *J. Org. Chem.* 71: 4878–4883.

153 Zhang, Q., Qu, D.-H., Wu, J. et al. (2013). *Langmuir* 29: 5345–5350.

154 Guan, W., Wang, G., Ding, J. et al. (2019). *Chem. Commun.* 55: 10788–10791.

155 Zhang, J., Huang, Y., Li, G., and Wei, Y. (2019). *Coord. Chem. Rev.* 378: 395–414.

156 Yu, G., Jie, K., and Huang, F. (2015). *Chem. Rev.* 115: 7240–7303.

157 Zhang, Y.-M., Liu, Y.-H., and Liu, Y. (2020). *Adv. Mater.* 32: 1806158.

158 Zhang, J. and Ma, P.X. (2009). *Angew. Chem. Int. Ed.* 48: 964–968.

159 Wang, J. and Jiang, M. (2006). *J. Am. Chem. Soc.* 128: 3703–3708.

160 Liu, L., Rui, L., Gao, Y., and Zhang, W. (2015). *Polym. Chem.* 6: 1817–1829.

161 Tao, W., Liu, Y., Jiang, B. et al. (2012). *J. Am. Chem. Soc.* 134: 762–764.

162 Böhm, I., Isenbügel, K., Ritter, H. et al. (2011). *Angew. Chem. Int. Ed.* 50: 7896–7899.

163 Liu, Y., Yu, C., Jin, H. et al. (2013). *J. Am. Chem. Soc.* 135: 4765–4770.

164 Park, C., Lee, I.H., Lee, S. et al. (2006). *Proc. Natl. Acad. Sci. U.S.A.* 103: 1199–1203.

165 Zhang, Z., Ding, J., Chen, X. et al. (2013). *Polym. Chem.* 4: 3265–3271.

166 Seidi, F., Shamsabadi, A.A., Amini, M. et al. (2019). *Polym. Chem.* 10: 3674–3711.

167 Kakuchi, T., Narumi, A., Miura, Y. et al. (2003). *Macromolecules* 36: 3909–3913.

168 Giacomelli, C., Schmidt, V., Putaux, J.-L. et al. (2009). *Biomacromolecules* 10: 449–453.

169 Yan, Q., Yuan, J., Cai, Z. et al. (2010). *J. Am. Chem. Soc.* 132: 9268–9270.

170 Harada, A., Takashima, Y., and Nakahata, M. (2014). *Acc. Chem. Res.* 47: 2128–2140.

171 Yan, Q., Xin, Y., Zhou, R. et al. (2011). *Chem. Commun.* 47: 9594–9596.

172 Stadermann, J., Komber, H., Erber, M. et al. (2011). *Macromolecules* 44: 3250–3259.

10

Supramolecular Polymers, Based on the Host–Guest Chemistry of Pillarenes

10.1 General Aspects

The utilization of non-covalent interactions to construct polymer-like materials – starting from small-molecule monomers – has evolved from a rather exotic field of research to an important component of today's chemistry [2–6]. Besides hydrogen-bonding [7] or ionic interactions [8], as well as metal-to-ligand complexation [9], the assembly of host–guest inclusion complexes represents one further powerful tool to prepare advanced supra(macro)molecular assemblies [10]. In this respect, various organic macrocycles have already been utilized over the last few decades; these macrocycles feature distinct topological structures and offer to accommodate a wide range of molecular guests. As outlined in Chapters 6–9, host–guest complexation of crown ethers [2, 11], cyclodextrins (CDs) [12], calixarenes [13], and cucurbiturils [14] is widely utilized in modern supramolecular chemistry and enables the facile formation of polymeric assemblies of diverse architectures and with tailor-made properties as well as sophisticated behaviors (e.g. molecular "muscles" or "machines"). In 2008, Ogoshi et al. introduced one additional class of macrocycles featuring the ability to accommodate guest molecules within their cavities: Pillar[*n*]arenes or – more commonly – pillarenes [1]. In general, pillar[*n*]arenes are composed of *n* hydroquinone units, which are bridged via methylene groups in their para-positions; from these, the D_5 symmetric pillar[5]arene (**1**) is the conformationally most stable representative, which is considerably more stable and rigid than, e.g. crown ethers or calixarenes (Figure 10.1). Pillarenes possess symmetrical, cylindrical architectures with two identical cavity gates and whose name refers to its pillar-like shape, which is composed of arene subunits. As a result, the close proximity of the electron-rich hydroquinone moieties allows electron-poor species to be efficiently bound within the cavities. This particular host–guest complexation behavior was already reported in the seminal report on pillarenes: host–guest complexes possessing a 1 : 1 stoichiometry were formed upon binding of a slim cationic methylviologen- or pyridinium-type guest (**2** and **3**) within the cavity of **1** (Figure 10.1); in contrast,

Supramolecular Polymers and Assemblies: From Synthesis to Properties and Applications, First Edition.
Ulrich S. Schubert, George R. Newkome, and Andreas Winter.

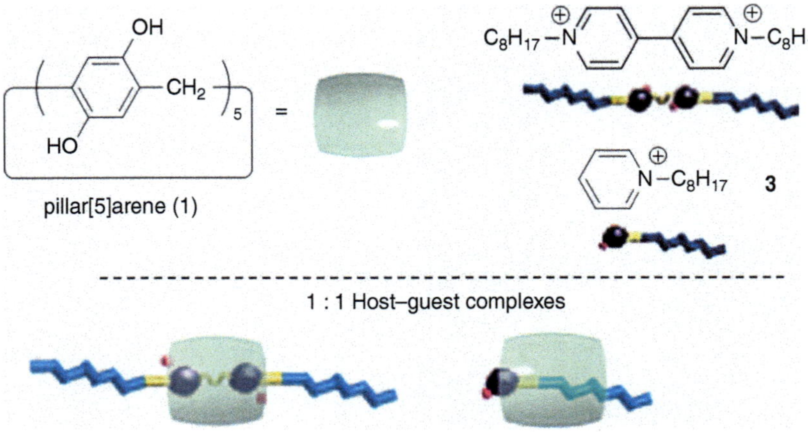

Figure 10.1 Schematic representation of pillar[5]arene (**1**) and two cationic guests (**2** and **3**) that formed host–guest complexes of 1 : 1 stoichiometry with **1**. Source: Ogoshi et al. [1]. © 2008 American Chemical Society.

bulky terminal groups (e.g. adamantyl moieties) prevented guest binding due to a mismatch within the cavity gate. For pillar[5]arene, a cavity diameter of 4.7 Å was calculated from the van-der-Waals radii of the atoms, thus being much smaller as for the next homolog pillar[6]arene (6.7 Å) [15]. Following the vocabulary used in Chapters 6–9, such 1 : 1 complexes have to be referred to as pseudorotaxanes; if the stopper units are introduced in order to prevent dissociation of the complexes, rotaxanes are obtained [11]. An overview of simple pillar[*n*]arene–guest complexes, including the corresponding association constants (K_a), has been given by Zhang and Zhao in a comprehensive review [16]. One further aspect is the ease of chemical modification (in comparison to other macrocyclic hosts, e.g. the cucurbiturils): Functionalization of pillar[*n*]arenes can occur via ether- or esterification of the OH-groups. Besides the modification of all available positions, the selective functionalization of only one or two positions can also be realized (such unsymmetrically functionalized species are often referred to as "copillarenes") [17]. This is of relevance when utilizing pillarenes as molecular sensors, e.g. for guest detection by fluorescence quenching (in this case, appropriate dyes have to be attached to the pillarene framework) or as monomers for the assembly of supramolecular polymers (see Section 8.3). Moreover, their solubility can be modulated over a wide range of organic solvents; thus, pillarenes are powerful complements to their water-soluble cucurbituril and cyclodextrin relatives, which possess similar cavity sizes [18].

Various review articles have summarized the methodologies to synthesize the various pillarenes, their solution, and solid-state structures, their host–guest chemistry as well as potential applications arising from the latter one [16, 18–24]. In the following, the utilization of pillarenes in the field of supramacromolecular chemistry, i.e. the self-assembly into poly(pseudo)rotaxanes and related structures will be discussed.

10.2 Host–Guest Complexation Between Pillarenes and Linear Polymers

A broad range of bis-paraquats and bis-pyridinium as well as bis-imidazolium salts has been employed as a guest for the threading into **1**, thus generating diverse pseudorotaxanes [16, 25]. The scope of this basic approach was extended by Ogoshi et al. toward viologen oligomers (**4**, Figure 10.2). Threading of these extended molecular guests into the cavity of **1** generated the first supramacromolecular pillarene-based materials [26]. These authors investigated the migration of the host along the axis of the guest, as a function of the solvent (a faster migration rate was observed in acetone/acetonitrile mixtures than in dimethylsulfoxide [DMSO]). Moreover, a dependency of the strength of the host–guest interaction on the length of the alkyl spacer was found: according to nuclear magnetic resonance (NMR) spectroscopic studies, guest **4a** was found to bind stronger than **4b**.

One application of pillarene-based complexation in the field of conducting polymers was also reported by the Ogoshi group [27]. From the family of conducting polymers, polyaniline (PANI) represents one of the most intensively studied materials due to its notable properties, such as manifold and controlled redox chemistry, environmental stability, and possibility for reversible doping [28]. Basically, four forms of PANI are known and all display distinct photophysical and electrochemical properties. The fully reduced form, the so-called leucoemeraldine base, is characteristically transparent; whereas, the fully oxidized species, referred to as pernigraniline base, has a dark violet coloration. The half-oxidized, blue-colored emeraldine base form of PANI can be protonated to give the green emeraldine salt, which is electrically conducting. The reversible interconversion between these different species can be achieved either electrically or chemically (i.e. by treatment with an appropriate reducing or oxidizing agent). The resulting color changes are of importance for sensing applications and can even be followed by the human eye. The emeraldine base of PANI ($M_w = 20\,000\,g\,mol^{-1}$) was used as polymer guest for the pseudorotaxane formation with **1** [27]. Remarkably, the initial blue color of the assembly vanished with time due to the formation of the colorless leucoemeraldine form (Figure 10.3). The reduction of the polymer was rationalized by a redox reaction with the hydroquinone moieties of the surrounding host molecule (in control experiments, the half-oxidized PANI was reduced with simple hydroquinone in a much less efficient manner). Moreover, the analogous pseudorotaxanes, derived

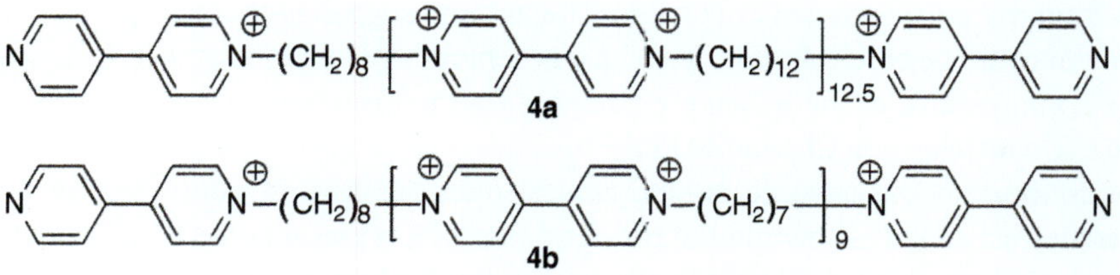

4a

4b

Figure 10.2 Schematic representation of the viologen oligomers **4** used as macromolecular guests.

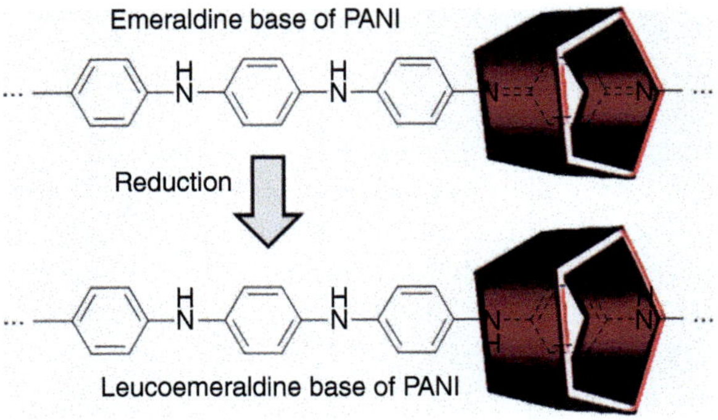

Figure 10.3 Schematic representation of the pillar[5]arene-based pseudorotaxane of PANI's emeraldine base and its reduction to the leucoemeraldine species.

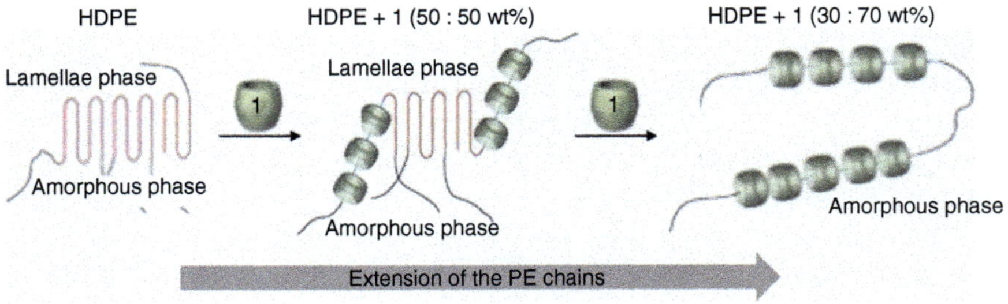

Figure 10.4 Schematic representation of the chain extension of polyethylene induced by the threading of **1** onto the chains. Source: Ogoshi et al. [29]. © 2013 Springer Nature.

from alternative macrocyclic hosts, such as CDs or cucurbiturils, and the emeraldine base retained their blue color due to the absence of reducing groups within the host materials (see Chapter 9).

Since **1** is known for its ability to give host–guest inclusion complexes even with linear alkanes, Ogoshi et al. investigated the pseudorotaxane formation using high-density polyethylene (HDPE), as the macromolecular guest [29]. Since HDPE is barely soluble in organic solvents, the self-assembly had to be performed in the melt. As a result, from the threading of **1** onto the polymer chains, a chain extension in concert with an increasing of the melting temperature (T_m) of the material from 126 to 152 °C could be observed (Figure 10.4). This change became readily apparent by a phase transition from the molten to the solid state, upon the addition of **1** to HDPE at 140 °C. At the same temperature, the phase transition could be reversed by adding 1,4-dibromobutane, as a competitive guest, which induced the dethreading of **1** from the polyethylene (PE) chains. The authors suggested that this particular behavior could be of relevance to develop a new molding and processing techniques for PE, comparable to the polymer processing via their intermediate complexation with cyclodextrins (see Chapter 9) [30].

In contrast to the cationic guests, neutral ones are less efficiently bound by pillarenes (as already shown in the previous example, representative K_a values are listed in Cragg and Sharma [19]). To circumvent this limitation, when aiming for the synthesis of rotaxanes with nonionic polymers as axis/thread, liquid pillar[5]arenes were used as hosts [31]. The rotaxanes **5** (Figure 10.5a) were obtained in a two-step

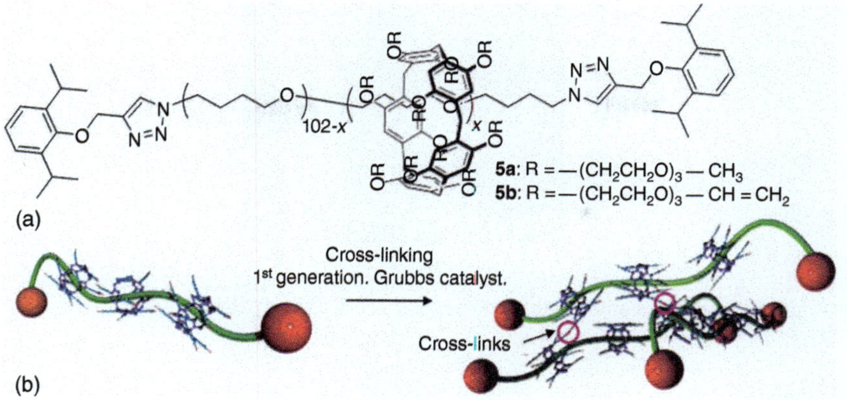

Figure 10.5 (a) Schematic representation of the pillar[5]arene-based polymer-containing rotaxanes **5**. (b) Representation of the Ru-catalyzed cross-linking of **5b** into a topological polymer gel. Source: Ogoshi et al. [31]. © 2014 The Royal Chemical Society.

procedure involving (i) pseudorotaxane formation and (ii) end capping of the pending azido groups of the polytetrahydrofuran guest with an alkyne stopper in a Cu(I)-catalyzed azide-alkyne cycloaddition reaction (CuCAAC, the so-called "click" reaction). In the case of **5b**, the reactive vinyl moieties on the pillarene framework were used for cross-linking via an olefin metathesis reaction (Figure 10.5b). Thusly, the resultant topological gel exhibited swelling behavior in organic solvents with medium polarity.

10.3 Supramolecular Polymers, Derived from Pillarene-based Host–Guest Interactions

Basically, linear supramolecular polymers can be assembled from appropriate monomers via directional and strong non-covalent interactions. The assemblies, formed under these conditions, have shown chain-like behavior in solution as well as in the solid state [32]. As outlined in Chapters 6–9, this methodology can be applied to a variety of cavity-containing macrocyclic host materials and, in the context of this chapter, also to pillarenes [22]. Again, complementary host and guest monomers are required that can assemble in a polyassociative process into linear supramolecular polymers. For this purpose, heteroditopic AB-type monomers can be used to form homopolymers. Alternatively, equimolar mixtures of heterocomplementary, homoditopic AA- and BB-type monomers will give alternating heteropolymers via two-component associative processes. Noteworthy, the broad range of nonlinear supramolecular polymers can also be assembled involving pillarene-based host–guest chemistry (hyperbranched or graft/comb polymers, polymer networks, dendritic, and star-shaped architectures). As for the other types of host–guest inclusion complexes – supramolecular polymerization will always compete with the formation of discrete species of lower molar masses: self-inclusion complexes as well as dimerized or (macro)cyclized oligomers.

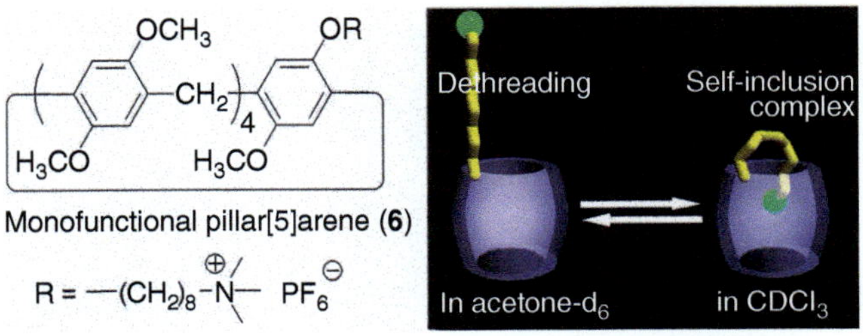

Figure 10.6 Schematic representation of the host–guest conjugate **6** and its reversible formation of a self-inclusion complex. Source: Guan et al. [33]. © 2011 The Royal Chemical Society.

Monofunctional host–guest conjugate (**6**), for example, exhibited such a self-inclusion behavior, i.e. threading of the trimethylammonium-bearing tail into the cavity of the pillarene host (Figure 10.6) [34]. Cation–π interactions facilitated this process in weakly polar solvents, such as $CHCl_3$; however, dethreading readily occurred in a more polar environment (e.g. in acetone); thus, the structure of the molecule could be modulated simply via the solvent's polarity. One further example of such self-inclusion complexes was given by Guan et al., who observed the fluoride-mediated isomerization of such assemblies [33].

The formation of supramolecular double-threaded dimers, in equilibrium with the self-inclusion complexes, was reported for the mono-functionalized pillarenes **7a** and **7b** Figure 10.7a) [35, 36]. The reversible association of these derivatives could be controlled by adjusting the environment's polarity (i.e. the solvent composition) or the temperature as well as by offering competing guests. The pillarene derivative **7c** featuring a long aliphatic tail exhibited a different self-assembly behavior [37]: At high concentrations or in the solid state, a linear supramolecular polymer was formed [$(\mathbf{7c})_n$, Figure 10.7b]. These authors rationalized this supramolecular polymerization to be driven mainly by CH–π interactions; however, owning to the weak nature of these polymers, very high concentrations (e.g. c. 500 mM for **7c**) were required to assemble polymers of high molar masses. Detailed NMR spectroscopic and viscosity measurements were conducted to study the supramolecular polymerization in solution; moreover, rod-like fibers of $(\mathbf{7c})_n$ were fabricated from highly concentrated solutions and studied by scanning electron microscopy (SEM). The regularly shaped fibers revealed an average diameter of 9.5 mm and confirmed the formation of a stable, even processable, supramolecular polymer.

As a conclusion from the former example, supramolecular polymerization of monofunctional pillarene monomers can be facilitated when the strength of the host–guest interaction is increased. In line with this argumentation, Stoddard and coworkers employed the pillarene **8** bearing an appending MV^{2+} moiety (MV^{2+}: methylviologen or viologen/paraquat) [38]. Due to the significantly stronger character of the cation–π interactions (compared to CH–π interactions), supramolecular polymerization of **8** readily occurred above the concentration limit of c. 20 mM; whereas, at even higher concentrations (i.e. above 25 mM), a fibrous organogel was obtained. This was evident from optical as well as SEM microscopy

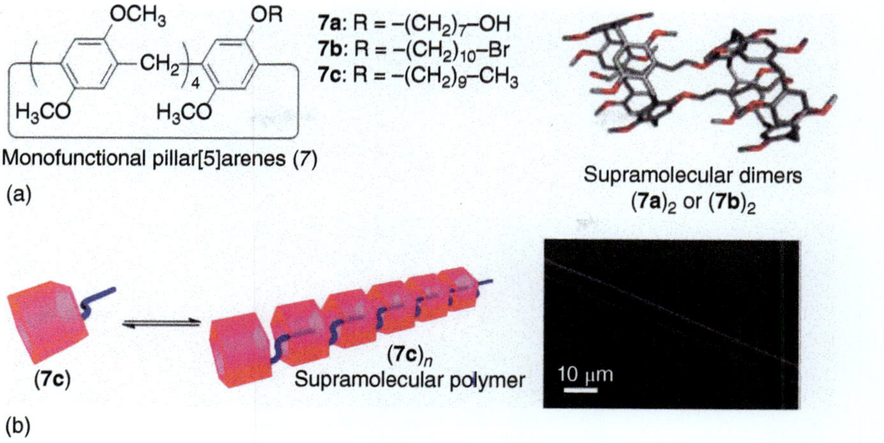

Figure 10.7 Schematic representation of the monofunctional pillar[5]arenes **7** that assembled into supramolecular dimers (**7a** and **7b**) or a supramolecular polymer (**7c**). (a) Ball-and-stick model of such a supramolecular dimer. (b) Representative SEM image of a gold-coated fiber of $(\mathbf{7c})_n$. Source: Liu et al. [35]. © 2012 American Chemical Society.

imaging and X-ray powder diffraction (XRD) measurements (Figure 10.8). The gelation process was rationalized by the entanglement of polymer chains and trapping of water molecules within the intertwined structures. Diverse related supramolecular polymers were also prepared from pillarene derivatives with an ammonium-functionalized side chain [39–43]. All of these polymers featured a rich stimuli-responsive behavior: their assembly/disassembly could be triggered by changes in pH, presence of certain anions (e.g. chloride), and polarity of the solvent. As an example, the solvent polarity was shown to induce a reversible contraction–extension process of the supramolecular polymer when the solvent was switched between $CHCl_3$ and DMSO [42]. Remarkably, electrospinning – as a convenient and versatile technique to fabricate functional nanofibers [44] – could be applied to these supramolecular polymers generating nanofibers with diameters of up to 800 nm (according to SEM imaging) [40].

The self-assembly of host–guest conjugates into polymers or discrete species can be controlled and enhanced via the presence of appropriate functional groups. In this respect, the photo-responsive supramolecular self-assembly, as reported by Wang et al., represents a versatile example [45]. These authors utilized a stiff stilbene derivative, as photo-active moiety, within the arm of the mono-functionalized pillarene **9** (Figure 10.9). An equilibrium between the self-inclusion complex and the dimeric species $(\mathbf{9})_2$ was observed for the Z-isomer of **9**. Upon photo-irradiation, the Z-isomer was transformed into the E-form, which readily formed a supramolecular polymer (for geometric reasons, the E-form could not form any self-inclusion complexes). The observed low molar mass was attributed to the rather weak CH–π interactions within the guests' cavities. However, upon protonation of the terminal imidazole moieties strong binding via cation–π interaction became enabled, thus leading to a supramolecular polymer of high molar mass.

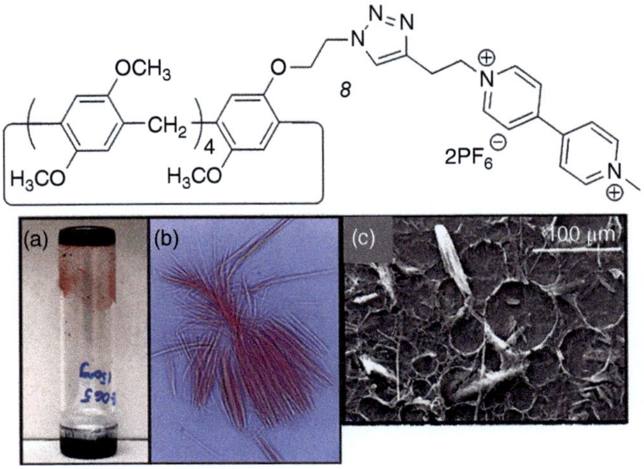

Figure 10.8 Schematic representation of the host–guest conjugate **8**. (a) Image of the supramolecular gel $(\mathbf{8})_n$; (b) optical microscopy image (500×) of aggregates of $(\mathbf{8})_n$; (c) SEM image of the gel in CH_2Cl_2. Source: Strutt et al. [38]. © 2012 The Royal Chemical Society.

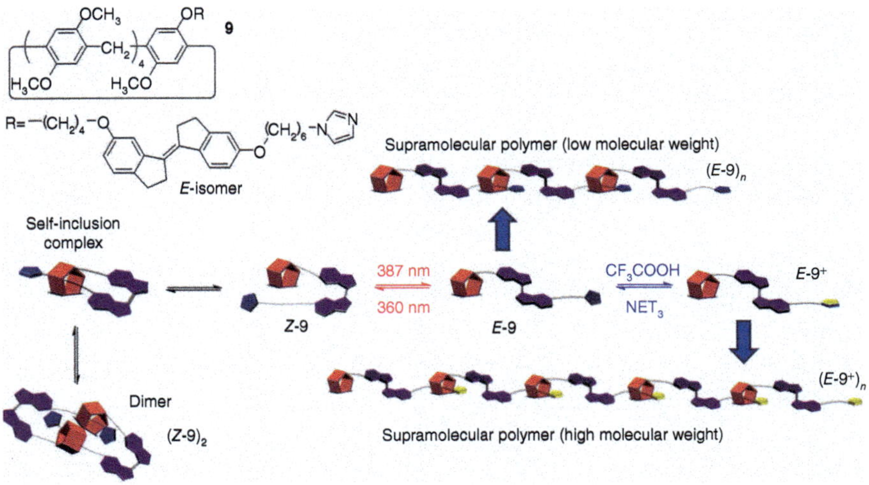

Figure 10.9 Schematic representation of the photo-switchable host–guest conjugate **9** and its self-assembly into supramolecular architectures. Source: Wang et al. [45]. © 2014 The Royal Chemical Society.

The Stoddard group incorporated a metalloporphyrin unit between the two binding sites of the host–guest conjugate **10** [46]. Due to the presence of the planar π-conjugated porphyrin entities, π–π stacking interactions of the initial linear supramolecular polymer strands generated large bundles as well as discrete fibers (Figure 10.10). These bundles entangled into an elastic organogel, which was analyzed by rheology measurements. An excellent example for a supramolecular self-sorting polymerization was published by Ogoshi et al. [47]: Two pillarenes of different cavity sizes and two selective guest moieties were

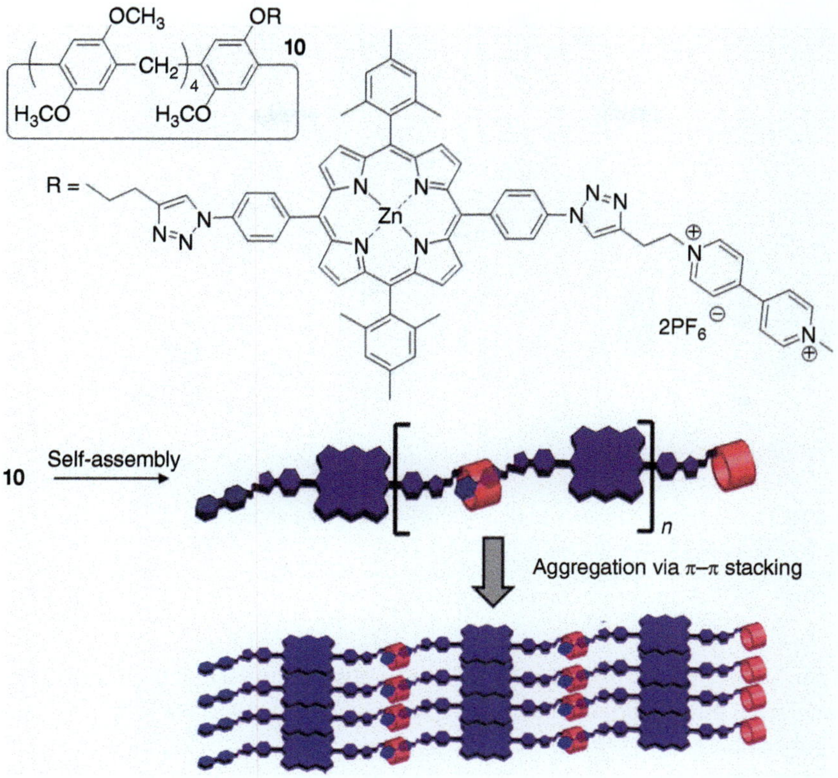

Figure 10.10 Schematic representation of the host–guest conjugate **10** that assembled into a linear supramolecular polymer and, subsequently into aggregates via π–π stacking interactions. Source: Fathalla et al. [46]. © 2015 The Royal Chemical Society.

used in this study. The smaller pillar[5]arene was equipped with a bulky cationic 1,4-diazabicyclo[2.2.2]octane (DABCO) moiety – a potent guest for pillar[6]arene, which is too bulky to match with the smaller pillar[5]arenes' cavity (thus, self-inclusion or homo-association could be excluded). Analogously, the pillar[6]arene was bearing a small pyridinium head – optimized in size to match with the small pillar[5]arene's cavity. The mixing of these two heteroditopic monomers induced a self-sorting process and, finally, gave a linear supramolecular polymer in which the pillar[5]arene and pillar[6]arene moieties were placed along the backbone in a strictly alternating fashion.

Utilizing fluorescent moieties within the molecular structure of a homoditopic monomer, either of AA- or BB-type, allows for the assembly of supramolecular polymers with rich photophysical properties. As a first example in this respect, Song et al. used a host–host conjugate, in which the two pillar[5]arene sites were linked via a central 9,10-distyryl-anthracene dye [48]. The supramolecular polymer, obtained from the self-assembly with a ditopic neutral guest, exhibited a bright yellow emission that was enhanced significantly when, e.g. compared to a host-only solution (here, a typical though less efficient aggregation-induced emission [AIE] behavior

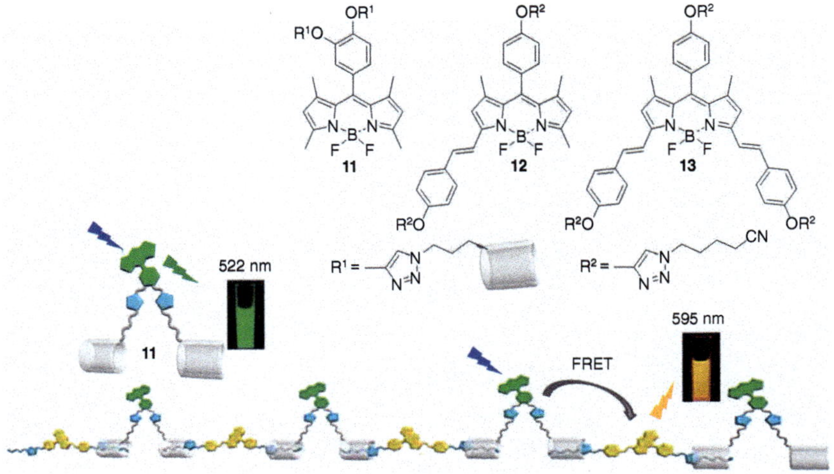

Figure 10.11 Schematic representation of the BODIPY-containing host (**11**) and guest molecules (**12** and **13**). For the supramolecular polymer $(\mathbf{11}\cdots\mathbf{12})_n$, FRET from the donor dye to the acceptor one was observed. Source: Meng et al. [49]. Figure reproduced with kind permission; © 2015 The Royal Chemical Society.

could be observed). A variety of control experiments confirmed that the fluorescence enhancement directly originated from the formation of the rigid-linear supramolecular polymer.

Against the background of artificial photosynthesis, Wang et al. focused on boron-dipyrromethene (BODIPY) dyes, as light-harvesting components, due to their high absorptivity and tunability regarding their emission color (BODIPY)[49]. For this purpose, the BODIPY-containing host (**11**) and guest molecules (**12** and **13**) were prepared (Figure 10.11). The BODIPY dye of **11** exhibited a green emission (λ_{fl} = 522 nm) upon excitation with visible light (λ_{ex} = 470–490 nm). Within the supramolecular polymer comprising monomers **11** and **12**, Förster-type resonance energy transfer (FRET) occurred leading to the emission of yellow light (λ_{fl} = 595 nm) originating from the dye of **12**, when the dye **11** was excited (Figure 10.11). The FRET process became possible due to the close proximity of the two different chromophores within the supramolecular polymer. Similarly, the supramolecular network of **11** and **13** exhibited red emission from the BODIPY dye of **13** (λ_{fl} = 669 nm).

Besides the previous examples where the host–guest conjugates utilized for supramolecular polymerization can be defined as AB-type monomers, supramolecular polymerizations employing AA- and BB-type monomers (i.e. a host–host and a guest–guest conjugate) in an equimolar ratio can also be found in the literature. In a rather straightforward approach, an AA-type host monomer, comprising two pillar[5]arene macrocycles and a central photo-responsive azobenzene moiety, was combined with a bis-pyridinium BB-type guest molecule to afford a supramolecular polymer of high molar mass (Degree-of-polymerization [DP] of c. 470) [50]. The DP could be decreased significantly by photochemically switching

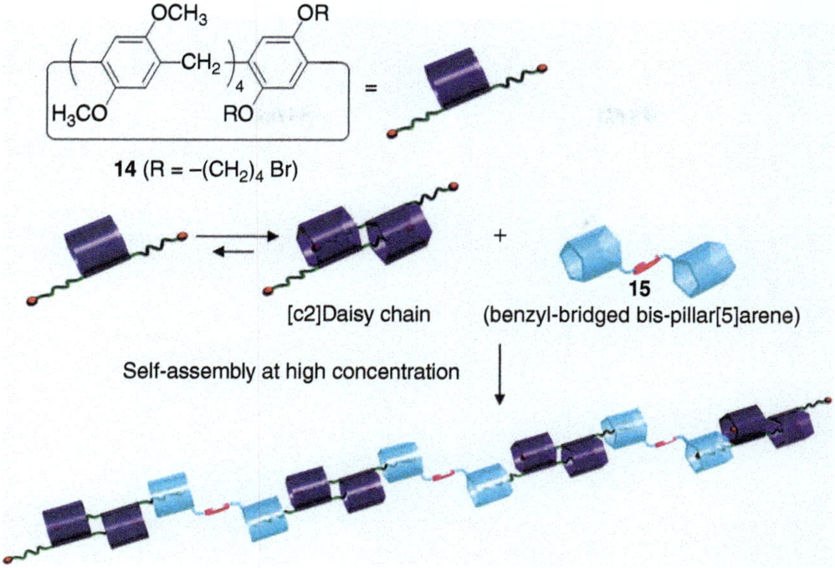

Figure 10.12 Schematic representation of the two-step self-assembly of the pillarene-based building blocks **14** and **15** into a supramolecular polymer. Source: Wang et al. [51]. © 2013 The Royal Chemical Society.

the trans-form of the azobenzene moiety to the cis-isomer – thus due to steric hindrance, partial depolymerization was observed (DP of c. 25); re-isomerization back to the trans-form led to an increase of the DP back to the initial value – hence, this light-induced polymerization/depolymerization process revealed a very high degree-of-reversibility.

In a related approach, Wang et al. reported on a supramolecular polymer featuring [*c*2]daisy chains (a definition is given in Chapter 6) and bis-pillar[5]arenes in a strictly alternating fashion (Figure 10.12) [51]. The authors first assembled a [*c*2]daisy-chain-type pseudorotaxane from the heterotritopic AB_2-type co-pillarene **14**. This double-threaded dimer featured two pending side chains, as the vacant recognition sites, which were further used for the intermolecular host–guest complexation with the bis-pillarene derivative **15**. The supramolecular polymerization was studied by NMR; the dynamic light scattering (DLS) measurements confirmed the formation of a supramolecular polymer with high molar mass (a hydrodynamic radius of c. 200 nm was observed for a 150 mN solution of $(\mathbf{14})_2$ and **15** [in a 1 : 1 ratio] in $CHCl_3$).

As pointed out in Chapters 3 and 4, supramolecular chemistry can be employed to assemble polymers with fluorescence properties, which can be used in opto-electronic applications. Similar to Meijer's work on the reversible self-assembly of fluorescent components due to quadruple H-bonding interactions of the ureidopyrimidinone (Upy) moieties [53], Zhang, Huang, and coworkers synthesized the pillar[5]arene-based ditopic host **16** as well as the corresponding imidiazole-functionalized guests **17** (Figure 10.13) [52]. The supramolecular polymerization of **16** and **17a** gave a highly fluorescent polymer (emission quantum

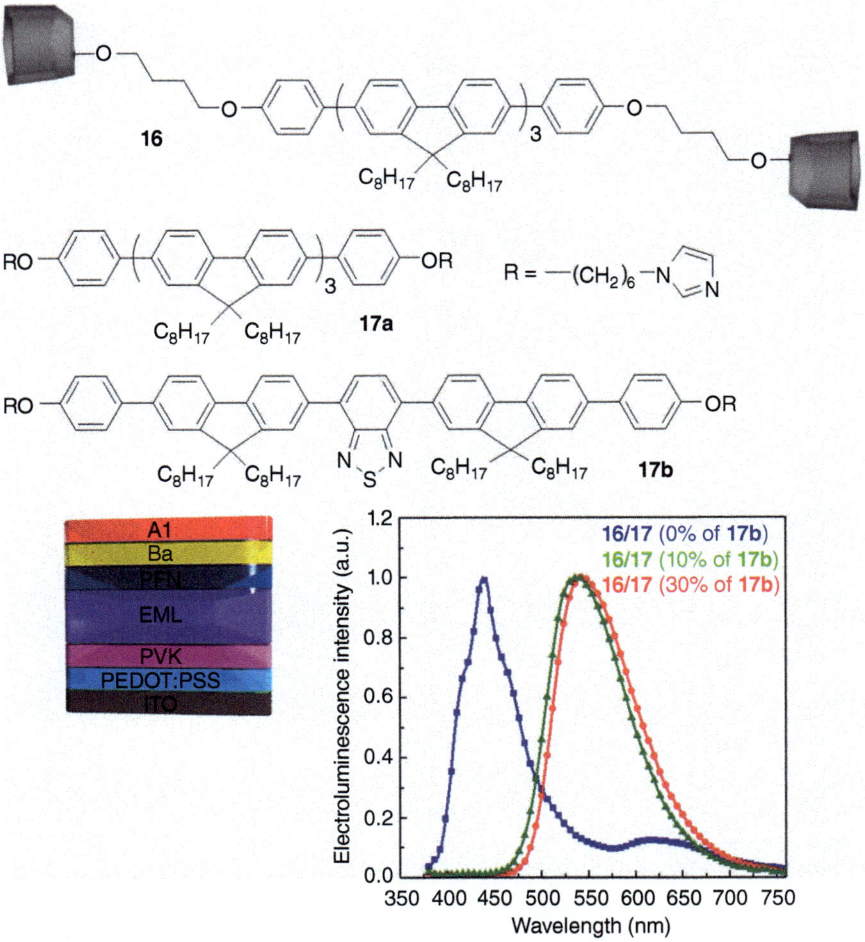

Figure 10.13 Schematic representation of the fluorescent host **16** and guests **17** which were employed, as monomers, to assemble linear supramolecular polymers. The device configuration, containing these materials, as emissive layer (EML), as well as the obtained electroluminescence spectra are also depicted. Source: Yang et al. [52]. © 2017 American Chemical Society.

yield of c. 82%); the emission color thereof could be shifted from blue to green by a gradual doping with **17b**. The supramolecular polymers were utilized, as emissive layer, in photoluminescent devices, which showed a maximum luminance efficiency in the range of 5 cd A^{-1}.

Hu et al. reported on a dynamic polyrotaxane that was prepared in a two-step self-assembly process [54]. First, an α,ω-diaminoalkane was threaded into the cavity of **1**; then, this pseudorotaxane was locked by attaching Upy moieties to the NH_2-groups (Figure 10.14). Due to the ability of Upy to dimerize via quadruple H-bonding interactions (see Chapter 3), the supramolecular polymerization into a polyrotaxane could be achieved at high concentrations. The self-assembly process was studied by NMR and viscometry measurements. The dynamic nature of the

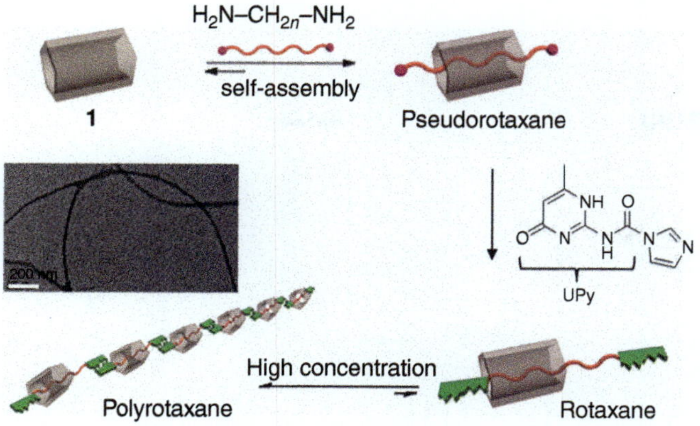

Figure 10.14 Schematic representation of the formation of a dynamic polyrotaxane via a two-step self-assembly process. A TEM microscopy image of the supramolecular polymer obtained from 1,4-diaminobutane ($n = 4$) is also shown. Source: Hu et al. [54]. Figure reproduced with kind permission; © 2012 American Chemical Society.

supramolecular polymer was verified by the highly reversible depolymerization upon dilution.

Recently, Huang and coworkers demonstrated that supramolecular polymers can also be assembled from the highly directional halogen-bonding (X-bonding) interactions between 4-pyridyl-functionalized pillar[5]arene **18** and 1,2,4,5-tetrafluoro-3,6-diiodobenzene [55]. A stable polypseudorotaxane was prepared by threading *n*-hexane molecules into the pill[5]arene cavities along the linear polymer chain (Figure 10.15). This material was characterized in solution by 1D/2D NMR spectroscopy and, remarkably, even in the solid state by XRD analysis. The fibrous crystalline supramolecular polymers had a needle-like morphology, as revealed by SEM imaging.

Dynamic covalent chemistry (DCC) [57] in concert with pillarenes' host–guest chemistry was employed by Xu et al. to prepare a double-dynamic polymer (Figure 10.16) [56]. First, a [*c*2]daisy chain with two pending anthracene moieties was assembled from the pillar[5]arene **19** and a BB-type bis-imidazolium guest. The photochemical dimerization of the pending anthracene moieties of the [*c*2]daisy chain in a [2+2]-cycloaddition gave a polyrotaxane-like polymer (according to a "supramonomer polymerization" approach, see Section 1.4). The heating of this polymer, either in solution or as a gel, enabled the depolymerization via two pathways (depending of the extent of thermal treatment): disassembly of only the host–guest inclusion complexes or of both components – the complexes and the photo-dimerized anthracene sites. Remarkably, the polymer structure could be reformed upon cooling and/or irradiation. One further example for a degradable supramolecular polymer, though not in a reversible fashion, was given by the Yao group by incorporating a diselenide moiety with the aliphatic bridge between two pillar[5]arene units [58]. At high concentrations, this ditopic host assembled into a supramolecular polymer with a neutral ditopic guest. The diselenide unit could

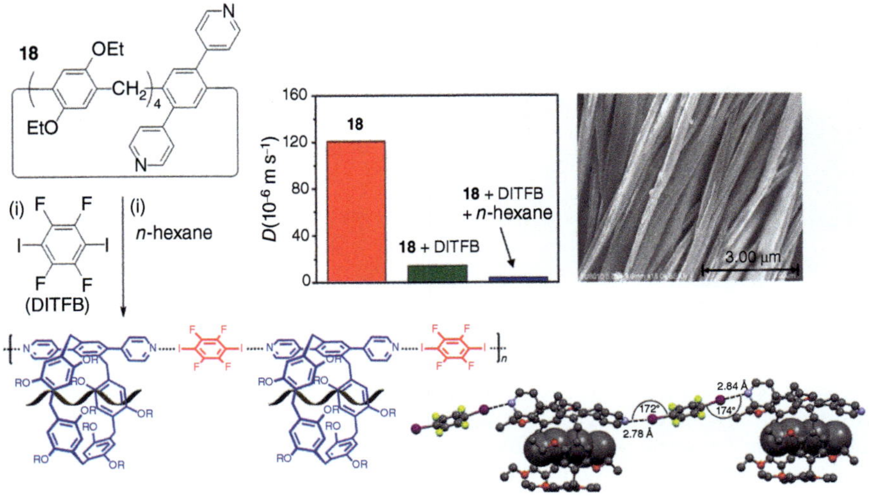

Figure 10.15 Schematic representation of the two-step self-assembly of a polypseudorotaxane containing X-bonding and host–guest interactions within the main and the side chains, respectively. The solid-state structure of the assembly, as determined by XRD analysis is depicted. A comparison of the diffusion coefficients (*D*), as determined by DOSY measurements as well as a SEM image of the materials are also shown. Source: Liu et al. [55]. Figure reproduced with kind permission; © 2018 Wiley-VCH.

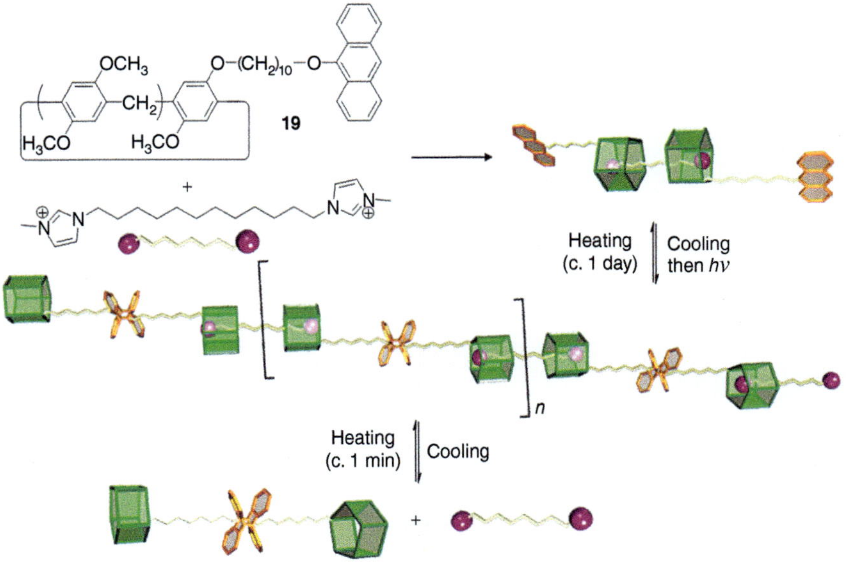

Figure 10.16 Schematic representation of the formation of a double-dynamic polymer, based on supramolecular host–guest interactions and dynamic covalent chemistry. Source: Xu et al. [56]. © 2013 American Chemical Society.

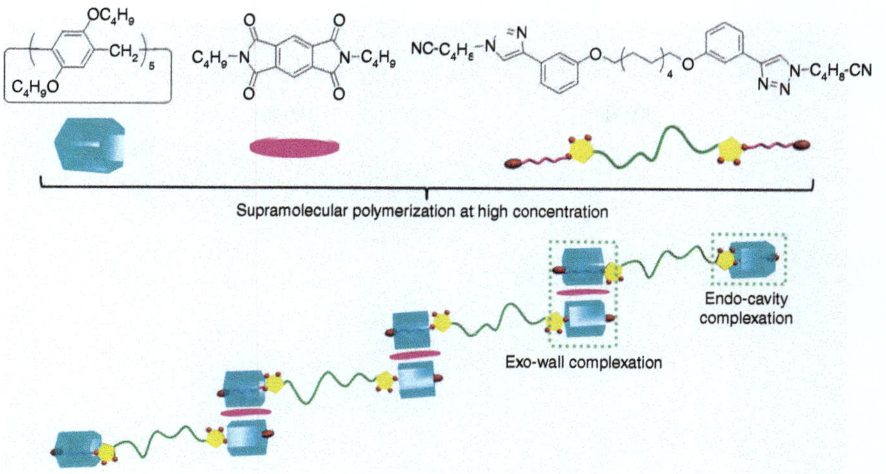

Figure 10.17 Schematic representation of a supramolecular polymer, based on an exo-wall and endo-cavity complexation. Source: Wang et al. [59]. © 2015 The Royal Chemical Society.

be cleaved by γ-ray irradiation or oxidatively, e.g. by adding peroxides or reduced glutathione.

A remarkable three-component supramolecular polymer featuring a pillarene-based host–guest complexation (referred to as an endo-cavity inclusion) and exo-wall interactions was published by Li and coworkers (Figure 10.17) [59]. The binding of molecules to the outside of the molecular structure represents a rather exotic motif in supramolecular chemistry. It was shown that a sandwich-type charge–transfer complex of 1 : 2 stoichiometry can be formed from an electron-rich pillar[5]arene and large electron-poor dyes, such as *N*,*N*′-bisalkylpyromellitic diimide. If such a complexation is conducted at high concentrations and in the presence of a neutral ditopic guest, a supramolecular polymer is formed.

A completely different type of architecture was reported by Stoddard et al. starting from the amino-functionalized pillar[5]arene **20** (Figure 10.18). The subsequent microwave-assisted condensation with terephthaldehyde afforded the oxazole-annulated pillarene **21**. This building block, in turn, was used further for the synthesis of a flexible column of pillar[5]arenes (up to the 9-mer) [60]; however, the host–guest chemistry of these columnar oligo-pillarenes has, to the best of our knowledge, not (yet) been investigated further.

10.4 Hyperbranched and Cross-linked Assemblies

As for the other types of supramolecular polymers that rely on host–guest interactions (e.g. crown ethers binding, calixarenes or CDs), macromolecular architectures beyond linear assemblies have been derived from the binding of guest molecules within the pillarenes' cavities: hyperbranched, dendritic, graft/comb copolymers,

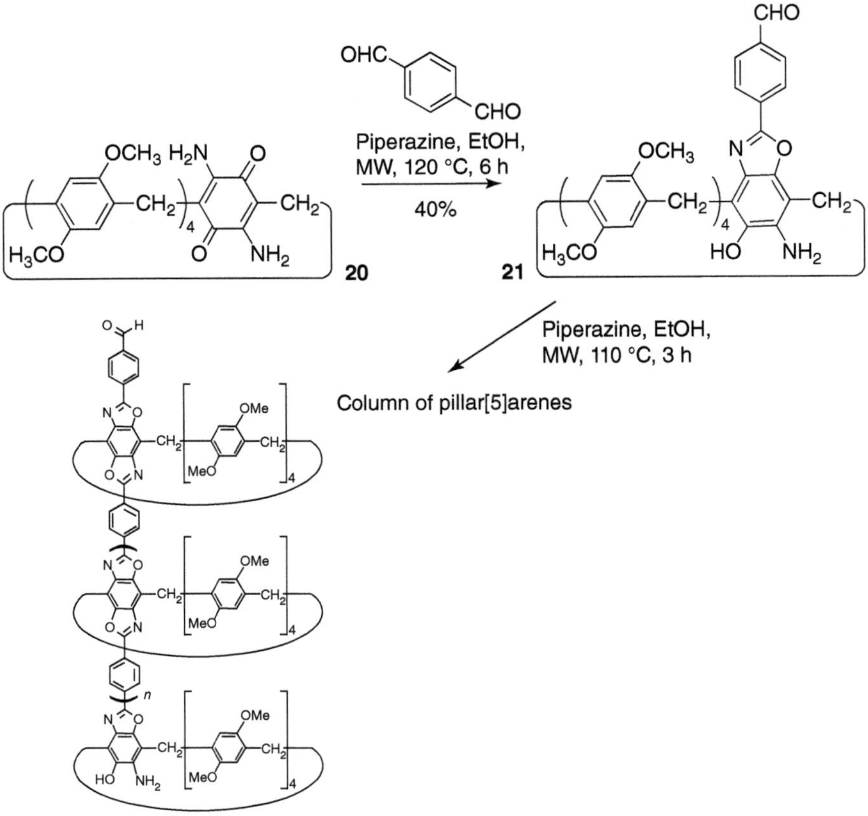

Figure 10.18 Schematic representation of the synthesis of columnar oligo-pillarenes according to Stoddard et al. ($n = 0$–9). Source: Strutt et al. [60]. © 2014 John Wiley and Sons.

and polymer networks [22]. Some selected examples will be discussed in the following.

A neutral hyperbranched polymer was assembled from the AB_2-type co-pillarene **22** at high concentration in a $CHCl_3$ solution (Figure 10.19) [61]. Thereby, a supramolecular polymerization was facilitated by a high K_a value of $1.2 \times 10^4\,M^{-1}$ for binding of the guest within the pillar[5]arenes' cavities. The resultant material was investigated by NMR, viscometry, DLS, and transmission electron microscopy (TEM). The latter two techniques verified the presence of large, spherical nanoparticles – thus confirming that a supramolecular polymer of high molar mass was formed (Figure 10.19). In the presence of 1,4-dicyanobutane, as a strong competitor for binding, depolymerization of $(\mathbf{19})_n$ was observed (a pseudorotaxane of **19** with a 1,4-dicyanobutane axis was formed instead).

The formation of a supramolecular polymeric network using the homoditopic host **11** and the homotritopic guest **13** has already been mentioned above (Figure 10.11) [49]. One further example for the self-assembly into such an A_2/B_3-type network was reported by Li et al. [63]. A molar mass of 127 kDa was calculated for this system in $CHCl_3$. In the same context, the orthogonal combination, i.e. starting from a

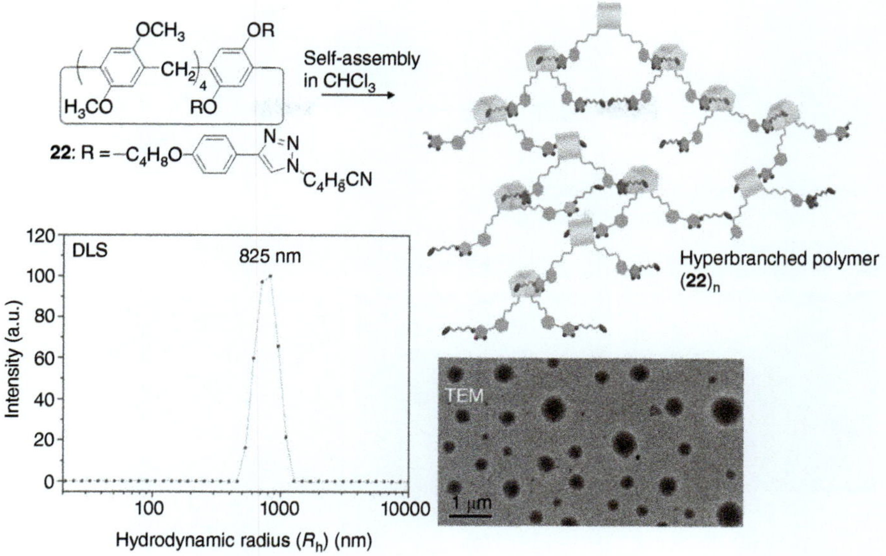

Figure 10.19 Schematic representation of the self-assembly of **22** into a hyperbranched supramolecular polymer. The distribution of the hydrodynamic diameter (R_h) according to DLS as well as a representative TEM image of the obtained nanoparticles are also shown. Source: Xiaoyang et al. [61]. Figure reproduced with kind permission; © 2013 John Wiley and Sons.

homotritopic host and a homoditopic guest (A_3/B_2-type) was also investigated [62]. The dynamic supramolecular polymer exhibited an interesting reversible multidimensional transformation from 0D (vesicular structures) via 1D (tubular objects) and 2D (layers) to 3D (stacked layers) architectures with increasing concentration (Figure 10.20).

In a related work, a ditopic AA-type host and a tetratopic B_4-guest were combined to afford a reversible supramolecular network [64]; the same type of material was also derived from the alternative combination of an A_4-type host with a BB-type guest [65, 66]. In the latter case, a tetraphenylethene-centered tetra-pillarene was used. The central tetraphenylethylene (TPE)-type dye has a well-known chromophore with AIE properties [67]. Due to formation of a multidimensional assembly, offering a rigid environment, the fluorescence was significantly enhanced. Finally, an A_6/B_2-type combination of multitopic host and guest molecules needs to be mentioned in this context [68].

Cross-linked supramolecular polymers represent one further type of topology to be briefly introduced within this section. Wang and coworkers utilized pillarene-based host–guest complexation in combination with quadruple H-bonding for the assembly of a supramolecular cross-linked polymer [69]. This work is an extension of their previous report on linear polymers featuring the same two types of non-covalent interactions (Figure 10.14) [54]. A linear supramolecular polymer was initially assembled by the dimerization of the self-complementary Upy moieties attached to a pillar[5]arene host (**23**); this polymer $(\mathbf{23})_n$ was subsequently cross-linked by adding

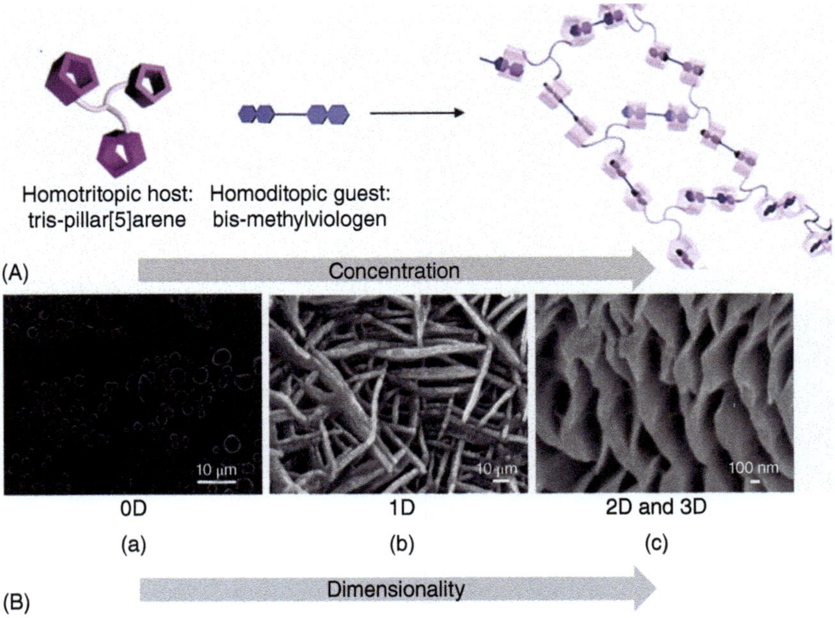

Figure 10.20 (A) Schematic representation of the formation of a dynamic supramolecular polymer network. (B) The TEM images of the polymer at certain concentrations are also shown [(a–c) 1, 2, and 5 mM]. Source: Zhang et al. [62]. Figure reproduced with kind permission; © 2013 The Royal Chemical Society.

an alkyl-bridged homoditopic guest (**24**) (Figure 10.21). In contrast to polymer $(\mathbf{23})_n$ and the cross-linked material incorporating **23**, for which no homogeneous films could be fabricated (using polycarbonate of $M_w = 19\,000\,\mathrm{g\,mol^{-1}}$, as the matrix), exploiting a poly(ethylene glycol)-bridged ditopic host enabled the formation of a yellow, translucent film (the yellow color originated from charge–transfer [CT] interactions between the electron-rich host and the electron-poor MV^{2+}-type guest entities).

Organic polymers equipped with either pillarenes or designated guest moieties in their side chains are of importance when aiming for supramolecular, cross-linked materials [22]. Basically, such polymers could also be employed to construct more advanced comb- or graft-like copolymer architectures – however, examples for this cannot (yet) be found in the literature.

A redox-responsive supramolecular network was formed by making use of the pillar[6]arene-ferrocenium recognition [70]. The Wang group prepared two copolymers with pending ferrocene and pillar[6]arene moieties, respectively. Due to the very weak binding of ferrocene (Fc) within the cavity of pillar[6]arene, no polymeric network could be formed. After oxidation, the ferrocenium cation (Fc^+), however, featured a much stronger host–guest interaction with the host and, thus, gave a gel-like material with viscoelastic properties. It was further shown that redox switching allowed a reversible transformation of the polymer mixture from a cross-linked to a non-cross-linked state.

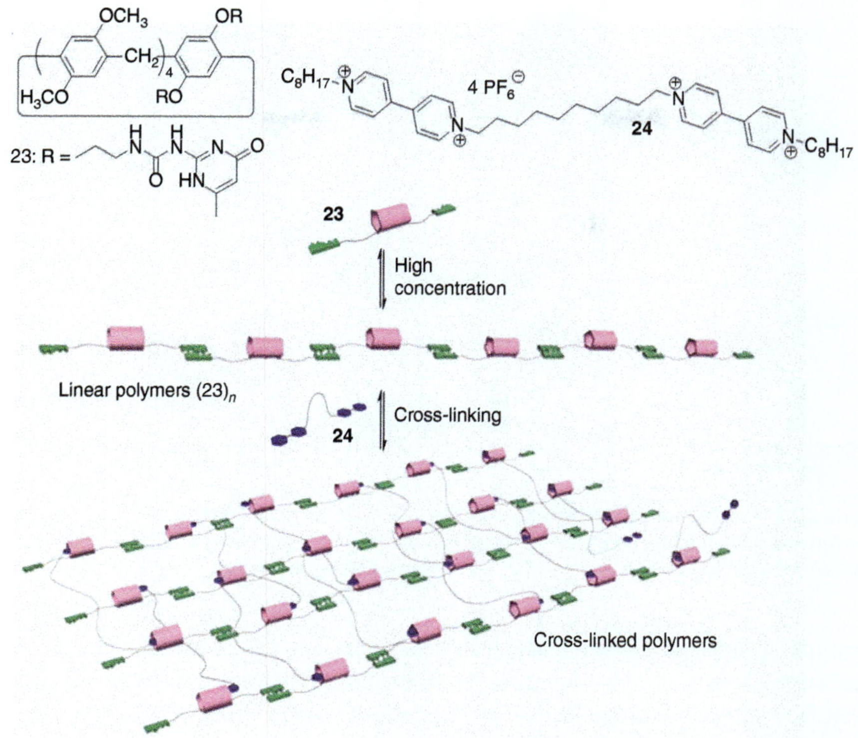

Figure 10.21 Schematic representation of the two-step self-assembly of a cross-linked supramolecular polymer. Source: Hu et al. [69]. © 2013 The Royal Chemical Society.

In another example, a conjugated polymer, i.e. of the poly(phenylene ethylene)-type, with two pillar[5]arene moieties on every second phenyl ring, was prepared by using a Sonogashira cross-coupling polymerization [71, 72]. It was shown that the fluorescence of the π-conjugated material **25** was quenched upon binding of cationic alkyl-pyrazinium guest molecules – the latter served as acceptors in the light-induced electron transfer from the donor-type polymer (Figure 10.22). The fluorescence could be restored by adding halide anions, whereby the recovery occurred in the order $Cl^- > Br^- > I^-$. The hard chloride ions strongly interacted with the pyrazinium cations due to ion pairing, thereby withdrawing them from the pillarenes' cavities.

10.5 Supramolecular Assemblies, Based on Amphiphilic Pillar[5]arenes

Basically, the functionalization of pillar[*n*]arenes can be achieved much more easily when compared to other common macrocyclic hosts [22]. The broad synthetic toolbox applicable to pillarenes also allows synthesizing amphiphilic pillar[*n*]arene-based materials featuring hydrophobic and hydrophilic groups. Both

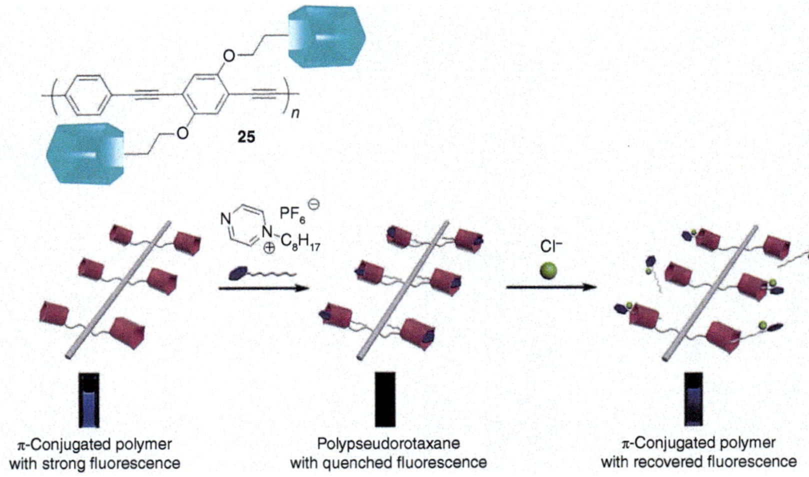

Figure 10.22 Schematic representation of the PPE-type polymer **25** and an anion-responsive polypseudorotaxane prepared thereof. Source: Sun et al. [71]. Figure reproduced with kind permission; © 2013 The Royal Chemical Society.

amphiphilic hosts and their host–guest complexes have been found to assemble into various types of nanostructures in polar and apolar solvents. Though such assemblies are not supramolecular polymers in a narrow sense of the definition, they still represent ensembles that have been formed via supramolecular interactions from small-molecule species. Recently, a detailed review article has been devoted to such pillarene-based, self-assembled amphiphiles (PSAs) regarding their preparation, characterization, properties, and potential applications, in particular in the broad field of biomedicine [23]. In the following, a short overview concerning such materials will be given using selected examples.

For instance, Yao et al. prepared the amphiphilic pillar[5]arene **26** exhibiting NH_2-groups at the termini of the pending arms – five each on the hydrophilic and hydrophobic side [73, 74]. This amphiphile **26** exhibited a rich self-assembly behavior and a variety of nanostructures could be identified by DLS measurements, TEM, and atomic force microscopy (AFM) imaging as well as UV/vis and infrared spectroscopy (Figure 10.23): Initially, bilayer vesicles were formed in an aqueous media. Some of these were ordered into necklace-like structures when coming in a close proximity to each other. Simultaneously, other vesicles fused into larger vesicles and gave sheet-like structures upon rupture. Finally, multiwalled microtubes were assembled – either by fusion of the necklace-like species or by roll-up of the sheet-like assemblies driven by H-bonding interactions. The entire process was found to take about four months to complete! In subsequent work, vesicular and microtubular structures were derived from amphiphilic pillar[5]arenes, similar to **26**, but the length of the aliphatic chains on the hydrophobic side as well as the nature of the polar head group of the hydrophilic portion were varied [75]. Vesicle formation was also observed for pillar[5]arenes comprising a hydrophobic side with five *n*-alkyl moieties and a hydrophilic side with five highly polar oligo(ethylene

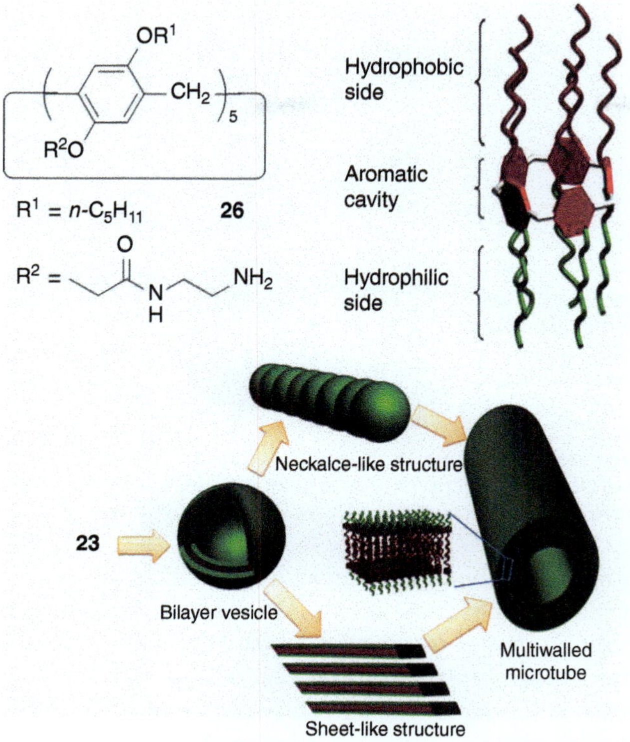

Figure 10.23 Schematic representation of the amphiphilic pillar[5]arene **26** and the nanostructures assembled thereof. Source: Yao et al. [73]. © 2012 American Chemical Society.

glycol) chains [76, 77]. These "biocompatible" pillarene-based vesicles were able to deliver dyes (e.g. rhodamine-B or fluorescein) for bioimaging applications into cells [76] or to take-up magnetic iron-oxide nanoparticles, thus generating hybrid magneto-responsive supramolecular vesicles [77].

The amphiphilic pillar[5]arene **27** assembled into stimuli-responsive, reverse multi-lamellar giant vesicles or an adaptive gel (in $CHCl_3$ or in a water/tetrahydrofuran [THF] mixture, respectively) [78]. DLS measurements and various microscopy imaging techniques were applied to study the formation of these assemblies; thereby, the influence of environmental conditions and the feedback to external forces were of interest (Figure 10.24). Besides a response to physical treatment (i.e. heat, sonication or stirring), disassembly was induced by KPF_6 and reassembly was achieved by adding benzo-18-crown-6, as a scavenger for the potassium ions – both processes occurred independent of solvent composition. Also binding of an appropriate competitive guest into the cavity led to a dissociation of the supramolecular assemblies, thus forming a small-molecule rotaxane. An amphiphilic pillar[6]arene was found to form enzyme-responsive micelles in aqueous medium [79]. In the presence of adenosine triphosphate (ATP), as guest, a transformation into double-layered vesicles occurred. The initial morphology could be restored upon treatment with an

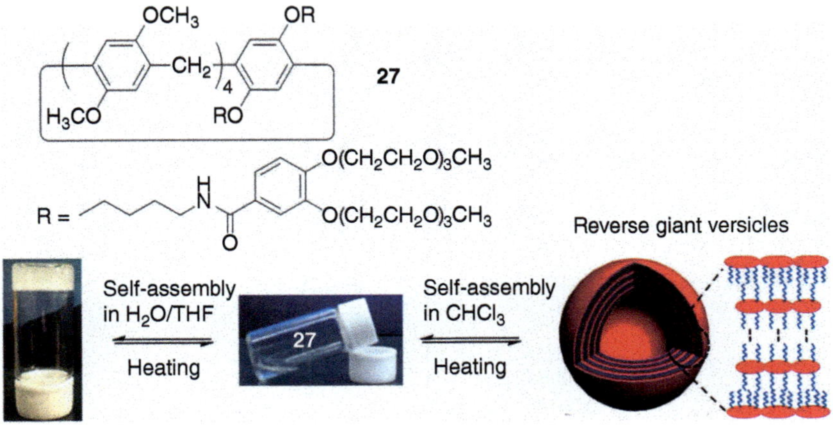

Figure 10.24 Schematic representation of the amphiphilic pillar[5]arene **27** and its reversible self-assembly into a gel-like material or reverse giant vesicles. Source: Gao et al. [78]. Figure reproduced with kind permission; © 2013 The Royal Chemical Society.

ATP-hydrolyzing enzyme. In a different approach, an amphiphilic pillar[5]arene equipped with five Fc^+ moieties gave redox-responsive vesicles [80]. This particular system was successfully tested as a glutathione-responsive carrier in the context of drug delivery and gene transfection in various types of cancer cells.

As an alternative approach to utilizing amphiphilic pillarenes for the self-assembly into a range of different nanostructures, supramolecular amphiphiles can also be obtained from the host–guest complexation of appropriate building blocks. From the broad range of such assemblies reported in literature [22, 23], the photo-switchable system, published by the Huang group, is highlighted here [81]. Due to the increased cavity size of pillar[6]arenes compared with their smaller five-membered homologues, azobenzene-type guests, via their stretched E-configuration only, can be threaded into the host. In contrast to the spherical aggregates formed by amphiphile E-**28** in $CHCl_3$, the host–guest complex of E-**28** and **29** only generated irregularly shaped aggregates (Figure 10.25). Upon photo-irradiation, isomerization of **28** from the E- to the Z-form occurred – accompanied by the dethreading of the guest from the host. The resulting mixture of Z-**28** and **29** was then reassembled into large vesicle-like aggregates. The entire process could be reversed by photochemically switching back to the E-isomer. On the basis of the same concept, a structural change on the nanoscale between solid nanoparticles (formed by the self-assembly of E-configured azobenzene amphiphiles, such as **28**) and vesicles (formed by a water-soluble pillar[6]arene and the Z-**28**) could be induced by photochemical isomerization of the azo-group [83]. The photo-induced disassembly of supramolecular nanostructures was employed to develop various nanocarriers for potential drug-delivery applications [82, 84, 85]. For example, the combination of a hydrophilic pillar[6]arene (bearing 12 carboxylate moieties) and a cationic azobenzene-centered surfactant generated vesicles that could be loaded with mitoxantrone (MTZ), an antineoplastic agent widely used in cancer treatment [82]. It was shown in *in vitro* experiments that stability under physiological conditions

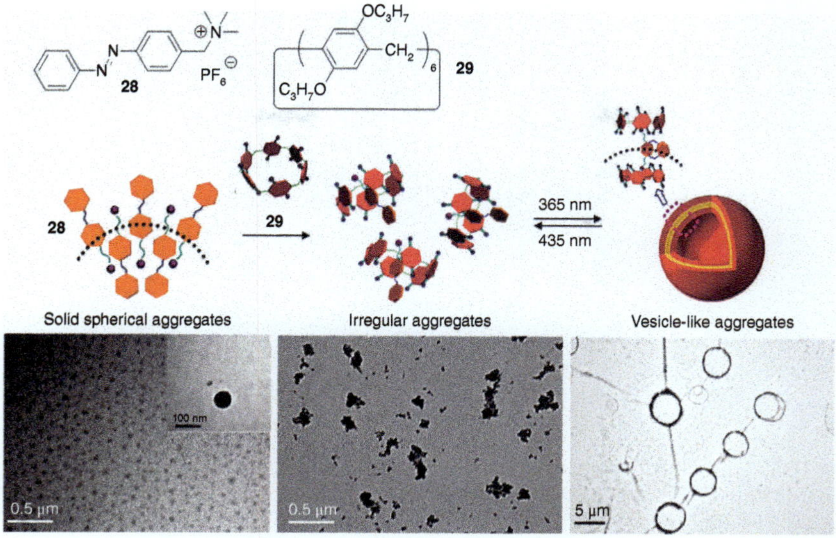

Figure 10.25 Schematic representation of the reversible vesicle formation via supramolecular interactions between pillar[6]arene **29** and the photo-switchable guest **28**. Source: Yu et al. [81]. Figure reproduced with kind permission; © 2012 American Chemical Society.

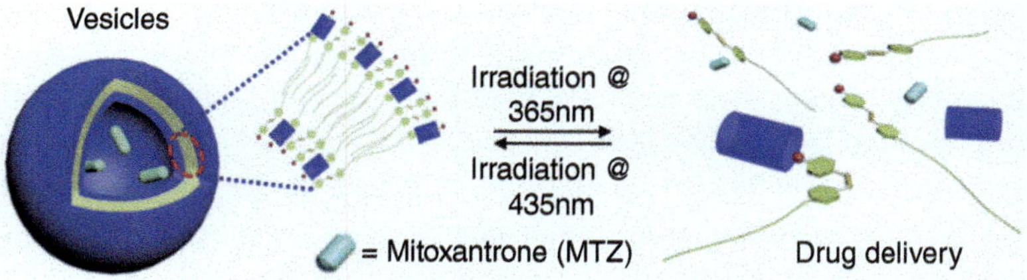

Figure 10.26 Schematic representation of drug-loaded vesicles exhibiting a light-induced release. Source: Hu et al. [82]. © 2015 John Wiley and Sons.

was a given and rapid release of the incapsulated drug could be induced either by external irradiation with UV light (Figure 10.26) or by adding an acid (thus mimicking the environment of cancer cells). Cytotoxicity tests and flow cytometric analysis revealed that the activity of MTZ against cancer cells was retained but the cytotoxicity against normal cells was reduced significantly (i.e. comparing the performance of the UV-released drug vs. a direct application).

Besides the encapsulation of drugs into supramolecular nanostructures, prodrugs having a recognition site for the binding to (water-soluble) pillarenes have also been considered with respect to the assembly of supramolecular amphiphiles with biomedical relevance (e.g. therapy or bioimaging) [76, 87]. In particular, guest molecules comprising the anticancer drug doxorubicin (DOX) have been employed to some extent. For this purpose, the DOX moiety was equipped with MV^{2+} [88] or a pyridinium unit [86, 89] at the end of a flexible tail. Cao et al. assembled well-defined supramolecular nanoparticles by combining the DOX-containing guests **30** and the

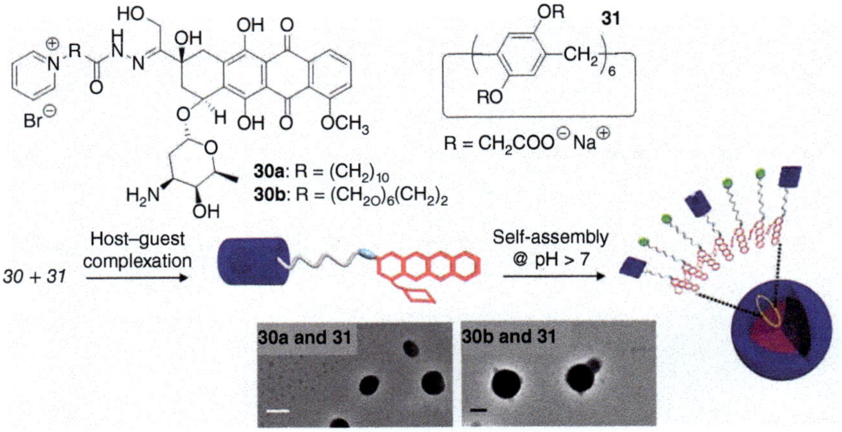

Figure 10.27 Schematic representation of the self-assembly of **30** and **31** into vesicular nanoparticles; the TEM images show representative nanoparticles observed in both cases (scale bar: 200 nm). Source: Cao et al. [86]. Figure reproduced with kind permission; © 2015 American Chemical Society.

water-soluble pillar[6]arene **31**; TEM imaging and DLS measurements were used to characterize these species (Figure 10.27) [86]. The nanoparticles were stable in the neutral pH regime but readily disassembled at a pH value of 5.5 (mimicking the environment in cancer cells); simultaneously, the acid-labile hydrazine bond was cleaved, thereby releasing the DOX drug. Performing intracellular location experiments allowed the authors to conclude that an efficient take-up by cancer cells led to an accumulation of DOX, in particular for SKOV3-type cancer cells. Cytotoxicity experiments indicated an efficient inhibition of the proliferation of cancer cells and, thus, a reasonable antitumor activity.

For an overview regarding other stimuli-responsive pillarene-based amphiphilic assemblies, in particular targeting the controlled release of drugs, the reader is referred to the reviews by Ogoshi et al. [22] as well as Zhang et al. [23].

10.6 Concluding Remarks

In 2008, Ogoshi introduced pillarenes as new macrocyclic hosts for the supramolecular complexation of guest molecules. Are pillarenes – from today's point-of-view – just “one more” macrocyclic species from which supramolecular materials, in particular supramolecular polymers, can be assembled via host–guest interactions? A range of such host materials, e.g. crown ethers, cyclodextrins, or calixarenes, have been known for a much longer period of time and have proven their versatility in this regard. The clear answer to this is “no.” The structural features, i.e. their rigid symmetrical shape, the electron-rich nature of the cavity, and the ease of (un)symmetrical functionalization, offer distinct advantages over the more established host materials [2, 22]. Both neutral and cationic guests can efficiently be bound via CH–π and cation–π interaction, respectively.

Various types of supramacromolecular architectures, based on pillarenes' host–guest chemistry, have already been reported: linear, hyperbranched, and cross-linked polymers have to be highlighted in this context. These materials exhibit a broad stimuli-responsive behavior that, perspectively, might be useful for applications in optical sensing, drug delivery, bioimaging, and/or materials science. Moreover, a diversity of responsive nanostructures has been derived from amphiphilic building blocks. Such responsive supramolecular amphiphiles are difficult to realize when using the other common macrocyclic hosts.

In summary, pillar[*n*]arenes have experienced a rapid development from their initial structural description to an important template in modern supramolecular chemistry that can (at least) compete with the established members of the macrocyclic host family.

References

1 Ogoshi, T., Kanai, S., Fujinami, S. et al. (2008). *J. Am. Chem. Soc.* 130: 5022–5023.
2 Yang, L., Tan, X., Wang, Z., and Zhang, X. (2015). *Chem. Rev.* 115: 7196–7239.
3 Winter, A., Hager, M.D., and Schubert, U.S. (2012). Supramolecular polymers. In: *Polymer Science: A Comprehensive Review*, vol. 5 (eds. H.-W. Schmidt and M. Ueda), 269–310. Amsterdam: Elsevier BV.
4 Brunsveld, L., Folmer, B.J.B., and Meijer, E.W. (2000). *MRS Bull.* 25: 49–53.
5 Brunsveld, L., Folmer, B.J.B., Meijer, E.W., and Sijbesma, R.P. (2001). *Chem. Rev.* 101: 4071–4098.
6 de Greef, T.F.A. and Meijer, E.W. (2010). *Aust. J. Chem.* 63: 596–598.
7 Binder, W.H. and Zirbs, R. (2007). *Adv. Polym. Sci.* 207: 1–78.
8 Faul, C.F.J. and Antonietti, M. (2003). *Adv. Mater.* 15: 673–683.
9 Winter, A. and Schubert, U.S. (2016). *Chem. Soc. Rev.* 45: 5311–5357.
10 Wei, P., Yan, X., and Huang, F. (2015). *Chem. Soc. Rev.* 44: 815–832.
11 Zheng, B., Wang, F., Dong, S.-Y., and Huang, F.-H. (2012). *Chem. Soc. Rev.* 41: 1621–1636.
12 Harada, A., Takashima, Y., and Yamaguchi, H. (2009). *Chem. Soc. Rev.* 38: 875–882.
13 Guo, D.-S. and Liu, Y. (2012). *Chem. Soc. Rev.* 41: 5907–5921.
14 Barrow, S.J., Kasera, S., Rowland, M.J. et al. (2015). *Chem. Rev.* 115: 12320–12406.
15 Yu, G., Jie, K., and Huang, F. (2015). *Chem. Rev.* 115: 7240–7303.
16 Zhang, H. and Zhao, Y. (2013). *Chem Eur. J.* 19: 16862–16879.
17 Strutt, N.L., Forgan, R.S., Spruell, J.M. et al. (2011). *J. Am. Chem. Soc.* 133: 5668–5671.
18 Xue, M., Yang, Y., Chi, X. et al. (2012). *Acc. Chem. Res.* 45: 1294–1308.
19 Cragg, P.J. and Sharma, K. (2012). *Chem. Soc. Rev.* 41: 597–607.
20 Ogoshi, T. (2012). *J. Inclusion Phenom.Macrocyclic Chem.* 72: 247–262.
21 Meier, H. and Cao, D. (2013). *Nachr. Chem.* 61: 408–410.

22 Ogoshi, T., Yamagishi, T.-a., and Nakamoto, Y. (2016). *Chem. Rev.* 116: 7937–8002.
23 Zhang, H., Liu, Z., and Zhao, Y. (2018). *Chem. Soc. Rev.* 47: 5491–5528.
24 Li, H., Yang, Y., Xu, F. et al. (2019). *Chem. Commun.* 55: 271–285.
25 Li, C., Xu, Q., Li, J. et al. (2010). *Org. Biomol. Chem.* 8: 1568–1576.
26 Ogoshi, T., Nishida, Y., Yamagishi, T.-a., and Nakamoto, Y. (2010). *Macromolecules* 43: 3145–3147.
27 Ogoshi, T., Hasegawa, Y., Aoki, T. et al. (2011). *Macromolecules* 44: 7639–7644.
28 Syed, A.A. and Dinesan, M.K. (1991). *Talanta* 38: 815–837.
29 Ogoshi, T., Kayama, H., Aoki, T. et al. (2013). *Polym. J.* 46: 77–81.
30 Gurarslan, A., Joijode, A., Shen, J. et al. (2017). *Polymers* 9: 673.
31 Ogoshi, T., Aoki, T., Ueda, S. et al. (2014). *Chem. Commun.* 50: 6607–6609.
32 Dong, S., Zheng, B., Wang, F., and Huang, F. (2014). *Acc. Chem. Res.* 47: 1982–1994.
33 Guan, Y., Liu, P., Deng, C. et al. (2014). *Org. Biomol. Chem.* 12: 1079–1089.
34 Ogoshi, T., Demachi, K., Kitajima, K., and Yamagishi, T.-a. (2011). *Chem. Commun.* 47: 7164–7166.
35 Liu, L., Wang, L., Liu, C. et al. (2012). *J. Organomet. Chem.* 77: 9413–9417.
36 Zhang, Z., Yu, G., Han, C. et al. (2011). *Org. Lett.* 13: 4818–4821.
37 Zhang, Z., Luo, Y., Chen, J. et al. (2011). *Angew. Chem. Int. Ed.* 50: 1397–1401.
38 Strutt, N.L., Zhang, H., Giesener, M.A. et al. (2012). *Chem. Commun.* 48: 1647–1649.
39 Han, C., Xia, B., Chen, J. et al. (2013). *RSC Adv.* 3: 16089–16094.
40 Wang, K., Wang, C.-Y., Wang, Y. et al. (2013). *Chem. Commun.* 49: 10528–10530.
41 Shi, B., Xia, D., and Yao, Y. (2014). *Chem. Commun.* 50: 13932–13935.
42 Gao, L., Zhang, Z., Zheng, B., and Huang, F. (2014). *Polym. Chem.* 5: 5734–5739.
43 Xia, B., Zheng, B., Han, C. et al. (2013). *Polym. Chem.* 4: 2019–2024.
44 Agarwal, S., Greiner, A., and Wendorff, J.H. (2013). *Prog. Polym. Sci.* 38: 963–991.
45 Wang, Y., Xu, J.-F., Chen, Y.-Z. et al. (2014). *Chem. Commun.* 50: 7001–7003.
46 Fathalla, M., Strutt, N.L., Sampath, S. et al. (2015). *Chem. Commun.* 51: 10455–10458.
47 Ogoshi, T., Kayama, H., Yamafuji, D. et al. (2012). *Chem. Sci.* 3: 3221–3226.
48 Song, N., Chen, D.-X., Xia, M.-C. et al. (2015). *Chem. Commun.* 51: 5526–5529.
49 Meng, L.-B., Li, D., Xiong, S. et al. (2015). *Chem. Commun.* 51: 4643–4646.
50 Ogoshi, T., Yoshikoshi, K., Aoki, T., and Yamagishi, T.-a. (2013). *Chem. Commun.* 49: 8785–8787.
51 Wang, X., Han, K., Li, J. et al. (2013). *Polym. Chem.* 4: 3998–4003.
52 Yang, X., Cai, W., Dong, S. et al. (2017). *ACS Macro Lett.* 6: 647–651.
53 Abbel, R., Grenier, C., Pouderoijen, M.J. et al. (2009). *J. Am. Chem. Soc.* 131: 833–843.
54 Hu, X.-Y., Wu, X., Duan, Q. et al. (2012). *Org. Lett.* 14: 4826–4829.
55 Liu, P., Li, Z., Shi, B. et al. (2018). *Chem. Eur. J.* 24: 4264–4267.
56 Xu, J.-F., Chen, Y.-Z., Wu, L.-Z. et al. (2013). *Org. Lett.* 15: 6148–6151.

57 Jin, Y., Yu, C., Denman, R.J., and Zhang, W. (2013). *Chem. Soc. Rev.* 42: 6634–6654.
58 Zhou, Y., Jie, K., Shi, B., and Yao, Y. (2015). *Chem. Commun.* 51: 11112–11114.
59 Wang, S., Wang, Y., Chen, Z. et al. (2015). *Chem. Commun.* 51: 3434–3437.
60 Strutt, N.L., Zhang, H., Schneebeli, S.T., and Stoddart, J.F. (2014). *Chem. Eur. J.* 20: 10996–11004.
61 Xiaoyang, W., Hongmei, D., Jian, L. et al. (2013). *Macromol. Rapid Commun.* 34: 1856–1862.
62 Zhang, H., Nguyen, K.T., Ma, X. et al. (2013). *Org. Biomol. Chem.* 11: 2070–2074.
63 Li, C., Han, K., Li, J. et al. (2013). *Chem. Eur. J.* 19: 11892–11897.
64 Sun, N., Xiao, X., and Jiang, J. (2015). *Polym. Chem.* 6: 5015–5020.
65 Song, N., Chen, D.-X., Qiu, Y.-C. et al. (2014). *Chem. Commun.* 50: 8231–8234.
66 Wu, J., Sun, S., Feng, X. et al. (2014). *Chem. Commun.* 50: 9122–9125.
67 Mei, J., Leung, N.L.C., Kwok, R.T.K. et al. (2015). *Chem. Rev.* 115: 11718–11940.
68 Li, Z.-Y., Zhang, Y., Zhang, C.-W. et al. (2014). *J. Am. Chem. Soc.* 136: 8577–8589.
69 Hu, X.-Y., Wu, X., Wang, S. et al. (2013). *Polym. Chem.* 4: 4292–4297.
70 Xia, W., Ni, M., Yao, C. et al. (2015). *Macromolecules* 48: 4403–4409.
71 Sun, S., Hu, X.-Y., Chen, D. et al. (2013). *Polym. Chem.* 4: 2224–2229.
72 Sun, S., Shi, J.-B., Dong, Y.-P. et al. (2013). *Chin. Chem. Lett.* 24: 987–992.
73 Yao, Y., Xue, M., Chen, J. et al. (2012). *J. Am. Chem. Soc.* 134: 15712–15715.
74 Yao, Y., Xue, M., Zhang, Z. et al. (2013). *Chem. Sci.* 4: 3667–3672.
75 Yao, Y., Wei, P., Yue, S. et al. (2014). *RSV Adv.* 4: 6042–6047.
76 Zhang, H., Ma, X., Nguyen, K.T., and Zhao, Y. (2013). *ACS Nano* 7: 7853–7863.
77 Zhou, J., Chen, M., and Diao, G. (2014). *ACS Appl. Mater. Interfaces* 6: 18538–18542.
78 Gao, L., Zheng, B., Yao, Y., and Huang, F. (2013). *Soft Matter* 9: 7314–7319.
79 Zhou, J., Chen, M., and Diao, G. (2014). *Chem. Commun.* 50: 11954–11956.
80 Chang, Y., Yang, K., Wei, P. et al. (2014). *Angew. Chem. Int. Ed.* 53: 13126–13130.
81 Yu, G., Han, C., Zhang, Z. et al. (2012). *J. Am. Chem. Soc.* 134: 8711–8717.
82 Hu, X.-Y., Jia, K., Cao, Y. et al. (2015). *Chem. Eur. J.* 21: 1208–1220.
83 Xia, D., Yu, G., Li, J., and Huang, F. (2014). *Chem. Commun.* 50: 3606–3608.
84 Tong, Z., Zhou, J., Huang, R. et al. (2017). *J. Polym. Sci., Part A: Polym. Chem.* 55: 2477–2482.
85 Xia, D., Shangguan, L., Xue, M., and Shi, B. (2016). *New J. Chem.* 40: 9890–9894.
86 Cao, Y., Li, Y., Hu, X.-Y. et al. (2015). *Chem. Mater.* 27: 1110–1119.
87 Zhang, H., Ma, X., Nguyen, K.T. et al. (2014). *ChemPlusChem* 79: 462–469.
88 Wang, Y., Du, J., Wang, Y. et al. (2015). *Chem. Commun.* 51: 2999–3002.
89 Cao, Y., Zou, X., Xiong, S. et al. (2015). *Chin. J. Chem.* 33: 329–334.

11

Supramolecular Polymers, Formed by Orthogonal Non-covalent Interactions*

11.1 Introduction

One of the main differences between "conventional" organic polymers and supramolecular polymers is the reversible and stimuli-responsive character of the latter, arising from the non-covalent interactions that their structures rely on [1–3]. In recent years, the level of complexity has been increased significantly to tailor the properties of such polymers with respect to their proposed application: optoelectronics, photovoltaics, drug-delivery systems, porous membranes, catalysts, self-healing materials, etc. [3–11]. Besides the general classification of supramolecular polymers into "main chain" and "side chain" species [12], they can also be discussed with respect to the type of non-covalent interactions upon which they were constructed: hydrophobic, ionic/electrostatic, π–π stacking or hydrogen-bonding (H-bonding) interactions, metal-to-ligand coordination, host–guest complexation, etc. [3, 13, 14]; each of these exhibits special characteristics, such as bond strength, selectivity of binding, and degree-of-directionality.

Whereas, most of the supramolecular polymers reported so far simply feature one of these secondary interactions, Nature teaches how multiple non-covalent interactions can efficiently be utilized [15]: for instance, the folding of the covalent building blocks of polypeptide chains into proteins is a consequence of cooperative H-bonding, hydrophobic, and/or π–π stacking interactions. Inspired by these natural systems, chemists faced the challenge to also fabricate artificial supramolecular architectures, based on multiple supramolecular interactions. In general, the multiple non-covalent interactions involved in a single supramolecular architecture should be orthogonal to each other. The definition of "orthogonality" in this concern was coined by Weck and coworkers [16]: "Orthogonal supramolecular interactions should not interact with each other. The self-assembly into supramolecular polymers, driven by the interplay of selected orthogonal non-covalent interactions,

* Parts of this chapter are reproduced from *Chem. Commun.* **2007**, 155-162 (DOI: 10.1039/b706855g).

Supramolecular Polymers and Assemblies: From Synthesis to Properties and Applications, First Edition.
Ulrich S. Schubert, George R. Newkome, and Andreas Winter.

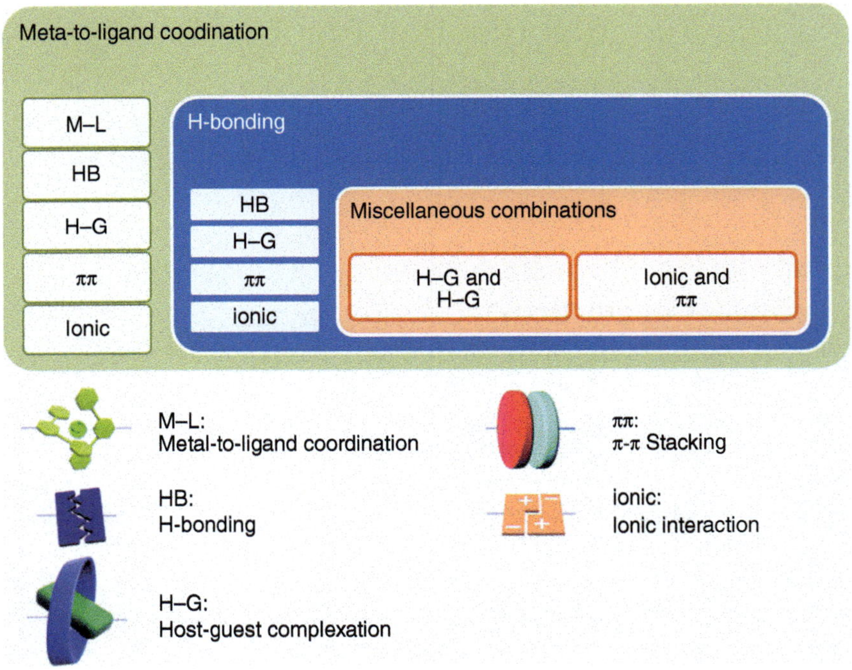

Figure 11.1 The important combinations of orthogonal supramolecular interactions. Source: Li et al. [17]. © 2012 The Royal Chemical Society.

enables the preparation of complicated, hierarchically ordered materials with novel structures and properties [17]. Moreover, the responsiveness of such materials can be extended even to multiple external stimuli, which thereby can pave the way toward novel materials, featuring a high degree-of-tunability and -reversibility."

Today, the concept of orthogonality is of utmost importance in all branches of synthetic chemistry – organic, macromolecular, and supramolecular – and always relates to the function of a material [18]. Supramolecular polymers, based on multiple, i.e. orthogonal, non-covalent interactions, have been surveyed in various reviews [3, 17, 19–22]. Considering the literature cited therein, three classes of supramolecular polymers can be identified from which the first two have up-to-now experienced the greatest attention (Figure 11.1):

1) Metal-to-ligand coordination in combination with others,
2) H-bonding interactions in combinations with others and, finally,
3) Miscellaneous combinations (i.e. various combinations of non-covalent interactions excluding metal-to-ligand coordination and H-bonding interactions).

In addition, biomimetic systems have been introduced rather recently, using the high selectivity for binding in proteins recognition sites (thus in an orthogonal manner) [23, 24].

11.2 Orthogonal Combinations of Supramolecular Interactions Involving Metal-to-Ligand Coordination

Complexes of transition metal ions with chelating ligands do not only feature a broad range of geometries, they also offer a distinct optoelectronic or magnetic property that can be employed to synthesize inorganic–organic hybrid materials for potential applications in, e.g. catalysis, optoelectronics, gas storage, or memory devices. In general, the typically directional metal-to-ligand coordination features high binding constants (K_a) and, at least to some extent, a reversibility-of-binding. The properties of such metal–ligand complexes can easily be tuned by variation of the components, i.e. the metal ions and its (chelating) ligands. Due to these aspects, metal-to-ligand coordination has been successfully introduced into supramolecular chemistry enabling the synthesis of a broad range of architectures; the field of metallo-supramolecular polymers has been discussed in Chapter 4 [10, 25–30]. In most of these cases, only one single type of metal-to-ligand coordination is employed to assemble a metallo-supramolecular system; however, different metal ions show a high preference in binding to different ligands; thus, certain kinds of coordination geometries will be obtained. The interplay of different types of metal-to-ligand coordination can result in complex supramolecular architectures, driven by the specificity of the self-assembly process. This strategy has not only been applied to obtain advanced, discrete 2D and 3D structures [31, 32] but also metallopolymers with alternating or random order of the incorporated types of metal-to-ligand moieties [9, 33–35]. Some examples have been already introduced in Chapter 4 and two representative metallo-supramolecular polymers will be highlighted here.

Yu et al. reported on the chiral "self-supported" polymer catalyst **1** [9], which was synthesized in a two-step self-assembly process by selective coordination of Fe(II) and Rh(I) centers to the terpyridine (tpy) and Feringa's MonoPhos sites [36], respectively (Figure 11.2a) [9]. Thereby, the Fe(II) centers acted as "supramolecular adhesive" linking the catalytically active Rh(I) sites. The performance of **1** in a heterogeneous asymmetric hydrogenation was comparable with that of the established homogeneous Rh(I)-MonoPhos system (Figure 11.2b). Advantageously, the incorporation of the catalyst into a kinetically stable (metallo-) supramolecular polymer allowed the efficient recycling: the catalyst could be reused in more than 10 cycles without losing its activity and enantioselectivity.

The Higuchi and Kurth groups prepared a variety of metallo-supramolecular polymers, based on bis-terpyridine ligands, for potential applications in electrochromic devices (see Chapter 4) [38]. To realize a multicolor electrochromic behavior, two different redox-active metal ions, i.e. Fe(II) and Ru(II), were incorporated in an alternating fashion via a two-step, self-assembly, approach [37]. Thin polymer films were fabricated by spin coating, and the electrochemical and electrochromic properties of these samples were investigated (Figure 11.3): a three-color scheme with high optical contrast ratios (68% and 37% at 508 and 585 nm, respectively) and very fast response times (only a few seconds) were observed. The coloration efficiencies (η)

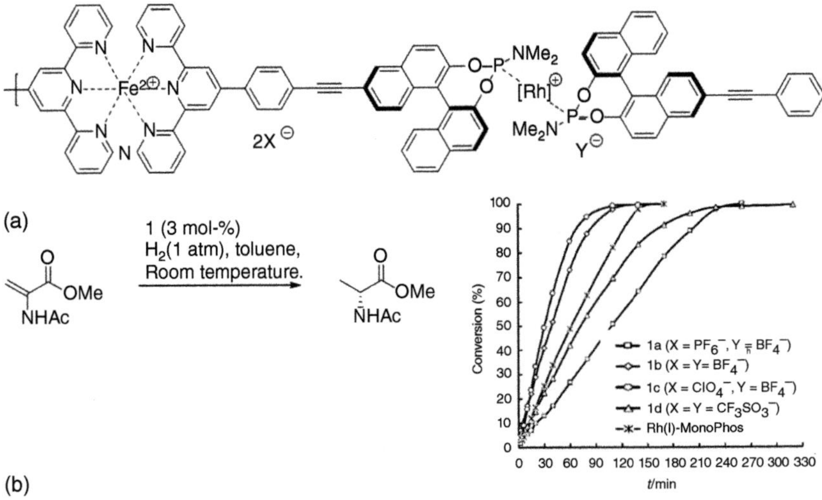

Figure 11.2 (a) Schematic representation of the metallo-supramolecular polymer **1**; (b) performance of **1** in an asymmetric hydrogenation reaction compared with Rh(I)-MonoPhos, as the benchmark. Source: Yu et al. [9]. © 2010 John Wiley and Sons.

of the metallo-supramolecular polymer were 242.1 and 188.2 $cm^2\ C^{-1}$ at the aforementioned wavelengths.

The combination of metal-to-ligand coordination and H-bonding interactions, as orthogonal pair of non-covalent interactions, was pioneered by the Schubert group [40]. The self-complementary UPy (UPy: 2-ureido-pyrimidinone) system is known to dimerize in nonpolar solvents, a wide range of supramolecular polymers, assembled from this particular interaction, is known from the literature ($K_a = 6 \times 10^7\ M^{-1}$ in $CHCl_3$, see also Section 3.2) [2, 3, 41]. The addition of a divalent transition metal ion to the heteroditopic ligand **3a** induced the supramolecular polymerization by forming terpyridine bis-complexes (Figure 11.4). The polymeric nature of the material was assured by ^{1}H NMR spectroscopy (i.e. signals arising from the H-bonded protons of the UPy-dimer were observed in $CDCl_3$ at between 10.2 and 13.1 ppm), UV/vis titration experiments as well as viscosity measurements. However, the low solubility of the resultant polymer hindered the investigation of the proposed underlying ring-chain equilibrium (Section 1.3.2) and to determine the degree-of-polymerization (DP). These limitations could be overcome by introducing a macromolecular spacer (in this study, poly[ε-caprolactone], PCl was used) between the two supramolecular binding sites (**2b**) [39]. Concentration-dependent viscometry measurements revealed the characteristic behavior of a reversible supramolecular polymers, a critical polymerization concentration (cpc) of 0.25 mM was estimated for **2b**. A good film-forming behavior of $\{[Fe(\mathbf{2b}\cdots\mathbf{2b})](PF_6)_2\}_n$ was observed; rheology confirmed the improved properties in comparison with pristine **2b**. The multiple-stimuli responsiveness of **2b** was studied: H-bonding could be "switched off" by changing either the temperature or solvent conditions; whereas, the terpyridine bis-complex could selectively be opened via electrochemistry or by adding a strongly competing ligand, e.g. HEEDTA (*h*ydroxy*e*thyl*e*thylene-*d*iamine*t*ri*a*cetic

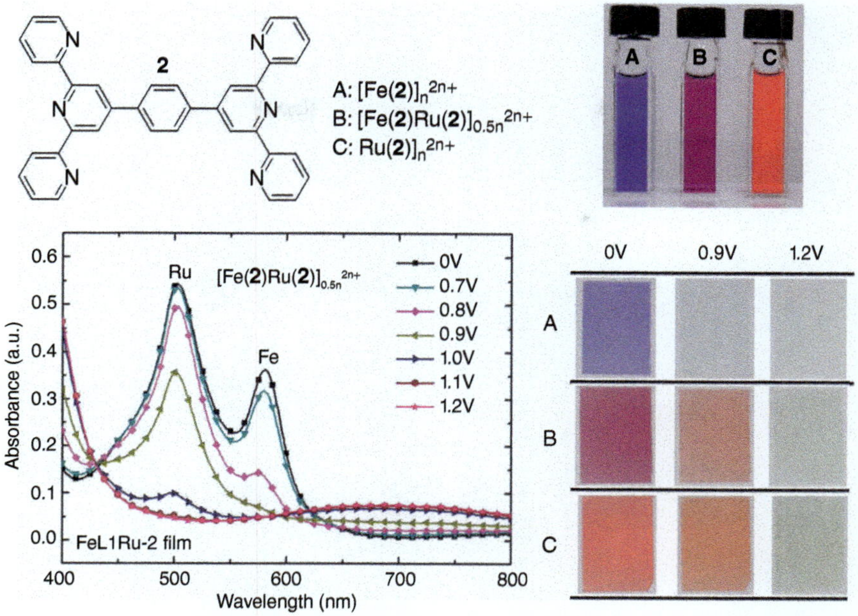

Figure 11.3 Schematic representation of the bis-terpyridine ligand **2** that was used for the formation of the homometallic and heterobimetallic metallopolymers comprising Fe(II) and/or Ru(II) ions. The polymers exhibited pronounced characteristics in their photophysical properties as revealed by their coloration that could be switched electrochemically by applying an external voltage. Source: Hu et al. [37]. Figure reproduced with kind permission; © 2013 The Royal Chemical Society.

acid). The authors concluded that the combination of properties, arising from the presence of the two different supramolecular arrays, was essential for the stimuli-responsive character of the material. In continuation of this work, the same group introduced a heteroditopic initiator for the nitroxide-mediate polymerization (NMP) [42]. The NMP of styrene directly gave the heterotelechelic polymer **4**; size-exclusion chromatography (SEC) of **4** showed a low dispersity (*Đ*) with a molar mass (M_n) of 6800 g mol^{-1}. Similar to the self-assembly of **2**, macroligand **4** was polymerized into a high-molar mass material by addition of divalent transition metal ions. Diffusion-ordered NMR spectroscopy (DOSY) verified the integration of both types of non-covalent interactions within the supramolecular polymer: the same relative diffusion coefficient was found for all resonances arising from the tpy, polystyrene, and UPy fragments $\{[Fe(\mathbf{3}\cdots\mathbf{3})](PF_6)_2\}_n$ (see also Section 12.8) [43].

The chiral Rh(I)-MonoPhos catalyst, which was already "immobilized" by incorporation into the supramolecular polymer **1** via two types of metal-to-ligand coordination (*vide supra*), was also combined with UPy, as a self-complementary H-bonding moiety. The resultant coordination polymer revealed, similar to **1**, an excellent reactivity and enantioselectivity, as a reusable, heterogeneous hydrogenation catalyst [44].

The Hirsch group utilized the same metal-to-ligand binding site as that of Schubert et al., i.e. a tpy moiety, but replaced the self-complementary UPy system

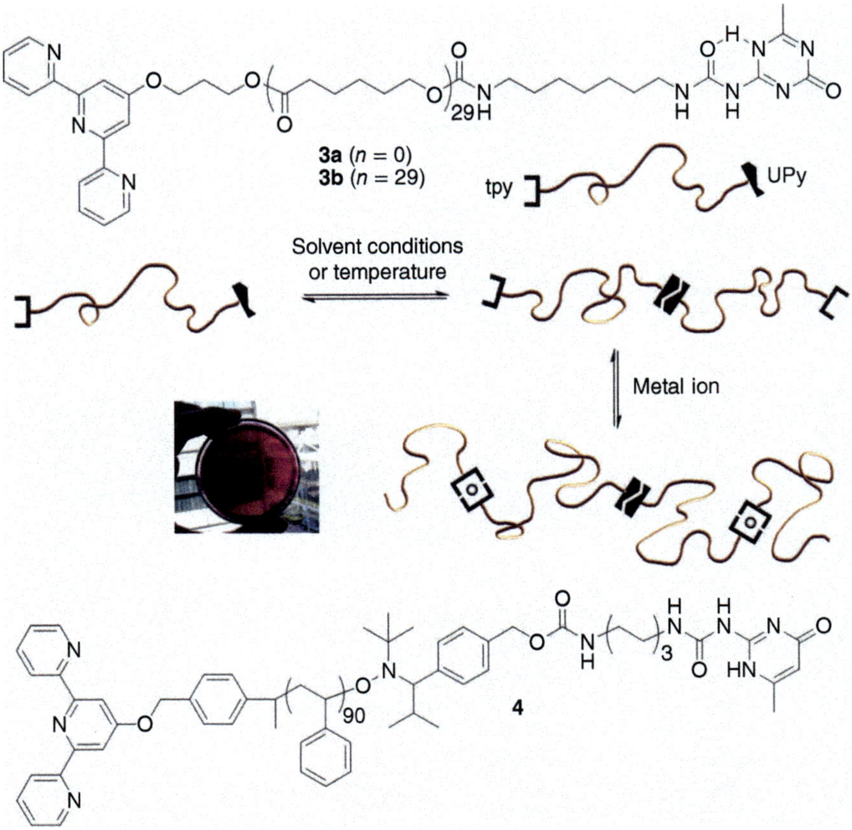

Figure 11.4 Schematic representation of the heteroditopic macroligands **3** and **4**, as monomers for the preparation of multi-stimuli-responsive supramolecular polymers of high molar mass. Source: Hofmeier et al. [39]. © 2005 American Chemical Society.

by a heterocomplementary Hamilton receptor, that was capable of forming strong sextuple H-bonds with cyanurates (see also Section 3.2) [46]. To construct a supramolecular polymer containing two orthogonal non-covalent interactions, **5** was self-assembled with the homoditopic cyanurate derivative **6** and Fe(II) ions in a 2 : 1 : 1 ratio (Figure 11.5) [45]. The reversibility of the H-bonding was confirmed by a sharp decrease in viscosity upon addition of a monotopic cyanurate derivative. In a similar approach, but using a Ru(II)-bipyridine tris-complex, as core unit, instead a supramolecular dendron could be prepared [47]. This example of using orthogonal non-covalent interactions indicates the high potential of this methodology in the design of advanced supramolecular structures with a high degree-of-hierarchy.

Finally, the Weck group combined metal-to-ligand coordination with H-bonding interactions to obtain random copolymers, where the supramolecular binding sites were located in the pendant side chains [16]. Ring-opening metathesis polymerization (ROMP) of two norbornene monomers and subsequent treatment with pyridine and *N*-butylthymine gave **7** (Figure 11.6); whereby, the anchoring groups

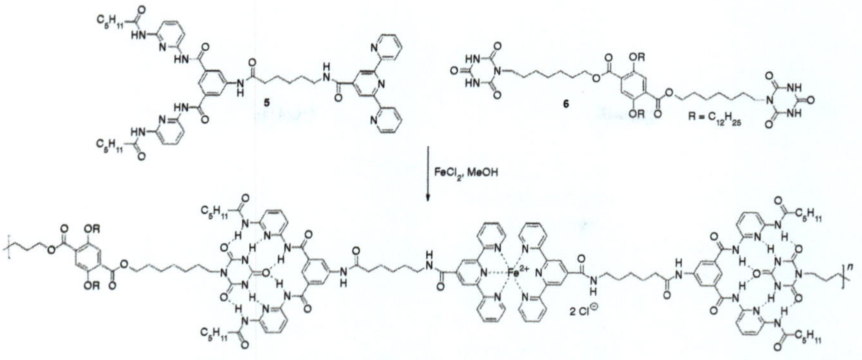

Figure 11.5 Schematic representation of the self-assembly of **5** and **6** in presence of Fe(II) ions. Source: Grimm et al. [45]. © 2011 John Wiley and Sons.

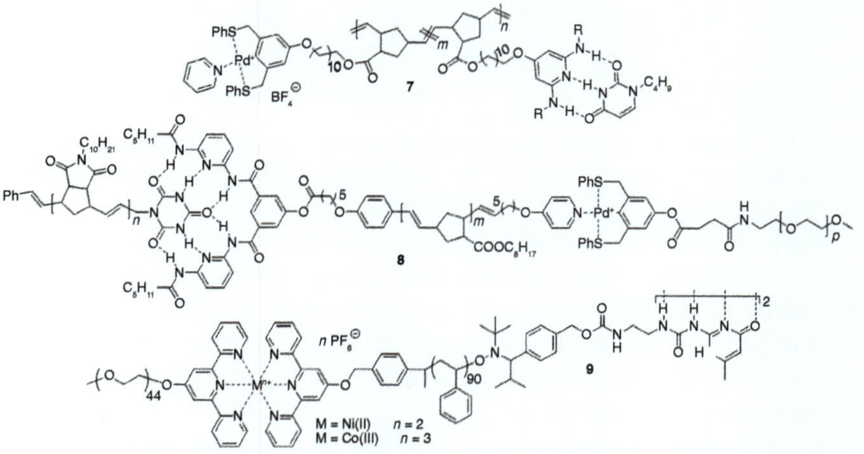

Figure 11.6 Schematic representation of supramolecular block copolymers **7–9**, based on metal-to-ligand coordination and H-bonding interactions. Source: Mansfeld et al. [48]. © 2013 The Royal Chemical Society.

in the side chains exhibited good selectivity and independence from each other. The scope of ROMP was later extended toward supramolecular ABC-type triblock copolymers (such as **8**, **6**) using a Ru(II)-based initiator that was functionalized unsymmetrically with the two recognition units [49]. In this context, the formation of the AB_2A-type quasi-triblock copolymer **8** is also noteworthy: Mansfeld et al. formed a heteroleptic Ni(II) bis-complex via an one-pot, two-step assembly of tpy-functionalized poly(ethylene glycol) and **4**; the supramolecular polymer was obtained via dimerization of the UPy end groups of **4** (Figure 11.6) [48]. Thus, copolymers having two types of orthogonal non-covalent interactions within the backbone or in the side chain can be achieved by the cooperation of established polymer synthesis and asymmetric supramolecular chemistry.

A different type of supramolecular polymer was obtained from the macromolecular building block **10** ($M_{n,SEC}$ = 4.7 kDa, $Đ$ = 1.19), which assembled into

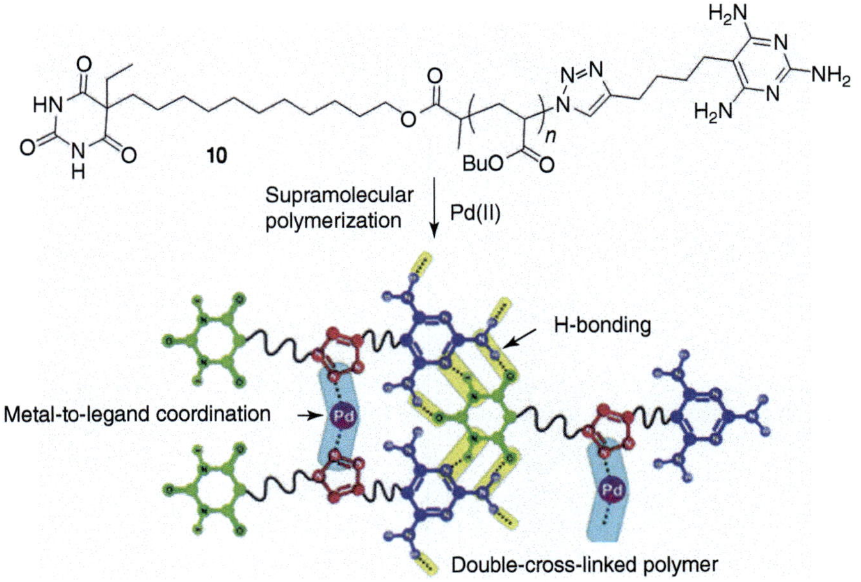

Figure 11.7 Schematic representation of the self-assembly of **10** and Pd(II) ions into a double-cross-linked polymer. Source: Chen et al. [50]. © 2019 The Royal Chemical Society.

polymer network via association of its complementary H-bonding moieties [50]. Double-cross-linking of the polymer chains occurred in the presence of Pd(II) ions (Figure 11.7). This process was accompanied with a drastic change in the morphology: the H-bonded network gave homogeneous films; whereas, with the increasing content of Pd(II) ions, a change to disc-like nano-objects and, eventually, large aggregates were observed.

The combination of metal-to-ligand coordination with ionic interactions, i.e. based on Coulombic attraction, has also been employed to a significant extend. The Yashima group reported a double-stranded supramolecular polymer with controlled helicity, thereby mimicking multistranded biomacromolecules (e.g. DNA or collagen) [51]. Addition of *cis*-$PtPh_2(DMSO)_2$ to an amidinium-carboxylate-bridged ligand (**11**) gave the one-handed helical polymer with square-planar Pt(II) complexes within the polymer's main chains (Figure 11.8a). 1D- and 2D-NMR, atomic force microscopy (AFM) imaging as well as dynamic light scattering (DLS) measurements were used to investigate the supramolecular polymer. According to the DLS measurements, a DP of c. 20 was reached (hydrodynamic radius: $R_H = 9.5$ nm, Figure 11.8b); cyclic dichroism (CD) spectroscopy disclosed the one-handed helical structure of the supramolecular polymer strands.

Why is the incorporation of ionic interactions, as a type of non-covalent binding, into supramolecular polymers advantageous? H-Bonding interactions give, in many cases, a strong and directional binding; however, one is limited to nonpolar solvents, such as $CHCl_3$. In polar solvents, binding is weakened due a competing effect of the solvent itself. In contrast to this, charged binding sites being capable of forming strong and directional ionic interactions can circumvent this limitation.

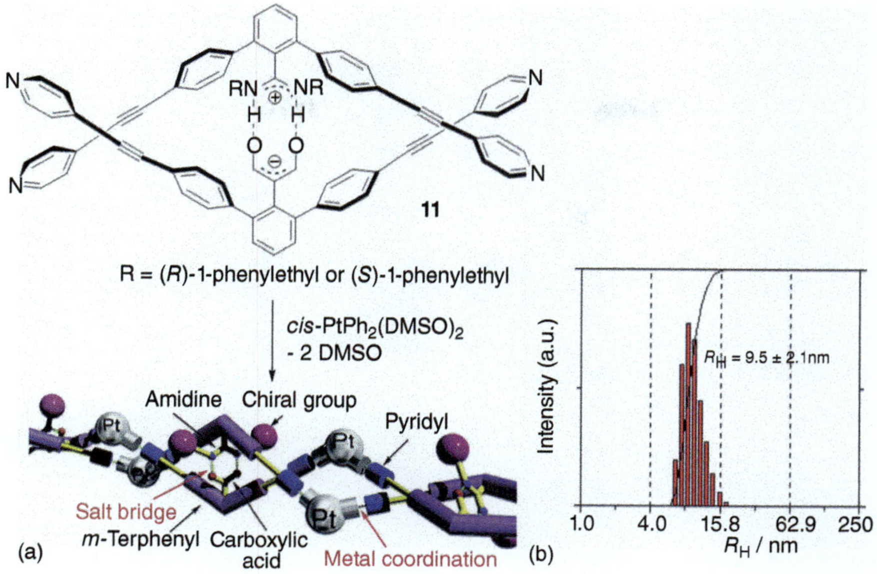

Figure 11.8 (a) Schematic representation of the formation of a double-stranded helical supramolecular polymer via metal-to-ligand coordination and ionic interactions. (b) Representation of the size distribution of the polymer (data derived from DLS measurements). Source: Ikeda et al. [51]. © 2006 American Chemical Society.

The Schmuck group utilized the heteroditopic building block **12** featuring a tpy unit, as the designated binding site for metal ions, and a zwitterionic guanidinio-carbonyl pyrrole carboxylate motif, as a self-complementary system for Coulombic binding (Figure 11.9, see also Chapter 2) [52, 53]. Strong association of both units ($K_a > 10^8\ M^{-1}$) was observed in DMSO, thus allowing the growth of linear supramolecular polymer chains of reasonable length even in highly polar solvents. Viscometry suggested an underlying ring-chain-mediated polymerization mechanism with a cpc of 0.35 mM. The subsequent assembly of the individual polymer chains into globular aggregates was studied by UV/vis absorption spectroscopy, transmission electron microscopy (TEM) imaging, and light/neutron scattering experiments. It was shown further that the association state could be modulated reversibly via the addition (or removal) of transition metal ions or by adjusting the pH value of the solution (Figure 11.9): low-molar mass species (i.e. monomer **12**, ion-paired dimer **12**···**12**, tpy bis-complex $[M(\mathbf{12})_2]^{2+}$, and a high-molar mass polymer $\{[(\mathbf{12})M(\mathbf{12})]^{2+}\}_n$) can be formed selectively by applying different external stimuli.

As detailed in Chapter 9, cyclodextrins (CDs) are a well-known host component in supramolecular chemistry and are, due to their solubility in water and nontoxicity, of considerable interest with respect to biological and pharmaceutical applications (e.g. drug or gene delivery, tissue engineering, etc.) [55–57] and are also utilized in polymer science (i.e. polymer processing) [58]. A linear supramolecular polypseudorotaxane combining metal-to-ligand coordination

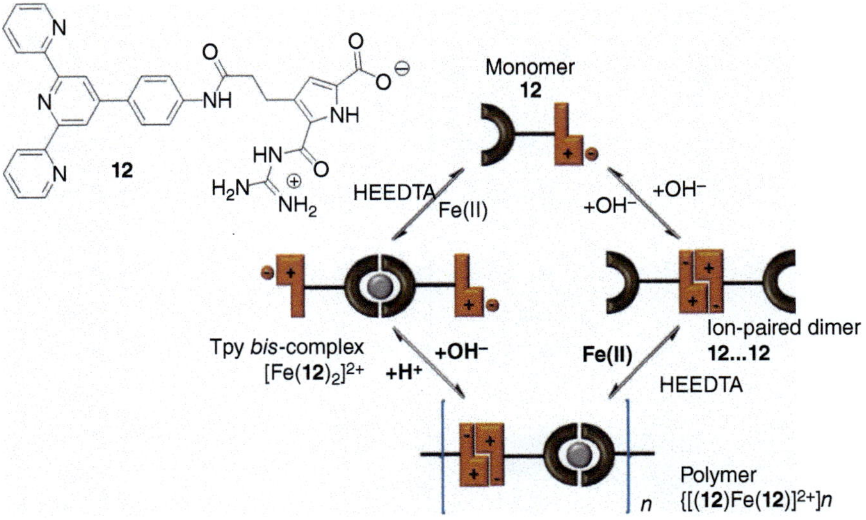

Figure 11.9 Schematic representation of the heteroditopic building block **12** and its use in the formation of a multi-stimuli-responsive supramolecular polymer. Source: Gröger et al. [52]. © 2011 American Chemical Society.

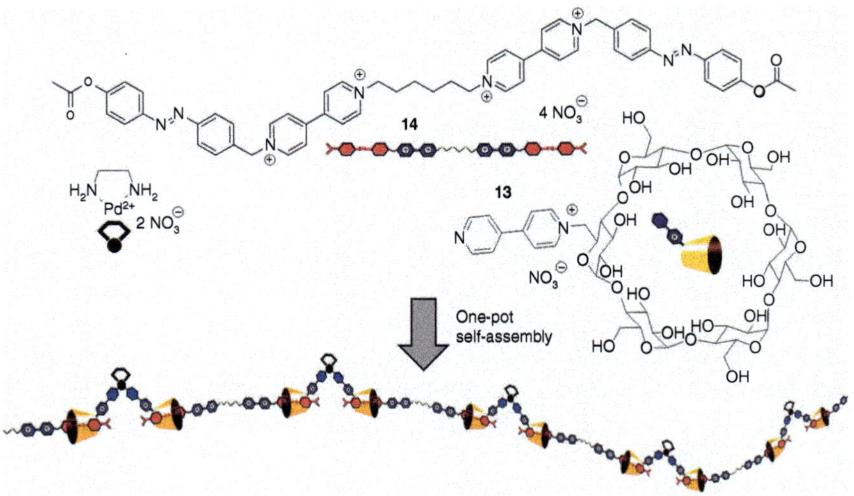

Figure 11.10 Schematic representation of a supramolecular pseudorotaxane with Pd(II) complexes in the main chain. Source: Zhu et al. [54]. © 2011 American Chemical Society.

and host–guest interactions, i.e. making use of CD inclusion complexes, was reported by Zhu et al. [54]. For this purpose, the functionalized β-CD unit **13**, the bis-azobenzene derivative **14** (the two viologen units of **14** were acting as hydrophilic electropositive barriers), and a Pd(II) precursor complex were self-assembled in water (Figure 11.10). It was further shown by viscometry that photoisomerization of the azobenzene units upon irradiation with UV light (λ = 345 nm) enabled a partially reversible depolymerization (the

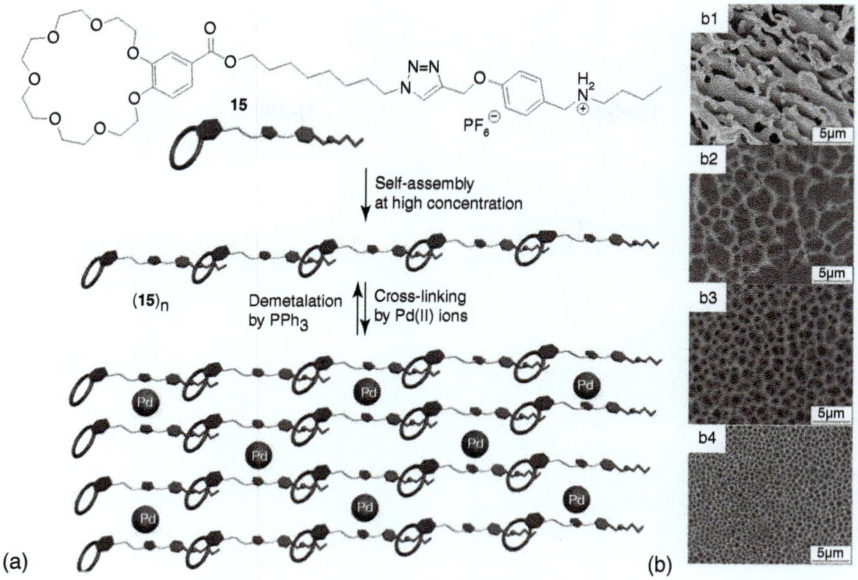

Figure 11.11 (a) Schematic representation of the two-step, self-assembly of **15** into a cross-linked supramolecular polymer; (b) representative SEM images of $(\mathbf{15})_n$ with increasing amount of Pd(II) ions (top to bottom: 0%, 20%, 60%, and 100%, respectively). Source: (a) Yan et al. [59]. © 2011 John Wiley and Sons, (b) Yan et al. [59]. Figure reproduced with kind permission; © 2012 Wiley-VCH.

trans-isomer was converted into the *cis*-form that was released from the CD cavity).

In a different approach, heteroditopic monomers, such as **15**, were applied to synthesize supramolecular polymers, based on crown ether recognition (for the host–guest chemistry of crown ethers, see Chapter 6) [59]. The monomers were prepared by Cu(I)-catalyzed azide-alkyne "click" (CuCAAC) reaction. As a result, a 1,4-disubstituted 1,2,3-triazole unit, as the designated binding site for metal ions [61], was incorporated into the pending arm of each monomer. The triazoles were utilized for cross-linking with Pd(II) ions subsequent to the supramolecular polymerization of, e.g. **15**, at high concentrations (Figure 11.11a). The addition of Pd(II) ions and their removal by PPh_3 afforded a reversible transition between the gelated and solubilized state; for $(\mathbf{15})_n$, a critical gelation concentration (cgc) of 30 mM was determined [59]. In addition, this gel was responsive to changes in temperature and pH value as well as to potassium ions. The pore size of the gel could be adjusted via the ratio between Pd(II) ions and polymer $(\mathbf{15})_n$ (Figure 11.11b); the material could be fabricated into shape-persistent, free-standing, yet elastic objects. These properties were attributed to a synergistic effect of the dynamic and reversible host–guest interaction and the mechanically robust metallo-supramolecular cross-linking. The same orthogonal combination of supramolecular interactions, was utilized by Stang and coworkers to assemble a multifunctional supramolecular gel comprising metallo-supramolecular cages, which were linked by host–guest interactions, i.e. crown ether recognition (Figure 11.12) [60]. The fluorescent material exhibited a

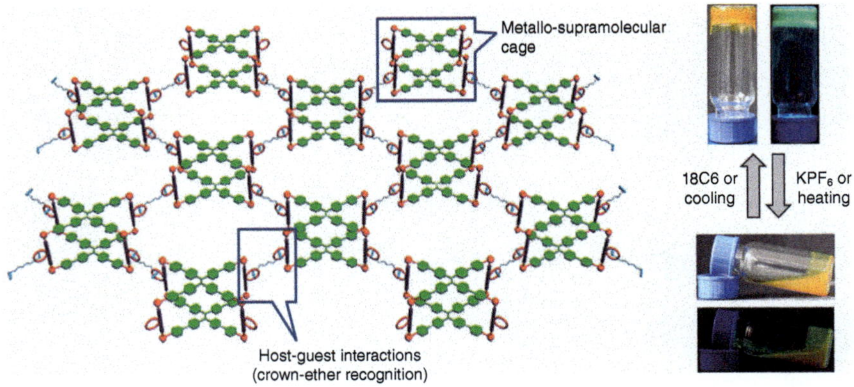

Figure 11.12 Schematic representation of the supramolecular gel formed by the cross-linking of metallocages via crown ether recognition. The material exhibited a multi-stimuli-responsive behavior (e.g. switchability between a gel-like and liquid state). Source: Lu et al. [60]. Figure reproduced with kind permission; © 2018 American Chemical Society.

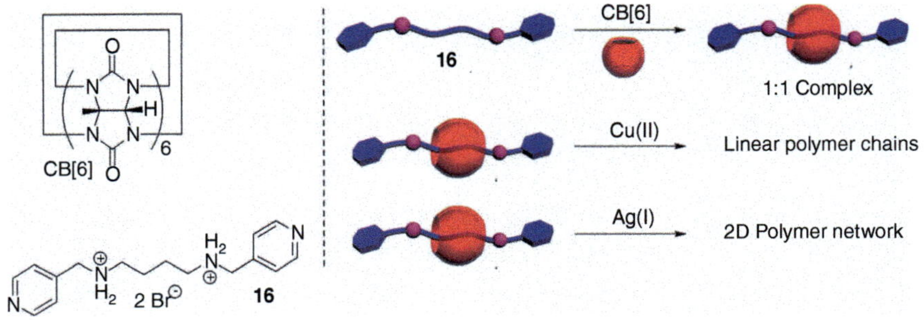

Figure 11.13 Schematic representation of the formation of the 1 : 1 complex **16**@CB[6] and its metallo-supramolecular polymerization into linear chains or a 2D polymer networks when Cu(II) or Ag(I) ions, respectively, are added. Source: Whang et al. [62]. © 1996 American Chemical Society.

rich stimuli-responsive and self-healing behavior due to the dynamic nature of the supramolecular interactions.

In Chapter 7, the application of cucurbiturils (CB[n]s) as macrocyclic hosts for the assembly of supramolecular polymers via host–guest complexation has been detailed. As for other types of host-guest interactions, also CB[n]-binding has *inter alia* been combined with metal-to-ligand coordination to obtain advanced supramolecular polymeric structures [17, 64, 65]. The earliest examples in this respect were published by Kim who, for instance, assembled **16** into the cavity of CB[6] [66]. The thusly formed complex of 1 : 1 stoichiometry formed zig-zag-shaped polymer chains when Cu(II) ions were added; the supramolecular polymerization was rationalized by selective *cis*-coordination of the pyridine sites of **16**@CB[6] to the metal centers (Figure 11.13) [62]. In the case of Ag(I) ions, a different architecture was obtained in which a supramolecular 2D network of high structural

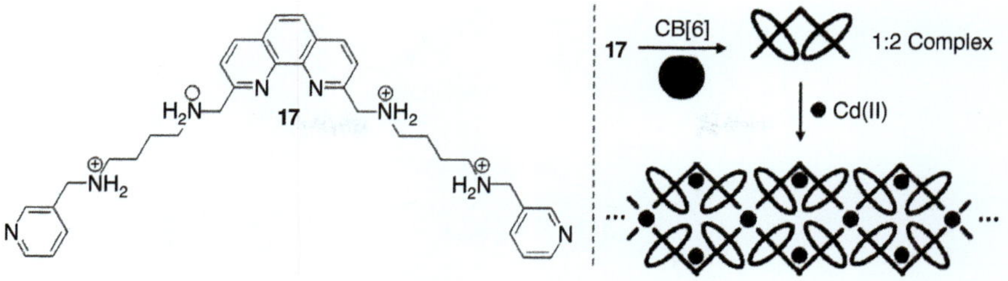

Figure 11.14 Schematic representation of the two-step assembly of **16–17** with CB[6] and Cd(II) ions into a double-chained supramolecular 1D polymer. Source: Redrawn from Park et al. [63].

regularity was obtained due to a different binding mode of the Ag(I) ions compared with the Cu(II) ones [67]. These reports already reflect the key synthetic strategy toward supramolecular polymers featuring CB[*n*]-based host–guest interactions and metal-to-ligand coordination: the host–guest complex has to be assembled first, supramolecular polymerization is achieved in the second step by adding an appropriate (transition) metal ion [66]. In other words, the formation of supramolecular polymer chains results from metal-to-ligand coordination and the host–guest complexes within the repeating units play mainly a structural role in terms of bringing rigidity into the otherwise flexible assembly. In a related study, the homoditopic building block **17** was combined with CB[6] into an 1 : 2 complex [63]. This, in turn, generated the double-chained or zig-zag-shaped supramolecular polymers on treatment with Cd(II) or Co(II) ions, respectively (Figure 11.14). More recently, Hu and coworkers assembled a supramolecular polymer gel by, first, forming a pseudo[2]rotaxane from a dimethoxypillar[5]arene, as host, and an amide-bridged bis-pyridine, as guest [68]. The metallo-supramolecular polymerization of this assembly gave a 1D polymer which, driven by intrachain H-bonding interactions, eventually generated a supramolecular gel in highly concentrated solutions. Pillar[5]arene-based host–guest interactions were also employed by Xia and coworkers for the supramolecular cross-linking of a linear metallo-supramolecular polymer, based on Ag(I)–pyridine bis-complexes [69]. This material was stimuli responsive, i.e. a reversible sol–gel transition could be induced by heating and cooling as well as by adding I^- ions, as competitive ligand for the Ag(I) centers.

As pointed out in Chapter 4, complexation of transition metal ions by tpy ligands in a 1 : 2 fashion represents one of the most important examples for metallo-supramolecular complexation. The Zhang group combined this particular interaction with the supramolecular chemistry of CB[*n*]s. However, the order of the assembly steps was inverted compared with Kim's pioneering work [70]: first, the Fe(II) bis-complex $[Fe(\mathbf{17})_2]^{2+}$ was formed and, subsequently, dimerized due to π–π stacking interactions of the naphthalene residues within the cavities of the added CB[8] molecules (Figure 11.15a). The authors attributed the increased water solubility to the presence of the cationic complexes, which acted as a rigid-rod linkage between the naphthalene units. Joseph et al. investigated this concept of supramolecular polymerization further and identified strong H-bonding interactions between the tpy ligands and the CB[8]'s carbonyl groups (independent of the

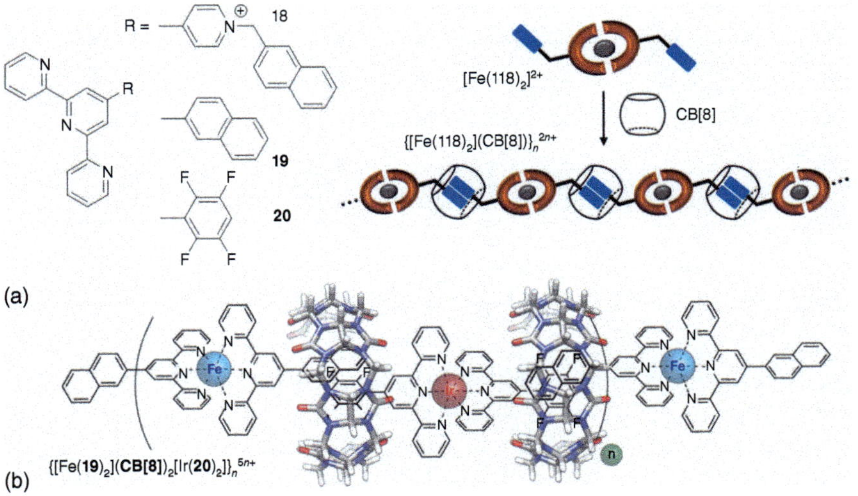

Figure 11.15 Schematic representation of the typ ligands **18–20**, as monomers, for the two-step supramolecular polymerization: (a) homometallic polymer, based on homotrimeric host–guest complexes, and (b) heterobimetallic polymer, based on heterotrimeric host–guest complexes. Source: (a) Liu et al. [70]. © 2013 The Royal Chemical Society, (b) Joseph et al. [71]. © 2014 American Chemical Society.

coordinated transition metal ion and of the substituent R in 4′-position of the tpy **18–20**, Figure 11.15a) [71]. The thermodynamics of the supramolecular polymerization were analyzed and a favorable enthalpic contribution (c. 20.2 kcal mol^{-1}) in concert with a highly detrimental entropic one (c. −10.2 kcal mol^{-1}) was identified. These authors also studied the one-pot, two-step self-assembly of tpy **19** and CB[8] in the presence of Fe(II) or Ir(III) ions. According to DOSY measurements, DPs of 11 and 54 were reached for these two metal ions, respectively (1 mM in water). A similar behavior was observed when using ligand **20** instead (in this case, the DPs were estimated to be 27 and 15, respectively). Moreover, a remarkable supramolecular self-sorting behavior was found: the equimolar mixture of $\{[Fe(\mathbf{19})_2](CB[8])\}_n^{2n+}$ and $\{[Ir(\mathbf{20})_2](CB[8])\}_n^{3n+}$ reassembled into a strictly alternating heterobimetallic polymer (Figure 11.15b). This process was rationalized by the strong quadrupole–quadrupole interactions between the naphthyl and tetrafluorophenyl moieties within CB[8]'s cavity (a behavior that is reminiscent of the long-known benzene–perfluorobenzene pairing [72]).

11.3 Orthogonal Combinations of Supramolecular Interactions Involving *H*-Bonding

The linkage of building blocks via (directional) multiple-hydrogen bonding represents one of the key strategies in modern supramolecular chemistry (seen in Chapter 3). The selectivity of H-bonding, as exemplified by DNA's base pairing in Nature, can be applied to construct supramolecular polymers. As a proof-of-concept,

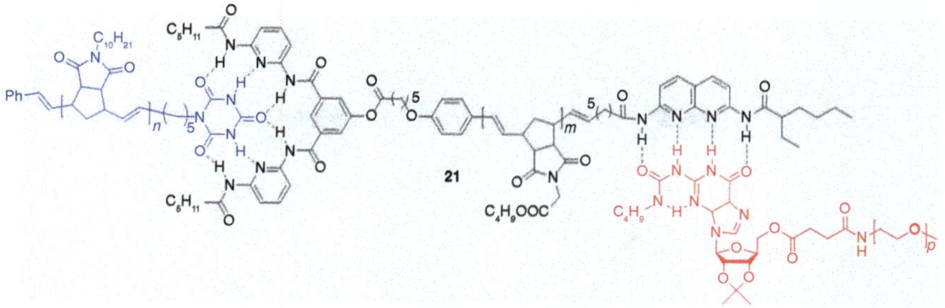

Figure 11.16 Schematic representation of supramolecular triblock copolymer **21**. Source: Yang et al. [73]. © 2010 American Chemical Society.

the Weck group prepared random and block copolymers, bearing thymine and cyanuric acid recognition sites at their chain ends, by ROMP; a highly dynamic, self-assembly process was observed, when the complementary building blocks with a diaminopyridine unit and a Hamilton receptor, respectively, were added [74]. In a more sophisticated approach, a supramolecular ABC-type triblock copolymer (**21**) was reported where the blocks were linked via pairing of a Hamilton receptor with a cyanurate on the one hand and 2,7-diamino-1,8-naphthyridine with ureidoguanosine on the other (Figure 11.16) [73]. The supramolecular self-sorting efficiency of the involved H-bonding entities was reflected by the fact that the assembly of **21** occurred in a one-pot process.

Nieuwenhuizen et al. investigated how UPy-functionalized molecules with additional urea or urethane moieties in the pendant side chain self-assembled in various solvents [75]. Lateral aggregation due to urea/urethane H-bonding induced a 1D-stacking of the UPy-dimerized molecules (Figure 11.17). It was shown by concentration-dependent ^{1}H NMR and DOSY spectra, that the aromatic units in the side chain stabilized the rather weak array by π–π stacking interactions. In apolar solvents, such as *n*-heptane, the formation of large helical assemblies was confirmed by temperature-dependent CD spectroscopy (chirality of **22** was arising from stereocenters in the alkyl substituents on the UPy and benzene rings).

The concept was extended from small building blocks to homoditopic polymeric unimers [77]: poly(ethylene-*co*-butylene) (PEB) was modified with UPy moieties at both termini via urea (U) linkages; the unimers of the general formula UPy-U-PEB-U-UPy (**23**) formed fibrillar aggregates with fiber lengths of c. 7 nm. Replacing the hard PEB of **23** by the soft poly(dimethylsiloxane) (PDMS) changed the mode-of-aggregation [78]: phase separation occurred due to the incompatibility of the PDMS block with the hard H-bonding array; crystallization, as a result from π–π stacking of the UPy-dimers, led to fiber formation. With increasing length of the PDMS chain, the morphology could be changed from the fibrous to a spherical state. Substituents on the pyrimidinone moiety were found to strongly influence the aggregation mechanism of **23** [76]: necessarily, the 5-position had to be unsubstituted (presumably a requirement to enable interchain interactions); for the 6-position either a small size (i.e. $-CH_3$) or a regular stereochemical constitution (i.e. enantiomerically pure alkyl groups) of the substituent

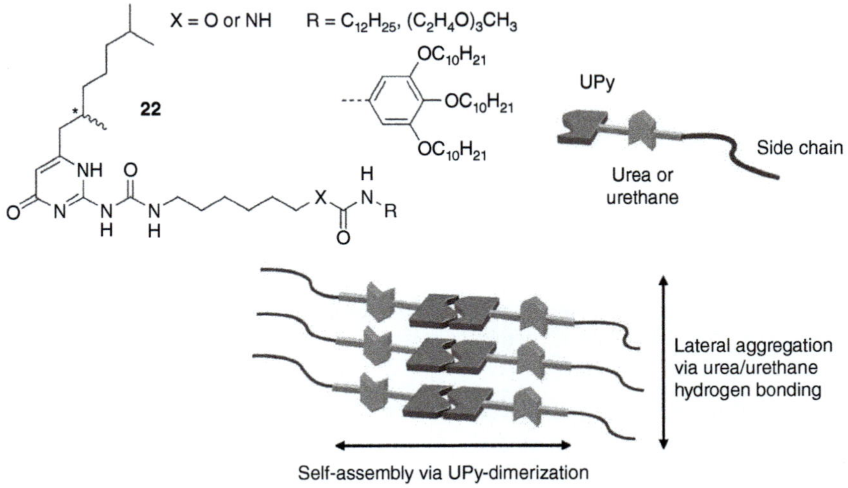

Figure 11.17 Schematic representation of the formation of 1D-stacks from **22** by synergistical H-bonding interactions. Source: Nieuwenhuizen et al. [75]. © 2010 John Wiley and Sons.

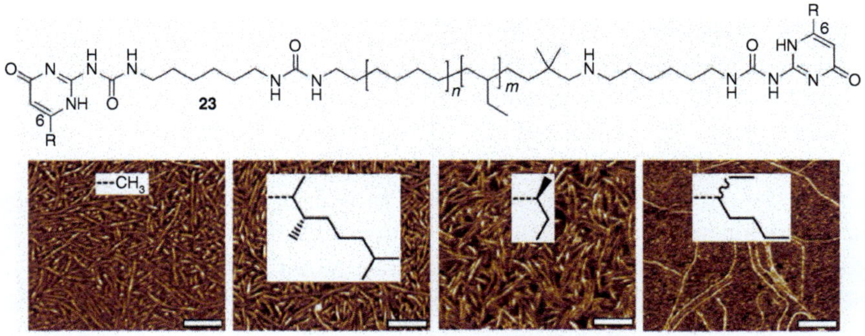

Figure 11.18 Schematic representation of homoditopic polymer **23**. The AFM imaging revealed fiber formation as a result from substituents in 6-position (scale bar: 100 nm). Source: Appel et al. [76]. Figure reproduced with kind permission; © 2011 American Chemical Society.

were reported to be essential with respect to the crystallization rate and degree (Figure 11.18).

Two monotopic polymers of low molar mass (**24**: M_n = 3900 g mol^{-1}; **25**: M_n = 6200 g mol^{-1}) were prepared as viscous, sticky oils; in contrast, the heteroditopic material **26** (M_n = 4000 g mol^{-1}) appeared as an elastic material. Both **24** and **25** self-assembled via H-bonding as shown in Figure 11.19 (the aggregation behavior of **24** was detailed above; **25** formed helical, columnar aggregates [80]). The addition of **26**, as compatibilizer, to a mixture of **24** and **25** led to cross-linking of the individual assemblies into a solid material with elastomeric properties (according to rheology measurements) [79]. Thus, the bulk properties of a binary

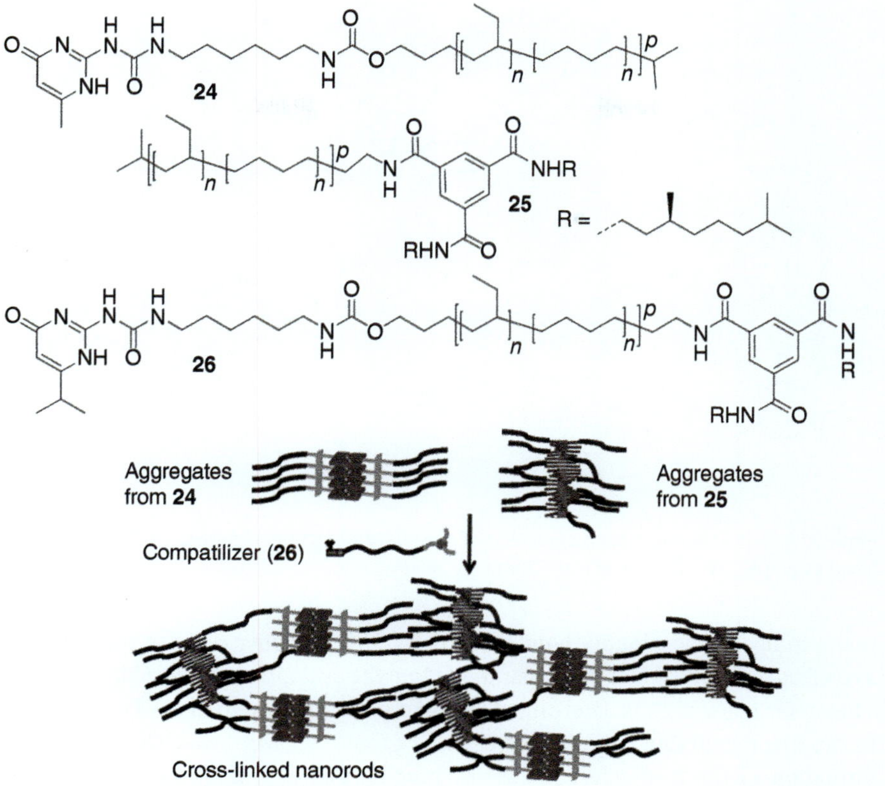

Figure 11.19 Schematic representation of polymers **24–26** and their self-assembly behavior. Source: Mes et al. [79]. © 2011 American Chemical Society.

orthogonal polymer blend can be significantly enhanced by adding an appropriate supramolecular compatibilizer.

The combination of H-bonding interactions with metal-to-ligand coordination has already been discussed above (see Section 11.2). Besides this important interplay of orthogonal non-covalent interactions, H-bonding interactions have also been combined with other types of supramolecular forces: host–guest complexation as well as π–π stacking, and ionic interactions.

About 15 years ago, the first structurally simple, discrete supramolecular assemblies, based on H-bonding and host–guest complexation, were reported. For example, Tokunaga and Seo reported on the UPy-dimerization of a crown ether derivative; a pseudo[2]rotaxane dimer was obtained by inclusion of a bis-benzylammonium species [82]. In a different approach, a UPy-heterodimer was obtained by the reassembly of two UPy-based homodimers at low concentration: driven by the formation of an intramolecular inclusion complex, the homodimers were converted into a heterodimer assembly [83]. Shao et al. showed the self-assembly of an UPy-bridged strapped metalloporphyrin cleft with a bipyridine-containing cyclophane derivative and a rigid fumaramide [81]. This advanced supramolecular assembly (**27**) featured a quadruple H-bonding unit, a

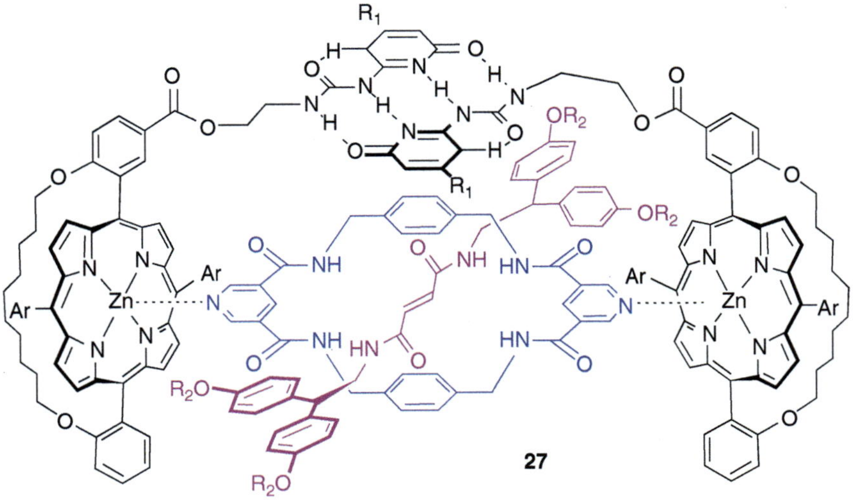

Figure 11.20 Schematic representation of the supramolecular architecture **27**. Source: Shao et al. [81]. © 2004 American Chemical Society.

metal-to-ligand coordinative bond [to link the Zn(II)-porphyrin and the cyclophane] as well as the inclusion of the fumaramide into the cyclophane cavity (Figure 11.20). The molar mass of **27** as determined by vapor pressure osmometry (VPO) was close to the theoretical one [M_{exp} = (3900 ± 400) g mol^{-1} vs. M_{theo} = 4368 g mol^{-1}]; the formation of the host–guest inclusion complex was monitored by ^{1}H NMR and UV/vis titration experiments, revealing K_a value of c. 10^4 M^{-1}.

A ring-chain equilibrium, based on H-bonding and host–guest interactions, was reported by Xiao et al. [84]: at high concentrations, a 1,5-dioxonaphthalene and a hydroquinone derivative, both bearing two UPy moieties at the end of pending side chains, self-assembled into linear supramolecular polymers due to quadruple H-bonding interactions. In the presence of a macrocyclic bis(paraquat-*p*-phenylene) system, the equilibrium could be pushed toward the chain-like assemblies (over a wide concentration range), formation of an interlocked strong donor–acceptor [2]catenane was identified as the driving force for this process. This example shows how supramolecular ring-chain equilibria can be triggered by introducing an orthogonal non-covalent interaction, i.e. host–guest complexation in this case.

The same group also showed that heteroditopic monomers of low molar mass can be self-assembled into supramolecular polymers of high molar mass (though only at relatively high concentrations) [85]: classical UPy-dimerization and paraquat recognition by a crown ether unit were employed as orthogonal supramolecular interactions in this particular example (Figure 11.21). DLS measurements gave a hydrodynamic radius (R_h) of 31 nm for an equimolar mixture of **28** and **29** in a $CHCl_3$/MeCN mixture. ^{1}H NMR spectroscopy was applied to determine the DP according to Carother's equation (see Section 1.2): at a concentration of 433 mM, a DP of 38.5 was calculated – corresponding to a molar mass of 659 000 g mol^{-1}.

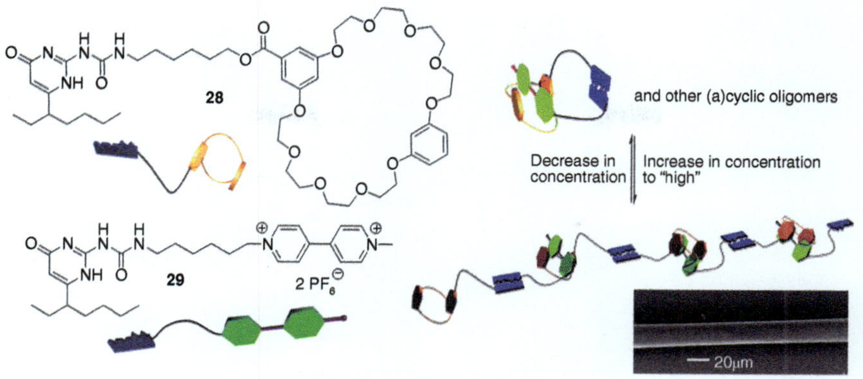

Figure 11.21 Schematic representation of monomers **28** and **29** as well as their self-assembly into a linear supramolecular polymer; the SEM image of a fiber grown from (**28**···**29**)$_n$ is also shown. Source: Li et al. [85]. Figure reproduced with kind permission; © 2011 The Royal Chemical Society.

Fibers could be drawn from these solutions, thus giving direct physical proof for the formation of a material of high molar mass (a typical SEM image of such a fiber is shown in Figure 11.21). In general, fiber formation is only possible for polymers of high molar mass [86].

The Wang group prepared a polypseudorotaxane from two homoditopic monomers: the bis-UPy derivative **30** and bis-paraquat **31** [87]. Monomer **30** gave a linear supramolecular polymer (due to the formation of UPy dimers) which, in solution, could be cross-linked by addition of **31** (Figure 11.22). Due to the threading of the paraquat moieties into the cavities of the crown ether units, a polypseudorotaxane network was formed. Thereby, the viscosity of the solution could be adjusted via the amount of added **31**, i.e. the cross-linking density.

The star-shaped tris-porphyrin derivatives **32** were investigated by the Elemans group [88, 89]. Self-assembly in alkane solvents (e.g. *n*-hexane or cyclohexane) occurred by concerted intermolecular H-bonding and π–π stacking interactions of the amide bonds and the porphyrin arrays, respectively. Under highly dilute conditions (c. 10^{-5} M), these interactions afforded large aggregates of columnar triplex helical structure (Figure 11.23a); however, this type of stacking could only be detected if the aliphatic side chains were chiral (**32b**) – no stable helical aggregates could be observed in the absence of chirality (**32a**). Temperature-dependent UV/vis absorption and CD spectroscopy suggested a selective face-to-face-type packing of the porphyrin rings of **32b** in *n*-hexane at room temperature; whereas, a combination of head-to-tail- and face-to-face-type ordering was observed in the same solvent at elevated temperatures. The size of the columnar architectures was estimated from DLS measurements (length of c. 300 nm, width of 5.1 nm) and allowed them to calculate a molar mass of 1.8×10^6 g mol^{-1} [89]. In $CHCl_3$, as a good universal solvent for **32**, no aggregation could be observed up to 0.2 mM. When deposited from solution onto mica substrates, equidistant lines of well-defined fibers were formed (Figure 11.23b); in contrast, only randomly oriented fibers were obtained from an alkane solution of preorganized **32** [88]. This processing

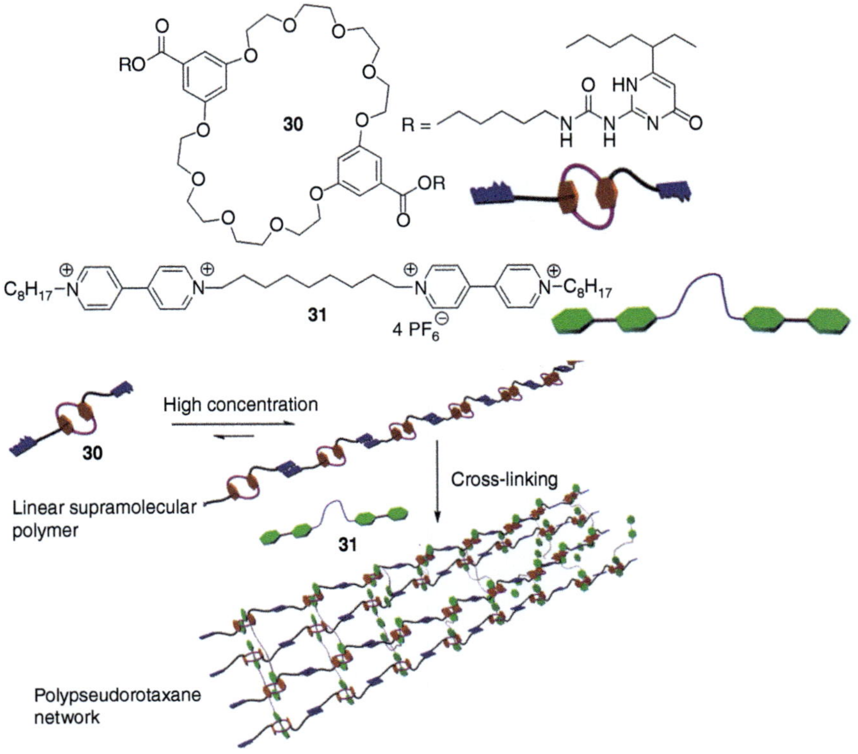

Figure 11.22 Schematic representation of the self-assembly of **30** and **31** into a pseudorotaxane network. Source: Li et al. [87]. © 2011 The Royal Chemical Society.

approach is of interest in view of surface patterning of small organic molecules, e.g. for optoelectronic applications, avoiding any demanding surface-patterning step (e.g. by lithography).

Ajayaghosh et al. utilized the short *oligo*(*p*-phenylene-ethynylene) (OPE) derivative **33** – bearing hydroxymethyl moieties at both ends – for the concentration-dependent fabrication of morphologies: from nanoparticles and microspheres to giant fibrillar structures [90]. Self-assembly was driven by concerted supramolecular aggregation, based on π–π stacking, H-bonding and, to some extent, also van-der-Waals interactions (Figure 11.24a). Though single H-bonding interactions are intrinsically weak (see Section 3.2.1.1), they were found to be crucial for the assembly of **33**. In a control experiment, the OPE segment missing the OH functionalities did not assemble into larger aggregates. AFM imaging showed that the size of the nanospheres increased with the concentration of **33**: at a concentration of 10^{-6} M, the size of the assemblies was below 100 nm; whereas, a size of c. 10 μm was observed at a concentration of 10^{-4} M. Above the cgc of 3.4×10^{-3} M, further assembly of the aggregates into extended superstructures, which eventually coagulated, was observed (Figure 11.24b).

The orthogonal supramolecular grafting onto a random copolymer bearing recognition sites for Coulombic and triple H-bonding interactions was studied by Nair and

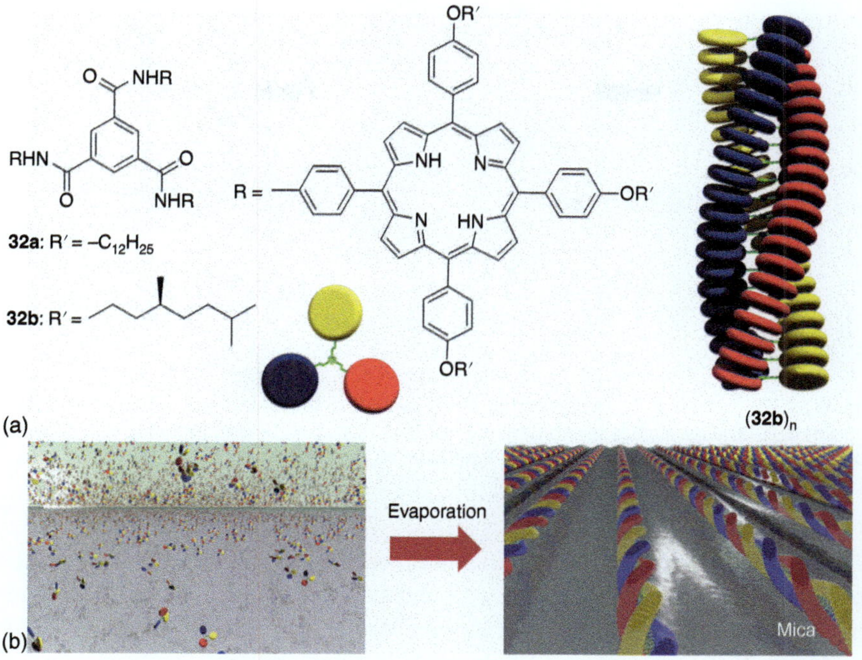

Figure 11.23 (a) Schematic representation of the self-assembly of a star-shaped molecule **32** into a helical columnar triplex; (b) schematic representation of the fabrication of an ordered array of columnar lines on a mica substrate. Source: van Hameren et al. [88]. © 2008 American Chemical Society.

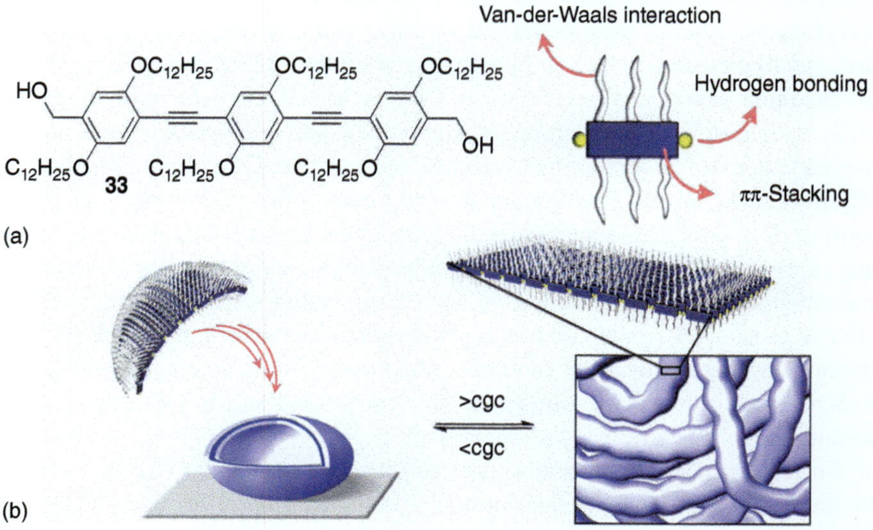

Figure 11.24 (a) Schematic representation of OPE **33** and (b) its concentration-dependent self-assembly into nanoparticles, microspheres or giant fibrillar structures. Source: Ajayaghosh et al. [90]. © 2006 John Wiley and Sons.

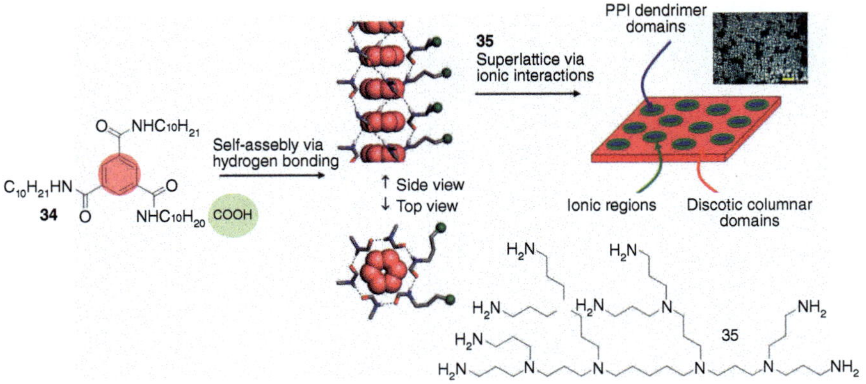

Figure 11.25 Schematic representation of the two-step, self-assembly of **34** and **35** into a supramolecular superlattice; the polarizing optical microscope (POM) image is also shown (**34**-to-**35** ratio of 8 : 1, 100 °C). Source: Fitié et al. [92]. © 2008 American Chemical Society.

Weck [91]. Similar to the investigation of other types of side-chain functionalized copolymers (*vide supra*), the random copolymer was prepared by ROMP. No interference of the recognition sites upon supramolecular grafting was found by ^{1}H NMR and infrared (IR) spectroscopy. While the H-bonding array was kept constant, i.e. *N*-butylthymine was attached to polymer-bound 2,6-diaminopyridine, the anions coordinated to the cationic side chains were varied over a wide range (e.g. iodide as well as organic carboxylates, sulfonates, and phenolates).

Sijbesma and coworkers reported on a nanocomposite featuring polar, dendrimer-like channels, which arose from columnar stacks with liquid crystalline (LC) properties [92]. The material was prepared in a two-step assembly approach: first, intramolecular H-bonding interactions of discotic **34** led to the formation of LC columnar mesophases; secondly, ionic interactions between the carboxyl-groups of **34** with the free amino-groups of the star-shaped oligo(propyleneimine) **35**, perpendicular to the pre-formed columns, gave the supramolecular superlattice (Figure 11.25). Worth mentioning, the columnar array of **34** was – in sharp contrast to most discotic LC materials with large aromatic systems – not relying on π–π stacking interactions [1, 93]. The highest stability of the superlattice was observed at an 8 : 1 ratio of **34** and **35**, providing a stoichiometric ratio of –COOH and $-NH_2$ units. Such supramolecular superlattices might be of relevance – although only after chemical modification – for potential applications as size- or polar-selective membranes/barriers as well as for the templating of surfaces [17].

Jamadar and Das very recently reported on a supramolecular polymer formed by the self-assembly of the benzene-1,3,5-tricarboxamide **36** by an interplay of H-bonding and π–π stacking interactions [94]. The outward-pointing pyridine moieties were employed, as acceptors, for the binding of halogen-bond donors, such as the tetrafluorobenzene derivatives **37**, in water (Figure 11.26). Due to the amphiphilic nature of the supramolecular architecture, spherical micelles, or 1D

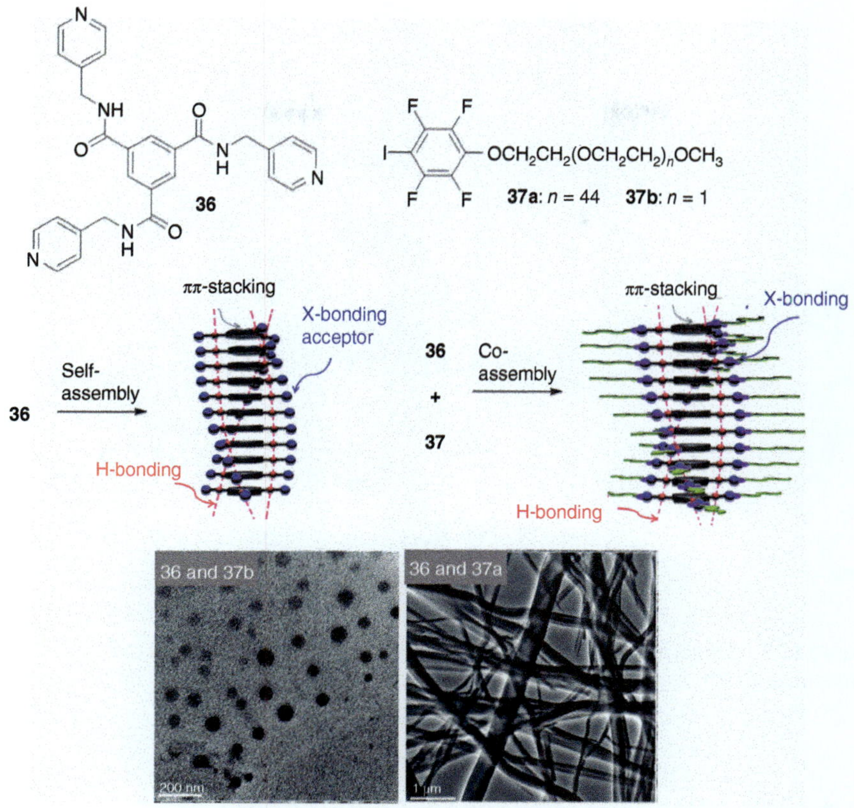

Figure 11.26 Schematic representation of the self-assembly of **36** into supramolecular polymers and grafted polymers (in the absence and presence of **37**, respectively). Representative TEM images of the spherical micelles and 1D fibers, obtained from the coassembly of **36** with **37a** and **37b**, respectively, are also shown. Source: Jamadar and Das [94]. © 2020 The Royal Chemical Society.

fibers were obtained for the combination of **36** with **37a** and **37b**, respectively. This study demonstrates that H-bonding and X-bonding can, under appropriate solvent conditions (i.e. in highly polar media, where X-bonding is strongly favored over H-bonding), be used to assemble hierarchically structured supramolecular architectures. Noteworthy, X-bonding has, up to now, mainly been employed to assemble supramolecular polymers in the solid state [95].

The sulfonamide-containing oligopeptides **38** were studied by Cai and coworkers regarding their self-assembly behavior [96]. Folding of **38** gave 4_{13} helices which, in turn, assembled via intramolecular H-bonding into chain-extended polypeptides. These 1D supramolecular polymers afforded – driven by an interplay between C–X···π, C–X···X–C, and X-bonding – a 3D architecture (Figure 11.27). This study demonstrated that X-bonding represents a versatile tool not only for the formation of supramolecular polymers but also in the design of biomimetic structures, such as artificial peptide foldamers [97].

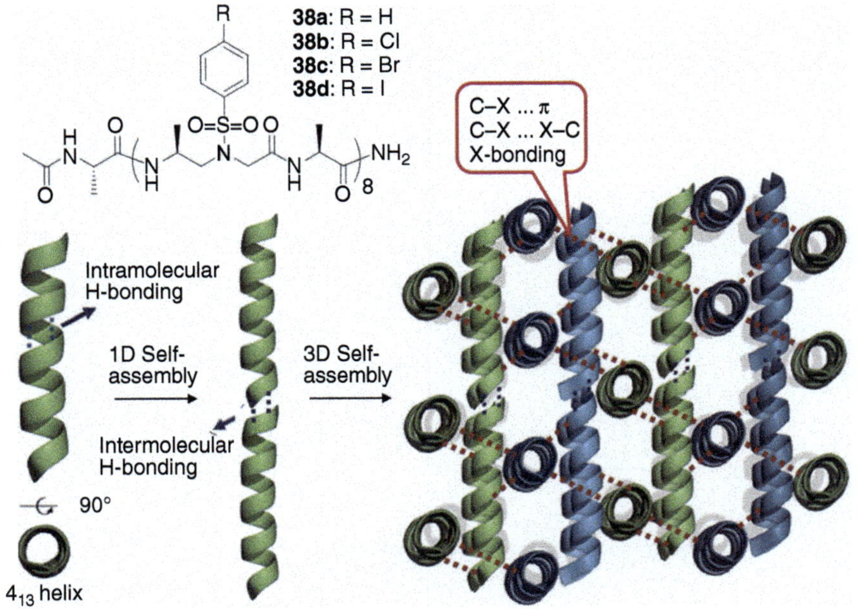

Figure 11.27 Schematic representation of the hierarchical self-assembly of the sulfonamide-functionalized oligopeptide **38**. Source: Teng et al. [96]. © 2019 John Wiley and Sons.

11.4 Miscellaneous Orthogonal Combinations of Supramolecular Interactions

When excluding metal-to-ligand coordination and H-bonding interactions, very few combinations of orthogonal non-covalent interactions, that have been realized into supramolecular polymers, remain. The complexation of guest molecules by crown ethers, as host materials, is one of the earlier examples for supramolecular chemistry, introduced by C.J. Pedersen already in 1967 [99]. In general, crown ether moieties feature a size-selectivity for the interaction with guest molecules – selectivity can principally be achieved via the size of the macrocycle and the number of *O*-atoms incorporated. Taking this basic doctrine of host–guest complexation, based on crown ether chemistry, orthogonality can be achieved when different crown ether units are combined with different guest molecules (see also Chapter 6) [17, 100]. Two small-molecule heteroditopic building blocks (**35** and **36**) were designed that the crown ether unit of **35** was matched to the guest residue of **36** (and vice versa) [98]. These particular pairs of host–guest systems were introduced by Stoddard and coworkers already in the 1990s [101, 102]. Supramolecular self-sorting of **35** and **36** gave a supramolecular alternating copolymer (Figure 11.28). The supramolecular polymerization followed a ring-chain-mediated mechanism (see Section 1.3.2); the rather low critical concentration of c. 40 mM was attributed to the flexible linkers of **35** and **36**. According to Carothers' equation (see Section 1.3) at a

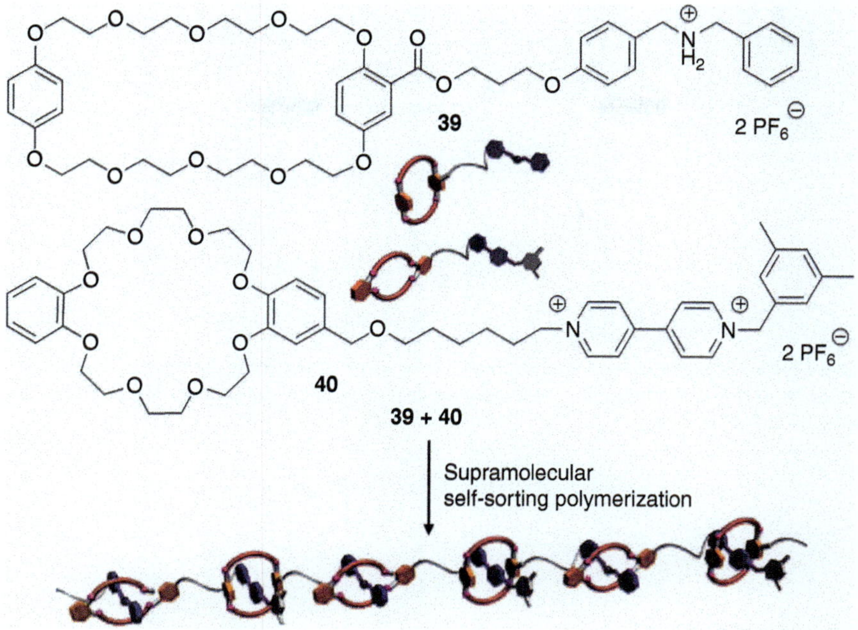

Figure 11.28 Schematic representation of the supramolecular polymerization of **39** and **40** due to orthogonal host–guest complexation. Source: Wang et al. [98]. © 2008 American Chemical Society.

concentration of 650 mM, the DP was 56.6 – corresponding to a molar mass as high as 1 260 000 g mol^{-1}.

Furthermore, a linear main-chain polypseudorotaxane with a supramolecular backbone was prepared from three molecular building blocks [103]. First, the host–guest interaction between two homoditopic monomers, i.e. a tris-(crown ether) and a bis-paraquat derivative gave a linear supramolecular polymer. Thereby, the paraquat units were complexed selectively by the terminal larger crown ethers moieties; whereas, the central smaller cavity remained empty. The final polypseudorotaxane was than formed by inclusion of the dibenzylammonium salt (Figure 11.29).

So far, supramolecular polymers, mainly based on ionic/Coulombic interactions, have hardly been reported (one rare example is the self-assembly of a zwitterionic pyrrole derivative, as shown by Schmuck [105], see Section 2.3.1). The typically low selectivity and directionality of this type of non-covalent interaction hinder a broad application in this respect. Faul and Antonietti pointed out that ionic self-assembly (ISA) is often accompanied by cooperative binding (for details on this mechanism, see Section 1.3.3), i.e. once a first bond is formed, it stimulates further binding until the final self-assembled structure is obtained [106]. Moreover, ionic interaction can be enhanced by additional non-covalent binding (e.g. metal-to-ligand coordination, H-bonding or π–π stacking). Examples for the first two combinations have already been discussed in Section 11.2 and 11.3. Gröhn et al. prepared supramolecular

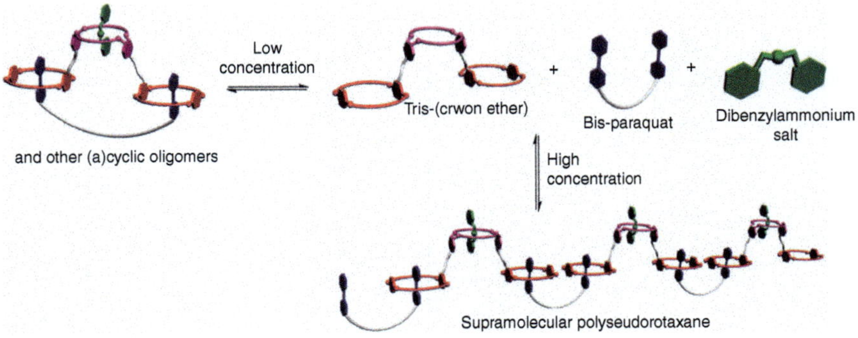

Figure 11.29 Schematic representation of the polypseudorotaxane formation via an orthogonal three-component self-assembly process. Source: Wang et al. [103]. © 2009 The Royal Chemical Society.

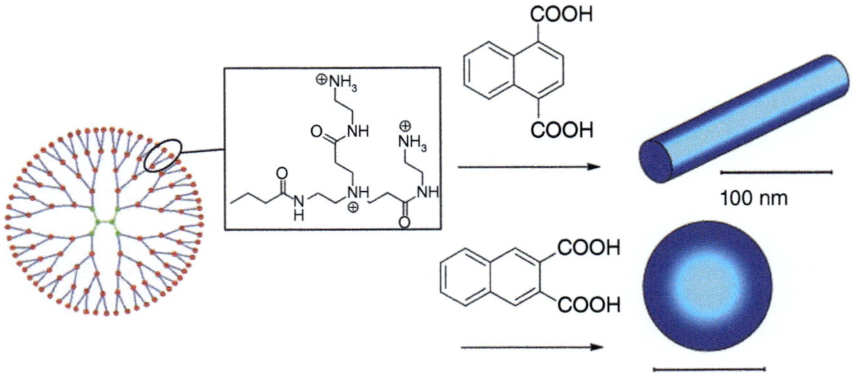

Figure 11.30 Schematic representation of the self-assembly of a polycationic dendron with naphthalene dicarboxylic acids into cylindrical and spherical aggregates. Source: Redrawn from Gröhn et al. [104]. © 2019 John Wiley and Sons.

polymers starting from a dendritic poly(amidoamine), as macromolecular "multication," and aromatic oligocarboxylates [104] or oligosulfonates [107, 108]. Due to a interplay of ionic and π–π stacking interactions, spherical or cylindrical aggregates were obtained (Figure 11.30); thereby, the morphology could simply be controlled via the geometry of the aromatic counterion (i.e. allowing or preventing the formation of supporting π–π stacks) [104]. Adjusting the pH value allowed reversible disassemble and reassemble to give the spherical aggregates. The encapsulation of a fluorescent dye, showing no significant supramolecular interaction(s) with either of the building blocks, into the vesicles was confirmed by confocal laser scanning microscopy (CLSM) [108]. Thus, these spherical assemblies might be of interest for (pharmacological) catch-and-release applications in polar medium.

Wang et al. utilized an interplay of ionic and π–π stacking interactions for the directed self-assembly of the oppositely charged porphyrins **41** and **42** into nanotubular architectures (in water at a pH value <7, Figure 11.31) [109]. The nanotubes were found to be sensitive toward light: on exposure to light, the structures

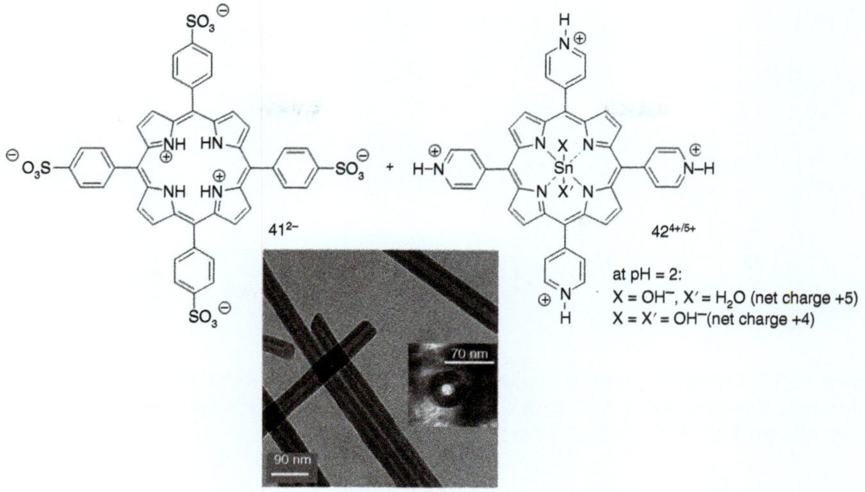

Figure 11.31 Schematic representation of the pair of oppositely charged porphyrins (**41**$^{2-}$ and **42**$^{4+/5+}$); a representative TEM image of the resultant nanotubes is also shown (inset: a tube trapped in a vertical orientation by a thick mat of tubes). Source: Wang et al. [109]. Figure reproduced with kind permission; © 2004 American Chemical Society.

transformed into much smaller and rod-like species; remarkably the nanotubes slowly reformed in the dark. This light-responsive behavior was rationalized by a light-induced electron transfer between the porphyrin moieties, which not only disturbed the charge balance, but also softened the tube walls. Perylene bisimides (PBIs) and related dyes have attracted considerable interest due to their ability to assemble into supramolecular polymers via π–π stacking interactions (see Chapter 5) [2]; this supramolecular polymerization has recently been confirmed to be of living character [111, 112]. The self-assembly of the PBI derivative **43**, equipped with two quaternary ammonium units, and the fourfold negatively charged Cu(II)-phthalocyanine **44** in water was shown by Guan et al. [110]. The supramolecular polymerization of these components gave an 1D structure, which was believed to be a twisted triple helix in which a combination of ionic and π–π stacking interactions accounted for its stability (Figure 11.32).

One further example for the application of a rylene dye, i.e. a naphthalene bisimide (NBI) **41**, in a two-component self-assembly was reported by Schmuck and coworkers (Figure 11.33) [113]. In the first step, the 2 : 1 mixture of **45** and a flexible, tetra-cationic guanidiniocarbonyl pyrrole derivative **46** gave a supramolecular complex, based on ionic interactions. Driven by π–π stacking interactions, this 1 : 2 species assembled further into a supramolecular 3D network that was investigated by DLS, NMR, fluorescence titration experiments, viscosity measurements as well as AFM and TEM imaging. At high concentrations (i.e. $c \geq 70$ mM), the polymeric network transformed into a gel. This sol–gel-phase transition could be reversed by applying external stimuli: either heating/cooling (transition temperature of 56 °C) or changing the pH value allowed it to modulate the material's state in solution.

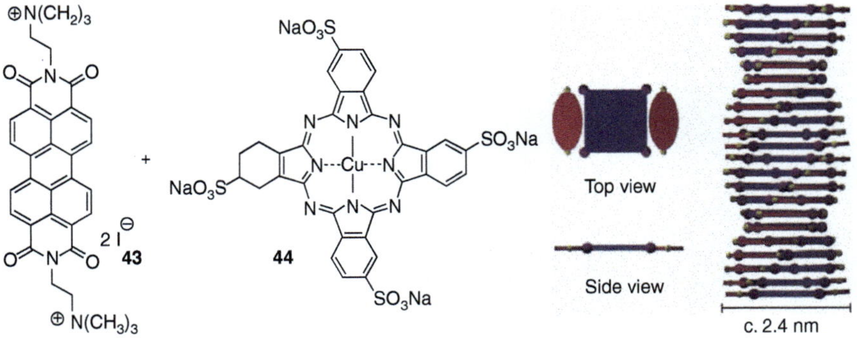

Figure 11.32 Schematic representation of the oppositely charged PBI (**43**) and phthalocyanine dyes (**44**), which gave a supramolecular helically stacked superstructure. Source: Guan et al. [110]. © 2005 John Wiley and Sons.

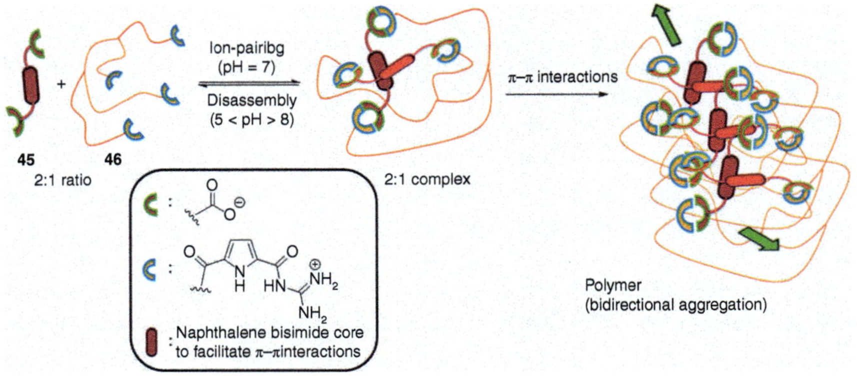

Figure 11.33 Schematic representation of the proposed model of the two-step, self-assembly of the carboxy-functionalized NBI **45** and the tetra-cationic guanidiniocarbonyl pyrrole **46**. Source: Samanta et al. [113]. © 2005 John Wiley and Sons.

Beyond the combination of cucurbiturils supramolecular chemistry with metal-to-ligand complexation (see Section 11.2), also various other dynamic/reversible types of bonding have been investigated: thiol-disulfide [115], diselenides [116], and cyclodextrin host–guest complexation [114]. Concerning the latter, Song et al. utilized the difunctional monomer **47** featuring a diazobenzene moiety and a short peptide residue, as the termini. In the presence of CB[8], dimerization of the peptide occurred giving rise to a heterotrimeric 2 : 1 complex. When treated with the β-CD dimer **48**, a supramolecular polymer was formed instantaneously (Figure 11.34a). According to asymmetric flow field flow fractionation (AF4), as a versatile technique to investigate supramolecular polymers [3, 30], a molar mass of 5.4×10^4 g mol^{-1} and a dispersity (Đ) of c. 1.4 was reached (Figure 11.34b). This particular analytical method is explained further in Chapter 12.

Finally, the self-sorting supramolecular polymerization of three different heterodifunctional components, such as **49–51**, in Haino's study, has to be mentioned. This approach, mimicking the formation of natural biopolymers, such as DNA or

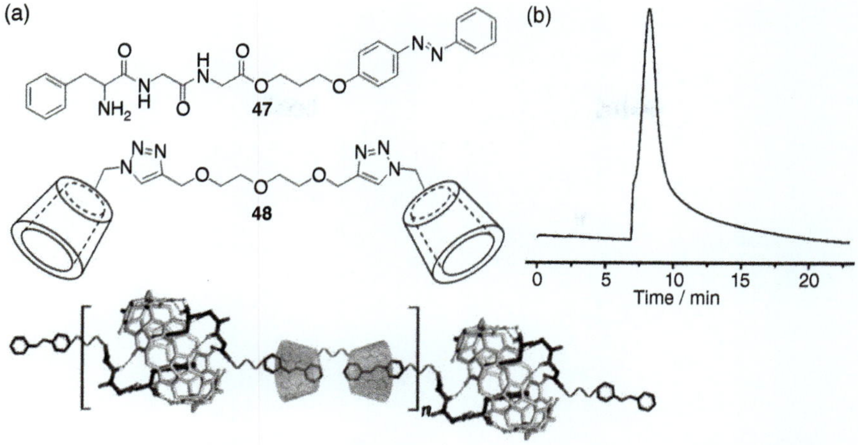

Figure 11.34 Schematic representation of the functionalized monomers **47** and **48**. (a) Schematic representation of the supramolecular polymer formed by **47**, CB[8], and **48** in a directional two-step assembly approach. (b) The AF4 elution curve of the supramolecular polymer (water, as eluent, UV detector). Source: Song et al. [114]. © 2014 The Royal Chemical Society.

proteins, allowed one to assemble sequence-controlled supramolecular copolymers, based on three orthogonal non-covalent interactions [117]. When mixed in a 1 : 1 : 1 ratio, the three monomers formed a regular ABC-type supramolecular copolymer due to a remarkable self-sorting process (DP of c. 200 at a concentration of 10 mM in 1,2-dichloroethane, Figure 11.35). The high K_a values of the respective interactions represented the major driving force for this efficient process ($K_a = 3.1 \times 10^4$ M^{-1} for **49/51**, $K_a = 7.3 \times 10^5$ M^{-1} for **49/50** and $K_a = 1.5 \times 10^4$ M^{-1} for **50/51**, respectively, in 1,2-dichloroethane at 298 K).

It was shown by Ghosh and coworker. that NBIs can form alternate donor–acceptor stacks with perylene derivatives (see also Section 5.3) [119]; high K_a values were found for this type of π–π stacking in a wide range of solvents. Shi and coworkers. combined this charge–transfer (CT) interaction with the host–guest chemistry of pillar[5]arenes to assemble a linear supramolecular polymer, which was characterized by 1D/2D NMR spectroscopy and viscosity measurements (Figure 11.36) [118]. Moreover, rod-shaped fibers were fabricated from a highly concentrated solution of the monomers, thus corroborating the existence of extended supramolecular polymer chains.

11.5 Biomimetic Orthogonal Self-Assembly: Protein Recognition

A fundamentally different approach is the usage of biological building blocks in the preparation of artificial supramolecular assemblies. In principle, binding of such "natural" systems, such as enzymes, relies on defined combinations of typical

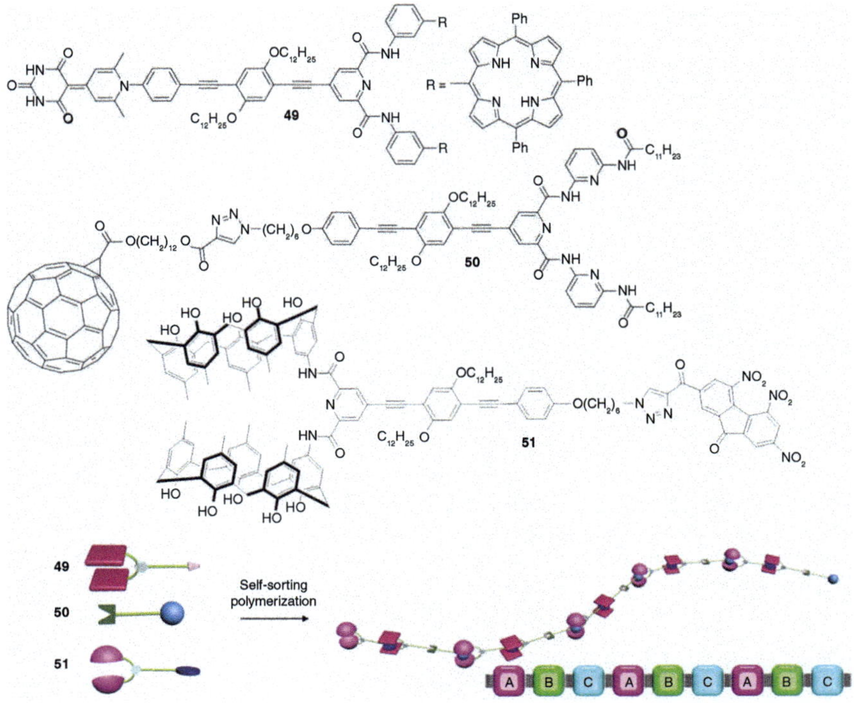

Figure 11.35 Schematic representation of the heteroditopic monomers **49–51** used for the sequence-controlled supramolecular copolymerization. Source: Hirao et al. [117]. © 2017 Springer Nature.

non-covalent interactions: metal-to-ligand-coordination as well as H-bonding, ionic, and π–π stacking interactions. The main advantages of introducing such entities into supramolecular polymers are their biocompatibility and biodegradability; furthermore, the ability to introduce biometric functions, which are orthogonal to those offered by the "traditional" non-covalent interactions is of relevance in this context [120, 121]. The Meijer group studied the self-assembly of a heteroditopic oligo(ethylene glycol) derivative; the interaction between the two functional groups, i.e. a ribonuclease *S*-peptide and *S*-protein, gave an enzyme-bridged supramolecular polymer in which the heme–protein recognition was utilized [122]. Moreover, it was shown that also such a supramolecular polymerization can be described by the principle models, as outlined in Chapter 1 (a ring-chain-mediated mechanism in this particular case).

Hayashi and coworkers externally modified myoglobin (Mb), an O_2-storage hemoprotein, by binding a heme moiety to the Cys^{125} residue on the protein's surface [23]. The external heme and protein matrix underwent a number of interprotein reactions to eventually form a fibrous supramolecular assembly (thereby, the naturally protein-bound heme was released throughout the process). The supramolecular polymer featured a remarkable chemical stability and could be converted into an enormous protein clusters by covalent cross-linking.

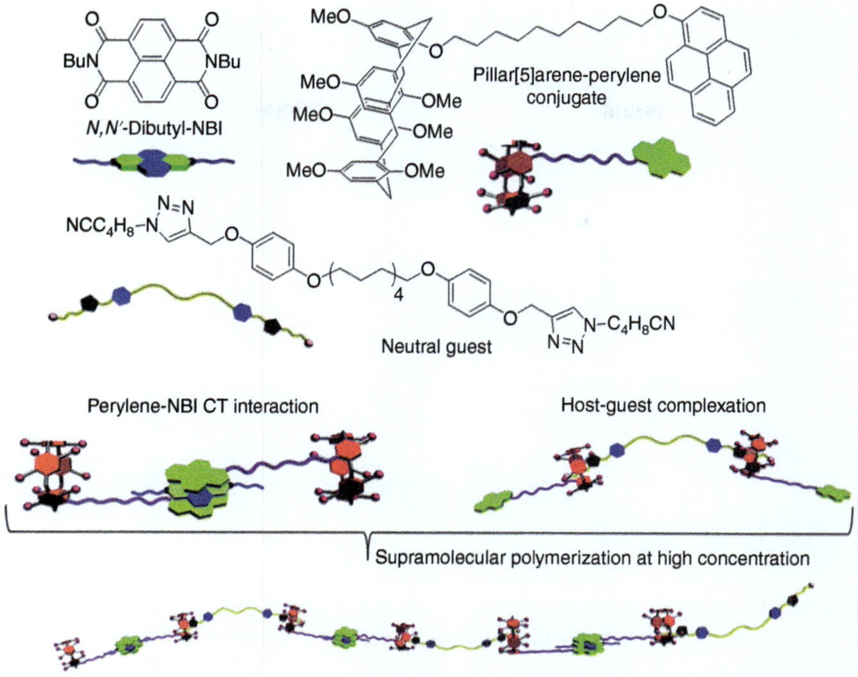

Figure 11.36 Schematic representation of the formation of a supramolecular polymer due to CT interactions and pillar[5]arene-based host–guest complexation. Source: Chen et al. [118]. © 2016 The Royal Chemical Society.

Remarkably, the nano-biomaterial even retained the oxygen-storage function of native myoglobin. The same group also developed a strategy to assemble a supramolecular polymer based on two "natural" orthogonal binding modes (Figure 11.37a) [24]: the heme-bis(biotin) conjugate **52** was bound selectively to the free sites of a myoglobin-dimer via its heme moiety; subsequently supramolecular polymerization was achieved by binding the biotin residues to streptavidin (SAv, the presence of four potential binding sites per SAv was confirmed by a biotin-4-fluorescein assay). AFM imagining was applied to further characterize the supramolecular material (Figure 11.37b). Flexible [Mb-**52**-SAv-**52**-Mb]$_n$ fibers were observed; the heights of the objects ranged from 3 to 5 nm, thus fitting to the expected sizes of SAv (c. 5 nm) and Mb (c. 3–4 nm). Due to the high kinetic stability of the biomimetic non-covalent interactions, the 1D protein copolymer could even be investigated by SEC (Figure 11.37c). In agreement with a polyaddition-like mechanism, a broad molar mass distribution was detected. Upon reduction of the disulfide bridges of the Mb-dimers, the SEC signal was shifted to faster elution times; the narrow distribution was assigned to the monodisperse [Mb-**52**-SAv-**52**-Mb] fragment.

These same authors also prepared a version of the cytochrome b_{562} H63C mutant (Cyt^{H63C}) to which a heme moiety was attached via a flexible linker. The conjugates **53** gave, at a pH value of 7.3, supramolecular polymers due to intramolecular

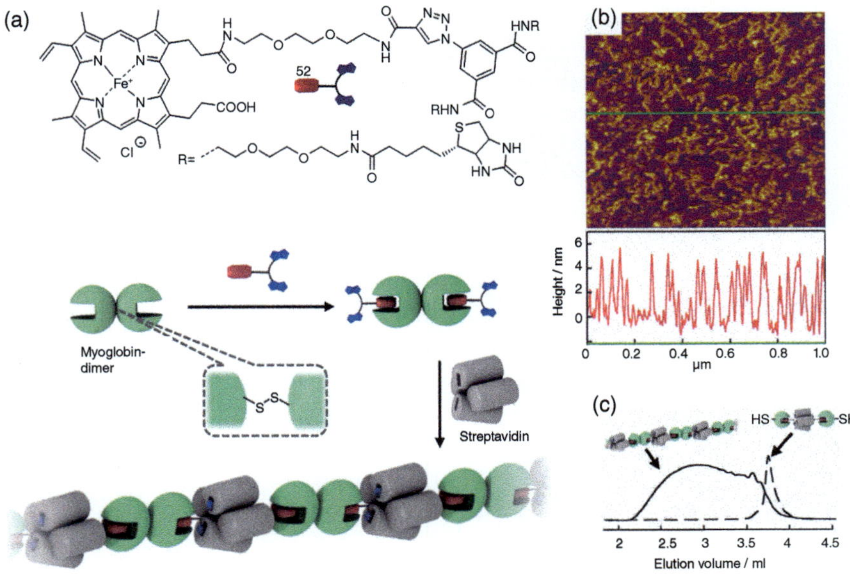

Figure 11.37 (a) Schematic representation of the self-assembly of the a myoglobin-dimer and streptavidin with heme-bis(biotin) conjugate **52**; (b) AFM height image of the supramolecular polymer on a modified mica substrate (100 mM K_3PO_4 buffer, pH value of 7); (c) SEC traces (UV/vis absorption detector) of the supramolecular polymer (solid line) and the fragment obtained after cleavage of the disulfide bridges (dashed line). Source: Oohara et al. [24]. Figure reproduced with kind permission; © 2012 Wiley-VCH.

heme–Cyt^{H36C} interactions (Figure 11.38a) [123]. In the case of short spacers, the assemblies were stabilized by intraprotein H-bonding interactions. The size of the assemblies was estimated by AFM imaging; according to these measurements, the fibers had a length of c. 350 nm, corresponding to the >100-mer protein. Remarkably, SEC analysis of the supramolecular polymers could be performed (proteins covering a wide molar mass range were used, as references, Figure 11.38b). More recently, rigid spacer units were also employed for the heme–Cyt^{H36C} conjugation [124]. The conjugate containing a diazo moiety was temperature responsive, i.e. the thermodynamically stable linear assembly could be converted by heating to 65 °C into a kinetically trapped micellar species for which a diameter in the range of c. 15 nm was determined by DLS measurements (Figure 11.39). This change in the supramolecular structure was attributed to a denaturation of the protein, which was reversed within 96 hours at 25 °C (thus, the initial fiber-like architecture was restored). The coassembly of a heme–protein and a Zn(II)-porphyrin–protein conjugate gave a supramolecular polymer, similar to $(\mathbf{53})_n$, possessing a random distribution of Zn(II)- and Fe(II)-containing porphyrins [125]. These systems showed a single-step, electron-transfer process in which an electron was transferred from the photoexcited Zn(II) sites to the Fe(II) ones (rate constants for the charge-separation and charge-recombination step were determined as 320–600 s^{-1} and 580–930 s^{-1}, respectively).

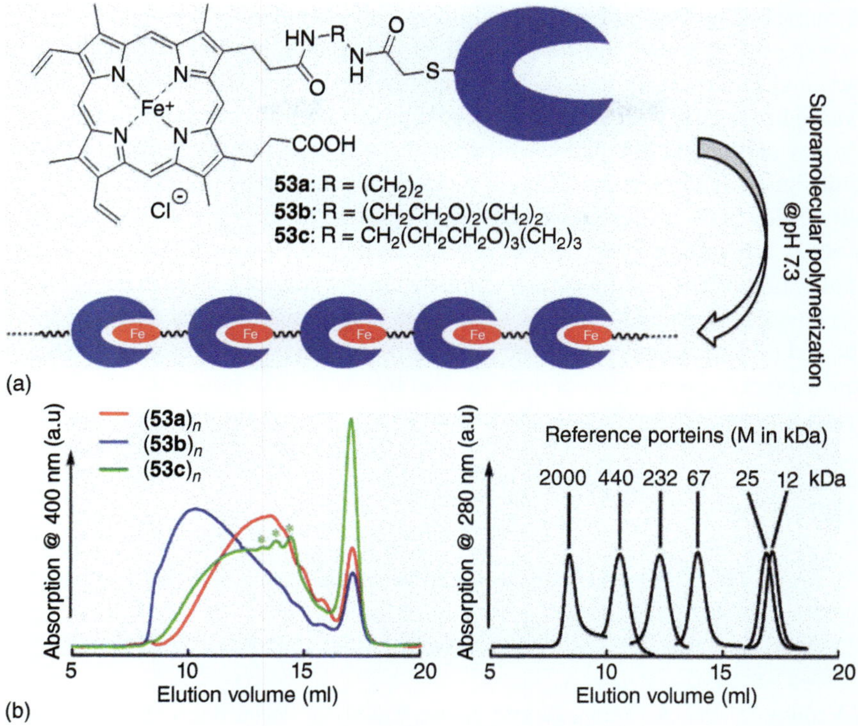

Figure 11.38 (a) Schematic representation of the formation of the supramolecular polymers $(\mathbf{53})_n$, based on protein recognition. (b) SEC analysis of $(\mathbf{53})_n$ using reference proteins, as standards. Source: Kitagishi et al. [123]. © 2007 American Chemical Society.

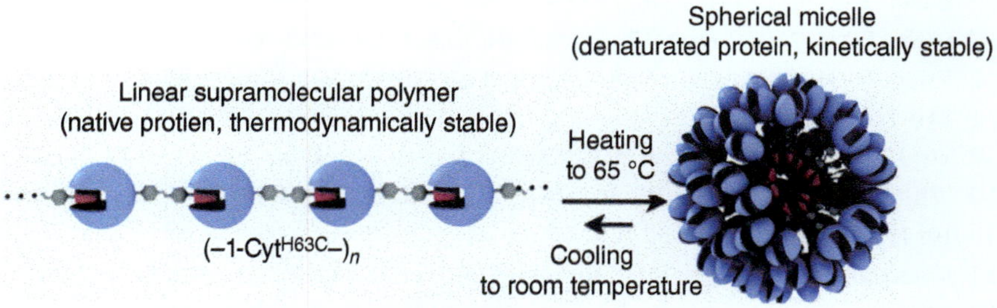

Figure 11.39 Schematic representation of the reversible switching between a thermodynamically and a kinetically stable architecture (linear polymer chain vs. spherical micelle). Source: Oohora et al. [124]. © 2017 The Royal Chemical Society.

11.6 Concluding Remarks

Supramolecular materials constructed from different, in particular orthogonal, non-covalent interactions are gaining increasing interest due to their intrinsic reversibility of binding. These nature-inspired architectures offer a wide range of properties that cannot (or only hardly) be realized with conventional "organic" polymers: self-healing and multistimuli-responsive behaviors have to be named in this respect. The proper choices of (small-molecule) building blocks equipped with functional groups being able to form supramolecular bonds allow one to design advanced materials with such macroscopic properties and precision. Thereby, the interplay of different interactions determines the architecture formed, the different architectures determine the different macroscopic properties [17]. Using orthogonality in supra(macro)molecular chemistry is still a rather young field of research; however, remarkable progress has already been made – as highlighted in this chapter. Recently, X-bonding and biomimetic interactions (e.g. protein recognition) have emerged as an add-on to the established non-covalent interactions; these are expected to significantly widen the scope in supra(macro)molecular chemistry. A more balanced combination of non-covalent interactions is, however, still required to realize materials that do not only impact science but eventually daily's life.

References

1 Brunsveld, L., Folmer, B.J.B., Meijer, E.W., and Sijbesma, R.P. (2001). *Chem. Rev.* 101: 4071–4098.
2 de Greef, T.F.A., Smulders, M.M.J., Wolffs, M. et al. (2009). *Chem. Rev.* 109: 5687–5754.
3 Yang, L., Tan, X., Wang, Z., and Zhang, X. (2015). *Chem. Rev.* 115: 7196–7239.
4 Aida, T., Meijer, E.W., and Stupp, S.I. (2012). *Science* 335: 813–817.
5 Cordier, P., Tournilhac, F., Soulié-Ziakovic, C., and Leibler, L. (2008). *Nature* 451: 977–980.
6 Hoeben, F.J.M., Jonkheijm, P., Meijer, E.W., and Schnenning, A.P.H.J. (2005). *Chem. Rev.* 105: 1491–1546.
7 Lehn, J.-M. (2002). *Proc. Natl. Acad. Sci. U.S.A.* 99: 4763–4768.
8 Krieg, E., Weissman, H., Shirman, E. et al. (2011). *Nat. Nanotechnol.* 6: 141–146.
9 Yu, L.-T., Wang, Z., Wu, J. et al. (2010). *Angew. Chem. Int. Ed.* 49: 3627–3630.
10 Whittle, G.R., Hager, M.D., Schubert, U.S., and Manners, I. (2011). *Nat. Mater.* 10: 176–188.
11 Korevaar, P.A., George, S.J., Markvoort, A.J. et al. (2012). *Nature* 481: 492–496.
12 Ciferri, A. (2005). *Supramolecular Polymers*. New York, NY: Taylor & Francis.
13 Winter, A., Hager, M.D., and Schubert, U.S. (2012). Supramolecular polymers. In: *Polymer Science: A Comprehensive Review*, vol. 5 (eds. H.-W. Schmidt and M. Ueda), 269–310. Amsterdam: Elsevier BV.
14 Zhang, X. and Wang, C. (2011). *Chem. Soc. Rev.* 40: 94–101.
15 Lehn, J.-M. (1988). *Angew. Chem. Int. Ed. Engl.* 27: 89–112.
16 Pollino, J.M., Stubbs, L.P., and Weck, M. (2004). *J. Am. Chem. Soc.* 126: 563–567.
17 Li, S.-L., Xiao, T.-X., Lin, C., and Wang, L.-Y. (2012). *Chem. Soc. Rev.* 41: 5950–5968.
18 Wong, C.-H. and Zimmerman, S.C. (2013). *Chem. Commun.* 49: 1679–1695.

19 Hofmeier, H. and Schubert, U.S. (2005). *Chem. Commun.*: 2423–2432.

20 Hoogenboom, R., Fournier, D., and Schubert, U.S. (2008). *Chem. Commun.*: 155–162.

21 Xiao, T., Zhou, L., Sun, X.-Q. et al. (2020). *Chin. Chem. Lett.* 31: 1–9.

22 Stupp, S.I., Clemons, T.D., Carrow, J.K. et al. (2020). *Isr. J. Chem.* 60: 124–131.

23 Oohara, K., Onoda, A., Kitagishi, H. et al. (2011). *Chem. Sci.* 2: 1033–1038.

24 Oohara, K., Burazerovic, S., Onoda, A. et al. (2012). *Angew. Chem. Int. Ed.* 51: 3818–3821.

25 Chiper, M., Hoogenboom, R., and Schubert, U.S. (2009). *Macromol. Rapid Commun.* 30: 565–578.

26 Wild, A., Winter, A., Schlütter, F., and Schubert, U.S. (2011). *Chem. Soc. Rev.* 40: 1459–1511.

27 Shunmugam, R., Gabriel, G.J., Aamer, K.A., and Tew, G.N. (2010). *Macromol. Rapid Commun.* 31: 784–793.

28 Dobrawa, R. and Würthner, F. (2005). *J. Polym. Sci., Part A: Polym. Chem.* 43: 4981–4995.

29 Lindoy, L.F., Park, K.-M., and Lee, S.S. (2013). *Chem. Soc. Rev.* 42: 1713–1727.

30 Winter, A. and Schubert, U.S. (2016). *Chem. Soc. Rev.* 45: 5311–5357.

31 De, S., Mahata, K., and Schmittel, M. (2010). *Chem. Soc. Rev.* 39: 1555–1575.

32 Chakraborty, S. and Newkome, G.R. (2018). *Chem. Soc. Rev.* 47: 3991–4016.

33 Zhou, G.-J., He, Y., Yao, B. et al. (2010). *Chem. Asian J.* 5: 2405–2414.

34 Wild, A.W., Hager, M.D., Görls, H., and Schubert, U.S. (2012). *Macromol. Rapid Commun.* 33: 517–521.

35 Stulz, E., Ng, Y.-F., Scott, S.M., and Sanders, J.K.M. (2002). *Chem. Commun.*: 524–525.

36 Erre, G., Enthaler, S., Junge, K. et al. (2008). *Coord. Chem. Rev.* 252: 471–491.

37 Hu, C.-W., Sato, T., Zhang, J. et al. (2013). *J. Mater. Chem. C* 1: 3408–3413.

38 Higuchi, M. (2009). *Polym. J.* 41: 511–520.

39 Hofmeier, H., Hoogenboom, R., Wouters, M.E.L., and Schubert, U.S. (2005). *J. Am. Chem. Soc.* 127: 2913–2921.

40 Hofmeier, H., El-ghayoury, A., Schenning, P.H.J., and Schubert, U.S. (2004). *Chem. Commun.*: 318–319.

41 Binder, W.H. and Zirbs, R. (2007). *Adv. Polym. Sci.* 207: 1–78.

42 Mansfeld, U., Hager, M.D., Hoogenboom, R. et al. (2009). *Chem. Commun.* 23: 3386–3388.

43 Mansfeld, U., Winter, A., Hager, M.D. et al. (2013). *Polym. Chem.* 4: 113–123.

44 Shi, L., Wang, X.-W., Sandoval, C.A. et al. (2006). *Angew. Chem. Int. Ed.* 45: 4108–4112.

45 Grimm, F., Ulm, N., Gröhn, F. et al. (2011). *Chem. Eur. J.* 17: 9478–9488.

46 Chang, S.-K. and Hamilton, A.D. (1988). *J. Am. Chem. Soc.* 110: 1318–1319.

47 Grimm, F., Hartnagel, K., Wessendorf, F., and Hirsch, A. (2009). *Chem. Commun.*: 1331–1333.

48 Mansfeld, U., Winter, A., Hager, M.D. et al. (2013). *Polym. Chem.* 4: 3177–3181.

49 Ambade, A.V., Yang, S.-K., and Weck, M. (2009). *Angew. Chem. Int. Ed.* 48: 2894–2898.

50 Chen, S., Wang, K., Geng, Z. et al. (2019). *Polym. Chem.* 10: 4740–4745.
51 Ikeda, M., Tanaka, Y., Hasegawa, T. et al. (2006). *J. Am. Chem. Soc.* 128: 6806–6807.
52 Gröger, G., Meyer-Zaika, W., Böttcher, C. et al. (2011). *J. Am. Chem. Soc.* 133: 8961–8971.
53 Gröger, G., Stepanenko, V., Würthner, F., and Schmuck, C. (2009). *Chem. Commun.*: 698–700.
54 Zhu, L.-L., Lu, M.-Q., Zhang, Q.-W., and Tian, H. (2011). *Macromolecules* 44: 4092–4097.
55 Del Valle, E.M.M. (2004). *Process Biochem.* 39: 1033–1046.
56 Harada, A., Takashima, Y., and Yamaguchi, H. (2009). *Chem. Soc. Rev.* 38: 875–882.
57 Li, J. and Loh, X.-J. (2008). *Adv. Drug Delivery Rev.* 60: 1000–1017.
58 Gurarslan, A., Joijode, A., Shen, J. et al. (2017). *Polymers* 9: 673.
59 Yan, X.-H., Xu, D.-H., Chi, X.-D. et al. (2012). *Adv. Mater.* 24: 362–369.
60 Lu, C., Zhang, M., Tang, D. et al. (2018). *J. Am. Chem. Soc.* 140: 7674–7680.
61 Crowley, J.D. and McMorran, D.A. (2012). *Top. Heterocycl. Chem.* 28: 31–83.
62 Whang, D., Jeon, Y.-M., Heo, J., and Kim, K. (1996). *J. Am. Chem. Soc.* 118: 11333–11334.
63 Park, K.-M., Lee, E., Roh, S.-G., and Kim, K. (2004). *Bull. Korean Chem. Soc.* 25: 1711–1713.
64 Wei, P., Yan, X., and Huang, F. (2015). *Chem. Soc. Rev.* 44: 815–832.
65 Barrow, S.J., Kasera, S., Rowland, M.J. et al. (2015). *Chem. Rev.* 115: 12320–12406.
66 Kim, K. (2002). *Chem. Soc. Rev.* 31: 96–107.
67 Whang, D. and Kim, K. (1997). *J. Am. Chem. Soc.* 119: 451–452.
68 Wu, X., Yu, Y., Gao, L. et al. (2016). *Org. Chem. Front.* 3: 966–970.
69 Liu, Y., Shangguan, L., Wang, H. et al. (2017). *Polym. Chem.* 8: 3783–3787.
70 Liu, Y., Huang, Z., Tan, X. et al. (2013). *Chem. Commun.* 49: 5766–5768.
71 Joseph, R., Nkrumah, A., Clark, R.J., and Masson, E. (2014). *J. Am. Chem. Soc.* 136: 6602–6607.
72 Patrick, C.R. and Prosser, G.S. (1960). *Nature* 187: 1021.
73 Yang, S.-K., Ambade, A.V., and Weck, M. (2010). *J. Am. Chem. Soc.* 132: 1637–1645.
74 Burd, C. and Weck, M. (2005). *Macromoelcules* 38: 7225–7230.
75 Nieuwenhuizen, M.M.L., de Greef, T.F.A., van der Bruggen, R.L.J. et al. (2010). *Chem. Eur. J.* 16: 1061–1612.
76 Appel, W.P.J., Portale, G., Wisse, E. et al. (2011). *Macromolecules* 44: 6776–6784.
77 Kautz, H., van Beek, D.J.M., Sijbesma, R.P., and Meijer, E.W. (2006). *Macromolecules* 39: 4265–4267.
78 Botterhuis, N.E., van Beek, D.J.M., van Gemert, G.M.L. et al. (2008). *J. Polym. Sci., Part A: Polym. Chem.* 46: 3877–3885.
79 Mes, T., Koenigs, M.M.E., Scalfani, V.F. et al. (2011). *ACS Macro Lett.* 1: 105–109.

80 Stals, P.M.J., Smulders, M.M.J., Martín-Rapún, R. et al. (2009). *Chem. Eur. J.* 15: 2071–2080.

81 Shao, X.-B., Jiang, X.-K., Zhao, X. et al. (2004). *J. Org. Chem.* 69: 899–907.

82 Tokunage, Y. and Seo, T. (2002). *Chem. Commun.*: 970–971.

83 Wang, X.-Z., Li, X.-Q., Shao, X.-B. et al. (2003). *Chem. Eur. J.* 9: 2904–2913.

84 Xiao, T.-X., Li, S.-L., Zhang, Y.-J. et al. (2012). *Chem. Sci.* 3: 1417–1421.

85 Li, S.-L., Xiao, T.-X., Wu, Y.-F. et al. (2011). *Chem. Commun.* 47: 6903–6905.

86 Gibson, H.W., Yamaguchi, N., and Jones, J.W. (2003). *J. Am. Chem. Soc.* 125: 3522–3533.

87 Li, S.-L., Xiao, T.-X., Hu, B.-J. et al. (2011). *Chem. Commun.* 47: 10755–10757.

88 van Hameren, R., van Buul, A.M., Castriciano, M.A. et al. (2008). *Nano Lett.* 8: 253–259.

89 van Hameren, R., Schön, P., van Buul, A.M. et al. (2006). *Science* 314: 1433–1436.

90 Ajayaghosh, A., Varghese, R., Praveen, V.K., and Mahesh, S. (2006). *Angew. Chem. Int. Ed.* 45: 3261–3264.

91 Nair, K.P. and Weck, M. (2007). *Macromolecules* 40: 211–219.

92 Fitié, C.F., Tomatsu, I., Byelov, D. et al. (2008). *Chem. Mater.* 20: 2394–2404.

93 Chen, S.-S., Tian, L., and Zhao, D.-H. (2010). *Acta Phys. Chim. Sin.* 26: 1124–1134.

94 Jamadar, A. and Das, A. (2020). *Polym. Chem.* 11: 385–392.

95 Zapata, F., González, L., Caballero, A. et al. (2018). *J. Am. Chem. Soc.* 140: 2041–2045.

96 Teng, P., Gray, G.M., Zheng, M. et al. (2019). *Angew. Chem. Int. Ed.* 58: 7778–7782.

97 Pizzi, A., Pigliacelli, C., Bergamaschi, G. et al. (2020). *Coord. Chem. Rev.* 411: 213242.

98 Wang, F., Han, C.-Y., He, Q.-Z. et al. (2008). *J. Am. Chem. Soc.* 130: 11254–11255.

99 Pedersen, C.J. (1967). *J. Am. Chem. Soc.* 89: 7017–7036.

100 Zheng, B., Wang, F., Dong, S.-Y., and Huang, F.-H. (2012). *Chem. Soc. Rev.* 41: 1621–1636.

101 Ashton, P.R., Philp, D., Reddington, M.V. et al. (1991). *J. Chem. Soc., Chem. Commun.*: 1680–1683.

102 Ashton, P.R., Chrystal, E.J.T., Glink, P.T. et al. (1996). *Chem. Eur. J.* 2: 709–728.

103 Wang, F., Zheng, B., Zhu, K.-L. et al. (2009). *Chem. Commun.*: 4375–4377.

104 Gröhn, F., Klein, K., and Brand, S. (2008). *Chem. Eur. J.* 14: 6886–6869.

105 Schmuck, C. (2001). *Tetrahedron* 57: 3063–3067.

106 Faul, C.F.J. and Antonietti, M. (2003). *Adv. Mater.* 15: 673–683.

107 Willerich, I. and Gröhn, F. (2008). *Chem. Eur. J.* 14: 9112–9116.

108 Gröhn, F., Klein, K., and Koynov, K. (2010). *Macromol. Rapid Commun.* 31: 75–80.

109 Wang, Z.-C., Medforth, C.J., and Shelnutt, J.A. (2004). *J. Am. Chem. Soc.* 126: 15954–15955.

110 Guan, Y., Yu, S.-H., Antonietti, M. et al. (2005). *Chem. Eur. J.* 11: 1305–1311.

111 Ogi, S., Stepanenko, V., Sugiyasu, K. et al. (2015). *J. Am. Chem. Soc.* 137: 3300–3307.
112 Wagner, W., Wehner, M., Stepanenko, V. et al. (2017). *Angew. Chem. Int. Ed.* 56: 16008–16012.
113 Samanta, K., Ehlers, M., and Schmuck, C. (2016). *Chem. Eur. J.* 22: 15242–15247.
114 Song, Q., Li, F., Tan, X. et al. (2014). *Polym. Chem.* 5: 5895–5899.
115 Yang, H., Bai, Y., Yu, B. et al. (2014). *Polym. Chem.* 5: 6439–6443.
116 Tan, X., Yang, L., Huang, Z. et al. (2015). *Polym. Chem.* 6: 681–685.
117 Hirao, T., Kudo, H., Amimoto, T., and Haino, T. (2017). *Nat. Commun.* 8: 634.
118 Chen, P., Mondal, J.H., Zhou, Y. et al. (2016). *Polym. Chem.* 7: 5221–5225.
119 Das, A. and Ghosh, S. (2014). *Angew. Chem. Int. Ed.* 53: 1092–1097.
120 Ringler, P. and Schulz, G.E. (2003). *Science* 302: 106–109.
121 Reynhout, I.C., Cornelissen, J.J.L.M., and Nolte, R.J.M. (2007). *J. Am. Chem. Soc.* 129: 2327–2332.
122 Bastings, M.M.C., de Greef, T.F.A., van Dongen, J.L.J. et al. (2010). *Chem. Sci.* 1: 79–88.
123 Kitagishi, H., Oohora, K., Yamaguchi, H. et al. (2007). *J. Am. Chem. Soc.* 129: 10326–10327.
124 Oohora, K., Onuma, Y., Tanaka, Y. et al. (2017). *Chem. Commun.* 53: 6879–6882.
125 Kajihara, R., Oohora, K., and Hayashi, T. (2019). *J. Inorg. Biochem.* 193: 42–51.

12

Characterization of Supramolecular Polymers*

12.1 Introduction

In general, supramolecular assemblies – polymers as well as small-molecule assemblies – are linked via (directed) non-covalent interactions [1–9]. The non-covalence in bonding gives rise to novel properties – beyond and, at times, even complementary to those associated with "conventional" systems [10]. For instance, small changes in temperature can result in significant viscosity changes (i.e. thermal responsiveness); thus, supramolecular polymers are often much easier to be processed then "traditional" polymers. The reversibility of the non-covalent interactions is a key feature that makes them potential candidates, e.g. for recyclable and self-healing materials [11–13].

Concerning these supramolecular polymers – based, in particular, on H-bonding and host–guest interactions or metal-to-ligand coordination – two main aspects are in the focus of current research: understanding the mechanisms for self-assembly [3] and bringing supramolecular materials "into action" [i.e. development of potential (new or improved) applications] [14]. For both, a reliable characterization of the supramolecular material is of utmost importance. The molar mass of supramolecular polymers, e.g. represents important information – only at a high degrees-of-polymerizations (DPs) can polymer-like properties be identified.

However, the dynamic nature of the non-covalent bonding, being responsible for the unique properties of supramolecular polymers, is an intrinsic limitation concerning the characterization of these materials. The self-assembly process is strongly dependent on parameters, such as: the solvent, concentration, temperature, composition, etc. Even a slight stimulus from the environment might lead to drastic changes in the ultimate supramolecular architecture. To overcome this restriction, attempts have been made to capture supramolecular assemblies via postchemical stabilization [15–17]. However, this strategy is certainly not generally applicable, and characterization remains a major challenge: many well-established analytical tools that can be applied to covalent (macro)molecules do not work for non-covalent systems

* Parts of this chapter are reproduced from *Chem. Soc. Rev.* **2016**, *45*, 5311–5357 (DOI: 10.1039/c6cs00182c) by permission of The Royal Chemical Society.

Supramolecular Polymers and Assemblies: From Synthesis to Properties and Applications, First Edition.
Ulrich S. Schubert, George R. Newkome, and Andreas Winter.

(or only to a limited extent). However, a certain analytical method will only provide some information on a supramolecular assembly. As a result, combinations of (ideally independent) characterization techniques are often required to comprehensively investigate the molecular structure/composition of supramolecular architectures. Some spectroscopic methods, e.g. vibration spectroscopy (i.e. IR or Raman), are rather unspecific and will only be applied for basic structural characterization.

12.2 Estimation of the Molar Mass from the Theories of Supramolecular Polymer Science

The basic concepts for supramolecular polymerization have already been detailed in Chapter 1 [3, 18]. Essentially, the self-assembly process can be described as a thermodynamic equilibrium. Considering the simplest case, i.e. the isodesmic association (see Section 1.3.1), the DP can be estimated from Eq. (12.1) [2]. As shown in Figure 12.1, high DP values are obtained either for high binding constants (K_a) or high monomer concentrations (c). For instance, a DP of c. 100 will be reached at a concentration of c. 0.05 M and a K_a value of 4.5×10^5 M^{-1} [19].

$$DP = (K_a \cdot c)^{1/2} \tag{12.1}$$

DP: degree-of-polymerization, K_a: association constant, c: total monomer concentration.

Measurement of the K_a value represents an important issue when studying supramolecular assemblies in solution [20]. The strength of non-covalent interactions in supramolecular systems can be quantified by titrating a solution of a guest to that of a host, thereby monitoring the changes of a certain physical property/measure. In particular, spectroscopic (e.g. NMR, UV/vis absorption or emission) or calorimetric techniques can be applied for this purpose. Though such titration experiments are, at least on first glance, straightforward to perform, some crucial aspects still need to be considered to obtain reliable results. According to Thordarson, these are as follows [20]: first of all, utilizing non-linear regression methods, instead of linear ones, will enable a more accurate quantification of the binding parameters (thereby, using advanced data analysis methods, such as global analysis, is important); second, the stoichiometry of the particular interaction needs to be known (molecular-modeling studies can help to establish these). The spectroscopic (or calorimetric) method must be carefully selected, and the concentrations of the host and guest have to be adjusted. Finally, the uncertainties and confidence intervals for the obtained results need to be considered. Briefly, the scope and limitations of the most important titration techniques will be evaluated. It has to be emphasized that all common titration methods, in particular the spectroscopic ones, are sensitive to dilution and temperature effects as well as to the presence of impurities in either of the used solutions.

^{1}H NMR (and, to a minor extent, ^{13}C NMR or ^{19}F NMR) spectroscopy can not only quantify host–guest interactions but also offer versatile information on the mode and stoichiometry of the interaction. Modern NMR spectrometers allow

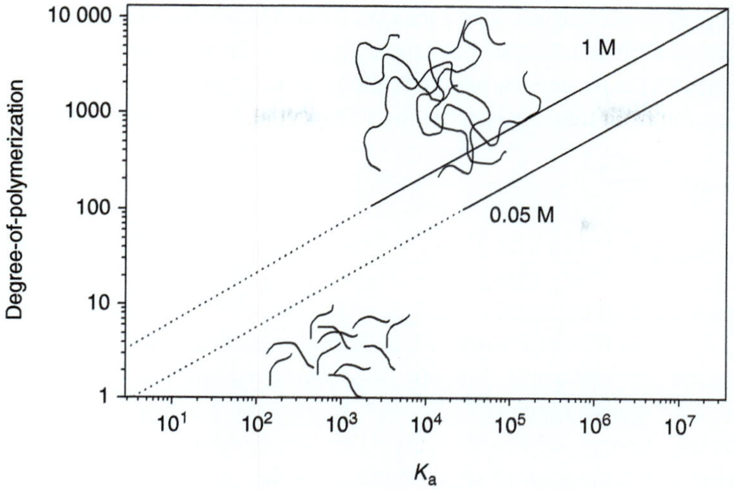

Figure 12.1 Theoretical plot of the degree-of-polymerization (DP) vs. association constant (K_a in M^{-1}) for a supramolecular polymerization (two different concentrations, isodesmic self-assembly model). Source: Brunsveld et al. [2]. © 2001 American Chemical Society.

for the measurement of high-quality spectra even in the sub-mM concentration regime (routinely as low as 10^{-4} M); thus, NMR spectroscopy should be applicable for supramolecular systems with K_a values in the range of $10^6\,M^{-1}$ (however, the commonly stated limit for NMR titration experiments is still at a K_a value of $10^5\,M^{-1}$) [21]. Moreover, the relative exchange rate within the supramolecular system, i.e. the relationship between K_a and the kinetic rates [Eq. (12.2)] as well as the timescale of the NMR experiment must also be considered.

$$K_a = \frac{k_1}{k_{-1}} = \frac{[HG]}{[H] \cdot [G]} \tag{12.2}$$

k_1 and k_{-1}: forward and backward rate constant, respectively, [H]: concentration of the host, [G]: concentration of the guest, [HG]: concentration of the host–guest complex.

Two different scenarios can be distinguished when dealing with NMR titration experiments, and these, in turn, represent the limiting factors: the system under investigation is in either the fast or slow exchange region. Upon interaction of a host with a guest, the chemical shifts experience certain shifts ($\Delta\delta$). These $\Delta\delta$ values as well as the K_a value, along with the associated ratio of rate constants (i.e. k_1/k_{-1}), represent a function of the free host's mole fraction (f_H) divided by k_{-1} [Eq. (12.3a)]; note, that this is equivalent to the undermentioned sum-of-inverses of the emission lifetimes of host–guest assemblies ($1/\tau_{HG}$) and the free host ($1/\tau_H$). Feeney et al. demonstrated that this rather simple assumption is valid only when the system is in the fast exchange region; otherwise, Eq. (12.3b) becomes relevant [22]. Equation (12.3b) also suggests that it can be of avail to focus only on resonances in titration experiments that are in the fast exchange regime, even if they only experience relatively small $\Delta\delta$ values. On the contrary, resonances featuring large $\Delta\delta$

values might appear broad due to a slow exchange (thus, the accurate integration of the signals becomes difficult). For fast-exchanging systems, the limit for performing reliable NMR titration experiments, hence, remains at a K_a value of $10^5\ M^{-1}$ (one prominent exception from this was reported by Wilcox et al. [23]).

$$\Delta\delta = \delta_{\Delta HG} \cdot \left(\frac{[HG]}{[H]_0}\right) \tag{12.3a}$$

$$2\pi \cdot |\nu_H - \nu_{HG}| \ll \left(\frac{1}{\tau_H} + \frac{1}{\tau_{HG}}\right) \tag{12.3b}$$

$\Delta\delta$: complexation-induced chemical shift, $\delta_{\Delta HG} = \delta_{HG} - \delta_H$; [HG]: concentration of the complex; $[H]_0$: initial host concentration; ν_H and ν_{HG}: resonance frequencies (in Hz) of the host and the host–guest complex, respectively; τ_H and τ_{HG}: lifetimes of the host and the host–guest complex.

Besides NMR spectroscopy, UV/vis absorption titration experiments represent the second most common technique to study supramolecular interactions. Basically, host concentrations as low as 10^{-7} M can be utilized (e.g. for porphyrin derivatives) – this corresponds to K_a values up to $10^9\ M^{-1}$ for simple 1 : 1 adducts (an in-depth evaluation on how concentration and K_a influence the feasibility of such titration experiments is given in Ref. [20]). However, it has to be noted that, for $K_d > [H]_0$ ($K_d = K_a^{-1}$), to obtain reliable values for the probability of binding (i.e. p values in the range of 0.2–0.8), an significant excess of $[G]_0$ is always required (see Figure 12.2) [24]. In this particular case, as many data points as possible should be collected while adding 1–50 eq. of guest molecules. If the $[H]_0/K_d$ quotient is ≥100, it is no longer necessary to accurately know $[H]_0$. As shown in Figure 12.3, in the example of a system with a K_a value of $100\ M^{-1}$, the binding isotherms for initial host concentrations ($[H]_0$) of 10^{-5} and 10^{-6} M are almost identical. Moreover, the concentrations must be chosen properly to allow the relevant UV/vis absorption band(s) to be analyzed according to Lambert–Beer's law [Eq. (12.4)]. To simplify the analysis of the spectra, the added guest should ideally not show any absorption bands in the relevant region where the host or host–guest complex absorbs the light. This is the case, for example, in a titration experiment in which transition

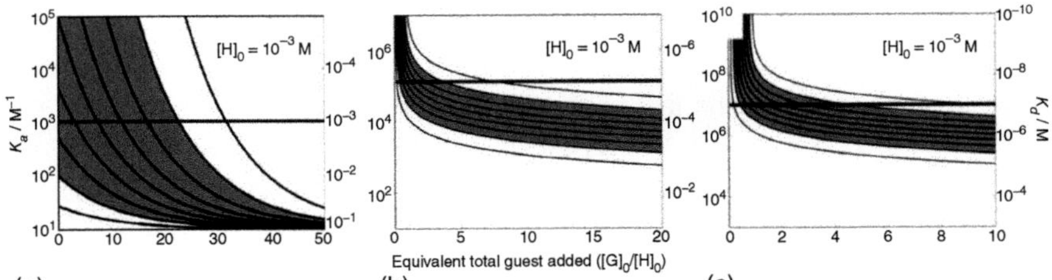

Figure 12.2 Representative plots of the probability of binding (p) according to Weber [24] as a function of the total guest concentration ($[G]_0$) and the K_a value (left axis) as well as the dissociation constant (K_d, right axis) at three different host concentrations [(a)–(c): $[H]_0 = 10^{-3}$ M, 10^{-5} M, and 10^{-7} M]. The thick horizontal bar indicates where $[H]_0$ is equal to K_d and K_a^{-1}. The contour lines represent 0.1 units of p (the p range from 0.2 to 0.8 is shaded gray. Source: Thordarson [20]. © 2011 Royal Society of Chemistry.

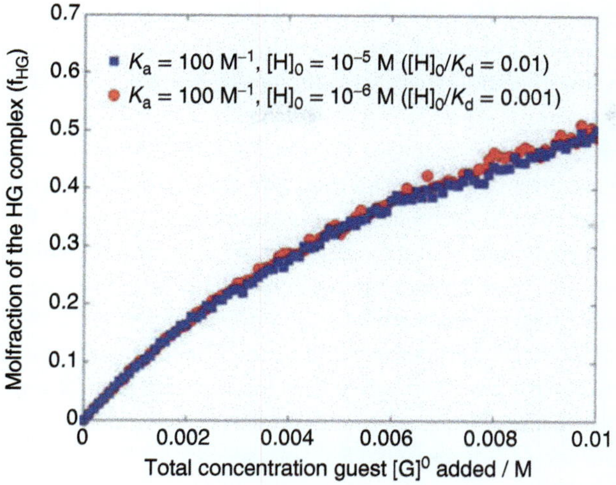

Figure 12.3 Simulated binding isotherms (1 : 1 complexation, $K_a = 100\,M^{-1}$) for $[H]_0 = 10^{-6}$ M (red curve) and 10^{-5} M (blue curve); a random error of 2% was included in the simulation. Source: Thordarson [20]. © 2011 Royal Society of Chemistry.

metal ions are titrated to an organic ligand (i.e. metal-to-ligand coordination). As important, addition of the guest should induce a distinct change in the UV/vis absorption behavior (ΔA_{obs}); such a change (increase or decrease) in the molar absorptivity (ε) upon binding, independent of the actual type of the interaction, represents a fundamental assumption when performing UV/vis titration measurements [Eq. (12.5)] [25–27]. The same basic model, though it needs to be modified, also applies for more complicated systems, such as the supramolecular association of individual components into oligomers or even polymers [28].

$$A = d \cdot c \cdot \varepsilon \tag{12.4}$$

A: absorbance ($A < 1$); d: length of path (in cm); c: concentration; and ε: molar absorptivity.

$$\begin{aligned} &\Delta A_{obs} = \varepsilon_{\Delta HG} \cdot ([HG]) \\ &\text{1:1 adduct “}\mathbf{HG}\text{”} \\ &\Delta A_{obs} = \frac{\varepsilon_{\Delta HG}[H]_0K_1[G] + \varepsilon_{\Delta HG_2}[H]_0K_1K_2[G]^2}{1 + K_1[G] + K_1K_2[G]^2} \\ &\Delta A_{obs} = \frac{\varepsilon_{\Delta HG}[G]_0K_1[H] + 2\varepsilon_{\Delta H_2G}[G]_0K_1K_2[H]^2}{1 + K_1[H] + K_1K_2[H]^2} \\ &\text{1:2 adduct “}\mathbf{HG_2}\text{” 1:2 adduct “}\mathbf{H_2G}\text{”} \end{aligned} \tag{12.5}$$

$\Delta A_{obs} = A_{obs} - A[H]_0$; ε_X: molar absorptivity of the species X; $\varepsilon_{\Delta HG} = \varepsilon_{HG} - \varepsilon_H$; $\varepsilon_{\Delta HG_2} = \varepsilon_{HG_2} - \varepsilon_H$; $\varepsilon_{\Delta H_2G} = \varepsilon_{HG} - 2\varepsilon_H$; K_1 and K_2: first and second stepwise association constants; and $[X]_{(0)}$: (initial) concentration of species X.

If working under highly dilute conditions, particular diligence is required to still offer a high accuracy of the host's concentration under the titration experiment conditions (e.g. cautions weighing in, usage of stock solutions). If the host concentration becomes too low to precisely quantify it (e.g. when only a small amount of material is available), this value becomes an unknown parameter to be dealt with in the data analysis, i.e. the fitting process. However, for large $[H]_0/K_d$ ratios (above 100), this restriction is no longer valid – at least in the case of the interaction of the components in a 1 : 1 fashion; the precise $[H]_0$ value must not be known necessarily (Figure 12.3).

Third, fluorescence titration represents maybe the most sensitive technique (at least when compared with the prior two methods): routinely, this analytical tool can be applied accurately even at sub-μM down to nM concentrations. Therefore, fluorescence spectroscopy is of particular virtue for determining very high K_a values (i.e. $>10^6\ M^{-1}$). Ideally, fluorescence titration experiments should be performed in the low-concentration regime, where the absorbance (A) at the employed excitation wavelength (λ_{ex}) is below 0.05. At higher concentrations, the linear dependency of the fluorescence response (F) on the absorbed light will be lost (inner filter effects are also known to contribute to the loss of linearity in such a concentration range) [29]. At low concentrations, the observed fluorescence (F_{obs}) of a species can be expressed by Eq. (12.6) [29].

$$F_{obs} = I_0 \cdot \varphi \cdot \varepsilon \cdot d \cdot [X] = k_x \cdot [X] \tag{12.6}$$

F_{obs}: observed fluorescence; I_0: intensity of excitation light; φ: fluorescence quantum yield; d: path length (in cm); ε: molar absorptivity; k_X: proportionality constant for species X; and [X]: concentration of species X.

The application of fluorescence titration to determine K_a is of particular relevance when only one of the species in solution is fluorescent. For example, a host (or guest) molecule alone is nonemitting and fluorescence is "turned on" upon its interaction with the guest (or host); alternatively, the emission of free host (or guest) molecules might be "turned off" when the corresponding binding partner is added [20]. The latter might be caused by either a static or a dynamic (collisional) quenching. The Stern–Volmer relation allows to quantify dynamic quenching processes according to Eq. (12.7a); practically, the fluorescence ratio (F_0/F) is plotted, as a function of the quencher's concentration ([Q]). In an ideal case, the static quenching process (for an 1 : 1 ratio of interacting molecules) can be expressed by basically the same relation (Eq. (12.7b)); however, both mechanisms are known contribute to an observed fluorescence quenching, and thus, the analysis of the obtained titration data is, in praxis, not as straightforward.

$$\frac{F_0}{F} = 1 + K_{SV} \cdot [Q] \tag{12.7a}$$

$$\frac{F_0}{F} = 1 + K_a \cdot [Q] \tag{12.7b}$$

F_0: initial fluorescence intensity; F: measured florescence intensity; K_{SV}: Stern–Volmer constant; [Q]: quencher concentration; and K_a: association constant.

Supramolecular fluorescence titration experiments can be conducted on the basis of three different scenarios. For clarity reasons, only host–guest interactions of 1 : 1 stoichiometry will be evaluated in the following (details on fluorescence titration studies on 1 : 2- and 2 : 1-type systems are available elsewhere [20]). In the first and most straightforward case, only the formed supramolecular species is fluorescent (i.e. the complexation "switches on" the fluorescence). The observed fluorescence of the host–guest complex HG can then simply be described via Eq. (12.8a). If both the host and the host–guest complex are emissive, both contributions have to be taken into account: the emission of the free host [Eq. (12.8a)] and the emission after the addition of the guest [Eq. (12.8b)] [26]. Due to contributions from the dynamic quenching of the host's emission, k_H^0 and k_{H} are not essentially equal [20]. Often, the two equations are combined into Eq. (12.8c) [26].

Fluorescence of the pure host

$$F_0 = k_H^0 \cdot [\mathrm{H}]_0 \tag{12.8a}$$

Fluorescence after the addition of guest

$$F_{\mathrm{obs}} = k_{\mathrm{H}} \cdot [\mathrm{H}] + k_{\mathrm{HG}} \cdot [\mathrm{HG}] \tag{12.8b}$$

$$\frac{F_{\mathrm{obs}}}{F_0} = \frac{k_{\mathrm{H}}/k_{\mathrm{H}}^0 + (k_{\mathrm{HG}}/k_{\mathrm{H}}^0) \cdot K_{\mathrm{a}} \cdot [\mathrm{G}]}{1 + K_{\mathrm{a}} \cdot [\mathrm{G}]} \tag{12.8c}$$

In the second case, the assembled host–guest system is no longer fluorescent (thereby, contributions from dynamic quenching are excluded, i.e. $k_H^0 = k_{\mathrm{H}}$). Then, Eq. (12.8c) is simplified to the classic Sterm–Volmer relation [Eq. (12.8a)] describing static quenching [26].

Excluding again any dynamic quenching but having a residual emission of the host–guest complex (i.e. $k_{\mathrm{HG}} \neq 0$), Eq. (12.9) is obtained. Apparently, fluorescence and UV/vis titration experiments can be analyzed on a common set of fundamental equations. The same holds true also for host–guest systems of other stoichiometries – as long as the assumption $k_H^0 = k_{\mathrm{H}}$ holds true [20].

$$\Delta F_{\mathrm{obs}} = k_{\Delta\mathrm{HG}} \cdot [\mathrm{HG}] \tag{12.9}$$

$\Delta F_{\mathrm{obs}} = F_{\mathrm{obs}} - F_0$: observed change in fluorescence; $k_{\Delta\mathrm{HG}} = k_{\mathrm{HG}} - k_{\mathrm{H}}$: change in the proportionality constants between complex and host.

The general aspects, discussed earlier for the three most important titration techniques, can be translated making, at least in theory, any type of spectroscopy a potential tool to investigate supramolecular binding. As the only requirement, a detectable change of a property in course of the titration experiment needs to occur. Beyond spectroscopy, also calorimetric means can be utilized for the very same purpose. Basically, calorimetry relies on the measurement of the enthalpy (H) increase upon addition of a host to a guest. In an isothermal titration calorimeter (ITC), as the most commonly utilized setup, the heat (Q) formed/absorbed during the binding process is detected. After the elimination of dilution effects, Q can simply be expressed by the molar enthalpy (ΔH) multiplied by the number of host–guest complex formed (in moles), i.e. the product of [HG] and volume (V). Thus, Eq. (12.10a) is derived for a simple complex of an 1 : 1 stoichiometry (for the analogous 1 : 2- and 2 : 1-type systems, the respective equations are deduced in Thordarson's review [20]). The main advantage of calorimetric titration studies is that, according to the laws of thermodynamics, the changes in the free energy (ΔG) and the entropy of the system (ΔS) can also readily be obtained [Eq. (12.10b)].

$$Q = \Delta H_{\mathrm{HG}} \cdot V \cdot [\mathrm{HG}] \tag{12.10a}$$

$$\Delta G = -R \cdot T \cdot \ln K_{\mathrm{a}} = \Delta H - T \cdot \Delta S \tag{12.10b}$$

Q: heat; ΔH_{HG}: molar enthalpy of complex formation; V: volume; ΔG: change in free energy; ΔS: change in entropy; R: gas constant; and T: temperature (in K).

12.3 Size-Exclusion Chromatography

Size-exclusion chromatography (SEC) represents one of the most powerful tools in polymer science to determine the molar mass and dispersity (Đ) of a (conventional) polymer in solution. Basically, the solubilized polymer chains are separated with respect to their molecular size, which can be determined by the retention time. Utilizing known standard samples [e.g. polystyrene, poly(ethylene glycol) (PEG), or poly(methyl methacrylate) (PMMA)], relative molar masses and dispersities can be estimated in a fast and reliable way [30]. However, it must be emphasized that the examination of samples via SEC is accompanied by a considerable dilution of the sample. As pointed out earlier, in reversible and dynamic systems – as supramolecular assemblies intrinsically are – the DP will always feature a certain concentration dependency. Thus, significant tailing of the SEC trace, indicative of (at least partial) dissociation, is a common phenomenon. For this reason, SEC is generally not the most favorite tool to derive the molar mass of supramolecular polymers. However, in special cases where a sufficient kinetic stability of the supramolecular bonds is given, as for some systems based on metal-to-ligand coordination or multiple hydrogen-bonding interactions, SEC can be applied [31–35]. The only crucial requirement is that the involved non-covalent bonds exhibit very slow association/disassociation kinetics.

Regarding metallopolymers, i.e. supramolecular polymers formed by metal-to-ligand coordination, three examples shall be highlighted briefly. The Kobuke group studied the formation of the supramolecular porphyrin arrays $[\mathbf{1}]_n$ due to intramolecular Zn(II)–imidazolyl interactions [35]. The broad distribution observed in the SEC was compared with the narrow distributions of commercial polystyrenes of different M_{n} values; a DP in the range from 80 to 400 was calculated from these data (Figure 12.4a). Paulusse and Sijbesma investigated the stimuli-responsive behavior of their Pd(II)- and Pt(II)-containing polymers by SEC [36]: In this work, the dissociation of the supramolecular polymer chains, induced by sonication, was verified to be a highly reversible process.

Thirdly, the metallo-supramolecular polymerization of ditopic bis-terpyridine ligands in the presence of transition metal ions has to be named. Schubert and coworkers showed that the binding of such ligands (e.g. **2**) to Ru(II) centers gave a

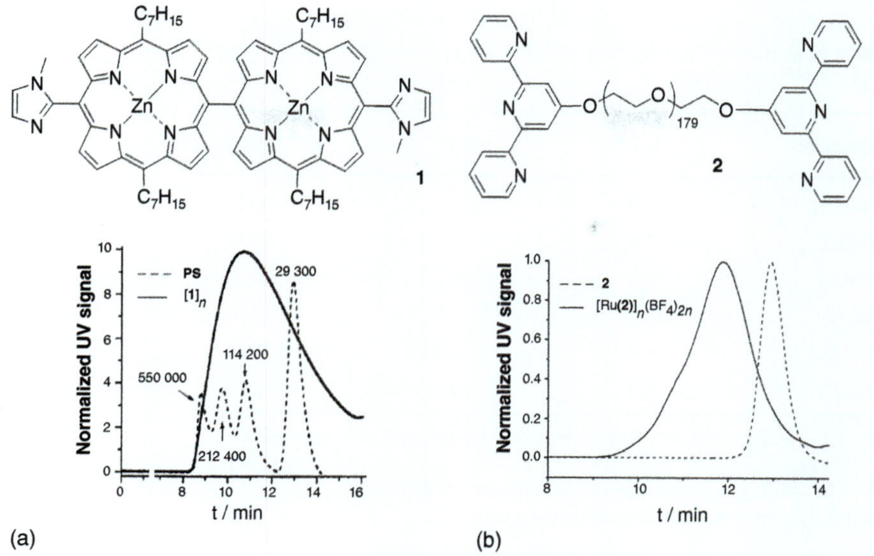

Figure 12.4 (a) SEC traces of $[\mathbf{1}]_n$ and polystyrene references (eluent: $CHCl_3$); (b) SEC traces of **2** and $\{[Ru(\mathbf{2})](BF_4)_2\}_n$ (eluent: DMAc, containing 5.5 mM NH_4PF_6). Source: (a) Ogawa and Kobuke [35]. © 2000 John Wiley & Sons, (b) Meier et al. [32]. © 2006 American Chemical Society.

kinetically stable polymer that could be analyzed by SEC (Figure 12.4b) [31, 32]. The formation of a supramolecular polymer of high molar mass was concluded from a pronounced shift of the SEC trace to shorter elution times, i.e. higher M_n values. A linear PMMA calibration was utilized to estimate a M_n value of 1.38×10^5 g mol^{-1} ($Đ = 1.55$), corresponding to a DP of c. 15. It was found to be crucial to add an appropriate conducting salt, such as NH_4PF_6, to the eluent to improve the quality of the measurements by suppressing the tailing of the SEC trace. According to the Irving–Williams series [39], the stability of high-spin complexes of the divalent first row transition metal ions follows the general order: Mn(II) < Fe(II) < Co(II) < Ni(II) < Cu(II) > Zn(II). Thus, the corresponding complexes exhibit different stability constants (K values), where K_1 represents the terpyridine mono-complex (i.e. 1 : 1 ratio of ligand and metal ion) and K_2 the analogous bis-complex (2 : 1 ratio) – these values were derived from calorimetric measurements (see also Section 12.2). In all cases, the complexation followed a pseudo-first-order kinetic behavior, as determined by a stopped-flow method (Table 12.1) [37]. Applying ITC enabled one to determine the thermodynamic parameters, such as the binding enthalpies of the complexes $[M(tpy)_2]^{2+}$ [38]. Taking these data into account, Schubert and coworkers prepared metallopolymers from ligand **2** and the moderately binding Ni(II), Co(II), and Fe(II) ions. The SEC analysis of these supramolecular polymers revealed a sufficient stability only in the case of the Ni(II)-containing example; in the other two cases, considerable dissociation was observed [9, 40].

Table 12.1 Stability constants and binding enthalpies of various $[M(tpy)_2]^{2+}$ complexes.

Metal ion	$\lg K_1$ [a]	$\lg K_2$ [a]	ΔH^0 (kJ mol^{-1}) [b]
Mn(II)	4.4	n.a.	n.a.
Fe(II)	7.1	13.8	−79.9
Co(II)	8.4	9.9	−61.5
Ni(II)	10.7	11.1	−66.9
Cu(II)	n.a.	n.a.	−92.2
Zn(II)	6.7	5.2	−60.7

a) The stability constants were determined by a stopped-flow procedure in water.
b) The binding enthalpies were determined by ITC in MeCN.
Source: Holyer et al. [37], Dobrawa et al. [38].

Supramolecular dendrimers, based on multiple H-bonding interactions, were reported by Zimmerman et al. [33]; the molar masses, as determined by SEC vs. polystyrene standards, were deviating by 20% from the theoretical ones. At this point, a general limitation in applying SEC for the determination of molar masses of supramolecular polymers needs to be addressed [41]: the lack of suitable calibration methods. Even if a supramolecular material is stable under SEC conditions, the obtained molar mass (distribution) cannot be related to the "real" values due to missing appropriate reference values. The available standards (*vide supra*) can be applied for organic polymers, and as a rule-of-thumb, the more similar a material is to the used standard, the more reliable the determined M_n and M_w values are. Since supramolecular polymers – in particular the charged, metal-containing systems – often behave very different on the column than, e.g. polystyrene or PEG, an SEC analysis of such materials remains always a crude estimation only.

Mechanical bonding is another type of interaction that allows the formation of supramolecular polymers [43]. In contrast to the previous examples, in such systems, the non-covalent bonding is supported by topological interlocking. As a result, the DP is no longer concentration-dependent and does not decrease upon SEC analysis. The most prominent representatives of this class of supramolecular materials are polycatenanes and polyrotaxanes (e.g. see Chapters 6–10). For instance, the interlocked poly(daisy chain) **3**, featuring a pH-responsive behavior, was prepared by Stoddard and coworkers (Figure 12.5) [42]. An SEC setup equipped with multiangle light scattering (MALS) detector gave an M_n value of c. 33 000 g mol^{-1} ($Đ$ = 1.85) for the assembly. Similarly, the polymerization of a host–guest complex, assembled from a crown ether and viologen derivative, by low-temperature esterification, gave a poly(daisy chain) that could be studied by SEC (M_n = 64 000 g mol^{-1}, $Đ$ = 1.5, DP ≈ 45) [44]. A stable polyrotaxane was prepared by threading α-cyclodextrin (α-CD) moieties onto a PEG chain; the

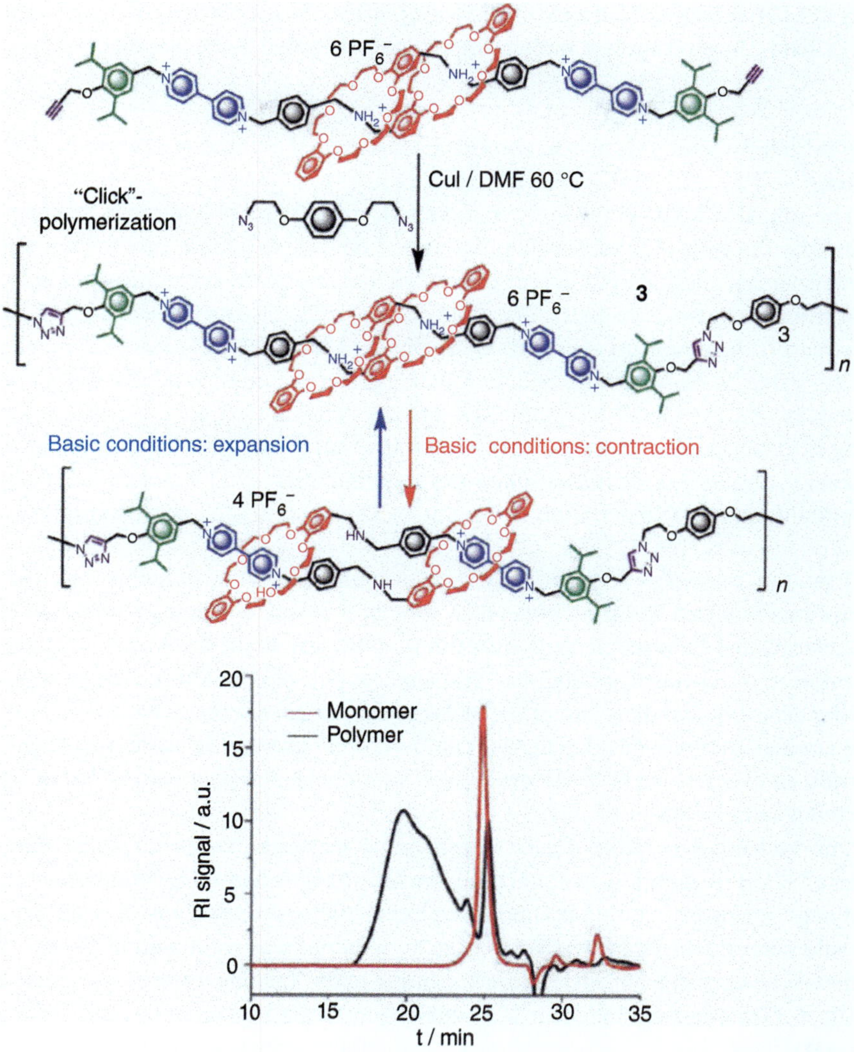

Figure 12.5 Schematic representation of the synthesis of the poly(daisy chain) **3**. The SEC traces of **3** and of the utilized monomer are also shown (eluent: DMF containing 0.2 M NH_4PF_6). Source: Fang et al. [42]. © 2009 American Chemical Society.

obtained α-CD-containing pseudopolyrotaxane was then end-capped with bulky pyrene units to prevent dethreading [45].

12.4 Viscometry

The determination of molar masses of polymers by viscosity measurements represents one of the oldest methods available. The empirical Mark–Houwink equation

[Eq. (12.11)] relates the intrinsic viscosity ($[\eta]$) to the viscosity-averaged molar mass (M_η) that is, in good approximation, equal to the M_w value of the polymer.

$$[\eta] = K \cdot (M_\eta)^\alpha \tag{12.11}$$

$[\eta]$: intrinsic viscosity; K and α: empirical constants; M_η: viscosity-averaged molar mass.

The so-called Mark–Houwens parameters (K and α) of a polymer can be determined on the basis of suitable calibration experiments with a number of fractions of low dispersity. Then, Eq. (12.11) can be applied to calculate the molar mass of a polymer from the intrinsic viscosity. The same holds principally true for supramolecular polymers if the empirical constants are available. This, however, already limits the scope of the Mark–Houwink equation with respect to supramolecular polymers: suitable covalent models for these dynamic systems are required to obtain the empirical constants. Otherwise, only a qualitative analysis is possible. For instance, viscosity titration experiments allow the determination of the stoichiometry in supramolecular polymers from the point of equivalence (i.e. the maximum in relative viscosity) [46, 47].

The self-assembly of poly(dimethylsiloxane)s **4** via association of their acid-functionalized termini generated a supramolecular polymer that could be investigated by viscometry (Figure 12.6a) [48]. To obtain the Mark–Houwink parameters, a covalent analog was prepared, as a model polymer. The analysis of the experimental data, in particular for $(\mathbf{4a})_n$, revealed a good agreement with the simulated curves (Figure 12.6b). The deviation observed for **4b/c** was mainly attributed to a reduced fit of the structure of supramolecular polymers to that of the model polymer.

The Meijer group utilized a monofunctional molecule, as the chain-stopping agent [49]. The addition of this compound to the solution of a supramolecular polymer, formed by multiple H-bonding interactions (see Chapter 3), revealed a drastic decrease in viscosity. The DP of the material was estimated to be $2/x$ (x denotes the stopper's mole fraction). The fitting of the analytical viscosity data gave a DP of 700, corresponding to an M_w value of 500 000 g mol^{-1} (at a concentration of 40 mM).

Beyond the determination of molar masses, viscometry can also be applied to gain insight into supramolecular polymerizations – even in a quantitative way [50–53]. In particular, supramolecular polymerizations, following a ring-chain-mediated mechanism, are of interest, where the critical polymerization concentration (cpc) needs to be determined (for a detailed view on this type of supramolecular polymerization, see Section 1.3.2 and Ref. [3]). In short, the cpc represents the limit in such a supramolecular polymerization, where above linear polymer chains and below macrocyclic species dominate. The viscosities of these two species correlate differently with concentration; thus, the cpc can easily be read out from the viscosity–concentration plots. As an example, the self-assembly of the homoditopic crown ether (**5**) and bis-paraquat derivatives (**6**), as reported by Huang et al. follows this mechanism (Figure 12.7) [50]. The bilogarithmic plot shows a linear relationship in the low- as well as high-concentration regime;

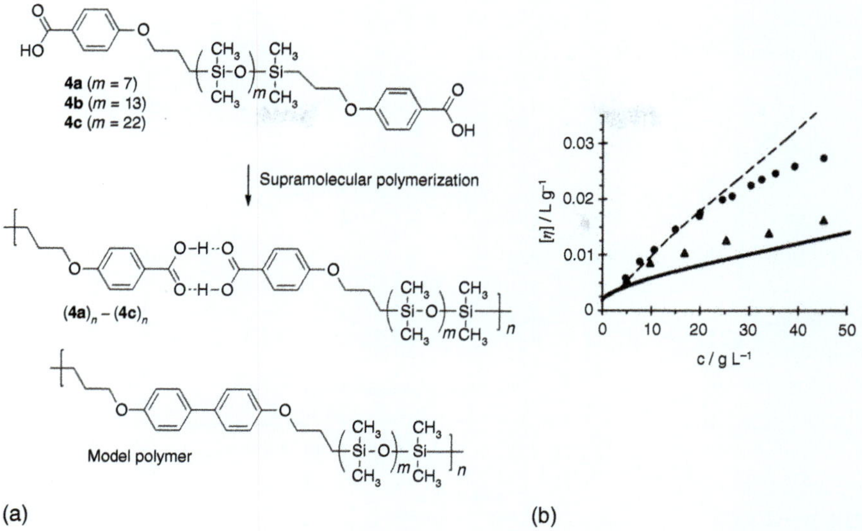

Figure 12.6 (a) Schematic representation of the formation of the supramolecular polymers $(\mathbf{4})_n$. (b) Reduced viscosity of $(\mathbf{4a})_n$ in CH_2Cl_2 (▲x025B2;) and CCl_4 (●x025CF;), the corresponding calculated curves are also shown (dashed and solid line, respectively). Source: Abed et al. [48]. © 2001 Elsevier.

the intercept of the linear fits is defined as the cpc (in this particular example: 80 mM).

12.5 Light Scattering

The determination of both molecular size and morphology of samples by light scattering techniques is well established in polymer science. Basically, there are two different methods available: static light scattering (SLS) and dynamic light scattering (DLS) [54].

The former is highly suited for measuring the molar mass of macromolecular compounds. Analysis of the scattering data by double extrapolation of the concentration (c) and the measurement angle (θ) to zero in a so-called Zimm-plot representation gives access to the M_w value. Moreover, the radius-of-gyration (R_g) and the second virial coefficient (A_2) can be derived from such an analysis (Eq. (12.12)) [55]. However, this approach is only valid for macromolecules with a size of $<\lambda/20$.

$$\frac{K \cdot c}{\Delta R(\theta, c)} = \frac{1}{M_W} \cdot \left[1 + \frac{16\pi^2}{3\lambda^2} \cdot R_g \sin^2\left(\frac{\theta}{2}\right)\right] + 2 \cdot A_2 \cdot c \frac{K \cdot c}{\Delta R(\theta \rightarrow, c \rightarrow 0)} = \frac{1}{M_W} \tag{12.12}$$

K: optical constant; c: concentration; $\Delta R(\theta,c)$: Rayleigh ratio; R_g: radius-of-gyration; θ: angle of the scattered light, A_2: second virial coefficient.

Covalent polymers that are not prone to aggregation generally exhibit a linear behavior; for supramolecular polymers, on the contrary, often nonlinear plots are

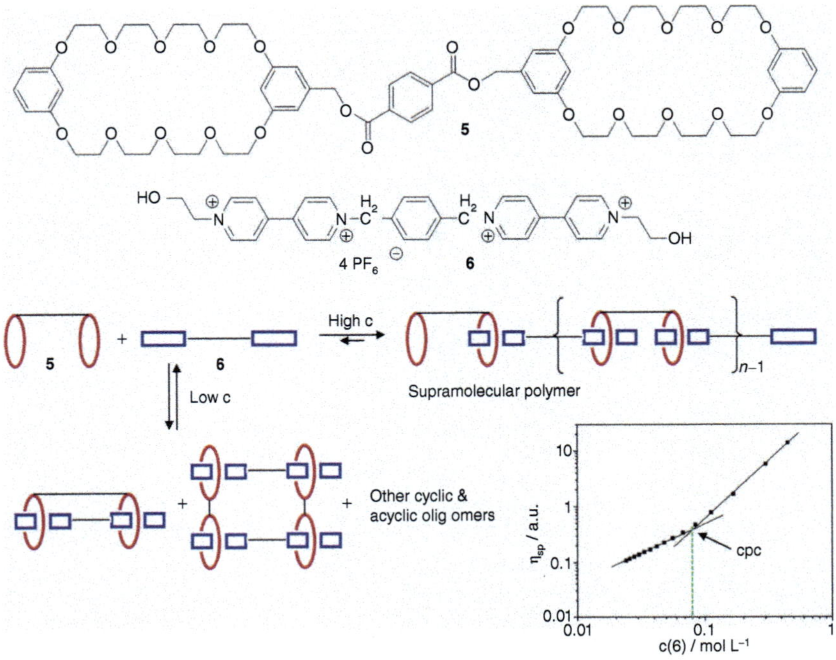

Figure 12.7 Schematic representation of the ring-chain-mediated polymerization of **5** and **6**. The bilogrithmic viscosity–concentration plot allowed the determination of the cpc value. Source: Huang et al. [50]. © 2007 American Chemical Society.

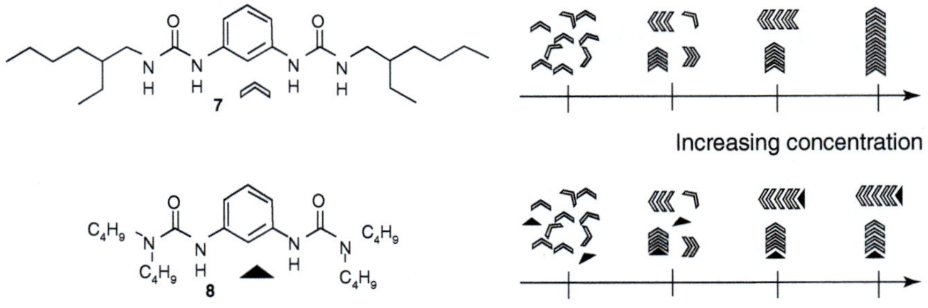

Figure 12.8 Schematic representation of the supramolecular polymerization of monomer **7** in the absence (a) and presence of the chain-stopping agent **8** (b). Source: Lortie et al. [56]. © 2005 American Chemical Society.

obtained, since the concentration dependency of the molar mass results in the variation of A_2. Thus, extrapolating the concentration to zero is not straightforward. In order to circumvent this issue, a chain-stopping agent may be utilized [56, 57]. The self-assembly of monomer **7**, for example, can be controlled by the addition of a chain-stopping agent (**8**): According to the SLS data, the M_w and R_g values were decreasing significantly; whereas, the A_2 value remained constant [56]. As a result, the experimental M_w value was exceeding the theoretical value, by only a factor of 9–13 (Figure 12.8).

In the special case of a supramolecular polymerization occurring in a highly dilute regime, the effect of A_2 can (almost) be neglected. This effect was demonstrated via the example of the self-assembly of Zn(II)-containing metalloporphyrins with pyridine derivatives, as shown by Chen et al. [58]: In toluene solution, the SLS experiments revealed an M_w value larger than 4.2×10^6 g mol^{-1} (DP $\geq$ 460).

The second technique, DLS, is of particular relevance for analyzing the size distribution of small particles, which, for example, originate from the aggregation of (supramolecular) polymer chains [54, 59]. Besides this application that has been adopted for almost all types of supramolecular polymers, other phenomena have been investigated. Representatively, a contribution by the Weck group shall be named [60]: the supramolecular polymerization of Pd(II)-pincer complexes ions with bis-pyridyl macroligands was facilitated by adding $AgBF_4$, eventually leading to a significant increase in particle sizes.

12.6 Vapor Pressure Osmometry

In general, the vapor pressure of a solution is lower than that of the pure solvent (under the same experimental conditions). This simple principle represents the basis for vapor pressure osmometry (VPO). According to Raoult's law, the molar mass (the M_n value for polymers) can be related to the vapor pressure [Eq. (12.13)] [61]. This colligative property has been utilized to investigate not only simple organic molecules but also polymers and, to some extent, even their supramolecular counterparts. In the latter two cases, the number-averaged molar mass, the M_n value, is obtained.

$$M_n = \frac{K_{VPO} \cdot c}{\Delta T} \tag{12.13}$$

K_{VPO}: experimentally determined constant; c: concentration; and ΔT: difference in temperature.

Convincing examples for the applicability of VPO in the field of supramolecular polymer science were reported by the Harada group [62–64]. The supramolecular polymer assembled from the homoditopic β-cyclodextrin (**9**) and the adamantyl derivatives (**10**) in a 1 : 1 stoichiometry was studied by his technique. An increase of M_n with increasing concentration was observed (i.e. 6×10^4 g mol^{-1} at 5 mM and 1×10^5 g mol^{-1} at 20 mM, see Figure 12.9) [62]. In a control experiment, the molar mass of the bis-β-CD ligand alone was, as expected, independent of the concentration.

As a considerable limitation, VPO cannot be utilized for the analysis of all different types of supramolecular polymers. For a reliable determination of the molar mass, the precise amount of the substance (n) in the solution to be investigated must be known. However, supramolecular polymers, based on metal-to-ligand coordination, often carry, due to their intrinsic salt-like structure, counterions that also need to be considered. But what is the degree-of-dissociation? The two possible extreme scenarios are as follows: full dissociation of the initial material (n_0) into the multiple-charged polymer and the associated counterions or the formation of

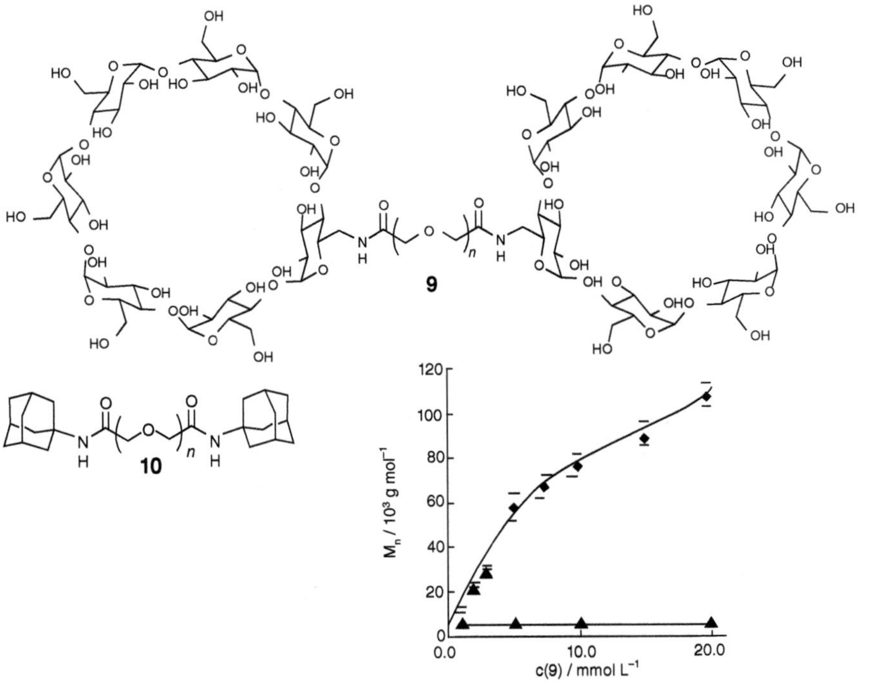

Figure 12.9 Schematic representation of the β-cyclodextrin-dimer (**9**) and the adamantyl-dimer (**10**). The evolution of the M_n of the supramolecular polymer $(\mathbf{9}\cdots\mathbf{10})_n$ as a function of the β-CD-dimer concentration (determined by VPO) is also shown. Source: Hasegawa et al. [62]. © 2005 American Chemical Society.

close ion pairs (i.e. no dissociation). In the first case, the resulting amount n is the sum over all molecules (i.e. polymer chains and anions); in the second case, n is simply equal to n_0. However, in a "real" metallopolymer, the situation will be at some point between these two extremes, and therefore, VPO cannot be utilized for such materials.

12.7 Analytical Ultracentrifugation

Analytical ultracentrifugation (AUC) was developed as a technique for the characterization of biomolecules, in particular proteins, almost 80 years ago [65]. The new generations of ultracentrifuges, introduced by Beckman in the 1990s, initiated a new period of theoretical and applied research in this field. Today, AUC represents a highly versatile tool for the elucidation of the solution structure and conformation of proteins and protein complexes as well as of synthetic macromolecules [66, 67]. The theoretical background to AUC, the methods, and typical applications have been detailed in several books [68–72] and review articles [73–77]. Basically, the instrument consists of a high-speed centrifuge, a rotor with several cell compartments, and an optical system that can measure the concentration gradients of macromolecules under centrifugal force (e.g. UV/vis absorption or interference measurements). Studying the concentration distributions under a centrifugal field gives access to a broad range of information [66]: molecular structure and conformation as well as secondary (i.e. non-covalent) interactions; the parameters one can derive include the molar mass, the sedimentation, and the diffusion coefficient as well as binding affinities and virial coefficients. Among others, AUC can be used to investigate the sedimentation of species via velocity and equilibrium measurements. In velocity experiments, a large number of intermediate profiles is recorded during the sedimentation process. In contrast, only the sedimentation diffusion equilibrium profile is analyzed for the latter method.

Sedimentation equilibrium measurements are typically performed under idealized conditions to suppress any wide-ranging interactions between sedimenting species or the macroscopic internal field, which might result from spatial separation of solubilized ions and settling species of opposite charge. Preferably, such experiments are performed at very low concentrations and, if the sedimenting material is charged, at relatively high ionic strengths (e.g. metallo-supramolecular polymers). As an independent method, AUC allows one to directly determine the average molar mass(es), the association state, the K_a values [73, 78], and even the dispersity [79]. A variety of parameters can be obtained from sedimentation velocity measurements: the sedimentation coefficient, the molar mass, and the size distribution (the latter only in the case of spherical species) [80], as well as the average diffusion and friction coefficients [78]. However, some systems (e.g. colloids and polyelectrolytes) are prone to be unstable in the presence of high salt concentrations; therefore, special nonideal theories for the sediment formation have to be employed [81]. In order to derive any of the aforementioned parameters, numerical methods have to be used. In particular, Lamm's equation [82] – describing the sedimentation profiles of monodisperse species – represents the cornerstone in analyzing AUC experiments [Eq. (12.14)] [71, 72, 74]. The first and second terms on the right-hand side are proportional to D and $S\omega^2$, respectively, and describe the competition between diffusion and sedimentation (D and S denote the diffusion and sedimentation coefficients, respectively): Sedimentation strives for concentrating the solute near the outer radius of the rotating cell; whereas, diffusion aims at equalizing the solute concentration throughout the cell.

$$\frac{\partial c}{\partial t} = D\left[\left(\frac{\partial^2 c}{\partial r^2}\right) + \frac{1}{r}\left(\frac{\partial c}{\partial r}\right)\right] - S\omega^2\left[r\frac{\partial c}{\partial r} + 2c\right] \tag{12.14}$$

c: concentration; t: time; r: radius of the cell; D: diffusion coefficient; S: sedimentation coefficient; and ω: rotor angular velocity.

Thereby, S is defined as $v/(\omega^2\ r)$, where v and $\omega^2\ r$ represent the sedimentation velocity and the centrifugal acceleration, respectively. At infinite dilution, S can be expressed by Eq. (12.15a); for noninteracting, ideally settling species, the Svedberg equation simply relates S and D to the molar mass [M, Eq. (12.15b)] [71, 72, 74]. A range of mathematical approaches to determine the S and M values (and, in the case of polydisperse species, also their distributions and averages) has been published. Comfortably, the SEDFIT program, developed by Schuck at the National Institute of

Health (NIH), offers finite-element solutions of Lamm's equation via a direct fitting of the velocity data [83] (www.analyticalultracentrifugation.com).

$$S = \frac{M(1 - \overline{v} \cdot \rho)}{N_A \cdot f} \quad (12.15a)$$

$$S = \frac{M(1 - \overline{v} \cdot \rho)}{R \cdot T} \cdot D \quad (12.15b)$$

N_A: Avogadro's number; ρ: solvent density; $\overline{v}$: partial specific volume; M: molar mass; f: friction coefficient; T: absolute temperature; and R: gas constant.

Going beyond the classical application of AUC, which is the investigation of biomacromolecules, the groups of D. Schubert and U.S. Schubert demonstrated that also metallo-supramolecular materials, discrete assemblies (e.g. grid-like assemblies) [78, 84–86] as well as polymers can be characterized by the means of AUC. Sedimentation velocity and equilibrium experiments as well as transient state analysis have been performed in this context. For example, the aforementioned metallo-supramolecular polymer $\{[Ru(\mathbf{2})](PF_6)_2\}_n$ was studied by sedimentation experiments (Figure 12.4b). The estimated M_n value, derived from the exponential fitting of the raw data, was compared with those determined by two further independent methods, i.e. by SEC and viscometry [31]. All these techniques gave an M_n value of c. 130 000 g mol^{-1}, corresponding to a DP of about 15. Thus, AUC represents a powerful add-on to determine the molar mass of macromolecular materials (in addition to SEC, NMR spectroscopy, viscometry, etc.). Moreover, similar main-chain metallo-supramolecular polymers, composed of rigid-rod π-conjugated ligands, were also investigated [47]. The basic challenges in applying AUC for such highly charged assemblies (e.g. determination of $\overline{v}$, avoidance of nonideal sedimentation) were addressed by Tziatiois et al. [84]. It has been shown by the so-called buoyant-density method that the $\overline{v}$ value is solvent-dependent and biased by the salt concentration. The primary charge effects, contributing to nonideal behavior of the sample, can often be suppressed by adding salt (though, a correction for the secondary charge effect might still be necessary) [67].

Water-soluble π-conjugated polyrotaxanes represent one further example for supramolecular polymers whose molar masses were estimated by AUC experiments (Figure 12.10) [87]. Michels et al. showed that neither gel electrophoreses, nor small-angle neutron scattering (SANS) nor mass spectrometry (MS) was suited for this purpose. Though, SEC could be applied for some examples, very high ionic strengths were generally required to reduce the strong electrostatic interactions between the polyelectrolyte-type supramolecular chains.

12.8 NMR Spectroscopy

The molar mass (M_n) of polymers can, in general, be determined by NMR spectroscopic end-group analysis: the M_n value can simply be calculated from the integral ratio of the protons of the end groups and the polymer backbone (if they are well separated from the other signals). Concerning supramolecular polymers, also the

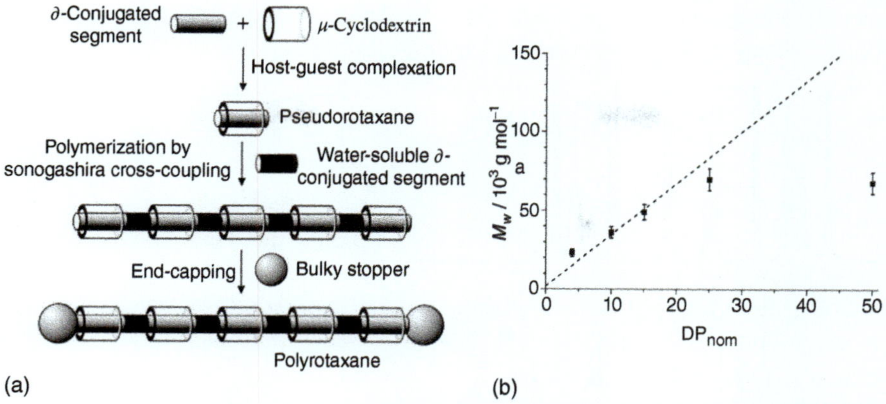

Figure 12.10 (a) Schematic representation of the stepwise synthesis of a water-soluble polyrotaxane. (b) Plot of the experimental M_w according to AUC as a function of the theoretical DP (DP_{nom}, the nonideal behavior at high DP values was attributed to accidental chain termination and fractionation upon purification. Source: From Michels [87]. © 2003 John Wiley & Sons.

DP can be estimated by NMR spectroscopy, though some assumptions have to be made. Quantitative interacting of a host with its guest (or between metal ions and organic ligands) will result in a change in the chemical shift ($\Delta\delta$) and, at a certain monomer concentration (c), to $\Delta\delta_c$. From this, the DP can be obtained according to Eq. (12.16) [88].

$$DP = \frac{1}{(1-p)} \quad \text{with} \quad p = \frac{\Delta\delta_c}{\Delta\delta} \tag{12.16}$$

$\Delta\delta$: change of the chemical shift at full complexation; $\Delta\delta_c$: change of the chemical shift at a certain monomer concentration.

For example, Higuchi and coworkers determined the DP and the molar mass of the Pt(II)-containing polymer **11** by applying this method (Figure 12.11) [89]. Due to the interaction of the Pt(II) ion with the coordinated ligands, significant downfield shifts of the signals assigned to the terpyridine (tpy) as well as to the pyridine (py) moieties could be observed. Taking the low-intensity signals of the metal-free binding sites, a DP of 40 was calculated for **11** – corresponding to an M_n of c. 30 000 g mol^{-1}.

Moreover, supramolecular polymerizations typically follow step-growth polymerization pathways and, thus, exhibit rather broad $Đ$ values. In order to study the supramolecular polymerization, one can perform concentration-dependent NMR spectroscopy. With increasing concentration of the monomer, the signals in the NMR spectra become broadened, indicative of the presence of supramolecular polymers [90, 91].

Besides the commonly applied NMR spectroscopic end-group analysis, diffusion-ordered NMR spectroscopy (DOSY) represents a state-of-the-art technique in order to characterize supramolecular polymers, in general [93]. The evaluation of the diffusion coefficient (D) allows one to carefully compare the sizes of supramolecular polymers, at least in a qualitative manner. For example, the self-assembly of the heteroditopic molecule **12a** via charge–transfer (CT) interactions of

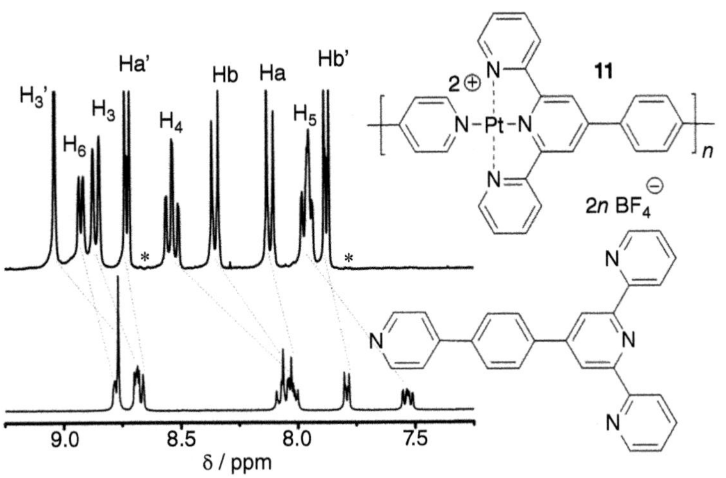

Figure 12.11 Comparison of the ^{1}H NMR spectra of the metallo-supramolecular polymer **11** and the corresponding metal-free ligand (both recorded in DMSO-d_6). In the spectrum of **11**, the signals assigned to the end groups are asterisked. Source: Brunsveld et al. [89]. © 2015 American Chemical Society.

the electron-rich porphyrin-based molecular tweezers (head group) and the electron-poor 4,5,7-trinitrofluorenone (TNF, tail group) was studied by Haino et al. [92]. DOSY measurements revealed that the diffusion behavior of the model compound **12b**, missing the TNF-tail, was independent of the concentration ($D \approx 3.02 \times 10^{-10}\ m^2\ s^{-1}$): whereas, **12a** exhibited a strong dependency: D decreased with increasing concentration (Figure 12.12). Thus, supramolecular interactions could be excluded for **12b** and the formation of a supramolecular polymer of high molar mass, at high concentrations, was confirmed for **12a**. Since all aggregates were assumed to be hydrodynamically spherical, the authors calculated the average size of $(\mathbf{12a})_n$ according to the Stokes–Einstein equation; the size of the polymer was found to be c. 660 times that of the monomer (in fair approximation, this value was equal to the DP).

This example highlights the potential of DOSY, as analytical tool, to characterize supramolecular polymers. A similar study was performed by Li et al. [94]: the self-assembly of a crown ether derivative bearing two pendant self-complementary UPy moieties (UPy: ureidopyrimidinone) gave a supramolecular polymer, based on quadruple H-bonding interactions (for further details on this type of supramolecular polymers, see Chapter 3). Subsequently, the polymer chains were cross-linked into pseudorotaxane networks by the addition of a bis-paraquat derivative. Due to the supramolecular cross-linking, the diffusion coefficient was decreased significantly compared with that of the linear supramolecular polymers. Related to this, the formation of supramolecular quasi-triblock copolymers, mediated by cucurbit[8]uril, as macrocyclic host for the encapsulation of two molecular guests, was demonstrated by the Scherman group [95]. The ternary complex, formed within the hydrophobic cavity of the host, typically consists of an electron-deficient paraquat derivative and an electron-rich aromatic system (e.g. naphthol) [96].

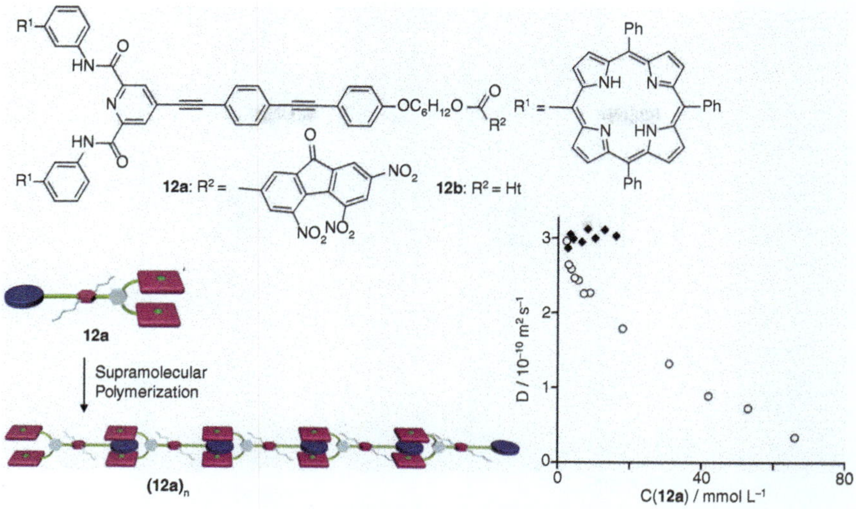

Figure 12.12 Schematic representation of the supramolecular polymerization of **12a**. The concentration dependency of the diffusion coefficient (*D*) is also shown. Source: Haino et al. [92]. © 2012 John Wiley & Sons.

The formation of a supramolecular polymer, based on two different types of non-covalent interactions, i.e. metal-to-ligand coordination in concert with H-bonding, was reported by Mansfeld et al. [97]. DOSY measurements confirmed the presence of the two types of supramolecular bonds within the chain-extended supramolecular polymer: the same relative diffusion coefficients were observed for all signals of $\{[(\mathbf{13})Fe(\mathbf{13})](PF_6)_2\}_n$ (Figure 12.13Ba). To examine the orthogonal responsiveness of the supramolecular entities, trifluoroacetic acid (TFA) was added to the chain-extended polymer. As expected, the quadruple H-bonding of the UPy-dimer was disrupted without affecting the more pH robust metal-to-ligand bonds (Figure 12.13Bb): the disappearance of the signals arising from H-bonding in the ^{1}H NMR spectrum in concert with a significant change in the diffusion behavior confirmed the selective responsiveness. For two different molecular species, the ratio of diffusion coefficients is inversely proportional to the square or cubic root of the corresponding ratio of molar masses [Eq. (12.17)]. Thus, on the basis of calibration curves, the diffusion behavior observed in DOSY experiments can be related to the molar mass [98].

$$\sqrt[3]{\frac{M_j}{M_i}} \leq \frac{D_i}{D_j} \leq \sqrt{\frac{M_j}{M_i}} \qquad (12.17)$$

$D_{i,j}$: diffusion coefficients and $M_{i,j}$: molar masses.

In this context, it is worth noting that DOSY measurements can also be performed for other nuclei than the "traditional" ^{1}H one: Paulusse and Sijbesma applied ^{31}P-DOSY experiments for studying their metallo-supramolecular polymers (formed from bis-phosphinoalkane ligands and $PdCl_2$) [34].

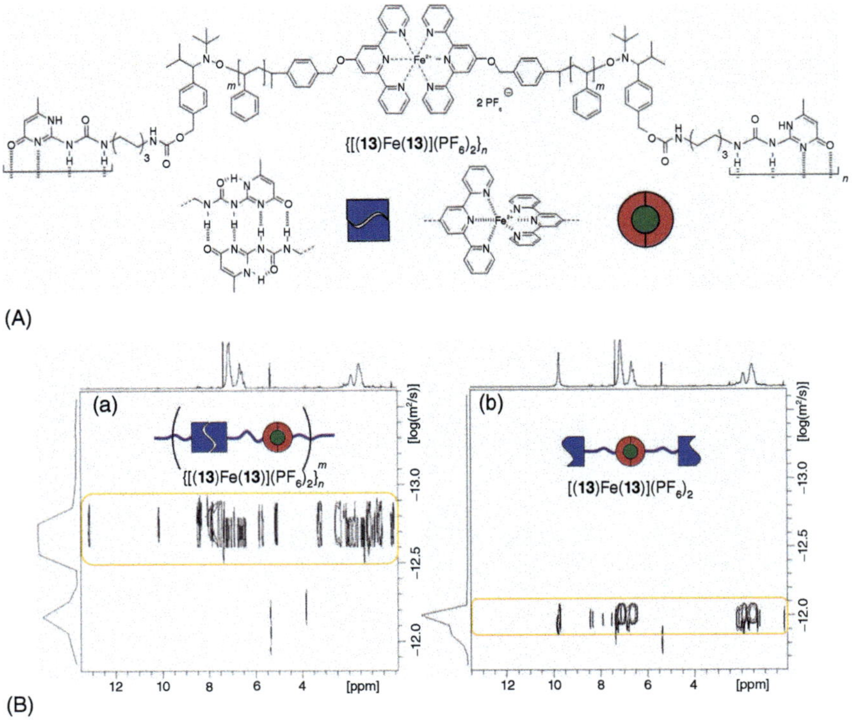

Figure 12.13 (A) Schematic representation of the chain-extended supramolecular polymer $\{[(\mathbf{13})Fe(\mathbf{13})](PF_6)_2\}_n$. (B) DOSY of $\{[(\mathbf{13})Fe(\mathbf{13})](PF_6)_2\}_n$ in the absence (a) and presence of TFA (b). Source: Mansfeld et al. [97]. © 2013 The Royal Society of Chemistry.

A broad range of sophisticated NMR spectroscopic methods, in solution as well as in the solid state, are known today and has already been employed in the field of (supra-)macromolecular chemistry [100]. One representative example utilizing solid-state NMR spectroscopy under magic-angle spinning (MAS), i.e. ^{1}H double-quantum (DQ) NMR spectroscopy, with respect to a supramolecular polymer, based on H-bonding interactions, shall finally be named. This particular method allows one to combine the sensitivity of conventional ^{1}H NMR spectroscopy (e.g. with respect to the strength of H-bonding interactions) with a quantitative view on dipole–dipole interactions between protons. Measuring 2D ^{1}H DQ-MAS NMR spectra allows, therefore, to extract detailed information on the arrangement of H-bonding arrays and even allows one to estimate the proton–proton distances. Linear supramolecular polymers, based on quadruple H-bonding interactions, have already been detailed in Chapter 3. The UPy moiety is known for its ability to irreversibly tautomerize from the keto- to the enol-form upon thermal treatment (Figure 12.14). Sijbesma and coworkers employed ^{1}H DQ-NMR spectroscopy to elucidate this structural transformation in the bulk material in the real time [99]. The kinetics of the process were analyzed, which gave an Arrhenius activation energy of c. 145 kJ mol^{-1}.

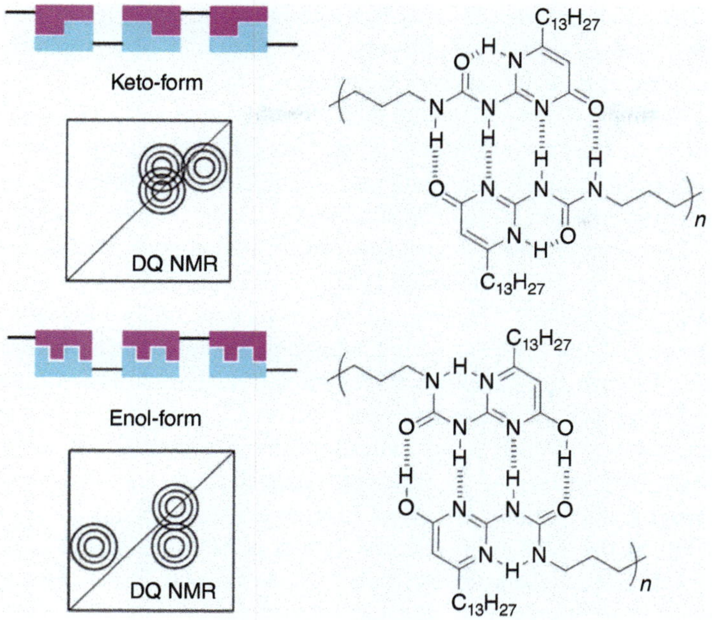

Figure 12.14 Analysis of a H-bonded supramolecular polymer in the keto- and the enol-form by ^{1}H DQ-MAS NMR spectroscopy. Source: Schnell et al. [99]. © 2002 The Royal Chemical Society.

12.9 Mass Spectrometry

The paradigm that MS is a rough and destructive method was overthrown in the 1980s when the soft methods for ionization, such as electrospray ionization (ESI) [101] and matrix-assisted laser desorption/ionization (MALDI) [102], were introduced. Thus, the application of MS in analyzing the molar mass of biomacromolecules and synthetic polymers was limited, for a long time, by the intrinsically nonexistent volatility and often poor thermal instability of such these materials [103]. Today, MALDI and ESI time-of-flight (TOF) MS represent powerful tools for the structural investigation of synthetic polymers (i.e. chemical composition and end-group analysis) [104, 105]. However, MALDI-TOF is difficult to apply for the analysis of advanced supramolecular materials, such as polymers, due to their poor stability under the required experimental conditions. Thus, only fragments or short oligomers can be detected in most cases, making MALDI-TOF MS an unsuited method to characterize supramolecular polymers so far [106–108].

Basically, the same limitations also hold true for ESI-TOF MS, though substantial progress has been made in this field, now enabling the fast and precise analysis of complex supramolecular assemblies of defined structure (i.e. discrete and stoichiometric species, not polymers yet). Wesdemiotis and coworkers established a multidimensional MS setup by coupling ESI to ion mobility mass spectrometry (IM-MS) and energy-resolved (gradient) tandem mass spectrometry (gMS^2) [109].

Thereby, the soft ESI conditions prevent (or at least minimize) undesired fragmentation of the assembly; the subsequent IM unit enables the separation of supramolecular ions by their charge and collision cross section (the latter being a function of size and shape). The thusly separated ions are then identified in the orthogonal MS dimension by measuring their *m/z* values and isotope patterns. Finally, the gMS^2 system gives insight into the preferred fragmentation and dissociation pathways of the assemblies. In particular, 2D and 3D metallo-supramolecular assemblies of defined architectures, based on terpyridine bis-complexes, were studied by the Newkome and Wesdemiotis groups via this methodology [109].

Beyond MALDI- and ESI-TOF MS, two alternative methodologies have been developed to circumvent the problem of fragmentation: coldspray (CS) ionization and Fourier-transform ion cyclotron resonance (FT-ICR) MS. The former technique basically employs an ESI source that is cooled by liquid nitrogen to reduce the internal energy gained upon ionization of the sample; as a result, the degree-of-fragmentation can be significantly reduced [110, 111]. However, commercial setups are not yet available, thus making CS-MS difficult to apply. Second, FT-ICR MS, as the method offering the highest possible mass resolution, can be utilized to detect metallo-supramolecular ions with high mass accuracy and well-resolved isotope patterns [112]; as a major drawback, isomers/conformers having exactly the same mass cannot be resolved.

Besides all of the limitations in analyzing supramolecular polymers by MS, one example where qualitatively good mass spectra were obtained shall be given. Miyauchi et al. investigated the self-assembly of the heteroditopic building block **14** via host–guest interactions (i.e. intermolecular inclusion of the *para*-(*tert*-butoxyamino)cinnamoylamino moieties into the cavities of the α-cyclodextrin rings (see also Chapter 6) [63]. Analysis via turbo ion spray TOF MS allowed the detection of assemblies up to the 14 mer (Figure 12.15).

12.10 Microscopy Imaging

The utilization of scanning probe microscopy (SPM) and electron microscopy allows one to investigate the morphology and shape of supramolecular architectures over various scales. From the various SPM methods, scanning tunneling microscopy (STM) is capable of imaging a sample at the atomic scale – the requirement for this being a rigid molecular structure of sufficiently large size [113]. In general, imaging is performed at low concentrations in order to avoid any aggregation and/or interloping of individual chains.

12.10.1 Scanning Probe Microscopy

Different types of supramolecular polymers have been studied by STM [114–119]. One representative example in this respect is the material that was obtained from the self-assembly of the cyclodextrin-Tb(III) array **15** with C_{60} (Figure 12.16a) [114]. STM imaging of the polymer, when deposited on a HOPG surface (HOPG:

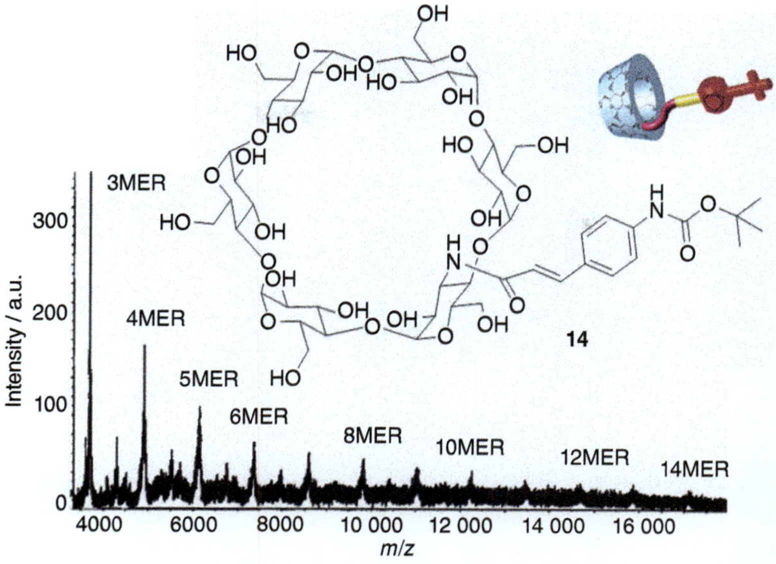

Figure 12.15 Positive turbo ion spray time-of-flight MS of **14**. Source: Miyauchi et al. [63]. © 2005 American Chemical Society.

highly ordered pyrolytic graphite), clearly revealed the fine structure of the chain-like assembly (Figure 12.16b); the width and height of the observed structure corresponded with those deduced from a computed model (MM2 calculations, Figure 12.16c).

The structure of a supramolecular, biocomponent polymer (**16**, Figure 12.17a) was investigated by the Lehn group. The material, based on highly directive sextuple hydrogen-bonding interactions, was formed at the liquid–graphite interface [117]. The monolayer of **16** formed by physisorption on the surface displayed a polycrystalline structure, which was stable over time (Figure 12.17c). The observed packing mode reflected the expected arrangement of a Hamilton receptor and a cyanurate (Figure 12.17b, see also Section 3.2.1.5). An area of $5.5 \pm 0.6\,\mathrm{nm}^2$ per dimer was calculated from the unit cell parameters, corresponding to one molecule of each component within each unit cell.

As a highly specialized technique, ultrahigh-vacuum low-temperature STM (UHV-LT-STM) imaging has been introduced, to image structures with atomic resolution via the direct measurement of electronic states; the STM images are typically acquired at 6 K on a clean Au(III) surface. This method has recently been applied, for example, to visualize single metallo-supramolecular chains [120]. The stepwise assembly procedure (see Chapter 4), using Fe(II) and Ru(II) ions, gave linear chains containing loops in every repeat unit. Using UHV-LT-STM in combination with atomic-scale scanning-tunneling spectroscopy (STS) allowed one to discriminate between two different topologies (arising from intra- and intermolecular coordination, Figure 12.18).

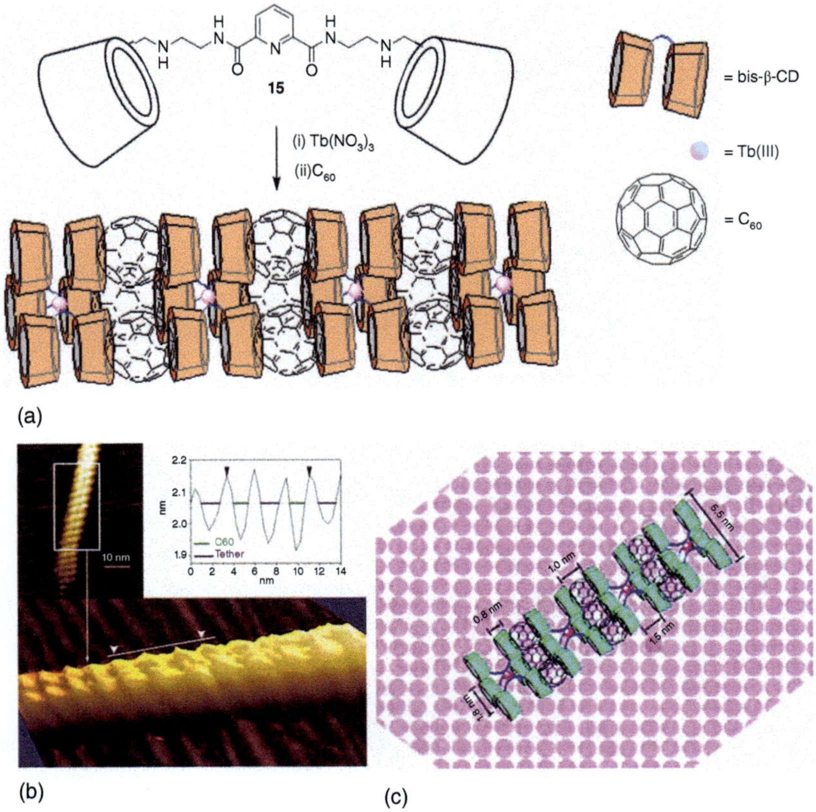

Figure 12.16 (a) Schematic representation of the self-assembly of **15** and C_{60} into a supramolecular polymer; (b) STM image of the polymer deposited onto a HOPG surface (the inset shows the corresponding line profile; (c) schematic representation of the polymer with structural parameters. Source: (a, c) Liu et al. [114]. © 2006 American Chemical Society, (b) Figure reproduced with kind permission; © 2006 American Chemical Society.

In a similar fashion, the shape-persistent metallo-supramolecular nanostructures, prepared via sophisticated stoichiometric self-assembly processes, were analyzed with these two methods. Starting from a fractal polymer, i.e. the molecular Sierpiński hexagonal gasket, a broad range of such assemblies have been prepared making use of multistep directional and one-step self-sorting procedures [123]. The Sierpiński motif, containing Fe(II) and Ru(II) ions, was visualized by UHV-LT-STM imaging: a line of such gaskets settled on a ridge on the Au(111) surface; for each gasket, a precise hexagonal pattern with 12 ± 1 nm diameter and 0.8 nm height was determined (Figure 12.19) [121]. Moreover, tunneling conductance spectra showed a single molecule at 6 K with a 1 eV energy gap. Very recently, Zhang and coworkers reported the synthesis of cyclic mixed-metal nanostructures under kinetic and thermodynamic control [124]. Going far beyond the structural characterization, all eight possible positional isomers were identified on a Ag(111) substrate by UHV-LT-STM imaging and STS. These authors also

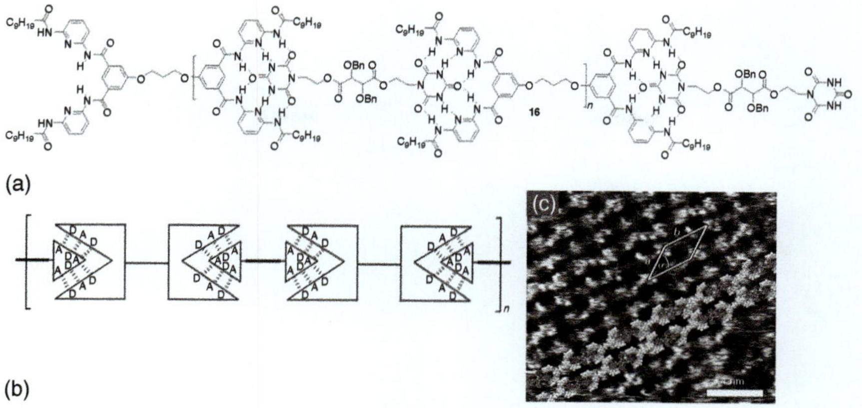

Figure 12.17 (a, b) Schematic representation of the supramolecular polymer **16** and the proposed binding motif. (c) STM image of a monolayer of **16** at the liquid–graphite interface (self-assembled from a 1,3,5-trichlorobenzene solution, the scale bar is 4 nm). Source: (a, b) Ciesielski et al. [117]. Figure reproduced with kind permission; © 2009 John Wiley & Sons, (c). Ciesielski et al. [117]. © 2009 Wiley-VCH.

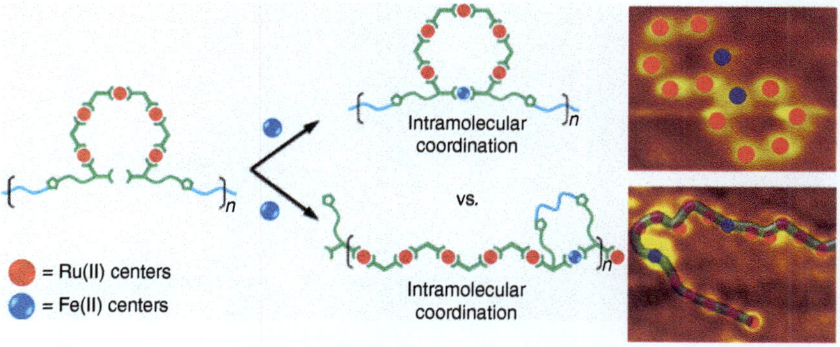

Figure 12.18 Schematic representation of the formation of loops within metallo-supramolecular polymers. The corresponding UHV-LT-STM images are also depicted. Source: Li et al. [120]. © 2020 American Chemical Society.

generated Zn(II)-containing supramolecular "Kandinsky circles" and "spiderwebs" by applying efficient self-sorting strategies (Figure 12.20) [122]. These structures, as well as a 20-nm-wide metallo-supramolecular hexagonal grid [125], were characterized *inter-alia* by UHV-LT-STM imaging and STS. These studies show how advanced microscopy-imaging techniques can be utilized to accurately determine the structure and dimensions of supramolecular assemblies.

Atomic force microscopy (AFM) represents one further high-resolution type of SPM (in the sub-nm regime) that has also been employed to image the morphology of supramolecular sample surfaces and can yield a 3D surface profile [127–133]. In contrast to STM imaging, also nonconductive surfaces can be investigated by this technique [134]. The Kurth group applied AFM as an analytical tool to determine the shape and length of rigid-linear metallopolymers that were spin-coated

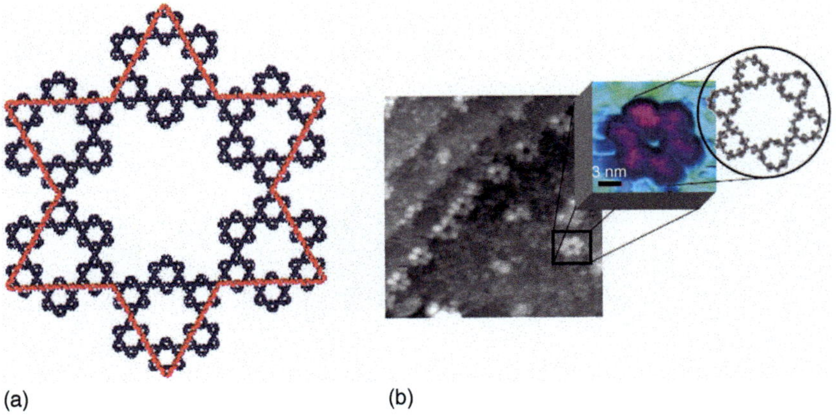

Figure 12.19 (a) Schematic representation of a metallo-supramolecular Sierpiński hexagonal gasket. (b) Visualization of the assembly by UHV-LT-STM imaging. Source: Newkome et al. [121]. Figure reproduced with kind permission; © 2006 AAAS.

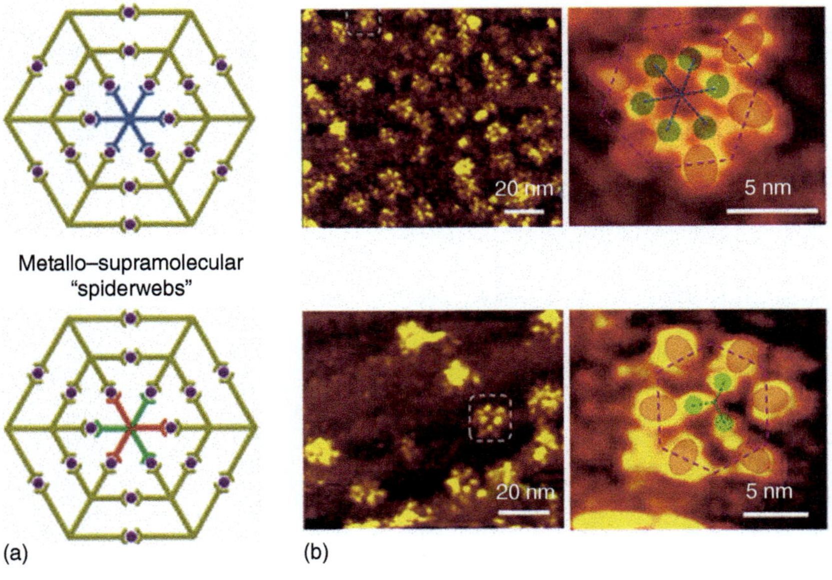

Figure 12.20 (a) Schematic representation of two metallo-supramolecular "spiderwebs." (b) The corresponding UHV-LT-STM images are also depicted. Source: Wang et al. [122]. Figure reproduced with kind permission; © 2019 American Chemical Society.

onto mica substrates [126]. For metallopolymer **17**, the DP could be adjusted via the ligand-to-Fe(II) ratio. The increase of the DP with increasing ratio was monitored by AFM imaging: at ratios of 0.6, 0.85, and 0.98, the estimated length of the rods was 150, 320, and 460 nm, respectively (Figure 12.21a–c). The height of the rods was 1.2–1.6 nm, thus being close to the value derived from X-ray scattering experiments (c. 1.2 nm) [135]. In a control experiment, metallopolymer **18**, based on a flexible bis-terpyridine ligand, was investigated, indicating the presence of smaller, discrete

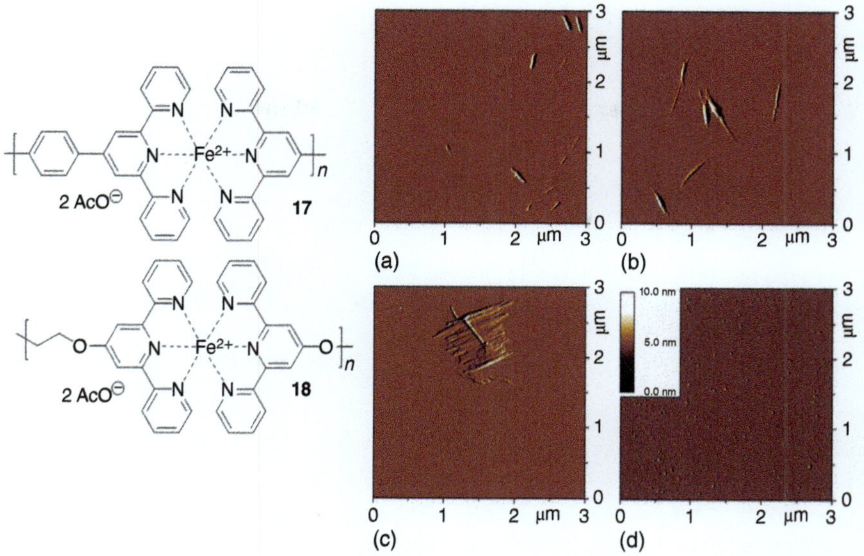

Figure 12.21 AFM images of spin-coated solutions of **17** (a–c) and **18** (d) on mica substrates; for **17**, various ligand-to-metal ratios were utilized (a: 0.6, b: 0.85, c: 0.98). Source: Schwarz et al. [126]. Figure reproduced with kind permission, © 2010 American Chemical Society.

assemblies ($d \approx 6$ nm, Figure 12.21d) – presumably rings with three to four repeating units.

Beyond its utilization as microscopy imaging method, AFM also allows the measurement of inter- and intramolecular interaction forces with pN resolution [134]. When used for 1D-force measurements, this particular version of AFM is referred to as single-molecule force spectroscopy (SMFS) or sometimes molecular force probe (MFP). AFM on the one hand and SMFS on the other hand have to be distinguished from each other: When applying it in the imaging mode, the AFM cantilever scans the surface of a sample in a line-by-line fashion; whereas, in SMFS experiments, the cantilever is lifted perpendicular to the sample's surface [136, 137]. The motion of the cantilever is operated by a piezo element, thus offering a sub-nm resolution. Displacement of the tip from the surface is directly recorded by a capacitor or a linear-voltage differential transformer; as a result, this method enables one to obtain force-extension curves of single molecules with remarkably high resolution. Many examples are known from the literature where SMFS has been applied [19]: entropic/enthalpic elasticity of single polymer strands, force-induced conformational transitions, melting or unzipping force of double-stranded DNA, interactions of macromolecules and small molecules, interfacial conformation, and desorption force of macromolecules.

The scope of SMFS has also been extended to (metallo-)supramolecular polymers. Kudera et al. measured the binding strength of Ru(II)-terpyridine bis-complexes at the molecular level [138]. For this purpose, the authors assembled a Ru(II)-terpyridine complexes, as the metallo-supramolecular linkage between the

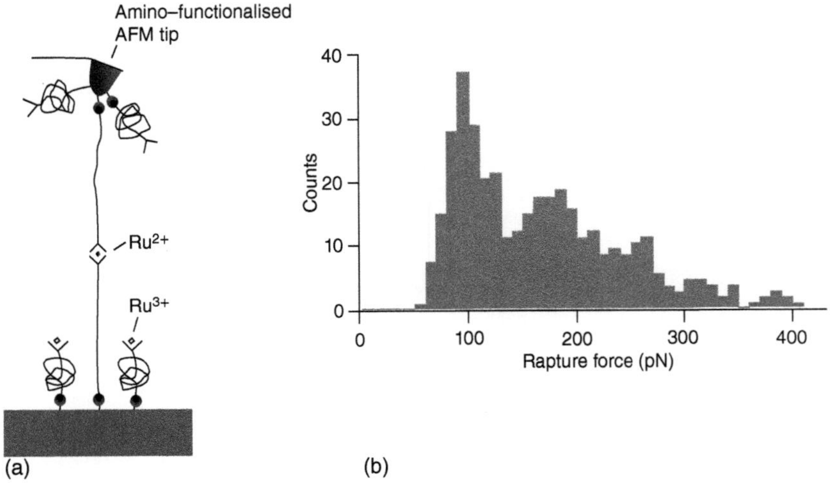

Figure 12.22 (a) Schematic representation of the experimental setup; (b) histogram of the bound rapture forces at a pulling velocity of 118 nm s^{-1}. Source: Kudera et al. [138]. © 2003 John Wiley & Sons.

PEG chains that were pending from the surface and an AFM tip (Figure 12.22a). The force-extension curves were measured by slowly lifting the cantilever up to the point of the rupture. A binding force of c. 95 pN, at a force loading of 1 nN s^{-1}, was measured (Figure 12.22b); the histogram of the bond rupture featured two weaker peaks (at 171 and 253 pN). Based on Monte Carlo simulations, the latter were assigned to the simultaneous rupture of two or three complexes, respectively.

In a different approach, the Vansco group studied a supramolecular polymer, based on quadrupole H-bonding interactions, by SMFS [139]. Like the previous example, both the AFM tip and the substrate were functionalized with UPy moieties. In the presence of a solution of a homoditopic monomer with two UPy end groups, a supramolecular polymer bridging the tip and substrate was formed (supramolecular polymers, based on the UPy–UPy dimerization, are discussed in Chapter 3). A maximum stretching length of c. 150 nm could be achieved, corresponding to a DP of ≥15. The addition of dimethylsulfoxide (DMSO), as a strong competitor for the H-bonding interaction, induced the disassembly of the supramolecular polymer chain. Related to the former two studies, also other types of supramolecular polymers have been analyzed by SMFS: supramolecular polymers, based on Pd(II)-pincer complexes [140], oligonucleotide strands [141] and host-enhanced CT interactions (i.e. making use of the host–guest chemistry of cucurbit[8]urils) [142].

12.10.2 Electron Microscopy

Two types of electron microscopy are commonly used for the investigation of supramolecular materials: transmission electron microscopy (TEM) and scanning

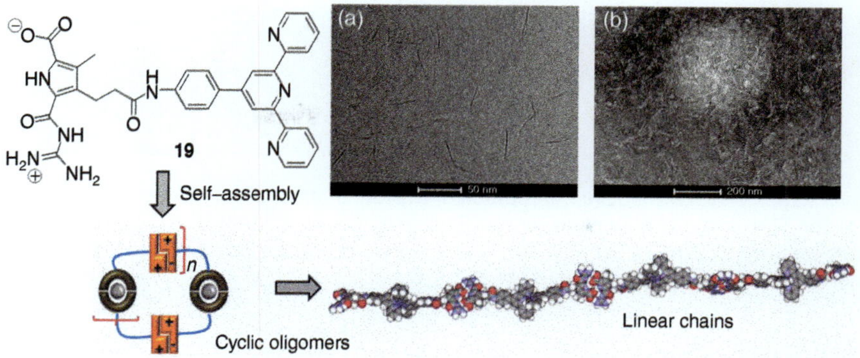

Figure 12.23 Schematic representation of the self-assembly of **19** into a linear supramolecular polymer, based on two different types of noncovalent interactions; the cryo-TEM images of $\{[Fe(\mathbf{19})_2]Cl_2\}_n$ are also shown (a: no staining, b: negative staining with uranyl acetate). Source: Gröger et al. [143]. Figure reproduced with kind permission; © 2011 American Chemical Society.

electron microscopy (SEM). Typically, TEM is utilized for the visualization of aggregates formed by (supramolecular) polymers [13, 40, 144–148]. The Schmuck group reported on the self-assembly of the heteroditopic building block **19** via metal-to-ligand coordination (i.e. complexation of Fe(II) ions by the tridentate tpy ligand) and dimerization of a self-complementary zwitterion [143] (Figure 12.23). A linear supramolecular polymer could be evidenced by various analytical methods (also confirming the presence of cyclic oligomers, as intermediates). Cryo-TEM, a technique giving direct structural data from vitrified solutions, showed linear molecules in aqueous solution with a length of 30–100 nm (Figure 12.23a, a length of 50 nm corresponds to a DP of c. 13). Due to the presence of metal centers, generally a good contrast to the image can be achieved. Hydrophobic interactions between the polymer strands led to the aggregation of single polymers into larger assemblies. Negative staining with uranyl acetate enhanced this process giving globular clusters with diameters ranging from 200 to 600 nm (Figure 12.23b).

The Newkome group prepared a water-soluble hexameric metallo-macrocycle that formed, as a multiple cation, an ion pair with the negative sites of oxidized single-wall carbon nanotubes (ox-SWCNTs) [150]. The composite material was characterized by TEM: the increase in diameter to c. 100 nm – in comparison to commercial ox-SWCNT – was assigned to the assembly of the orthogonally charged species via electrostatic interaction. Schmittel and coworkers constructed a dynamic "polynanocage" in a two-step procedure (Figure 12.24a) [149]: first, the heteroleptic phenanthroline (HETPHEN) methodology was applied to assemble a trisphenanthroline ligand and a linear bis(pyridinecarbaldehyde) derivative with Cu(I) ions into a nanoprism (HETPHEN: *het*eroleptic *phen*anthroline complexation [151]); the six aldehyde entities (each three on the top and on the bottom of the cage) were subsequently reacted with *p*-phenylenediamine to generate the polymeric material, based on Schiff base bindings. The formation of a polymer-like assembly

Figure 12.24 (a) Schematic representation of the formation of a "polynanocage"; (b and c) AFM and TEM image of the polynanocage. Source: (a) Fan et al. [149]. © 2012 American Chemical Society; (b, c) Fan et al. [149]. Figure reproduced with kind permission; © 2012 American Chemical Society.

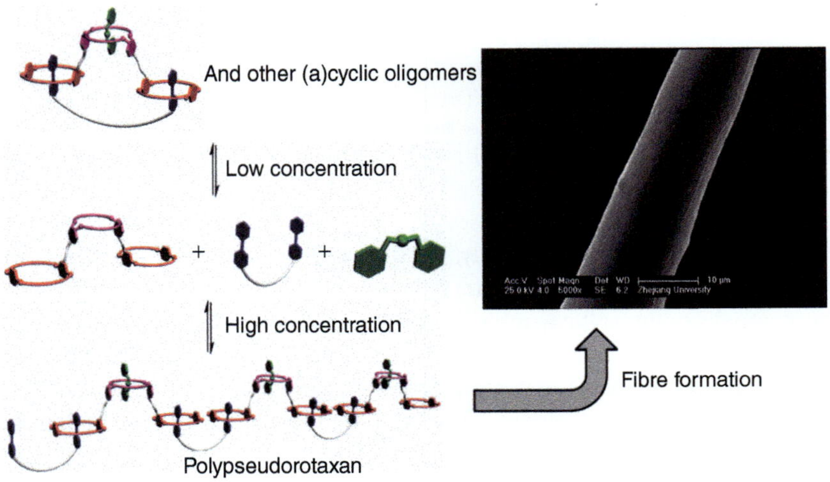

Figure 12.25 Schematic representation of the formation of a pseudorotaxane of high molar mass that could be processed into rod-like fibers. Source: Wang et al. [152]. Figure reproduced with kind permission; © 2009 Royal Society of Chemistry.

was confirmed by a broad range of characterization tools, including AFM and TEM imaging (Figure 12.24b/c).

SEM represents the second important type of electron microscopy; this particular method allows to image supramolecular materials of μm sizes (or even beyond). Wang et al., for example, prepared rod-shaped fibers of regular shape from a highly concentrated solution of a crown ether-based polypseudorotaxane (Figure 12.25) [152]. Generally, fiber formation is indicative of the high molar mass of a polymer – thus, SEM was applied as an indirect proof of the high DP of the self-assembly process. More recently, similar studies employing SEM for analyzing supramolecular polymers, based on CT interactions [153], or Ru(II)-based metallopolymers (obtained by electropolymerization) [154] have been published.

12.11 Small/Wide-Angle X-Ray Scattering

From the established X-ray diffraction (XRD) methods available, mainly two techniques are commonly used to analyze the structure of polymers in the solid

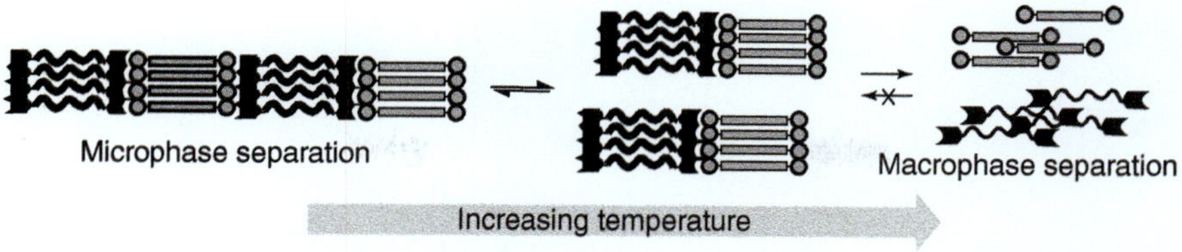

Figure 12.26 Schematic representation of supramolecular diblock copolymers, based on multiple H-bonding interactions and their temperature-dependent microphase-separation behavior [the two different representations denote a polyisobutylene and a poly(ether ketone) constituent block, respectively]. Source: Binder and Zirbs [145]. © 2006 Springer.

state: wide- and small-angle X-ray scattering measurements [wide-angle X-ray scattering (WAXS) and small-angle X-ray scattering (SAXS), respectively]. The former represents a versatile technique for evaluating the crystalline structure, the disorder, and orientation of a material; whereas, the morphology of semicrystalline polymers can be investigated by SAXS [155]. Consequently, a broad range of macromolecular materials have been subjected to solid-state analysis by WAXS and SAXS – ranging from block copolymers to charged polymers, e.g. ionomers [156]. A proper analysis of the scattering data allows one to conclude the macroscopic morphology of a polymer, i.e. the (semi)crystalline, mesomorphic, or amorphous structure.

Diblock copolymers can be obtained either by covalently linking the constituent blocks or by utilizing supramolecular interactions for the same purpose. Examples for making use of the latter methods can, in particular, be found for H-bonding interactions (Chapter 3) and metal-to-ligand-coordination (Chapter 4). For example, Binder et al. investigated supramolecular polymers in which the phase-separating blocks, i.e. poly(etherketone) (PEK) and poly(isobutylene) (PIB), were linked via triple [157] or sextuple H-bonding interactions [145]. The microphase separation of the immiscible blocks was confirmed by TEM imaging and SAXS investigations (Figure 12.26). Furthermore, the strength of the employed H-bonding interaction was found to strongly impact the microphase-separation behavior: the sextuple hydrogen-bonding array offered a stronger binding compared to the analogous system, based on triple H-bonding interactions – even up to temperatures of 80 °C, thus beyond the glass-transition temperature (T_g) of the PEK as the higher melting constituent.

In a similar fashion, Al-Hussein et al. compared the melt morphology of a Ru(II)-containing diblock copolymer with a hard polystyrene and a soft PEG block to that of a conventional diblock copolymer missing the central terpyridine bis-complex [158, 159]. Due to the presence of the charged complexes, aggregation occurred over a wide temperature regime – as revealed by a broad signal in the SAXS curve.

Rowan and coworkers investigated the chain-extended polymers, obtained from the self-assembly of macroligand **20** with Zn(II) ions [13]. The SAXS measurements of the thin spin-coated films revealed a microphase-separated lamellar morphology in which the Zn(II) complexes formed a hard phase physically cross-linking the

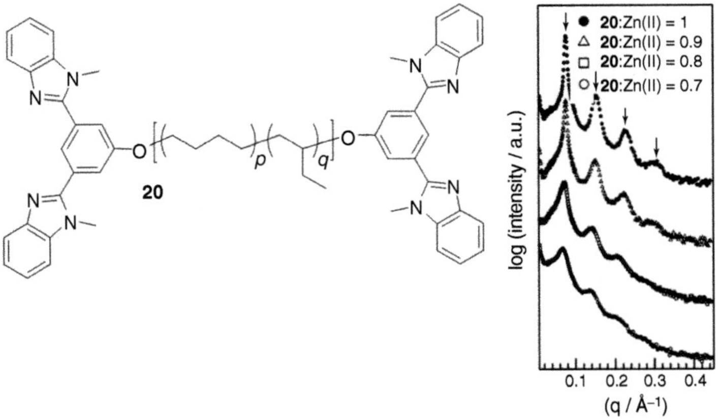

Figure 12.27 SAXS data for the metallo-supramolecular polymer obtained from macroligand **20** with Zn(II) ions [SAXS curves at various **20** to Zn(II) ratios are shown]. Source: Burnworth et al. [13]. © 2011 Nature Publishing Group.

poly(ethylene-*co*-butylene) soft domains. The SAXS data showed strong Bragg diffraction maxima at integer multiples of the scattering vector magnitude of the primary diffraction peak, being characteristic for well-ordered layered morphologies (Figure 12.27; the periodicity of the lamellar increased with decreasing Zn(II)/macroligand ratios).

12.12 X-Ray Crystallography

In contrast to organic and inorganic molecules, for which a structural characterization by XRD analysis represents a highly common technique, its application to polymers and, in particular supramolecular ones, is extremely rare. This limitation can be attributed to the intrinsic properties of synthetic polymers that might be highly crystalline, semicrystalline, microcrystalline, or amorphous; it might even be possible to observe these in a single polymer sample. Whereas, XRD analysis represents a powerful analytical tool in understanding the structure and function of biopolymers (e.g. DNA and proteins) [160]. For structurally "more defined" molecules, single crystals suited for XRD analysis are comparably easy to obtain, and thus, crystallography can readily be performed – this also holds true for monodisperse supramolecular assemblies, e.g. 2D and 3D architectures, based on metal-to-ligand coordination [161–163]. Regarding crystalline supramolecular polymers, which have been analyzed by XRD, some examples will be given in the following.

The threading of CD rings onto preformed polymers is an important application with respect to polymer processing (see Section 9.2.1) [165]. For many of these polypseudorotaxanes, the formation of channel-like crystalline structures was reported [7, 166]. The structural characterization by XRD analysis was, for example, performed for the β-CD-containing polypseudorotaxanes, derived of the nonionic surfactant Igepal CO-520 [167], poly(trimethylene oxide) (P3MO)

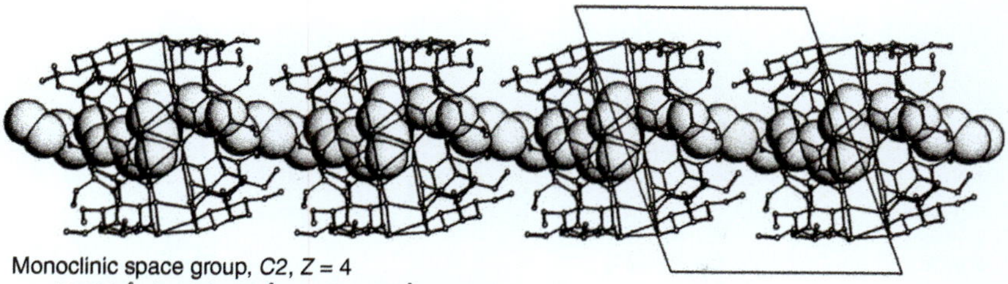

Figure 12.28 Representation of the solid-state structure of the β-CD–PEG polypseudorotaxane, with the PEG chain reaching through four unit cells of β-CD dimers. The H-bonding interactions between pairs of head-to-head connected CDs are shown (view along the *b*-axis, hydrogen atoms are omitted for clarity). Source: Udachin et al. [164]. © 2000 American Chemical Society.

[168], poly(propylene glycol) (PPG) [168], and PEG [164]. In the latter case, the structure is composed of a PEG chain on which β-CD dimers were threaded in a head-to-head-type fashion (a ratio of 1 : 3 with respect to CD rings and the monomeric $-C_2H_4O-$ units of the polymer were found, Figure 12.28). The PEG chain exhibited a highly disordered conformation within the channels formed by the threaded CD moieties held together by H-bonding interactions.

Also, for other types of host–guest interactions, examples for crystalline supramolecular polymers can be found. Huang and coworkers obtained single crystals of the supramolecular polymer $(\mathbf{21})_n$ based on the polyassociation of the alkyl-functionalized pillar[5]arene **21**, as an AB-type monomer (Figure 12.29) [169]. In a similar fashion, Colquhoun and coworkers reported on the single-crystal structure of a polypseudorotaxane of the donor–acceptor type [170].

For many coordination polymers, the solid-state structure has been reported; however, for most of these materials, the definition of supramolecular polymers cannot be applied (i.e. the polymer structure must be retained also in solution). This point of view excludes metal-organic frameworks (MOFs) [171] as well as the purely organic ones (i.e. relying on ionic and/or H-bonding interactions only) [172].

Though in very few cases, the single-crystal structure of supramolecular polymers could be determined, in most cases small-molecule model assemblies/complexes have to be formed, crystallized, and analyzed accordingly by XRD measurements to conclude on the overall structure of the related supramolecular polymer.

12.13 Small-Angle Neutron Scattering

SANS represents another classic technique, in particular suited for determining the M_w value and the radius-of-gyration (R_g) of polymers in solution (a few studies have also dealt with the determination of samples dispersities). The complex fundamentals of the method have been summarized elsewhere [174], and only some basic aspects are addressed here. Under dilute conditions, i.e. at polymer concentrations well below the critical concentration for overlap of the coils, scattering

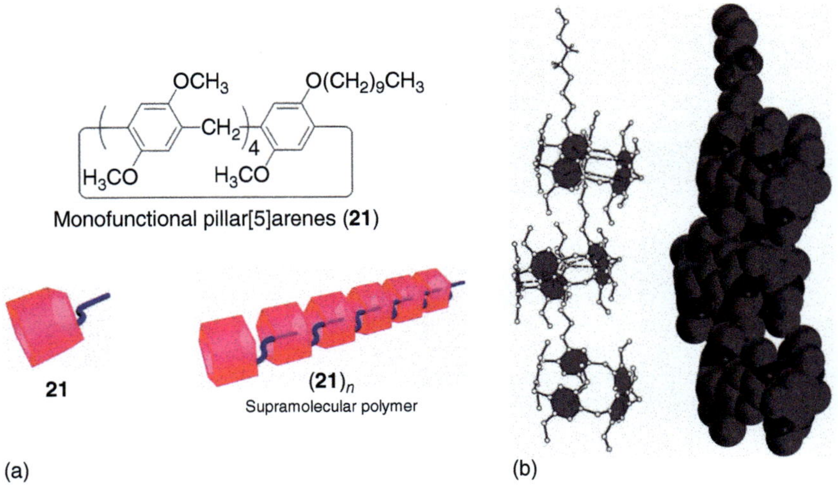

Figure 12.29 (a) Schematic representation of the supramolecular polymer formed by the polyassociation of the AB-type monomer **21**. (b) Two views of the 1D arrangement of $(\mathbf{21})_n$ (only the hydrogen atoms involved in the C–H···π interactions are shown, all other hydrogen atoms were omitted for clarity). Source: Zhang et al. [169]. © 2011 John Wiley & Sons.

will hardly be influenced by intermolecular interference. In this case, the scattered intensity (I) vs. the scattering vector (Q) can be discussed in three different regions (Figure 12.30). At low Q values, the neutron radiation is probing large distances, therefore the scattering is sensitive to the overall dimension of the randomly coiled polymer chain; this is referred to as the Guinier region and I appears almost as a plateau in the bilogarithmic plot. The intermediate region features a negative slope. The probed distances correspond to the interior of the polymer coil, and thus, the scattering is sensitive to chain statistics. For a Gaussian chain, as expected in a theta solvent (i.e. under conditions where the polymer coils behave as ideal chains), the scattering decays as Q^{-2}, and for a self-avoiding walk, as expected in a good solvent, it decays as Q^{-1} [175]. The crossover between the Guinier and the intermediate region occurs at $Q \sim R_g^{-1}$; however, the precision of the crossover is typically too low to allow a reliable determination of R_g.

The local structure of the macromolecule is then reflected in the third region (the so-called Porod region). For thin rigid-rod polymer chains, the scattering will decay as Q^{-1}. The crossover between the intermediate region and the Porod region occurs at $Q \sim L^{-1}$ (L: persistence length of the polymer); again, this is too imprecise to generate reliable values for L. Thus, in practice, the Guinier region is better suited to extract information on a polymer's structure from the scattering data [173].

Whittell et al. showed the example of a series of almost monodisperse poly(ferrocenyldimethylsilane)s that SANS can be applied to determine the R_g, M_w, and $Đ$ values of metallopolymers in solution [173]. The good agreement of the estimated data with those obtained from a parallel SEC analysis indicated the high potential of SANS in characterizing polymers, including supramolecular ones. This implies in particular for those systems where other methods cannot be

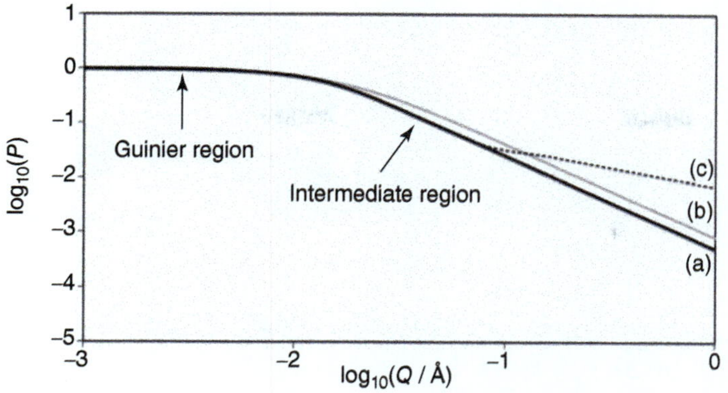

Figure 12.30 The form factor (P) for two polymers whose chains follow self-avoiding walk statistics and which have a R_g of 100 Å. The Guinier and intermediate regions can be clearly seen. For (a) the chains are perfectly monodisperse ($Đ$ of 1.0) and for (b) the $Đ$ value is large (i.e. $M_w/M_n = 6.4$). The dashed line (c) qualitatively indicates the effect of a persistence length (L) of c. 10 Å, giving a third region with a slope of −1. Source: Whittell et al. [173]. © 2013 John Wiley & Sons.

applied due to column interactions (SEC), paramagnetism (NMR spectroscopy), strong absorption of visible light (light scattering), or dilution/dissociation effects (SEC or AUC).

For example, cucurbit[8]uril-based supramolecular polymers were studied by SANS measurements [176]. Zhang and coworkers confirmed the formation of supramolecular polymers of high molar masses in water; however, the derived Porod exponent of c. 2.5 indicated that the polymer chains were collapsed into coils in this environment [177].

Regarding supramolecular polymers, assembled via H-bonding, the study by Brás et al. is worth mentioning [178]. The authors showed that the supramolecular polymerization of two homoditopic monomers could be described by random walk statistics. The SANS data were analyzed using a random phase formalism, which is typically applied to pure polymer systems. Both in solution and in the melt, linear assemblies could be observed, and the presence of macrocyclic species or micellar aggregates could be excluded.

Terech and coworkers prepared a set of metallo-supramolecular polymers, based on the multitopic cyclam-centered bis-terpyridine ligand **22** (Figure 12.31a) [179]. Various parameters were identified as crucial for the formation of viscoelastic gels (i.e. nature of the metal and counter ion, stoichiometry, concentration, solvent polarity). Moreover, the Co(II)-containing metallopolymers revealed a reversible gel-to-liquid transition when the metal centers were oxidized to the +III state (Figure 12.31b). SANS, as a noninvasive technique, was applied to identify the supramolecular structure in the gelated state. Basically, swollen gels can also be probed with SANS on the basis of the elastic deviation of a neutron beam and the analysis of the large-scale fluctuations of the neutron scattering length density. However, when applying SANS to molecular gels, one has to consider the special features of self-assembled networks, such as the presence of plain fibers, bundles,

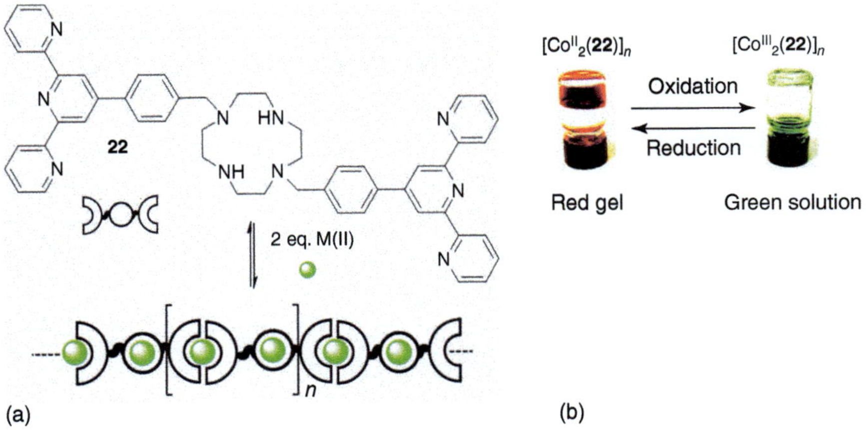

Figure 12.31 (a) Schematic representation of the metallo-supramolecular polymerization of **21** with various divalent metal ions [e.g. Co(II), Ni(II)]. (b) The Co(II)-containing polymer showed an electrochemically driven switchability between the gel and sol states. Source: (a) Gasnier et al. [179]. © 2009 American Chemical Society, (b) Gasnier et al. [179]. Figure reproduced with kind permission; © 2009 American Chemical Society.

and (to some extent) organized domains. The authors concluded from SANS and rheology measurements that linear rigid-rod metallo-supramolecular polymers were formed and any cross-linking involving metal-to-ligand interactions could be excluded. Thus, the 3D network could only be established on the basis of weak fiber–fiber interactions being responsible for the highly shear-sensitive rheological behavior. It has to be pointed out that coordination of metal ions by the chelating cyclam units played a crucial role with respect to the 1D chain growth.

12.14 Asymmetric Flow Field-Flow Fractionation

In 1987, Wahlund and Giddings introduced asymmetric flow field-flow fractionation (AF4), which is today one of most important field-flow fractionation techniques [180]. The in-line connection of an AF4 setup to a detector (e.g. ultraviolet, differential refraction, or light scattering) gives versatile information of the sample under investigation: molar mass, dispersity, size, and shape/conformation. Since no stationary phase is required, AF4 represents a mild separation and characterization method, almost circumventing nonspecific interactions and shear forces, thus enabling a broad separation range from nano- up to micrometers – far beyond the scope of other common separation techniques (Figure 12.32a). These features make AF4 a powerful and versatile tool in view of characterizing macromolecules as well as supramolecular assemblies [181–183]. The separation in AF4, a typical setup is depicted in Figure 12.32b, is just achieved by an eluent flow through a channel, where a perpendicular flow force is applied. The channel itself consists of two joined, spacer-separated plates. The bottom one, which is porous, is covered by a semipermeable membrane acting as an accumulation wall. Due to the cutoff

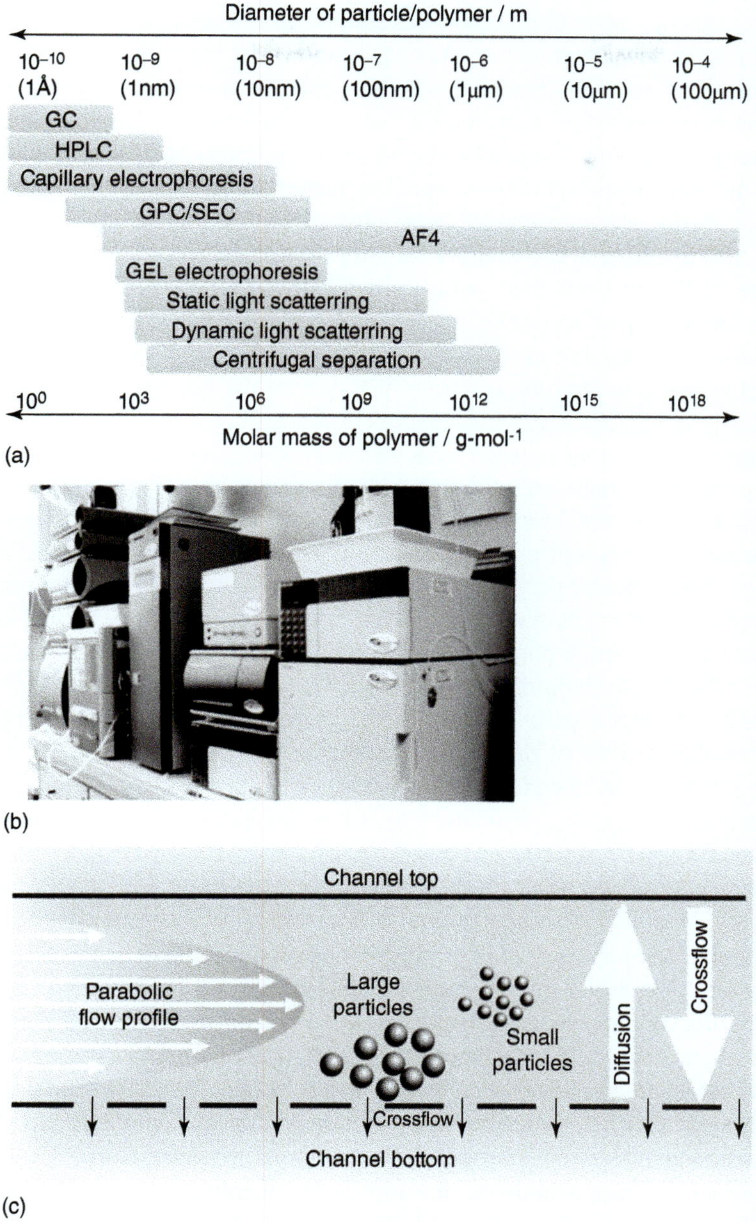

Figure 12.32 (a) Comparison of the molar mass and size detection/separation range for different techniques; (b) Picture of a typical AF4 setup (www.schubert-group.uni-jena.de); (c) Schematic representation of the separation field in a AF4 channel (www.nanolytics.de). Source: (a) Yang et al. [177]. © 2015 American Chemical Society, (b) U.S. Schubert, (c) © Nanolytics.

(i.e. typically 5–10 kDa), the membrane is permeable only for the solvent and small molecules but impermeable for most polymers or nanoparticles. Therefore, the sample is kept in the channel and flow-transported to the channel outlet. The principle of a field-flow fractionation channel is shown in Figure 12.32c (www .nanolytics.de). The parabolic eluent flow profile within the channel leads to a slower movement closer to the channel walls than in the central part of the channel. Simultaneously, a cross-flow is applied while the sample is transported by the eluent pushing the macromolecules/particles toward the membrane on the bottom. During this process, diffusion acts as counterforce, eventually establishing an equilibrium distribution of the sample in the channel. As a result, smaller molecules/particles having higher diffusion coefficients will be found in the channel center. Thus, the samples are separated according to their size: smaller molecules are eluted before the larger species, due to the differences in the eluent flow velocity (this separation is opposite to that known from SEC, where the large molecules are eluted first).

When coupled to different detector systems (*vide supra*), AF4 can be utilized to determine the molar mass and dispersity of supramolecular polymers. Advantageously, AF4 allows the investigation even of complex heterogeneous and polydisperse mixtures without prior fractionation. Though the versatility of AF4 in analyzing supramolecular polymers has already been documented in literature [177], some crucial limitations in view of studying supramolecular polymers need to be addressed: supramolecular polymers, based on labile non-covalent interactions, are prone to dissociation upon dilution – as discussed earlier for the application of SEC, this is also problematic in AF4. Therefore, reliable results will only be obtained from the analysis of those materials that retain their macromolecular structure in solution – independent of concentration. Moreover, AF4 is a highly specialized technique requiring particular parameters for analyzing a polymer sample (i.e. optimized solvent composition and eluent flow, etc.). Generalized methods have, up to now, not been established – at least not in view of supramolecular polymers.

As an example for applying AF4 in studying a supramolecular polymerization, Huang et al. employed the tetracationic building block **24** for the two-step self-assembly of a supramolecular polymer, based on the size-selective binding of cucurbit[7]uril and cucurbit[8]uril (Figure 12.33a) [184]. Due to self-sorting of the components, the same type of material was formed – independent of the order of the assembly steps. The highest molar mass was observed at an equimolar ratio of the three components ($M_W = 97\,000\,\text{g mol}^{-1}$, $Đ = 1.5$); with increasing CB[8] : CB[7] ratio, a decrease of the molar mass, due to depolymerization, became evident (Figure 12.33b).

12.15 Taylor Dispersion Analysis

The aforementioned X-ray or light scattering methods represent appropriate tools for the characterization of multichain aggregates; the same holds true for DOSY and

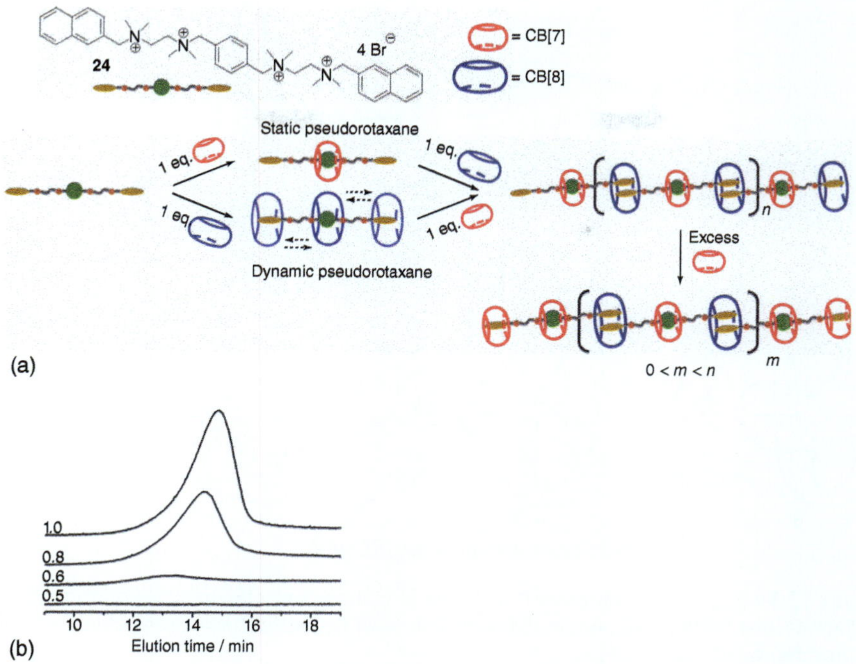

Figure 12.33 (a) Schematic representation of the supramolecular self-sorting polymerization of **24** with two different cucurbituril derivatives. (b) Representation of the ratio-dependent AF4 elution curves obtained by the MALS detector (the labeling of the curves indicates the applied CB[8]-to-CB[7] ratios). Source: Huang et al. [184]. © John Wiley & Sons.

SEC measurements (the latter when coupled to a viscometer). However, the application of all these techniques to characterize single-chain nanoparticles (SCNPs) remains a challenge [186]. SCNPs are relatively young class of nanoparticles, having diameters >30 nm, which can be formed, e.g. by the intramolecular cross-linking of polymers via metal-to-ligand complexation within their ligand-equipped side chains (Figure 12.34) [185]. The limitations in analyzing such assemblies can, at least to some extent, be circumvent by utilizing the so-called Taylor dispersion analysis (TDA), which allows one to determine the hydrodynamic radii over some orders of magnitude in a highly accurate fashion. Thus, small molecules, proteins, macromolecules, nanoparticles, and their self-assembly processes in solution can be studied by TDA [187].

Though the fundamentals of this approach were already introduced by Taylor in the 1950s, it's application in the field of polymer analytics emerged rather recently [189]. Basically, three independent phenomena are employed for the particle size determination by TDA – the optical extinction, the translational diffusion, and the sheer-enhanced dispersion of nanoparticles under a steady laminar flow [190]. According to Lambert–Beer's law [Eq. (12.4)], the particle's UV/vis absorption scales linearly with the size, i.e. the hydrodynamic radius (even when some larger aggregates are present) [191]. As elaborated by Balog and coworkers, the SCNP's

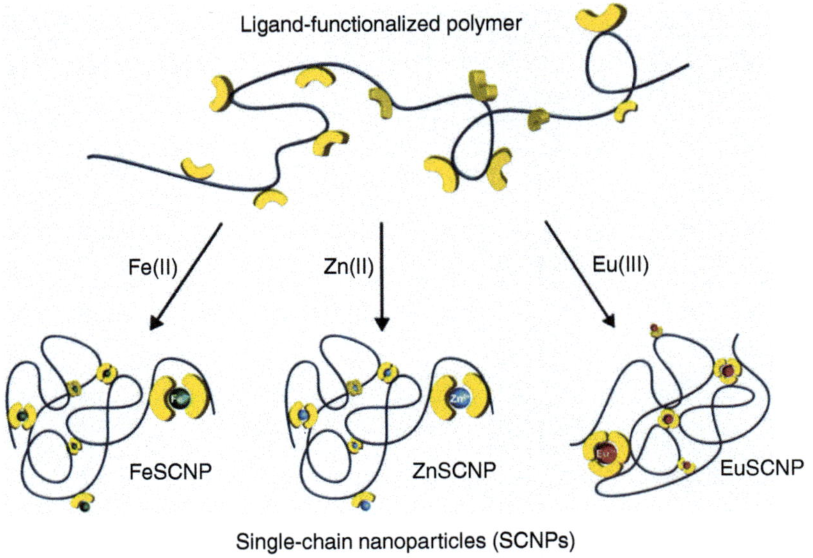

Figure 12.34 Schematic representation of the formation of single-chain nanoparticles by intramolecular metallo-supramolecular crosslinking of ligand-functionalized polymers. Source: Neumann et al. [185].

translational self-diffusion and sheer-enhanced dispersion, under a steady laminar flow through a microfluidic channel, are analyzed by recording the absorption profile at a certain wavelength (Figure 12.35a) [188]. In such an experiment, the combination of convection and diffusion effects results in changes of the width of the absorbing particles' concentration band within the channel. The hydrodynamic radii are finally obtained by fitting the measurement results to the Stokes–Einstein equation.

Very recently, Schrettl and coworkers applied TDA to their intramolecularly cross-linked polymers (Figure 12.34) [185]. To suppress the intermolecular cross-linking and favor the SCNP formation, the metal-to-ligand complexation of the pending tridentate binding sites was conducted in a highly dilute concentration regime. The thus-obtained TDA profiles for three different types of metal ions and the calculated hydrodynamic radii are summarized in Figure 12.35b. The Eu(III) ions prefer tris-tridentate coordination geometry, and thus, the SCNPs appeared more compact than those obtained with Fe(II) or Zn(II) ions. It has also been demonstrated that a polydispersity analysis, based on TDA measurements, is possible [192, 193]; however, this strategy has not yet been applied to supramolecular materials.

12.16 They Are Very Complex Structures but Totally Timely ...

The question of what the structure of a molecule is like should, ideally, always be answered with a counter question: what is the analytical technique under

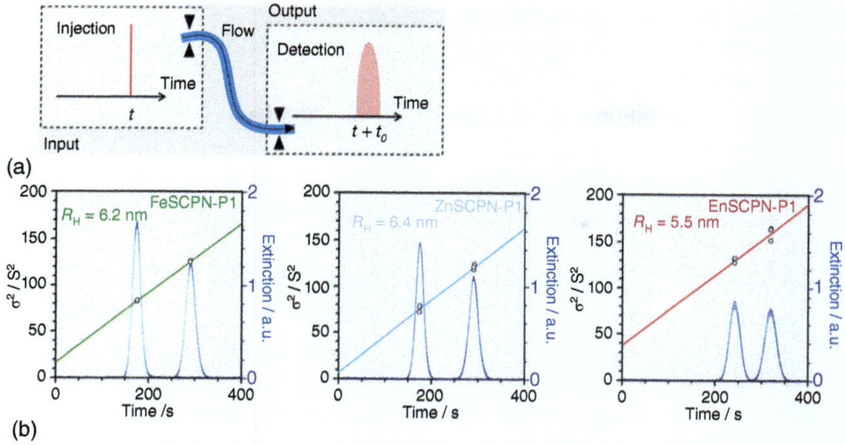

Figure 12.35 (a) Idealized representation of a TDA measurement in which the particles travel from the injection point to the detector with a residence time (t_0); in this way, the particles are dispersed by a combination of pressure-driven laminar flow and translational diffusion. (b) TDA profile of three different SCNPs. Source: (a) Lemal et al. [188]. © 2019 American Chemical Society, (b) Neumann [185].

consideration? For conventional organic molecules, including polymers, this might be rather hairsplitting on a first glance due to the intrinsic shape persistency of the structure arising from their solely covalent bonds. However, also such molecules might form aggregates under certain conditions or exist in different conformers or tautomers, polymers might be solubilized in an extended form or be dispersed as a collapsed random coil. For supramolecular polymers exhibiting some degree-of-reversibility of the underlying non-covalent interactions, eventually leading to a partially dynamic behavior, the choice of an appropriate analytical method to characterize the material is no longer straightforward.

The modern analytical techniques allow one to investigate supramolecular polymers or their aggregates regarding the molecular structure (i.e. molar mass, dispersity, etc.). However, combinations of independent methods are often required to draw a rather precise picture of the supramolecular polymers structure. Owing to the diversity of non-covalent interactions that enable the assembly of such polymers (i.e. H-bonding, metal-to-ligand coordination, CT interactions, or host–guest complexation), universal methods, e.g. determining the DP are still difficult to establish.

References

1 Lehn, J.-M. (2002). *Polym. Int.* 51: 825–839.
2 Brunsveld, L., Folmer, B.J.B., Meijer, E.W., and Sijbesma, R.P. (2001). *Chem. Rev.* 101: 4071–4098.
3 de Greef, T.F.A., Smulders, M.M.J., Wolffs, M. et al. (2009). *Chem. Rev.* 109: 5687–5754.
4 Fouquey, C., Lehn, J.-M., and Levelut, A.-M. (1990). *Adv. Mater.* 2: 254–257.

5 Fox, J.D. and Rowan, S.J. (2009). *Macromolecules* 42: 6823–6835.
6 Zheng, B., Wang, F., Dong, S.-Y., and Huang, F.-H. (2012). *Chem. Soc. Rev.* 41: 1621–1636.
7 Harada, A., Takashima, Y., and Yamaguchi, H. (2009). *Chem. Soc. Rev.* 38: 875–882.
8 Yang, S.K., Ambade, A.V., and Weck, M. (2011). *Chem. Soc. Rev.* 40: 129–137.
9 Chiper, M., Hoogenboom, R., and Schubert, U.S. (2009). *Macromol. Rapid Commun.* 30: 565–578.
10 de Greef, T.F.A. and Meijer, E.W. (2008). *Nature* 453: 171–173.
11 Hager, M.D., Greil, P., Leyens, C. et al. (2010). *Adv. Mater.* 22: 5424–5430.
12 Cordier, P., Tournilhac, F., Soulié-Ziakovic, C., and Leibler, L. (2008). *Nature* 451: 977–980.
13 Burnworth, M., Tang, L., Kumpfer, J.R. et al. (2011). *Nature* 472: 334–337.
14 Aida, T., Meijer, E.W., and Stupp, S.I. (2012). *Science* 335: 813–817.
15 Masuda, M., Jonkheijm, P., Sijbesma, R.P., and Meijer, E.W. (2003). *J. Am. Chem. Soc.* 125: 15935–15940.
16 Wilson, J.N., Masuda, M., Sijbesma, R.P., and Meijer, E.W. (2005). *Angew. Chem. Int. Ed.* 44: 2275–2279.
17 Moreau, J.J.E., Pichon, B.P., Man, M.W.-C. et al. (2003). *Angew. Chem. Int. Ed.* 43: 203–206.
18 Martin, R.B. (1996). *Chem. Rev.* 96: 3043–3064.
19 Liu, Y., Wang, Z., and Zhang, X. (2012). *Chem. Soc. Rev.* 41: 5922–5932.
20 Thordarson, P. (2011). *Chem. Soc. Rev.* 40: 1305–1323.
21 Wilcox, C.S. (1991). Design, synthesis, and evaluation of an efficacious functional group dyad. methods and limitations in the use of NMR for measuring host-guest interactions. In: *Frontiers in Supramolecular Organic Chemistry and Photochemistry* (eds. H.-J. Schneider and H. Dürr), 123–144. Weinheim: Wiley-VCH.
22 Feeney, J., Batchelor, J.G., Albrand, J.P., and Roberts, G.C.K. (1979). *J. Magn. Reson.* 33: 519–529.
23 Wilcox, C.S., Adrian, J.C. Jr.,, Webb, T.H., and Zawacki, F.J. (1992). *J. Am. Chem. Soc.* 114: 10189–10197.
24 Weber, G. (1965). The binding of small molecules to proteins. In: *Molecular Biophysics* (eds. B. Pullman and M. Weissbluth), 369–397. New York: Academic Press.
25 Connors, K.A. (1987). *Binding Constants*. New York: Wiley.
26 Connors, K.A. (1996). Measurement of cyclodextrin complex stability constants. In: *Comprehensive Supramolecular Chemistry*, vol. 3 (eds. J.L. Atwood, J.E.D. Davies, D.D. MacNicol and F. Vögtle), 205–241. Oxford: Pergamon Press.
27 Tsukube, H., Furuta, H., Odani, A. et al. (1996). Determination of stability constants. In: *Comprehensive Supramolecular Cheimistry*, vol. 8 (eds. J.L. Atwood, J.E.D. Davies, D.D. MacNicol and F. Vögtle), 425–482. Oxford: Pergamon Press.
28 Webb, J.E.A., Crossley, M.J., Turner, P., and Thordarson, P. (2007). *J. Am. Chem. Soc.* 129: 7155–7162.
29 Parker, C.A. and Rees, W.T. (1962). *Analyst* 87: 83–111.

30 Kostanski, L.K., Keller, D.M., and Hamielec, A.E. (2004). *J. Biochem. Biophys. Methods* 58: 159–186.

31 Meier, M.A.R., Hofmeier, H., Abeln, C.H. et al. (2006). *e-Polymers* 6: article ID 16 (7 pages).

32 Meier, M.A.R., Wouters, D., Ott, C. et al. (2006). *Macromolecules* 39: 1569–1576.

33 Zimmerman, S.C., Zeng, F.-W., Reichert, D.E.C., and Kolotuchin, S.V. (1996). *Science* 271: 1095–1098.

34 Paulusse, J.M.J. and Sijbesma, R.P. (2003). *Chem. Commun.*: 1494–1495.

35 Ogawa, K. and Kobuke, Y. (2000). *Angew. Chem. Int. Ed.* 39: 4070–4073.

36 Paulusse, J.M.J. and Sijbesma, R.P. (2008). *Chem. Commun.*: 4416–4418.

37 Holyer, R.H., Hubbard, C.D., Kettle, S.F.A., and Wilkins, R.G. (1966). *Inorg. Chem.* 5: 622–625.

38 Dobrawa, R., Ballester, P., Saha-Möller, C.R., and Würthner, F. (2006). Thermodynamics of 2,2′:6′,2″-terpyridine-metal ion complexation. In: *Metal-Containing and Metallosupramolecular Polymers and Materials*, vol. 928 (eds. U.S. Schubert, G.R. Newkome and I. Manners), 43–62. Washington, DC: American Chemical Society.

39 Irving, H. and Williams, R.J.P. (1953). *J. Chem. Soc.*: 3192–3210.

40 Chiper, M., Meier, M.A.R., Wouters, D. et al. (2008). *Macromolecules* 41: 2771–2777.

41 Winter, A. and Schubert, U.S. (2016). *Chem. Soc. Rev.* 45: 5311–5357.

42 Fang, L., Hmadeh, M., Wu, J. et al. (2009). *J. Am. Chem. Soc.* 131: 7126–7134.

43 Raymo, F.M. and Stoddard, J.F. (1999). *Chem. Rev.* 99: 1643–1664.

44 Zhang, M.-M., Li, S.-J., Dong, S.-Y. et al. (2011). *Macromolecules* 44: 9629–9634.

45 Jarroux, N., Guégan, P., Cheradame, H., and Auvray, L. (2005). *J. Phys. Chem. B* 109: 23816–23822.

46 Winter, A., Friebe, C., Chiper, M. et al. (2009). *J. Polym. Sci., Part A: Polym. Chem.* 47: 4083–4098.

47 Wild, A., Schlütter, F., Pavlov, G.M. et al. (2010). *Macromol. Rapid Commun.* 31: 868–874.

48 Abed, S., Boileau, S., and Bouteiller, L. (2001). *Polymer* 42: 8613–8619.

49 Sijbesma, R.P., Beijer, F.H., Brunsveld, L. et al. (1997). *Science* 278: 1601–1604.

50 Huang, F.-H., Nagvekar, D.S., Zhou, X.-C., and Gibson, H.W. (2007). *Macromolecules* 40: 3561–3567.

51 Niu, Z.-B., Huang, F.-H., and Gibson, H.W. (2011). *J. Am. Chem. Soc.* 133: 2836–2839.

52 Scherman, O.A., Ligthart, G.B.W., Sijbesma, R.P., and Meijer, E.W. (2006). *Angew. Chem. Int. Ed.* 45: 2072–2076.

53 Folmer, B.J.B., Cavini, E., Sijbesma, R.P., and Meijer, E.W. (1998). *Chem. Commun.*: 1847–1848.

54 Berne, B.J. and Pecore, R. (2000). *Dynamic Light Scattering: With Applications to Chemistry, Biology and Physics*. Toronto: General Publishing Company Ltd.

55 Zimm, B.H. (1945). *J. Chem. Phys.* 13: 141–145.

56 Lortie, F., Boileau, S., Bouteiller, L. et al. (2005). *Macromolecules* 38: 5283–5287.

57 Knoben, W., Besseling, N.A.M., and Cohen Stuart, M.A. (2006). *Macromolecules* 39: 2643–2653.

58 Chen, S.-G., Yu, Y., Zhao, X. et al. (2011). *J. Am. Chem. Soc.* 133: 11124–11127.

59 Braun, J., Renggli, K., Razumovitch, J., and Vebert, C. (2012). Dynamic light scattering in supramolecular materials chemistry. In: *Supramolecular Chemistry: From Molecules to Nanomaterials*, vol. 2 (eds. P.A. Gale and J.W. Steed), 411–425. Hoboken, NJ: Wiley.

60 Yang, S.K., Ambade, A.V., and Weck, M. (2009). *Chem. Eur. J.* 15: 6605–6611.

61 Guggenheim, E.A. (1937). *Trans. Faraday Soc.* 33: 151–156.

62 Hasegawa, M., Miyauchi, M., Takashima, Y. et al. (2005). *Macromolecules* 38: 3724–3730.

63 Miyauchi, M., Takashima, Y., Yamaguchi, H., and Harada, A. (2005). *J. Am. Chem. Soc.* 127: 2984–2989.

64 Ohga, K., Takashima, Y., Takahashi, S. et al. (2005). *Macromolecules* 38: 5897–5904.

65 Svedberg, T. and Pedersen, K.O. (1940). *The Ultracentrifuge*. Oxford: Calderon Press.

66 Liu, J., Andya, J.D., and Shire, S.J. (2006). *APPS J.* 8: E580–E589.

67 Rasa, M. and Schubert, U.S. (2006). *Soft Matter* 2: 561–572.

68 Schachman, H.K. (1959). *Ultracentrifugation in Biochemistry*. New York: Academic Press.

69 Harding, S.E., Rowe, A.J., and Horton, J.C. (eds.) (1992). *Analytical Ultracentrifugation in Biochemistry and Polymer Science*. Cambridge: Royal Society of Chemistry.

70 Schuster, T.M. and Laue, T.M. (eds.) (1994). *Modern Analytical Ultracentrifugation: Acquisition and Interpretation of Data for Biological and Synthetic Polymer Systems*. Boston, MA: Birkhäuser.

71 Scott, D.J., Harding, S.E., and Rowe, A.J. (eds.) (2005). *Analytical Ultracentrifugation: Techniques and Methods*. Cambridge: Royal Society of Chemistry.

72 Mächtle, W. and Börger, L. (2006). *Analytical Ultracentrifugation of Polymers and Nanoparticles*. Berlin: Springer-Verlag.

73 Hensley, P. (1996). *Structure* 4: 367–373.

74 Lebowitz, J., Lewis, M.S., and Schuck, P. (2002). *Protein Sci.* 11: 2067–2079.

75 Cole, J. and Hansen, J. (1999). *J. Biomol. Tech.* 10: 163–176.

76 Laue, T.M. and Stafford, W.F. III, (1999). *Annu. Rev. Biophys. Biomol. Struct.* 28: 75–100.

77 Pavlov, G.M., Perevyazko, I.Y., Okatova, O.V., and Schubert, U.S. (2011). *Methods* 54: 124–135.

78 Schubert, D., Tziatzios, C., Schuck, P., and Schubert, U.S. (1999). *Chem. Eur. J.* 5: 1377–1383.

79 Rasa, M., Tziatzios, C., Lohmeijer, B.G.G. et al. (2006). *Prog. Colloid Polym. Sci.* 131: 165–171.

80 Cölfen, H. and Völkel, A. (2004). *Prog. Colloid Polym. Sci.* 127: 31–47.

81 Dhont, J.K.G. (1996). *An Introduction to Dynamics in Colloids*. Amsterdam: Elsevier B.V.

82 Lamm, O. (1929). *Ark. Mat. Astron. Fys.* 21B: 1–4.

83 Schuck, P. (2000). *Biophys. J.* 78: 1606–1619.

84 Tziatzios, C., Durchschlag, H., Weidl, C.H. et al. (2002). *Synthetic Macromolecules with Higher Structural Order*, vol. 812 (ed. I.M. Khan), 185–200. Washington, DC: ACS Publishing.

85 Tziatzios, C., Durchschlag, H., Sell, B. et al. (1999). *Prog. Colloid Polym. Sci.* 113: 114–120.

86 Happ, B., Pavlov, G.M., Altuntas, E. et al. (2011). *Chem. Asian J.* 6: 873–880.

87 Michels, J.J., O'Connell, M.J., Taylor, P.N. et al. (2003). *Chem. Eur. J.* 9: 6167–6176.

88 Gibson, H.W., Yamaguchi, N., and Jones, J.W. (2003). *J. Am. Chem. Soc.* 125: 3522–3533.

89 Chakraborty, C., Pandey, R.K., Hossain, M.D. et al. (2015). *ACS Appl. Mater. Interfaces* 7: 19034–19042.

90 Yan, X.-H., Zhou, M., Chen, J.-H. et al. (2011). *Chem. Commun.* 47: 7086–7088.

91 Chen, Y.-Y., Tao, Y.-T., and Lin, H.-C. (2006). *Macromolecules* 39: 8559–8566.

92 Haino, T., Watanabe, A., Hirao, T., and Ikeda, T. (2012). *Angew. Chem. Int. Ed.* 51: 1479–1476.

93 Cohen, Y., Avram, L., and Frish, L. (2005). *Angew. Chem. Int. Ed.* 44: 520–554.

94 Li, S.-L., Xiao, T.-X., Hu, B.-J. et al. (2011). *Chem. Commun.* 47: 10755–10757.

95 Zayed, J.M., Biedermann, F., Rauwald, U., and Scherman, O.A. (2010). *Polym. Chem.* 1: 1434–1436.

96 Rauwald, U., Biedermann, F., Deroo, S. et al. (2010). *J. Phys. Chem. B* 114: 8606–8615.

97 Mansfeld, U., Winter, A., Hager, M.D. et al. (2013). *Polym. Chem.* 4: 113–123.

98 Timmermann, P., Weidmann, J.-L., Jolliffe, K.A. et al. (2000). *J. Chem. Soc., Perkin Trans. 2*: 2077–20889.

99 Schnell, I., Langer, B., Söntjens, S.H.M. et al. (2002). *Phys. Chem. Chem. Phys.* 4: 3750–3758.

100 Spiess, H.W. (2017). *Macromolecules* 50: 1761–1777.

101 Fenn, J.B., Mann, M., Meng, C.-K. et al. (2005). *Mass Spectrom. Rev.* 9: 37–70.

102 Karas, M., Bahr, U., and Gießmann, U. (1991). *Mass Spectrom. Rev.* 10: 335–357.

103 Nielen, M.W.F. (1999). *Mass Spectrom. Rev.* 18: 309–3344.

104 Gruendling, T., Weidner, S., Falkenhagen, J., and Barner-Kowollik, C. (2010). *Polym. Chem.* 1: 599–617.

105 Barner-Kowollik, C., Gruendling, T., Falkenhagen, J., and Weidner, S. (eds.) (2012). *Mass Spectrometry in Polymer Science*. Weinheim: Wiley-VCH.

106 Huang, F.-H. and Gibson, H.W. (2005). *Chem. Commun.*: 1696–1698.

107 Huang, F.-H. and Gibson, H.W. (2004). *J. Am. Chem. Soc.* 126: 14738–14739.

108 Herbert, D.E., Gilroy, J.B., Chan, W.-Y. et al. (2009). *J. Am. Chem. Soc.* 131: 14958–14968.

109 Guo, K., Guo, Z., Ludlow, J.M. et al. (2015). *Macromol. Rapid Commun.* 36: 1539–1552.

110 Sakamoto, S., Fujita, M., Kim, K., and Yamaguchi, K. (2000). *Tetrahedron* 56: 955–964.

111 Yamaguchi, K. (2003). *J. Mass Spectrom.* 38: 473–490.

112 Engeser, M., Rang, A., Ferrer, M. et al. (2006). *Int. J. Mass spectrom.* 255–256: 185–194.

113 Stroscio, J.A. and Kaiser, W.J. (eds.) (1993). *Scanning Tunneling Microscopy.* London: Academic Press Ltd.

114 Liu, Y., Chen, G.-S., Chen, Y. et al. (2006). *Nano Lett.* 6: 2196–2200.

115 Fathalla, M., Neuberger, A., Li, S.-C. et al. (2010). *J. Am. Chem. Soc.* 132: 9966–9967.

116 Miyauchi, M., Hoshino, T., Yamaguchi, H. et al. (2005). *J. Am. Chem. Soc.* 127: 2034–2035.

117 Ciesielski, A., Schaeffer, G., Petitjean, A. et al. (2009). *Angew. Chem. Int. Ed.* 48: 2039–2043.

118 Vonau, F., Aubel, D., Bouteiller, L. et al. (2007). *Phys. Rev. Lett.* 99: 086103 (086104 pages).

119 Stepanenko, V., Stocker, M., Müller, P. et al. (2009). *J. Mater. Chem.* 19: 6816–6826.

120 Li, Z., Li, Y., Zhao, Y. et al. (2020). *J. Am. Chem. Soc.* 142: 6196–6205.

121 Newkome, G.R., Wang, P., Moorefield, C.N. et al. (2006). *Science* 312: 1782–1785.

122 Wang, H., Li, Y., Yu, H. et al. (2019). *J. Am. Chem. Soc.* 141: 13187–13195.

123 Chakraborty, S. and Newkome, G.R. (2018). *Chem. Soc. Rev.* 47: 3991–4016.

124 Wang, L., Song, B., Li, Y. et al. (2020). *J. Am. Chem. Soc.* 142: 9809–9817.

125 Zhang, Z., Li, Y., Song, B. et al. (2020). *Nat. Chem.* 12: 468–474.

126 Schwarz, G., Bodenthin, Y., Geue, T. et al. (2010). *Macromolecules* 43: 494–500.

127 Song, B., Wu, G.-L., Wang, Z.-Q., and Zhang, X. (2009). *Langmuir* 25: 13306–13310.

128 Zhu, L.-L., Lu, M.-Q., Zhang, Q.-W., and Tian, H. (2011). *Macromolecules* 44: 4092–4097.

129 El-ghayoury, A., Schenning, P.H.J., van Hal, P.A. et al. (2001). *Angew. Chem. Int. Ed.* 40: 3660–3663.

130 Dobrawa, R., Lysetska, M., Ballester, P. et al. (2005). *Macromolecules* 38: 1315–1325.

131 Fernández, G., Pérez, E.M., Sánchez, L., and Martín, N. (2008). *Angew. Chem. Int. Ed.* 47: 1094–1097.

132 Hofmeier, H., Schmatloch, S., Wouters, D., and Schubert, U.S. (2003). *Macromol. Chem. Phys.* 204: 2197–2203.

133 Messmore, B.W., Hulvat, J.F., Sone, E.D., and Stupp, S.I. (2004). *J. Am. Chem. Soc.* 126: 14452–14458.

134 Neuman, K.C. and Nagy, A. (2008). *Nat. Methods* 5: 491–505.

135 Meister, A., Förster, G., Thünemann, A.F., and Kurth, D.G. (2003). *ChemPhysChem* 4: 1095–1100.

136 Rief, M., Gautel, M., Oesterhelt, F. et al. (1997). *Science* 276: 1109–1112.

137 Rief, M., Oesterhelt, F., Heymann, B., and Gaub, H.E. (1997). *Science* 275: 1295–1297.

138 Kudera, M., Eschbaumer, C., Gaub, H.E., and Schubert, U.S. (2003). *Adv. Funct. Mater.* 13: 615–620.

139 Zou, S., Schönherr, H., and Vancso, G.J. (2005). *Angew. Chem. Int. Ed.* 44: 956–959.

140 Kersey, F.R., Yount, W.C., and Craig, S.L. (2006). *J. Am. Chem. Soc.* 128: 3886–3887.

141 Kersey, F.R., Lee, G.-R., Marszalek, P., and Craig, S.L. (2004). *J. Am. Chem. Soc.* 126: 3038–3039.

142 Liu, Y., Yu, Y., Gao, J. et al. (2010). *Angew. Chem. Int. Ed.* 49: 6576–6579.

143 Gröger, G., Meyer-Zaika, W., Böttcher, C. et al. (2011). *J. Am. Chem. Soc.* 133: 8961–8971.

144 Ruokolainen, J., Mäkinen, R., Torkkeli, M. et al. (1998). *Science* 280: 557–560.

145 Binder, W.H. and Zirbs, R. (2007). *Adv. Polym. Sci.* 207: 1–78.

146 Yao, S., Beginn, U., Gress, T. et al. (2004). *J. Am. Chem. Soc.* 126: 8336–8348.

147 Würthner, F., Kaiser, T.E., and Saha-Möller, C.R. (2011). *Angew. Chem. Int. Ed.* 50: 3376–3410.

148 Hoeben, F.J.M., Jonkheijm, P., Meijer, E.W., and Schnenning, A.P.H.J. (2005). *Chem. Rev.* 105: 1491–1546.

149 Fan, J., Lal Saha, M., Song, B. et al. (2012). *J. Am. Chem. Soc.* 134: 150–153.

150 Wang, P., Moorefield, C.N., Li, S. et al. (2007). *J. Mater. Chem.* 17: 3023–3029.

151 De, S., Mahata, K., and Schmittel, M. (2010). *Chem. Soc. Rev.* 39: 1555–1575.

152 Wang, F., Zheng, B., Zhu, K.-L. et al. (2009). *Chem. Commun.*: 4375–4377.

153 Chi, X.-D., Xu, D.-H., Yan, X.-H. et al. (2013). *Polym. Chem.* 4: 2767–2772.

154 Venkatanarayanan, A., Spehar-Deleze, A.-M., Dennany, L. et al. (2008). *Langmuir* 24: 11233–11238.

155 Murthy, N.S. (2016). X-ray diffraction from polymers. In: *Polymer Morphology* (ed. Q. Guo), 14–36. Hoboken, NJ: Wiley.

156 Chu, B. and Hsiao, B.S. (2001). *Chem. Rev.* 101: 1727–1762.

157 Binder, W.H., Kunz, M.J., and Ingolic, E. (2004). *J. Polym. Sci., Part A: Polym. Chem.* 42: 162–172.

158 Al-Hussein, M., Lohmeijer, B.G.G., Schubert, U.S., and de Jeu, W.H. (2003). *Macromolecules* 36: 9281–9284.

159 Al-Hussein, M., de Jeu, W.H., Lohmeijer, B.G.G., and Schubert, U.S. (2005). *Macromolecules* 38: 2832–2836.

160 Parker, M.W. (2003). *J. Biol. Phys.* 29: 341–362.

161 Chakraborty, S., Endres, K.J., Bera, R. et al. (2018). *Dalton Trans.* 47: 14189–14194.

162 Xie, T.-Z., Guo, K., Guo, Z. et al. (2015). *Angew. Chem. Int. Ed.* 54: 9224–9229.

163 Chakraborty, S., Hong, W., Endres, K.J. et al. (2017). *J. Am. Chem. Soc.* 139: 3012–3020.

164 Udachin, K.A., Wilson, L.D., and Ripmeester, J.A. (2000). *J. Am. Chem. Soc.* 122: 12375–12376.

165 Gurarslan, A., Joijode, A., Shen, J. et al. (2017). *Polymers* 9: 673.

166 Wenz, G., Han, B.-H., and Müller, A. (2006). *Chem. Rev.* 106: 782–817.

167 Guerrero-Martínez, A., Ávila, D., Martínez-Casado, F.J. et al. (2010). *J. Phys. Chem. B* 114: 11489–11495.

168 Kamitori, S., Matsuzaka, O., Kondo, S. et al. (2000). *Macromolecules* 33: 1500–1502.

169 Zhang, Z., Luo, Y., Chen, J. et al. (2011). *Angew. Chem. Int. Ed.* 50: 1397–1401.

170 Zhu, Z., Li, H., Liu, Z. et al. (2012). *Angew. Chem. Int. Ed.* 51: 7231–7235.

171 Zhou, H.-C., Long, J.R., and Yaghi, O.M. (2012). *Chem. Rev.* 112: 673–674.

172 Kitagawa, S., Kitaura, R., and Noro, S.-i. (2004). *Angew. Chem. Int. Ed.* 43: 2334–2375.

173 Whittell, G.R., Gilroy, J.B., Grillo, I. et al. (2013). *J. Polym. Sci., Part A: Polym. Chem.* 51: 4011–4020.

174 Higgins, J.S. and Benoit, H.C. (1997). *Polymers and Neutron Scattering*. Oxford: Calderon Press.

175 Friese, V.A. and Kurth, D.G. (2008). *Coord. Chem. Rev.* 252: 199–211.

176 del Barrio, J., Horton, P.N., Lairez, D. et al. (2013). *J. Am. Chem. Soc.* 135: 11760–11763.

177 Yang, L., Tan, X., Wang, Z., and Zhang, X. (2015). *Chem. Rev.* 115: 7196–7239.

178 Brás, A.R., Hövelmann, C.H., Antonius, W. et al. (2013). *Macromolecules* 46: 9446–9454.

179 Gasnier, A., Royal, G., and Terech, P. (2009). *Langmuir* 25: 8751–8762.

180 Wahlund, K.G. and Giddings, J.C. (1987). *Anal. Chem.* 59: 1332–1339.

181 Wagner, M., Holzschuh, S., Traeger, A. et al. (2014). *Anal. Chem.* 86: 5201–5210.

182 Podzimek, S. (2011). Asymmetric flow field flow fractionation. In: *Light Scattering, Size Exclusion Chromatography and Asymmetric Flow Field Flow Fractionation*, 259–305. Weinheim: Wiley-VCH.

183 Messaud, F.A., Sanderson, R.D., Runyon, J.R. et al. (2009). *Prog. Polym. Sci.* 34: 351–368.

184 Huang, Z., Yang, L., Liu, Y. et al. (2014). *Angew. Chem. Int. Ed.* 53: 5351–5355.

185 Neumann, L.N., Urban, D.A., Lemal, P. et al. (2020). *Polym. Chem.* 11: 586–592.

186 Blasco, E., Tuten, B.T., Frisch, H. et al. (2017). *Polym. Chem.* 8: 5845–5851.

187 Taladriz-Blanco, P., Rothen-Rutishauser, B., Petri-Fink, A., and Balog, S. (2019). *Anal. Chem.* 91: 9946–9951.

188 Lemal, P., Petri-Fink, A., and Balog, S. (2019). *Anal. Chem.* 91: 1217–1221.

189 Taylor, G.I. (1953). *Proc. R. Soc. London, Ser. A* 219: 186–203.

190 Balog, S., Urban, D.A., Milosevic, A.M. et al. (2017). *J. Nanopart. Res.* 19: 287.

191 Urban, D.A., Milosevic, A.M., Bossert, D. et al. (2018). *Colloid Interface Sci. Commun.* 22: 29–33.

192 Balog, S. (2018). *Anal. Chem.* 90: 4258–4262.

193 Cipelletti, L., Biron, J.-P., Martin, M., and Cottet, H. (2014). *Anal. Chem.* 86: 6471–6478.

Index

a

AA-and BB-type monomers 4, 221, 232, 241, 245, 253, 279, 314–322, 325, 327, 338, 369, 373–374, 391, 396
AAA···DDD systems 60
A_2-and B_3-type monomers 219
AB_2-type co-pillarene 397, 402
AB_2A-type quasi-triblock copolymer 421
AB-type diblock copolymers 90–91, 100, 292
AB-type monomers 4, 9, 14, 128, 130, 164, 219, 231, 232, 241–242, 244–245, 247–248, 259, 279–280, 311–314, 325–327, 338, 369–373, 391, 396, 487–488
A-frame 127
acid-initiated polymerization of isocyanides 19
Adair's two-site binding model 227
adamantyl derivatives 467
ADDA-DAAD-type hetero-associate 90, 97
adenosine triphosphate (ATP) 288, 407–408
Ag(I)–pyridine bis-complexes 427
aggregation-caused quenching (ACQ) 331
aggregation-induced emission (AIE) 247–248, 290, 330–333, 395, 403
alkane solvents 433
alkyl-bridged bis-Napy derivative 92
alkyl-bridged homoditopic guest 404
alkyl-functionalized pillar[5]arene 487
A–[M]–A homopolymers 140–141, 147
ambipolar conductive materials 218
amide-bridged bis-pyridine 427
amidinium-carboxylate-bridged ligand 422
amido-bridged melamine dimers 73
3-aminobenzamidines 32
amino-functionalized pillar[5]arene 401
1,1′-(6-amino-1,3,5-triazine-2,4-diyl)diurea unit 85
amphiphilic donor-acceptor molecule 211
amphiphilic pillar[5]arene 405–410
amphiphilic pillar[n]arene-based materials 405
analytical ultracentrifugation (AUC) 12, 141, 366, 468–471, 476, 489
anti-cooperative supramolecular polymerization 16–18, 20, 106
anti-/gauche-conformations 86
apparent binding constants 231
arene π-complexation 120–121, 172–173, 176
aromatic π-electronic surface 195
aromatic charge-transfer (CT) interactions 209–213
aromatic cycloheptatrienyl (Ch^+) cation 216
aromatic diisocyanide derivatives 130
aromatic oligocarboxylates 440
Arrhenius activation energy 474
artificial chlorophyll 70
artificial peptide foldamers 437

Supramolecular Polymers and Assemblies: From Synthesis to Properties and Applications, First Edition.
Ulrich S. Schubert, George R. Newkome, and Andreas Winter.

aryl-bridged metallopolymers 173
association constants 5–6, 14, 16–18, 30, 34, 60, 76, 240, 252, 303, 315, 388, 454–455, 457–458
associative polymers 67
asymmetric flow field flow fractionation (AF4) 442, 490–492
atom-transfer radical polymerization (ATRP) 90, 101
atomic force microscopy (AFM) 41, 45, 49, 51, 72, 75, 92, 104, 143–145, 198, 202, 222, 230–231, 279–280, 283–286, 288, 317, 322, 336, 358, 406, 422, 430, 434, 441, 445–446, 479–482, 484
atomic-scale scanning-tunneling spectroscopy (STS) 477
Au(I)-containing metallopolymers 131–132, 161
Au(I)-containing monomers 162

b

barbituric acid 103
bay-positions 69, 205–206
BB-type bis-imidazolium guest 317, 399
β-CD dimer 373–374, 442, 468, 487
benzamidinium cation 32
benzene-1,3,5-tricarboxamide 436
benzene-perfluorobenzene pairing 428
benzo-18-crown-6 407
benzoic acid 65, 67
bidirectional chain-growth mechanism 120
binding constants 12, 14, 30, 43, 202, 226, 231, 276, 281, 283, 285, 361, 417, 454
biocompatible pillarene-based vesicles 407
biocomponent polymer 477
bio macromolecular building blocks 291–295
biomimetic orthogonal self-assembly 443–447
biomimetic systems 416
biopolymers 291, 376, 442, 486
bio-supramolecular polymerizations 18
biotin-4-fluorescein assay 445
bipyridine-containing cyclophane derivative 431
4,4′-bipyridine derivatives 65
bis-azobenzene derivative 424
bis-benzylammonium species 431
bis-calix[5]arene 229, 314–315
1,1′-bis(chlorozinc)-functionalized ferrocene 174
bis-exTTF molecular tweezers 229–231
bis-imidazolium salts 389
2,6-bis(1-methyl-1*H*-benzo[d|i]imidazol-2-yl)pyridine (mebip) 149
bis(η^6-naphthalene)chromium 176
bis-paraquat derivatives 389, 433, 439, 464, 472
bisphosphine and ancillary diisocyanide ligands 127
bis-pillarene derivative 397
bis-pyridine derivative 225
bis-pyridinium 389
bis-pyridinium BB-type guest molecule 396
bis-pyridyl macroligands 467
1,3-bis(pyrrol-2-yl)-1,3-diketone derivative 215
1,3-bis(pyrrol-2-yl)-1,3-diketone difluoroboron complexes 215
2,6-bis(pyrrol-2-yl)-4-nitrophenols 215
bis-terpyridine ligands 138, 147, 149, 152–154, 156, 169, 417, 460
bis-UPy derivative 433
bis-UPy-functionalized dithienylethylene derivative 94
bis-UPy-functionalized poly(ethylene glycol) (PEG) 88
bis-UPy-functionalized PTHF 89, 92
block copolymers 37–39, 48, 63, 81, 140–141, 145, 252, 292, 356, 359
bolaamphiphiles 82–84, 330
bolaform surfactants 84
bolaphiles 84
boron-centered derivative 208
boron-containing derivatives 165

boron-dipyrromethene (BODIPY) 165, 396
bottleneck effect 18
bovine serum albumin (BSA) 358
Bragg diffraction 486
building blocks 1, 29–30, 32, 35, 37–42, 48, 106, 119, 128, 130, 132, 164, 168, 172, 195, 220, 221, 231, 232, 240, 252–253, 261–263, 279, 291–295, 308, 316–317, 350, 364, 421, 423–424, 427–429, 438–440, 443, 476, 492
  small-molecule 42–46, 241–252, 448

c

calix[4]arenes 34–35, 103, 277, 301, 303–304, 307–312, 318–319, 323, 332–334
calix[n]arenes 35, 52, 228, 301–303, 322–328, 335
calixarenes 387, 401
  -based supramolecular polycaps 302, 304–307, 309
  supramacromolecular chemistry 229
calixarene-induced aggregation (CIA) 330, 333, 335
calix[4]pyrrole receptors 226, 322
carbene–metal complexes 132
carbon nanotubes (CNTs) 229, 354, 358, 483
carbon-to-metal bonds 160–162, 165
carboxylates 30, 33, 36, 130, 168, 267, 325, 408, 423
Carother's equation 5, 432
catechol bis-complexes 168
catenanes 227
cationic alkynylplatinum(II) terpyridine complexes 220
CB[*n*]-based host-guest interactions 427
[c2]daisy-chain-type pseudorotaxane 397
α-CD-containing pseudopolyrotaxane 463
β-CD-containing polypseudorotaxanes 486
central TPE-type dye 403
chain-extended polymers 63, 93, 101, 140, 166, 210, 252, 254, 291, 295, 473, 485
chain-extended supramolecular block copolymers 99
chain-extended supramolecular polymer 92, 101, 252–253, 291, 473–474
charge-transfer (CT) complexes 279, 401
chemical-vapor deposition (CVD) 71
chiral self-supported polymer catalyst 417
chiral Rh(I)-MonoPhos catalyst 419
cholinesterase (ChE) 334
circular dichroism (CD) 49, 65, 156, 204, 314, 353, 371
cis-$PtPh_2(DMSO)_2$ 422
classic Sterm–Volmer relation 459
classical polymers 3, 80
Co(II)-containing metallopolymers 489
coacervates 50
coiled fibers 222
coldspray (CS) ionization 476
coloration efficiencies 152, 156, 417–418
columnar triplex helical structure 433
columnar-type stacking of molecules 195
commercial polybutadiene 66
Commision Internationale l'Éclairage (CIE) 155–156, 170
concentration-dependent FT-IR absorption measurements 67
concentration-dependent kinetic measurements 18
concentration-dependent viscometry measurements 418
conductive/conjugated polymers 353
conventional polycondensation reactions 4, 19
cooperative nucleation–elongation mechanism 80, 162
cooperative supramolecular polymerization 6, 97, 130
  acid-initiated polymerization of isocyanides 19

cooperative supramolecular polymerization (*contd.*)
  amphiphilic perylene derivatives 20
  complex thermodynamics of 17
  covalent world 20
  downhill 17, 18
  (electro)static interactions 18
  NEP 16
  nucleated 17
  oligomer formation features 18
  powers 18
  radical polymerization 19
coordination nanosheet (CONASH) 133, 155
coordination polymers 118, 124, 126–128, 130, 133–159, 164, 165, 167, 169, 171, 302, 419, 487
coordinative bonding 120–123, 176
Cope-type reaction 94
copillarenes 388
core-shell micelles 146, 376
Cotton effect 76
Coulombic interactions 29, 37, 439
covalent living polymerizations 20
covalent polymers 3, 23, 36, 87, 90, 107, 118–119, 248, 280, 309, 465
covalent-supramolecular polymer (CSP) 248–249
covalent/ionic bonding 159–172
critical micellization concentration (cmc) 267, 335
critical polymerization concentration (cpc) 8, 86, 224, 245, 325, 418, 464
cross-linked supramolecular polymers 316, 403, 405, 425
crown ethers 387, 401
  aspects 239–240
  based polypseudorotaxane 484
  dendritic cell 253
  derivative 472
  hyperbranched polymer 253–258
  macromolecular building blocks 252–253
  poly(pseudo)rotaxanes 263–266
  pseudorotaxanes to rotaxanes 258–263
  recognition 425
  small-molecule building blocks 241–252
  supramolecular amphiphiles 266–269
cryogenic transmission electron microscopy (cryo-TEM) 210, 322, 333, 483
C_3-symmetric diphenylisoxazoles 200
Cu(I)-catalyzed azide-alkyne cycloaddition reaction (CuAAC) 261, 391, 425
Cu(II)-containing chain-extended polymers 166
cucurbit[7]uril 492
cucurbit[8]uril 22, 210, 239, 472, 482, 489, 492
cucurbit[8]uril-based supramolecular polymers 489
cucurbiturils (CB[n]s) 275, 387, 426
  aspects 275–276
  CB[n]s with small organic guest molecules 276–277
  supramolecular polymers 278
    bio macromolecular building blocks 291–295
    small organic molecules 278–291
cucurbiturils supramolecular chemistry 442
cyanurates 73, 79, 98, 103, 310, 420, 429, 477
cyclic eight-residue peptide 105, 106
cyclic mixed-metal nanostructures 478
cyclic oligomers 67, 224, 241, 245, 250, 282, 301, 307, 317, 319–320, 370, 373, 483
cyclic peptide scaffolds 104
cyclodextrin-Tb(III) array 476
cyclodextrins (CDs) 263, 275, 387, 388, 423
  based on CD-recognition 368–375
  defined 343

host-guest complexation 442
linear polypseudorotaxanes 345–347
moieties 309
polypseudorotaxanes 347–348
conductive/conjugated polymers 353–355
inorganic polymers 353
linear block copolymers 356–359
poly(ethylene glycol) 347
polyamides and polyurethanes 350–351
polyamines 355–356
polyesters 349–350
polyethers 348–349
polymeric architectures 356
polyolefins 352–353
polyrotaxanes 353, 362–364
side-chain polymers 360–362
star-shaped polymers 359–360
pre-assembled pseudorotaxanes 364–368
structures 343
supramolecular polymerization 368–375
supramolecular polymers
AA-and BB-type monomers 373–374
AA-type monomers 374–375
diblock copolymers 375–379
α-cyclodextrin (α-CD) moieties 462
α-cyclodextrin rings 476
γ-cyclodextrin (γ-CD) 227, 228
cyclohexane 197, 433
cyclometallating binding 163
cytochrome b_{562} H63C mutant (Cyt^{H63C}) 445
cytosine 81, 84

d

DADA-type fashion 221, 223
Dainton–Ivin equation 9
daisy chains 241–243, 259–261, 370, 397, 399, 462
D-alanine-based melamine monomer 76
D–A-type supramolecular polymer 209
D–A-type interaction 231
1D cofacial supramolecular polymer 199
DDAA-type unit tautomerizes 61
degree-of-polymerization (DP) 40, 126, 221, 243, 278, 347, 396
delivery of doxorubicin (DOX) 292
dendrimers 231–232, 255, 257–258, 263, 292, 376, 462
dendritic poly(amidoamine) 440
depolymerization process 221, 397
3D hyperbranched polymer 241, 253
9,10-dialkoxyanthracene 73
2,7-diamido-1,8-naphthyridine unit (Napy) 62
α,ω-diaminoalkane 398
2,7-diamino-1,8-naphthyridine 429
diaminopyridine unit 429
diaminopyridines 57, 73, 102, 429, 436
2,6-diaminopyridines 73, 436
2,6-diaminotriazine 81
2,6-diamino-1,3,5-triazine 73, 81
1,4-diazabicyclo[2.2.2]octane 128, 335, 395
diazobenzene moiety 94, 442
diazobenzene tethers 73
dibenzylammonium salt 256–257, 261, 439
1,3-dibenzylguanidinium 30
diblock copolymers 42, 90–91, 100, 292, 375–379, 485
1,2-dichloroethane 326, 443
1,4-dicyanobutane 402
Diels–Alder reaction 221, 364
diffusion-ordered NMR spectroscopy (DOSY) 45, 169, 202, 279, 316, 419, 429, 471
diffusion-ordered spectroscopy (DOSY) 87, 221, 282, 284, 288, 292, 371, 400, 428, 472–473, 492
measurements 87
difunctional monomer 442
β-diketonate derivatives 165
β-diketonate moiety 166
diketopyrrolopyrroles (DPPs) 71–72, 200–201, 331

dimerization of carboxylic acids 67
dimerization process 30
dimerize via quadruple H-bonding interactions 61–62, 398
dimerized/(macro)cyclized oligomers 391
dimethoxypillar[5]arene 427
dinuclear pincer-type precursor complex 163–164
α,ω-dinucleobase-functionalized PTHF 84
1,5-dioxonaphthalene 432
diphenylphenanthrene (DPP) 331
3,5-diphenylisoxazole moieties 199
3,5-diphenylisoxazoles 199
2,6-diphenylpyridine spacer 220
dipole–dipole-type interaction 57
directly interacting monomers 4
diselenides 399, 442
9,10-distyryl-anthracene dye 395
1,4-disubstituted 1,2,3-triazole unit 425
ditopic bidentate ligands 133–137
ditopic bis-terpyridine ligands 460
ditopic molecular tweezers 226
ditopic monodentate ligands 123
  bisphosphine and ancillary diisocyanide ligands 127
  chelate effect 123
  heterobimetallic polymer 123
  perylenebisimide 130
  polymer-analogues to cisplatin 127
  $[Pt_3(\mu_3\text{-}CO)(\mu\text{-}dppm)_3]^{2+}$ precursor 126
  heterobimetallic polymer 124
  sonication-induced dissociation 125
  zig-zag-shaped metallo-supramolecular polymers 130
  4-pyridyl-substituted porphyrin derivative 131
  Au(I)-containing metallopolymers 132
  metallopolymer 127
  metallopolymers containing ditopic NHC-type ligands 133
  polymer-analogues to cisplatin 128
  porphyrin rings 128
  Shish-Kebab-type metallo-supramolecular polymers 129
  sonication-induced dissociation 124
  zig-zag-shaped metallo-supramolecular polymers 130
ditopic pyridyl-functionalized cruciform-shaped ligand 164
ditopic tridentate ligands
  bis-terpyridines 138, 139
  bis-tridentate ligands with Ru(II) ions 150, 151
  ABA-type triblock copolymers 147, 149
  chiral Ru(II)-and Fe(II)-containing metallopolymers 157
  core-shell-corona morphologies 147, 148
  diblock copolymer 140, 141
  electrostatic self-assembly processes 153
  hydrophobic PS core and the hydrophilic PEG corona 143, 144
  LCST temperature 142
  macroligands 143, 147
  mixing triangle of 156
  Ru(II)-containing chain-extended polymer 140
  "scorpionate"-type metallopolymer $[Fe(47)]_n$ 158
  terminal metal-binding site 141, 142
  terpyridine *bis*-complexes 143
  tpy-containing diblock copolymer 145, 148
  transition and rare-earth metal ions 149, 150
  transition metal ions 137, 138
  tridentate 2,6-bis(1*H*-pyrazol-1-yl)pyridine units 156
  Zn(II) ions and π-conjugated bis-terpyridine ligands 153, 154
diverse factors 196
diverse pseudorotaxanes 389

DNA 1, 58, 104, 227, 280, 347, 371, 428, 442, 481, 486
dodecyltrimethylammonium cations 47
double-chained/zig-zag-shaped supramolecular polymers 426, 427
doubled-stranded supramolecular polymer 422
double dynamers 102
double-dynamic polymer 102–103, 399–400
double hydrogen-bonding interactions 66–72
downhill cooperative supramolecular polymerizations 17, 18
doxorubicin (DOX) 292–294, 409–410
doxorubicin hydrochloride (DOX) 329
1D protein copolymer 445
1D supramolecular polymers 155, 250, 437
D_5 symmetric pillar[5]arene 387
dynamic "polynanocage," 483–484
dynamic covalent chemistry (DCC) 262–263, 399–400
dynamic light scattering (DLS) 40, 44–46, 75, 94, 100–101, 210, 280, 286, 292, 294, 312–314, 321, 327–329, 333–334, 361, 374, 377, 397, 402–403, 406–407, 410, 422–423, 432–433, 441, 446, 465, 467
dynamic polyrotaxane 398–399

e

EDTA (EDTA:ethylenediamine-tetraacetic acid) 46, 145
effective molarity (EM) 13–14
electro-spray ionization (ESI) 284, 475–476
(electro)static interactions 18
electrochromism 136, 152, 155
electron microscopy 73, 482–484
electron paramagnetic resonance (EPR) spectroscopy 286
electron-rich pillar[5]arene 401
electrospray ionization (ESI) 475–476
electrospray ionization mass spectrometry (ESI MS) 35, 284
electrostatic interactions 20, 30, 40, 42, 49–50, 52, 60, 80, 196, 203, 213, 267, 311, 331, 470, 483
enantiomerically pure alkyl groups 429–430
endo-cavity inclusion 401
enzyme-bridged supramolecular polymer 444
Ercolani's ring-chain model 14
ESI time-of-flight (TOF) MS 475
Eu(III) ions 149, 166, 169–170, 494
exo-ditopic macrocyclic ligand 135
exTTF pincers 231

f

face-to-face-type orientation 195
Fe(II) bis-complex $[Fe(17)_2]^{2+}$ 427
"Ferrari" pigment 71, 200
ferrocene (Fc) 157, 170, 172–175, 404
ferrocene-containing polymers 173
fill factors (FF) 200
flexible $[Mb\text{-}52\text{-}SAv\text{-}52\text{-}Mb]_n$ fibers 445
Flory–Huggins theory 10
Flory's "principle of equal reactivity," 8 19
9-fluorenyllithium 19
fluorescein 407
fluorescence titration 316, 441, 458
foldamer approach 211
Fourier-transform infrared (FT-IR) 67–68, 104–105, 337, 350, 352
Fourier-transform ion cyclotron resonance (FT-ICR) mass spectrometry 476
fractal polymer 478
free association model 10–11
free-radical polymerization (FRP) 266
of vinylferrocene 172
Fukuzawa's bis-calix[5]arenes 229
fumaramide 431–432
functionalized β-CD unit 424

g

Gibbs free energy 2–3, 22, 69, 203
(gradient) tandem mass spectrometry (gMS^2) 475–476
green emeraldine salt 389
grid-like assemblies 470
Grotthuss-type mechanism 170

h

half-oxidized, blue-colored emeraldine base 389
halogen bonding 2, 64, 399, 448
Hamilton receptor 98–99, 420, 429, 477
hard and soft acids and bases (HSAB) 171
hard polystyrene 485
^{1}H double-quantum (DQ) NMR spectroscopy 474
helix-helix diblock copolymer 100
heme-bis(biotin) conjugate 445–446
heme-protein recognition 444
heme–Cyt^{H36C} conjugation 446
heme–protein 444, 446
hetero-complementary Hamilton receptor 420
heteroatom-containing ligand 171
heteroatom-containing triangulene 208
heterobimetallic coordination polymer 124, 171
heterobimetallic polymer 123, 428
heterocomplementary, homoditopic AA-and BB-type monomers 391
heterodifunctional components 442
heteroditopic AB-type monomers 14, 231, 241, 391
heteroditopic building block 231, 423–424, 438, 476, 483
heteroditopic macroligands 420
heteroditopic material 430
heteroditopic molecule 471
heteroditopic monomers 12, 44, 218, 221–222, 225, 229–230, 244, 259, 314, 374, 395, 425, 432, 444
heteroditopic oligo(ethylene glycol) derivative 13, 444
heterogeneous hydrogenation catalyst 419
heterogeneous nucleation 18, 23
heteroleptic Ru(II)–typ bis-complexes 143
heterotelechelic polymer 101, 419
heterotritopic AB_2-type co-pillarene 241, 397
hexa-*peri*-substituted HBCs 197–198
high binding constants 281, 361, 417, 454
high-density polyethylene (HDPE) 390
highly-ordered pyrolytic graphite (HOPG) 365, 476–478
Hill coefficients 232
^{1}H NMR spectroscopy 30, 31, 33, 35, 43–44, 77, 85, 93, 97–98, 106, 337, 348, 418, 432, 454
Hofmeister effect 36
Hogsteen-type base-pairing interaction 82
homo-and heteroditopic bolaamphiphiles 82
homo-and heterometallic polydecker assemblies 176
homo-or heterotelechelic monomers via H-bonding interactions 63–64
homoditopic β-cyclodextrin 467
homoditopic AA-and BB-type monomers 4, 241, 374, 391
homoditopic building block 262, 427
homoditopic components 221
homoditopic crown-ether 464
homoditopic cyanurate derivative 420
homoditopic host 262, 318, 369, 402
homoditopic macromonomers 93
homoditopic monomers 4, 218, 223, 245–246, 285, 314, 373, 395, 433, 439, 482, 489
homoditopic polymeric unimers 429
homogeneous Rh(I)-MonoPhos system 417
homolog pillar[6]arene 388

homotelechelic bis-UPy-functionalized polymers 87–89
homotelechelic monomers 73, 86–87, 92
homotritopic guest 402
homotropic allosteric binding process 227
honeycomb PBI-melamine network 78
host-enhanced charge-transfer interactions 482
host-guest complexation 2, 4, 119, 140, 195, 226, 250, 255, 277, 281, 295, 301, 356, 374, 387, 389–391, 397, 401, 403, 408, 415, 426, 431–432, 438–439, 442, 445, 495
host-guest interactions
  AA and BB-type monomers 314, 316, 317, 319–322
  AB-type monomers 312–314
host-stabilized charge-transfer (HSCT) interaction 210
hydrogen-bonding (H-bonding) interactions 4, 428
  aprotic and non-polar solvents 58
  classification of 58, 59
  dimerize via quadruple H-bonding interactions 62
  heterodimeric assemblies via quadruple H-bonding 62
  interactions 29
  N-alkyl-cyanurates *via* sextuple H-bonding 62, 63
  vs. non-covalent interactions 57
  single H-bonding motifs 59
  supramolecular polymers
    double hydrogen-bonding interactions 66–72
    multiple H-bonding interactions 104
    parameters 64
    quadruple H-bonding interactions 84
    sextuple H-bonding interactions 98
    single H-bonding unit 65
    triple H-bonding interactions 72
  triple H-bonding motifs 58, 60, 61
  two-centered H-bonding motifs 59
  types of 57, 58
  UPy-moiety 62
hydrogen-evolution reaction (HER) 172
hydrophilic pillar[6]arene 408
hydrophobic effect 195, 203–204
hydroquinone derivative 432
hydroxyethylethylene-diaminetriacetic acid (HEEDTA) 418–419
hyperbranched and cross-linked assemblies 401–405, 411
hyperbranched supramolecular polymer 231, 232, 403

i

Ibidapo's classification 37
imide-melamine H-bonding interaction 75
imidiazole-functionalized guests 397
inorganic-organic hybrid materials 417
interlocked poly(daisy chain) 462
interlocked strong donor-acceptor [2]catenane 432
intermolecular carboxylate-to-guanidinium binding 32
intermolecular non-covalent interactions 2
intra-protein H-bonding interactions 446
intramolecular heme–Cyt^{H36C} interactions 446
intramolecular Zn(II)-imidazolyl interactions 460
ion mobility mass spectrometry (IM-MS) 475
ionic interactions 2, 4, 21, 29–52, 119, 353, 387, 422, 431, 436, 439, 441
ionic self-assembly (ISA) 213, 439
ionic/covalent bonding 120, 165, 176
ionomers 37, 485
isocyanate-terminated (macro)molecules 96
isodesmic association 454

isodesmic supramolecular polymerization (IDP) 6, 7, 98, 197, 227
  bifunctional monomer 9
  concentration or temperature of 8
  energy of 8
  intermolecular equilibrium constant 7, 8
  multi-stage open association mechanism 6
  non-covalent linkage 11
  temperature-dependency 10, 11
  transition line 9
isomeric dipyridones 69
isotactic polypropylene (iPP) 91
isothermal titration calorimetry (ITC) 30, 34, 63, 169, 227, 276, 284, 290, 295, 357, 371, 459, 461

j

Jacobson–Stockmayer parameters 67

l

L-alanine-derived melamine 76
Lambert–Beer's law 456, 493
Lamm's equation 469–470
latent dyes 72
LC polymers 65
LC supramolecular polymers 74, 82
Lehn's concept of double dynamic supramolecular polymers 102
leucoemeraldine base 389–390
light scattering 66, 281, 357–358, 374, 465, 490, 492
light-emitting diodes (LEDs) 95, 154, 164
linear bis(pyridinecarbaldehyde) derivative 483
linear main-chain polypseudorotaxane 439
linear metallo-supramolecular polymers
  coordinative bonding 159–172
    ditopic bidentate ligands 133–137
    ditopic monodentate ligands 123–133
    ditopic tridentate ligands 137–159
    pyridine-based ligands 122, 123
  metal-arene interactions 172–176
linear polymers 1, 4, 6, 63, 67, 85, 138, 166, 241, 264, 284, 288, 309, 318–320, 325, 345–356, 371, 399, 403, 464
  pillarenes and 389–391
linear polypseudorotaxanes 241–252, 344–356
linear rigid-rod metallo-supramolecular polymers 161, 490
linear supramolecular polymers 65, 67, 75, 98, 99, 253, 391, 403, 483
  π-electronic interactions 219
linear supramolecular polypseudorotaxane 423
liquid crystalline (LC) 305
  materials 216
  morphology 47
  properties 436
living crystallization 20
living supramolecular polymerization 20, 23
lower critical solution temperature (LCST) 142, 291, 376
Lu@C82 metallofullerenes 78
lyotropic LC phase 73

m

macrocyclic bis(paraquat-*p*-phenylene) system 432
macroligand 140–141, 143, 145–147, 157, 256–257, 419–420, 467, 485–486
macromolecules interactions 1
magic-angle spinning (MAS) 474
main-chain metallo-supramolecular polymers 470
Mark–Houwens parameters 464
Mark–Houwink equation 463
mass spectrometry (MS) 11, 35, 41, 106, 231, 255, 259, 284, 367, 475

matrix-assisted laser desorption/ionization (MALDI) 231, 242, 284, 475
matrix-assisted laser desorption/ionization time-of-flight (MALDI-TOF) 255, 284, 475
mechanical bonding 462
melamine-bearing PBIs 79
memory effect 66, 151
merocyanines 79, 196, 208
metal-arene interactions 172–176
metal-centered chirality 133, 138
metal-containing polymers 117–121, 177
metal-metal-to-ligand charge-transfer (MMLCT) process 218
metal-organic coordination networks 118
metal-organic frameworks (MOFs) 118, 487
metal-to-carbon bonds 159
metal-to-ligand complexation 295, 387, 442, 493–494
metal-to-ligand coordination 11, 29, 36, 49, 57, 76, 107, 117, 119, 147, 157, 176, 195, 213, 224, 232, 252, 262, 280–281, 318, 415–428, 495
  hydrogen bonding 2–4
metallo-macrocyclic assemblies 134
metallophilic interactions 20, 64, 162, 218
metallopolymers 21, 118–119, 121–139, 151–159, 161–177, 417, 419, 460–461, 468, 479–480, 488–489
metalloporphyrin-containing polymer 129
metallo-supramolecular cages 425
metallo-supramolecular polymers 131, 168, 417, 469
  bidirectional chain-growth mechanism 120
  definition 118
  ligand's binding sites 120
  linear 121–176
  synthesis of 121
(metallo-)supramolecular polymerizations 160
metathesis reaction 160, 214, 391
methylcyclohexane 75–76, 199, 204, 227, 325–326
methylenediphenyl-4,4′-diisocyanate (MDI) 351, 364
methylviologen-/pyridinium-type guest 387
micellar gels 147
mitoxantrone (MTZ) 408–409
molar enthalpy (ΔH) 459–460
molar mass estimation 454
molecular force probe (MFP) 481
molecular loop 320
molecular Sierpiński hexagonal gasket 478
molecular tweezers 213, 219–221, 224, 228, 229, 315
molecular tweezers/guest recognition motifs 212
monodisperse poly(ferrocenyl-dimethylsilane)s 488
monodisperse supramolecular assemblies 486
monofunctional host-guest conjugate 392
mono-functionalized pillarenes 392, 393
monofunctional pillarene monomers 392
monomer 1, 312–322, 369–375, 433
monomer anion 19
monomethyl ethers (MPEGs) 141, 147–148
Monte Carlo simulations 15, 482
more defined molecules 486
MPEG–[Ru]–PS diblock copolymers 143
multi-angle light scattering (MALS) detector 462
multi-stranded biomacromolecules 422
multifunctional supramolecular gel 425
multiple H-bonding interactions 58, 62, 64, 98, 104–107, 303, 462, 464, 485

multiple-stimuli responsive supramolecular polymers 420, 424
multitopic cyclam-centered bis-terpyridine ligand 489
myoglobin (Mb) 444–446

n
N-alkyl-cyanurates *via* sextuple H-bonding 62, 63
n-alkyltrimethylammonium surfactants 47
NAND logic gate 227
nano-biomaterial 445
naphthalene bisimide (NBI) derivative 196, 210, 441
natural anion binding systems 30
natural biopolymers 442–443
naturally protein-bound heme 444
N-butylthymine 420, 436
N-dodecyl cyanuric acid 74, 75
negatively charged Cu(II)-phthalocyanine 441
n-heptane 429
N-heterocyclic carbenes (NHCs) 132
n-hexane 399–400, 433
NH_2-functionalized poly(N-isopropyl acrylamide) (PNIPAM-NH_2) 39
Nigishi-type cross-coupling 174
$[Ni(mnt)_2]^{2-}$ 136
20-nm-wide metallo-supramolecular hexagonal grid 479
nitroaromatics 157, 219
nitroxide-mediated polymerization (NMP) 252, 353, 419
N-methylpyrrolidone (NMP) 252, 353, 419
N,N'-bisalkylpyromellitic diimide 401
11′-(N,N,N-trimethylammonium)-undecanylisobutyrate group 40
non-covalent interactions 1, 2
  degree-of-reversibility 5
  hydrogen-bonding interactions 57
  intermolecular 2
  strength and directionality of 3
  supramonomer 20
non-covalently linked monomers 2
non-trivial thermo-rheological behavior 87
nonionic surfactant Igepal CO-520 486
norbornene monomers 420
nuclear magnetic resonance (NMR) spectroscopy 11, 30, 63, 162, 243, 276, 389, 470–475
nucleated supramolecular polymerization 17–18
nucleation-elongation mechanism 17, 80, 162
nucleation-elongation polymerization (NEP) model 16
nucleobase triphosphates 21
nucleobase-pairing approach 84
nucleobase-pairing interactions 82

o
O_2-storage hemoprotein 444
oligo(ethylene glycol) chains 97, 138, 197, 407
oligofluorene (OF) 94
oligometallocene 174–175
oligonuclear metallo-supramolecular block copolymers 147
oligonucleotide strands 482
oligo(phenylene-ethynylene (OPE) 82–83, 434–435
oligo(phenylene ethynylene)-based Pd(II)-pyridyl complexes 20
oligo(phenylene-vinylene) (OPV) 94–95, 204–205
oligosulfonates 440
oppositely charged porphyrins 440–441
order-disorder transitions (ODTs) 37–38
organic field-effect transistors (OFETs) 72
organic polymers 117, 353, 404, 415, 462
organometallic polymers 159, 161, 172
organometallics 220
organoplatinum(II)-containing molecular tweezers 221

orthogonal combinations of supramolecular interactions
biomimetic orthogonal self-assembly 443–447
involving H-bonding 428–438
involving metal-to-ligand coordination 417–428
miscellaneous 438–443
orthogonality 415–416, 438, 448
outward-pointing pyridine moieties 436
oxazole-annulated pillarene 401
oxidized single-wall carbon nanotubes (ox-SWCNTs) 483

p

para-ethynyl-substituted Pt(II) pincer 164
para-(*tert*-butoxyamino)cinnamoylamino moieties 476
particle-in-a-sphere model 13
pathway selection 6
P-centered triangulenes 208–209
p-diisocyanoarenes 161
Pd(II) ions 244, 247, 422, 425
Pd(II)-and Pt(II)-containing polymers 132, 460
Pd(II)-containing metallo-supramolecular polymers 163
Pd(II)-pincer complexes 164, 467, 482
pentafulvalene dianion 173
pernigraniline base 389
perylene bisimides (PBIs) 69–72, 75–79, 94–95, 130, 196, 200–208, 327, 441
phenyl C_{61} butyric acid (PCBA) 231
phenyl-[6,6]-C_{61}-butyric acid methyl ester (PCBM) 95
photo-induced energy-/electron-transfer processes 227
phenyl 4-(*N*-alkylamino)benzoate 19
photochromic diazobenzene 94
photo-responsive azobenzene moiety 396
photo-responsive supramolecular self-assembly 393
P3HT-[Zn]-MPEG diblock copolymers 143
physical cross-linking 87
physical origin classification 2
π–π-stacked aggregates 196
π–π-stacking interaction 2, 69–71
attractive and repulsive 195, 196
charge–transfer to π-electronic ion-pairing interactions 213–218
columnar supramolecular polymers
amphiphilic HBCs 2 197, 198
boron-centered derivative 208, 209
charge–transfer or cation–anion π-interactions 197
dipole-dipole interactions 199
discotic amphiphile in aqueous medium 199
DPPs 199, 200
heteroatom-containing triangulene 208
hexa-*peri*-substituted hexabenzocoronenes 198
IDP 197
LC phase 202
PBIs dyes 200, 202–206, 208
P-centered triangulenes 208
unimers 197
donor-acceptor-type charge-transfer interactions 209–213
linear supramolecular polymers 218–232
π-conjugated anion receptors and anions 215
π-conjugated chromophores 94–95, 154, 207, 220
π-conjugated material 197, 405
π-conjugated metallopolymer 125, 129
π-conjugated molecules 196, 199, 208
π-conjugated segments 149, 154
π-conjugated systems 58, 79, 98, 195–197, 210, 215, 353
π-electron anion 216
receptors and anions 215

π-electron ion pairs
  aromatic cycloheptatrienyl (Ch^+) cation 216
  columnar assemblies 214
  cycloheptatrienyl (Ch^+) cation 217
  cyclopentadienyl anions 217
  metathesis 214
  self-assembly modes of 214
  square-planar Pt(II) complexes 218
  $TATA^+$ cation 217
π-extended analogs (exTTFs) 229
pillar[5]arene-based ditopic host 397
pillar[5]arene-based pseudorotaxane 390
pillar[5]arene macrocycles 396
pillar[n]arenes 387–388, 405
pillar[n]arene-guest complexes 388
pillarene-based host-guest interactions
  hyperbranched and cross-linked assemblies 401–405
  and linear polymers 389–391
  supramolecular assemblies, based on amphiphilic pillar[5]arenes 405–410
  supramolecular polymers derived from 391–401
pillarene-based self-assembled amphiphiles (PSAs) 406
pincers 132, 141, 163–165, 211–212, 220–222, 227, 231
planar π-conjugated porphyrin entities 394
planar π-electronic surface 226
planar metalloporphyrins 127
point-of-nucleation 17
poly(acrylic acid) (PAA) 47–48, 378
poly(butylene succinate) (PBS) 350
poly(ϵ-caprolactone) (PCl) 349
  -based heteroditopic macromonomers 253
  monomers 258
poly(dimethyl-aminoethyl methacrylate) (PDMAEMA) 41, 292
poly(dimethylsiloxane) (PDMS) 67–68, 90, 337, 353, 429, 464
poly(D,L-lactide) (PLA) 292–294
poly(etherketone) (PEK) 80, 81, 99, 485
poly(ethylene glycol) (PEG) 38, 267, 347, 460, 462
  bridged ditopic host 404
  chains 462, 482
poly(ethylene imine) (PEI) 47, 52, 152–153, 347, 355
poly(ethylene terephthalate) (PET) 350
poly(ethylene-*co*-butylene) (PEB) 129, 429
  soft domains 486
poly(ethylene-*co*-propylene) (PEP) 91
poly(3-hydroxypropionate) (P3HP) 350
poly(isobutylene) (PIB) 38, 80–81, 99, 346, 352, 485
poly(metal acetylide)s 159–161
poly(metal arylide)s 159, 161
poly(methyl acrylate) (PMA) 38, 266
poly(methyl methacrylate) (PMMA) 100, 253, 257, 316, 460, 461
poly(methylvinyl ether) (PMVE) 349
poly(*N*-isopropylacrylamide) (PNIPAM) 291, 294, 376
poly(*N*-isopropylacrylamide) (tpy-PNIPAM) 142
poly(*N*,*N*-dimethylacrylamide) (PDMA) 291
poly(phenylene-ethynylene) polymers (PPEs) 49, 262
poly(propylene glycol) (PPG) 487
poly(propylene imine) (PPI) 255
poly(propylene oxide) (PPO) 81–82, 349, 357–359, 367–368
poly(pseudo)rotaxanes 240–241, 263–264, 266, 343, 388
poly(tetrahydrofuran) (PTHF) segments 82, 84
poly(trimethylene oxide) (P3MO) 486–487
poly[(μ-η^6,η^6-naphthalene)chromium] 176
poly(4-vinylbenzyltrimethylammonium chloride) 50
polyaddition reactions 2, 4, 9, 133

polyamides 350–351
polyamines 355–356
polyaniline (PANI) 353–354, 365, 389–390
polycaprolactone 166
polycaps 302–309
polycationic poly(*N*,*N*-dimethyl-aminoethyl methacrylate) (PDMAEMA) block 41
polycondensation reactions 2, 4, 19, 138, 364
polycyclic aromatic hydrocarbons (PAHs) 196, 220
polydecker sandwich complexes 174
polydecker-type architectures 174
polyelectrolyte complexes (PECs) 49–52
polyelectrolyte-amphiphile complexes (PACs) 153
polyelectrolyte-surfactant complexes 36, 46, 47
polyesters 87, 165–167, 264, 349–350, 357, 364
polyethers 87, 348–349, 352, 360
polyisobutylene (PIB) 38, 80–81, 99, 346, 352, 485
polyisoprene (PI) 37, 292, 352
polymer networks 37, 39, 64–66, 82, 87, 89, 106, 149, 224–225, 249, 285, 290–291, 307, 315, 391, 402, 404, 422, 426
polymer-bound 2,6-diaminopyridine 436
polymeric micelles 2, 332
polymerization
  pre-assembled pseudorotaxanes 364–368
  techniques 1
  transition line 9
polymerization-induced phase separation (PIPS) 89
polymerize techniques 1
polymetallocenes 174–176
  structures 172
polymethacrylates 90, 360
polyolefins 352–353
polyoxometalates (POMs) 39–42, 202–203, 375–376, 436
polyplatinynes 159, 160
polypseudorotaxanes 399, 439, 486
  conductive/conjugated polymers 353–355
  linear block copolymers 356
  poly(ethylene glycol) 347–348
  polyamides and polyurethanes 350–352
  polyamines 355–356
  polyesters 349–350
  polyethers 348–349
  polymeric architectures 356
  polyolefins 352–353
  polyrotaxanes 353, 362–364
  side-chain polymers 360–362
  star-shaped polymers 359–360
polyrotaxane-like polymer 399
polyrotaxanes 262, 265–266, 337, 346, 347, 349, 351, 353, 362–368, 375–376, 462, 471
polystyrenes (PSs) 37, 252, 460, 462
polyurethanes 264–265, 350, 359, 364
Porod region 488
porphyrins 220
  -based molecular tweezers 221, 223
  molecules 221
power-conversion efficiencies (PCE) 200
protein recognition 443–448
proteins 36, 50, 88, 104, 285, 415–416, 443, 446–447, 468, 486, 493
proton-transfer processes 38
pseudo[2]rotaxane 427
pseudo[2]rotaxane dimer 431
pseudorotaxanes 16, 239–252, 336–337, 343–389
Pt(II)-containing polymer 460, 471
$[Pt_3(\mu_3\text{-}CO)(\mu\text{-}dppm)_3]^{2+}$ precursor 126
Puddephatt's synthesis 161
pyrene guest moiety 221
pyridine (py) 420, 471
  dicarboxamide spacer 222
  -based ligands 122, 123
4-pyridyl moiety 131

4-pyridyl-functionalized pillar[5]arene 399
4-pyridyl-substituted porphyrin derivative 131

q

quadruple H-bonding interactions
  K_a values 85
  anti-gauche-conformations 86
  bis-UPy-functionalized π-conjugated chromophores 95–96
  bis-UPy-functionalized biocompatible poly(ε-caprolactam) 88
  1,3,5-triamino-2,4,6-triazine derivatives 96–97
  AB-type diblock copolymers 90
  fiber-or mesh-like structures 88
  heterodifunctional H-bonding array 97
  heteroditopic H-bonding units 94
  homoditopic macromonomers 93
  light-emitting diodes (LEDs) 95
  non-self-complementary monomers 93
  PCBM molecules 95
  photoreversible formation 94–95
  supramolecular diblock copolymers 92
  UPy-and Napy-functionalized homopolymers 90–91
  UPy-containing poly(ε-caprolactone) (PCl) 88
  UPy-functionalized homotelechelic monomers 85–87
  UPy/1,1′-(6-amino-1,3,5-triazine-2, 4-diyl)diurea unit 85
quadrupole H-bonding interactions 482
quadrupole-TOF (QTOF) 284
quasi-star block copolymer architecture 39

r

radius-of-gyration 465, 487
random-coil polymers 13
ratio-of-diffusion-coefficients 473
redox-responsive supramolecular network 404
reversibility-of-binding 3, 220, 417, 447
reversible addition-fragmentation chain-transfer (RAFT) 41, 42, 99, 101, 291
  polymerization of styrene 41
reversible cross-linking *via* H-bonds 64
reversible supramolecular polymerization 94, 222
reversible thermoresponsive properties 66
rhodamine-B 152, 377, 379, 407
ribonuclease S-peptide 444
rigid-rod π-conjugated ligands 470
ring-chain equilibrium polymerization 15, 16, 68, 85, 86
ring-chain equilibrium supramolecular polymerizations 67
ring-chain mediated supramolecular polymerization 6, 12–16, 23, 43, 279, 325
  intra-and intermolecular association 13
  intramolecular reaction 12
  pseudorotaxanes 16
  spacer's rigidity 15
  supramolecular world 16
ring-opening metathesis polymerization (ROMP) 12, 16, 248, 420, 421, 429, 436
ring-opening polymerizations (ROPs) 1, 16, 23, 121, 177
rod-shaped fibers 443, 484
rotaxanes 227, 239, 258–263, 343–344, 390
Ru(II)-based metallopolymers 484
Ru(II)-bipyridine tris-complex 420
Ru(II)-containing chain-extended polymer 140
Ru(II)-containing diblock copolymer 485
Ru(II)-tpy bis-complexes 143
Ru(II)-terpyridine bis-complexes 481

$\{[Ru(2)](PF_6)_2\}_n$ 470
rylene dye 441

S
salen 167, 168
salicylaldimine units 167
salicylaldimine-type ligands 167
scanning electron microscopy (SEM) 41, 74, 76, 79, 88, 90, 166, 199, 200, 205, 227, 248, 252, 316, 322, 324, 325, 328, 392–394, 399, 400, 425, 433, 483, 484
scanning probe microscopy (SPM) 476–482
scanning-force microscopy (SFM) 310
scattered intensity 488
scattering vector 486, 488
schlieren texture 305, 306
second virial coefficient 465
SEDFIT program 469
sedimentation coefficient 469
sedimentation equilibrium measurements 469
self-assembled monolayer (SAM) 78
self-complementary monomers 4
self-complementary UPy system 418, 419
"self-stoppered" effect 93
sequence-controlled supramolecular copolymers 443
sextuple H-bonding interactions
  calix[4]arenes 103
  diblock copolymers 104
  double dynamers 102, 103
  Hamilton-receptor 98
  Hamilton-receptor and barbiturates 99, 100
  helix-helix diblock copolymer 100
  heterotelechelic polymer 101, 102
  linear supramolecular polymer 98–99
  pseudo-block copolymers 99
shape-persistent metallo-supramolecular nanostructures 478
Shish–Kebab-type metallo-supramolecular polymers 129, 130
Shish–Kebab-type metallopolymers 126, 130
Shish–Kebab-type supramolecular polymerization 128
side-chain functionalized copolymers 436
side-chain polymers 39, 360–362
Sierpiński motif 478
single-chain nanoparticles (SCNPs) 493–495
single-molecule force spectroscopy (SMFS) 280, 284, 285, 481, 482
single-walled carbon nanotubes (SWCNTs) 354, 483
site-selective complexation 356–361
site-selective polypseudorotaxane 357–361
size-exclusion chromatography (SEC) 11, 41, 107, 125, 131–132, 164–166, 255, 316, 364, 419, 445–447, 460–463, 488–489, 492–493
small angle neutron scattering (SANS) 45, 281, 283, 366, 371, 372, 470, 487–490
small angle X-ray scattering (SAXS) measurements 37, 77, 82, 92, 100, 281, 374, 485–490
small-molecule heteroditopic building blocks 438
small/wide angle X-ray scattering 484–486
soft poly(ethylene glycol) block 485
solvent effects 58
solvophobic forces 195
sonication-induced dissociation 124, 125
Sonogashira cross-coupling polymerization 262, 405
S-protein 444
square-planar coordination geometry 218
square-planar Pt(II) complexes 218, 422
star-shaped oligo(propyleneimine) 436
star-shaped polymers 40, 41, 256, 357, 359–360

star-shaped tris-porphyrin derivatives 433
static light scattering (SLS) 317, 375, 465–467
Stern–Volmer relation 458
stiff stilbene derivative 393
stimuli-controlled supramolecular polymerization 21
stimuli-responsive pillarene-based amphiphilic assemblies 410
stimuli-responsiveness 213, 227, 267, 418
Stokes–Einstein equation 472, 494
streptavidin 445, 446
structural monomer classification 2, 4
sulfonamide-containing oligopeptides 437
sulforhodamine B 152, 220
superconducting quantum interference device (SQUID) 156
supramacromolecular pillarene-based materials 389
supramolecular $(AB)_n$-type multiblock copolymer chains 37
supramolecular ABC-type triblock copolymers 421, 429
supramolecular adhesive 417
supramolecular amphiphiles 266, 327
supramolecular assemblies, based on amphiphilic pillar[5]arenes 405–410
supramolecular assemblies, on ionic interactions 35
- binding modes and discrete model assemblies 30–35
- polyelectrolyte complexes 49–52
- polymers, as building blocks 37–43
- polymers, in side chain 46–49, 51
- small molecules, as building blocks 42–46

supramolecular block copolymers 38, 48, 81, 99, 141, 142, 146, 147, 232, 292, 421
supramolecular cruciform 164
supramolecular dendrimers 255, 257, 258, 462
supramolecular diblock copolymers 90, 92, 104, 253, 268, 292, 375–379, 485
supramolecular droplets 89
supramolecular fluorescence titration experiments 458
supramolecular graft-block copolymers 37
supramolecular ladder-type polymers 65, 66
supramolecular polymer featuring [c2]daisy chains 397
"supramonomer polymerization" approach 399
- (partially) reversible character 5
- AA-and BB-type monomers 4
- AB-type monomers 4, 369–373
- cooperative 6, 16–20
- definition 1
- degree of polymerization 5
- degree-of-reversibility 5
- depolymerization 5
- homoditopic monomers A–A and B–B 4
- hydrogen-bonding or ionic interactions 4
- IDP 6, 7
- kinetic control 21–22
- ring-chain mediated supramolecular polymerization 6, 12–16
- self-complementary monomers 4
- stimuli-controlled 21
- thermodynamic equilibrium state 6
- traditional equilibrium state 6
- types of 5

supramolecular polymers
- analytical ultracentrifugation (AUC) 468–470
- architectures 302
- aspects 301–302
- asymmetric flow field-flow fractionation 490–492
- based on calixarenes 327–337
- Calix[4]pyrroles 322–327

calixarene scaffolds 309–311
calixarene-based 302–309
cucurbiturils 278
small organic molecules 278–291
definition of 2
electron microscopy 482–484
host-guest interactions
AA-and BB-type monomers 314–322
AB-type monomers 312–314
light scattering 465–467
mass spectrometry 475–476
molar mass estimation 454–460
NMR spectroscopy 471
scanning probe microscopy 476–482
self-assembly process 453
size-exclusion chromatography 460–463
small-angle neutron scattering (SANS) 487–490
small/wide angle X-ray scattering 484–486
Taylor dispersion analysis 492–494
unique properties of 453
vapor pressure osmometry 467–468
viscometry 463–465
X-ray crystallography 486–487
supramolecular quasi-triblock copolymers 472
supramolecular rubber 39, 40, 106
supramonomer 20
Svedberg equation 469
synthetic macromolecules 468

t

taco complexes 240
tailor-made metallopolymers 121
"tailor made" supramolecular polymers 3
Taylor dispersion analysis (TDA) 492–495
telechelic poly(dimethylsiloxane)s 67
terminal larger crown-ethers moieties 439
terpyridine (tpy) 218, 417, 471
terpyridine bis-complexes 138, 141, 146, 418, 476, 481, 485
2,2′:6′,2″-terpyridine (tpy) ligands 218
terpyridine mono-complex 461
1,2,4,5-tetrafluoro-3,6-diiodobenzene 399
1,2,4,5-tetrakis(diphenylphosphino) benzene 135
tetra(*n*-butyl)ammonium (TBA) cation 216, 217, 324, 325
tetra(*n*-butyl)ammonium iodide (TBAI) 324, 325
7,7,8,8-tetracyanoquinodimethane (TCNQ) 222, 226
4,5,7-trinitrofluorenone (TNF) 212, 213, 222, 472
4,8,12-trioxa-4,8,12,12c-tetrahydrodibenzo [cd|i,*mn*}]pyrene ($TOTA^+$) 216
tetra-meso-NO_2-substituted porphyrin derivative 131
tetra-cationic guanidiniocarbonyl pyrrole derivative 441
7,7,8,8-tetracyanoquinodimethane (TCNQ) 222, 226
tetrakis(guanidiniumcarbonyl pyrrole) cation 45–46
tetracarboxylate 46
tetracationic building block 492
tetraethylammonium iodide (TEAI) 324, 325
tetrafluoro-*p*-benzoquinone 226
tetrafluorobenzene derivatives 436
tetrahydrofuran (THF) 16, 19, 49, 143, 197, 347, 391, 407
tetraphenylethene-centered tetra-pillarene 403
tetraphenylethylene (TPE) 247, 289–291, 330, 403
tetrathiafulvalenes (TTFs) 220, 226, 229, 288, 323
tetrathiolate-type ligand 171
thermodynamic classification 2
thermodynamically-controlled step-growth polymerizations 14
thermoreversible gelation process 199

thin polymer films 177, 417
thiol-disulfide 442
thiolate-type ligand systems 171
threading-followed-by-stoppering 257–259
three-point-charges 195
thymine 59, 60, 73, 81–83, 102, 420, 429, 436
tpy-containing diblock copolymer 145
tpy-functionalized poly(ethylene glycol) 142, 421
traditional equilibrium state 6
traditional polymer chemistry 8
traditional polymer synthesis 121
transmission electron microscopy (TEM) 39–42, 45, 76–77, 84, 91–92, 144, 210, 268, 283, 322, 375–379, 402, 406, 410, 423, 482
transoid-and cisoid-conformer 94
travelling wave ion-mobility (TWIM) 284
trialkoxysilyl functionalities 89
1,3,5-triamino-2,4,6-triazine derivatives 96, 97, 103
trifluoroacetic acid (TFA) 473, 474
1,3,5-trinitrobenzene (TNB) 219, 226, 227, 323
2,4,7-trinitrofluorenone (TNF) 212, 213, 222, 472
4,5,7-trinitrofluorenone (TNF) 212, 213, 222, 472
2,4,6-trinitrotoluene (TNT) 219
triorganotin units 170
4,8,12-trioxa-4,8,12,12c-tetrahydrodibenzo [cd|i,*mn*}]pyrene (TOTA$^+$) 216
triple H-bonding interactions
  α,ω-dinucleobase-functionalized PTHF 84
  N-dodecyl cyanuric acid 74–75
  Ag-layered Si substrate 78
  chain-extended supramolecular block-copolymer 81
  cyanurates or barbiturates 73
  heterotelechelic PBI 77
  homotelechelic monomers via 73
  homotelechelic PPOs 81–82
  honeycomb network 78
  imide-melamine 75
  LC supramolecular polymer 73–74
  linear tape-like and cyclic rosette-like shapes 75
  melamine-bearing PBIs 79
  melamine/cyanurate 73, 75
  nucleobase-pairing interactions 82, 84
  OPE fragments 82, 83
  ribbon-or rope-like aggregates 79
  toroids and linear fibrils 80
tris-bidentate complexes 133, 134
tris-(crown ether) 439
tris-guanidinium 30
tris(8-hydroxyquinolinato)aluminum (Alq_3) 166, 167
trisphenanthroline ligand 483
tris(pyrazol-1-yl)borate (Tp) moieties 157
tris-tridentate coordination geometry 494
type-A polydecker assemblies 174, 175
type-B polydecker architectures 174
type-C polymetallocene 176
α,ω-type surfactants 84

u

ultrahigh-vacuum low-temperature STM (UHV-LT-STM) imaging 477–480
UPy fragments $[Fe(3...3)](PF_6)_2]_n$ 419
UPy-U-PEB-U-UPy 429
uracil 57, 73
uracil derivatives 57
uradiazoles 66
urea/urethane H-bonding 429
ureidoguanosine (UG) 92, 429
ureidopyrimidinone (UPy) 397, 472
  moieties 397–398
  and Napy-functionalized polyacrylates 90
  based homodimers 431
  bridged strapped metalloporphyrin cleft 431

containing poly(ε-caprolactone) (PCl) 88, 349, 418
equipped macromolecular components 88, 89
functionalized amphiphiles 89
functionalized homotelechelic monomers 85–87
functionalized peptide recognition sequences 88
functionalized π-conjugated chromophores 94, 96
heterodimer 431
terminated diblock copolymer 92
2-ureido-pyrimidinone (UPy) 418

V
van der Schoot's model 10, 11
van-der-Waals interactions 10, 333, 345, 352, 434
van't-Hoff data analysis 197, 202
vapor pressure osmometry (VPO) 69, 202, 264, 369–371, 373, 374, 432, 467–468
viologen oligomers 389
virtual molar mass 3, 68, 98, 108
viscoelastic gels 489
viscometry 67, 86, 93, 168, 169, 247, 253, 255, 291, 292, 307, 317, 318, 399, 402, 418, 423, 424, 463–465, 470

W
water-soluble π-conjugated polyrotaxanes 470
water-soluble cucurbituril 388
water-soluble hexameric metallo-macrocycle 483
water-soluble pillar[6]arene 408, 410
water-soluble pillarenes 409
Watson–Crick-or Hogsteen-type fashion 73
white-light emitting heterobimetallic supramolecular polymer 170, 171
Whitesides' concept of peripheral crowding 75
wide angle X-ray scattering (WAXS) measurements 100, 484–486
worm-like chain model 13

X
X-ray crystallography 106, 486–487
X-ray diffraction (XRD) methods 49, 65, 72, 77, 126, 202, 216–219, 227, 286, 290, 336, 345, 347, 351, 361, 370, 393, 399–400, 484, 486–487

Z
zig-zag-shaped metallo-supramolecular coordination polymer 130
zig-zag-shaped metallo-supramolecular polymers 130
zig-zag-shaped polymer chains 426
Zimm-plot representation 465
Zn(II)-and Fe(II)-containing porphyrins 446
Zn(II)-containing metalloporphyrins 467
Zn(II)-containing porphyrin-PBI-porphyrin triad 130
Zn(II)-containing supramolecular "Kandinsky circles," 479 480
Zn(II)-dipyrromethane dyes 165
Zn(II)-porphyrin–protein conjugate 446
Zn(II)–pyridine interaction 122
zwitterionic guanidiniocarbonyl pyrrole carboxylate motif 423
zwitterionic pyrrole derivative 30, 439